DATE DUE FOR RETURN

24. JUN. 1983

5 MAY 1995

?? JUN. 1984

26 SEP 1995

31/4/86

7 OCT 1996

20 JUN 1986

30 Sept 1986 (JH)

26 JUN 1987

31.12.87 DD

WORD PROBLEMS

STUDIES IN LOGIC

AND

THE FOUNDATIONS OF MATHEMATICS

VOLUME 71

Editors

A. HEYTING, *Amsterdam*
H. J. KEISLER, *Madison*
A. MOSTOWSKI, *Warszawa*
A. ROBINSON, *New Haven*
P. SUPPES, *Stanford*

Advisory Editorial Board

Y. BAR-HILLEL, *Jerusalem*
K. L. DE BOUVÈRE, *Santa Clara*
H. HERMES, *Freiburg i. Br.*
J. HINTIKKA, *Helsinki*
J. C. SHEPHERDSON, *Bristol*
E. P. SPECKER, *Zürich*

NORTH-HOLLAND PUBLISHING COMPANY
AMSTERDAM • LONDON

WORD PROBLEMS

Decision Problems and the Burnside Problem in Group Theory

Edited by

W. W. BOONE
University of Illinois
Urbana

F. B. CANNONITO
University of California
Irvine

R. C. LYNDON
University of Michigan
Ann Arbor

1973

NORTH-HOLLAND PUBLISHING COMPANY
AMSTERDAM • LONDON

Library of Congress Catalog Card Number: 70-146190
North-Holland ISBN: 0 7204 2271 X

PRINTED IN THE NETHERLANDS

Dedicated to the memory of
our friend and colleague
Hanna Neumann (1914–1971)

LIST OF CONTENTS

Introduction ix

S. AANDERAA: A proof of Higman's embedding theorem, using Britton extension of groups 1

S.I. ADJAN: Burnside groups of odd exponent and irreducible systems of group identities 19

S. BACHMUTH, H.Y. MOCHIZUKI and D.W. WALKUP: Construction of a non-solvable group of exponent 5 39

J.L. BRITTON: The existence of infinite Burnside groups 67

F.S. CANNONITO: The algebraic invariance of the word problem in groups 349

F.B. CANNONITO and B.W. GATTERDAM: The computability of group constructions, part I 365

D.J. COLLINS: The word, power and order problems in finitely presented groups 401

R.W. GATTERDAM: The Higman theorem for primitive-recursive groups – A preliminary report 421

W. HAKEN: Connections between topological and group theoretical decision problems 427

S. LIPSCHUTZ: On the word problems and T-fourth-groups 443

J. McCOOL and A. PIETROWSKI: On a conjecture of W. Magnus 453

R. McKENZIE and R.J. THOMPSON: An elementary construction of unsolvable word problems in group theory 457

T.G. McLAUGHLIN: A non-enumerability theorem for infinite classes of finite structures 479

C.F. MILLER III: Some connections between Hilbert's 10th problem and the theory of groups 483

C.F. MILLER III: Decision problems in algebraic classes of groups (a survey) 507

A.W. MOSTOWSKI: Uniform algorithms for deciding group-theoretic problems 525

B.H. NEUMANN: The isomorphism problem for algebraically closed groups 553

H. SCHIEK: Equations over groups 563

P.E. SCHUPP: A survey of small cancellation theory 569

D. TAMARI: The associativity problem for monoids and the word problem for semigroups and groups 591

L. WOS and G. ROBINSON: Maximal models and refutation completeness: semidecision procedures in automatic theorem proving 609

Problems 641

INTRODUCTION

This volume grew out of the conference "Decision Problems in Group Theory", (CODEP) held at the University of California, Irvine, in September 1969, with the support of the United States National Science Foundation, the Army Research Office, Durham, and the Air Force Office of Scientific Research (Air Force Systems Command). These same organizations supported the preparation of this volume.

Although cutting broadly across logic, group theory and other fields of algebra, topology and computer science, the mathematics discussed in the conference and this book has, we would argue, an inherent unity of purpose and method. The unifying aspects are the common concern with algorithms and the common use of a certain kind of combinatorial analysis. This analysis had been used in logic for a long time, particularly in the study of proofs, before it then spread – in the very development of the field we are here presenting – to the study of "word problems" in groups and semigroups. To us it seems unlikely that the argument of Novikov and Adjan for the Burnside Problem, published in 1968 and whose central ideas are explained in the present volume, or the argument of Britton for the same result in the present volume, could have been discovered by a "pure" group theorist!

All this is not to say that the field of word problems did not exist before this confluence of group theory with logic. Indeed, the origin of the field may be traced back to Max Dehn's fundamental problems posed in 1911. At that time Dehn was studying the fundamental groups of compact manifolds of genus at least 2. His concern was with the existence of algorithms to determine if given paths were contractible to a point, if given paths were homotopic, and if given spaces were homotopy equivalent. In group-theoretic terms these decision problems are, respectively, the word (or identity) problem, the conjugacy (or transformation) problem, and the isomorphism problem – now all known to be unsolvable in general for finitely presented abstract groups. Dehn did solve the word problem corresponding to the case of closed 2-manifolds of genus at least 2. Further, Wilhelm Magnus' proof of the Freiheitssatz, and his related solution of the word problem for groups given by one defining relation, also appeared in the 1930's before the union with logic. We should mention, too, that early on in the century, as an independent line of development, Axel Thue had formulated the word problem for finitely presented semi-groups – or, as one now says, Thue systems – and solved various special cases of this general problem.

But negative results, unsolvability results in group theory, were impossible before the union with logic, since the very idea of an algorithmically unsolvable problem was lacking. This lack was overcome in 1935–1936 by Alonzo Church and, independently, in 1936 by A.M. Turing when they gave equivalent precise mathematical definitions of the intuitive notion of algorithm. All these definitions, known today as "Church's Thesis", led to Church's negative solution of the decision problem for first-order arithmetic; and, subsequently, to independent negative solutions by Church and Turing to Hilbert's Entscheidungsproblem for pure predicate logic. Hindsight shows that virtually all unsolvability results in mathematics are, in the final analysis, a translation of such classical results into a new setting.*

The propitious event, to our minds, constructing the bridge from logic to algebra occurred in 1946–1947. For then Emil Post and A.A. Markov, independently, showed the word problem for Thue systems unsolvable. This result was the first unsolvability result outside the foundations and it marked off the field of word problems in much its present form. Ultimately this bridging would lead to the situation that, while one still had positive and negative results regarding the existence of algorithms, the logic and algebra are so merged in theorems and proofs that it is impossible to separate one from the other.

* Other formalizations of the intuitive notion of effective process are due to Gödel (based on a suggestion of Herbrand), Kleene, Post, and A.A. Markov. That all these definitions have turned out to be equivalent is often taken as evidence for the correctness of Church's Thesis.

This book is intended not only as a volume for the expert but also as a vehicle for entry into the field at various levels of knowledge and into various specialties. We have included many survey articles; the Britton proof and the Wos-Robinson proof are entirely self-contained.

The foremost reference we can mention to obtain the necessary background for this volume is the following: Joseph J. Rotman, *The theory of groups, an introduction* (Allyn and Bacon). The second edition (1973), particularly the last chapter, of Rotman's text gives virtually all the background material from logic (recursive function theoretic), needed to get into the present volume. Indeed, the only sizable gap for the reader would then be the Friedberg–Muchnik Theorem on the existence of "in between" recursively enumerable degrees – and this can be found in several standard logic texts.

Our next most important reference is: Wilhelm Magnus, Abraham Karrass and Donald Solitar, *Combinatorial group theory* (Interscience, New York, 1966). Referring to the text, the reader will be able to follow or look in at a deeper level.

An historical account of the field is given in: William W. Boone, *The theory of decision problems in group theory: a survey*; this is available on tape from the American Mathematical Society and is to appear in the Bulletin of the American Mathematical Society.

The following three items are parallel to parts of this volume:

William W. Boone, Wolfgang Haken and Valentin Poénaru, *On recursively unsolvable problems in topology and their classification,* in: *Contributions to Mathematical Logic* (North-Holland, Amsterdam, 1968).

Roger C. Lyndon and Paul Schupp, *Geometric group theory* (tentative title), Ergebnisse der Mathematik (Springer, Berlin, to appear).

Charles F. Miller, III, *On group-theoretic decision problems and their classification,* Annals of Mathematic Studies (Princeton Univ. Press, Princeton, N.J., 1971).

We do not attempt to attribute authorship to the problems listed in this volume and we wish here to thank all who have so generously helped in the preparation of this volume by their contribution to this very interesting problem set.

The editors also wish to acknowledge their gratitude to the North-Holland Publishing Company and particularly to Einar Fredriksson for the considerable help given to us during the preparation of this work. Similarly Jens Mennicke's mathematical help in carefully checking Britton's proof was so important that he deserves some such title as "honorary editor", which we hereby confer. Our thanks, too, to Peter X. Sarapuka for his careful translation of Adjan's article from the original Russian. And, finally, we thank the contributors, without whose efforts and cooperation, this volume, of course, could not exist. To them belongs the lion's share of credit for whatever merits this volume may have.

William W. Boone
Frank B. Cannonito
Roger C. Lyndon

A PROOF OF HIGMAN'S EMBEDDING THEOREM USING BRITTON EXTENSIONS OF GROUPS

Stål AANDERAA
University of Oslo

§1. Introduction

The aim of this paper is to give a new proof of the following theorem due to Higman (see Higman [5, Theorem 1, p. 456]):

Theorem 1. *A finitely generated group can be embedded in a finitely presented group iff it is recursively presented.*

Shoenfield [9, pp. 321–326] presents a variant of Higman's proof.

As pointed out by Higman [5], one half of Theorem 1 is trivial. Hence, from now on we shall concentrate on the non-trivial half of the theorem, which we shall state as a lemma:

Lemma 1. *Every finitely generated and recursively presented group can be embedded in a finitely presented group.*

Definition 1. A *positive group presentation* is a presentation $(S \mid D)$, where each relation in D is of the form $w = 1$ for some positive word over S.

Note that every group presentation $(S \mid D)$ can be turned into a positive presentation of the same group by replacing S by $S' = \{s' \mid s \in S\} \cup S$ and D by D', where D' is obtained from D by replacing each occurrence of s^{-1} by s' and by adding the relation $ss' = 1$ for every s in S.

Definition 2. A group is *recursively presented* if it has a positive presentation of the form

$$(u_1, u_2, \ldots, u_m \mid w = 1, w \in E),$$

where E is a recursive enumerable set of positive words on $u_1, u_2, \ldots, u_m$.

§2. Outline of the proof of Lemma 1

In Section 3 we shall start with some remarks on Britton extensions and Britton's Lemma. Then in Section 4 we shall outline the construction of the Turing machine Z_E and the corresponding semigroup T_E. In Section 5 we shall study the Boone-Britton group G (see Britton [3, pp. 22–23]). Our final aim is to prove Lemma 11 which is the only lemma in Section 5 which is used in Section 6. Lemma 11 corresponds to Lemma 5.1 of Higman [5, p. 473], and to Lemma 18 of Shoenfield [9, p. 335]. Lemma 11 shows that the group G_E so to speak semicomputes or recognizes the enumerable set of words E.

In Section 5 we first form the free product of G_E and the group R, which we want to embed. Then we construct three Britton extensions, which by using a finite number of new generators and new defining relations transfer the information given in group G_E about E to generate relations which make all but a finite number of defining relations in $G * R$ superfluous.

§3. Remarks on Britton extensions

We shall assume that the reader is familiar with Britton's Lemma, which is the main tool in this paper (see Britton [3, Lemma 4, p. 20]). We shall not state the lemma here since the reader, in order to understand the details of Section 4, must have the paper of Britton before him anyhow. We would only like to define some new concepts and to make some remarks.

Definition 3. A presentation $(S^* | D^*)$ is a *Britton extension* of $(S|D)$ with stable letters P, if the following two conditions are satisfied:

(1) $(S^* | D^*)$ has stable letters P and corresponding basis $(S|D)$.

(2) The isomorphism condition is satisfied.

All unexplained concepts are as in [3]. We shall not always state explicitly what the basis and the stable letters consist of when it should be obvious from the context.

Moreover, we shall not always define the mapping which is used to verify that the isomorphism condition holds. Note that the isomorphism condition is satisfied iff for all $v \in V$ and all pairs of words w, w', where w is a word on the set $\{A_i | i \in J_v\}$ and w' is the corresponding word on $\{B_i | i \in J_v\}$, then $w = 1$ in $(S|D)$ iff $w' = 1$ in $(S|D)$. We may also verify the isomorphism condition by proving that $A(v)$ and $B(v)$ have equivalent presentations, where A_i corresponds to B_i for $i \in J_v$.

We shall sometimes deviate a little from Britton's notation in order to avoid subsubscripts, and subsuperscripts. Instead of denoting the stable letters by p_v $(v \in V)$ or $\{p_v | v \in V\}$ we shall use the set P. Then p without subscript may be used to denote an arbitrary element of P. We shall often use the same letter to denote the group and its presentation. Thus $G_1 = (S_1 | D_1)$ means that G_1 may be used both to denote the group defined by the presentation $(S_1 | D_1)$ and the presentation $(S_1 | D_1)$ itself. We shall use $(G_1; S_2 | D_2)$ as an abbreviation for $(S_1 \cup S_2 | D_1, D_2)$ if $G_1 = (S_1 | D_1)$. $G_2 = (G_1; S_2 | D_2)$ is called an extension of the presentation G_1. Note that in most cases used in this paper G_1 is embedded in $(G_1; S_2 | D_2)$, but in some cases this is not true. We shall not explicitly define the homomorphism or isomorphism intended when the reader should have no difficulty in finding out the mapping we have in mind which will fit into the proof. Talking about the homomorphism (or isomorphism) of the group $G_1 = (S_1 | D_1)$ into $G_2 = (S_2 | D_2)$ when $S_1 \cap S_2 \neq \emptyset$, we often mean that the mapping

$$\varphi(s) = \begin{cases} s \text{ if } s \in S_1 \cap S_2, \\ 1 \text{ if } s \in S_1 \setminus S_2 \end{cases}$$

can be extended to a homomorphism of G_1 into G_2.

Definition 4. Let W be a word. We shall say that W is trivially (or freely) reduced iff W contains no subword of the form $s^{-1}s$ or ss^{-1}.

Definition 5. Let $(S^* | D^*)$ be a Britton extension of $(S|D)$ with stable letters P. A *pinch* is a word of one of the following forms:
(i) $p^{-1}Cp$, where C is a word on S, and $C \in A(p)$ and $p \in P$; or
(ii) pCp^{-1}, where C is a word on S, and $C \in B(p)$ and $p \in P$.

Definition 6. Given $(S^* | D^*)$ with basis $(S|D)$ and stable letters P, then a word on S^* is *p-reduced* (for a fixed $p \in P$) iff W contains no pinches of the forms $p^{-1}Cp$ or pCp^{-1} as subwords. Moreover, W is p-reduced iff W is p-reduced for all $p \in P$.

Boone [2, p. 57] defines the p-reduction of a word. We shall use p-reduction in a special case.

Definition 7. Let $(S^* | D^*)$ be a Britton extension of $(S|D)$ with stable letters P. Let $p \in P$ and suppose all relations in D^* involving p are of the form $p^{-1}Xp = X$. Let W and W' be words on S^*. Then W' is a *p-reduction* of W iff W contains a pinch $p^{-e}Cp^e$ $(e = \pm 1)$ and W' is the result of replacing $p^{-e}Cp^e$ by C.

Definition 8. A set X is called a *free set of generators in* G if the subgroup F generated by X is free on X.

Lemma 2. *Let G^* be a Britton extension of G with stable letters P. Let F be a free subgroup of G which is free on the set X. Let H be a subgroup of G such that $F \cap H = \{1\}$. Let $P_1 \subset P$ and suppose $F \cap A(p) = F \cap B(p) = \{1\}$ for all $p \in P_1$. Let F^* be the subgroup of G^* generated by $X \cup P_1$. Then F^* is a free group on the set $X \cup P_1$, and $F^* \cap H = \{1\}$. Moreover suppose that H' is a group* *generated by* $Y = \{h_i p_i g_i | 1 \leqslant i \leqslant 1\}$, *where* $h_i, g_i \in G$, $p_i \in P$, $p_i \notin P_1$ *and* $p_i \neq p_j$ *if* $i \neq j$ $(i = 1, 2, ..., l; j = 1, 2, ..., l)$, *then* H' *is a free group on* Y *and* $F^* \cap H' = \{1\}$.

Proof. Let w be a trivially reduced word on $P_1 \cup X$, and suppose $w = h$ in G^* for some $h \in H$. We shall prove that this implies $w \equiv 1$, which implies that F^* is free on the set $X \cup P_1$ and that $F^* \cap H = \{1\}$. Suppose that w is a word which involves at least one stable letter $p \in P_1$. Since $wh^{-1} = 1$ in G^*, and since p does not occur in h, we have that w contains a pinch of the form $p^{-\epsilon} C p^{\epsilon}$ $(\epsilon = \pm 1)$, where $p \in P_1$ and $C \in A(p) \cap F = \{1\}$ or $C \neq B(p) \cap F = \{1\}$. Hence $C = 1$ and $p^{-\epsilon} C p^{\epsilon} = 1$, which contradicts the fact that w is trivially reduced. Hence w involves no letter $p \in P_1$. That is, $w \in F$, since w is a word on X. But $F \cap H = \{1\}$ which shows that $w \equiv 1$ and $h = 1$ in G^*.

To prove the last part of the lemma, let w be trivially reduced word on $X \cup P_1$ and let u be a trivially reduced word on Y, and let $w = u$ in G^*. Then $w^{-1}u = 1$ in G^*, and exactly as before we can prove that w is a word on X and if u contains a pinch of the form $p_i^{-1} C p_i$ then u contains a subword of the form $g_i^{-1} p_i^{-1} h_i^{-1} h_i p_i g_i$, which contradicts the assumption that u is trivially reduced. Moreover, by a similar argument we can prove that u contains no pinch of the form $p_i C p_i^{-1}$. Hence $u \equiv 1$ and $w = 1$ in G^*. This proves the last part of the lemma.

§4. The Turing machine and the corresponding semigroups

The following discussion leading to Lemma 4 presupposes familiarity with Turing machines.

We shall use the convention that s_0 is the blank square.

Definition 9. T is a Turing machine, its alphabet is the set $\{s_0, s, ..., s_{M-1}\}$ of all s-letters occurring in its quadruples. If w is a positive word on the non-blank symbols $\{s_1, s_2, ..., s_{M-1}\}$, then T *accepts* w if there is a computation of T beginning with $q_1 w$. (We use "computation of T" in the sense of Davis [4, Definition 1.9, p. 7].)

Informally, we regard the machine T as being started in the state q_1 scanning the leftmost symbol in w, and T *accepts* w iff it eventually stops. We shall also call w an *input* to the Turing machine T in the case we are studying the (finite or infinite) sequence of instantaneous descriptions $q_1 w \to \alpha_2 \to \alpha_3 \to \cdots$. We shall say that a set of positive words is *recursively enumerable* iff the set of its Gödel numbers is recursively enumerable.

Lemma 3. *Let E be a recursively enumerable set of positive words on the letters $\{a_1, a_2, ..., a_m\}$. Then there exists a Turing machine Z_E with alphabet $S' = \{s_0, s_1, s_2, ..., s_{M-1}\}$ and states $Q' = \{q_1, q_2, ..., q_{N-2}\}$ such that*

(1) $\{a_1, a_2, ..., a_m\} \subset \{s_1, s_2, ..., s_{M-1}\}$;

(2) *if w is a positive word on the letters $\{a_1, a_2, ..., a_m\}$, then the Turing machine Z_E accepts w iff $w \in E$.*

Remark. Note that from now on E is a recursively enumerable set of words on $\{a_1, a_2, ..., a_m\}$, while in Definition 2, E was a recursively enumerable set of positive words on $\{u_1, u_2, ..., u_m\}$.

Proof of Lemma 3. Let ω be the set of non-negative integers. Numbers are represented by a sequence of symbols 1 on its tape. An input is numerical iff it consists of a sequence of 1's, i.e. non-negative integers. A subset A of ω is recursively enumerable (abbreviated r.e.) iff $A = \emptyset$ or A = range (f) for some recursive function f. A basic theorem on r.e. sets says that A is r.e. iff A = domain(ψ) for some partial recursive function ψ. Since Turing computable is the same as recursive, we have that A is r.e. iff there exists a Turing machine which accepts the numerical input n iff $n \in A$. Since a Turing machine can compute the Gödel number of a word w given as input, we obtain Lemma 3.

By Post's construction (see [6]), or better by Davis modification of Post's construction, we obtain a semigroup $T = T_E$ which corresponds to the r.e. set E. T is generated by $S \cup Q$, where $S = S' \cup \{s_M\}$ and $Q = Q' \cup \{q_0, q_{N-1}, q_N\}$. We shall write q for q_0.

A word Σ on the letters $S \cup Q$ is *special* if either $\Sigma \equiv q_0 \equiv q$ or $\Sigma \equiv wq_jw'$ for some $j \in \{0, 1, 2, ..., N\}$ and for some positive words w and w' on $S = \{s_0, s_1, ..., s_M\}$.

Lemma 4. *Let E be a recursively enumerable set of words on the letters* $\{a_1, a_2, ..., a_m\}$. *Then there exists a semigroup* $T = T_E$ *generated by the set* $S \cup Q$ ($S = \{s_0, s_1, ..., s_M\}$, $Q = \{q_0, q_1, ..., q_N\}$) *and defining relations*

$$\Sigma_i = \Gamma_i \ (i = 1, 2, ..., P),$$

where each Σ_i *and* Γ_i *is a special word, such that for each positive word w on the letters* $\{a_1, a_2, ..., a_m\}$ *we have that*

(*) $hq_1wh = q$ *in* T *iff* $w \in E$.

Moreover,

(**) *if* $W = q$ *in* T, *then* $W \equiv q$ *or* $W \equiv h\beta q_j \beta' h$, *where* $j \in \{0, 1, 2, ..., N\}$ *and* β *and* β' *are positive words on* $\{s_0, s_1, ..., s_{M-1}\}$.

§5. Properties of the Boone-Britton group

By Britton's modification of Boone's construction we obtain a group $G = G_E$ such that the following lemma holds.

Lemma 5. *Let E be a r.e. set of positive words on* $\{a_1, a_2, ..., a_m\}$. *Then there exists a finitely presented group* $G = G_E$ *such that*

(1) G_E *is generated by* $S \cup Q \cup \{r_1, r_2, ..., r_P, l_1, l_2, ..., l_P, t, k\}$.

(2) $\{a_1, a_2, ..., a_m\} \subset \{s_1, s_2, ..., s_M\}$.

(3) *If* Σ *is a special word then*

$$k^{-1}\Sigma^{-1}t\Sigma k = \Sigma^{-1}t\Sigma \text{ in } G_E \text{ iff } \Sigma = q \text{ in } T_E.$$

(4) *If w is a positive word on* $\{a_1, a_2, ..., a_m\}$ *then*

$$k^{-1}(hq_1wh)^{-1}t(hq_1wh)k = (hq_1wh)^{-1}t(hq_1wh) \text{ iff } w \in E.$$

It turns out that we need to know more about Boone's group. In order to prove Lemma 11, we start with the following lemma:

Lemma 6. *The subgroup of G generated by the set* $S \cup Q \cup \{t\}$ *is free on this set.*

Proof. The detail of the proof is left to the reader, since by using Lemma 2 the proof becomes easy, although somewhat lengthy since many details have to be checked.

We first prove that $S \cup \{p_0, ..., p_N\}$ is a free set of generators in G_3, by repeated application of Lemma 2. Then we prove that the group generated by $S \cup \{p_0, p_1, ..., p_N, z\}$ in H_2 is free on this set (see Britton [3, p. 24–25]). Hence $S \cup Q$ is a free set of generators in G_2. By application of Lemma 2 again, we complete the proof of Lemma 6.

Lemma 7. *The subgroup of G generated by the set* $S \cup Q \cup \{k\}$ *is free on this set.*

Proof. Let

$$G_1' = (G_2, k \mid k^{-1} x k = x,\ k^{-1} r_i k = r_i,\ i = 1, 2, ..., P).$$

Then we have that

$$G = (G_1, t \mid t^{-1} y t = y,\ t^{-1} l_i t = l_i,\ t^{-1} (qkq^{-1}) t =$$

$$= qkq^{-1},\ i = 1, 2, ..., P).$$

Hence G_1' is a Britton extension of G_2 with stable letter k and G is a Britton extension of G_1 with stable letter t. Hence Lemma 7 can be proved in the same way as Lemma 6. This completes the proof of Lemma 7.

By modification of Britton's proof we can prove:

Lemma 8. *Let Δ be a word on $s_0, s_1, ..., s_M, q, q_0, q_1, ..., q_N$ and let Δ be trivially reduced. Then $k^{-1}\Delta^{-1}t\Delta k = \Delta^{-1}t\Delta$ implies that Δ is a special word.*

Proof. First we shall show that Δ contains exactly one occurrence of a symbol from the alphabet $\{q_0, q_1, ..., q_N\}$. Exactly as in [3, eq. (4), p. 25], we can prove that

(4′) $$\Delta = N_0 z^{i_1} N_1 z^{i_2} \cdots N_{b-1} z^{i_b} N_b = LpzR \text{ in } H_2,$$

where $N_0, N_1, ..., N_b$ are words on $s_0, s_1, ..., s_M, p_0, p_1, ..., p_N$, and where L is a word on $l_1, l_2, ..., l_P, y$ and where R is a word on $r_1, r_2, ..., r_P, x$.

Let K be the group generated by $s_0, s_1, ..., s_M, p_0, p_1, ..., p_N$. Then $K \cap A(z) = 1$ and $K \cap B(z) = 1$ by Lemma 2 (see the proof of Lemma 6). Since (4′) implies

(4″) $$R^{-1}z^{-1}p^{-1}L^{-1}N_0 z^{i_1} N_1 z^{i_2} \cdots N_{b-1} z^{i_b} N_b = 1,$$

we have by Britton's Lemma that $p^{-1}L^{-1}N_0 \in A(z)$. Hence for some $C \in B(z)$ we have

(4‴) $$R^{-1}CN_1 z^{i_2} \cdots N_{b-1} z^{i_b} N_b = 1.$$

But since $K \cap A(z) = 1$ and $K \cap B(z) = 1$, since K, $A(z)$ and $B(z)$ are free groups, and Δ is freely reduced, we have that $b = 1$ and $i_1 = 1$. Hence we have that

(4⁗) $$\Delta = Fq_sG = Fp_szG = LpzR \text{ in } H_2,$$

where F and G are words on $s_0, s_1, ..., s_M$. It remains to prove that F and G are positive words. To prove this we change Lemma 8 of [3, p. 26] to read as follows:

Lemma 8′. *Let F and G be reduced words on $s_0, s_1, ..., s_M$ and let $\Delta \equiv Fq_sG$. If $v_1, ..., v_n$ and $e_1, ..., e_n$ and words L (in l_i, y), R (in r_i, x) exist such that* (6) *and* (7) *hold, then Δ is a positive word* *(and hence a special word), moreover, in the semigroup T, Δ can be transformed into q by a sequence of at most n elementary transformations.*

Proof. The proof is by induction on n as before. The first place where we have to change the proof is in part (s) of [3, p. 27]. Change the second paragraph of part (s) to read as follows:

"The reader is reminded that, in (8), F and H_i are trivially reduced words (possibly empty) in the s_b, and H_i is even a positive word; and Y is a word in y only. We shall prove that F has the form $F \equiv UH_{v_1}$ (i.e., the form UH_i), for some reduced word U. Now reduced words W, F', H_i' in the s_b certainly exist such that W and H_i' are positive and such that $F \equiv F'W$, $H_i \equiv H_i'W$, and suppose W has maximal length, so it is sufficient to prove that $H_i' \equiv 1$. Assume that $H_i' \not\equiv 1$."

The rest of part (s) may be left almost unchanged. We shall here quote some basic facts and add some new facts:

Since $\Delta \equiv Fq_s G \equiv UH_i q_{g_i} K_i V$, the word Δ can be transformed into the word $\Delta^* \equiv UF_i q_{s_i} G_i V$ by one elementary transformation, as before. By (6*) and (7*) of [3, p. 29], we obtain that $\Delta^* \equiv F^* q_{s_i} G^* \equiv UF_i q_{s_i} G_i V$ is a special word, and Δ^* can be transformed into q in at most $n-1$ steps. Hence U and V are positive words, and since H_i and K_i are positive words, we obtain that $UH_i q_{g_i} K_i V \equiv \Delta$ is a special word; moreover, Δ can be transformed into q in at most n steps. When $e_1 = -1$, only slight changes have to be made as before. This completes the proof of the revised Lemma 8′.

Hence our Lemma 8 is proved.

Remark. A shorter proof of our Lemma 8 may be possible by applying [2, Lemma 26, p. 74] and some other results in [2].

Lemma 9. *Let $\Delta_1, \Delta_2, \Delta_3, \Delta_4$ be reduced words on $s_0, s_1, \ldots, s_M, q, q_0, q_1, \ldots, q_N$. Suppose $k^{-1}\Delta_1^{-1} t\Delta_2 k = \Delta_3^{-1} t\Delta_4$, in the group G_E. Then $\Delta_1 \equiv \Delta_2 \equiv \Delta_3 \equiv \Delta_4$ and Δ_1 is a special word.*

Proof. Since the group obtained by adding the relation $k = 1$ to G is isomorphic to G_1, the group obtained from G by deleting k as a generator and deleting all relations containing k, we have that $k^{-1}\Delta_1^{-1}t\Delta_2 k = \Delta_3^{-1}t\Delta_4$ implies $\Delta_1^{-1}t\Delta_2 = \Delta_3^{-1}t\Delta_4$. Hence $\Delta_1 \equiv \Delta_3$ and $\Delta_2 \equiv \Delta_4$ since the letters in $\Delta_1^{-1}t\Delta_2$ and $\Delta_3^{-1}t\Delta_4$ form a free set of generators, by Lemma 7. In the same way, by adding the relation $t = 1$ to G which is isomorphic to the group G_1' obtained by deleting t from the generators and deleting all relations in the presentation of G which contain t, we obtain that $k^{-1}\Delta_1^{-1}t\Delta_2 k = \Delta_3^{-1}t\Delta_4$ and $t = 1$ implies $k^{-1}\Delta_1^{-1}\Delta_2 k = \Delta_3^{-1}\Delta_4$. Hence $\Delta_1 \equiv \Delta_2$ and $\Delta_3 \equiv \Delta_4$. Thus $\Delta_1 \equiv \Delta_2 \equiv \Delta_3 \equiv \Delta_4$ and $k^{-1}\Delta_1^{-1}t\Delta_1 k = \Delta_1^{-1}t\Delta_1$, hence Δ_1 is a special word by Lemma 8. This proves Lemma 9.

Lemma 10. *Let H_1 be the subgroup of $G = G_E$ generated by*

(1) $S \cup Q \cup \{t, k\}$.

Then H has the presentation

(2) $(S \cup Q \cup \{t, k\} \mid k^{-1}\Sigma^{-1}t\Sigma k = \Sigma^{-1}t\Sigma,$

where Σ is a special word and $\Sigma = q$ in T).

Proof. Let H_1' be a group having the presentation above. Let w be a word on the letters (1). Suppose $w = 1$ in H_1', then $w = 1$ in H_1 by [3, Lemma 7, p. 23]. Suppose $w = 1$ in H_1 and $w \neq 1$ in H_1'. We shall prove that this leads to a contradiction. We may choose w to have as few occurrences of k as possible. From Lemmas 6 and 7 it follows that w must contain at least two occurrences of t and two occurrences of k.

Since $w = 1$ in H_1 we also have $w = 1$ in G. By Britton's Lemma, we have that w must contain a pinch of the form $k^{-\epsilon}Ck^{\epsilon}$ ($\epsilon = \pm 1$), where $C \in A(k) = B(k)$. Hence $k^{-\epsilon}Ck^{\epsilon} = C$ in H_1. Suppose $k^{-\epsilon}Ck^{\epsilon} = C$ is true in H_1'. Then $w = w'$ in H_1', where w' is obtained from w by replacing $k^{-\epsilon}Ck^{\epsilon}$ in w by C. Then $w' = 1$ in H_1 and $w' = w \neq 1$ in H_1'. Since w' contains a smaller number of occurrences of k than w, this contradicts the fact that w is minimal. Hence $k^{-1}Ck = C$ in H_1 but $k^{-1}Ck \neq C$ in H_1'.

Let $w'' = k^{-1}CkC^{-1}$. Then $w'' = 1$ in H_1 and $w'' \neq 1$ in H_1'; moreover, w'' contains just two occurrences of k. The group G of Britton (see [3, p. 28]) is obtained from G_2 by a Britton extension as follows:

$$G_1 = (G_2; t \mid t^{-1} l_i t = l_i,\ t^{-1} y t = y,\ i = 1, 2, ..., P),$$

$$G = (G_1; k \mid k^{-1} r_i k = r_i,\ k^{-1} x k = x,\ k^{-1}(q^{-1} t q) k = q^{-1} t q,$$

$$i = 1, 2, ..., P).$$

Here t is first added as stable letter and then k is added as stable letter in the second extension. But we may as well obtain G from G_2 by first adding k as stable letter and then t in the next extension:

$$G_1' = (G_2; k \mid k^{-1} r_i k = r_i,\ k^{-1} x k = x,\ i = 1, 2, ..., P),$$

$$G \ = (G_1; t \mid t^{-1} l_i t = l_i,\ t^{-1} y t = y,\ t^{-1}(q k q^{-1}) t = q k q^{-1},$$

$$i = 1, 2, ..., P).$$

Among the words w on the letters $S \cup Q \cup \{t, q\}$ choose a w such that

(1) $w = 1$ in H_1;
(2) $w \neq 1$ in H_1';
(3) w contains just two occurrences of k;
(4) w contains as few occurrences of t as possible.

We can now prove that w contains just two occurrences of t by using the fact that G is a Britton extension of G_1'. The proof is analogous to the proof above. Hence w contains no more than two occurrences of k and two occurrences of t. By using Britton's lemma and Lemma 9 it is easy to prove that $w = 1$ in H_1' also. Hence we have a contradiction which proves Lemma 10.

Lemma 11. *Let* $k_0 = hkh^{-1}$ *and let* $t_0 = q_1^{-1} h^{-1} t h q_1$. *Then the subgroup of* G *generated by* $a_1, a_2, ..., a_m, t_0, k_0$ *has the presentation:*

$$(a_1, a_2, \ldots, a_m, t_0, k_0 \mid k_0^{-1}(w^{-1} t_0 w) k_0 = w^{-1} t_0 w,$$

$$w \text{ a positive word on } \{a_1, \ldots, a_m\} \text{ and } w \in E).$$

Proof. $hq_1wh = q$ in $T = T_E$ by Lemma 4, and we obtain Lemma 11 from Lemma 10, since $k_0^{-1}w^{-1}t_0wk_0 = w^{-1}t_0w$ iff

$$k^{-1}(hq_1wh)^{-1}t(hq_1wh)k = (hq_1wh)^{-1}t(hq_1wh).$$

Note that our Lemma 11 corresponds to [5, Lemma 5.1, p. 473] and to [9, Lemma 18, p. 335].

§6. Completion of the proof of Lemma 1

Given a word w on the letters $\{a_1, a_2, \ldots, a_m\}$, let w_u be the result of replacing all occurrences of a_i in w by u_i for all $i = 1, 2, \ldots, m$ simultaneously. Moreover w_b is defined to be a word on the letters $\{b_1, b_2, \ldots, b_m\}$ in the same way. (Example: $(a_1a_2a_3^{-1})_u = u_1u_2u_3^{-1}$, and $(a_1a_2a_3^{-1})_b = b_1b_2b_3^{-1}$.) Let

$$R = (u_1, u_2, \ldots, u_m \mid w_u = 1, w \in E).$$

Let G_4 be the free product of $G = G_E$ and R, i.e.,

$$G_4 = G * R.$$

We shall now consider a sequence of extensions of G_4 and prove that we have a sequence of Britton extensions:

$$G_5 = (G_4; b_1, b_2, \ldots, b_m \mid b_i^{-1}u_jb_i = u_j,\ b_i^{-1}a_jb_i = a_j,$$

$$b_i^{-1}k_0b_i = k_0u_i^{-1}, \text{ for all } i, j = 1, 2, \ldots, m),$$

$$G_6 = (G_5; d \mid d^{-1}k_0d = k_0,\ d^{-1}a_ib_id = a_i, \text{ for all } i = 1, 2, \ldots, m),$$

$$G_7 = (G_6; p \mid p^{-1}t_0p = t_0d,\ p^{-1}k_0p = k_0,\ p^{-1}a_ip = a_i,$$

$$\text{for all } i = 1, 2, \ldots, m).$$

Lemma 12. *The subgroups* $\langle a_1, \ldots, a_m, k_0 \rangle_4$ *and* $\langle a_1, \ldots, a_m, t_0 \rangle_4$ *of* G_4 *are free on the displayed generating sets.*

Remark. The subscript 4 indicates that the subgroups are computed in G_4, i.e., only the relations of G_4 are used.

Proof of Lemma 12. The Lemma is an immediate consequence of Lemma 11.

Lemma 13. G_5 *is a Britton extension of* G_4.

Proof. By Lemma 12, we have

$$A(b_i) = \langle u_1, \ldots, u_m, a_1, \ldots, a_m, k_0 \rangle_4$$

$$= \langle u_1, \ldots, u_m, a_1, \ldots, a_m \rangle_4 * \langle k_0 \rangle_4 .$$

Hence the mapping $\varphi_i(u_j) = u_j$, $\varphi_i(a_j) = a_j$ and $\varphi_i(k_0) = k_0 u_i^{-1}$ may be extended to a homomorphism of $A(b_i)$ onto $B(b_i)$. Since $B(b_i) = A(b_i)$ and the map $\psi_i(u_j)$ given by $\psi_i(u_j) = u_j$, $\varphi_i(a_j) = a_j$ and $\psi_i(k_0) = k_0 u_i$ can be extended to a homomorphism of $B(b_i)$ into $A(b_i)$, and since the extension of ψ_i is the inverse of the extension φ_i we have that φ_i is an isomorphism. This shows that the isomorphism condition holds and Lemma 13 is proved.

Lemma 14. G_6 *is a Britton extension of* G_5.

Proof. Let $G_5' = (G_5 \mid b_i = u_i = 1)$. Then G_5' is isomorphic to G_4. Since $\langle a_1, a_2, \ldots, a_m, k_0 \rangle_4$ is a free group on the displayed letters, $\langle a_1, a_2. \ldots, a_m, k_0 \rangle$ and $\langle a_1 b_1, a_2 b_2, \ldots, a_m b_m, k_0 \rangle$ are also free groups on the displayed letters in G_5' and then also in G_5. This makes it easy to prove that the isomorphism condition holds. The rest of the proof of Lemma 14 is left to the reader.

Definition 10. Let G_4', G_5', G_6', G_7' be the result of deleting all the relations of the form $w_u = 1$, $w \in E$, in the presentations of G_4, G_5, G_6, G_7, respectively.

Lemma 15. *Let the presentation of H be an extension of the presentation of* G'_6. *Let w be a word on the letters* $a_1, a_2, ..., a_m$. *Then* $w_u = 1$ *in H iff* $k_0^{-1} w_b k_0 = w_b$ *in H.*

Proof. Since $b_i^{-1} k_0 b_i = k_0 u_i^{-1}$ is a defining relation in G'_5, and hence also in G'_6 and H, we have that $k_0^{-1} b_i k_0 = b_i u_i$. Since b_i and u_j commute, we also have $k_0^{-1} w_b k_0 = w_b w_u$. Hence if $w_u = 1$ in H, then $k_0^{-1} w_0 k_0 = w_b$ in H. Moreover, if $k_0^{-1} w_b k_0 = w_b$, then $w_b = w_b w_u$, hence $w_u = 1$. This proves Lemma 15.

Lemma 16. *Let the presentation of H be an extension of the presentation of* G'_6. *Let w be a word on the letters* $a_1, a_2, ..., a_m$, *and suppose* $k_0^{-1} w^{-1} t_0 w k_0 = w^{-1} t_0 w$ *in H. Then* $k_0^{-1} w^{-1} t_0 d w k_0 = w^{-1} t_0 d w$ *in H if* $w_u = 1$ *in H.*

Proof. Since a_i and b_j commute in H, and since $d a_i = a_i b_i d$ in H, we have that

(8) $$w^{-1} t_0 d w = w^{-1} t_0 w w_b d.$$

Here k_0 commutes with d, and we have supposed that k_0 commutes with $w^{-1} t_0 w$. Hence

$$
\begin{array}{lcl}
k_0^{-1} w^{-1} t_0 d w k_0 & = & w^{-1} t_0 d w, \\
 & \Updownarrow & \text{by (8)} \\
k_0^{-1} (w^{-1} t_0 w) w_b d k_0 & = & w^{-1} t_0 w w_b d, \\
 & \Updownarrow & \text{by the hypothesis} \\
(w^{-1} t_0 w) k_0^{-1} w_b k_0 d & = & (w^{-1} t_0 w) w_b d, \\
 & \Updownarrow & \text{by cancellation} \\
k_0^{-1} w_b k_0 & = & w_b, \\
 & \Updownarrow & \text{by Lemma 15} \\
w_u & = & 1.
\end{array}
$$

This proves Lemma 16.

Lemma 17. *G_7 is a Britton extension of G_6.*

Proof. The group $A(p)$ has the presentation

(9) $(a_1, a_2, \ldots, a_m, t_0, k_0 \mid k_0^{-1} w^{-1} t_0 w k_0 = w^{-1} t_0 w,\ w \in E)$,

by Lemma 11, since G is embedded in G_6.

In order to prove that the isomorphism condition holds, it is sufficient to prove that the group $B(p) = \langle a_1, a_2, \ldots, a_m, t_d, k_0 \rangle_6$ (where $t_d = t_0 d$) has the presentation

$$B' = (a_1, a_2, \ldots, a_m, t_d, k_0 \mid k_0^{-1} w^{-1} t_d w k_0 = w^{-1} t_d w,\ w \in E).$$

Let $w \in E$. Then $k_0^{-1} w^{-1} t_0 w k_0 = w^{-1} t_0 w$ in G_6 by Lemma 11 and $w_b = 1$ in G_6 by definition. Since the presentation of G_6 is an extension of the presentation of G_6', we have by Lemma 16 that $k_0^{-1} w^{-1} t_d w k_0 = w^{-1} t_d w$ in G_6 and hence also in $B(p)$. Let W be a word on the generators of B'. Suppose $W = 1$ in B'. Then $W = 1$ in $B(p)$ since $B(p)$ satisfies the defining relation of B'. Suppose $W = 1$ in $B(p)$. Let

$$G_6'' = (G_6 \mid u_i = 1, b_i = 1, d = 1, t_d = t_0).$$

Then G_6'' is a homomorphic image of G_6, and G'' is isomorphic to G. Hence $W = 1$ in G_6''. Let W' be the result of replacing t_d by t_0 in W. Then $W' = 1$ in G_6'', but W' is a word on the generators of G and G is isomorphic to G_6''. Hence $W' = 1$ in G and so $W' = 1$ in $A(p)$. But $A(p)$ and B' are isomorphic by definition. Hence $W = 1$ in B'. This proves Lemma 17.

Lemma 18. *G_7' is isomorphic to G_7, G_7' is finitely presented and R is embedded in G_7'.*

Proof. Recall that G_7' is the result of deleting all the relations of the form $w_u = 1$ $(w \in E)$ in G_7. We shall prove that all the relations

$w_u = 1$ $(w \in E)$ can be proved in G'_7. Let $w \in E$. Then

(10) $$k_0^{-1} w^{-1} t_0 w k_0 = w^{-1} t_0 w_0.$$

Hence

(11) $$p^{-1}(k_0^{-1} w^{-1} t_0 w k_0) p = p^{-1}(w^{-1} t_0 w_0) p,$$

(12) $$k_0^{-1} w^{-1} t_0 d w k_0 = w^{-1} t_0 d w,$$

since p commutes with all letters in (10) except t_0 and $p^{-1} t_0 p = t_0 d$.

The presentation of G'_7 is an extension of the presentation of G'_6. Hence by (10), (12) and Lemma 16 we have $w_u = 1$ in G'_7, as desired. Hence G'_7 and G_7 are isomorphic and G'_7 is obviously finitely presented. Moreover, since G_7 is obtained from G_4 by a sequence of Britton extensions, we have that R is embedded in G_7. Hence R is embedded in G'_7. This completes the proof of Lemma 18.

Since R was an arbitrary recursively presented group, this also proves Lemma 1.

Acknowledgements

In July 1970, I wrote an outline of this paper. I am very grateful to Professor W.W. Boone, Dr. C.F. Miller, III, and Dr. D.J. Collins for proposing some simplifications and for encouraging me to write down my work and get it published. I am especially grateful to Professor W.W. Boone; without his encouragement this paper would hardly have been written. I would also like to thank Professor J.J. Rotman. He has given me an early draft of chapter 12 of the second edition of his book on group theory. I have benefited from his exposition in certain places.

References

[1] W.W. Boone, The word problem, Ann. Math. 70 (1959) 207–265.
[2] W.W. Boone, Word problems and recursively enumerable degrees of unsolvability. A sequel on finitely presented groups, Ann. Math. 84 (1966) 49–84.
[3] J.L. Britton, The word problem, Ann. Math. 77 (1963) 16–32.
[4] M. Davis, Computability and unsolvability (McGraw-Hill, New York, 1958).
[5] G. Higman, Subgroups of finitely presented groups, Proc. Roy. Soc. London A 262 (1961) 455–475.
[6] E.L. Post, Recursive unsolvability of a problem of Thue, J. Symbolic Logic 12 (1947) 1–11.
[7] H. Rogers, Jr., Theory of recursive functions and effective computability (McGraw-Hill, New York, 1967),
[8] J.J. Rotman, The theory of groups: An introduction (Allyn and Bacon, 1965; second ed. 1973).
[9] J.R. Shoenfield, Mathematical logic (Addison-Wesley, Reading, Mass., 1967).

BURNSIDE GROUPS OF ODD EXPONENT AND IRREDUCIBLE SYSTEMS OF GROUP IDENTITIES

S.I. ADJAN

Steklov Institute of Mathematics, Moscow

In 1902 Burnside [1] proposed the following problem: "Is every group finite when it has a finite number of generators and satisfies the identity relation

(1) $x^n = 1$?"

This question was answered in the negative in [4]. It is shown in [4] that, for arbitrary $m > 1$ and arbitrary odd $n \geqslant 4381$, there is an infinite group $\Gamma(m, n)$ on m generators satisfying (1). Prior to this, affirmative answers to the problem proposed by Burnside were obtained for $n = 3$ [1], $n = 4$ [2], and $n = 6$ [3].

In order to construct $\Gamma(m, n)$ in [4], a certain classification of periodic words on the group alphabet

(2) $a_1, a_2, \ldots, a_m, a_1^{-1}, a_2^{-1}, \ldots, a_m^{-1}$

is introduced and a theory for transformations of words, which is consistent with (1), is constructed with a fixed odd $n \geqslant 4381$. Affirmative answers to the word and conjugacy problems for free Burnside groups with odd $n \geqslant 4381$ were obtained by the same writers on the basis of this theory in the sequels [5] and [6]. The existence of infinite groups in which all abelian subgroups are finite is proved in [6].

By using a somewhat altered version of the theory constructed in [4], the writer of the present paper proved in [7] and [9] that the infinite system of group identities.

(3) $$(x^{kn}y^{kn}x^{-kn}y^{-kn})^n = 1,$$

where $n \geqslant 4381$ is a fixed odd number and k ranges over all the prime numbers, is irreducible, that is, no identity in this system can be derived from the others in this system.

From this it can be deduced that there is a group which is defined by two generators and a recursively enumerable set of identity relations of form (3) and has an unsolvable word problem.

The present paper sets forth the fundamentals of the theory of word transformations constructed in [4]* and the ideas of the proofs of the results obtained in [4,5,6 and 7].

In addition to the odd number $n \geqslant 4381$, we also set by definition some auxiliary numerical parameters which will be used in our determinations:

$$p = 18, \quad p_1 = 37, \quad g_1 = 2p + 2p_1 + 13,$$

$$g_2 = g_1 + 2p_1, g_3 = g_2 + 2p_1, \quad g = g_3 + 2p_1 + 1.$$

We consider words on (2). The empty word is denoted by 1. Then length of the word X is denoted by $\partial(X)$. The word that is the inverse of X is denoted by X^{-1}. If A is a nonempty word and t a positive integer, A^t denotes the result of writing t copies of A one after another. For example, A^3 is AAA.

Letter-for-letter identity of two words X and Y is denoted by $X \mathrel{\overset{\circ}{=}} Y$. We say that word E occurs in word X if there are words P and Q such that $X \mathrel{\overset{\circ}{=}} PEQ$. In this case, if P (resp. Q) is empty, we say that E is an end (resp. start) of X. The same word E can occur in different places in a given word X. In order to distinguish different occurrences of a word E in a given word X, we shall use the symbol $*$, placing it before the first letter and after the last letter of the occurrence of E that is being considered. If $X = PEQ$, where X, P, E and Q are words on (2), then the word $P * E * Q$ shows us a particular occurrence of E in X. A word of the form

* The condition $\partial(X) \geqslant 2p\partial(A)$ is omitted from the statements of Lemmata 36 and 41.1 of Chapter I in [4].

$P * E * Q$ is called an occurrence in PEQ and E is called the base of this occurrence. Occurrences are denoted by the letters V and W, with various indices or without them.

We denote equality by definition by the symbol $\rightleftharpoons$. Let $V \rightleftharpoons P * E * Q$ and $W \rightleftharpoons R * D * S$ be two occurrences in the same word, that is, $PEQ \overset{\circ}{=} RDS$. If $\partial(P) < \partial(R)$ and $\partial(Q) > \partial(S)$, we say that V lies to the left of W and we write $V < W$. If $\partial(P) \leqslant \partial(R)$ and $\partial(Q) \leqslant \partial(S)$, we say that W is contained in V. If $P \overset{\circ}{=} R$ and $\partial(Q) \geqslant \partial(S)$, we say that W is a start of V or that V starts with W. If $\partial(P) \leqslant \partial(R)$ and $Q \overset{\circ}{=} S$, we say that W is an end of V or that V ends with W.

Let $V \rightleftharpoons P * E * Q$ and $W \rightleftharpoons R * E * S$ be respective occurrences of the same word E in words X and Y. If an occurrence $V_1 \rightleftharpoons P_1 * C * Q_1$ is contained in V, then there are words A and B such that $P_1 \overset{\circ}{=} PA$, $E \overset{\circ}{=} ACB$ and $Q_1 \overset{\circ}{=} BQ$. Then the occurrence $RA * C * BS$ in Y is denoted by $\phi(V_1, V; W)$.

A word X on (2) is called uncancellable if no words of the form $a_i a_i^{-1}$ or $a_i^{-1} a_i$ occur in it. Uncancellable words of the form $A^t A_1$, where A_1 is a start of A and $t > 1$, are called periodic words with period A. An occurrence $P * E * Q$ in a periodic word $A^t A_1$ is called an occurrence interior relative to the period A, if $\partial(P) \geqslant 8\partial(A)$ and $\partial(Q) \geqslant 8\partial(A)$. Two occurrences $P * E * Q$ and $R * D * S$ in a periodic word $A^t A_1$ are said to correspond in phase relative to the period A if $\partial(R) - \partial(P) = \partial(Q) - \partial(S) = k\partial(A)$, for some integer k. This means that $R * D * S$ is obtained from $P * E * Q$ as a result of a shift to the left or right by an integral number of periods A. In this case we have $E \overset{\circ}{=} D$.

The following fundamental concepts of the theory under consideration are introduced by means of joint induction on a natural parameter α, called the rank.

(1) Kernel of rank α. Reduced and minimized words of rank α. By Π_α and M_α we mean the set of all reduced words of rank α and the set of all minimized words of rank α, respectively. If $X \in M_\alpha$ and a word E occurs in X, then $E \in \Pi_\alpha$.

(2) The relation $X \overset{\alpha}{\simeq} Y$, called an equivalence of rank α, is defined for X and Y both in Π_α. It is reflexive, symmetric and transitive. The function $W = f_\alpha(V, Y)$ is uniquely defined for an arbitra-

ry kernel V of rank α of a reduced word X of rank α if $X \overset{\alpha}{\sim} Y$ and $Y \in \Pi_\alpha$, and its value W is a kernel of rank α of the word Y. If $W = f_\alpha(V, Y)$, then $V = f_\alpha(W, X)$.

(3) The operation $[X, Y]_\alpha = Z$, called a *close-up* of rank α, is defined for arbitrary words X and Y both in Π_α, and its value Z also is in Π_α.

We first define these concepts for $\alpha = 0$.

If X is a word on (2), every occurrence $P * E * Q$ in X, where E is one of the letters in (2), is called a kernel of rank 0 of the word X. By a reduced word of rank 0 we mean any uncancellable word on (2). It is called a minimized word of rank 0, that is, $M_0 = \Pi_0$.

Two words X and Y in Π_0 are called equivalent in rank 0 if $X \overset{\circ}{=} Y$. We write $X \overset{\circ}{\sim} Y$ in this case. If $X \overset{\circ}{\sim} Y$ and V is a kernel of rank 0 of X, then by definition we set $f_0(V, Y) = V$.

A close-up of rank 0 for $X, Y \in \Pi_0$ is defined in the following manner: $[X, Y]_0 = Z$ if and only if there are words X_1, Y_1 and T such that $X \overset{\circ}{=} X_1 T$, $Y \overset{\circ}{=} T^{-1} Y_1$, $X_1 Y_1 \overset{\circ}{=} Z$ and $Z \in \Pi_0$.

Now let us assume that $\alpha > 0$ and that all the concepts listed in paragraphs (1) – (3) are defined for $\alpha - 1$. Then we define all of these concepts for α. Together with these concepts, we also define a number of other concepts for $\alpha - 1$ and α which are no less important for the theory under consideration than the ones given in paragraphs (1)–(3). We do not include them in the initial listing only because they are either derived from the enumerated concepts or they are meaningful only for $\alpha > 0$.

In defining these or other concepts of our theory, we shall at the same time cite some of their characteristic properties in order, as far as possible, to make the meaning of these properties understandable for the reader. Some of these properties follow easily from the definition, while the proofs of others require painstaking analysis of the concepts introduced.

An occurrence W in $X \in \Pi_{\alpha-1}$ is called a regular occurrence of rank $\alpha - 1$ if it starts and ends with supporting kernels of rank $\alpha - 1$ of X. Let us assume that $X, Y \in \Pi_{\alpha-1}$, $X \overset{\alpha-1}{\sim} Y$ and W is a regular occurrence of rank $\alpha - 1$ in X. If a kernel V_1 of rank $\alpha - 1$ is a start of W and a kernel V_2 of rank $\alpha - 1$ is an end of W, then $f_{\alpha-1}(W, Y)$ denotes an occurrence in Y for which $f_{\alpha-1}(V_1, Y)$ is a start and $f_{\alpha-1}(V_2, Y)$ is an end. From $W_1 = f_{\alpha-1}(W, Y)$ we obtain $W = f_{\alpha-1}(W_1, X)$ since this is true for kernels of rank $\alpha - 1$ by the induction hypothesis.

Let us assume that $V \doteq P * E * Q$ and $W \doteq R * C * S$ are regular occurrences of rank $\alpha - 1$ in X, $Y \in \Pi_{\alpha-1}$, respectively. We say that W and V are mutually normalized in rank $\alpha - 1$ if the following conditions are satisfied: for arbitrary $Z \in \Pi_{\alpha-1}$, where $Z \overset{\alpha-1}{\sim} X$ and $f_{\alpha-1}(V, Z) = P_1 * D * Q_1$, there is a $Z_1 \in \Pi_{\alpha-1}$ such that $Z_1 \overset{\alpha-1}{\sim} Y$ and $f_{\alpha-1}(W, Z_1) = R_1 * D * S_1$, for some R_1 and S_1; conversely, for any such word Z_1 there is a $Z \in \Pi_{\alpha-1}$ such that $Z \overset{\alpha-1}{\sim} (V, Z) = P_1 * D * Q_1$, for some P_1 and Q_1.

It is evident from the definitions given that two occurrences $P * E * Q$ and $R * D * S$ respectively in X, $Y \in \Pi_\alpha$ are mutually normalized in rank 0 if and only if $E \overset{\circ}{=} D$.

The relation of being mutually normalized in rank $\alpha - 1$ is reflexive, symmetric and transitive. If $X \overset{\alpha-1}{\sim} Y$ and W is a regular occurrence of rank $\alpha - 1$ in X, than W and the occurrence $f_{\alpha-1}(W, Y)$ are mutually normalized in rank $\alpha - 1$.

A periodic word $X \doteq A^t A_1$ with a period A is called a periodic word of rank α if $X \in \Pi_{\alpha-1}$, $\partial(X) \geqslant 2p\partial(A)$, and there is a kernel of rank $\alpha - 1$ of X interior relative to the period A. It follows from properties of the concepts introduced by us that if two occurrences V and W in a periodic word X of rank α that are interior relative to the period A correspond to each other in phase relative to the period A and if one of them is a kernel (resp. regular occurrence) of rank $\alpha - 1$, then the other is also a kernel (resp. regular occurrence) of rank $\alpha - 1$, moreover, in this case V and W are mutually normalized in rank $\alpha - 1$. This assertion is obvious for $\alpha = 1$.

A word $Y \in \Pi_{\alpha-1}$ is called an integral word of rank α if it is equivalent in rank $\alpha - 1$ to some periodic word X of rank α. In this case X is called an original periodic word for Y. In particular, every integral word of rank 1 is letter-by-letter equal to its original periodic word of rank 1. In the general case an integral word need not be periodic. In virtue of the reflexivity of the relation $X \overset{\alpha-1}{\sim} Y$, every periodic word of rank α is an integral word of rank α.

Every kernel V of rank $\alpha - 1$ of a periodic word X of rank α that is interior relative to a period A is called a supporting kernel: A

kernel V of rank $\alpha - 1$ of an integral word Y of rank α is called a supporting kernel if for an original periodic word X of rank α the kernel $f_{\alpha-1}(V, X)$ is a supporting kernel of X. Supporting kernels V and W of an integral word Y of rank α are said to correspond in phase relative to an original periodic X if and only if the supporting kernels $f_{\alpha-1}(V, X)$ and $f_{\alpha-1}(W, X)$ of X correspond to each other in phase relative to a period A of X.

If a regular occurrence $P * E * Q$ of rank $\alpha - 1$ in an integral word Y of rank α starts and ends with supporting kernels and contains at least three supporting kernels which correspond to each other in phase relative to an original periodic word X, then the word E is called a semi-integral word of rank α. Then, $P * E * Q$ is called a generating occurrence of the semi-integral word E of rank α. By the number of segments of a semi-integral word E of rank α we mean the maximal number of supporting kernels that are contained in its generating occurrence and which correspond to each other in phase relative to the original periodic word X.

We say that A is a minimal period in rank $\alpha - 1$ of a periodic word $X \doteqdot A^t A_1$ of rank α if, for any generating occurrence $W \doteqdot P * E * Q$ in X and for any generating occurrence $W_1 \doteqdot R * E * S$ that is mutually normalized with it and is in an arbitrary integral word Y_1 of rank α that has an original periodic word X_1 with a period B, the following condition is fulfilled: if two supporting kernels V_1 and V_2 of Y_1 are contained in W_1 and correspond to each other relative to X_1, then the occurrences $\phi(V_1, W_1; W)$ and $\phi(V_2, W_1; W)$ are supporting kernels of X and correspond to each other in phase relative to A.

A periodic word X with a period A minimal in rank $\alpha - 1$ is called an elementary periodic word of rank α if for every occurrence $W \doteqdot P * E * Q$ in X that is interior relative to A, where E is a semi-integral word of rank α generated by a generating occurrence $W_1 \doteqdot R * E * S$ in an integral word Y_1 of rank α with an original periodic word X_1 which has a period B minimal in rank $\alpha - 1$ and contains at least p segments, the following condition is fulfilled: if W and W_1 are mutually normalized in rank $\alpha - 1$, then any two supporting kernels V_1 and V_2 of Y_1 that are contained in W_1 correspond to each other in phase relative to X_1 if and only if the occurrences $\phi(V_1, W_1; W)$ and $\phi(V_2, W_1; W)$ are supporting kernels of X and correspond to each other in phase relative to A.

If X is an elementary periodic word of rank α, then every integral word Y of rank α obtained with original word X will be called an integral elementary word of rank α, and every semi-integral word E of rank α generated by an occurrence $P * E * Q$ in Y will be called an elementary word of rank α generated by $P * E * Q$. If the number of segments of elementary word E is not less than the natural number l, we say that E is an elementary l-power of rank α.

Let $W_1 \rightleftharpoons R * E * S$ be an occurrence in a word $Z \in \Pi_{\alpha-1}$ of an elementary word E of rank α generated by an occurrence $W \rightleftharpoons P * E * Q$ in an integral word Y of rank α. If V_1 and V_2 are phase-correspondent supporting kernels of generating occurrence W, then the occurrences $\phi(V_1, W; W_1)$ and $\phi(V_2, W; W_1)$ in Z are called supporting kernels that correspond in phase relative to W. If W_1 and W are mutually normalized in rank $\alpha-1$, then W_1 is called a normalized occurrence of the elementary word E in the word Z.

If W_1 is normalized, then every one of its supporting kernels is a kernel of rank $\alpha-1$ of Z. Generally speaking, different generating occurrences of the same semi-integral word E of rank α may have different phase-correspondences of their supporting kernels. However, if E is an elementary p-power of rank α, then any two of its generating occurrences W and W' are mutually normalized in rank $\alpha-1$ and the phase-correspondence of their supporting kernels are transformed from one to another by the function $\phi(V, W; W')$. Therefore, the question of whether an occurrence of an elementary p-power E of rank α in $Z \in \Pi_{\alpha-1}$ is normalized, the phase-correspondence of its supporting kernels and, consequently, also the number of its segments, are all independent of the choice of W.

If $R \overset{\circ}{=} R_1 v$ and $W_2 \rightleftharpoons R_1 * vE * S$ is a normalized occurrence of the elementary word vE of rank α, then W_2 is called a normalized continuation of W_1 to the left. Moreover, if the word v is non-empty, we say W_1 is continuable to the left in normalized fashion.

We also define in a similar fashion the normalized continuation to the right $R * Eu * S$ and the normalized continuation in both directions $R_1 * vEu * S_1$. A normalized occurrence W_1 of an elementary word E is called maximal if it is not continuable in normalized fashion either to the left or to the right. It is obvious that for every normalized occurrence W_1 of an elementary word of rank α in $Z \in \Pi_{\alpha-1}$ its maximal normalized extension is uniquely defined. Two normalized occurrences W_1 and W_2 of the respective elementary words E_1 and E_2 of rank α in the same word Z are called compatible if there is a common normalized continuation for both of these occurrences.

The compatibility relation is symmetric and transitive. Two compatible occurrences of elementary words have the same maximal normalized continuation. If two normalized occurrences of elementary p-power of rank α in $Z \in \Pi_{\alpha-1}$ are not compatible, then neither is contained in the other, i.e., one lies to the left of the other.

Every normalized occurrence $W \rightleftharpoons R * E * S$ of an elementary word E of rank α in $Z \in \Pi_{\alpha-1}$ is a regular occurrence of rank $\alpha-1$. Moreover, if $Z \overset{\alpha-1}{\approx} Z_1$ and $Z_1 \in \Pi_{\alpha-1}$, then the occurrence $f_{\alpha-1}(W, Z_1)$ in Z_1 (cf. par. (2)) is a normalized occurrence of some elementary word of rank α which contains exactly the same number of segments as E.

If $P * E * Q$ is any occurrence in a word $Z \in \Pi_{\alpha-1}$ of an elementary word E of rank α that contains $l \geqslant p$ segments and $E \overset{\circ}{=} uE_1v$, where $Pu * E_1 * vQ$ starts with the supporting kernel that is seventh from the left and corresponds in phase to the starting supporting kernel of $P * E * Q$ and ends with the supporting kernel that is seventh from the right and corresponds in phase to the ending supporting kernel of $P * E * Q$, then $Pu * E_1 * vQ$ is a normalized occurrence of an elementary $(l-12)$-power of rank α.

Let $W \rightleftharpoons P * B^tB_1 * Q$ be a normalized occurrence of an elementary p-power B^tB_1 of rank α with period B in a word $X \in M_{\alpha-1}$, where $B \overset{\circ}{=} B_1B_2$. If there are words $Y \in \Pi_{\alpha-1}, D, P_1$, Q_1, T_1 and T_2 such that

$$(B^{-1})^{n-t-1} B_2^{-1} \overset{\alpha}{\sim}{}^{1} Y,\ Y \mathrel{\overline{\underline{\circ}}} T_1^{-1} D T_2^{-1},\ P \mathrel{\overline{\underline{\circ}}} P_1 T_1,$$
$$Q \mathrel{\overline{\underline{\circ}}} T_2 Q_1, P_1 D Q_1 \in \Pi_{\alpha-1},$$

then the transition

(4) $\qquad PEQ \rightarrow P_1 D Q_1$

is called a simple reversal of the occurrence W. Moreover, if the maximal normalized continuation of W contains at least $l \geqslant p$ segments and the occurrence $P_1 * D * Q_1$ contain a normalized occurrence W_1 of an elementary l-power of rank α generated with an original periodic word with period B^{-1}, then transition (4) is called a simple l-reversal of rank α of the occurrence W. Moreover, we say that the occurrence W and W_1 corresponds to each other in individuality in reversal (4). A regular occurrence V of rank $\alpha - 1$ contained in an occurrence $W_2 \rightleftharpoons * P_1 * T_1 EQ$ is called stable in the reversal (4) if V is mutually normalized in rank $\alpha - 1$ with an occurrence $V' \rightleftharpoons \phi(V, W_2; W_3)$ in the word $P_1 D Q_1$, where $W_3 \rightleftharpoons * P_1 * DQ_1$. Moreover, V' is called the image of occurrence V in the reversal (4). In a similar fashion, we define the stability of the regular occurrence of rank α contained in $PET_2 * Q *$ and its image in reversal (4). If V is a normalized occurrence of an elementary p-power of rank α that is stable in (4), then we say that V and its image V' in reversal (4) correspond to each other in individuality in this reversal.

Let $W \rightleftharpoons P * E * Q$ be a normalized occurrence of an elementary p-power E of rank α in a word $X \in \Pi_{\alpha-1}$. The transition $X \rightarrow Z$, where $Z \in \Pi_{\alpha-1}$, is called an l-reversal of rank α of the occurrence W if there are words $X_1 \in M_{\alpha-1}$ and $Z_1 \in \Pi_\alpha$ such that $X \overset{\alpha}{\sim}{}^{1} X_1$, $Z \overset{\alpha}{\sim}{}^{1} Z_1$ and the transition $X_1 \rightarrow Z_1$ is a simple l-reversal of rank α of the occurrence $f_{\alpha-1}(W, X_1)$ in X_1. We say that a regular occurrence V of rank $\alpha - 1$ in the word X is stable in the reversal $X \rightarrow Z$ and that the occurrence V_1 in the word Z is its image if the occurrence $f_{\alpha-1}(V_1, Z_1)$ is the image of the occurrence $f_{\alpha-1}(V, X_1)$ in the simple reversal $X_1 \rightarrow Z_1$. In this case we write $f_{x \rightarrow z}(V) = V_1$. Normalized occurrences W_1 and W_2 of elementary

p-powers of rank α in the words X and Z, respectively, are said to correspond in individuality in the reversal $X \to Z$ if the occurrences $f_{\alpha-1}(W_1, X_1)$ and $f_{\alpha-1}(W_2, Z_1)$ correspond in individuality in the simple reversal $X_1 \to Z_1$. Moreover, if W_1' and W_2' are normalized occurrences of elementary p-powers of rank α in the words X and Z and are compatible with W_1 and W_2, respectively, then they also are said to correspond in individuality in the reversal $X \to Z$.

If $l > r \geqslant p$, then every l-reversal of rank α is an r-reversal of rank α. If $l \geqslant p$ and if a maximal normalized continuation of an occurrence W of an elementary p-power of rank α in a word $X \in \Pi_{\alpha-1}$ contains at least l and at most $n-l-28$ segments, then an l-reversal of occurrence W can be made. If two normalized occurrences W_1 and W_2 of elementary p-power of rank α in a word $X \in \Pi_{\alpha-1}$ are compatible, then every l-reversal of W_1 is an l-reversal of W_2. If a transition $X \to Z$ is an l-reversal of an occurrence W and an occurrence W_1 in the word Z corresponds to W in individuality in this reversal, then the transition $Z \to X$ is an l-reversal of the occurrence W_1.

Assume the transition $X \to Z$ is a g_1-reversal of rank α of the occurrence W_1, while $W \doteq R * E * S$ is a normalized and not compatible with W_1 occurrence in X of an elementary l-power E of rank α, where $l \geqslant 2p_1 + p$. If $W_1 < W$, then the end W' of W that starts with the $(p_1 + 1)$–th supporting kernel from the left which corresponds in phase to the starting supporting kernel of W is a normalized occurrence of an elementary $(l - p_1)$-power of rank α which is compatible with W and stable in the reversal $X \to Z$. A similar statement holds for $W < W_1$. If $V_1 < V < W$ (or $W < V < V_1$), where V and V_1 are regular occurrences of rank $\alpha - 1$ and V is stable in a p-reversal of W, then V_1 is also stable in a reversal of W.

Let N_α denote the set of all words $X \in \Pi_{\alpha-1}$ such that X does not contain any normalized occurrences of elementary $(n - g + 3p_1)$-powers of rank α. Let W_1 and W_2 be two normalized occurrences of elementary g_1-powers of rank α in $X \in N_\alpha$ that are not mutually compatible. We say that W_1 and W_2 adjoin each other if there is no normalized occurrence of an elementary p-power of rank α between them which is stable in a reversal of either one of them.

It is easy to construct on the three symbols 1, 2 and 3 an infinite sequence

(5) $c_1, c_2, \ldots, c_i, \ldots$

such that, for arbitrary i, the word $c_1 c_2 \ldots c_i$ cannot be represented in the form $REES$, with non-empty E (cf. [8]). Let us choose some such sequence (5).

Let $V_0 \rightleftharpoons P * A_0 * Q$ be a normalized occurrence of an elementary g-power A_0 of rank α in a word $X \in N_\alpha$. The set of occurrences in X

(6) $$V_i \rightleftharpoons PA_0 u_1 A_1 u_2 \ldots u_i * A_i * u_{i+1} \ldots A_{k-1} u_k A_k Q_1,$$

where $i = 1, 2, \ldots, k$ and $k \geqslant 1$, is called a right cascade of the occurrence V_0 if the following conditions are fulfilled:

(a) For $0 < i < k$, each V_i is a normalized occurrence of an elementary g_{c_i}-power of rank α where c_i is the i-th term in (5) and a maximal normalized right continuation of V_i contains less than $g_{c_i} + p_1$ segments. Occurrence V_k is a normalized occurrence of an elementary $(g_{c_k} + p_1)$-power of rank α.

(b) The occurrences V_i and V_{i+1} adjoin each other for $0 \leqslant i < k$.

(c) For $0 < i \leqslant k$, every V_i is stable in the reversals of V_{i-1} and V_{i+1}, and no normalized continuation of V_i, differing from it, has this property.

We call occurrence V_1 the first element of cascade (6). The left cascade of occurrence V_0 is defined in a similar fashion.

An occurrence V_0 is said to be real on the right (left) if either V_0 does not have a right (left) cascade or V_0 is stable in the reversal of the first element of its right (left) cascade. If V_0 is real both on the right and on the left, we say that V_0 is a real occurrence of rank α.

Let the transition $X \to Y$ be a g-reversal of a normalized occurrence V of an elementary p-power of rank α in a word $X \in N_\alpha$ and let $Y \in N_\alpha$. We call this reversal $X \to Y$ a real reversal of rank α if there are real occurrences V_1 and W_1 of rank α in the words X and Y, respectively, such that V_1 is compatible with V and W_1 corresponds to V in individuality in this reversal. An occurrence V is said to be a really active occurrence of rank α if it has a real reversal of rank α.

A normalized occurrence V of an elementary p-power of rank α in a word $X \in N_\alpha$ is called a kernel of rank α of the word X if it is stable in the reversal of any really active occurrence of rank α in the word X that is not compatible with it and no normalized continuation of V possesses this property if the continuation differs from V. A kernel V of rank α is called a really active or really inactive kernel, depending on whether it is a really active occurrence or not.

If a kernel V contains less than t segments, its maximal normalized continuation contains less than $t + 2p_1$ segments. Every normalized occurrence of an elementary $(p + 2p_1)$-power of rank α in a word $X \in N_\alpha$ is compatible with some kernel of rank α of X. If two kernels V_1 and V_2 of rank α in the word X are different, then they are not compatible and neither of them is contained in the other, i.e., either $V_1 < V_2$ or $V_2 < V_1$. If a kernel V of rank α of a word X is really active, its maximal normalized continuation does not intersect with other kernels of X.

If $X \rightarrow Y$ is a real reversal of rank α, then each kernel V of rank α of the word X corresponds in individuality in this reversal to a unique kernel V' of rank α of the word Y, moreover the kernels V and V' of the words X and Y, respectively, also correspond to each other in individuality in the real reversal $Y \rightarrow X$. We write this correspondence as $V' = f_\alpha(V, Y)$ or $V = f_\alpha(V', X)$.

A word X is called a reduced word of rank α if $X \in N_\alpha$ and every kernel of rank α of X contains less than $n - g - 7p_1 - 103$ segments. Moreover, if $X \in M_{\alpha - 1}$ and each kernel of rank α contains less than $(n + 3)/2$ segments, then the word X is called a word minimized in rank α. By Π_α (respectively, M_α) is denoted the set of all reduced (respectively, minimized) words of rank α.

A word occurring in a reduced word of rank α need not itself be reduced. However, no elementary $(n + \frac{3}{2} + 2p_1 + 12)$-powers of rank α occur in X if $X \in M_\alpha$. Hence, $PEQ \in M_\alpha$ implies $E \in \Pi_\alpha$. The relations $\Pi_\alpha \subset \Pi_{\alpha - 1}$ and $M_\alpha \subset M_{\alpha - 1}$ follow immediately from the definitions.

Two reduced words X and Y of rank α are called equivalent in rank α if either $X \overset{\alpha - 1}{\sim} Y$ or there is a sequence of real reversals of rank α

$$(7) \qquad X \to X_1 \to X_2 \to \cdots \to X_i \to X_{i+1} \to \cdots \to X_\lambda \to Y.$$

Here the words X_i need not be reduced, but they are all in the set N_α. The equivalence in rank α of the words X and Y is denoted by $X \overset{\alpha}{\approx} Y$. Obviously, the relation $X \overset{\alpha}{\approx} Y$ is reflexive, symmetric and transitive. For every word $X \in \Pi_\alpha$ there is a $Y \in M_\alpha$ such that $X \overset{\alpha}{\approx} Y$.

Let $X, Y \in \Pi_\alpha$ and $X \overset{\alpha}{\approx} Y$. If $X \overset{\alpha-1}{\approx} Y$, then an occurrence V in X is a kernel of rank α if and only if the occurrence $f_{\alpha-1}(V, Y)$ is a kernel of rank α in Y. In this case the single-valued function mapping all the kernels of rank α of X into kernels of Y is said to be equal by definition to the function $f_{\alpha-1}(V, Y)$. In the case when the words X and Y are not equivalent in rank $\alpha - 1$ and a sequence of real reversals (7) exists, the function $f_\alpha(V, Y)$ is defined as a composition of the mappings $f_\alpha(V, Y)$ that were described above for one real reversal of rank α. It is obvious that if $X \overset{\alpha}{\approx} Y$, then $V_1 = f_\alpha(V, Y)$ implies $V = f_\alpha(V_1, X)$.

A word $Z \in \Pi_\alpha$ is called the result of a close-up of rank α of the reduced words X and Y of rank α if there are words $X_1 \in \Pi_\alpha$ and $Y_1 \in \Pi_\alpha$ such that $X \overset{\alpha}{\approx} X_1$, $Y \overset{\alpha}{\approx} Y_1$ and $Z = [X_1, Y_1]_0$. In this case we write $Z = [X, Y]_\alpha$. This operation is defined for any pair of reduced words X and Y of rank α and is independent of the choice of the intermediate words up to equivalence in rank α, i.e., if $X_1 \overset{\alpha}{\approx} X_2$ and $Y_1 \overset{\alpha}{\approx} Y_2$, then $[X_1, Y_1]_\alpha \overset{\alpha}{\approx} [X_2, Y_2]_\alpha$. The operation $[X, Y]_\alpha$ is associative.

With this, the inductive definition of all the basic concepts of the theory under consideration is completed. Now we can define the desired group $\Gamma(m, n)$.

A word X on (2) is called an absolutely reduced word of rank 0 if $X \in \Pi_0$ and no elementary p-power of rank 1 occurs in X. A word $X \in \Pi_\alpha$, if $\alpha > 0$, is called an absolutely reduced word of rank α if there is a normalized occurrence in X of some elementary p-power of rank α and there are no normalized occurrences in X of elementary p-powers of rank $\alpha + 1$. The set of all absolutely reduced words of rank α is denoted by $\mathfrak{A}_\alpha$. It follows easily from the definitions we have introduced that, for arbitrary i and j, $\mathfrak{A}_i \subset \Pi_j$ and if $X \in \mathfrak{A}_i$, then $X^{-1} \in \mathfrak{A}_i$. Two words X and Y from the set

$\mathfrak{A} \rightleftharpoons \bigcup_{i=0}^{\infty} \mathfrak{A}_i$ are called equivalent if $X \overset{\alpha}{\approx} Y$ for some α. The equivalence of the words X and Y is denoted by $X \sim Y$. Let $\mathfrak{B}$ denote the set of all the classes of equivalent words that are formed when $\mathfrak{A}$ is partitioned by the relation $X \sim Y$. If $X \in \mathfrak{A}$, then an element of $\mathfrak{B}$ containing the word X is denoted by $\{X\}$.

The operation of multiplication in $\mathfrak{B}$ is introduced in the following manner. By the product of elements u and v in $\mathfrak{B}$ is meant an element $w \in \mathfrak{B}$ such that for some representatives of these classes, $X \in u$, $Y \in v$ and $Z \in w$, the following relations hold: $X \in \mathfrak{A}_\alpha$, $Y \in \mathfrak{A}_\beta$ and $Z = [X, Y]_\gamma$, where $\gamma = \max(\alpha, \beta)$. The product of the elements u and v defined in this way is denoted by $u \circ v$. The operation $u \circ v$ is uniquely defined for any two elements of $\mathfrak{B}$ and is associative. Moreover, the set $\mathfrak{B}$ together with this operation constitute a group for which the elements $\{a_1\}, \{a_2\}, \ldots, \{a_m\}$ form a set of generators, while the element $\{1\}$ is the identity element. The inverse of $\{X\}$, if $X \in \mathfrak{A}$, is the element $\{X^{-1}\}$. This group is denoted by $\Gamma(m, n)$. The identity relations (1) are fulfilled in it, where n is a fixed odd number chosen by us. The proof of this is quite involved.

It is simple to show that $\Gamma(m, n)$ is infinite and we can do this here. It follows from the definition of a kernel of rank α that, for $\alpha > 0$, the base of any kernel of rank α is an elementary p-power of rank $\alpha + 1$ is an elementary p-power of rank α. Hence, for arbitrary $\alpha > 0$, an elementary p-power of rank α contains an occurrence of an elementary p-power of rank 1, i.e., a periodic word containing at least $p - 1 = 17$ complete periods. Assume $X \in \mathfrak{A}_0$. Then X does not have any kernels of rank α, for arbitrary $\alpha > 0$. Hence, for arbitrary $\alpha > 0$, a real reversal $X \to X_1$ of rank α is impossible. This means, for arbitrary $\alpha > 0$ and $Y \in \mathfrak{A}$, that $X \overset{\alpha}{\approx} Y$ implies $X \overset{\alpha-1}{\approx} Y$. Thus, $X \sim Y$ implies $X \overset{\circ}{=} Y$, i.e., the class $\{X\}$ consists of the one word X. In particular, the generators $\{a_i\}$ and the unit element $\{1\}$ of our group also are of this kind. Thus, it is sufficient to show that $\mathfrak{A}_0$ is infinite. But this follows trivially from the fact that $m > 1$ and the existence of a sequence on two symbols a_1 and a_2 constructed as in [8]

$$a_{i_1}, a_{i_2}, \ldots, a_{i_j}, \ldots$$

for which, for arbitrary j, the word $a_{i_1} a_{i_2} \ldots a_{i_j}$ cannot be represented in the form $PEEEQ$, with non-empty E.

Let $B(m, n)$ be a free Burnside group with generators (2) and identity relations (1). For an arbitrary word X on alphabet (2) a word Y can be actually displayed such that $Y \in \mathfrak{A}$ and $X = Y$ in $B(m, n)$. Two words X and Y from $\mathfrak{A}$ are equal in $B(m, n)$ if and only if $X \sim Y$. It follows immediately from these two assertions that the correspondence

$$a_i \text{---} \{ a_i \}$$

induces an isomorphism of $B(m, n)$ onto $\Gamma(m, n)$. It follows from $X \in \mathfrak{A}_i$, $X \sim Y$ and $Y \in \mathfrak{A}$ that $\partial(Y) < \pi(i, \partial(X))$, where $\pi(i, j)$ is some recursive function. Therefore we have an algorithm that solves the word problem for $B(m, n)$.

Let $B(m, n, \alpha)$, for $\alpha > 0$, denote a group defined by the generators (2) and all defining relations of the form

$$A^n = 1,$$

where A^n is an elementary word, with period A, of rank $\leqslant \alpha$. If two words X and Y are equal in $B(m, n)$, then they are equal in $B(m, n, \alpha)$, if $\alpha = \partial(X) + \partial(Y)$. If $B(m, n)$ can be defined by means of a finite number of defining relations, then, for some $\alpha > 0$, every equality in $B(m, n)$ is also fulfilled in $B(m, n, \alpha)$. However, it is easy to construct an elementary word C^n, with period C, of rank $\alpha + 1$, that is not equal to 1 in $B(m, n, \alpha)$. Consequently, identity relation (1) can not be replaced by a finite number of defining relations, when we have odd $n \geqslant 4381$. All the results stated in the last two paragraphs are given with complete proofs in [5].

The following fundamental lemma is proved in [6] by using the theory described above and the isomorphism between $B(m, n)$ and $\Gamma(m, n)$. If $AB = BC$ and if $A \neq 1$ in $B(m, n)$, then there is a word E such that $B = E$ in $B(m, n)$ and $\partial(E) \leqslant n \cdot \pi(j, \pi(j, j))$, where $j = \partial(A) + \partial(C)$. This provides us with an algorithm that solves the conjugacy problem for $B(m, n)$, when we have odd $n \geqslant 4381$. It follows from this lemma that the centralizer of any non-unit element of $B(m, n)$ contains a finite number of elements. Thus, for $m > 1$ and odd $n \geqslant 4381$, $B(m, n)$ is an example of an infinite group in which all of the abelian subgroups are finite.

We consider in [7] and [9] a generalization of the theory of word transformation that is described above, in order to prove that the system of group identities (3) is irreducible when $n = 4381$ and k ranges over all the prime numbers. In order to construct this new theory, in addition to using all of the concepts in [4], we also define by induction on the rank α the concept of an admissible word of rank α which is associated with a given set of prime numbers K.

If, as a result of replacing the variables x and y on the left-hand side of (3) by reduced words A and B of rank $\alpha - 1$ and interpreting the operation of multiplication as a close-up of rank $\alpha - 1$, a word is obtained within the brackets which is equivalent in rank $\alpha - 1$ to the result of a close-up of rank $\alpha - 1$ of the form $[T, C^s, T^{-1}]_{\alpha-1}$, where $s > 0$, then we say that the word C is generated by the substitution in rank $\alpha - 1$ of the pair of words (A, B) in (3).

Let K be some set of prime numbers. A periodic word $C^t C_1$ of rank α with period C minimal in rank $\alpha - 1$ is said to be admissible in rank $\alpha - 1$ relative to the set K by a periodic word of rank α if there are words A and B such that the word C is generated by the substitution in some rank $\beta < \alpha$ of the pair of words (A, B) in some relation (3), for $k \in K$. In particular, a periodic word $C^t C_1$ of rank 1 with period C is admissible in rank 0 if and only if some word of the form

$$A^{kn} B^{kn} A^{-kn} B^{-kn},$$

where $k \in K$ and where A and B are words on (2), is conjugate to some word C^s in the free group, where $s > 0$. Integral, semi-integral and elementary words of rank α obtained with original periodic words of rank α that are admissible in rank $\alpha - 1$ are also said to be admissible in rank $\alpha - 1$.

Reversals of occurrences of admissible elementary words of rank α are called admissible reversals. Only admissible reversals of rank α are considered in the new theory. It is necessary to consider only normalized occurrences of admissible elementary words of rank α in defining a cascade of rank α of a real or really active occurrence of rank α. A normalized occurrence of an elementary word of rank α which is not admissible is not really active, no matter how many segments it may contain. Hence, the conditions concerning the number of segments that are stated in the definition of the sets of words N_α, Π_α and M_α apply only to occurrences of admissible elementary words of rank α. In particular, words belonging to these sets may contain occurrences of kernels of rank α having more than n segments, but their bases will not be admissible elementary words of rank α and these kernels will not be really active.

The proofs in the new theory of all the assertions made in [4] remain basically the same. Only some of these assertions need to be reformulated by taking into account the remarks made above. For example, an admissible elementary p-power of rank $\alpha + 1$ always contains occurrences of elementary p-powers of rank α (the bases of its supporting kernels), but it can not contain occurrences of admissible elementary p-powers of rank α. If an elementary word E of rank α, generated by some original periodic word of rank α with period C, does not contain an occurrence of any admissible elementary $2p$-power of rank $\beta < \alpha$, then E is a periodic word with period C. Hence, if a word $X \in \Pi_\alpha$ does not contain an occurrence of any admissible periodic $2p$-power of rank 1 and if $X \overset{\alpha}{\approx} Y$, then $X \overset{\circ}{\equiv} Y$.

In a manner similar to that in which the group $\Gamma(m, n)$ was defined above, a group $\Gamma(m, n, K)$, which also depends on the choice of the set K, is constructed on the basis of the concepts of the new theory. Moreover, the definition of the set $\mathfrak{A}_\alpha$ must involve normalized occurrences of admissible elementary p-powers. The assertion that, for $k \in K$, all identities (3) are fulfilled in $\Gamma(m, n, K)$ is proved in a manner similar to that for Lemma 22 of Chapter VII in [4]. Moreover, important use is made of the following basic property of the concept of admissibility: if a word $C \in \Pi_{\alpha-1}$ is generated by the substitution in rank $\alpha > 0$ of a pair of words (A, B) in some relation (3) and if C does not contain an occurrence of any admissible elementary $2p$-power of rank α, then C can be generated by the substitution in rank $\alpha - 1$ of some pair of words

(A_1, B_1) in the same relation (3). From this property it follows that if C does not contain an occurrence of any periodic $2p$-power of rank 1 that is admissible in rank 0 and if the periodic word $C^t C_1$ of rank α with period C is admissible in rank $\alpha - 1$, then it is also admissible in rank 0. Since a word

$$(8) \qquad a_1^{kn}\, a_2^{kn}\, a_1^{-kn}\, a_2^{-kn},$$

for $k \notin K$, does not contain an occurrence of any periodic p-power of rank 1 that is admissible in rank 0, then (8) is in the set $\mathfrak{A}_0$ of the new theory and is not equivalent to any word different from it. This means that (3) is not fulfilled in $\Gamma(m, n, K)$ when $k \notin K$.

Thus, identity relation (3) is fulfilled in the group $\Gamma(m, n, K)$ that we constructed if and only if $k \in K$.

If we take as K the set of all prime numbers not equal to a given prime number l, then we find that relation (3) for $k = l$ does not follow from the other identities in system (3).

The fact that the system of group identities (3) is irreducible implies immediately that a continuum exists of systems of group identities of the form (3) that are not pairwise equivalent. A proof that the set of all group varieties is a continuum which is based on a completely different idea has been obtained by A.Ju. Ol'shanskii (Ol'shanskii's result is published in the journal Izvestija AN SSSR, Ser. Mat. 1970, V. 34, No2). He defines group varieties not by using systems of identities, but by means of systems of finite groups belonging to these varieties. Since the set of all finite systems of group identities is denumerable, this result implies the existence of a group variety that is not defined by any finite system of identities. However, it has not been possible to actually exhibit such a variety on the basis of Ol shanskiĭ's proof.

If an identity has been derived from some infinite system of identities, then only a finite number of identities from this system could have been used in the derivation. Hence, it follows from the fact that the system of identities (3) is irreducible that this system (and every one of its infinite subsystems) is not equivalent to any finite system of identities, i.e., a group variety defined by (3) does not have a finite basis.

Let S be some recursively enumerable and undecidable set of prime numbers, while $\Gamma(S)$ is a group defined by two generators and the identity relations (3) for $k \in S$. Every relation in $\Gamma(S)$ is an identity relation. Moreover, for given n, (3) is fulfilled in $\Gamma(S)$ if and only if $k \in S$. Consequently $\Gamma(S)$ has an unsolvable word problem.

References

[1] Burnside, W., On an unsettled question in the theory of discontinuous groups, Quart, J. Pure Appl. Math. 33 (1902) 230-238.

[2] Sanov, I.N., (Solution of Burnside's problem for exponent 4) Rešenie problemy Bernsaĭda dlja pokazatelja 4, Leningrad. Gos. Univ. Učen. Zap. Ser. Mat. Nauk. 10 (1940) 166-170. (Russian) MR2, 212.

[3] Hall, Marshall, Jr., Solution of the Burnside problem for exponent 6, Proc. Nat. Acad. Sci. USA 43 (1957) 751-753. MR 19, 728.

[4] Novikov, P.S. and S.I. Adjan, (Infinite periodic groups) O beskonečnyh periodičeskih gruppah, I, II, III, Izv. Akad. Nauk SSSR Ser. Mat. 32 (1968) 212-244, 251-524, 709-731 = Math. USSR Izv. 2 (1968) 209-236, 241-479, 665-685. MR 39 # 1532 a-b-c.

[5] Novikov, P.S. and S.I. Adjan, (Defining relations and the world problem for free periodic groups of odd order) Opredeljajuščie sootnošenija i problema toždestva dlja svobodnyh periodičeskih grupp nečetnogo porjadka, Izv. Akad. Nauk SSSR Ser. Mat. 32 (1968) 971-979 = Math. USSR Izv. 2 (1968) 935-942.

[6] Novikov, P.S. and S.I. Adjan, (On abelian subgroups and the conjugacy problem in free periodic groups of odd order) O kommutativnyh podgruppah i probleme soprjažennosti v svobodnyh periodičeskih gruppah nečetnogo porjadka, Izv. Akad. Nauk SSSR Ser. Mat. 32 (1968) 1176-1190 = Math. USSR Izv. 2 (1968) 1131-1144. MR 38 # 2197.

[7] Adjan, S.I., (Infinite irreducible systems of group identities) Beskonečnye neprivodimye sistemy gruppovyh tozdestv, Dokl. Akad. Nauk SSSR 190 (1970) 499-501 = Sov. Math. Dokl. 11 (1970), No1, 114-115.

[8] Aršon, S.E., (Proof of the existence of n-valued infinite asymmetric sequences) Dokazatel'stvo suščestvovanija n-značnyh beskonečnyh asimmetričnyh posledovatel'nosteĭ, Mat. Sb. (N.S.) 2 (44) (1937) 769-779. (Russian).

[9] Adjan, S.I., (Infinite irreducible systems of group identities) Beskonečnye neprivodimye sistemy gruppovyh toždestv, Izv. Akad. Nauk SSSR Ser. Mat. 34 (1970) 715-734 = Math. USSR Izv. 4 (1970)

CONSTRUCTION OF A NON-SOLVABLE GROUP OF EXPONENT 5

S. BACHMUTH, H.Y. MOCHIZUKI and
D.W. WALKUP
University of California, Santa Barbara, California,
Boeing Research Laboratory, Seattle, Washington

1. Introduction

Let R be the free, associative, noncommutative ring of characteristic 5 without unit element generated by the countable set of indeterminates $x_1, x_2, \ldots$. Let L be the Lie ring imbedded in R generated by $x_1, x_2, \ldots$, where addition in L is the same as in R and multiplication in L is commutation $[x, y] = xy - yx$ in R. Let I be the ideal of R generated by all cubes $x^3, x \in L$, and let H be the ideal of R generated by all elements $g(x, y) = x^2y - 3xyx + 3yx^2$, $x, y \in L$. A major part of this paper, §3, is devoted to proving the following theorem.

Theorem 1. *The ring R/H is not nilpotent. In particular, $x_1{}^2 x_2{}^2 \ldots x_n{}^2 \not\equiv 0 \pmod{H}$ for any $n \geqslant 1$.*

Since $g(x, x) = x^3$, it is obvious that $H \supset I$ and hence Theorem 1 implies that the ring R/I is not nilpotent either. Theorem 1 also implies:

Corollary 1. *There exists a Lie ring L' of characteristic 5 which satisfies the third Engel condition, i.e., $[[[x, y], y], y] = 0$ for all $x, y \in L'$, which is not nilpotent as a Lie ring.*

Since this corollary is not needed for our group theory applications below, we will not indicate a proof here. A proof that Corollary 1 and the first part of Theorem 1 are actually equivalent has been given by Walkup in his thesis [9].

In §4 we use Theorem 1 and the tools developed in §3 to establish:

Theorem 2. *There exists a group of exponent* 5 *which is locally solvable but not solvable.*

A theorem of Kostrikin [6] states that for any finite cardinal m and any prime p there is a largest finite group of exponent p with at most m generators. This result and Theorem 2 taken together imply:

Corollary 2. *There exists a variety of groups which is a locally solvable and locally finite variety but not a solvable variety.*

Other consequences of Theorem 2 relevant to the restricted Burnside problem may be inferred from the results of G. Higman [5] and Kostrikin [6].

For the sake of completeness we mention the following strengthening of a result of Higgins and Heineken (see Bruck [2] or Bachmuth, Mochizuki and Walkup [1]). A proof can be found in [9].

Theorem 3. *Suppose the ring R' has characteristic prime to* 2, 3, and 5 (*i.e.*, $30x = 0$ *implies* $x = 0$). *Suppose further that R' is generated by elements $x_1, x_2, \ldots$. Let L' be the Lie ring embedded in R' generated by $x_1, x_2, \ldots$ and let I' be the ideal of R' generated by the cubes $x^3, x \in L'$. Then R'/I' is nilpotent of index at most* 9.

2. Preliminaries

In this section we introduce some useful elementary concepts, and in Lemma 1 below we apply the familiar process of linearization to obtain a set of elements which span the ideal H as a subspace of R.

A natural basis for R, considered as a vector space over the field Z_5 of integers modulo 5, is provided by the set of all *monomials* $M = x_{i_1} x_{i_2} \ldots x_{i_m}$, $m \geqslant 1$. The integer m is the *degree* of M. Unless otherwise indicated, when we write a product of the form PMQ, M a monomial, we include the case in which M is the "empty" monomial of degree 0, that is, we include the product PQ as one of the cases. The analogues of the monomials in R are the *commutators* in L, which may be defined inductively as follows: Any indeterminate is a commutator of degree 1, and if C_1 and C_2 are commutators of degree d_1 and d_2 respectively, then $C = [C_1, C_2] = C_1 C_2 - C_2 C_1$ is a commutator of degree $d = d_1 + d_2$. It is easy to see that the commutators span the Lie ring L, considered as a vector space over Z_5.

Given any monomial $M = x_{i_1} x_{i_2} \ldots x_{i_m}$ in R, we denote by R_M the subspace of R spanned by all monomials M' (including M) obtainable from M by rearranging the order of the indeterminates in the product. In the special case $M = x_1{}^2 x_2{}^2 \ldots x_n{}^2$ we will write R_n for R_M. Of course the ideal H is a subspace of R, and we write $H_M = R_M \cap H$ and $H_n = R_n \cap H$.

We shall say that an ideal J of the ring R is an *I-substitution ideal* if $x \in J$ implies $x' \in J$, where x' is obtained from x by substituting any (not necessarily distinct) *indeterminates* systematically for the indeterminates which figure in x. Similarly, we shall say that J is a *Lie substitution ideal* if $x \in J$ implies $x' \in J$, where x' is obtained from x by substituting any elements of L for the indeterminates of x. Clearly the ideals I and H defined in the introduction are Lie substitution ideals, and any Lie substitution ideal is obviously a substitution ideal.

Lemma 1. *For any monomial M, the subspace $H_M = R_M \cap H$ of R_M is spanned by the polynomials of the form*

$$M_1 h(C_1, C_2, C_3) M_2, \tag{2.1}$$

where h is the polynomial function defined by

$$h(x, y, z) = xyz + zyx + 2yzx + 2yxz,$$

M_1 and M_2 are (possibly empty) monomials, and C_1, C_2, C_3 are *commutators such that $M_1 C_1 C_2 C_3 M_2$ is in R_M.*

Proof. By its definition as an ideal of R, H is spanned by the elements of the form $P_1 g(L_1, L_2)P_2$, where P_1, P_2 are arbitrary polynomials from R and L_1, L_2 are arbitrary elements from the Lie ring L. It is readily computed that

$$\begin{aligned} 2h(x, x, y) - h(x, y, x) &= 6x^2 y + 2xyx - 2yx^2 \\ &\equiv x^2 y - 3xyx + 3yx^2 \pmod 5 \\ &\equiv g(x, y) \pmod 5 \end{aligned}$$

and

$$\begin{aligned} g(y, x) + g(z, x) + g(x, z) + g(y, z) - g(y+z, x) - g(x+y, z) \\ = -4xyz - 4zyx + 2yzx + 2yxz \\ \equiv xyz + zyx + 2yzx + 2yxz \pmod 5 \\ \equiv h(x, y, z) \pmod 5. \end{aligned}$$

It follows that H is also spanned by the elements of the form $P_1 h(L_1, L_2, L_3)P_2$. Since every element P_i of R is a sum of monomials, since every element L_i of L is a sum of commutators, and since h is linear in each of its arguments, it is clear that H is spanned by elements of the form (2.1). The restriction to H_M is obvious.

Remark. Since R has characteristic 5, the above congruences modulo 5 are superfluous statements. However, in the above lemma and in later places, we often add this extra comment to emphasize the role played by characteristic 5.

3. Proof of Theorem 1

In this section we prove the following strengthened version of Theorem 1. Some of the machinery developed here will also be used in the proof of Theorem 2 in §4.

Theorem 1′. *For each* $n \geqslant 1$, $H_n = R_n \cap H$ *is a proper subspace of* R_n, *and* $x_1{}^2 x_2{}^2 \ldots x_n{}^2$ *together with* H_n *span* R_n.

The proof of Theorem 1′ may be divided into three main steps. Let H_n^0 be the subspace of H_n spanned by generators of H_n of the form $M_1 h(x, y, z) M_2$ where x, y, and z are indeterminates. In the first step we define a Collection Algorithm which enables us to use induction on n. In the second step (Lemma 2), we use this Collection Algorithm to define a linear map ϕ from R_n into itself with kernel H_n^0 and image spanned by $x_1{}^2 \ldots x_n{}^2$. In particular, H_n^0 and $x_1{}^2 \ldots x_n{}^2$ span R_n, so that Theorem 1′ holds with H_n^0 in place of H_n. In the final step (Lemma 3) we show that $H_n = H_n^0$.

For any (nonempty) monomial M let H_M^0 be the subspace of $H_M = R_M \cap H$ spanned by the polynomials of the form $M_1 h(x, y, z) M_2$, where M_1, M_2 are (possibly empty) monomials and x, y, z are indeterminates such that $M_1 xyz M_2$ is in R_M. The relation $h(x, y, z) \equiv 0$ modulo H implies many others. We indicate below six of the simplest ones. Each of the six expressions in (3.1) is equivalent to 0 modulo H for any x, y, z in L. If x, y, z are restricted to being indeterminates, each of these expressions, when pre- and post-multiplied by appropriate monomials M_1 and M_2, becomes a member H_M^0.

$$
(3.1)\qquad
\begin{aligned}
h(x, y, z) &= xyz + zyx + 2yxz + 2yzx \\
x^2 y - yx^2 &\equiv 2h(x, x, y) - 2h(x, y, x) \pmod 5 \\
xyx + 2yx^2 &\equiv -2h(x, y, x) \pmod 5 \\
2x^2 y + xyx &\equiv -h(x, x, y) - h(x, y, x) \pmod 5 \\
\bar{h}(x, y, z) &= zyx + xyz + 2xzy + 2zxy \\
&\equiv -2h(x, y, z) - h(x, z, y) - h(y, x, z) \pmod 5
\end{aligned}
$$

$$S(x, y, z) = xyz + xzy + yxz + zxy + yzx + zyx$$
$$= \tfrac{1}{2}\{h(x, y, z) + \bar{h}(x, y, z)\}.$$

Notice that $\bar{h}(x, y, z)$ is the polynomial obtained from $h(x, y, z)$ by reversing the order of the factors in each term.

If Theorem 1′ is true, then for each polynomial P in R_n there exists a unique polynomial $\varphi(P)$ of the form $\beta x_1{}^2 x_2{}^2 \ldots x_n{}^2$ (β not always zero) such that $P \equiv \varphi(P)$ (mod H_n). In the process of proving Theorem 1′ we will construct an explicit algorithm for computing $\varphi(P)$ for any P in R_n. A principle ingredient in this algorithm is the following:

Collection algorithm. An integer $n \geqslant 1$, and a polynomial P in some R_{M^*} are given. It is assumed that M^* is of degree at most 2 in the indeterminate x_n. A polynomial $\theta_n(P)$ is sought such that $P \equiv \theta_n(P)$ (mod $H^0_{M^*}$) and such that the occurrences of the indeterminate x_n in $\theta_n(P)$ are restricted to the two right-most positions in each term.

(1) Set Q equal to P.

(2) Choose any term βM of Q, $\beta \not\equiv 0$ (mod 5), in which x_n appears other than in the last two positions. If no such term can be found, go to step 6. Otherwise, go to step 3.

(3) Determine (possibly empty) monomials M_1 and M_2 such that the monomial M chosen in step 2 has the form $M_1 x_n x_j x_k M_2$, where x_j and x_k are indeterminates not necessarily distinct from x_n and M_1 does not involve x_n.

(4) Perform one of the following operations on Q.

(a) If $j = n \neq k$, replace βM by $\beta M_1 x_k x_n^2 M_2$.

(b) If $j \neq n = k$, replace βM by $-2\beta M_1 x_k x_n^2 M_2$.

(c) If $j \neq n \neq k$, replace βM by

$$-\beta M_1\, x_k\, x_j\, x_n\, M_2 - 2\beta M_1\, x_j\, x_k\, x_n\, M_2 - 2\beta M_1\, x_j\, x_n\, x_k\, M_2.$$

(5) Collect like terms of Q, reduce all coefficients modulo 5, and go to step 2.

(6) Define $\theta_n(P)$ to be Q and stop.

Proposition 1. *For any n, any R_{M^*}, and any $P \in R_{M^*}$, the Collection Algorithm is well-formed, terminates in a finite number of steps, and yields a unique polynomial $\theta_n(P)$ independent of the order in which terms are chosen in step 2. Moreover θ_n is a linear map of R_{M^*} into itself and $P \equiv \theta_n(P)$ modulo $H^0_{M^*}$ for all P in R_{M^*}.*

Proof. The proposition is readily derived from the following five observations: (i) Given any M satisfying the conditions in step 2, the representation $M = M_1 x_n x_j x_k M_2$ defined in step 3 is unique. (ii) Given the representation in step 3, exactly one of the conditions listed in step 4 must hold. (iii) Each of the operations given in step 4 preserves the property that $Q \in R_{M^*}$. (iv) Each of the operations given in step 4 replaces βM by a polynomial, in each of whose terms the occurrences of x_n are farther to the right. (v) Each of the operations given in step 4 is equivalent to adding elements of the form $M_1 f M_2$ to Q, where each f has the form of one of the expressions listed in (3.1).

Proposition 2. *Suppose n and P satisfy the restrictions stated in the Collection Algorithm and P can be factored in the form $P_1 P_2 P_3$, (respectively $P_2 P_3$ or $P_1 P_2$), where P_1 does not involve x_n. Then each P_i is an element of some R_{M_i}, n and P_i satisfy the restrictions of the Collection Algorithm, and*

$$\theta_n(P) = \theta_n \{ P_1 \cdot \theta_n(P_2) \cdot P_3 \}$$

(respectively

$$\theta_n(P) = \theta_n \{\theta_n(P_2) \cdot P_3 \}$$

or

$$\theta_n(P) = P_1 \cdot \theta_n(P_2)).$$

Lemma 2. *For each $n \geq 1$, H_n^0 is a proper subspace of R_n and $x_1{}^2 x_2{}^2 \ldots x_n{}^2$ together with H_n^0 span R_n.*

Proof. Define a map φ_n of R_n into itself by induction on n as follows: φ_1 is the identity map of R_1 onto itself, and if $n \geqslant 2$ then $\varphi_n(P) = \varphi_{n-1}(P')x_n{}^2$, where $\theta_n(P) = P'x_n{}^2$. Obviously φ_n is a linear map of R_n onto the subspace of dimension 1 spanned by $x_1{}^2x_2{}^2 \ldots x_n{}^2$ and $\varphi_n(P) \equiv P$ modulo H_n^0. In order to complete the proof of the lemma if suffices to show that φ_n takes H_n^0 into zero. This is immediate for $n = 1$, since H_1^0 is trivially zero by Lemma 1. Thus the present lemma will follow by induction on n if we show that θ_n takes H_n^0 into $H_{n-1}^0x_n{}^2$. To do this it suffices, in view of the definition of H_n^0 to show that

$$\theta_n\{M_1h(x,y,z)M_2\} = Qx_n{}^2 \qquad (3.2)$$
$$Q \equiv 0 \pmod{H_{n-1}^0}$$

for any monomials M_1, M_2 and (not necessarily distinct) indeterminates x, y, z such that M_1xyzM_2 is in R_n. We distinguish 12 cases depending on the location of the two occurrences of x_n in $M_1h(x,y,z)M_2$. In what follows N_i denotes a (possibly empty) monomial and x, y, z, s, t, etc., denote indeterminates distinct from x_n but not necessarily distinct from each other. Since h is symmetric in the first and third argument, it is unnecessary to list additional cases involving $h(x,y,x_n)$, etc.

(1) $N_1h(x,y,z)N_2x_nN_3x_nN_4$

(2) $N_1x_nN_2x_nN_3h(x,y,z)N_4$

(3) $N_1h(x_n,x_n,z)N_2$

(4) $N_1h(x_n,y,x_n)N_2$

(5) $N_1x_nN_2h(x_n,y,z)N_3$

(6) $N_1x_nN_2h(x,x_n,z)N_3$

(7) $N_1h(x_n,y,z)x_nN_2$

(8) $N_1h(x_n,y,z)tN_2x_nN_3$

(9) $N_1h(x,x_n,z)x_nN_2$

(10) $$N_1 h(x, x_n, z) t N_2 x_n N_3$$

(11) $$N_1 x_n N_2 h(x, y, z) x_n N_3$$

(12) $$N_1 x_n N_2 h(x, y, z) t N_3 x_n N_4$$

Consider Case 1. By Proposition 2

$$\begin{aligned}\theta_n \{N_1 h(x, y, z) N_2 x_n N_3 x_n N_4\} &= N_1 h(x,y,z) N_2 \cdot \theta_n \{x_n N_3 x_n N_4\} \\ &= N_1 h(x,y,z) N_2 P x_n{}^2,\end{aligned}$$

which is obviously an element of $H^0_{n-1} x_n{}^2$. This takes care of the first, and easiest case. Consider Case 2. Again using Proposition 2, we have

$$\begin{aligned}\theta_n \{N_1 x_n N_2 x_n N_3 h(x,y,z) N_4\} &= \theta_n \{N_1 \cdot \theta_n \{x_n N_2 x_n\} \cdot N_3 h(x,y,z) N_4\} \\ &= \theta_n \{N_1 P x_n{}^2 N_3 h(x, y, z) N_4\} \\ &= N_1 P N_3 h(x, y, z) N_4 x_n{}^2\end{aligned}$$

which is again obviously an element of $H^0_{n-1} x_n{}^2$. (Here we have made use of the observation that $\theta_n \{x_n{}^2 Q\} = Q x_n{}^2$. Note that this is an observation about the action of the Collection Algorithm, and not just an observation about squares of indeterminates commuting modulo H^0_M!) Now consider one of the most complex cases, Case 12. Again by Proposition 2, we have

$$\theta_n \{N_1 x_n N_2 h(x,y,z) t N_3 x_n N_4\} = \theta_n \{N_1 \cdot \theta_n \{x_n N_2\} h(x,y,z) t N_3 x_n N_4\}.$$

From the description of the Collection Algorithm it is obvious that

$$\theta_n \{x_n N_2\} = P_0 x_n + \Sigma_{i \geq 1} P_i x_n s_i,$$

where each s_i is an indeterminate. (If the degree of N_2 is 0 or 1, then $\theta_n \{x_n N_2\} = x_n$ or $x_n s$ respectively.) Actually, a close examination of the Collection Algorithm will disclose that any s_i, if present at all, must be the last indeterminate of N_2. Thus, after another application of Proposition 2, we have

$$\begin{aligned}(3.3)\qquad &\theta_n\{N_1x_nN_2h(x,y,z)tN_3x_nN_4\} = \\ &= \theta_n\{N_1P_0\cdot\theta_n\{x_nh(x,y,z)t\}N_3x_nN_4\} \\ &+ \theta_n\{N_1P_1\cdot\theta_n\{x_nsh(x,y,z)t\}N_3x_nN_4\}.\end{aligned}$$

By the same reasoning

$$\theta_n\{x_nh(x,y,z)t\} = Q_{00}x_n + Q_{01}x_nt$$

$$\theta_n\{x_nsh(x,y,z)t\} = Q_{10}x_n + Q_{11}x_nt.$$

Substituting this back into (3.3) and applying Proposition 2 again, we obtain

$$\begin{aligned}&\theta_n\{N_1x_nN_2h(x,y,z)tN_3x_nN_4\} = \\ &= N_1P_0Q_{00}P_0'N_4x_n^2 + N_1P_0Q_{01}P_1'N_4x_n^2 \\ &+ N_1P_1Q_{10}P_0'N_4x_n^2 + N_1P_1Q_{11}P_1'N_4x_n^2,\end{aligned}$$

where $\theta_n\{x_nN_3x_n\} = P_0'x_n^2$ and $\theta_n\{x_ntN_3x_n\} = P_1'x_n^2$. In order to prove (3.2) for Case 12 it will suffice to prove each Q_{ij} lies in the appropriate H_M^0. Now it may happen that certain of the indeterminates s, x, y, z, t are not distinct. However, an examination will show that the Collection Algorithm depends in no way upon the identity of any indeterminate other than x_n. Thus the proof of (3.2) for Case 12 will follow from a proof of the propositions:

$$(3.4)\quad\begin{aligned}&(12a)\quad \theta_5\{x_5h(x_1,x_2,x_3)x_4\} = Q_{00}x_5 + Q_{01}x_5x_4, \\ &\qquad Q_{00} \in H^0_{x_1x_2x_3x_4},\ Q_{01} \in H^0_{x_1x_2x_3}; \\ &(12b)\quad \theta_6\{x_6x_1h(x_2,x_3,x_4)x_5\} = Q_{10}x_6 + Q_{11}x_6x_5, \\ &\qquad Q_{10} \in H^0_{x_1x_2x_3x_4x_5},\ Q_{11} \in H^0_{x_1x_2x_3x_4}.\end{aligned}$$

In similar fashion the proof of the remaining nine cases reduces to the proof of the following propositions. In some cases it turns out in practice that the application of the Collection Algorithm yields 0 directly. These cases have been indicated.

(3.5)

(3) $\theta_2\{h(x_2, x_2, x_1)\} = 0$

(4) $\theta_2\{h(x_2, x_1, x_2)\} = 0$

(5a) $\theta_3\{x_3 h(x_3, x_1, x_2)\} = 0$

(5b) $\theta_4\{x_4 x_1 h(x_4, x_2, x_3)\} = Qx_4{}^2,\ Q \in H^0_{x_1x_2x_3}$

(6a) $\theta_3\{x_3 h(x_1, x_3, x_2)\} = 0$

(6b) $\theta_4\{x_4 x_1 h(x_2, x_4, x_3)\} = Qx_4{}^2,\ Q \in H^0_{x_1x_2x_3}$

(7) $\theta_3\{h(x_3, x_1, x_2)x_3\} = 0$

(8) $\theta_4\{h(x_4, x_1, x_2)x_3\} = 0$

(9) $\theta_3\{h(x_1, x_3, x_2)x_3\} = 0$

(10) $\theta_4\{h(x_1, x_4, x_2)x_3\} = Q_0 x_4 + Q_1 x_4 x_3,$
$Q_0 \in H^0_{x_1x_2x_3},\ Q_1 = 0$

(11a) $\theta_4\{x_4 h(x_1, x_2, x_3)x_4\} = Qx_4{}^2,\ Q \in H^0_{x_1x_2x_3}$

(11b) $\theta_5\{x_5 x_1 h(x_2, x_3, x_4)x_5\} = Qx_5{}^2,\ Q \in H^0_{x_1x_2x_3x_4}.$

Each of these is a proposition about an explicit finite space R_M of dimension at most 5! over Z_5. A FORTRAN program which the authors have used to verify these propositions is reproduced in the Appendix. The algorithm employed by the program is explained in §5.

Remark. Characteristic 5 is crucial in the proof that φ_n maps H_n^0 into zero. If division by 5 were possible, (3.2) would no longer be valid. In fact if we replace characteristic 5 by characteristic 7, and use identity $g(x, y) = 0$, then $x^2y^2 = 0$ (cf. Higgins [4], Theorem 3).

Proposition 3. *If $M = M_1 M_2$ is any monomial in R_n, then $M_1 M_2 \equiv M_2 M_1$ modulo H_n^0.*

Proof. The proposition is trivial for $n = 1$. Assume therefore that $n \geqslant 2$ and the proposition has been proved for smaller values of n. Clearly it will suffice to prove $M = yM_0 \equiv M_0 y \pmod{H_n^0}$, where y is an indeterminate. Since $n \geqslant 2$ there is some indeterminate $x_j \neq y$ such that M has the form $M = yN_1 x_j N_2 x_j N_3$. Applying Propositions 1 and 2 we have

$$yM_0 \equiv \theta_j(yM_0) = yN_1 PN_3 x_j^2 \pmod{H_n^0}$$
$$M_0 y \equiv \theta_j(M_0 y) = N_1 PN_3 y x_j^2 \pmod{H_n^0},$$

where $\theta_j(x_j N_2 x_j) = Px_j^2$. From the inductive hypothesis it follows that $yN_1 PN_3 = N_1 PN_3 y \pmod{H_{n-1}^0}$. Thus $yM_0 \equiv M_0 y \pmod{H_n^0}$ as required.

The proof of Lemma 3 will involve repeated applications of the simple observation contained in the following proposition.

Proposition 4. *Suppose $P(t_1, t_2, \ldots t_m)$ is a polynomial in $H^0_{t_1 t_2 \ldots t_m}$. Suppose further that $P(C_1, \ldots, C_m)$ is a polynomial in R_n, where each C_i is a commutator of degree d_i, $1 \leqslant i \leqslant m$, and the sum of the degrees of any three of these commutators is less than d. Suppose finally that $M_1 h(C, C', C'')M_2 \in R_n$ implies $M_1 h(C, C', C'')M_2 \in H_n^0$ whenever the sum of the degrees of the commutators C, C', C'' is less than d. Then $P(C_1, \ldots, C_m) \in H_n^0$.*

Proof. By the first hypothesis, $P(t_1, \ldots, t_m)$ is the sum of polynomials of the form $N_1 h(t_i, t_j, t_k)N_2$. Hence $P(C_1, \ldots, C_m)$ is the sum of polynomials of the form $P_1 h(C_i, C_j, C_k)P_2$, each of which is in H_n^0 by the third hypothesis and the assumption concerning the degree of the commutators C_i.

We can now prove the following lemma, which, combined with Lemma 2 proves Theorem 1′.

Lemma 3. $H_n = H_n^0$.

Proof. We must show that every generator (2.1) of H_n lies in H_n^0. In view of Proposition 3, we may restrict attention to generators of the form

$$G = M_1 h(C', C'', C''').$$

We prove the assertion $G \in H_n^0$ by induction on the pair of parameters $d \geqslant 3$ and $n \geqslant 1$, where $d = d_1 + d_2 + d_3$ is the sum of the degrees of the commutators C', C'', C'''. Clearly the assertion is trivially true for any n if $d = 3$ and vacuously true for any d if $n = 1$. Consider, therefore, any generator G with $d > 3$ and $n > 1$ and assume the inductive hypothesis that the assertion holds when either d or n are smaller. We distinguish four cases.

Case I. C'' has degree at least 2. Then the G has the form $G = M_1 h(C_1, [C_2, C_3], C_4)$. Note that no three of the commutators C_1, C_2, C_3, C_4 and the commutators of degree 1 (indeterminates) in M_1 have a sum of degrees equal to d. The desired conclusion $G \equiv 0 \pmod{H_n^0}$ follows from Proposition 4, the inductive hypothesis in d, and the identity

$$h(w, [x, y], z) \equiv 0 \pmod{H_{wxyz}^0}. \tag{3.6}$$

This identity and the identities (3.7), (3.9), (3.10) and (3.11) below can all be verified by the method described in § 5.

Case II. Either C' or C''' is of degree at least 3. Then G has the form $G = M_1 h([[C_1, C_2], C_3], C_4, C_5)$. Applying Proposition 4, the inductive hypothesis in d, and the identity

$$[[x, y], z] \equiv -2(xz + zx)y + 2(yz + zy)x \pmod{H_{xyz}^0} \tag{3.7}$$

we get

$$\begin{aligned} G \equiv & -2M_1 h((C_1 C_3 + C_3 C_1)C_2, C_4, C_5) + \\ & + 2M_1 h((C_2 C_3 + C_3 C_2)C_1, C_4, C_5) \pmod{H_n^0}. \end{aligned} \tag{3.8}$$

Applying Proposition 4, the inductive hypothesis in d, and the identity

$$x(yz + zy) \equiv (yz + zy)x \pmod{H^0_{xyz}} \tag{3.9}$$

repeatedly we may commute $(C_1 C_3 + C_3 C_1)$ in the first term of (3.8) to the left until it is outside the influence of h. Applying the same process to the second term as well, we have

$$G \equiv P_1 h(C_2, C_4, C_5) + P_2 h(C_1, C_4, C_5) \pmod{H^0_n},$$

which is equivalent to 0 modulo H^0_n by the inductive hypothesis on d.

Case III. G has one of the forms $M_1 h([x, z], y, y)$, $M_1 h([x,y], y, z)$, or $M_1 h([x, y], y, [z, w])$. This case is dismissed immediately by the identities

$$\begin{aligned} &h([x, z], y, y) \equiv 0 \pmod{H^0_{xy^2z}} \\ &h([x, y], y, z) \equiv 0 \pmod{H^0_{xy^2z}}. \end{aligned} \tag{3.10}$$

For $M_1 h([x, y], y, [z, w])$ we use the second identity, induction on d, and Proposition 4.

Case IV. $G = N_1 y N_2 h([w, x], y, C_1)$. Of course, by Proposition 2 and the argument preceding (3.3),

$$G \equiv N_1 P_0 y h([w, x], y, C_1) + N_1 P_1 y s h([w, x], y, C_1) \pmod{H^0_n}.$$

By Proposition 4, the inductive hypothesis in d, and the identities

$$\begin{aligned} &y h([w, x], y, z) \equiv 0 && \pmod{H^0_{y^2wxz}} \\ &y s h([w, x], y, z) \equiv -y^2 h([w, x], s, z) && \pmod{H^0_{y^2swxz}} \end{aligned} \tag{3.11}$$

we have

$$G \equiv -N_1 P_1 y^2 h([w, x], s, C_1) \equiv -y^2 N_1 P_1 h([w, x], s, C_1) \pmod{H^0_n}.$$

But by the inductive hypothesis on n, $N_1 P_1 h([w, x], s, C_1) \equiv 0$ (mod $H^0_{M'}$), where $N_1 N_2 w x C_1 \in R_{M'}$. Thus $G \equiv 0$ modulo H^0_n. This completes the proof of the lemma.

4. A non-solvable group of exponent 5

Assume now that R has an identity. Let G be the multiplicative semigroup of $R/(H + U)$ generated by all elements of the form $e_i = 1 + x_i$, where U is the ideal of R spanned by all monomials with at least one indeterminate factor repeated. It is readily seen that $e_i^4 \equiv 1 - x_i$ (mod U) is an inverse of e_i in G. Hence G is a group. We defer until the end of this section a proof that G has exponent 5. We show now that G is not solvable.

Consider the group commutator $(e_1, e_2) = e_1^{-1} e_2^{-1} e_1 e_2$. We have

$$\begin{aligned}(e_1, e_2) &\equiv (1 - x_1)(1 - x_2)(1 + x_1)(1 + x_2)\\ &\equiv 1 + [x_1, x_2] - x_1^2 + x_1 x_2 x_1 + \ldots\\ &\equiv 1 + [x_1, x_2] \qquad \text{(mod } U\text{)}.\end{aligned}$$

Moreover

$$(e_1, e_2)^{-1} = (e_2, e_1) \equiv 1 - [x_1, x_2] \quad \text{(mod } U\text{)}.$$

And in general any (compound) group commutator formed exclusively from the generators e_i is equivalent modulo U to 1 plus the corresponding ring commutator in the indeterminates x_i. In particular $1 + C_s$ is a member of the s-fold derived group $G^{(s)}$ of G, where C_s is any member of the following sequence of ring commutators:

$$C_0 = x_1$$
$$C_1 = [x_1, x_2]$$
$$C_2 = [[x_1, x_2], [x_3, x_4]]$$

$$C_3 = [[[x_1, x_2], [x_3, x_4]], [[x_5, x_6], [x_7, x_8]]]$$

etc.

Now suppose, in order to reach a contradiction, that G is solvable, i.e., $C_s \equiv 0 \pmod{H + U}$ for some s. (We may suppose $s \geqslant 3$.) Then obviously $C_s \equiv 0 \pmod{H}$, and it follows by substitution of indeterminates that $C_s{}^* \equiv 0 \pmod{H}$, where

$$C_2{}^* = [[x_1, x_2], [x_1, x_3]]$$
$$C_3{}^* = [[[x_1, x_2], [x_1, x_3]], [[x_4, x_2], [x_4, x_5]]]$$
$$C_4{}^* = [C_3{}^*, [[[x_6, x_7], [x_6, x_3]], [[x_8, x_7], [x_8, x_9]]]]$$

etc.

Using the methods in §5 it is easy to verify the identity

(4.1) $$[[x, y], [x, z]] \equiv 2x^2 [y, z] \pmod{H^0_{x^2 yz}}.$$

By repeated application of this identity we have

$$C_2{}^* \equiv +2x_1{}^2 [x_2, x_3] \qquad \pmod{H}$$
$$C_3{}^* \equiv [+2x_1{}^2 [x_2, x_3], +2x_4{}^2 [x_2, x_5]]$$
$$\equiv 4x_1{}^2 x_4{}^2 [[x_2, x_3], [x_2, x_5]]$$
$$\equiv -2x_1{}^2 x_2{}^2 x_4{}^2 [x_3, x_5] \qquad \pmod{H}$$

and in general, if $s \geqslant 3$,

$$C_s{}^* \equiv -2x_1{}^2 x_2{}^2 \ldots x_t{}^2 x_{t+2}^2 \ldots x_{2t}^2 [x_{t+1}, x_{2t+1}] \pmod{H}$$
$$t = 2^{s-2}.$$

By an application of the identity

(4.2) $$[x_1, x_2] x_1 x_2 \equiv 2x_1{}^2 x_2{}^2 \pmod{H_2^0}$$

it follows that

$$0 \equiv C_s{}^* x_{t+1} x_{2t+1} \equiv +x_1{}^2 x_2{}^2 \ldots x_{2t+1}^2 \pmod{H_{2t+1}}.$$

Since this contradicts Theorem 1′, we conclude that G is not solvable.

It remains to be shown that G has exponent 5. Every element of G has the form

$$\begin{aligned} g &= (1 + x_{i_1})(1 + x_{i_2}) \ldots (1 + x_{i_n}) \\ &= 1 + P(x_{i_1}, x_{i_2}, \ldots, x_{i_n}) \end{aligned}$$

for some n. Now

$$g^5 = (1 + P)^5 \equiv 1 + P^5 \pmod{5}. \tag{4.3}$$

We shall show that $P^3 \equiv 0$ modulo $H + U$. To do so we need to introduce a family of polynomials T_3 described by Bruck in [2].

For each $n \geqslant 1$ the polynomial $T_3(x_1, \ldots, x_n)$ is defined to be the multilinear part (homogeneous of total degree n and of degree one in each of $x_1, \ldots, x_n$) of the polynomial

$$\{P(x_1, \ldots, x_n)\}^3 = \{(1 + x_1)(1 + x_2) \ldots (1 + x_n) - 1\}^3.$$

It is observed on p. S2.1 of [2] that

$$\begin{aligned} T_3(\ldots, y_i, y_{i+1}, \ldots) &= T_3(\ldots, y_{i+1}, y_i, \ldots) + \\ &\quad + T_3(\ldots, [y_i, y_{i+1}], \ldots). \end{aligned} \tag{4.4}$$

(Note that the second term on the right has fewer arguments.) It is readily computed that $T_3(x_1)$ and $T_3(x_1, x_2)$ are zero and $T_3(x_1, x_2, x_3)$ is just the symmetric function $S(x_1, x_2, x_3)$ already mentioned in (3.1). Let $\mathcal{I}$ denote the ideal of R spanned by elements of the form

$$M_1 T_3(x_{i_1}, \ldots, x_{i_n}) M_2$$

for $n \geqslant 3$. It is apparent from (4.4) and the definition given in § 2 that $\mathcal{G}$, as well as H, is a Lie substitution ideal.

Lemma 4. $\mathcal{G} \subset H$.

Proof. Since $\mathcal{G}$ and H are both Lie substitution ideals, it will suffice to show that the following proposition holds for all n:

> If $y_1, ..., y_n$ are commutators in $x_1, ..., x_m$ such that $y_1 y_2 ... y_n \in R_{x_1 x_2 ... x_m}$, then $T_3(y_1, ..., y_n) \equiv 0 \pmod{H}$.

This proposition is true trivially for $n \leqslant 2$. Moreover, since H is a Lie substitution ideal, it suffices to prove the proposition for a fixed but arbitrary $n \geqslant 3$ in the special case each of the commutators y_i is just the indeterminate x_i, assuming the inductive hypothesis that the proposition is true (without restriction on the degrees of the commutators) for smaller values of n. Now by (4.4) and the inductive hypothesis, $T_3(x_1, ..., x_n)$ is equivalent modulo H_M to the same expression with the indeterminates permuted in any way. Thus

$$
\begin{aligned}
T_3(x_1, ..., x_n) &\equiv 6T_3(x_1, ..., x_n) \\
&\equiv S^*\{T_3(x_1, ..., x_n)\} \pmod{H_M},
\end{aligned}
$$

where $S^*\{P\}$ denotes the sum of P and the five other polynomials obtained from P by permuting the three indeterminates x_1, x_2, and x_3. Clearly, $S^*\{T_3(x_1, ..., x_n)\}$ is the sum of polynomials of the form $Q = S^*\{N_1 x_1 N_2 x_2 N_3 x_3 N_4\}$. From the properties of the Collection Algorithm, including Proposition 2, it is clear that each Q is equivalent modulo H_M to a sum of expressions of the form

$$
\begin{array}{ll}
S^*\{N' x_1 x_s x_2 x_t x_3 N''\} & S^*\{N' x_1 x_2 x_t x_3 N''\} \\
S^*\{N' x_1 x_s x_2 x_3 N''\} & S^*\{N' x_1 x_2 x_3 N''\}.
\end{array}
$$

But each of these is in H_M as a consequence of identities

(4.5) $S^*\{x_1x_4x_2x_5x_3\} \equiv 0 \pmod{H^0_{x_1x_2x_3x_4x_5}}$

etc.

which can be verified by the methods of § 5. Thus each Q and hence $T_3(x_1, ..., x_n)$ is in H_M. This completes the proof of the lemma.

We may now complete the proof that G has exponent 5. In view of (4.3) and the fact that $H + U$ is an I-substitution ideal, it suffices to show

$$P^3 = \{(1+x_1)(1+x_2)...(1+x_n)-1\}^3 \equiv 0 \pmod{H+U}.$$

Certain of the terms in the expansion of P^3 are in U. It is not very difficult to see that the remaining terms may be regrouped in the form

$$\Sigma T_3(x)$$

where the sum is taken over all (increasing) subsequences

$$x = (x_{j_1}, x_{j_2}, ..., x_{j_s})$$

of $(x_1, ..., x_n)$, for $3 \leqslant s \leqslant n$. Thus $P^3 \in \mathcal{T} + U$. By Lemma 4, $P^3 \in H + U$, which is what we proposed to show.

Remarks. We have given a more or less direct proof that G is not solvable. Other proofs are possible. Thus a variation on the proof given above will show that G is not nilpotent. Then by a theorem of Tobin [8] the group G of prime exponent can not be solvable. (Actually Tobin speaks only of finitely generated groups, but the proofs can be adapted to the general case.) Yet another approach, which the authors used originally [1], is to construct a group $\mathcal{G}$ which is the split extension of an abelian normal subgroup A by G. Nilpotency of $\mathcal{G}$ implies $x_1x_2... x_n \equiv 0 \pmod{H+U}$, which contradicts Theorem 1. Thus $\mathcal{G}$ is not nilpotent. By the theorem of Tobin, $\mathcal{G}$ is not solvable, and consequently G is not solvable.

5. A reduction algorithm

In §3 and §4 we have made use of more than 20 identities which hold modulo H_M^0 in various spaces R_M. In this section we present an algorithm for automating the verification of these identities. We do not know if the Reduction Algorithm given below is capable of verifying every valid identity in R_M modulo H_M^0 for any M, but we do not need so strong a result.

As a minor convenience, let us make the following definition: We say that a monomial $M = x_{i_1} x_{i_2} \dots x_{i_m}$ has *Property A* if $i_l < i_{l+2}$ for $1 \leqslant l \leqslant m-2$. Note that a monomial can not have Property A if it has degree 3 or more in any one indeterminate.

Reduction Algorithm. A polynomial P in some R_{M*} is given. It is desired to produce a polynomial $\psi(P)$ such that $P \equiv \psi(P)$ modulo H_{M*}^0 and such that each monomial of $\psi(P)$ has Property A.

(1) Set Q equal to P.

(2) Choose any term βM of Q, $\beta \neq 0$ (mod 5), which does not have Property A. If no such term can be found, go to step 6. Otherwise, go to step 3.

(3) Determine (possibly empty) monomials M_1 and M_2 such that the monomial M chosen in step 2 has the form $M_1 x_i x_j x_k M_2$, where $i \geqslant k$, but $M_1 x_i x_j$ has Property A.

(4) Perform one of the following operations on Q.

(a) If $i = j = k$, delete βM from Q.

(b) If $i = j > k$, replace βM by $\beta M_1 x_k x_i^2 M_2$.

(c) If $i > j = k$, replace βM by $\beta M_1 x_k^2 x_i M_2$.

(d) If $i = k < j$, replace βM by $-2\beta M_1 x_i^2 x_j M_2$.

(e) If $i = k > j$, replace βM by $-2\beta M_1 x_j x_i^2 M_2$.

(f) If $i > j > k$ or $i > k > j$, replace βM by

$$-\beta M_1 x_k x_j x_i M_2 - 2\beta M_1 x_j x_k x_i M_2 - 2\beta M_1 x_j x_i x_k M_2 .$$

(g) If $j > i > k$, replace βM by

$$-\beta M_1 x_k x_j x_i M_2 - 2\beta M_1 x_k x_i x_j M_2 - 2\beta M_1 x_i x_k x_j M_2 .$$

(5) Collect like terms of Q, reduce all coefficients modulo 5, and go to step 2.
(6) Set $\psi(P)$ equal to Q and stop.

Proposition 5. *For any M^* and any $P \in R_{M^*}$, the Reduction Algorithm is well-formed, terminates in a finite number of steps, and yields a unique polynomial $\psi(P)$ independent of the order in which terms are chosen in step 2. Moreover ψ is a linear map of R_{M^*} into itself, $P \equiv \psi(P)$ modulo $H^0_{M^*}$, and each term of $\psi(P)$ has Property A.*

Proof. The proof is essentially the same as for Proposition 1, except that observation (iv) must be replaced by: (iv′) Each of the operations given in step 4 either deletes βM or replaces it by a polynomial, each of whose terms is lexicographically less than βM. (The term $\beta' M'$ is lexicographically less than βM if $M = M_1 x_i M_2$, $M' = M_1 x_j M_2'$, and $i > j$.)

As mentioned above, we do not know whether the Reduction Algorithm will reduce every element of H^0_M to the zero polynomial. However, any element of R_M which does reduce to zero must be in H^0_M. We have found in practice that each of the elements of H^0_M corresponding to the identities used in §3 and §4, as well as all others experimented with in the course of finding reasonably direct proofs, were reduced to zero by the Reduction Algorithm. Nevertheless, it is conceivable that the Reduction Algorithm might fail to completely reduce some of the expressions listed below if the indeterminates were renumbered. It is for this reason that we take the trouble to list them explicitly.

(3.6) $h(x_1, [x_2, x_3], x_4) \equiv 0$

(3.7) $[[x_1, x_2], x_3] + 2(x_1x_3 + x_3x_1)x_2 - 2(x_2x_3 + x_3x_2)x_1 \equiv 0$

(3.9) $x_1(x_2x_3 + x_3x_2) - (x_2x_3 + x_3x_2)x_1 \equiv 0$

(3.10a) $h([x_1, x_2], x_3, x_3) \equiv 0$

(3.10b) $h([x_1, x_2], x_2, x_3) \equiv 0$

(3.11a) $x_1 h([x_2, x_3], x_1, x_4) \equiv 0$

(3.11b) $x_1 x_2 h([x_3, x_4], x_1, x_5) + x_1{}^2 h([x_3, x_4], x_2, x_5) \equiv 0$

(4.1) $[[x_1, x_2], [x_1, x_3]] - 2x_1{}^2[x_2, x_3] \equiv 0$

(4.2) $[x_1, x_2]x_1 x_2 - 2x_1{}^2 x_2{}^2 \equiv 0$

(4.5a) $S^*\{x_1 x_4 x_2 x_5 x_3\} \equiv 0$

(4.5b) $S^*\{x_1 x_4 x_2 x_3\} \equiv 0$

(4.5c) $S^*\{x_1 x_2 x_4 x_3\} \equiv 0$

(4.5d) $S^*\{x_1 x_2 x_3\} = S(x_1, x_2, x_3) \equiv 0$

We conclude with an observation concerning the Collection Algorithm and Lemma 3. As is shown in the final example in the Appendix, $\psi h([x_1, x_2], x_3, x_4)$ is not zero. As already noted, this does not immediately prove that $h([x_1, x_2], x_3, x_4)$ is not an element of $H^0_{x_1 x_2 x_3 x_4}$. However, a tedious but straightforward hand computation shows that the six monomials in $R_{x_1 x_2 x_3 x_4}$ with Property A form a basis for $R_{x_1 x_2 x_3 x_4}$ modulo $H^0_{x_1 x_2 x_3 x_4}$. It follows that $h([x_1, x_2], x_3, x_4)$ is not an element of $H^0_{x_1 x_2 x_3 x_4}$ and consequently $H^0_{x_1 x_2 x_3 x_4} \neq H_{x_1 x_2 x_3 x_4}$, in contrast with the result in Lemma 3. This observation points up the importance of Case IV and the induction on n in Lemma 3.

Appendix

Reproduced below for the convenience of the reader is a FORTRAN program which the authors have used to verify all the assertions in (3.4) and (3.5) and the identities listed at the end of §5. The program output for typical cases (and the more difficult cases) is also included.

We remark that the verification of most of the cases in (3.5) and at the end of §5 can be done routinely by hand. Indeed, the original proof of the authors as outlined in [1] did not rely on a computer at all, either directly or indirectly as an experimental tool.

The proofs presented here achieve a reduction in complexity and overall length at the modest expense of verifying a few complicated identities.

```
      COMMON INDET(10,200), ICOEF(200), ITEMP(10), ITCOEF
      COMMON LENGTH, N, ITERM, LASTRM
      LOGICAL COLECT, REDUCE
C           LENGTH        =  NUMBER OF FACTORS IN EACH TERM.
C           N             =  INDETERMINATE TO BE COLLECTED.  (IF  N = 0
C                              ONLY THE REDUCTION ALGORITHM IS TO BE
C                              PERFORMED.)
C           LASTRM        =  TOTAL NUMBER OF TERMS CURENTLY IN  Q.
C           ICOEF(I)      =  COEFFICIENT OF  ITH  TERM OF POLYNOMIAL  Q.
C           INDET(L,I)    =  SUBSCRIPT OF  LTH  INDETERMINATE FACTOR
C                              OF  ITH  TERM OF THE POLYNOMIAL  Q.
C           ITERM         =  TERM OF  Q  CURENTLY BEING EXAMINED.
C                              (ALL PREVIOUS TERMS HAVE BEEN
C                              COLLECTED/REDUCED.)
   10 CALL INPUT
      LIMIT = LENGTH - 2
      COLECT = .FALSE.
      REDUCE = .TRUE.
      IF(N.LE.0) GO TO 90
      COLECT = .TRUE.
      REDUCE = .FALSE.
   90 IF(LIMIT.LE.0) GO TO 200
      ITERM = 0
  100 ITERM = ITERM + 1
      IF(ITERM.GT.LASTRM) GO TO 200
C     FIND FIRST POSITION IN TERM WHERE OPERATION IS POSSIBLE.
      DO 110 L = 1,LIMIT
      I = INDET(L,ITERM)
      J = INDET(L+1,ITERM)
      K = INDET(L+2,ITERM)
      IF(COLECT.AND.I.EQ.N  .OR.
     1     REDUCE.AND.I.GE.K.AND.J.NE.N.AND.K.NE.N) GO TO 115
  110 CONTINUE
C     NO OPERATION POSSIBLE ON THIS TERM.
      GO TO 100
C 115 COPY ORIGINAL TERM AND DELETE FROM POLYNOMIAL.
  115 ITCOEF = ICOEF(ITERM)
      DO 120 LL = 1,LENGTH
  120 ITEMP(LL) = INDET(LL,ITERM)
      IDEL = ITERM
      CALL DELETE(IDEL)
      IF(REDUCE) GO TO 130
C     DETERMINE WHICH COLLECTION OPERATION TO PERFORM.
      IF(J.EQ.N.AND.K.NE.N) GO TO 140
      IF(J.NE.N.AND.K.EQ.N) GO TO 170
      IF(J.NE.N.AND.K.NE.N) GO TO 180
      STOP 1
C 130 DETERMINE WHICH REDUCTION OPERATION TO PERFORM.
  130 IF(I.EQ.J.AND.J.EQ.K) GO TO 100
      IF(I.EQ.J.AND.J.GT.K) GO TO 140
      IF(I.GT.J.AND.J.EQ.K) GO TO 150
      IF(I.EQ.K.AND.K.LT.J) GO TO 160
      IF(I.EQ.K.AND.K.GT.J) GO TO 170
      IF(I.GT.J.AND.J.GT.K  .OR.  I.GT.K.AND.K.GT.J) GO TO 180
      IF(J.GT.I.AND.I.GT.K) GO TO 190
      STOP 2
```

```
C 140 REPLACE  XXY  BY  YXX.
  140 ITEMP(L) = K
      ITEMP(L+2) = I
      CALL ADTERM
      GO TO 100
C 150 REPLACE  XYY  BY  YYX .
  150 ITEMP(L) = K
      ITEMP(L+2) = I
      CALL ADTERM
      GO TO 100
C 160 REPLACE  XYX  BY  -2XXY
  160 ITCOEF = MOD(-2*ITCOEF,5)
      ITEMP(L+2) = J
      ITEMP(L+1) = K
      CALL ADTERM
      GO TO 100
C 170 REPLACE  XYX  BY  -2YXX .
  170 ITCOEF = MOD(-2*ITCOEF,5)
      ITEMP(L) = J
      ITEMP(L+1) = I
      CALL ADTERM
      GO TO 100
C180 REPLACE  XYZ  BY  -ZYX -2YZX -2YXZ .
  180 ITCOEF = MOD(-ITCOEF,5)
      ITEMP(L) = K
      ITEMP(L+2) = I
      CALL ADTERM
      ITCOEF = MOD(2*ITCOEF,5)
      ITEMP(L) = J
      ITEMP(L+1) = K
      CALL ADTERM
      ITEMP(L+2) = K
      ITEMP(L+1) = I
      CALL ADTERM
      GO TO 100
C 190 REPLACE  XYZ  BY  -ZYX -2ZXY -2XZY .
  190 ITCOEF = MOD(-ITCOEF,5)
      ITEMP(L) = K
      ITEMP(L+2) = I
      CALL ADTERM
      ITCOEF = MOD(2*ITCOEF,5)
      ITEMP(L+2) = J
      ITEMP(L+1) = I
      CALL ADTERM
      ITEMP(L+1) = K
      ITEMP(L) = I
      CALL ADTERM
      GO TO 100
C 200 COLLECTION OR REDUCTION ALGORITHM COMPLETED.
  200 IF(REDUCE) GO TO 250
      WRITE(6,201)
  201 FORMAT(/,' RESULT OF COLLECTION ALGORITHM ...',/)
      CALL OUTPUT
C     IF POLYNOMIAL IS ZERO, GO TO NEXT CASE.
      IF(LASTRM.EQ.0) GO TO 10
      REDUCE = .TRUE.
      COLECT = .FALSE.
      GO TO 90
  250 WRITE(6,251)
  251 FORMAT(/,' RESULT OF REDUCTION ALGORITHM ...',/)
      CALL OUTPUT
      GO TO 10
      END
```

```
      SUBROUTINE INPUT
      COMMON INDET(10,200), ICOEF(200), ITEMP(10), ITCOEF
      COMMON LENGTH, N, ITERM, LASTRM
      DIMENSION TITLE(20)
      READ(5,100) TITLE
  100 FORMAT(20A4)
      WRITE(6,100) TITLE
      READ(5,101) LENGTH, N, LASTRM
  101 FORMAT(3I5)
      WRITE(6,102) LENGTH, N, LASTRM
  102 FORMAT(/,' LENGTH ='I3,' N =',I3,', NO. OF TERMS ='I5,/)
      DO 10 I = 1,LASTRM
   10 READ(5,103) ICOEF(I),(INDET(LL,I),LL=1,LENGTH)
  103 FORMAT(2X,I2,2X,10I2)
      CALL OUTPUT
      RETURN
      END

      SUBROUTINE OUTPUT
      COMMON INDET(10,200), ICOEF(200), ITEMP(10), ITCOEF
      COMMON LENGTH, N, ITERM, LASTRM
      DATA X/' X  '/
      IF(LASTRM.GT.0) GO TO 10
      WRITE(6,101)
      RETURN
   10 DO 20 I = 1,LASTRM
   20 WRITE(6,102) ICOEF(I),(X,INDET(LL,I),LL=1,LENGTH)
      RETURN
  101 FORMAT('      POLYNOMIAL IS ZERO.')
  102 FORMAT(5X,I2,10(A2,I1))
      END

      SUBROUTINE DELETE(/IDEL/)
      COMMON INDET(10,200), ICOEF(200), ITEMP(10), ITCOEF
      COMMON LENGTH, N, ITERM, LASTRM
C     DELETE TERM NUMBER IDEL FROM  Q.  ADJUST  LASTRM  AND  ITERM.
      DO 10 I = IDEL,LASTRM
      ICOEF(I) = ICOEF(I+1)
      DO 10 LL = 1,LENGTH
   10 INDET(LL,I) = INDET(LL,I+1)
      IF(ITERM.GE.IDEL) ITERM = ITERM - 1
      LASTRM = LASTRM - 1
      RETURN
      END


      SUBROUTINE ADTERM
C     ADDS TERM IN ITEMP TO THE POLYNOMIAL.
      COMMON INDET(10,200), ICOEF(200), ITEMP(10), ITCOEF
      COMMON LENGTH, N, ITERM, LASTRM
      DO 20 I = 1,LASTRM
      DO 10 LL = 1,LENGTH
   10 IF(INDET(LL,I).NE.ITEMP(LL)) GO TO 20
C     TERM IN TEMP COMBINES WITH TERM NUMBER I OF THE POLYNOMIAL.
      ICOEF(I) = MOD(ICOEF(I) + ITCOEF,5)
      IF(ICOEF(I).NE.0) RETURN
C     TERMS CANCEL.  DELETE.
      CALL DELETE(I)
      RETURN
   20 CONTINUE
C     TERM DOES NOT COMBINE.  ADD NEW TERM AT END OF POLYNOMIAL.
      LASTRM = LASTRM + 1
      IF(LASTRM.GT.200) STOP 3
      ICOEF(LASTRM) = ITCOEF
      DO 30 LL = 1,LENGTH
```

```
30 INDET(LL,LASTRM) = ITEMP(LL)
   RETURN
   END
```

```
          X6X1H(X2,X3,X4)X5

LENGTH =   6 N =   6, NO. OF TERMS =     4

      1 X6 X1 X2 X3 X4 X5
      1 X6 X1 X4 X3 X2 X5
      2 X6 X1 X3 X2 X4 X5
      2 X6 X1 X3 X4 X2 X5

RESULT OF COLLECTION ALGORITHM ...

      1 X2 X1 X4 X3 X6 X5
     -2 X2 X1 X3 X4 X6 X5
     -2 X1 X2 X4 X3 X6 X5
      1 X4 X1 X2 X3 X6 X5
     -2 X4 X1 X3 X2 X6 X5
     -2 X1 X4 X2 X3 X6 X5
     -2 X3 X1 X4 X2 X6 X5
     -2 X3 X1 X2 X4 X6 X5
     -1 X1 X3 X4 X2 X6 X5
     -2 X2 X1 X3 X5 X4 X6
     -4 X2 X1 X3 X4 X5 X6
      1 X1 X2 X3 X5 X4 X6
     -3 X1 X2 X3 X4 X5 X6
     -2 X1 X2 X3 X4 X6 X5
     -2 X4 X1 X3 X5 X2 X6
     -4 X4 X1 X3 X2 X5 X6
      1 X1 X4 X3 X5 X2 X6
     -3 X1 X4 X3 X2 X5 X6
     -2 X1 X4 X3 X2 X6 X5
     -4 X3 X1 X2 X5 X4 X6
     -3 X3 X1 X2 X4 X5 X6
     -2 X1 X3 X2 X5 X4 X6
      1 X1 X3 X2 X4 X5 X6
     -4 X3 X1 X4 X5 X2 X6
     -3 X3 X1 X4 X2 X5 X6
     -2 X1 X3 X4 X5 X2 X6
      1 X1 X3 X4 X2 X5 X6
     -1 X1 X3 X2 X4 X6 X5

RESULT OF REDUCTION ALGORITHM ...

     POLYNOMIAL IS ZERO.
```

```
        X1X2H((X3,X4),X1,X5) + X1X1H((X3,X4),X2,X5)

LENGTH =  6 N =  0, NO. OF TERMS =    16

     1 X1 X2 X3 X4 X1 X5
     1 X1 X2 X5 X1 X3 X4
     2 X1 X2 X1 X3 X4 X5
     2 X1 X2 X1 X5 X3 X4
    -1 X1 X2 X4 X3 X1 X5
    -1 X1 X2 X5 X1 X4 X3
    -2 X1 X2 X1 X4 X3 X5
    -2 X1 X2 X1 X5 X4 X3
     1 X1 X1 X3 X4 X2 X5
     1 X1 X1 X5 X2 X3 X4
     2 X1 X1 X2 X3 X4 X5
     2 X1 X1 X2 X5 X3 X4
    -1 X1 X1 X4 X3 X2 X5
    -1 X1 X1 X5 X2 X4 X3
    -2 X1 X1 X2 X4 X3 X5
    -2 X1 X1 X2 X5 X4 X3

RESULT OF REDUCTION ALGORITHM ...

     POLYNOMIAL IS ZERO.


        H((X1,X2),X3,X4)

LENGTH =  4 N =  0, NO. OF TERMS =    8

     1 X1 X2 X3 X4
     1 X4 X3 X1 X2
     2 X3 X1 X2 X4
     2 X3 X4 X1 X2
    -1 X2 X1 X3 X4
    -1 X4 X3 X2 X1
    -2 X3 X2 X1 X4
    -2 X3 X4 X2 X1

RESULT OF REDUCTION ALGORITHM ...

    -4 X3 X1 X4 X2
    -4 X1 X3 X2 X4
     1 X1 X2 X4 X3
    -1 X2 X1 X4 X3
     1 X2 X1 X3 X4
     3 X1 X2 X3 X4
```

References

[1] S. Bachmuth, H.Y. Mochizuki, D.W. Walkup, A non-solvable group of exponent 5, Bull. Amer. Math. Soc. 76 (1970) 638–640.
[2] R.H. Bruck, Engel conditions in groups and related questions, Lecture notes, Australian National Univ., Canberra, Jan. 1963.
[3] H. Heineken, Endomorphismenringe und Engelsche Elemente, Arch. der Math. 13 (1962) 29-37.
[4] P.J. Higgins, Lie rings satisfying the Engel condition, Proc. Cambridge Philos. Soc. 50 (1954) 8-15.
[5] G. Higman, On finite groups of exponent 5, Proc. Cambridge Philos. Soc. 52 (1956) 381–390.
[6] A.I. Kostrikin, Solution of the restricted Burnside problem for the exponent 5, Izvestia Akad. Nauk. SSSR, 19 (1955) 233-244;
[7] A.I. Kostrikin, The Burnside problem, Izvestia Akad. Nauk. SSSR, Ser. Mat. 23 (1959) 3-34.
[8] S. Tobin, On a theorem of Baer and Higman, Canad. J. Math. 8 (1956) 263-270.
[9] D.W. Walkup, Lie rings satisfying Engel conditions, Thesis, Univ. of Wisconsin, 1963, 154 pages.

THE EXISTENCE OF INFINITE BURNSIDE GROUPS

J.L. BRITTON
University of Kent at Canterbury

Preface

This paper is a record of research work carried out at the University of Glasgow and the University of Kent at Canterbury; also during visits of about two months each to the University of Western Australia (1965) and the University of Illinois, Urbana (1968).

I first reported my ideas in a brief address at the University of Oxford (1964) and gave progress reports later at the International Conference on the Theory of Groups at Canberra (1965), at the International Congress of Mathematicians, Moscow (1966) and more fully at the University of Illinois, Urbana (1968). However I first reported on the final version to the Edinburgh Mathematical Society in Aberdeen (1970).

It was a personal disappointment that working out the technical details of the method known in principle in 1964, and unchanged since then, should have taken so much time.

I wish to thank the North-Holland Publishing Company and the Organizing Committee of the CODEP conference, W.W. Boone, F.B. Cannonito and R.C. Lyndon, for suggesting that I should present the full text in this volume rather than a report of my address to the conference. My apologies are due to the other contributors since the manuscript reached the publishers six months later than the date then promised. During this time I verified "Wright's Law" that the difficulties of presenting a research paper increase with the cube of its length.

I wish to thank the U.S. Army Research Office-Durham for a travel grant enabling me to attend the conference.

I owe special thanks to J. Mennicke for his valuable comments and for suggesting many improvements to the manuscript. I acknowledge with thanks the support received from the University of Bielefeld and the Deutsche Forschungsgemeinschaft.

Summary

The primary purpose of this paper is to prove, in Chapter II, that the Burnside Group B_d^e on d generators with exponent e ($d \geqslant 2$, $e > 0$) is infinite for all sufficiently large odd e.

This proof is independent of the proof of Novikov and Adyan [5].

The secondary purpose of this paper is to give in Chapter I a self-contained account of a new class of groups, called Generalized Tartakovskii Groups.

We contend that B_d^e is a Generalized Tartakovskii Group, for sufficiently large odd e.

No previous knowledge of the results of Tartakovskii [6] or related results of Greendlinger, Lipschutz, Lyndon, McCool, Schiek, Schupp and the author is assumed. Indeed the entire paper assumes little more that the concepts free product, free group, group given by generators and defining relations.

CHAPTER I

Generalized Tartakovskii Groups

§1. Introduction

The results of this chapter are presented in Theorems 1, 2, 3, 4, 5, 6.

For Chapter II, only the statements of Theorems 4, 5, 6 are required together, of course, with their supporting definitions (including 2.1, 2.8B, 3.1–3.5, 6.1–6.3, 6.6).

§2.

2.1. Notation. Let Π be a free product of groups $G_\gamma (\gamma \in \Gamma)$. An element X of Π is either the identity I or has normal form $a_1 a_2 \ldots a_n$, where $n \geqslant 1$ is the length $L(X)$ of X, $a_i \in G_{\gamma_i}$ $(i = 1, \ldots, n)$ and $\gamma_i \neq \gamma_{i+1}$ $(i = 1, \ldots, n-1)$. We use capital letters for elements of Π, and small letters for elements of Π having unit length. The G_γ are called the *constituent groups* of Π and $\equiv$ denotes equality in $\Pi \cdot x \sim y$ means that x, y belong to the same constituent group. We write In X, Fin X for a_1, a_n respectively.

The product of $X, Y \in \Pi$ is denoted by $X \cdot Y$ but we shall use the "dot convention" that writing $W_1 W_2 \cdots W_k$ instead of $W_1 \cdot W_2 \cdot \cdots \cdot W_k$ indicates that the length of the element is $\Sigma_{i=1}^k L(W_i)$.

Let $X \in \Pi$; if $L(X) \geqslant 2$ and In X, Fin X are in different constituent groups we say that X is *externally reduced* (E.R.).

If $L(X) \geqslant 2$ and In X, Fin X are not mutually inverse we say that X is externally *cancelled* (E.C.).

Let $X \equiv a_1 a_2 \cdots a_n$ be E.R. Then a cyclic arrangement (C.A.) of X is any element $a_i \cdots a_n a_1 \cdots a_{i-1}$ $(i = 1, ..., n)$.

By a split of an element x of length 1 we mean any conjugate $y^{-1} \cdot x \cdot y$ where $y \sim x$; a *split* of an E.C. element $X \equiv a_1 a_2 \cdots a_n$ is defined as follows: put $Y \equiv a_2 \cdots a_n \cdot a_1$ if $a_1 \sim a_n$, otherwise put $Y \equiv X$; thus Y is E.R., say $Y \equiv b_1 b_2 \cdots b_s$; then (i) any element

$$b_i'' b_{i+1} \cdots b_s b_1 \cdots b_{i-1} b_i',$$

$$\text{where } b_i = b_i' \cdot b_i''$$

is a split of X and (ii) any C.A. of Y is a split of X. Let $\mathrm{Sp}(X)$ be the subset of Π consisting of all splits of X; similarly we may write C.A.(X) if X is E.R.

Throughout the paper, we mean by a *word* the normal form of an element of Π; the *empty word* means I.

A *subword* of $W \in \Pi$ is an ordered triple A, B, C of elements of Π such that $W \equiv ABC$; the *value* of the subword is B. Distinct subwords may have the same value, but from now on we shall not distinguish notationally between a subword and its value.

If $W \equiv CD$ we call C a left subword of W and D a right subword of W. A left subword of W is sometimes denoted by W^{L} or W^{l}. Similarly W^{R} or W^{r} denotes a right subword. An arbitrary subword of W is sometimes denoted by W^{M} or W^{m}.

If $X, Y \in \Pi$ we write $\beta(X, Y)$, $\epsilon(X, Y)$ for the number of cancellations and amalgamations respectively in the product $X \cdot Y$. Thus $\epsilon(X, Y) = 0$ or 1 and $L(X, Y) = L(X) + L(Y) - 2\beta(X, Y) - \epsilon(X, Y)$. We may write $X \cdot Y \equiv X^{\mathrm{L}} c^{\epsilon} Y^{\mathrm{R}}$ where $L(X^{\mathrm{L}}) = L(X) - \beta - \epsilon$, $L(Y^{\mathrm{R}}) = L(Y) - \beta - \epsilon$, $\epsilon = \epsilon(X, Y)$, $\beta = \beta(X, Y)$.

By the *dual* of a statement we mean the new statement obtained by interchanging the words "left" and "right".

If $X \not\equiv I$ then an *external cancellation* is the operation of deleting In X and Fin X from X if these are mutually inverse and $L(X) \geqslant 2$. Denote by $E(X)$ the result of carrying out all external cancellations; it will be either E.C. or of length 1.

2.2. The relations $\approx$ 0 ··· 0. The relation $X \approx Y$ where $X, Y \in \Pi$ means Y is E.C. and X is a split of Y.

The relation $X \approx Y \circ Z$ where $X, Y, Z \in \Pi$ means Y, Z are E.C. and there exist X', Y', Z' such that

Y', Z' are splits of Y, Z,

$Y' \cdot Z' \not\equiv I$,

$E(Y' \cdot Z') \equiv X'$,

X is a split of X'.

Note that X has length 1 or is E.C.

(Informally, X arises from the E.C. words Y, Z by the following process. Take splits Y', Z', then externally cancel the word $Y' \cdot Z'$ and finally take a split. Similarly, below, X arises from $Y_1, \ldots, Y_n$ by repetition of this process.)

We consider certain formal expressions involving elements of Π, the symbol $\circ$ and brackets. These are defined inductively as follows. (i) Any element of Π is an admissible expression; (ii) if ξ and η are admissible expressions then so is $(\xi \circ \eta)$.

The relation $X \approx Y_1 \circ Y_2 \circ \cdots \circ Y_n$, where $X, Y_1, \ldots, Y_n \in \Pi$ ($n \geqslant 2$) and where $Y_1 \circ Y_2 \circ \cdots \circ Y_n$ has been assigned brackets making it an admissible expression, means the following. Where $Y_1 \circ Y_2 \circ \cdots \circ Y_n$ has the form $(Y_1 \circ \cdots \circ Y_r) \circ (Y_{r+1} \circ \cdots \circ Y_n)$, there exists A, B such that $A \approx Y_1 \circ \cdots \circ Y_r$, $B \approx Y_{r+1} \circ \cdots \circ Y_n$ and $X \approx A \circ B$. Note that the case $n = 2$ is consistent with the previous definition.

Clearly if $X \approx Y_1 \circ \cdots \circ Y_n$, then $X \equiv T_1^{-1} \cdot Y_1 \cdot T_1 \cdot \cdots T_n^{-1} \cdot Y_n \cdot T_n$ for some $T_1, \ldots, T_n$ in Π.

2.3. We shall consider a subset Ω' of Π satisfying the following conditions and the group $G = \Pi/[\Omega']$ where square brackets denote "normal subgroup generated by".

1. Each element of Ω' is E.R. and $\Omega' = L_1 \cup L_2 \cup \cdots$.

2. L_i is closed with respect to the operations cyclic arrangement and inverse, ($i = 1, 2, \ldots$).

3. The sets L_i are pairwise disjoint, and if L_j is empty then so is L_{j+1}.

2.4. Let $\Omega = \mathrm{Sp}(\Omega')$. For $n = 1, 2, 3, \ldots$, put $X \in C_n$ if there exist $Y_1, \ldots, Y_n$ in Ω such that $X \approx Y_1 \circ \cdots \circ Y_n$ for some bracketing. Thus $C_1 = \Omega$.

2.5. Definition. Let W be a relator, i.e., $W \in [\Omega']$ and let $W \not\equiv I$ so that there exists a representation

$$W \equiv T_1^{-1} \cdot W_1 \cdot T_1 \cdot \cdots \cdot T_n^{-1} \cdot W_n \cdot T_n$$

($T_i \in \Pi$, $W_i \in \Omega'$). The *preweight* of this representation, and of the sequence $W_1, ..., W_n$ means the sequence $(d_1, d_2, ...)$ where d_s is the number of $W_1, ..., W_n$ in L_s. $(d_1, d_2, ...) + (d_1', d_2', ...)$ means $(d_1 + d_1', d_2 + d_2', ...)$. Put $(d_1', d_2', ...) < (d_1, d_2, ...)$ if there exists j such that $d_j' < d_j$ and $k > j \Rightarrow d_k' = d_k$. This totally orders the preweights of the representation of W; it is a well-odering and the *weight* of W means the least preweight of any representation of W. It is denoted by $w(W)$.

If $W \in L_a$ then the weight of W is $\leqslant (0, ..., 0, 1, 0, 0, ...)$ where the 1 occurs in the ath place; if equality holds we call W irredundant.

2.6. Definition. $J \in \Pi$ is a *natural element* if there exist $W_1, ..., W_s$ in Ω' such that (i) $J \approx W_1 \circ \cdots \circ W_s$ with some bracketing and (ii) the weight of J equals the preweight of the sequence $W_1, ..., W_s$. (Each of $W_1, ..., W_s$ is then irredundant.)

2.7. Our first aim is to prove

Theorem 1. *If no natural element has length* 1 *then every non-empty relator has a natural element as a subword, hence has length at least* 2.

2.7A. Let $R_0 \not\equiv I$ be a relator and take a fixed representation

$$(1) \qquad R_0 \equiv \prod_{i=1}^{N} T_i^{-1} \cdot W_i \cdot T_i$$

of smallest preweight. Consider all representations of the form

$$(2) \qquad R_0 \equiv \prod_{i=1}^{h} S_i^{-1} \cdot A_i \cdot S_i$$

where

(3) $A_i \approx W_{\alpha(i,1)} \circ \cdots \circ W_{\alpha(i,q_i)}$.

$\{\alpha(i, 1), ..., \alpha(i, q_i)\}$ being a subset J_i of $J = \{1, 2, ..., N\}$.
It follows that $J = J_1 \cup \cdots \cup J_h$. (1) itself is of this form. Of all representations (2) take those with h least; of these take those with $\Sigma_{i=1}^{h} L(S_i^{-1} \cdot A_i \cdot S_i)$ least; finally of these choose one with least vector $(l_1, ..., l_h)$ where $l_i = L(S_i^{-1} \cdot A_i \cdot S_i)$ and $(l_1, ..., l_h) < (k_1, ..., k_h)$ means that for some i, $l_i < k_i$ and $j < i \Rightarrow l_j = k_j$. Let $M1$ means the choice of (1) and let M2, M3, M4 be the successive restrictions just made on the choice of (2).

Since if (3) holds for A_i it holds for any split of A_i, we may assume $S_i^{-1} A_i S_i$ to be written without dots.

If A_i is expressed in the form $\Pi_{j=1}^{t} U_j^{-1} \cdot V_j \cdot U_j$ $(V_j \in \Omega')$ then the preweight of $V_1, ..., V_t$ is at least that of $W_{\alpha(i,1)}, ..., W_{\alpha(i,q_i)}$

2.8. In the following lemmas we assume that in (2) each A_i has length $\geqslant 2$.

2.8A. Lemma. *The number β of cancellations and the number ϵ of amalgamations between the ith and $(i + 1)$th terms in* (2) *satisfy*

1. $\beta < \text{Max}(L(S_i), L(S_{i+1}))$,
2. $2\beta + \epsilon < l_{i+1}$,
3. $2\beta + \epsilon \leqslant l_i$.

2.8B. Note. If X, Y are two consecutive terms in (2) and $X \cdot Y \equiv X^{\text{L}} c^{\epsilon} Y^{\text{R}}$ (cf. 2.1) then 2.8A implies that $X^{\text{L}} \not\equiv I$ and $Y^{\text{R}} \not\equiv I$.

Proof of 2.8A. In this proof, abbreviate $L(X) < L(Y)$ to $X < Y$.

Write the terms in question $T^{-1}AT$, $S^{-1}BS \equiv E$.

1: Suppose $S \leqslant T$; (there is a dual proof if $T \leqslant S$). Suppose $\beta \geqslant T$. Then $E \equiv T^{-1}E_1$ say and

$$T^{-1}AT \cdot S^{-1}BS \equiv T^{-1}A \cdot E_1 \equiv E \cdot E_1^{-1} \cdot A \cdot E_1 \equiv S^{-1}BS \cdot (E_1^{-1} \cdot A \cdot E_1).$$

By M3, $2T + A \leqslant 2E_1 + A$, i.e., $T \leqslant E_1$, hence

$$T \leqslant \tfrac{1}{2}(T + E_1) = \tfrac{1}{2}E = S + \tfrac{1}{2}B < S + B.$$

Since $T^{-1}E_1 \equiv S^{-1}BS$ we have $T^{-1} \equiv S^{-1}B_1$, $B \equiv B_1B_2$, $E_1 \equiv B_2S$. Let $\hat{B} = B_2 \cdot B_1$. Then

$$T^{-1}AT \cdot S^{-1}BS \equiv S^{-1} \cdot B_1 \cdot A \cdot B_2 \cdot S \equiv S^{-1}B_1 \cdot A \cdot \hat{B} \cdot B_1^{-1}S.$$

A and B are E.C. $\hat{B}$ is a split of B. Now $A \cdot \hat{B} \not\equiv I$ by M1. Let $D = E(A \cdot \hat{B})$; thus $D \approx A \circ B$. Now $A \approx W_{\alpha(i,1)} \circ \cdots \circ W_{\alpha(i,q_i)}$ and $B \approx W_{\alpha(i+1,1)} \circ \cdots \circ W_{\alpha(i+1,q_{i+1})}$ hence

$$D \approx (W_{\alpha(i,1)} \circ \cdots \circ W_{\alpha(i,q_i)}) \circ (W_{\alpha(i+1,1)} \circ \cdots \circ W_{\alpha(i+1,q_{i+1})})$$

contradicting M2.

2: If $\beta \leqslant S$ then $2\beta + \epsilon \leqslant 2S + 1 < 2S + B = l_{i+1}$ so assume $\beta > S$. Now $\beta < \mathrm{Max}(S,T)$ so $\beta < T$, $T \equiv T_1X$, $L(X) = \beta$, $E \equiv X^{-1}E_1$, $T^{-1}AT \cdot S^{-1}BS \equiv X^{-1}T_1^{-1}AT_1 \cdot E_1 \equiv E \cdot M$ where $M \equiv E_1^{-1} \cdot T_1^{-1}AT_1 \cdot E_1$. By M3, $M \geqslant T^{-1}AT$. Now $T_1 \neq I$ so

$$2E_1 + 2T_1 + A - 2\epsilon \geqslant 2T + A = 2T_1 + 2X + A$$

Hence $E - \epsilon \geqslant X$ and $E = X + E_1 \geqslant 2X + \epsilon = 2\beta + \epsilon$.

If $E = 2\beta + \epsilon$ then $E = 2\beta + \epsilon < 2T + \epsilon < 2T + A$ and also $E = X + \epsilon$ hence $L(M) = L(T^{-1}AT)$ contrary to M4.

3: This is proved similarly to 2 except that now equality may occur.

2.8C. Lemma. *Let X, Y, Z be three consecutive terms in* (2). *Writing $X \cdot Y = X^{\mathrm{L}}c^{\epsilon}Y^{\mathrm{R}}$, $Y \cdot Z = Y^{\mathrm{L}}d^{\epsilon'}Z^{\mathrm{R}}$ where ϵ, ϵ' are the number of amalgamations in $X \cdot Y$, $Y \cdot Z$ respectively we have*

$X \cdot Y \cdot Z = X^{\mathrm{L}}HZ^{\mathrm{R}}$ for some H of length $\geqslant 1$ and

$\mathrm{In}\, H \sim \mathrm{In}(c^{\epsilon}Y^{\mathrm{R}})$, $\mathrm{Fin}\, H \sim \mathrm{Fin}(Y^{\mathrm{L}}d^{\epsilon'})$

Also H has the form $c^{\epsilon}Jd^{\epsilon'}$.

Proof. $X \cdot Y \cdot Z = X^{\mathrm{L}}c^{\epsilon}Y^{\mathrm{R}} \cdot Y^{-1} \cdot Y^{\mathrm{L}}d^{\epsilon'}Z^{\mathrm{R}}$.

Case 1. $L(Y^{\mathrm{R}}) + L(Y^{\mathrm{L}}) > L(Y)$.
Here $Y \equiv PQR$, $Y^{\mathrm{L}} \equiv PQ$, $Y^{\mathrm{R}} \equiv QR$ say, $Q \not\equiv I$. Hence

$$X \cdot Y \cdot Z = X^{\mathrm{L}} c^{\epsilon} Q \, d^{\epsilon'} Z^{\mathrm{R}}$$

without dots and the result follows.
Case 2. $L(Y^{\mathrm{R}}) + L(Y^{\mathrm{L}}) < L(Y)$
Here $L(Y) - \beta - \epsilon + L(Y) - \beta' - \epsilon' \leqslant L(Y) - 1$, so

$$(L(Y) - 2\beta - \epsilon) + (L(Y) - 2\beta' - \epsilon') \leqslant \epsilon + \epsilon' - 2$$

By the previous lemma $2\beta + \epsilon < L(Y)$ and $2\beta' + \epsilon' \leqslant L(Y)$. Hence $0 < \epsilon + \epsilon' - 2$, a contradiction.
Case 3. $L(Y^{\mathrm{R}}) + L(Y^{\mathrm{L}}) = L(Y)$.
Modifying the argument of Case 2 we find $0 < \epsilon + \epsilon'$. Now $Y^{\mathrm{L}} \neq I$; for $L(Y) - \beta' - \epsilon' = 0$ implies $2\beta' + \epsilon' \leqslant \beta' + \epsilon'$ hence $\beta' = 0$ and $L(Y) = \epsilon' \leqslant 1$, a contradiction; similarly $Y^{\mathrm{R}} \not\equiv I$. Thus $Y^{\mathrm{L}} = y_1 \cdots y_j$, $Y^{\mathrm{R}} = y_{j+1} \cdots y_p$, $Y = y_1 \cdots y_p$, $1 \leqslant j < p$. Now

$$X \cdot Y \cdot Z = X^{\mathrm{L}} c^{\epsilon} \cdot d^{\epsilon'} Z^{\mathrm{R}}$$

We show that the dot can be removed.
If $\epsilon = \epsilon' = 1$ then $c \sim y_j \not\sim y_{j+1} \sim d$.
If $\epsilon = 0$, $\epsilon' = 1$ then $\mathrm{Fin}\, X^{\mathrm{L}} \not\sim \mathrm{In}\, Y^{\mathrm{R}} = y_{j+1} \sim d$.
If $\epsilon = 1$, $\epsilon' = 0$ then $c \sim y_j = \mathrm{Fin}\, Y^{\mathrm{L}} \not\sim \mathrm{In}\, Z^{\mathrm{R}}$.
Let $H = c^{\epsilon} d^{\epsilon'}$. We show $\mathrm{In}\, H \sim \mathrm{In}(c^{\epsilon} Y^{\mathrm{R}})$; this is trivial if $\epsilon = 1$ so let $\epsilon = 0$. $\mathrm{In}\, H = d \sim y_{j+1} = \mathrm{In}\, Y^{\mathrm{R}}$. Similarly $\mathrm{Fin}\, H \sim \mathrm{Fin}(Y^{\mathrm{L}} d^{\epsilon'})$.

2.8D. Lemma. *Let* $K = B_i \cdot B_{i+1} \cdot \cdots \cdot B_j$ *be a product of at least three consecutive terms of* (2). *Let*

$$\beta' = \beta(B_i, B_{i+1}) \qquad \epsilon' = \epsilon(B_i, B_{i+1})$$

$$\beta = \beta(B_{j-1}, B_j) \qquad \epsilon = \epsilon(B_{j-1}, B_j)$$

Thus we may write

$$B_i \cdot B_{i+1} = B_i^{\mathrm{L}} c^{\epsilon'} B_{i+1}^{\mathrm{R}} \qquad B_{j-1} \cdot B_j = B_{j-1}^{\mathrm{L}} d^{\epsilon} B^{\mathrm{R}};$$

Then $K = B_i^{\mathrm{L}} c^{\epsilon'} Q\, d^{\epsilon} B_j^{\mathrm{R}}$.

Proof. This is true for three factors by the previous lemma. Assume it is true for $K = B_i \cdot \cdots \cdot B_j$; we prove it is true for $B_i \cdot \cdots \cdot B_j \cdot B_{j+1} = K \cdot B_{j+1}$.

Let $B_j \cdot B_{j+1} = B_j^{\mathrm{L}} f^{\epsilon''} B_{j+1}^{\mathrm{R}}$. Now $K = A d^{\epsilon} B_j^{\mathrm{R}}$ and $B_{j-1} \cdot B_j \cdot B_{j+1} = B_{j-1}^{\mathrm{L}} H B_j^{\mathrm{R}}$, $L(H) \geqslant 1$,

$$\mathrm{In}\, H \sim \mathrm{In}(d^{\epsilon} B_j^{\mathrm{R}}), \qquad \mathrm{Fin}\, H \sim \mathrm{Fin}(B_j^{\mathrm{L}} f^{\epsilon''}), \qquad H = d^{\epsilon} J f^{\epsilon''}$$

Hence

$$K \cdot B_{j+1} = A d^{\epsilon} B_j^{\mathrm{R}} \cdot (B_{j-1}^{\mathrm{L}} d^{\epsilon} B_j^{\mathrm{R}})^{-1} \cdot B_{j-1}^{\mathrm{L}} H B_{j+1}^{\mathrm{R}} = A \cdot H B_{j+1}^{\mathrm{R}}.$$

Now $\mathrm{Fin}\, A \nsim \mathrm{In}(d^{\epsilon} B_j^{\mathrm{R}}) \sim \mathrm{In}\, H$. Since $A = B_i^{\mathrm{L}} c^{\epsilon'} Q$ we have $B_i \cdots B_{j+1} = B_i^{\mathrm{L}} c^{\epsilon'} Q\, d^{\epsilon} J f^{\epsilon''} B_{j+1}^{\mathrm{R}}$.

2.9. Lemma. R_0 *contains some* A_i *as a subword.*

Proof. This is so if $h = 1$, so assume $h \geqslant 2$. Consider $X = S_i^{-1} A_i S_i$ where $L(S_i)$ is maximal. If $i = 1$ let $Y = S_2^{-1} A_2 S_2$. Then $\beta(X, Y) < L(S_1)$. Write $X \cdot Y = X^{\mathrm{L}} c^{\epsilon} Y^{\mathrm{R}}$. Then $R_0 = X^{\mathrm{L}} c^{\epsilon} D$. But X^{L} has the form $S_i^{-1} A_i P$, hence A_i is a subword of R_0.

Similarly, if $i = h$. Now let $1 < i < h$. Then

$$B_1 \cdot \cdots \cdot B_i = Q A_i S_i \qquad B_i \cdot \cdots \cdot B_n = S_i^{-1} A_i R.$$

Hence $R_0 \equiv Q A_i S_i \cdot (S_i^{-1} A_i S_i)^{-1} \cdot S_i^{-1} A_i R \equiv Q A_i R$ as required.

2.10. Proof of Theorem 1. This now follows since in (2) each of $A_1, \ldots, A_h$ is natural.

§3.

In this section we introduce some concepts we shall need.

3.1. Definition. Let $W_1, W_2, \ldots, W_n$ $(n \geqslant 1)$ be a sequence in Π of which not all the terms are I. Its *component sequence* and *patches* are as follows.

First define the component sequence of a non-empty word $W = x_1 x_2 \cdots x_m$ to be $x_1, x_2, \ldots, x_m$ and the component sequence of the word I to be the empty sequence. Then the component sequence of $W_1, \ldots, W_n$ is the result of writing in succession the component sequences of $W_1, \ldots, W_n$.

Denote it by $d_1, d_2, \ldots, d_p$. Then $p \geqslant 1$. It may be written $S_1, S_2, \ldots, S_q$ where S_i is a non-empty subsequence all of whose terms belong to the same constituent group G_{γ_i} and $\gamma_i \neq \gamma_{i+1}$ $(i = 1, \ldots, q-1)$. Call $S_1, S_2, \ldots, S_q$ *prepatches.*

If $q \geqslant 2$ and $\gamma_1 = \gamma_q$ the patches are $S_2, S_3, \ldots, S_{q-1}$ and $S_q S_1$. In the opposite case the patches are $S_1, S_2, \ldots, S_q$.

3.2. Definition. A *partition* is a sequence $W_1, \ldots, W_n$ $(n \geqslant 1)$ in Π such that either (i) all terms are I or (ii) the product of the terms in each patch is not I. It is *proper* if no W_i is I. (If $W = W_1 \cdot W_2 \cdot \cdots \cdot W_n$ we say $W_1, \ldots, W_n$ is a partition of W)

3.3. Note. Let W be a partition of one term. Then W is I, or $L(W) = 1$ or W is E.C.

3.4. A partition is *strong* if it has the form (ii) of 3.2 and the product of the terms in any non-empty subsequence of a patch is not I. This is the case if no patch has more than two terms. A one-termed partition is strong.

3.5. Let $W \in \Pi$. Then a *division* of W is a subsequence $W_1, \ldots, W_n$ whose product is W such that, if the component sequence $d_1, \ldots, d_p$ is non-empty and is factorized into $S_1, S_2, \ldots, S_r$ (S_i non-empty, all terms in $S_i \in G_{\gamma_i}$, $\gamma_i \neq \gamma_{i+1}$) then the product of the terms in each S_i is not I.

Each W_i is called a *segment* of W.

3.6. Lemma. *Let $W_1, \ldots, W_n$ be a division of W. Let $T = W_i \cdot W_{i+1}$. Then $W_1, \ldots, W_{i-1}, T, W_{i+2}, \ldots, W_n$ is a division of W. Similarly for partitions.*

Proof. Let $W_i = x_1 \cdots x_m$, $W_{i+1} = y_1 \cdots y_n$ (the lemma is trivial if W_i or W_{i+1} is I or if $x_m \sim y_1$ does not hold).

Case 1. Both W_i, W_{i+1} have length $\geqslant 2$. Then $x_m \cdot y_1 \neq I$. If there is an amalgamation the new component sequence is

$$\ldots, x_1, \ldots, (x_m y_1), y_2, \ldots, y_n, \ldots$$

if not the new component sequence is the same as the old.

Case 2. $m = 1, n \geqslant 2$ (or dually). Let $z_1, \ldots, z_\gamma, x_m, y_1$ $(r \geqslant 0)$ be the patch containing x_m, y_1 in the original sequence. Then the new patch is either $z_1, \ldots, z_r$ if there is cancellation (when $x_m y_1 = I$ and $r \neq 0$); $z_1, \ldots, z_r, (x_m y_1)$ if there is amalgamation; $z_1, \ldots, z_r, x_m, y_1$ if there is no amalgamation. Hence the product of terms in the patch is unchanged.

Case 3. $m = n = 1$. Again the product is unchanged.

3.7. Corollary. *Let $W_1, \ldots, W_n$ be a division of $W = x_1 \cdots x_q$. Then the product of the terms in S_j is x_j and the number of S_j is q.*

Proof. By 3.6 W is a division of W and when we combine two terms the product of terms in the jth patch is unchanged.

3.8. Corollary. *Any division of an* E.C. *word is a partition. Any division of a word of length* 1 *is a partition.*

3.9. Lemma. *Let $W_1, \ldots, W_n$ be a partition (division) of W and let $U_1, \ldots, U_q$ be a division of W_i. Then $W_1, \ldots, W_{i-1}, U_1, \ldots, U_q$, $W_{i+1}, \ldots, W_n$ is a partition (division) of W.*

Proof. Say $W_i = x_1 \cdots x_p$. Then the component sequence of $U_1, \ldots, U_q$ is $x_1^1 \cdots x_1^{\alpha_1} x_2^1 \cdots x_2^{\alpha_2} \cdots x_p^1 \cdots x_p^{\alpha_p}$ where $x_j = x_j^1 \cdot x_j^2 \cdot \cdots \cdot x_j^{\alpha_j}$. Hence the product of terms in any patch of $W_1, \ldots, W_n$ is preserved.

3.10. Lemma. *Let $R_1, \ldots, R_\rho$ and $S_1, \ldots, S_\sigma$ be two divisions of W, $W \not\equiv I$. Then there is a common subdivision, i.e., there is a division $U_1, \ldots, U_\mu$ of W and divisions $R_i^1, \ldots, R_i^{\beta_i}$ of R_i, $S_j', \ldots, S_j^{\gamma_j}$ of S_j such that $U_1, \ldots, U_\mu$ is $R_1^1, \ldots, R_1^{\beta_1}, \ldots, R_\rho^1, \ldots, R_\rho^{\beta_\rho}$ and also is $S_1^1, \ldots, S_1^{\gamma_1}, \ldots, S_\sigma^1, \ldots, S_\sigma^{\gamma_\sigma}$. Finally, $\mu \leqslant \rho + \sigma - 1$.*

Proof. Let $W = x_1 x_2 \cdots x_p$. Consider any x_j. Then the jth prepatch of $R_1, \ldots, R_\rho$ is such that the product of its terms $a_1, a_2, \ldots, a_m$ is x_j. Define c by $a_1 = h_1 \cdot h_2 \cdot \cdots \cdot h_{n-1} \cdot c$ where $b_1, b_2, \ldots, b_n$ is the jth prepatch of $S_1, S_2, \ldots, S_\sigma$. Then $b_n = c a_2 a_3 \cdots a_m$ since $b_1 \cdots b_{n-1} c a_2 \cdots a_m = x_j = b_1 b_2 \cdots b_n$. Consider the product

$$(2) \qquad x_j = b_1 \cdot b_2 \cdot \cdots \cdot b_{n-1} \cdot c \cdot a_2 \cdot a_3 \cdot \cdots \cdot a_n .$$

Note that any a_i $(i = 1, \ldots, m)$ is a subproduct and any b_k $(k = 1, \ldots, n)$ is a subproduct. Rewrite (2) as $x_j^1 x_j^2 \cdots x_j^{\beta_j}$. Now consider the product

$$(3) \qquad x_1^1 \cdots x_1^{\beta_1} x_2^1 \cdots x_2^{\beta_2} \cdots x_\rho' \cdots x_p^{\beta_p}$$

each of $R_1, \ldots, R_\rho$ is equal to the product of terms in a subproduct of (3) and similarly for each $S_1, \ldots, S_\sigma$. Thus (3) factorises into $\mu \leqslant \rho + \sigma - 1$ part products where if U_j is the product of terms in the jth part product $(j = 1, \ldots, \mu)$ then each R_i and each S_j is a part product of

$$(4) \qquad U_1 \cdot U_2 \cdot \cdots \cdot U_\mu .$$

No U_k can obtain both x_j^r, x_j^{r+1} otherwise by examining (2) it would follow that U_k properly contains the end point of an R_a or S_b.

Hence the subsequence of (3) which forms U_k is just the normal form of U_k. Hence $U_1, \ldots, U_k$ is a division of W.

Consider any R_l. We have, say, $R_l = U_i \cdot U_{i+1} \cdot \cdots \cdot U_p$. Now each prepatch P is the component sequence of $U_i, \ldots, U_p$ is clearly a subsequence of a prepatch P' in the component sequence of $U_1, \ldots, U_k$ in fact P is $R_l \cap P'$. If we show the product of the terms in P is not

I then $U_i, \ldots, U_p$ is a division of $U_i \cdot \cdots \cdot U_p$ i.e., of R_l. But a typical P' in (2) so a typical $R_l \cap P'$ is either $b_1, b_2, \ldots, b_{n-1}, c$ or a_k (some $k \geqslant 2$).

3.11. Note. Trivially if $W_1, \ldots, W_n$ is a partition then so is $W_{i+1}, W_{i+2}, \ldots, W_n, W_1, \ldots, W_i$ where $1 \leqslant i \leqslant n$.

3.12. Lemma. *Let $W_1, \ldots, W_n$ be a strong partition of W and let U be a split of W. Then there is a strong partition of U of the form*

$$W_i'', W_{i+1}, \ldots, W_n, W_1, W_2, \ldots, W_{i-1}, W_i'$$

where W_i', W_i'' is a division of W_i.

Proof. This is trivial if $U = W$ so let $U \neq W$. Then

$$W = a_1 a_2 \cdots a_m, \qquad R = a_1 a_2 \cdots a_{r-1}, \qquad S = Y a_{r+1} \cdots a_m$$

$$x \cdot Y = a_r, \qquad W = R \cdot S, \qquad U = S \cdot R, \qquad L(Y) \leqslant 1.$$

Let the component sequence of $W_1, \ldots, W_n$ be $d_1, d_2, \ldots, d_p$. Since a strong partition is a division, $a_r = d_g \cdot d_{g+1} \cdot \cdots \cdot d_h$ say, and $R = d_1 \cdot \cdots \cdot d_{g-1} \cdot x$, $S = Y \cdot d_{h+1} \cdot \cdots \cdot d_p$.

If for some f $x = d_g \cdot d_{g+1} \cdot \cdots \cdot d_f$, $g \leqslant f \leqslant h$, then

$$U = d_{f+1} \cdot \cdots \cdot d_h \cdot \cdots \cdot d_p \cdot d_1 \cdot \cdots \cdot d_{g-1} \cdot d_g \cdot \cdots \cdot d_f$$

and the result follows. If x cannot be written in the above form put $d_g = x \cdot z$. Then no non-trivial subproduct of $x \cdot z \cdot d_{g+1} \cdot \cdots \cdot d_h$ equals I and

$$U = z d_{g+1} \cdots d_h \cdots d_p d_1 \cdots d_{g-1} x.$$

Suppose d_g lies in W_i say: $W_i = d_k \cdots d_g \cdots d_l$. Put $W_i' = d_k \cdots d_{g-1} x$, $W_i'' = z d_{g+1} \cdots d_l$. The result follows.

3.13. Proposition. *Let $X, Y \in \Pi$, $X \cdot Y \neq I$, $X \neq I$, $Y \neq I$. Let Z be the* *result of cancelling the element $X \cdot Y$ externally. Then X, Y have the form*

1. $X = X'QE, \qquad Y = E^{-1}RV^{-1}ZVS^{-1}X'^{-1}$, *or*
2. $X = T^{-1}BQE, \quad Y = E^{-1}RCT$, *where* $Z = BSC$, *or*
3. $X = Y'^{-1}S^{-1}V^{-1}ZVQE, \qquad Y = E^{-1}RY'$.

Here either Q, R, S are all I, or all have length 1 and $S = Q \cdot R$.

Proof. $X \cdot Y = X'SY'$, $X = X'QE$, $Y = E^{-1}RY^{-1}$, where Q, R, S are all I or all have length 1 and belong to the same constituent group; $S = Q \cdot R$.

T exists such that $T^{-1}ZT = X'SY'$.

Case 1. $L(X'S) \leqslant L(T^{-1})$.

$T^{-1} = X'SV^{-1}$ say, $Y' = V^{-1}ZT = V^{-1}ZVS^{-1}X'^{-1}$.

Case 2. $L(T^{-1}) < L(X'S) \leqslant L(T^{-1}Z)$.

Here $L(T^{-1}) \leqslant L(X')$ so $X' = T^{-1}B$ say, and $BSY' = ZT$. Now $Z = BSC$ say, so $Y' = CT$.

Case 3. $L(X'S) > L(T^{-1}Z)$.

Here $L(X') \geqslant L(T^{-1}Z)$ so $X' = T^{-1}ZV$ say and $T = VSY'$.

3.14. Lemma. *Let X, Y be E.C. and $X \cdot Y \neq I$. Let Z be the result of cancelling the element $X \cdot Y$ externally. Then there are splits X_0, Y_0 of X, Y respectively such that*

1. X_0^{-1}, F^{-1}, Z, F *is a division of* Y_0 *for some* F, *or*
2. *elements L, M, N exist such that*
   - L, N^{-1} *is a division of* X_0
   - N, M *is a division of* Y_0
   - L, M *is a division of* Z, *or*
3. F^{-1}, Z, F, Y_0^{-1} *is a division of* X_0 *for some* F.

Proof. In Case 1 of 3.13 $X'^{-1}, E^{-1}RV^{-1}ZVS^{-1}$ is a division say of Y_0, hence Y_0 is a split of Y. Now $E^{-1}RV^{-1}$, Z, VS^{-1} is a division of $E^{-1}RV^{-1}ZVS^{-1}$ and $E^{-1}Q^{-1}, SV^{-1}$ is a division of $E^{-1}RV^{-1}$. Hence by 3.9 $X'^{-1}, E^{-1}Q^{-1}, SV^{-1}, Z, VS^{-1}$ is a division of Y_0 hence so is $X_0^{-1}, SV^{-1}, Z, VS^{-1}$ where $X_0 = QE \cdot X'$. This is a split of X.

In Case 2, BQE, T^{-1} is a division of X_0, a split of X. Now BQ, E is a division of BQE so BQ, E, T^{-1}, hence $BQ, (E \cdot T^{-1})$ is a division of X_0. Similarly $(T \cdot E^{-1}), RC$ is a division of Y_0, a split of Y. It remains to show that BQ, RC is a division of Z. This is so.

Case 3 is now obvious.

3.15. Definition. A *semischeme* of k rows on symbols $u_1, \ldots, u_a$ is an array

$$\begin{array}{cccc} x_{11} & x_{12} & \cdots & x_{1q_1} \\ \cdots & \cdots & \cdots & \cdots \\ x_{k1} & x_{k2} & \cdots & x_{kq_k} \\ r_1 & r_2 & \cdots & r_b \end{array}$$

consisting of k rows and a *result* $r_1 r_2 \cdots r_b$ where

1. $k \geqslant 1, q_i \geqslant 1$ $(i = 1, \ldots, k)$, $b \geqslant 0$;
2. each x_{ij} has the form u_r or u_r^{-1}; each r_i has the form u_r;
3. for each $s = 1, \ldots, a$ the number of occurrences of $u_s^{\pm 1}$ is two: either an occurrence of u_s in a row and an occurrence of u_s^{-1} in a row, or an occurrence of u_s in a row and u_s in the result; in the first case u_s and u_s^{-1} are of *type* C while in the second case both occurrences of u_s are of *type* R;

(The total number of entries is $q_1 + \cdots + q_k + b = 2a$.)

4. the order in which the rows are written does not matter; the terms in any row or in the result may be cyclically permuted.

3.16. Definition. Given a sequence $\tau_1, \tau_2, \ldots, \tau_p$ each term of which is a non-empty sequence of symbols each of the form $u_i^{\pm 1}$ $(i = 1, \ldots, a)$ an *elementary operation* means any one of the following.

E1. Replace some τ_i by a cyclic arrangement of τ_i.

E2. Cancel two adjacent mutually inverse symbols in some τ_i.

E3. For some j, k where $j \neq k$ delete τ_k and replace τ_j by $\tau_j \tau_k$.

Notation: If after a finite sequence of elementary operations $\tau_1, \tau_2, \ldots, \tau_p$ is tranformed into $\nu_1, \nu_2, \ldots, \nu_q$ say, write

$$\tau_1, \tau_2, \ldots, \tau_p \to \nu_1, \nu_2, \ldots, \nu_q.$$

Example: Let τ_1 be $d, a^{-1} b$; τ_2 be e, b^{-1}, c; τ_3 be f, c^{-1}, a; σ be e, d, f. Then $\tau_1, \tau_2, \tau_3 \to \sigma$.

3.17. Definition. A semischeme is *reducible* if the rows $\rho_1, \ldots, \rho_k$ can be transformed into the result σ by elementary operations, i.e.,

$$\rho_1, \ldots, \rho_k \to \sigma$$

Examples (compare with 3.14):

$$\begin{array}{llllll} & x & & l\,,n^{-1} & & f^{-1}, z, f, y^{-1} \\ (1) & x^{-1}, f^{-1}, z, f & (2) & n, m & (3) & y \\ & z & & l\,, m & & z \end{array}$$

The following are trivial.

3.18. If $\mu_1, \mu_2, \ldots, \mu_k \to \nu_1, \nu_2, \ldots, \nu_q$ then the sequence $\mu_1, \ldots, \mu_k$ is the union of q disjoint subsequences $\theta_1, \ldots, \theta_q$ such that $\theta_i \to \nu_i$ $(i = 1, \ldots, q)$.

3.19. If φ is a sequence of elements of a group let $p(\varphi)$ denote the product of these elements. If $\mu_1, \ldots, \mu_k \to \nu$ then $p(\nu) = \Pi_{i=1}^k T_i^{-1} \cdot p(\mu_i) \cdot T_i$ for some T_i in a free group.

3.20. Let $\mu_1, \ldots, \mu_k \to \nu$. Let A be a term of μ_1, say $\mu_1 = \xi, A, \eta$ and let A survive the sequence of elementary operations, i.e., assume A is not cancelled in any elementary operation; thus A is a term of ν, say $\nu = \xi', A, \eta'$. Then

$$p(\eta')\, p(\xi') = p(\eta)\, p(\xi) \prod_{i=2}^{k} T_i^{-1} \cdot p(\mu_i) \cdot T_i \quad \text{in a free group.}$$

3.21. Assume $\mu_1, \ldots, \mu_k \to \nu$ and μ_1 contains two terms A, B that survive the sequence of elementary operations; thus $\mu_1 = A, \xi, B, \eta$ and $\nu = A, \xi', B, \eta'$ say. Then $p(\xi') = p(\xi)\, \Pi_{i=2}^r T_i^{-1} p(\mu_i)\, T_i$ and $p(\eta') = p(\eta)\, \Pi_{i=r+1}^k T_i^{-1} p(\mu_i)\, T_i$ (in a free group) without loss of generality.

3.22. Proposition. *In a semischeme with rows $\rho_1, \ldots, \rho_k$ where two distinct rows have the form α, Z and Z^{-1}, β and the result is σ we have*

$$p(\sigma) = T^{-1} \cdot p(\alpha) \cdot p(\beta) \cdot T \cdot \prod_{i=3}^{k} T_i^{-1} \cdot p(\rho_i) \cdot T_i.$$

Proof. Z and Z^{-1} cancel in the reduction say at the sth elementary operation. Hence there exists $r < s$ such that after $r-1$ operations the two rows have the form ξ, Z, η and ζ, Z^{-1}, θ, and the rth operation brings them together giving $\xi Z \eta \zeta Z^{-1} \theta$ or $\zeta Z^{-1} \theta \xi Z \eta$; we consider the first case only. After the $(s-1)$th operation we have say $\lambda, Z^{\epsilon}, Z^{-\epsilon}, \mu$ $(\epsilon = \pm 1)$ and the next operation changes it to λ, μ. Suppose $\epsilon = -1$; the case $\epsilon = 1$ is similar. Then

$$p(\eta)\, p(\xi) = p(\alpha)\, \pi_1, \qquad p(\theta)\, p(\zeta) = p(\beta)\, \pi_2$$
$$1 = p(\theta)\, p(\xi)\, \pi_3, \quad p(\mu)\, p(\lambda) = p(\eta)\, p(\zeta)\, \pi_4$$

$$p(\sigma) = s^{-1} p(\lambda)\, p(\mu)\, s \cdot \pi_5,$$

hence $p(\sigma) = T^{-1} p(\alpha)\, p(\beta)\, T\, \pi'_1 \pi'_2 \pi'_3 \pi'_4 \pi'_5$ where π'_i is a conjugate of π_i $(i = 1, \ldots, 5)$.

3.23. Proposition. *E2 followed by E1 or E3 is equivalent to E1 or E3 followed by E2.*

Proof. Trivial.

3.24. Proposition. *If* $\rho_1, \rho_2, \ldots, \rho_m \to 1$ *(the empty sequence) and if* ρ'_1 *is* u, ρ_1 *then* $\rho_1^1, \rho_2, \ldots, \rho_m \to u$.

Conversely $\rho'_1, \rho_2, \ldots, \rho_m \to u$ *implies* $\rho_1, \rho_2, \ldots, \rho_m \to 1$.

Proof. The converse is trivial.

Now $\rho_1, \rho_2, \ldots, \rho_m \to \sigma_1, \sigma_2, \ldots, \sigma_b \to 1$, where $\sigma_1, \sigma_2, \ldots, \sigma_b$ arises from $\rho_1, \rho_2, \ldots, \rho_m$ by E1, E3 only and $\sigma_1, \sigma_2, \ldots, \sigma_b \to 1$ by E2 only (hence $b = 1$). Say $\rho_1, \rho_2, \ldots, \rho_r \to \sigma_1$. Then ρ_1 is $\xi_0, \xi_1, \ldots, \xi_s$ and σ_1 is a C.A. of (1) $\xi_0 \alpha_1 \xi_1 \cdots \alpha_s \xi_s$ where $\rho_2, \ldots, \rho_r \to \alpha_1, \ldots, \alpha_s$. Hence

$$
\begin{array}{llll}
\rho_1' & u\,\xi_0\,\xi_1 \cdots \xi_s & & u\,\xi_0\,\alpha_1\,\xi_1 \cdots \alpha_s\,\xi_s \\
\rho_2 & \longrightarrow & \alpha_1 \longrightarrow & \\
 & & & \\
\rho_r & & \alpha_s &
\end{array}
$$

Now σ_1 cancels to the empty word hence so does (1). Hence $u\,\xi_0\,\alpha_1\,\xi_1 \cdots \alpha_s\,\xi_s$ cancels to u.

3.25. Proposition. *If α, x and x^{-1}, β are two distinct rows ρ_1, ρ_2 of a reducible semischeme with rows $\rho_1, \rho_2, \ldots, \rho_k$ and result σ then $(\alpha, \beta), \rho_3, \ldots, \rho_k \to \sigma$.*

Proof. In the reduction

$$
\begin{array}{lllll}
\alpha x & \to & \alpha_2^1\, x\, \alpha_1^1 & \to & \alpha_2^1\, x\, \alpha_1^1\, \beta_2^1\, x^{-1} \text{ (or } \beta_2^1\, x^{-1}\, \beta_1^1\, \alpha_2^1\, x\, \alpha_1^1) \to \\
x^{-1}\beta & & \beta_2^1\, x^{-1}\, \beta_1^1 & & \\
 & \to & \delta\, x^{\epsilon}\, x^{-\epsilon} & \to & \delta \to \sigma.
\end{array}
$$

Hence $\alpha\beta \to \alpha_2^1\,\beta\,\alpha_1^1 \to (\alpha_1^1\,\alpha_2^1)\beta \to \beta_2^1(\alpha_1^1\,\alpha_2^1)\beta_1^1$. Also if $\epsilon = 1$,

$$\alpha_1^1\,\beta_2^1 \to 1, \qquad \beta_1^1\,\alpha_2^1 \to \delta$$

hence $\alpha_1^1\,\beta_2^1 \to 1$ and $\beta_2^1\,\alpha_1^1 \to 1$. By 3.24

$$u\beta_2^1\,\alpha_1^1 \to u, \qquad (\alpha_2^1\,\beta_1^1)\,\beta_2^1\,\alpha_1^1 \to (\alpha_2^1\,\beta_1^1) \to \beta_1^1\,\alpha_2^1 \to \delta.$$

Similarly if $\epsilon = -1$.

3.26. Replacing r_i by r_i, u_{a+1} in both row and result of a semischeme S gives a semischeme S' on $u_1, \ldots, u_a, u_{a+1}$. If S is reducible then so is S'.

3.27. Note. If $a \neq 1$, deleting both occurrences of any u_i in S yields a semischeme S'. If S is reducible then so is S'.

3.28. Note. In a semischeme S, delete r_1, r_2 from the result and, in the row containing r_1, replace r_1 by r_2^{-1}; then a semischeme S' is obtained. If S is reducible then so is S'.

3.29. Definition. Let $X_1, ..., X_a \in \Pi$ and let S be a semischeme on $u_1, ..., u_a$. The array obtained from S by replacing u_i^ϵ by X_i^ϵ ($\epsilon = 1$ or -1 $i = 1, ..., a$) is called a *scheme*.

3.30. Proposition. *Consider a reducible scheme such that the rows and the result are partitions. Let the result be $W_1, ..., W_n$ where $W_{i-1} \equiv Xy'$ $W_i \equiv y$ and $y' \sim y$. If we replace Xy' by A, y^{-1} in the row, where $A = Xy' \cdot y$, and replace W_{i-1}, W_i in the result by A then we obtain a reducible scheme whose rows and result are partitions.*

Proof. In the underlying semischeme let r_1, r_2 correspond to Xy' and y respectively. Replacing r_1 by r_1, u we obtain a reducible semischeme S' by 3.26. In S', delete u, r_2 from the result and, in the appropriate row, replace u by r_2^{-1} obtaining a reducible semischeme S'' by 3.28. In S'' replace r_1, r_2 by A, y and replace all the other symbols as in the original scheme. We obtain a reducible scheme whose rows and result are partitions.

3.30A. Note. If some row of the original scheme is a division, the corresponding row of the new scheme is a division.

3.31. Proposition. *Let $W_1, ..., W_n$ be a proper partition that is not strong. Then $n \geqslant 2$ and some W_i has length 1 and is in the same constituent group as Fin W_{i-1} and In W_{i+1}; (interpret W_0 to be W_n and W_{n+1} to be W_1).*

Proof. Some patch contains three or more terms hence contains d_{j-1}, d_j, d_{j+1} say. Then d_j is W_i, $d_{j-1} =$ Fin W_{i-1} and d_{j+1} is In W_{i+1}.

§4.

The purpose of this section is to prove Theorem 2 below. We remark informally that a geometrical interpretation is possible in which each row ρ becomes a region ρ' in the plane, each entry A in ρ corresponds to an edge A' of ρ' and if X, X^{-1} are of Type C (cf. 3.15) then the edges X', $(X^{-1})'$ coincide. (cf. Lyndon [3]).

Theorem 2. *Let Ω' satisfy* 1, 2, 3 *of* 2.3. *Let W be a natural element so that $W \approx W_1 \circ \cdots \circ W_k$ $(W_i \in \Omega')$, $k \geqslant 1$. Then there is a scheme satisfying the following conditions.*

1. (a) *It is a reducible scheme of k rows.*
   (b) *The total number of entries is* $\leqslant 6k-4$.
2. *The ith row is a partition of a split W_i' of W_i.*

3$'$. *The result is a proper strong partition of W.*

4. *No row has the form* $\ldots, X, X^{-1}, \ldots$ *where X and X^{-1} are partners in the underlying semischeme.*
5. *In the underlying semischeme if one row has the form* $\ldots, x^{\epsilon}, y^{\epsilon'}, \ldots$ *then no other row has the form* $\ldots, y^{-\epsilon'}, x^{-\epsilon}, \ldots$.
   No row has the form $\ldots, x^{\epsilon}, y^{\epsilon'}, \ldots, y^{-\epsilon'}, x^{-\epsilon}, \ldots$ $(\epsilon = \pm 1, \epsilon' = \pm 1)$.

Note. A supplement to Theorem 2 is given in 6.12.

4.1. Lemma. *If there is a scheme satisfying* 1(a), 2 *and also:*
3. *The result is a partition of a split of W.*
then we may assume that in 3. *the partition is proper and strong.*

Proof. Of all schemes satisfying the hypothesis choose one whose result $R_1, \ldots, R_n$ has the least number of terms. Then no R_j is I otherwise we could delete R_j and its partner $R_{j'}$. The result follows if $n = 1$ so let $n \geqslant 2$. If $R_1, \ldots, R_n$ were not strong we could obtain a contradiction by using 3.30, 3.31.

4.2. Lemma. *With the hypothesis of Lemma* 4.1, *there is a scheme satisfying* 1(a), 2 *and* 3$'$.

Proof. Take a scheme S satisfying 1(a), 2 and with result $R_1, \ldots, R_n$ a proper strong partition of W', a split of W. This determines a strong partition of W of the form $R_i'', \ldots, R_n, R_1, \ldots, R_i'$ where R_i', R_i'' is a division of R_i. If in S we replace R_i in both row and result by R_i', R_i'' we obtain S' satisfying 1a, 2 and

3″. The result *is a* strong partition of W.
If R_i' or R_i'' is I we may delete it from row and result to obtain 3′.

4.3. Lemma. *A scheme exists satisfying* 1, 2, 3′.

Proof. The proof is by induction on k.

$k = 1$. Here $W \approx W_1$, i.e., W is a split of W_1. For the required scheme take the single row W and result W.

For the induction step we have $W \approx X \circ Y$

$$X \approx W_1 \circ \cdots \circ W_a, \qquad Y \approx W_{a+1} \circ \cdots \circ W_k \qquad (1 \leqslant a < k).$$

Since W is natural, so are X and Y. Since $W \approx X \circ Y$, X and Y are E.C. There are splits X', Y' of X, Y such that $X' \cdot Y' \not\equiv I$, $Z = E(X' \cdot Y')$, W is a split of Z. By 3.14 there are splits X_0, Y_0 of X', Y', hence of X, Y such that 1, 2 or 3 of 3.14 hold.

Case 1. 1 holds, i.e., X_0^{-1}, F^{-1}, Z, F is a division of Y_0. Now $y_0 \approx W_{a+1} \circ \cdots \circ W_k$ so by induction hypothesis there is a reducible scheme S_2 of $k - a$ rows satisfying 2 for $i = a + 1, \ldots, k$ and whose result $Y_1, \ldots, Y_{m_2}$ is a proper strong partition of Y_0, hence a division of Y_0. Hence there are divisions

$\xi_a^{-1}, \ldots, \xi_1^{-1}$	of X_0^{-1}	$\varphi_b^{-1}, \ldots, \varphi_1^{-1}$	of F^{-1}
$Z_1, \ldots, Z_c$	of Z	$F_1, \ldots, F_d$	of F
$d(Y_i)$	of Y_i	$(i = 1, 2, \ldots, m_2)$	

such that

$$(1) \quad d(Y_1)d(Y_2) \cdots d(Y_{m_2}) = \xi_a^{-1} \cdots \xi_1^{-1} \varphi_b^{-1} \cdots \varphi_1^{-1} Z_1 \cdots Z_c F_1 \cdots F_d.$$

Also there is a reducible scheme S_1 of a rows satisfying 2 for $i = 1, \ldots, a$ and whose result $X_1, \ldots, X_{m_1}$ is a proper strong partition of X_0. Hence there are divisions $d(\xi_i)$ of ξ_i $(i = 1, \ldots, a)$ $d(X_j)$ of X_j $(j = 1, \ldots, m_1)$ such that

$$(2) \qquad d(\xi_1) \cdots d(\xi_a) = d(X_1) \cdots d(X_{m_1})$$

Also there are divisions $d(\varphi_i)$ of φ_i $(i = 1, ..., b)$ $d(F_j)$ of F_j $(j = 1, ..., d)$ such that

$$(3) \qquad d(F_1) \cdots d(F_d) = d(\varphi_1) \cdots d(\varphi_b)$$

Let S_1' arise from S_1 by replacing each X_j by $d(X_j)$. Then S_1' has a rows and is reducible and 2 still holds for $i = 1, ..., a$; the result is $d(\xi_1) \cdots d(\xi_a)$ $(= \rho$, say).

Let S_2' arise from S_2 by replacing each Y_i by $d(Y_i)$. Then S_2' is reducible, with $k - a$ rows and satisfies 2 for $i = a + 1, ..., k$. Its result is the right side of (1). Let S_2'' arise from S_2' by replacing ξ_i^{-1} by $d(\xi_i)^{-1}$, φ_i^{-1} by $d(\varphi_i)^{-1}$, F_j by $d(F_j)$. S_2'' is reducible with $k - a$ rows and satisfies 2 for $i = a + 1, ..., k$. Its result, σ say, is

$$d(\xi_a)^{-1} \cdots d(\xi_1)^{-1} d(\varphi_b)^{-1} \cdots d(\varphi_1)^{-1} Z_1 \cdots Z_c d(F_1) \cdots d(F_d).$$

Let S_3 arise by taking the rows of S_1' and S_2'' and taking for result $Z_1, ..., Z_c$. S_3 is a reducible scheme since

$$\rho, \sigma \to d(\varphi_b)^{-1}, ..., d(\varphi_1)^{-1}, Z_1, ..., Z_c, d(F_1), ..., d(F_d)$$

$$\to Z_1, ..., Z_c \text{ (by (3))}.$$

S_3 has k rows and satisfies 2 for $i = 1, ..., k$.

We next consider the number of entries in S_3. The number of terms on each side of (1) is $m_2 - 3$, the number for (2) is $m_1 + a - 1$ and the number for (3) is $b + d - 1$. From (1) $a+b+c+d = m_2+3$. The increase in the number of entries in the rows (and also the increase in the number of entries in the result)

from S_1 to S_1' is $a - 1$

from S_2 to S_2' is 3

from S_2' to S_2'' is $(m_1 - 1) + (2(b + d - 1) - b - d)$.

Let the total number of entries in the rows of S_1, S_2 be a_1, a_2 respectively. S_3 has c terms in the result so its total number of entries is

$$a_1 + a_2 + (a - 1) + 3 + (m_1 - 1 + b + d - 2) + (m_2 + 3 - a - b - d)$$
$$= (a_1 + m_1) + (a_2 + m_2) + 2 \leqslant (6a - 4) + (6(k - a) - 4) + 2 = 6k - 6.$$

The result of S_3 is a division of Z, which is E.C. or has length 1 since it is a split of W and $W \approx X \circ Y$. Thus the result is a partition of Z. Now Z is a split of W so as in 4.2 we can obtain from S_3 a scheme satisfying 1, 2, 3′ with at most $6k - 4$ entries.

Case 2. 2 of 3.14 holds.

We may take L, M to be a proper strong partition.

Schemes S_1, S_2 exist as in Case 1. By considering $X_1, \ldots, X_{m_1}$ and L, N^{-1} we see that there are divisions $d(X_i)$, $d(L)$ and $d(N^{-1})$ such that $d(X_1) \cdots d(X_{m_1}) = d(L)\, d(N^{-1})$. Say $d(N^{-1})$ is $N_p^{-1}, \ldots, N_1^{-1}$. Let S_1' arise from S_1 by replacing X_i by $d(X_i)$. Then S_1' has result $d(L)\, N_p^{-1} \cdot \ldots \cdot N_1^{-1}$. There are divisions $d(Y_i)$ $(i = 1, \ldots, m_2)$ and $d(N)$, $d(M)$ such that $d(Y_1) \cdots d(Y_{m_2}) = d(N)\, d(M)$. Let S_2' arise from S_2 by replacing each Y_i by $d(Y_i)$. Then its result is $T_1 \cdots T_q\, d(M)$ where $d(N)$ is $T_1, \ldots, T_q$. Considering $N_1, \ldots, N_p$ and $T_1, \ldots, T_q$ we see that there are divisions such that

$$(*) \qquad d(N_1) \cdots d(N_p) = d(T_1) \cdots d(T_q).$$

Let S_1'' arise from S_1' by replacing N_i^{-1} by $d(N_i)^{-1}$. Let S_2'' arise from S_2' by replacing T_i by $d(T_i)$. Now take the rows of S_1'' and of S_2'' and result $d(L)\, d(M)$ to obtain a reducible scheme of k rows satisfying 2 for $i = 1, \ldots, k$; it is reducible since by $(*)$

$$d(L)\, d(N_p)^{-1} \cdots d(N_1)^{-1},\ d(T_1) \cdots d(T_q)\, d(M) \to d(L)\, d(M).$$

The increases in the row entries from S_1, S_2, S_1', S_2' to S_1', S_2', S_1'', S_2'' respectively are $1, 1, q-1, p-1$ respectively. Now $d(L)$, $d(M)$ have $m_1 + 1 - p$, $m_2 + 1 - q$ terms respectively, hence the number of entries in the final scheme is $a_1 + m_1 + a_2 + m_2 + 2 \leqslant 6k - 6$ as before.

Since L, M is a division of Z, so is $d(L)\, d(M)$. As in Case 1 we find a scheme satisfying 1., 2., 3′.

Case 3. Part 3 of 3.14 holds.

This is similar to Case 1.

4.4. Lemma. *If there is a scheme S satisfying* 1, 2, 3′ *then there is a scheme satisfying* 1, 2, 3′, 4, 5.

Proof. If S does not satisfy 4 consider a row ..., X, X^{-1}, ... Since this is a partition $L(X) \leqslant 1$. If $L(X) = 1$ then X, X^{-1} are two terms of a patch, so deleting them does not affect the product of the terms in this patch; thus if we delete X, X^{-1} from the row it remains a partition; this is also the case if $X \equiv I$. But deleting X, X^{-1} does not destroy reducibility.

If S does not satisfy 5, combine X^{ϵ}, $Y^{\epsilon'}$ into a single element (their product); partitions are not destroyed. Now deleting y in the underlying semischeme preserves reducibility so the result follows.

4.5. Note. Theorem 2 now follows.

§5.

5.1. Lemma. *In a reducible semischeme with k rows* $\xi_1, ..., \xi_k$ *and result* $r_1, r_2, ..., r_n$, *if the first row* ξ_1 *has the form* $r_1 r_j \cdots$ *then* $j = 2$ *or* $r_2 r_3 \cdots r_{j-1}$ *is in a free group equal to*

$$\prod_{\nu=1}^{s} T_\nu^{-1} \cdot \eta_\nu \cdot T_\nu$$

where $\eta_1, ..., \eta_s$ *form a subset of the rows* $\xi_2, ..., \xi_k$.

Proof. Consider an elementary operation applied to $\tau_1, \tau_2, ..., \tau_q$ where τ_1 is a cyclic arrangement of r_1, α, r_j, β (α and β are possibly empty sequences) and none of $\alpha, \beta, \tau_2, ..., \tau_q$ contains r_1 or r_j. A cyclic arrangement of one of the new terms has the form $r_1 \alpha' r_j \beta'$ where either $\alpha = \alpha'$ in a free group or $\alpha = \alpha_1 \alpha_2$ $\alpha' = \alpha_1 T_w \alpha_2$ where $2 \leqslant w \leqslant q$. In the second case $\alpha' = \alpha(\alpha_2^{-1} T_w \alpha_2)$ in a free group. Initially α is empty, and after a finite sequence of elementary operations it becomes $r_2 r_3 \cdots r_{j-1}$ if $j \geqslant 3$.

5.2. Definition. Let α, X, β be a partition of a split of an element of Ω', in L_a say. Let $L(X) \geqslant 2$. Let q be an integer $\geqslant a$. We say X is *q-special in* α, X, β if

1. where X^{-1}, B is a partition of an element of L_b and $b \leqslant q$ then $w(A \cdot B) < \bar{a} + \bar{b}$, where $A = \beta \cdot \alpha$ and $\bar{a} = (0, ..., 0, 1, 0, ...)$ with 1 as the ath term, and similarly for $\bar{b}$.

2. if $a = 1$ then if X^{-1}, B is a partition of a split of $A \cdot X$ we have $A \cdot B \cdot \equiv I$.

We say X is *special in* α, X, β if it is q-special in α, X, β for all $q \geqslant a$.

5.3. We now make a further assumption about the subset Ω' of 2.3:

4. If $X_1, X_2, ..., X_u$ is a partition of an element of $Sp(\Omega')$ where $u \leqslant 5$ then at least one X_i is special with respect to $X_1, ..., X_u$.

5.4. Lemma. *Let* Ω' *satisfy* 1, 2, 3, 4. *If* $W \approx W_1 \circ \cdots \circ W_k$ *is natural, so that there is a scheme* S *satisfying* 1, 2, 3′, 4, 5 *then for every such* S *we have:*

6′. *no row of* S *has the form* α, X, β, X^{-1} *where* X, X^{-1} *is a cancelling pair in the underlying semischeme and* X *is special in this row.*

Note: The proof of this lemma is postponed to Section 6 (6.12A).

Theorem 3. *Let* Ω' *satisfy* 1, 2, 3, 4. *Every non-empty relator* R *has a segment* A *such that there is a partition* $Z_1, ..., Z_v$ *of an element of* $\mathrm{Sp}(\Omega')$ *where* $v \leqslant 4$ *and*

(i) Z_1 is A

(ii) Z_1 *is special but none of* $Z_2, ..., Z_v$ *are special* (*with respect to* $Z_1, ..., Z_v$).

Hence $L(R) \geqslant L(A) \geqslant 2$.

5.5. Proof that Lemma 5.4 implies Theorem 3. The least possible weight of R is $(1, 0, 0, ...)$. In this case $R \equiv U^{-1} W' U$ where $W' \in S(L_1)$, and the theorem is true in this case with $v = 1$. Now let the theorem be true for all non-empty relators R' with weight smaller than that of R.

Let

$$R = \prod_{i=1}^{k} T_i^{-1} \cdot W_i \cdot T_i \qquad (W_i \in \Omega')$$

where $W_1, \ldots, W_k$ has weight equal to the weight of R. Then

$$R \equiv \prod_{i=1}^{h} S_i^{-1} \cdot A_i \cdot S_i, \qquad A_i \approx W_{\alpha(i,1)} \circ \cdots \circ W_{\alpha(i,q_i)}.$$

If $h \geqslant 2$ then $q_i < k$ for all i so the theorem is true for all A_i, hence $L(A_i) \geqslant 2$. Hence by 2.9, R contains some A_i as a subword. Hence the theorem is true for R. Thus, changing the notation, it remains to prove that if $W \approx W_1 \circ \cdots \circ W_k$ is natural then the theorem is true for W.

There is a scheme S satisfying $1, 2, 3', 4, 5$. By 5.4 S satisfies $6'$. We may assume that the total number of terms is minimal. Let $X_1, \ldots, X_c$ be a row consisting entirely of terms of type C. Suppose $c \leqslant 5$. Then some X_i is special. By 6, the partner X_i^{-1} of X_i does not lie in the row $X_1, \ldots, X_c$. Thus we have two distinct rows α, X_i and X_i^{-1}, β. By the irredundancy of $W_1, \ldots, W_k$ (cf. 2.6) we have

$$w(\alpha \cdot \beta) < w(\alpha \cdot X_i) + w(X_i^{-1} \cdot \beta)$$

But $W = S^{-1} \alpha\beta S \Pi_{i=3}^{k} T_i^{-1} \cdot W_i \cdot T_i$, contradicting that the weight of W equals the weight of $W_1, \ldots, W_k$.

Hence $c > 5$. Let d_i rows have i terms ($i = 1, 2, \ldots, 5$) so that $q = k - \Sigma d_i$ rows have $\geqslant 6$ terms each. Then

$$r + d_1 + 2d_2 + 3d_3 + 4d_4 + 5d_5 + 6q \leqslant 6k - 4.$$

Hence $c \leqslant -4 + 5d_1 + 4d_2 + 3d_3 + 2d_4 + d_5$.

Let d_i^j rows have i terms of which j are of type R. Suppose (i) $d_i^1 = 0$ ($i = 1, 2, 3, 4$). Hence $d_1 = 0$. Every row with two terms has both of type R. The three-termed rows contain together $2d_3^2 + 3d_3^3$ terms of type R. Each row with four terms has at least two terms of type R. Each row with five terms has at least one of type R. Hence

$$2d_2 + (2d_3^2 + 3d_3^3) + 2d_4 + d_5 \leqslant r, \quad 0 \leqslant -4 + 2d_2^2 + d_3^2$$

If (i) is false there is a row $Y_1, Y_2, \ldots, Y_u, u \leqslant 4$, where Y_1 has type R and $Y_2, \ldots, Y_u$ have type C. None of $Y_2, \ldots, Y_u$ are special otherwise a contradiction would follow as before. By 4, Y_1 is special. Hence the theorem is true.

If (i) is true then some row r, s, δ has the terms r, s of type R so the result is either (a) $\alpha\, r\beta s\gamma$ or (b) $\alpha\, s\beta r\gamma$. Since $d_2^2 + d_3^2 \geqslant 2$ there are at least two such rows so we may assume that if (b) holds then not both α, γ are empty.

Assume (a). If β is non-empty then $p(\beta)$ is a product $\Pi_{i=1}^t T_i^{-1} \cdot \eta_i \cdot T_i$ where $t < k$ and $\eta_1, \ldots, \eta_t$ is a subset of $W_1, \ldots, W_k$. By induction the hypothesis β has a segment as in the theorem hence so does W (cf. 3.6, 3.9). Hence we may suppose β is empty. Then the terms r, s may be amalgamated to form a single term (in both the row and the result), decreasing the total number of terms.

Assume (b). Then $p(\gamma)\, p(\alpha) = \pi_1$ and $p(\beta) = p(\delta)\, \pi_2$

$$sp(\beta)r = sp(\delta)\pi_2 r = r^{-1}(rsp(\delta))r \cdot r^{-1}\pi_2 r$$

so that $sp(\beta)r$ is a product of less than k conjugates provided π_1 is nonempty, and the theorem follows as before. Now let π_1 be empty. Since the result is a proper strong partition α and γ are both empty, a contradiction.

§6.

In this section we introduce Generalized Tartakovskii Groups.

6.1. Definition. Let M be a non-empty subset of Π closed to C.A. and inverses. Let $\epsilon_0, \epsilon_1, \epsilon_2$ be non-negative real numbers. Then M has a *size function using* $\epsilon_0, \epsilon_1, \epsilon_2$ if to each pair $A, B \in \Pi$ such that $AB \in M$ there is defined a real number $s_{AB}(A)$ such that

1. $0 \leqslant s_{AB}(A) \leqslant 1$. If $A \equiv I$ then $s_{AB}(A) = 0$. If $B \equiv I$ then $s_{AB}(A) \geqslant 1 - \epsilon_0$.
2. $L(A) = 1 \Rightarrow s_{AB}(A) \leqslant \epsilon_2$
3. $s_{AB}(A) = s_{A^{-1}B^{-1}}(A^{-1})$

4. if $ABCD \in M$ then $s_{BCDA}(B) \leqslant s_{ABCD}(ABC)$
5. if $ABC \in M$ then $s_{ABC}(AB) \leqslant s_{ABC}(A) + s_{BCA}(B) + \epsilon_1$.

6.2. Notation. If $X, Y \in \pi$ then $X \cong Y$ means that $L(X) = L(Y)$ and (a) $X \equiv Y \equiv I$ or (b) $L(X) = 1$ and $X \sim Y$ or (c) $X \equiv x_1 \cdots x_n$ $Y \equiv y_1 \cdots y_n$ and $x_1 \sim y_1, x_n \sim y_n, x_i = y_i$ $(i = 2, 3, ..., n-1)$

6.3. Definition. A non-empty subset Ω' of Π satisfies the *Generalized Tartakovskii Conditions*, and $G = \Pi/[\Omega']$ is a *Generalized Tartakovskii Group* if subsets $L_1, L_2, ...$ of Π and real numbers $p, p_0, p_0', \epsilon_0, \epsilon_1, \epsilon_2$ exist such that the following conditions hold: put $G_i = \Pi/[L_1 \cup \cdots \cup L_i]$, $G_0 = \Pi$.

1. $\Omega' = L_1 \cup L_2 \cup \cdots$ and each element of Ω' is E.R.
2. The L_i are pairwise disjoint, and if L_j is empty then so is L_{j+1} $(j = 1, 2, ...)$
3. Each L_i is closed under C.A. and inverses
4. $1/p_0' > 0,\ 1/p \geqslant 1/p_0 \geqslant 1/p_0' + \epsilon_1 + \epsilon_2$
   $5(1/p + \epsilon_1) \leqslant 1 - \epsilon_0 + \epsilon_1$
5. There is a size function on Ω' using $\epsilon_0, \epsilon_1, \epsilon_2$.
6. If $AX_1 \in L_t, BX_2 \in L_s$ and $X_1 \cong X_2$, say $t \leqslant s$, then (a) and (b) below hold where P is defined by

$$AX_1 \equiv Ex, \quad BX_2 \equiv Fy, \quad P \equiv x \cdot E \cdot (y \cdot F)^{-1}$$

(and $L(x) = L(y) = 1$).

(a) If $t < s$ then
  (i) $S_{X_1A}(X_1) < 1/p$ or $P = U \cdot C \cdot U^{-1}$ in G_{t-1} for some $u \in \Pi$ and $C \in L_s$
and (ii) $s_{X_2B}(X_2) < 1/p_0'$.
(b) If $t = s$ then
  (i) $s_{X_1A}(X_1) < 1/p_0$ and $s_{X_2B}(X_2) < 1/p_0$
or (ii) $P = I$ in G_{t-1}.

6.4. Definition. A non-empty subset Ω' of Π satisfies the *Tartakovskii Conditions* and $G = \Pi/[\Omega']$ is a *Tartakovskii Group* if a real number p_0 exists such that

1. Ω' consists of E.R. elements and is closed under C.A. and inverses.

2. $5 \leqslant p_0 < L = \mathrm{Min}\, L(W)$ for $W \in \Omega'$.

3. If $AX_1, BX_2 \in \Omega'$ and $X_1 \cong X_2$ and P is defined by $AX_1 \equiv Ex$, $BX_2 \equiv Fy$, $P \equiv x \cdot E \cdot (y \cdot F)^{-1}$ $(L(x) = L(y) = 1)$ then

(i) $L(X_1)/L(X_1A) < 1/p_0$ and $L(X_2)/L(X_2B) < 1/p_0$

or (ii) $P = I$ in Π.

6.5. Proposition. *Any Tartakovskii Group is a Generalized Tartakovskii Group.*

Proof. Let Ω' satisfy the Tartakovskii Conditions. Put $\epsilon_2 = 1/L$, $\epsilon_1 = \epsilon_0 = 0$, $1/p = 1/p_0$, $1/p_0' = 1/p_0 - \epsilon_2$. Then $s_{AB}(A) = L(A)/L(AB)$ is a size function for Ω'. Put $\mathsf{L}_1 = \Omega'$ and take $\mathsf{L}_2, \mathsf{L}_3, \ldots$ empty.

6.6. Definition. Let $X_1, \ldots, X_m$ be a partition. The left (right) semiclosure of X_i is the product Y_i of those patches of which the last (first) term arises from X_i $(i = 1, \ldots, m)$.

6.7. Proposition. *Let Ω' satisfy the Generalized Tartakovskii Conditions. Let α, X be a partition of a split of an element of L_t. Let Y be the left semiclosure of X. Let X^{-1}, β be a partition of a split of an element of L_s. (i) Let $s_{YZ}(Y) \geqslant 1/p_0$. Then we have $t \leqslant s$; if $t = s$ then $w(\alpha \cdot \beta) < \bar{s} + \bar{t}$ and X is special in α, X. (Here YZ is the appropriate element of L_t.) (iii) If $s_{YZ}(Y) \geqslant 1/p_0'$ and $s_{YZ}(Y) > \epsilon_2$ then $t \leqslant s$ or $w(\alpha \cdot \beta) < \bar{s} + \bar{t}$.*

Proof. Since $1/p \geqslant 1/p_0 > \epsilon_2$ we have $s_{YZ}(Y) > \epsilon_2$, hence $L(Y) \geqslant 2$, in all cases. Hence $X \equiv x_1 x_2 \cdots x_n$, $n \geqslant 2$. Now (1) $Y \equiv \bar{x}_1 x_2 \cdots x_{n-1}$ if x_n is not a patch (of $\alpha \cdot X$) or (2) $Y \equiv \bar{x}_1 x_2 \cdots x_{n-1} x_n$ if x_n is a patch, where $\bar{x}_1 \sim x_1$. Consider β^{-1}, X. This is a split of an element J of L_s which either (3) contains a subword $T \equiv \hat{x}_1 x_2 \cdots x_{n-1} \hat{x}_n$, if x_1, x_n are in different patches, or (4) contains a subword $T \equiv \hat{x}_1 x_2 \cdots \hat{x}_{n-1}$ if x_1, x_n are in the same patch; here $\hat{x}_1 \sim x_1$, $x_n \sim \hat{x}_n$.

Case 1. (1) and (3) do not both hold; (2) and (4) do not both hold.

Then $Y \cong T$. Let (i) hold. Then $s_{YZ}(Y) \geqslant 1/p_0 \geqslant 1/p_0'$ hence $t \leqslant s$. If $t = s$ then "$P = I$ in G_{t-1}"; but P here is conjugate in Π to $\alpha \cdot \beta$ so $w(\alpha \cdot \beta) < t$. If (iii) holds then $t \leqslant s$. Assume (ii) holds. If $\bar{t} < s$ then $P = U \cdot C \cdot U^{-1}$ in G_{t-1}, $C \in L_s$, so $w(\alpha \cdot \beta) < \bar{s} + \bar{t}$. If $t = s$ then $w(\alpha \cdot \beta) < \bar{t} < \bar{s} + \bar{t}$.

Case 2. (2) and (4) both hold.

Since $x_1 \sim x_n$ we have $n \geqslant 3$. Now where $Y \equiv Y' x_n$ we have $Y' \cong T$. Moreover $T \in L_s$. Hence $s(T) \geqslant 1 - \epsilon_0 \geqslant 5/p + 4\epsilon_1 \geqslant 1/p \geqslant 1/p_0 \geqslant 1/p_0'$. Hence $s \leqslant t$. If $s = t$ then since $s(T) \geqslant 1/p_0$, $w(\alpha \cdot \beta) < \bar{t} < \bar{s} + \bar{t}$; if $s < t$ then since $s(T) \geqslant 1/p$, $w(\alpha \cdot \beta) < \bar{s} + \bar{t}$.

Case 3. (1) and (3) both hold.

Here $Y \equiv \bar{x}_1 \cdots x_{n-1}$ and $T \equiv \hat{x}_1 \cdots x_{n-1} \hat{x}_n$. Put $T' \equiv \hat{x}_1 \cdots x_{n-1}$. The argument of Case 1 now holds with T replaced by T'.

6.8. Proposition. *If Ω' satisfies the Generalized Tartakovskii Conditions it satisfies* 1, 2, 3, 4 *of* 2.3 *and* 5.3.

Proof. Trivially 1, 2, 3 hold. Let $X_1, \ldots, X_c$ be a partition of a split of $W \in L_t$ where $c \leqslant 5$. Let Y_i be the left semiclosure of X_i. Then $Y_1\, Y_2 \cdots Y_c$ has no dots and is a C.A. of W. Omitting certain subscripts we have

$$1 - \epsilon_0 \leqslant s(Y_1 \cdots Y_c) \leqslant s(Y_1) + \cdots + s(Y_c) + (c-1)\epsilon_1$$

Now $c(1/p + \epsilon_1) \leqslant 5(1/p + \epsilon_1) \leqslant 1 - \epsilon_0 + \epsilon_1$ so $1/p \leqslant (1 - \epsilon_0 - (c-1)\epsilon_1)/c$ hence $1/p \leqslant s(Y_j)$ for some j $(1 \leqslant j \leqslant c)$. By 6.7 X_j is special.

6.9. We shall continue the discussion of *Generalized* Tartakovskii Groups in Section 7. Our aim in this section is to prove Lemma 5.4.

6.9A. Proposition. *Let Ω' satisfy the Tartakovskii Conditions and assume* $(*)$ *no constituent group of* Π *has an element of order* 2. *Then* 5.4 *holds in this case.*

Note: It will appear later than $(*)$ is unnecessary (cf. 6.12A).

Proof. Let $W \approx W_1 \circ \cdots \circ W_k$ be natural and let S be a scheme satisfying 1, 2, 3′, 4, 5. Assume one row has the form α, X, β, X^{-1} where X is special. Let $A = \beta \cdot X^{-1} \cdot \alpha$, $B = \alpha \cdot X \cdot \beta$. Then X^{-1}, B and X, A are partitions so $w(A \cdot B) < (1, 0, 0, ...)$, i.e., $A \cdot B \equiv I$. Hence

$$X\beta X^{-1} \alpha \cdot \alpha X \beta X^{-1} \equiv I.$$

Let $C = X \cdot \beta \cdot X^{-1}$. Then $C \cdot \alpha = C^{-1} \cdot \alpha^{-1}$. Now $L(X) \geqslant 2$ and α, X, β, X^{-1} is a partition hence $\beta \not\equiv I$; similarly $\alpha \not\equiv I$. Hence $L(C) \geqslant 3$.

Suppose (†) there is no amalgamation in $C \cdot \alpha$ and $C^{-1} \cdot \alpha^{-1}$. Then $C \equiv C^{-1}$ so $C^2 \equiv I$ and by (∗) $C \equiv I$, a contradiction. Now let (†) be false. Where $\alpha = a_1 \cdots a_n$, $C \equiv c_1 \cdots c_m$ we have

$$c_1 \cdots c_{m-1} x a_2 \cdots a_n = c_m^{-1} \cdots c_2^{-1} y a_{n-1}^{-1} \cdots a_1^{-1}, \quad x = c_m \cdot a_1, \; y = c_1^{-1} \cdot a_n^{-1}$$

Hence $c_1 \cdots c_{m-1} = c_m^{-1} \cdots c_2^{-1}$. In particular $c_1 = c_m^{-1}$ so $c_1 \cdots c_{m-1} c_m = c_m^{-1} \cdots c_2^{-1} c_1^{-1}$. Thus $C \equiv C^{-1}$, yielding $C \equiv I$ as before.

6.10. Lemma. *Let F be the free group on $s_1, s_2, \ldots$. Let Ω_1 be a subset of F consisting of elements of the form $S_{v_1}^{\epsilon_1} S_{v_2}^{\epsilon_2} \cdots S_{v_j}^{\epsilon_j}$ $(\epsilon_i = \pm 1)$ where $j \geqslant 6$ and $v_i \neq v_{i+1}$ $(i = 1, 2, ..., j)$ where v_{j+1} means v_1. Let Ω_1 be closed under cyclic arrangements and inverses. Assume that if $U, V \in \Omega_1$ then either the number $\beta(U, V)$ of cancellations in $U \cdot V$ is at most one or $U \cdot V = 1$. Then every non-identity element U of $[\Omega_1]$ has a subword A such that $(\exists B)\, AB \in \Omega_1$, $L(B) \leqslant 3$.*

Note: Here we have temporarily adapted our notation and definitions to the case of free groups rather than free products.

Proof. This is a special case of e.g. Greendlinger [1] but we shall derive it from our present results.

Let $k \geqslant 15$ be an integer and let H be the free group on

$$s_{11}, \ldots, s_{1k}, s_{21}, s_{22}, \ldots, s_{2k}, \ldots$$

If W is a word in $s_1, s_2, \ldots$ let $\overline{W}$ be the word obtained from W

by replacing s_i by $s_{i1}s_{i2}\cdots s_{ik}$ $(i = 1, 2, ...)$. Let Ω_2 be the set of all $\overline{W}$ $(W \in \Omega_1)$ and let Ω_3 be the closure of Ω_2 under cyclic arrangements and inverses. Let $K = H/[\Omega_3]$. Then $K = H/[\Omega_2]$. Add new generator s_i and new defining relations $s_i = s_{i1}s_{i2}\cdots s_{ik}$. Then remove this defining relation with the generator s_{i1}. This shows that K is the free product of a free group with $F/[\Omega_1]$. Now $U = 1$ in $F/[\Omega_1]$ hence $U = 1$ in K and $\overline{U} = 1$ in K.

We show Ω_3 satisfies the Tartakovskii Conditions with $p_0 = 6k/(k+3)$. Now $5 \leqslant p_0 < 6k$. If $U, V \in \Omega_3$ we find $\beta(U, V) \leqslant k$ or $U \cdot V = 1$. Let $AX_1, BX_2 \in \Omega_3$ and $X_1 \cong X_2$, $L(X_1)/L(AX_1) \geqslant 1/p_0$. Then $L(X_1) \geqslant k+3$ so $X_1 = c_1Xd_1$, $X_2 = c_2Xd_2$, $L(X) \geqslant k+1$. Now $\beta(d_1Ac_1X, X^{-1}c_2^{-1}B^{-1}d_2^{-1}) > k$ so $d_1Ac_1X \cdot X^{-1}c_2^{-1}B^{-1}d_2^{-1} = 1$.

Since the constituent groups are infinite cyclic, 5.4 hence Theorem 3 are available (cf. 6.9A). Hence $\overline{U}$ contains a segment A such that there is a partition $A_1, Z_2, ..., Z_v$ of an element J of $S(\Omega_3)$ where $v \leqslant 4$ and none of $Z_2, ..., Z_v$ are special. Hence $L(Z_i) < k+3$. Some split J_1 of J is in Ω_2 so has the form $\overline{W}$ $(W \in \Omega_1)$. Let $\lambda = L(W)$. Then $\lambda k \leqslant L(A \cdot Z_2 \cdot \cdots \cdot Z_v) \leqslant L(A) + 3(k+2)$, so $L(A) \geqslant k(\lambda - 3) - 6$ where $A = cA'd$, A' is a subword of $\overline{U}$ and of an element $\overline{V}$ of Ω_2 and $L(A') \geqslant k(\lambda - 3) - 8$. But a cancellation of length a between $\overline{X}, \overline{Y}$ implies a cancellation of length b between suitable $C \cdot A$ of $\overline{X}, \overline{Y}$ where b is the least number such that $k|b$ and $a \leqslant b$. Hence $\overline{U}$ has a subword A'' of length $\geqslant k(\lambda - 3)$ where $A''B'' \in \Omega_2$ and $L(A''B'') = \lambda k$. Hence the result.

6.11. Corollary. *Let F be the free group on $s_1, s_2, ...$. To each s_i assign a positive integer p_i. If $X \equiv s_{i_1}^{\delta_1} \cdots s_{i_k}^{\delta_k}$ (with no cancellations possible), $\delta_i = \pm 1$, let $w(X) = p_{i_1} + \cdots + p_{i_k}$. Let Ω_1 be a subset of F, closed under C.A. and inverses, consisting of elements of the form*

$$Y \equiv s_{v_1}^{\epsilon_1} \cdots s_{v_j}^{\epsilon_j} \ (\epsilon_i = \pm 1), \qquad w(Y) \geqslant 6$$

where if $j \geqslant 2$ then $v_i \neq v_{i+1}$ $(i = 1, ..., j-1)$, $v_j \neq v_1$. Assume if $U, V \in \Omega_1$, $U \equiv U'X$, $V \equiv X^{-1}V'$ then $w(X) \leqslant 1$ or $U \cdot V = 1$. Then every non-identity element U of $[\Omega_1]$ has a subword A such that $(\exists B)\ AB \in \Omega_1$ and $w(B) \leqslant 3$.

Proof. Replace s_i by $t_{i1}t_{i2}\cdots t_{ip_i}$.

6.12. Supplement to Theorem 2. *Let* Ω' *satisfy* 1, 2, 3, 4. *If* $W \approx W_1 \circ \cdots \circ W_k$ *is natural, so that there is a scheme* S' *satisfying* 1, 2, 3′, 4, 5, *then for every such* S' *we have*

6. *no row of* S' *has the form* α, X, β, X^{-1} *where* X, X^{-1} *is of Type C in the underlying semischeme.*

Proof. If some row of S' has the form α, X, β, X^{-1} where X, X^{-1} is a pair of Type C call X, X^{-1} a pair of *Type* CS. Assume some Type CS pair exists; we shall obtain a contradiction. Consider any sequence of elementary operations transforming the rows of S' to its result and consider the first elementary operation in this sequence in which a Type CS pair say X, X^{-1} is cancelled:

$$\begin{array}{ccccccc}
\alpha, X, \beta, X^{-1} & & \alpha', X, X^{-1} & & \alpha' & & \beta \\
\rho_2 & \to \cdots \to & \sigma_2 & \to & \sigma_2 & \to \cdots \to \sigma.\ \text{Hence} & \rho_2 \to \cdots \to 1 \\
\vdots & & \vdots & & \vdots & & \vdots \\
\rho_k & & \sigma_v & & \sigma_v & & \rho_h
\end{array}$$

without loss of generality. Now none of $\beta, \rho_2, \ldots, \rho_h$ contain a term of Type R of Type CS. If some ρ_i $(i = 2, \ldots, h)$ has c terms where $c \leqslant 5$ then a contradiction would follow as in 5.5. Hence each of $\rho_2, \ldots, \rho_h$ has at least six terms. Let the symbols in $\rho_2, \ldots, \rho_h$ be $u_1, \ldots, u_a$. Then $\beta = 1$ in the group $(u_1, \ldots, u_a | \rho_2 = 1, \ldots, \rho_h = 1)$. By 6.11 β contains at least three consecutive terms from some C.A. of $\rho_i^{\pm 1}$ (for some $i = 2, \ldots, h$). This is a contradiction since S' satisfies 5.

6.12A. *Note*: Lemma 5.4 follows immediately from 6.12. Also by 5.5 we have that Theorem 3 holds. Finally we note that in 6.9A, the hypothesis $(*)$ is unnecessary.

§7.

The main result for Generalized Tartakovskii Groups is:

7.1. Theorem 4. *Let Ω' satisfy the Generalized Tartakovskii Conditions and let $I \not\equiv R \in [\Omega']$. Then R has a subword A such that there is an element $A'B$ of Ω' where $A \cong A'$ and*

$$s_{A'B}(A') > 1 - \epsilon_0 - 3(1/p + \epsilon_1).$$

Moreover $L(R) \geqslant L(A) \geqslant 2$.

If $R \in [\mathsf{L}_1 \cup \cdots \cup \mathsf{L}_q]$ and $A'B \in \mathsf{L}_q$ then

$$s_{A'B}(A') > 1 - \epsilon_0 - 3(1/p_0 + \epsilon_1).$$

Proof. By 6.8, Theorem 3 is available. Hence R has a segment A_1 (i.e., A_1 is a term of a division of R) and there is a partition $A_1, A_2, ..., A_v$ of $J \in S(\Omega')$ where $v \leqslant 4$ and none of $A_2, ..., A_v$ is special. Let Y_i be the left semiclosure of A_i. As in 6.8,

$$1 - \epsilon_0 \leqslant s(Y_1 Y_2 \cdots Y_v) \leqslant s(Y_1) + \cdots + s(Y_v) + (v-1)\epsilon_1.$$

Now if $s(Y_i) \geqslant 1/p$ then A_i is special (cf. 6.7); hence $1 - \epsilon_0 < s(Y_1) + 3(1/p + \epsilon_1)$. Therefore $L(Y_1) \geqslant 2$, otherwise $s(Y_1) \leqslant \epsilon_2 < 1/p$ and

$$1 < \epsilon_0 + 4(1/p + \epsilon_1) - \epsilon_1$$

$$< \epsilon_0 + (1 - \epsilon_0 + \epsilon_1) - \epsilon_1 = 1.$$

Hence $A_1 = x_1 x_2 \cdots x_n$, $n \geqslant 2$. Y_1 is (a) $\bar{x}_1 x_2 \cdots x_{n-1}$ or (b) $\bar{x}_1 x_2 \cdots \bar{x}_n$ and an appropriate E.R. split of J contains a subword (c) $A' \equiv \bar{x}_1 x_2 \cdots x_{n-1} \bar{x}_n$ or (d) $A' \equiv \bar{x}_1 x_2 \cdots x_{n-1}$. ((d) is the case if and only if x_1, x_n lie in the same patch.) Not both (b), (d) can hold. Hence $Y_1 \subset A'$. Now a subword of R has the form $\hat{x}_1 x_2 \cdots x_{n-1} \hat{x}_n$ of which a left subword $A \cong A'$. We may suppose $J \equiv A'B$ say. Then

$$1 - \epsilon_0 - 3(1/p + \epsilon_1) < s_{A'B}(Y_1) \leqslant s_{A'B}(A')$$

We have $L(A) = L(A') \geqslant L(Y_1) \geqslant 2$.

To prove the last part of the theorem, first note that if $\Omega'' = \mathsf{L}_1 \cup \cdots \cup \mathsf{L}_q$ then Ω'' satisfies the Generalized Tartakovskii Conditions; moreover if α, X, β is a partition of a split of an element of L_q and $L(X) \geqslant 2$ then, with respect to Ω'',

$$X \text{ is } q\text{-special in } \alpha, X, \beta \Leftrightarrow X \text{ is special in } \alpha, X, \beta.$$

Thus if $s(Y_i) \geqslant 1/p_0$ for some $i = 2, ..., v$ then by 6.7 A_i is q-special; hence A_i is special; but this is a contradiction. Thus

$$1 - \epsilon_0 < s(Y_1) + 3(1/p_0 + \epsilon_1)$$

and the required inequality follows.

§8.

In this section we discuss equations of the form $A \cdot B = I$ and $A \cdot B \cdot C = I$ in a Generalized Tartakovskii Group.

8.1. Theorem 5. *Let Ω' satisfy the Generalized Tartakovskii Conditions. Let $C = D$ in G_n where C, D are not both I. Assume that if S is $*$-contained in C or D, i.e., $S \cong S'$ for some subword S' of C or D, and if $ST \in \mathsf{L}_i$, $1 \leqslant i \leqslant n$ then*

$$s_{ST}(S) \leqslant \begin{cases} 1 - \epsilon_0 - 4(1/p + \epsilon_1) & \text{if } i < n \\ 1 - \epsilon_0 + 4(1/p_0 + \epsilon_1) & \text{if } i = n. \end{cases}$$

Then there are divisions $C = P_1, ..., P_m$, $D = Q_1, ..., Q_m$, and words $X_0, X_1, ..., X_m$ where $X_0 \equiv X_m \equiv I$ such that for $j = 1, 2, ..., m$.

1. *$I \neq P_j \equiv Q_j$, $X_j \equiv X_{j-1} \equiv I$. (In this case put $t_j = 0$.) or*
2. *$P_j^{-1}, X_{j-1}^{-1}, Q_j, X_j$ is a partition of a split of an element W_j of L_{t_j} where $1 \leqslant t_j \leqslant n$.*

t_j is called the type of P_j and of Q_j
Moreover in 2 if $\wedge$ denotes left semiclosure then, where

$$1/p_0'' = \begin{cases} 1/p_0' & \textit{if } \epsilon_2 < 1/p_0' \\ \epsilon_2 + 1/p_0' & \textit{otherwise.} \end{cases}$$

$$s(\hat{X}_{j-1}^{-1}) < 1/p_0'',\ 1/p_0,\ 1/p \text{ according as } t_j >, =, < t_{j-1}$$

$$s(\hat{X}_j) \quad < 1/p_0'',\ 1/p_0,\ 1/p \text{ according as } t_j >, =, < t_{j+1}.$$

Similarly for right semiclosure.

8.1A. Note. Since $1 - \epsilon_0 \leqslant s(\widehat{p_j^{-1}}) + s(\widehat{X_{j-1}^{-1}}) + s(\widehat{Q_j}) + s(\widehat{X_j}) + 3\epsilon_1$ we have

$$s(\widehat{P_j^{-1}}) + s(\widehat{Q_j}) > \begin{cases} 1 - \epsilon_0 - 2/p - 3\epsilon_1 & \text{if } t_j < n \\ 1 - \epsilon_0 - 2/p_0 - 3\epsilon_1 & \text{if } t_j = n. \end{cases}$$

Proof of 8.1. We may suppose $C \not\equiv D$. Let $D^{-1} \equiv T^{-1}B$, $C \equiv AT$ where T is taken maximally. Then $A \cdot B = I$. Next $A \cdot B \equiv A^{\mathrm{L}} c^{\epsilon} B^{\mathrm{R}}$ ($\epsilon = 0$ or 1), and contains a natural element N. N $*$-contains a subword of L_q' of size $> 1 - \epsilon_0 - 3(1/p + \epsilon_1)$ (if $q = n$ replace p by p_0). Hence $N \not\subset A^{\mathrm{L}}$ and $N \not\subset B^{\mathrm{R}}$. Thus there is a division of N of the form $A^{\mathrm{R}}, B^{\mathrm{L}}$. There is a scheme as in Theorem 2 whose result $R_1, \ldots, R_b$ is a proper strong partition of N, and such that no row has the form $R_i, R_{i+1}, \ldots$ ($i = 1, \ldots, b$). By 3.10 there is a division R_i', R_i'' of some R_i and divisions $R_1, \ldots, R_{i-1}, R_i'$ of A^{R}; $R_i'', R_{i+1}, \ldots, R_b$ of B^{L}. In the row containing R_i replace R_i by R_i', R_i''. Let G' be the group with generators the symbols of this scheme and defining relations the rows of this scheme. Let Θ consist of all C.A. and inverses of the defining relations.

The scheme satisfies 1a, 2, 3$'$, 4, 5 and no row has the form $\ldots, X, \ldots, X^{-1}, \ldots$. Also any row consisting of Type C terms has at least six terms. The defining relations satisfy $\beta(U, V) \leqslant 1$ or $U \cdot V \equiv 1$. Hence the result contains a subword E where $EF \in \Theta$ and $L(F) \leqslant 3$.

We first show that E contains R_i' or R_i''. If not then without loss of generality $E \subset R_1, ..., R_{i-1}$. E contains some R_j $(1 \leqslant j \leqslant i-1)$. Hence EF is a C.A. of the row containing R_j. Hence E does not contain R_{j-1} or R_{j+1}. Hence E is R_j. F contains no term of the result since $L(F) \leqslant 3$. This means that A' "contains more than" $1 - 3/p - 3\epsilon_1 - \epsilon_0$ of an element of Ω', contrary to hypothesis.

Next E contains both R_i', R_i''; for if it contains only one then the same contradiction occurs.

Hence $R_1 \cdots R_{i-1} F^{-1} R_{i+1} \cdots R_b = 1$ in the group obtained from G' by deleting the defining relation $EF = 1$, The left side could be empty but if not, no cancellation occurs, and it contains C where $CD \in \Theta$, $L(D) \leqslant 3$. Now $C \cap F \leqslant 1$ so C is not contained in any of

$$R_1 \cdots R_{i-1}, \qquad F^{-1}, \qquad R_{i+1} \cdots R_b$$

If $C \subset R_1 \cdots R_{i-1} F^{-1}$ then A' would "contain more than" $1 - 4/p - 4\epsilon_1$ of an element of Ω'. Hence $C \equiv R_{i-1} F^{-1} R_{i+1}$. Thus we see that the rows of the scheme are

$$R_i' \, R_i'' \, X_i$$

$$R_{i-1} X_i^{-1} R_{i+1} X_{i+1}$$

$$R_{i-2} X_{i+1}^{-1} R_{i+2} X_{i+2}$$

$$R_2 X_{b-2}^{-1} R_{b-1} X_{b-1}$$

$$R_1 X_{b-1}^{-1} R_b$$

With respect to Ω' consider any scheme as in Theorem 2 and let α, X and X^{-1}, β be two distinct rows, say the ith and jth rows respectively where X, X^{-1} is a pair of Type C. Say $W_i \in L_t$, $W_j \in L_s$. By the definition of a natural element (2.6) $w(\alpha \cdot \beta) \geqslant s + t$. Hence by 6.7 if Y is the left semiclosure of X in $\alpha \cdot X$ then

(i) $s(Y) \geqslant 1/p_0 \Rightarrow t < s$
(ii) $s(Y) < 1/p$
(iii) $s(Y) \geqslant 1/p_0'' \Rightarrow t \leqslant s$.

Thus $t > s$ implies $s(Y) < 1/p_0''$, $t = s$ implies $s(Y) < 1/p_0$ and $t < s$ implies $s(Y) < 1/p$.

Define A_1, B_1 by $A \equiv A_1 A^{\mathrm{R}}$, $B \equiv B^{\mathrm{L}} B_1$. Then

$$I = A \cdot B \equiv A_1 A^{\mathrm{R}} \cdot B^{\mathrm{L}} B_1 = A_1 \cdot N \cdot B_1 = A_1 \cdot B_1.$$

Either $A_1 \cdot B_1 \equiv I$ or $A_1 \cdot B_1$ contains a natural element N_1. The theorem now follows by a simple induction argument.

8.1B. Corollary. *In Theorem 5, if $L(C) \leqslant 2$ then $C \equiv D$.*

8.2. Theorem 6. *Let $A' \cdot B' \cdot C' = I$ in G_n where if S is $*$-contained in A', B' or C' and $ST \in \mathsf{L}_i$ $(i \leqslant n)$ then*

$$s_{ST}(S) \leqslant \begin{cases} 1 - 5/p - 5\epsilon_1 - \epsilon_0 & \text{if } i < n \\ 1 - 5/p_0 - 5\epsilon_1 - \epsilon_0 & \text{if } i = n \end{cases}$$

Then

1. *there are divisions A_1, A_2, A_3 of A'; B_1, B_2, B_3 of B'; C_1, C_2, C_3 of C'. Also $A_3 \cdot B_1 = X$, $B_3 \cdot C_1 = Y$, $C_3 \cdot A_1 = Z$ in G_n for some X, Y, Z.*

2. *Either (2a) A_2, X, B_2, Y, C_2, Z is a partition of a split of an element of Ω' or (2b) $A_2 \cdot X \cdot B_2 \cdot Y \cdot C_2 \cdot Z \equiv I$. Moreover if (2b) holds then either (i) A_2, B_2, C_2 all have length 1 and $X \equiv Y \equiv Z \equiv I$, or (ii) $A_2 \equiv B_2 \equiv C_2 \equiv I$.*

3. *For the equation $A_3 \cdot B_1 = X$ we have (unless $A_3 \equiv B_1 \equiv I$) divisions $A_3^{-1} \equiv P_1, \ldots, P_k$, $B_1 \equiv Q_1, \ldots, Q_k$ where for $i = 1, \ldots, k$ either $P_i \equiv Q_i$ and $X_{i-1} \equiv X_i \equiv I$, or $P_i^{-1} X_{i-1}^{-1} Q_i X_i$ is a partition of a split of an element of Ω'. Also $X_0 \equiv I$, $X_k^{-1} \equiv X$. Similarly for the other two equations.*

Proof. Take U maximal so that $(\exists V)\ A' \equiv A_1 U \quad B' \equiv V B_1 \quad U \cdot V = 1$ in G (possibly $U \equiv I$). V is uniquely determined by U; if $B' \equiv V' B_1'$, $U \cdot V' = 1$ then $V = V' \equiv VJ$ say and $J = I$. Now $B' \supset J$; if J were not empty B' $*$-contains E where $EF \in \Omega'$

$$s_{EF}(E) > 1 - 3/p - 3\epsilon_1.$$

Thus $A_1 \cdot B_1 \cdot C^1 = 1$. Take X maximal such that

$$(\exists Y)\, B_1 \equiv BX, \quad C' \equiv YC_2, \quad X \cdot Y = 1.$$

Then $A_1 \cdot B \cdot C_2 = 1$. Take S maximal such that

$$(\exists T)\, C_2 \equiv CS, \quad A_1 \equiv TA, \quad S \cdot T = 1.$$

Then

(1) $$A \cdot B \cdot C = 1$$

(2) $$A' \equiv TAU, \quad B' \equiv VBX, \quad C' \equiv YCS.$$

Note that if $A \equiv A''E$, $B \equiv FB''$, $E \cdot F = 1$ then $E \equiv F \equiv 1$ for $A' \equiv TA''EU$, $B' \equiv VFB''X$ hence $E \equiv I$; by the uniqueness $F \equiv 1$. Similarly for the pairs B, C and C, A.

Case 1. $A \equiv B \equiv C \equiv I$.

Here $A' \equiv TU$, $B' \equiv VX$, $C' \equiv YS$, $U \cdot V = X \cdot Y = S \cdot T = 1$.

Case 2. Exactly one or exactly two of A, B, C are I.

If two are I then the other $= I$ in G hence $\equiv I$.
If A is I then $B \cdot C = I$, hence $B \equiv C \equiv I$, a contradiction; similarly if B is I and if C is I.

Case 3. None of A, B, C is I.

Case 3a. All of A, B, C have length 1

$$A' \equiv TaU, \quad B' \equiv VbX, \quad C' \equiv YcS, \quad a \cdot b \cdot c \equiv I$$

$$U \cdot V = X \cdot Y = S \cdot T = 1.$$

Case 3b. Exactly two have length 1.

$$A \cdot b \cdot c = I, \qquad A = c^{-1}b^{-1} \quad \text{hence} \quad A \equiv c^{-1}b^{-1} \quad \text{(no dot).}$$

Thus we have the contradiction that cancellation occurs in $A \cdot b$.

Case 3c. At most one of A, B, C has length 1.

Note that if $A^{\mathrm{R}} \cdot B \cdot C^{\mathrm{L}} = 1$ then $C^{\mathrm{R}} \cdot A^{\mathrm{L}} = 1$. Hence $C^{\mathrm{R}} \equiv A^{\mathrm{L}} \equiv 1$, A^{R} is A and C^{L} is C.

Also note that A, B, C is a partition (this is so if all of A, B, C have length $\geqslant 2$ so there remains the case

$$a, b_1, ..., b_m, c_1, ..., c_n \qquad (m \geqslant 2, n \geqslant 2).$$

Now $a \cdot b_1 \not\equiv I$, $c_n \cdot a \not\equiv I$. Suppose $c_n \cdot a \cdot b_1 \equiv I$. Then by 8.1B $n = m = 1$ a contradiction.

Now $A \cdot B \cdot C$ is a non-empty relator so has a natural element as subword. N is either a subword of A, B or C or has the form $A^{\mathrm{R}} \cdot B^{\mathrm{L}}$ or $B^{\mathrm{R}} \cdot C^{\mathrm{L}}$ or $A^{\mathrm{R}} \cdot B \cdot C^{\mathrm{L}}$. Hence $N \equiv A \cdot B \cdot C$.

By Theorem 2 and 3.10 there is a scheme S satisfying 1a, 2, 4, 5. The result is a partition of N and has the form

(3) $\qquad A_1 \cdots A_r \; B_1 \cdots B_s \; C_1 \cdots C_t$

where $A_1 \cdots A_r$ is a division of A; similarly for B and C. No row is $\cdots X \cdots X^{-1} \cdots$ and any row consisting of Type C terms has $\geqslant 6$ terms. Let G^0 be the group with generators the symbols of this scheme and the rows as defining relations. Let Θ consist of all C.A. and inverses of the defining relations. If $U, V \in \Theta$ then $\beta(U, V) \leqslant 1$ or $U \cdot V \equiv 1$. Consider $A_1 \cdots A_r B_1 \cdots B_s$. If this contains E where $EF \in \Theta$ and $L(F) \leqslant 4$ then by an earlier argument $E \equiv A_r B_1$. Since $L(F) \leqslant 4$, F contains no result term. Consider (4) $A_1 \cdots A_{r-1} F^{-1} B_2 \cdots B_s$ noting that

$$A_1 \cdots A_{r-1} F^{-1} B_2 \cdots B_s C_1 \cdots C_t = 1$$

in G' where G' is obtained from G^0 by deleting the defining relator EF.

If (4) contains C where $CD \in \Theta$ and $L(D) \leqslant 4$, i.e., $L(C) \geqslant \lambda - 4$ then since $C \cap A_1 \cdots A_{r-1}$ has the form A_j and moreover $L(A_j) \leqslant \lambda - 6$ and since $F^{-1} \cap C \leqslant 1$ we have $C \equiv A_{r-1} F^{-1} B_2$. Consider (5) $A_1 \cdots A_{r-2} D^{-1} B_3 \cdots B_s$ and note that

$$A_1 \cdots A_{r-2} D^{-1} B_3 \cdots B_5 C_1 \cdots C_t = 1 \quad \text{in } G^2,$$

obtained from G' by deleting CD. Then by induction we have (6) $\Delta \cdot C_1 C_2 \cdots C_t = 1$ in G^i where Δ does not contain E where $EF \in \Theta$,

$L(F) \leqslant 4$ and *either* $i = 0$ and Δ is

$$A_1 \cdots A_r B_1 \cdots B_s \quad \textit{or} \quad \Delta \text{ is } A_1 \cdots A_{r-i} H^{-1} B_{i+1} \cdots B_s$$

and $L(H) \leqslant 4$. Combining these case $\Delta = A_1 \cdots A_{r-i} H^{-\delta_i} B_{i+1} \cdots B_s$ ($\delta_i = 0$ or 1). Consider $B_{i+1} \cdots B_s C_1 \cdots C_t$ and similarly remove subwords E where $EF \in \Theta$, $L(F) \leqslant 4$, as long as possible to obtain either

$$B_{i+1} \cdots B_s C_1 \cdots C_t \quad \text{or} \quad B_{i+1} \cdots B_{s-j} K^{-1} C_{j+1} \cdots C_t .$$

Combine these cases: $B_{i+1} \cdots B_{s-j} K^{-\epsilon_j} C_{j+1} \cdots C_t$ ($j = 0 \Rightarrow \epsilon_j = 0$, $j > 0 \Rightarrow \epsilon_j = 1$). Then

$$(7) \qquad A_1 \cdots A_{r-i} H^{-\delta_i} B_{i+1} \cdots B_{s-j} K^{-\epsilon_j} C_{j+1} \cdots C_t = 1 \quad \text{in } G^{i+j}$$

If $B_{i+1} \cdots B_{s-j}$ is empty there may be cancellation between $H^{-\delta_i}$ and $K^{-\epsilon_j}$; if so we obtain after cancellation

$$(8) \qquad \underline{A_1 \cdots A_{r-i} (H^{-\delta_i})^{\mathrm{L}} (K^{-\epsilon_j})^{\mathrm{R}}}\, C_{j+1} \cdots C_t = 1 \quad \text{in } G^{i+j}$$

Now in case (7) or (8) the leftmost underlined word "contains at most" $(\lambda - 5) + 1 = \lambda - 4$ and the same is true for the rightmost underlined word.

The left side of (7) or (8) is either empty or "contains at least" $\lambda - 3$ in which case $i + 1 = s - j$ and

$$(7') \qquad A_{r-i} H^{-\delta_i} B_{i+1} K^{-\epsilon_j} C_{j+1} U \in \Theta$$

or

$$(8') \qquad A_{r-1} (H^{-\delta_i})^{\mathrm{L}} (K^{-\epsilon_j})^{\mathrm{R}} C_{j+1} U \in \Theta$$

and

$$(7'') \qquad A_1 \cdots A_{r-i-1} U^{-1} C_{j+2} \cdots C_t = 1 \quad \text{in } G^{i+t+1}$$

in any case. In particular if (8′) holds

$$(8'') \qquad A_1 \cdots A_{r-i} (H^{-\delta_i})^{\mathrm{L}} (K^{-\epsilon_j})^{\mathrm{R}} C_{j+1} \cdots C_t = 1 .$$

Summarizing: If $B_{i+1} \cdots B_{s-j}$ is non-empty (so $i = s - j - 1$)

(9) $$A_{r-i+1} \cdots A_r B_1 \cdots B_i = H^{-\delta_i}$$

(10) $$B_{s-j+1} \cdots B_s C_1 \cdots C_j = K^{-\epsilon_j}$$

(11) $$C_{j+2} \cdots C_t A_1 \cdots A_{r-i-1} = U$$

and

$$A_{r-i} H^{-\delta_i} B_{s-j} K^{-\epsilon_j} C_{j+1} U \in \Theta$$

The right sides of (9), (10), (11) have length $\leqslant 1$.

If $B_{i+1} \cdots B_{s-j}$ is empty (hence $i = s - j$) (9), (10) hold

(12) $$C_{j+1} \cdots C_t A_1 \cdots A_{r-i} = W^{-1}, \qquad W \equiv (H^{-\delta_i})^{\mathrm{L}} (K^{-\epsilon_j})^{\mathrm{R}}$$

and

$$\emptyset \cdot H^{-\delta_i} \cdot \emptyset \cdot K^{-\epsilon_j} \cdot \emptyset \, [(H^{-\delta_i})^{\mathrm{L}} (K^{-\epsilon_j})^{\mathrm{R}}]^{-1} \equiv 1$$

The right sides of (9), (10), (12) have length $\leqslant 2$.

Finally in the case when the left side of (7) or (8) is empty we have (9) and (10) and $A_1 \cdots A_{r-i} \equiv 1$, $s - j + 1 - i - 1 = 0$,

$$C_{j+1} \cdots C_t \equiv 1\,, \qquad H^{-\delta_i} K^{-\epsilon_j} \equiv 1$$

$$C_{j+1} \cdots C_t A_1 \cdots A_{r-i} \equiv 1, \quad \emptyset \cdot H^{-\delta_i} \cdot \emptyset \cdot K^{-\epsilon_j} \cdot \emptyset \cdot 1 \equiv 1.$$

so that this case can be absorbed into a previous case.

CHAPTER II

Burnside Groups

§ 1.

1.1. Introduction

Let B_d^e denote the Burnside Group with d generators and exponent e ($d \geqslant 2$, $e > 0$). The argument which we shall give in this chapter proving that B_d^e is infinite for all sufficiently large odd e has the following formal structure.

In § 3 we state some axioms which are readily seen to be true when $n = 0$ (cf. 4.1). Assume they are true for some $n \geqslant 0$. The first axiom implies that $B_d^e = \Gamma(C_1, \ldots, C_n, J_n)$, where $C_1, \ldots, C_n$ and J_n are subsets of the free product

$$\Pi = \{a_1, \ldots, a_d \mid a_1^e = 1, \ldots, a_d^e = 1\}$$

and the notation $\Gamma(D)$, for a subset D of Π, means the group obtained from Π by adding the defining relations $Z^e = 1$ ($Z \in D$).

If J_n is empty one easily sees that B_d^e is infinite by using the Morse–Hedlund sequence (5.1, 5.2).

If J_n is not empty one shows that the axioms are true for $n + 1$; this occupies the main body of the chapter. Afterwards it is not difficult to show that $B_d^e = \Gamma(C_1, C_2, C_3, \ldots)$; the infiniteness follows. (cf. 5.3)

1.2. Remark. It was our intention to derive an explicit numerical value for A such that B_d^e is infinite for all odd $e \geqslant A$ ($d \geqslant 2$), but this aim was abandoned in the course of the proof because of the labour likely to be involved. However it will become clear to the reader that an explicit bound is obtainable from the method of proof, granted that the inequalities (which are almost all linear) are consistent.

1.3. *Note*: All considerations of consistency are postponed until Section 21, although the reader may prefer to consult Section 21 after reading each section (5, 6, ...).

§2.

From now on, Π denotes the free product

$$\{a_1, \ldots, a_d \mid a_1^e = 1, \ldots, a_d^e = 1\}$$

of $d \geqslant 2$ cyclic groups each of order e. (Our methods would in fact apply to an arbitrary free product provided that no constituent group has an element of order 2.) We adopt the following notation and definitions.

2.1. If $X, Y \in \Pi$ we say X is a C.A.I. of Y if X is a cyclic arrangement of Y or of its inverse Y^{-1}.

2.2. For any subset S of Π, if A is a subword of B and $B \in S$, we say A is a *subelement* of S.

2.3. Let X be an E.R. word. We say *X is a power* if $X = Y^s$ in Π for some $Y \in \Pi$ and $s \geqslant 2$. (Then Y is E.R. and $L(X) = s \cdot L(Y)$.)

2.4. If $W \equiv ABC$ and $A \not\equiv I$, $C \not\equiv I$ we say B is properly contained in W.

We sometimes write $S- \equiv T-$ to denote that one of S, T is a left subword of the other; (and dually).

2.5. A *point* of a word W may be defined as a left subword X of W; it is a left end point (L.E.P.) if $X \equiv I$, an interior point if $I \not\equiv X \not\equiv W$ a right end point (R.E.P.) if $X \equiv W$. A point of a subword of W determines a point of W in an onvious way. The point X is left of the point Y if $L(X) < L(Y)$ and we occasionally write $<$ to denote "left of" for points.

2.6. When we say "distinct subwords $S_1, S_2, ..., S_m$ of W" it will be assumed, unless otherwise stated, that for $i = 1, ..., m-1$

1. L.E.P. $S_i <$ L.E.P. S_{i+1}, or
2. L.E.P. $S_i =$ L.E.P. S_{i+1} and R.E.P. $S_i <$ R.E.P. S_{i+1}.

2.7. Let A, B be non-empty subwords of W. If neither is a subword of the other then either $W \equiv XAYBZ$ or $W \equiv XDEFZ$, where A is DE, B is EF and D, E, F are all non-empty; in the first case we say A, B are *disjoint* and in the second case we say A, B *overlap.* If A, B are disjoint and Y above is I we say A, B *touch.*

We say A *is left of* B if L.E.P. $A <$ L.E.P. B and R.E.P. $A <$ R.E.P. B. We do not write $A < B$ however.

2.8. The notation (A, B) where A, B are subwords of W means the subword of W generated by A, B, i.e., the smallest subword to contain both. Similarly for three or more subwords of W.

2.9. If $S_1, ..., S_m$ are subwords of W such that each S_i has a non-empty subword J_i such that J_i is disjoint from all $S_j, j \neq i$, then the largest such J_i is called the *kernel* of S_i with respect to $S_1, ..., S_m$. We also say that *kernels exist for* $S_1, ..., S_m$.

2.10. By a *parameter* we mean an explicitly given function of e (the exponent), not containing variables other than e. For example, $5e^{-4} + 6$. All parameters will be positive for all sufficiently large e.

2.11. Let X be E.R, and let $g > 0$ be any real number. Then $\delta_g(X)$ denotes a subword of some power of X or of X^{-1} of length $\leqslant geL(X)$ and $\epsilon_g(X)$ denotes a similar subword of length $> geL(X)$.

2.12. Cyclic (or circular) words. In Π we have (i) if $I \not\equiv W \in \Pi$ then W is a conjugate of J where J is E.R. or has length 1; (ii) if A, B are E.R. and conjugate then A is a cyclic arrangement of B.

A *weak cyclic word* is an equivalence class for the relation of conjugacy on Π.

A *cyclic word* is an equivalence class containing an E.R. element. Thus if X, Y are E.R. then $(X) = (Y)$ if and only if X is a C.A. of Y.

A *linearization* of a cyclic word C is an E.R. word X such that $(X) = C$.

A word, i.e., an element of Π may also be called a *linear word.*

2.13. A *weak incyclic word* is an equivalence class for the relation "X is a conjugate of Y or of Y^{-1}" on Π.

An *incyclic word* is such a class containing an E.R. element.

If X, Y are E.R. then $(X)' = (Y)'$ if and only if X is a C.A.I. of Y.

2.14. Let W be E.R. A *potential subword* of W is either

1. a subword A, B, C of W (cf. 2.1 of Chap. I), or
2. a subword X, Y, Z of W^2 such that

$$L(X) < L(W), \qquad L(Y) \leqslant L(W) \qquad L(Z) < L(W).$$

The *value* of the potential subword is B in case 1 and Y in case 2. In case 2 note that $W \equiv DEF$, $Y \equiv FD$, $D \not\equiv I$, $F \not\equiv I$ (also $X \equiv DE$, $Z \equiv EF$).

2.15. For each cyclic word C assume chosen a linearization Lin(C). By a *subword* of the cyclic word C we mean a potential subword of Lin(C).

In the sequel we shall not distinguish explicitly between a subword of a cyclic word and its value.

The concept of *kernel* may be applied to cyclic words (cf. 2.9).

2.16. **Translates.** Let X be E.R. and not a power. Let $m \geqslant 2$ and $ABC \equiv X^m$. By the first translate of B to the right we mean B' where $ABC \equiv A'B'C'$, $L(B) = L(B')$ and $L(A') = L(A) + L(X)$; it exists if and only if $L(AB) + L(X) \leqslant L(X^m)$; denote it by B^+. Dually, B^- means the first translate to the left. A translate of B is any subword obtained from B by applying operations $Y \to Y^+$, $Y \to Y^-$.

Let $S_1, ..., S_p$ be subwords of X^m for which kernels exist; we say the collection $S_1, ..., S_p$ is *closed to translation* if for each i:

if Y is a translate of S_i and $Y \subset (S_1, S_p)$ then Y is S_j (some j). The collection is *properly* closed to translation if in addition $S_1^+ \subset (S_1, S_p)$; in this case a unique $a > 0$ exists such that

$$S_i^+ = S_{i+a} \quad (i = 1, 2, ..., p-a), \quad 1 \leqslant p-a.$$

If X is E.R. and $m \geqslant 2$ the concept of translation can be extended to the cyclic word (X^m); if B is a subword then B^+ and B^- always exist.

2.16A. Proposition. *Let $S_1, ..., S_p$ be subwords of (X^m) for which kernels exist and assume that the translate of any S_i is an S_i. Note that $a = p/m$ is an integer. For any integer k define S_k to be S_r where $k \equiv r \pmod p$, $1 \leqslant r \leqslant p$. Then S_i^+ is S_{i+a} for all i. Let $Y \subset (X^m)$ and let y be the number of $S_1, ..., S_p$ contained in Y. Then*

$$|L(Y)/L(X) - y/a| < 3, \quad |L(Y)/L(X^m) - y/p| < 3/m.$$

Proof. Let $\sigma = y/a$, $\lambda = L(Y)/L(X)$ and $k = [\lambda]$ so that $k \leqslant \lambda < k+1$ and $\lambda \leqslant m$. Thus $L(Y) \geqslant kL(X)$. First suppose $k \geqslant 3$. Then $m \geqslant k \geqslant 3$. Hence each S_i has length $< 2L(X)$, otherwise $S_i^+ \subset S_i \cup S_i^{++}$. Hence $y \geqslant 1$. Let $S_{b+1}, ..., S_{b+y}$ be the S_i in Y. The subword between L.E.P. S_{b+1} and L.E.P. $S_{b+1+(k-1)a}$ has length $(k-1)L(X) \leqslant (\lambda - 1)L(X) = L(Y) - L(X)$ hence $S_{b-1+(k-1)a} \subset Y$. Thus $y \geqslant (k-1)a - 1 \geqslant (k-2)a$. Hence $\sigma \geqslant k - 2 > \lambda - 3$. The same is true if $k < 3$ since then $k \leqslant 2$ and $\lambda < 3 \leqslant 3 + \sigma$.

Conversely, let $[(y-1)/a] = k$; then $k \leqslant (y-1)/a < k+1$. If $y > 0$ Y contains $S_{b+1}, ..., S_{b+y}$ say. Now $y - 1 \geqslant ka$ and $k \geqslant 0$. The subword between L.E.P. S_{b+1} and S_{b+1+ka} has length $kL(X)$ so $L(Y) > kL(X)$ and $\lambda > k > (y-1)/a - 1 = \sigma - 1/a - 1 \geqslant \sigma - 2$. If $y = 0$ then $\sigma = 0$ so again $\lambda > \sigma - 2$. Thus $-2 < \lambda - \sigma < 3$ and the result follows.

2.17. Lemma. *Let S be a set with an equivalence relation $\sim$. Let*

$D = \{1, 2, ..., a + b - d\}$, *where a, b are positive integers and* $d = (a, b)$. *Let* $X_i \in S$ *for all* $i \in D$. *Let* $X_i \sim X_j$ *if* $i, j \in D$ *either* $i \equiv j$ (*mod* a) *or* $i \equiv j$ (*mod* b). *Then* $X_i \sim X_j$ *if* $i, j \in D$ *and* $i \equiv j$ (*mod* d).

Proof. Let $q \in \{1, 2, ..., d\}$. Put $a' = a/d$, $b' = b/d$ and $A_r = X_{q+(r-1)d}$ $(r = 1, 2, ..., a' + b' - 1$. $D' = \{1, 2, ..., a' + b' - 1\}$.

Let $i, j \in D'$ and $i \equiv j$ (mod a'); then $A_i \sim A_j$. Similarly, if $i, j \in D'$ and $i \equiv j$ (mod b') then $A_i \sim A_j$.

Without loss of generality $a \neq d$, $b \neq d$, $a \neq b$; hence $a' > b' > 1$. For any integer p let $Y_p = A_{\bar{p}}$ where $p \equiv \bar{p}$ (mod b') and $1 \leqslant \bar{p} \leqslant b'$.

Let $s \in D'$; then $Y_s \sim A_s$ since $A_{\bar{s}} \sim A_s$. Let k be any integer not divisible by b'. Then $k = xb' + y$, $1 \leqslant y < b'$. Hence $y + a' \in D'$ and $Y_k = Y_y = A_y \sim A_{y+a'} \sim Y_{y+a'} = Y_{k+a'}$.

Let b' divide h; then b' divides none of

$$a' + h,\ 2a' + h,\ ...,\ (b' - 1)a' + h$$

so $Y_{a'+h} \sim Y_{2a'+h} \sim \cdots \sim Y_{(b'-1)a'+h} \sim Y_{b'a'+h} = Y_h$. Hence for all integers k, $Y_k \sim Y_{k+a'}$. Also $Y_k \sim Y_{k+b'}$; hence $Y_k \sim Y_{k+1}$ since $(a', b') = 1$. Thus $A_k \sim A_{k+1}$ provided $k, k+1 \in D'$, and $A_1 \sim A_2 \sim \cdots \sim A_{a'+b'-1}$ and the result follows.

2.18. Lemma. *If* $X, Y \in \Pi$ *and* $XY \equiv YX$ *(where there are no dots) then there exist* $J \in \Pi$, $r \geqslant 0$, $s \geqslant 0$ *such that* $X \equiv J^r$, $Y \equiv J^s$.

Proof. This is easily proved by induction on $L(XY)$.

2.18A. Lemma. *Let X, Y be E.R. and* $X^{e'} \equiv UA$, $Y^{e''} \equiv UB$ *where* $e' \geqslant 1$, $e'' \geqslant 1$. *Let* $L(U) \geqslant L(X) + L(Y)$. *Then (1) a, b exist such that* $X^a \equiv Y^b$, $a \geqslant 1$, $b \geqslant 1$ *and (ii) a, b exist such that* $X \equiv J^b$, $Y \equiv J^a$ *for some* $J \in \Pi$.

Proof. We may suppose $L(X) \geqslant L(Y)$.

Case 1. $L(Y)$ divides $L(X)$. Since $L(U) \geqslant L(X)$ we have X is a left subword of U hence of $Y^{e''}$. Thus $X \equiv Y^b$, $b \geqslant 1$.

Case 2. Not Case 1. Let V be the left subword of U of length $L(X) + L(Y)$. Then $L(X) < L(V) \leqslant 2L(X)$ so V has the form XX^{L}

and $L(X^{\mathrm{L}}) = L(Y)$. The left subword of V of length $L(Y)$ is X^{L} and also Y. Thus $V \equiv (Y_1 Y_2)^a Y_1, a \geqslant 1$. Now X^{L} is $Y_2 Y_1$, Y is $Y_1 Y_2$, $Y_1 \equiv J^r$, $Y_2 \equiv J^s$, $r \geqslant 0, s \geqslant 0$. Therefore $X \equiv (Y_1 Y_2)^{a-1} Y_1 \equiv J^{(r+s)(a-1)+r}$, $Y \equiv J^{r+s}$, $Y_1 \not\equiv I$, $Y_2 \not\equiv I$ hence $J \not\equiv I$, $r \geqslant 1$, $s \geqslant 1$. Also $a \geqslant 2$ otherwise $L(X) = L(Y_1) < L(Y)$.

2.18B. Lemma. *Let C be E.R. and let s be a positive integer. Let $C^s \equiv XYC$. Then $X \equiv J^t$, $C \equiv J^u$ for some $J \in \Pi$, $t \geqslant 0$, $u \geqslant 0$. Thus if C is not a power* (cf. 2.3) $X \equiv C^t$.

Proof. $X \equiv C^v C_1$ where $v \geqslant 0$ and $C \equiv C_1 C_2$. Hence $XC \equiv C^v C_1 C_2 C_1$ so $C_1 C_2 \equiv C_2 C_1$. Therefore $C_1 \equiv J^r$, $C_2 \equiv J^{s'}$ and the result follows.

§3.

Let J_0 be the set of all E.R. words. For any subset C of Π let $\Gamma(C)$ be the group

$$\{a_1, ..., a_d \,|\, a_1^e = 1, ..., a_d^e = 1, Z^e = 1 \quad (Z \in C)\}$$

Then the Burnside Group B_d^e with d generators and exponent e is $\Gamma(J_0)$.

We shall now state some axioms†; immediately afterwards we shall begin to consider their meaning in an informal manner and discuss their validity when $n = 0, 1$.

Axiom 1. There are subsets $C_1, ..., C_n, J_0, J_1, ..., J_n$ of Π. $B_d^e = \Gamma(C_1, ..., C_n, J_n)$; $C_1, ..., C_n, J_n$ are pairwise disjoint; $C_1, ..., C_n$ are non-empty; $J_0 \supset J_1 \supset \cdots \supset J_n$; J_0 consists of all E.R. words: $C_i \subset J_{i-1}$ $(i = 1, ..., n)$.

Put $G_i = \Gamma(C_1, ..., C_i)$, $G_0 = \Pi$. If $X, Y \in \Pi$ write $X \underset{i}{=} Y$ if X equals Y in G_i.

Axiom 2. There are relations $\underset{i}{\sim}$ $(i = 0, 1, ..., n)$ on Π; $\underset{0}{\sim}$ is $\equiv$.

† The author hopes that the reader will bear with the unusual numbering of the axioms.

Axiom 3. There are relations $\underset{i}{\cong}$ $(i = 0, 1, ..., n)$ on the set of all cyclic words; $\underset{0}{\cong}$ is the equality relation ($\equiv$, say).

Axiom 4. $(X) \underset{i}{\cong} (Y)$ implies $X \underset{i}{=} U^{-1} \cdot Y \cdot U$ for some U.

3.1. Definition. Let $X, Y \in J_n$. Then $XE_n Y$ means $(\exists s, t)$ $(X^s) \underset{n}{\cong} (Y^{\pm t})$, $1 \leqslant s < t$.

$X\bar{\rho}_n Y$ means that some $\delta_f(X) \underset{n}{\sim}$ some $\epsilon_q(Y)$ (cf. 2.11); here f and q are parameters; they are regarded as given, and fixed from now on, even though their explicit definitions are not given at this moment. Similarly for all subsequent parameters. $((f, q))$

$X\bar{\sigma}_n Y$ means that some $\delta_f(X) \underset{n}{\sim}$ some $\epsilon_{q'}(Y)$ $((q'))$

$X\rho_n Y$ means $XE_n Y$ or $X\bar{\rho}_n Y$

$X\sigma_n Y$ means $XE_n Y$ or $X\bar{\sigma}_n Y$

$X >_n Y$ means there exist $X_0, X_1, ..., X_j$ in J_n $(j \geqslant 1)$ such that (i) X_0 is X, X_j is Y; (ii) $X_0 \alpha_1 X_1 \alpha_2 X_2 \cdots \alpha_j X_j$ where each α_i is one of $E_n, \bar{\rho}_n, \bar{\sigma}_n$; (iii) $(\exists i)\alpha_i = \bar{\sigma}_n \Rightarrow (\exists p)\alpha_{i+p} = \bar{\rho}_n, p > 0$. Since $q \geqslant q'$ (this will follow from the explicit definitions of q, q' when given), we have $X_0 \sigma_n X_1 \sigma_n X_2 \cdots \sigma_n X_{j-1} \rho_n X_j$. $((q \geqslant q'))$

3.2. Definition. For $k = 1, 2, 3, ...$ L_n^k consists of all cyclic arrangements of X^{ke} and of X^{-ke}, $X \in C_n$. L_n means L_n^1.

Axiom 5. Let $X_1, X_2, X_3, ...$ be an infinite sequence in J_n. Then not $X_i \sigma_n X_{i+1}$ $(i = 1, 2, ...,)$.

Axiom 6. If $0 < x \leqslant 1$, certain elements of Π are called *n-bounded* by x. Every element of Π is 0-bounded by x. If X is n-bounded by x then so is X^{-1} and every subword of X.

Informal remark. Usually, $\frac{1}{2} \leqslant x < 1$.

Axiom 7. $X \underset{n-1}{\sim} Y$ implies $X \underset{n}{\sim} Y$.

Axiom 8. $(X) \underset{n-1}{\cong} (Y)$ implies $(X) \underset{n}{\cong} (Y)$

3.3. Definition. A cyclic word is n-bounded by x if every linearization is n-bounded by x.

By the notation (n)-bounded by x we mean i-bounded by x $(i = 1, 2, ..., n)$; with $n - 1$ instead of n we write $((n - 1))$-bounded to avoid ambiguity.

Axiom 8a. If $X^e \neq I$ in G_n, X is E.R. and (X) is (n)-bounded by z, where $z \geqslant \frac{1}{2}$ then for any $t \geqslant 1$ (X^t) is (n)-bounded by $z + a_8$. Here a_8 is a parameter but z is a variable. $((a_8))$

Axiom 9. $X \in J_n$ if and only if $X^e \neq I$ in G_n, X is E.R. and (X) is (n)-bounded by str. (Here str is a parameter.) $((str))$

Put $str^+ = str + a_8$. $((str^+))$

Axiom 10. If X is (n)-bounded by b_∞ and $X \underset{n}{=} I$ then $X \equiv I$. $((b_\infty))$

Axiom 11. $\underset{n}{\cong}$ is an equivalence relation on the set of cyclic words (n)-bounded by b_2

$$(X) \underset{n}{\cong} (Y) \text{ implies } (X^k) \underset{n}{\cong} (Y^k)\ (k \geqslant 1). \qquad ((b_2))$$

Axiom 11a. If $W \in \Pi$ there exists W' such that $W' \underset{n}{=} W$ and W' is (n)-bounded by $b_{\min}$. $((b_{\min}))$

Axiom 11b. Let $W^e \neq I$ in G_n. Then there is an E.R. word K such that K is conjugate to W in G_n and (K) is (n)-bounded by $cb_{\min}$. $((cb_{\min}))$

Since $cb_{\min} \leqslant str$, any such K is in J_n. $((cb_{\min} \leqslant str))$

Axiom 12. Let $n \geqslant 1$. For any positive integer k there is a non-empty subset L_n^k of Π; its elements are E.R. and it is closed to C.A. and inverses. If $1 \leqslant r \leqslant n$, $1 \leqslant s \leqslant n$, $\mathsf{L}_r^k \cap \mathsf{L}_s^l \neq \emptyset$ then $r = s$ and $l = k$. $\mathsf{L}_1^k = L_1^k$. $\mathsf{L}_n = \mathsf{L}_n^1$. Also

$$[L_1 \cup \cdots \cup L_n] = [\mathsf{L}_1 \cup \cdots \cup \mathsf{L}_n].$$

Axiom 13. If $AB \in \mathsf{L}_i^k$ $(1 \leqslant i \leqslant n, k \geqslant 1)$ there is defined a real number $s_{AB}(A)$ such that $0 \leqslant s_{AB}(A) \leqslant k$; if $A \equiv I$ then $s_{AB}(A) = 0$; if $B \equiv I$ then $s_{AB}(A) \geqslant k - 2/e$; $s_{A^{-1}B^{-1}}(A^{-1}) = s_{AB}(A)$.

3.4. Definition. An element of Π is a *powerelement* of L_n if it belongs to some L_n^k; A is a *subpowerelement* of L_n if it is a subword of a powerelement of L_n. If A is a subpowerelement of L_n its n-size $s^n(A)$ is given by

$$s^n(A) = \text{Sup}\, s_{AB}(A),$$

taken over B, k such that $AB \in L_n^k$.

Axiom 14. $s^n(A)$ is finite.

Axiom 15. If $ABCD \in L_n^k$ then $s_{BCDA}(B) \leqslant s_{ABCD}(ABC)$.

Axiom 16. If $ABC \in L_n^k$ then

$$s_{ABC}(AB) \leqslant s_{ABC}(A) + s_{BCA}(B) + 2/e.$$

Axiom 16a. If $ABC \in L_n^k$ then

$$s_{ABC}(A) + s_{BCA}(B) \leqslant S_{ABC}(AB) + \epsilon_1^*. \qquad ((\epsilon_1^*))$$

Axiom 17. If $L(A) = 1$ then $s_{AB}(A) \leqslant \epsilon_2 = 1/2e$. $((\epsilon_2))$

Axiom 18. If $AB \in L_n^k$ and $s_{AB}(A) < h_0$ then $((h_0))$

$$(\exists C) AC \in L_n, \quad s_{AC}(A) = s_{AB}(A).$$

Axiom 19. If $AC \in L_n^l$, $s_{AC}(A) > r_0$ and $AB \in L_n^k$ then $s_{AC}(A) \leqslant s_{AB}(A) + \epsilon_3$. $((\epsilon_3))$

Hence if $s^n(A) > r_0$ and $AB \in L_n^k$ then $s^n(A) \leqslant s_{AB}(A) + \epsilon_3$. $((r_0))$

Axiom 20. Let $AX \in L_t$, $BX \in L_s$, $t \leqslant n$, $s \leqslant n$; say $t \leqslant s$; then (a) and (b) below hold.

(a) If $t < s$ then

(i) $s_{XA}(X) < r$ or $B \cdot A^{-1} = U \cdot C \cdot U^{-1}$ in G_{t-1} for some $U \in \Pi$, $C \in L_s$ $((r))$

and (ii) $s_{XB}(X) < r'_0$ $((r'_0))$

(b) If $t = s$ then

(i) $s_{XA}(X) < r_0$ and $s_{XB}(X) < r_0$

or (ii) $B \cdot A^{-1} = I$ in G_{t-1}.

Axiom 21. Let $n \geqslant 1$. Let $0 < x \leqslant 1$. Then W is n-bounded by x if and only if any n-subpowerelement X contained in W satisfies $s^n(X) < x$.

3.5. Definition. A is an *n-urelement* if it is a subpowerelement of L_n and $s^n(B) > r_0$ for every subpowerelement B of L_n containing A. A is *n-normal* if it is an n-urelement and a subelement of L_n.

Axiom 22. If $W \in \Pi$, each n-urelement S contained in W is contained in a unique maximum n-urelement $\bar{S}$ contained in W.

If X is E.R. and no C.A. of X is a subpowerelement of L_n then each n-urelement contained in (X) is contained in a unique maximal n-urelement contained in (X).

Axiom 23. If $AB \in L_n^k$ and $0 \leqslant \alpha < \beta < s_{AB}(A)$, and $\beta - \alpha > \epsilon_4$, then for some left subword A^L of A, $\alpha \leqslant s_{AB}(A^L) < \beta$. $((\epsilon_4))$

Axiom 24. Let ABC be a subelement of L_n, $s^n(A)$, $s^n(C) > s_{24}$ and A, C $((n-1))$-bounded by $b \leqslant b_{24}$. Then there is a subelement $A^L B' C^R$ of L_n, $((n-1))$-bounded by $b + q_{24}$ such that

$$ABC \underset{n-1}{=} A^L B' C^R$$

$$s^n(A^L) > s^n(A) - r_{24}, \quad s^n(C^R) > s^n(C) - r_{24},$$

$$|s^n(ABC) - s^n(A^L B' C^R)| < \epsilon_{24}, s^n(B') < s^n(B) + d_{24}.$$ $((s_{24}, b_{24}$ q_{24}, r_{24} $\epsilon_{24}, d_{24}))$

Axiom 25. Let A be an n-subelement of n-size $> s_{25}$, $A \underset{p}{\sim} Y$, $p < n$ and let A, Y be (p)-bounded by b_{25}. Then Y contains an n-subelement of n-size $> s^n(A) - r_{25}$. $((s_{25}, b_{25}, r_{25}))$

3.6. Definition. If X, Y are E.R. then $(X) \underset{n}{-} (Y)$ means that there are cyclic arrangements of X, Y that are equal in G_n (thus $\overline{n}$ is a relation on the set of all cyclic words).

Axiom 26'. If $(X) \overline{n} (Y)$ where both cyclic words are n-bounded by $b'_{26'}$ and $((n-1))$-bounded by $b_{26'}$, then any maximal n-normal subword U of (X), where $s^n(U) > c_{26'}$, determines a maximal n-normal subword Im U of (Y). $((b'_{26'}, b_{26'}, c_{26}))$

Axiom 26. Let (X) be $((n-1))$-bounded by s_0 and n-bounded by $1 - z_{26}$. Let $M_1, \ldots, M_k$ $(k \geqslant 1)$ be a set of maximal n-normal subwords each of n-size $\geqslant z > z_{26}$. $((s_0, z_{26}))$

If $k = 1$ assume (X) contains a maximal n-normal subword of n-size $> z_{26}$ different from M_1.

Then there are cyclic words X_1, X_2, X_3 $((n-1))$-bounded by $s_0 + s_{26}, s_0, s_0$ respectively and n-bounded by $1 - z + t_{26}$ such that $(X) \underset{n-1}{-} X_1 \underset{n}{-} X_2 \underset{n}{-} X_3$. $((s_{26}, t_{26}))$

If U is maximal n-normal in (X) of size $> w_{26}$ then there are maximal n-normal subwords U_1, U_2, U_3 of X_1, X_2, X_3 respectively such that Im $U = U_1$, Im $U_1 = U_2$, Im $U_2 = U_3$ and Im $U_1 = U$, Im $U_2 = U_1$, Im $U_3 = U_2$. Finally, $((w_{26}))$

if U is an M_j then $|s^n(U) + s^n(U_3) - 1| < r_{26}$

if U is not an M_j then $|s^n(U) - s^n(U_3)| < r_{26}$. $((r_{26}))$

We say that X_3 arises from (X) by *simultaneous replacement* of $M_1, \ldots, M_k$.

Axiom 26''. If V is maximal n-normal in X_3 of n-size $> v_{26}$ then V is of the form U_3 in Axiom 26; thus U exists in (X) of size $> w_{26}$ and U_1, U_2 exist as in Axiom 26. If V is maximal n-normal in X_2 of n-size $> v_{26}$ then V is of the form U_2 in Axiom 26. $((v_{26}))$

Axiom 26⁻. If X_3 arises from (X) by simultaneous replacement of $M_1, \ldots, M_k$ (using X_1, X_2) then X_3^{-1} arises from (X^{-1}) by simultaneous replacement of $M_1^{-1}, \ldots, M_k^{-1}$ (using X_1^{-1}, X_2^{-1}). If $U' \equiv U^{-1}$ is maximal n-normal in (X^{-1}) of size $> w_{26}$ then the corresponding words U'_1, U'_2, U'_3 of $X_1^{-1}, X_2^{-1}, X_3^{-1}$ are $U_1^{-1}, U_2^{-1}, U_3^{-1}$.

3.7. Definition. $(X) \underset{n}{\rightarrow} (Y)$ means that (X) is (n)-bounded by s_0

and either $(X) \equiv (Y)$ or (Y) arises from (X) by simultaneous replacement of some set of maximal n-normal subwords each of size $> r'$. $((r'))$

Thus $(X) \underset{n}{\rightarrow} (Y)$ implies $(X^{-1}) \underset{n}{\rightarrow} (Y^{-1})$.

Axiom 26A. Let $(RA) \underset{n}{\rightarrow} (SB)$, where R is max. n-normal of size $> r'$ and S is its (iterated) image as in Axiom 26. Let $ST \in \mathsf{L}_n$. Then there is a cyclic word H such that $(RA) \underset{n}{\rightarrow} H$ and any linearization of H is conjugate in G_{n-1} to $T^{-1} \cdot B$.

Axiom 26L. (Linear version of Axioms 26′, 26, 26″, 26A.)

(i) Axiom 26′ holds if X, Y are linear words and $X \underset{n}{=} Y$

(ii) Axiom 26 becomes: Let X be $((n-1))$-bounded by s_0 and n-bounded by $1 - z_{26}$. Let $M_1, \ldots, M_k$ $(k \geqslant 1)$ be a set of maximal n-normal subwords each of size $> z_{26}$. *Then* there exists Y, unique mod G_{n-1} $((n-1))$-bounded by s_0 and n-bounded by $1 - z_{26} + t_{26}$ such that $X \underset{n}{=} Y$. If U is maximal n-normal in X of n-size $> w_{26}$ then

if U is an M_j, $|s^n(U) + s^n(\operatorname{Im} U) - 1| < r_{26}$

if U is not an M_j, $|s^n(U) - s^n(\operatorname{Im} U)| < r_{26}$.

(iii) If V is maximal n-normal in Y of size $> v_{26}$ then V has the form $\operatorname{Im} U$ for some U.

(iv) If $A_1 R A_2 \underset{n}{\rightarrow} B_1 S B_2$, where $s^n(R) > r'$ and $S = \operatorname{Im} R$ and $ST \in \mathsf{L}_n$ then $(\exists H) A_1 R A_2 \underset{n}{\rightarrow} H \underset{n-1}{=} B_1 \cdot T^{-1} \cdot B_2$.

3.8. *Note*: The definition of $X \underset{n}{\rightarrow} Y$ for linear words is similar to 3.7.

Axiom 27. If X is an n-subpowerelement and $s^n(X) \geqslant \frac{1}{2}$ then X contains a subword of the form UUU, where U is E.R.

Axiom 28. If $(X) \underset{n}{\cong} (Y)$ where the cyclic words (X), (Y) are (n)-bounded by b_{28}, a_{28} respectively then $(X) \underset{n}{\rightarrow} \cdots \underset{1}{\rightarrow} (Y)$. Similarly for linear words: if $X \underset{n}{=} Y$ where X, Y are (n)-bounded by b_{28}, a_{28} respectively then $X \underset{n}{\rightarrow} \cdots \underset{1}{\rightarrow} Y$. $((b_{28}, a_{28}))$

If W is a cyclic or linear word (n)-bounded by b'_{25} then there exist cyclic or linear words respectively such that

$$W_n \underset{n}{\rightarrow} \cdots \underset{2}{\rightarrow} W_1 \underset{1}{\rightarrow} W_0, \qquad W_0 \equiv W. \qquad ((b'_{25}))$$

Axiom 28A. If $X \underset{n}{-} Y$, where X, Y are (n)-bounded by c_{28} and no linearization of X or Y is a subelement of $L_1 \cup \cdots \cup L_n$ then $X \underset{n}{\cong} Y$. $((c_{28}))$

Axiom 28B. Let X be a cyclic word, (n)-bounded by d_{28} and such that no linearization is a subelement of $L_1 \cup \cdots \cup L_n$. Then Y exists such that Y is (n)-bounded by *str*, $X \underset{n}{\cong} Y$ and if $X \equiv Z^s$ for some Z, s then $Y \equiv T^s$ for some T. $((d_{28}))$

Axiom 29. Let $X \in J_{n-1}$, $Y \in C_n$ and $X \underset{n-1}{>} Y$. Then either some $\delta_f(X)$ contains a subelement of L_n of size $> q_0$ or

$$(\exists s, t, U)\ X^s \underset{n-1}{=} U^{-1} \cdot Y^t \cdot U,\ U \in \Pi,\ 1 \leqslant s < |t|. \qquad ((q_0))$$

Axiom 30. Let $n \geqslant 2$. Let P be a subelement of L_n and $s^n(P) > s_{30}$. Let P be $((n-1))$-bounded by b_{30}. Then P contains subelements $X_1, \ldots, X_a$ of L_{n-1} as follows. $((s_{30}, b_{30}))$

1. There exists $PQ \in L_n$ where $s_{PQ}(P) > s^n(P) - 1/e^2$.
2. There is an integer $t > 0$ such that e divides t.
3. Either (a) each X_i is maximal $(n-1)$-normal in PQ of size $> r^*$, or (b) the X_i are disjoint and have size $> q^*$. If (b) holds then P contains no $(n-1)$-subelement of size $> c^*$. $((r^*, q^*, c^*))$
4. If $P' \subset P$ and those X_i contained in P' are $X_i, \ldots, X_j$ then

$$|(j+1-i)/t - s_{P'Q'}(P')| < \epsilon'_{30} \qquad ((\epsilon'_{30}))$$

where if $P \equiv XP'Y$ then $Q' \equiv YQX$. Also

$$|a/t - s_{PQ}(P)| < \epsilon'_{30} \text{ and } |a/t - s^n(P)| < \epsilon_{30}. \qquad ((\epsilon_{30}))$$

Axiom 31A. If $X \underset{n}{\sim} Y$ then certain pairs $E, K \in \Pi$ are called *end words for* $X \underset{n}{\sim} Y$. Also $X \underset{n}{=} E \cdot Y \cdot K$.

Axiom 31B. Let $T \underset{p}{\sim} Y$ with end words E, K, where $p \leqslant n-1$. Let T be an n-subpowerelement, $s^n(T) > s_{31}$ and let T, Y be (p)-bounded by $b \leqslant b_{31}$. Then there is a word $T^{\mathrm{L}} F Y^{\mathrm{R}}$ (p)-bounded by $b + q_{31}$

and equal in G_p to $T \cdot K^{-1}$ $(= E \cdot Y)$. Also Y^{R}, T^{L} are not subelements of $\mathsf{L}_1 \cup \cdots \cup \mathsf{L}_p$ and $s^n(T^{\mathrm{L}}) > s^n(T) - r_{31}$.
(Dually $Y^{\mathrm{L}} F T^{\mathrm{R}} = Y \cdot K = E^{-1} \cdot T$.) $((s_{31}, b_{31}, q_{31}, r_{31}))$

Axiom 31C. Let $U \underset{n}{\sim} V$ with end words E, K, where U, V are (n)-bounded by $b \leqslant f_{31}$. Let $U \equiv (S, T)$ where S, T are distinct maximal n-normal subwords of U of size $> g_{31}$. Similarly, let $V \equiv (S', T')$. Then there is a word $S^{\mathrm{L}} A T'^{\mathrm{R}}$, (n)-bounded by $b + h_{31}$ such that $S^{\mathrm{L}} A T'^{\mathrm{R}} \underset{n}{=} E \cdot V$. Moreover S^{L}, T'^{R} are n-normal words and $s^n(S^{\mathrm{L}}) > s^n(S) - j_{31}$, $s^n(T'^{\mathrm{R}}) > s^n(T') - j_{31}$. S^{L}, T'^{R} are contained in distinct maximal n-normal words S_1, T_1' and $s^n(S^{\mathrm{L}}) > s^n(S_1) - a_{31}$, $s^n(T'^{\mathrm{R}}) > s^n(T_1') - a_{31}$. $((f_{31}, g_{31}, h_{31}, j_{31}, a_{31}))$

Axiom 32′. L_n is contained in L_n.

Axiom 33. There is a subset $\mathrm{Rep}(L_n)$ of L_n. If AA' and $BB' \subset \mathrm{Rep}(L_n)$, $A \underset{n-1}{\sim} B$ and $s^n(A) > s_{33}$ then $A \equiv B$ and $A' \equiv B'$. $((s_{33}))$

Axiom 33′. If CAB is a subelement of L_n and $s^n(A) > t_{33}$, $((t_{33}))$ then there is a subelement $CA^{\mathrm{L}} EZFA^{\mathrm{R}} B$ of L_n equal to CAB in G_{n-1}, where Z is a subelement of $\mathrm{Rep}(L_n)$ and $s^n(Z) > s^n(A) - r_{33}$. Also $((r_{33}))$

1. A^{L} and A^{R} have n-size $> a_{33} \geqslant r_0'$ and $< w_{33}$. $((a_{33}, w_{33}))$
2. If A is $((n-1))$-bounded by $b \leqslant b_{33}$ then $A^{\mathrm{L}} EZFA^{\mathrm{R}}$ is $((n-1))$-bounded by $b + c_{33}$. $((b_{33}, c_{33}))$
3. E and F have n-size $< \epsilon_{33}$. $((\epsilon_{33}))$
4. $A^{\mathrm{L}} E$, FA^{R} have n-size $< q_{33}$. $((q_{33}))$

Axiom 33″. If ABC is a subelement of L_n and A, C are subelements of $\mathrm{Rep}(L_n)$ of size $> m_{33}$ then there is a subelement ADC of $\mathrm{Rep}(L_n)$ equal to ABC in G_{n-1}. If further $XABCY$ is a subpowerelement of L_n then so is $XADCY$ and

$$|s^n(XABCY) - s^n(XADCY)| < p_{33}. \qquad ((m_{33}, p_{33}))$$

Axiom $\overline{33}$. Let W be a subelement of L_n of size $> d_{33}$. Let W be $((n-1))$-bounded by f_{33} and $W \underset{n-1}{=} W^{\mathrm{L}} D W^{\mathrm{R}}$, where the right side is $((n-1))$-bounded by g_{33} and W^{L}, W^{R} have size $> h_{33}$. Then $W^{\mathrm{L}}DW^{\mathrm{R}}$ is a subelement of L_n. $((d_{33}, f_{33}, g_{33}, h_{33}))$

Axiom 33$^+$. Let $ABCD \in \mathsf{L}_n$ where A and C have n-size $> j_{33}$. As in Axiom 33′ let $A \underset{n-1}{=} A^{\mathrm{L}}EZFA^{\mathrm{R}}$, $C \underset{n-1}{=} C^{\mathrm{L}}E'Z'F'C^{\mathrm{R}}$. Let ZH, $Z'H' \in \mathrm{Rep}(L_n)$. Then ZH is a cyclic arrangement of $Z'H'$. $((j_{33}))$

Axiom 39. Let $U \equiv (Z_1, Z_2)$ be $((n-1))$-bounded by $b \leqslant b_{39}$ and n-bounded by b'_{39}, where Z_1, Z_2 are subelements of $\mathrm{Rep}(L_n)$ of size $> s_{39}$. Let Z_1, Z_2 be contained in distinct maximal n-normal subwords $\overline{Z}_1, \overline{Z}_2$ and let $s(Z_i) > s(\overline{Z}_i) - d'_{39}$ $(i = 1, 2)$. Let $Z_2T_2^{-1} \in \mathrm{Rep}(L_n)$. Then $U \cdot Z_2^{-1}T_2 \underset{n-1}{=} Z_1^{\mathrm{L}}HT_2^{\mathrm{R}}$, where the right side is $((n-1))$-bounded by $b + c_{39}$, and $s(Z_1^{\mathrm{L}}) > s(Z_1) - d_{39}$, $s(T_2^{\mathrm{R}}) > s(T_2) - d_{39}$. Also Z_1^{L}, T_2^{R} are contained in distinct maximal n-normal subwords $\overline{Z_1^{\mathrm{L}}}$, $\overline{T_2^{\mathrm{R}}}$ and $s(Z_1^{\mathrm{L}}) > s(\overline{Z_1^{\mathrm{L}}}) - f_{39}$, $s(T_2^{\mathrm{R}}) > s(\overline{T_2^{\mathrm{R}}}) - f_{39}$. Finally $Z_1^{\mathrm{L}}HT_2^{\mathrm{R}}$ is n-bounded by $\mathrm{Max}(b'_{39}, 1 - s(\overline{Z}_2)) + g_{39}$. $((b_{39}, b'_{39}, s_{39}, d'_{39}, c_{39}, d_{39}, f_{39}, g_{39}))$

Axiom 40. No element of L_n has the form $J^{-1}AJB$ where $s^n(J) > s_{40}$ and J is $((n-1))$-bounded by b_{40}. $((s_{40}, b_{40}))$

§4.

From now on we shall suppose that the statement of each axiom has been prefixed by "Let $n \geqslant 0$" or "Let $n \geqslant 1$" as follows:
$n \geqslant 0$, for Axioms 1, 2, 3, 4, 5, 6, 8a, 9, 10, 11, 11a, 11b, 31A.
$n \geqslant 1$, for all remaining axioms.

4.1. Lemma. *The axioms are true when $n = 0$.*

Proof. Axiom 1 reads "There is a subset J_0 of Π; $B_d^e = \Gamma(J_0)$ and J_0 consists of all E.R. words." This is so by the beginning of Section 3.

Axiom 5 will follow if we prove that $X \sigma_0 Y$ implies $L(X) > L(Y)$. If $X E_0 Y$ then $sL(X) = tL(Y) > sL(Y)$; if $X \overline{\sigma}_0 Y$ then $feL(X) > q'eL(Y) \geqslant feL(Y)$ since $q' \geqslant f$. $((q' \geqslant f))$

Axiom 11 reads "$\equiv$ is an equivalence relation on the set of all cyclic words." This is so. In Axiom 31A if $X \underset{0}{\sim} Y$, that is, $X \equiv Y$, the only pair of end words shall by definition be I, I.

4.2. *Note.* The remainder of Section 4 consists of an informal discussion of the axioms when $n = 1$. This is logically not necessary, but it is hoped that it will help the reader.

4.3. It is easy to show that $\underset{0}{>}$ is transitive and not reflexive, hence it determines a partial ordering on J_0 (e.g, take $n = 0$ in 5.5 below). Put X in C_1 if $X \in J_0$ and

$$Y \in J_0 \text{ implies not } X \underset{0}{>} Y.$$

Then $C_1 \neq \emptyset$. Let $k \geqslant 1$; put Y in L_1^k if Y is a C.A.I. of X^{ke} for some X in C_1. L_1 means L_1^1. Put $G_1 = \Gamma(C_1)$; thus $G = \Pi/[L_1]$.

Put $\mathsf{L}_1^k = L_1^k$ and $\mathsf{L}_1 = L_1$; if $AB \in \mathsf{L}_1^k$ we say, as in the axioms that A is a subpowerelement of L_1 or a 1-subpowerelement.

If $X \in \mathsf{L}_1^k \cap \mathsf{L}_1^l$ then $k = l$ (e.g., taken $n = 0$ in 5.15 below).

If $A, B \in L_1$ then $A \cdot B \equiv I$ or $\beta(A,B) < qL(B)$; to see this take $n = 0$ in 5.16 below. Thus G_1 satisfies the Tartakovskii conditions of Chapter I, since $q < 1/5$. $((q < 1/5))$

If $AB \in \mathsf{L}_1^k$ put $s_{AB}(A) = kL(A)/L(AB)$. Note that $L(AB) \geqslant 2ke$.

The following "size" axioms hold clearly: 13, 14, 15, 16, 16a, 17. Axiom 18 is true since $h_0 \leqslant 1$; Axiom 20 since $q \leqslant r_0$ and Axiom 23 since $\epsilon_4 \geqslant 1/2e$. Axiom 19 is true if $l = k = 1$ since $r_0 \geqslant q$; the general case can be reduced to this by considering a suitable A^{L} and using Axiom 18. $((h_0 \leqslant 1, q \leqslant r_0, \epsilon_4 \geqslant 1/2e))$

As in the axioms we have the concepts 1-urelement, 1-normal, 1-bounded (cf. Axiom 21). We prove axiom 22.

Consider a word ABC where B is a 1-urelement and AB, BC are 1-subpowerelements, say $ABX \equiv Z^{ke}$, $Z^e \in \mathsf{L}_1$, then $BXA \equiv Z_1^{ke}$, $Z_1^e \in \mathsf{L}_1$ and $BCY \equiv T^{he}$, $T^e \in \mathsf{L}_1$. Since $s(B) > r_0$ we obtain $Z_1 \equiv T$. Hence $BXABCY \equiv T^{(k+h)e}$. Thus ABC is a 1-urelement. Axiom 22 now follows. If A is a 1-subelement and $s(A) > r_0 + \epsilon_3$ then A is 1-normal (e.g. take $n = 1$ in 5.22 (iii)).

4.4. 1-Replacements. Consider a word XZY where Z is a 1-subelement, say $ZT^{-1} \in L_1$. Then $X \cdot T \cdot Y$ is said to arise from XZY by replacing Z; this concept is only useful if $s(Z) \geqslant r_0 + \epsilon_3$: then, $s_{ZT^{-1}}(Z) \geqslant r_0$ so by Axiom 20 if also $ZT'^{-1} \in L_1$ then $T \equiv T'$ so $X \cdot T \cdot Y \equiv X \cdot T' \cdot Y$.

4.4A. Proposition. *Let* $x_1 = r_0 + \epsilon_3 + \epsilon_2 + 4/e$, $x_2 = 1 - x_1$, $x_3 = r_0 + x_1$, $x_4 = 2\epsilon_3$. *Let XZ be 1-bounded by* $c \leqslant x_2$ *and let Z be maximal* 1-*normal in XZ. Let W′ be the result of replacing Z in XZ. Let* $ZT^{-1} \in L_1$. *Then* $W' \equiv X^L E T^R$ *where* $L(E) \leqslant 1$, $L(X^L) \geqslant L(X) - 1$, $L(T^R) \geqslant L(T) - 1$, T^R *is* 1-*normal and*

$$|s(\overline{T^R}) + s(Z) - 1| \leqslant x_4.$$

W′ is 1-*bounded by* $\mathrm{Max}\,(c + x_1 - \epsilon_3, 1 + x_4 - s(Z))$. *Finally, if V is maximal* 1-*normal in XZ (or W′) and not Z (or* $\overline{T^R}$*) and its size is* $y > x_3$ *then it determines a maximal* 1-*normal subword V′ of W′ (or XZ) where* $|s(V) - s(V')| \leqslant x_1$.

Proof. We have

$$s_{TZ-1}(T) = 1 - s_{ZT^{-1}}(Z) \geqslant 1 - s(Z) > 1 - c \geqslant 1 - x_2 > \epsilon_2$$

hence $L(T) \geqslant 2$, say $T \equiv aS$. Next

$$s_{SZ^{-1}a}(S) = s_{TZ^{-1}}(T) - 1/L(TZ^{-1}) > 1 - x_2 - 1/2e > r_0 + \epsilon_3$$

hence S is 1-normal, and so is T. Now $W' \equiv X \cdot aS$. No cancellation can occur since Z is maximal normal. If no amalgamation occurs, $W' \equiv XT$. Take $T^R \equiv T$, $X^L \equiv X$. $\overline{T^R}$ is T since there is no cancellation in $X \cdot Z$. Now

$$s(T) + s(Z) - 1 \leqslant s_{TZ^{-1}}(T) + s_{ZT^{-1}}(Z) + 2\epsilon_3 - 1 = 2\epsilon_3 \leqslant x_4$$

and

$$s(T) + s(Z) - 1 \geqslant 0 \geqslant -x_4.$$

If amalgamation occurs $X \equiv X'b$, $W' \equiv X'cS$, $c = b \cdot a$. Take $T^{\mathrm{R}} \equiv S$, $X^{\mathrm{L}} \equiv X'$. Then $\overline{T}^{\mathrm{R}}$ is S and

$$s(S) + s(Z) - 1 \geqslant -1/L(ZT^{-1}) \geqslant -1/2e > -x_4$$

$$s(S) + s(Z) - 1 \leqslant s_{sz^{-1}a}(S) + s_{ZT^{-1}}(Z) + 2\epsilon_3 - 1 =$$

$$= 2\epsilon_3 - 1/L(ZT^{-1}) \leqslant 2\epsilon_3 = x_4. \qquad ((\epsilon_3 > 1/4e))$$

Note that in any case $s(\overline{T}^{\mathrm{R}}) < 1 + x_4 - s(Z)$.

Suppose U is normal of size y but is necessarily maximal normal. $U \cap Z$ is not normal so it has size $\leqslant r_0 \leqslant x_3$ hence $U \not\subset Z$. Write $U \equiv U_1 U_2 U_3$ where $U_1 \equiv U \cap X^{\mathrm{L}}$, $U_3 \equiv U \cap Z$, $L(U_2) \leqslant 1$. Then

$$s(U) \leqslant s(U_1) + \epsilon_2 + r_0 + 4/e \quad \text{so} \quad s(U_1) > x_3 - \epsilon_2 - r_0 - 4/e \geqslant r_0 + \epsilon_3$$

hence U_1 is normal. Also $s(U_1) \geqslant y - \epsilon_2 - r_0 - 4/e = y - x_5$, say. Similarly if $U \subset W'$. Thus W' is 1-bounded by $\mathrm{Max}(c + x_5, 1 + x_4 - s(Z)) < 1$. If U above is now maximal normal then we have

$$y - x_5 \leqslant s(U_1) \leqslant s_J(U_1) + \epsilon_3 \leqslant s_{J'}(\overline{U}_1) + \epsilon_3 \quad (\text{for some } J, J')$$

so

$$s(\overline{U}_1) \geqslant y - x_5 - \epsilon_3 = y - x_1.$$

4.4B. Corollary. *Let $W \equiv (Z_1, Z_2)$ be 1-bounded by $c \leqslant 1 - 2r_0 - 3\epsilon_3 - 2\epsilon_2 - 8/e$ where Z_1, Z_2 are distinct maximal 1-normal of size $> 2r_0 + \epsilon_3 + \epsilon_2 + 4/e$. Let $Z_i T_i^{-1} \in \mathsf{L}_1$ $(i = 1, 2)$. Put $W'' \equiv T_1 Z_1^{-1} \cdot (Z_1, Z_2) \cdot Z_2^{-1} T_2$. Then $W'' \equiv (T_1'', T_2'')$ where T_1'', T_2'' are distinct maximal normal, $T''- \equiv T-$, $-T'' \equiv -T$,*

$$|s(Z_i) + s(T_i'') - 1| < r_0 + 3\epsilon_3 + \epsilon_2 + 4/e \quad (i = 1, 2)$$

and W'' is 1-bounded by $\mathrm{Max}(s(T_1''), s(T_2''), c + 2(r_0 + \epsilon_3 + 4/e))$.

Proof. Replace Z_1 in W to get $W' \equiv (T_1', Z_2')$, say. Then $|s(T_1') + s(Z_1) - 1| \leqslant x_4$ and W' is 1-bounded by $\mathrm{Max}(c + x_1 - \epsilon_3,$

$1 + x_4 - s(Z_1)) = c'$. Since $s(Z_2) > x_3$, $|s(Z_2) - s(Z_2')| \leqslant x_1$. Since $c + x_1 - \epsilon_3 \leqslant x_2$, $1 + x_4 - s(Z_1) \leqslant x_2$ we may replace Z_2' to get $\overline{W} \equiv (T_1'', T_2'')$ where $|s(Z_2') + s(T_2'') - 1| \leqslant x_4$ and $\overline{W}$ is 1-bounded by $\mathrm{Max}(c' + x_1 - \epsilon_3, 1 + x_4 - s(Z_2')) = c''$. Since $s(T_1') \geqslant 1 - s(Z_1) - x_4 > x_3$ we have $|s(T_1^1) - s(T_1'')| \leqslant x_1$. Thus $|s(Z_i) + s(T_i'') - 1| \leqslant x_1 + x_4$.

4.4C. Simultaneous 1-replacement. Take a cyclic word $W \equiv (Z_1 A_1 \cdots Z_s A_s)$ where $s \geqslant 2$ and the Z_i are 1-normal, say $Z_i T_i^{-1} \in L_1$. Let $J = \{\beta_1, \beta_2, ..., \beta_r\}$ be a non-empty subset of $1, 2, ..., s$. Let Q_i be T_i, if $i \in J$ and Z_i, otherwise. We wish to investigate the weak cyclic word $(Q_1 \cdot A_1 \cdot \cdots \cdot Q_S \cdot A_S)$ arising by simultaneous replacement of the Z_{β_j}. This investigation is easily reduced to 4.4A and 4.4B as follows; we give a sketch only.

We may write

$$W' \equiv \left(\prod_{i=1}^{s} Q_i \cdot A_i \cdot Q_{i+1} \cdot Q_{i+1}^{-1} \right)$$

where Q_{s+1} means Q_1 and the method is to show

(1) $Q_i \cdot A_i \cdot Q_{i+1} \equiv Q_i^{\mathrm{L}} B_i Q_{i+1}^{\mathrm{R}}$

(2) $L(Q_{i+1}^{\mathrm{R}}) + L(Q_{i+1}^{\mathrm{L}}) > L(Q_{i+1})$.

Now (2) implies say $Q_{i+1} \equiv DEF$, $Q_{i+1}^{\mathrm{R}} \equiv EF$, $Q_{i+1}^{\mathrm{L}} \equiv DE$ so $Q_{i+1}^{\mathrm{R}} \cdot Q_{i+1}^{-1} \cdot Q_{i+1}^{\mathrm{L}} \equiv E \equiv Q_{i+1}^{\mathrm{M}}$ say. Hence $W' \equiv (Q_1^{\mathrm{M}} B_1 Q_2^{\mathrm{M}} B_2 \cdots Q_s^{\mathrm{M}} B_s)$ which is cyclic.

To prove (1) is trivial if $i, i+1 \notin J$, while if $i, i+1 \in J$ we may use 4.4B. If $i \in J$, $i+1 \notin J$ then $Q_i \cdot A_i \cdot Q_{i+1} \equiv T_i \cdot A_i \cdot Z_{i+1}$ and we may use the dual of 4.4A.

If $r < s$ assume without loss of generality that Q_1 is Z_1. Then W, W' have the common subword Q_1^{M}; thus $W \underset{1}{-} W'$ (cf. 3.6).

If $r = s$ we show $W \underset{1}{-} W_0 \underset{1}{-} W'$ for some W_0. Put

$$W_0 \equiv \left(Z_1 \cdot A_1 \cdot \prod_{i=2}^{s} T_i \cdot A_i \right) \equiv (Z_1^{\mathrm{M}} C_1 T_2^{\mathrm{M}} C_2 \cdots T_s^{\mathrm{M}} C_s);$$

then $W \underset{1}{-} W_0$. Now let W_1 arise from W_0 by replacing Z_1^{M}. Then $W_0 - W_1$. But W_1 is W', hence $W \underset{1}{-} W_0 \underset{1}{-} W'$.

4.5. Suppose $X = Y$ in G_1. By 8.1 of Chapter I, if X and Y are suitably 1-bounded, there are divisions $P_1, ..., P_m$ of X and $Q_1, ..., Q_m$ of Y (we shall simplify this discussion by regarding terms in a division as subwords), and words $X_0, ..., X_m$ such that $X_0 \equiv X_m \equiv I$ and for each i either (i) $P_i \equiv Q_i, X_{i-1} \equiv X_i \equiv I$ and $t_i = 0$ or (ii) $P_i^{-1} X_{i-1}^{-1} Q_i X_i \in \mathsf{L}_1$ and $t_i = 1$.

If $t_i = 1$, P_i is 1-normal. Clearly if we simultaneously replace the P_i such that $t_i = 1$ in X we obtain Y.

Hence (or directly) a maximal normal subword of X or Y of suitably large size determines a maximal normal subword of Y or X, respectively.

4.6. We now consider the axioms in turn, for $n = 1$, and comment on those axioms not already covered by the preceding discussion.

For Axiom 1 we must first define J_1. If W is E.R. and $W^e \neq I$ in G_1 consider the set of all E.R. words conjugate to W in G_1 and choose an element W' from this set such that (W') has as small a 1-bound as possible, say $f(W)$. It can be shown that $\operatorname{Sup} f(W)$ is approximately $\frac{1}{2}$. Thus if $str \geqslant cb_{\min} \geqslant \operatorname{Sup} f(W)$ then Axiom 11b is true for $n = 1$ and J_1 may be defined as in Axiom 9. Axiom 1 now follows. In Axiom 2, $\underset{1}{\sim}$ may for the purposes of this informal discussion be taken to the equality in G_1. In Axiom 3, $\underset{1}{\cong}$ should be thought of as a transitive relation similar in meaning to "linearizations are equal in G_1" (which is not transitive).

Axiom 5 is the only axiom that we shall not attempt to justify here.

Next we prove Axiom 8a. Let $N \subset (X^t)$, $t \geqslant 2$, where $NT \equiv Y^e \in \mathsf{L}_1$ and $s(N) \geqslant \frac{1}{2}$.

If $L(N) \geqslant 2L(X)$ then by 2.18A, $X_1 \equiv J^\sigma$, $Y \equiv J^\tau$ for some C.A.I. X_1 of X. Now $\tau = 1$ by the definition of C_1. Hence $X_1^e \underset{1}{=} 1$, contradiction.

If $L(X) < L(N) < 2L(X)$ then $N \cap N^+$ is not normal, otherwise $N \cup N^+$ is normal and has length $> 2L(X)$ so the previous case applies. Let $N \equiv AB$ where $B = N \cap N^+$. Then $A \subset (X)$ so $s(N) < 2/e + z + r_0$ as required.

If $L(N) \leqslant L(X)$ then $s(N) < z$ as required.

Axiom 10 simply expresses the fact that in the Tartakovskii

Group G_1, $X \underset{1}{=} I$ implies that X contains a subelement of L_1 of large size.

Axioms 24, 25 are true, trivially. Axiom 26′ follows from 4.5. In Axiom 26A, H is obtained from (SB) by replacing S. Axiom 27 is obvious. Axiom 28 follows from 4.5, at least in the linear case.

Next we prove Axiom 29. We have $X \equiv X_0 \alpha_1 X_1 \cdots \alpha_j X_j \equiv Y$ as in 3.1. If $\alpha_i = E_0$ for all i we obtain $(X^a) \equiv (Y^b)$ for some a, b where $a < b$. Now consider the opposite case. Recall that $L(X_i) > L(X_{i+1})$.

We first show that if D is a word of the form $\delta_f(X_{i+1})$ then D also has the form $\delta_f(X_i)$. If $(X_i^s) \equiv (X_{i+1}^{\pm t})$ then, enlarging s, t if necessary, $D^{\pm 1}$ is a subword of the right side and hence of the left side. If $\delta_f(X_i) \equiv \epsilon_{q'}(X_{i+1})$ then $D^{\pm 1}$ is a subword of the right side since $q' \geqslant f + 1/e$. (($q' \geqslant f + 1/e$))

Next if C has the form $\epsilon_q(X_i)$ and if $(X_i^s) \equiv (X_{i+1}^{\pm t})$ then C has the form $\epsilon_q(X_{i+1})$.

Combining these two results we obtain $\delta_f(X) \equiv \epsilon_q(Y)$; but $\epsilon_q(Y)$ has size $> q$.

Axiom 31B is trivial. For Axiom 31C, assume $U \equiv (S, T) \underset{1}{=} (S', T') \equiv V$ where S, T, S', T' are maximal 1-normal. We indicate how to prove that there is a word $S^{\mathrm{L}} A T'^{\mathrm{R}} \underset{1}{=} U \underset{1}{=} V$.

Let us say that S has property X if S is $\bar{P}_i$, $t_i = 1$, that is, the maximal normal determined by P_i (cf. 4.5). If T does not have property X then $-T \equiv -T'$: take (S, T) for the required word. If S does not have X then $S- \equiv S'-$: take (S', T'). If S, T have X let U'' arise from (S, T) by replacing T; say $U'' \equiv (S'', T'')$. Then $S- \equiv S''-$ and $-T'' \equiv -T'$: take U'' for the required word.

In Axiom 33 define $\mathrm{Rep}(L_1)$ to be L_1. The next four axioms are trivial. Axiom 39 follows from 4.4A.

For Axiom 40, we obtain from Axiom 20 that

$$BJ^{-1}AJ \cdot J^{-1}AJB \equiv I$$

Hence $A \cdot A \equiv I$. Since e is odd, $A \equiv I$, hence cancellation occurs in $J^{-1}AJB$, a contradiction. ((e odd))

§5.

Take three symbols a, b, c and let $s_1 s_2 s_3 \cdots$ be a Morse–Hedlund sequence for a, b, c; i.e., each s_i is a, b or c and there do not exist $i, p > 0$ such that $s_i s_{i+1} \cdots s_{i+p-1}$ equals $s_{i+p} s_{i+p+1} \cdots s_{i+2p-1}$ cf. Leech [2]. In the free product $\Pi = (a_1, ..., a_d : a_1^e = \cdots = a_d^e = 1)$ $d \geqslant 2$, let $A \equiv a_1 a_2$, $B \equiv a_1 a_2^2$, $C \equiv a_1^2 a_2$. Let $S_{mn}(a, b, c)$ denote $s_{m+1} s_{m+2} \cdots s_n$ when $0 \leqslant m < n$ and put $X_{mn} \equiv S_{mn}(A, B, C)$.

5.1. Proposition. *X_{mn} does not have the form $UXXXV$ where $X \not\equiv I$.*

Proof. Suppose not. Then X is E.R. Say $X \equiv x_1 x_2 \cdots x_n$, $n \geqslant 2$. Let $Y \equiv x_2 \cdots x_n x_1$. Then X^3 contains Y^2. Either x_1 or x_2 has the form a_1^h; let Z be X or Y accordingly. Then $Z \equiv a_1^{h_1} a_2^{k_1} \cdots a_1^{h_s} a_2^{k_s}$, $n = 2s$. X_{mn} contains $ZZ \equiv u(A, B, C) v(A, B, C)$; since $u(A, B, C) \equiv v(A, B, C)$ we have $u(a, b, c) = v(a, b, c)$, a contradiction.

5.2. Theorem. *G_n is infinite. Thus if J_n is empty then B_d^e is infinite.*

Proof. It is sufficient to show that no two of $X_{01}, X_{02}, X_{03}, \ldots$ are equal in G_n. If this were not so, then $X_{rs} \underset{n}{=} I$ for some r, s $1 \leqslant r < s$. By Axiom 10 X_{rs} is not (n)-bounded by b_∞ so for some p, $1 \leqslant p \leqslant n$, X_{rs} contains S such that $s^p(S) \geqslant b_\infty \geqslant \frac{1}{2}$ (Assume $b_\infty \geqslant \frac{1}{2}$.) cf. Axiom 21. By Axiom 27 S contains UUU where U is E.R., in contradiction to 5.1.

5.3. Remark. The main effort of the remainder of the paper consists in showing that if the axioms hold for some n and if J_n is non-empty then the axioms hold for $n + 1$. Afterwards it will not be difficult to deduce that, if J_n is non-empty for all n, then $B_d^e = \Gamma(C_1, C_2, C_3, \ldots)$. We now prove that the infiniteness of B_d^e follows from this. If a finite group G is given by a finite number of generators and defining relations $R_i = 1$ $(i = 1, 2, 3, \ldots)$ then there exists h such that $R_1 = 1, \ldots, R_h = 1$ are defining relations for G. Thus the finiteness of B_d^e would imply $B_d^e = \Gamma(C_1, \ldots, C_n) = G_n$ for some n; hence B_d^e is infinite.

5.4. Proposition. *Let $X_1, X_2, X_3, \dots$ be an infinite sequence in J_n. Then not $X_1 >_n X_2 >_n X_3 > \cdots$.*

Proof. If $X \rho_n Y$ then $X \sigma_n Y$. ((Assume $q \geqslant q'$.))
The result now follows from Axiom 5.

5.5. Corollary. *$>_n$ determines a partial ordering on J_n.*

Proof. It is transitive, Suppose $X >_n X$; then $X >_n X >_n X >_n \cdots$, contrary to 5.4.

5.6. Definition. Put X in C_{n+1} if X is in J_n and

$$Y \in J_n \Rightarrow \text{not } X >_n Y.$$

5.7. Proposition.
(i) *C_{n+1} is non-empty.*
(ii) *If $X \in J_n \backslash C_{n+1}$ then $(\exists Y)\ Y \in C_{n+1}, X >_n Y$.*

Proof. Let $X \in J_n$. Either $X \in C_{n+1}$ or $X >_n X_1$ for some $X_1 \in J_n$. Either $X_1 \in C_{n+1}$ or $X_1 >_n X_2$ for some $X_2 \in J_n$. And so on. By 5.4 either $X \in C_{n+1}$ or $X >_n X_1 >_n \cdots >_n X_p \in C_{n+1}$ for some $p \geqslant 1$.

5.8. Proposition. *Let $Y \in C_{n+1}$, $n \geqslant 1$. Then $(\exists Z)\ Z \in C_n$, $Y >_{n-1} Z$.*

Proof. $C_{n+1} \subset J_n \subset J_{n-1}$. Since $Y \in J_n$, $Y \notin C_n$. Thus $Y \in J_{n-1} \backslash C_n$. Hence $Y >_{n-1} Z \in C_n$.

5.9. Proposition.
(i) *Let $Y \in C_{n+1}$. Then not $Y \rho_n Z$.*
(ii) *Let $Y \in C_i$, $1 \leqslant i \leqslant n$, $Z \in C_{n+1}$. Then not $Y \sigma_{i-1} Z$.*

Proof. (i) Suppose $Y \rho_n Z$. Then $Y >_n Z$, contrary to $Y \in C_{n+1}$.
(ii) Suppose $Y \sigma_{i-1} Z$. Then $Y, Z \in J_{i-1}$. Now $Z \subset J_n \subset J_i$ so $Z \notin C_i$. Thus $Z \in J_{i-1} \backslash C_i$. Hence $Z >_{i-1} T \in C_i$. Since $Y \in C_i$ we do not have $Y >_{i-1} T$, hence $Z \equiv Z_0 E_{i-1} Z_1 \cdots E_{i-1} Z_k \equiv T$, say. If $0 \leqslant j < k$ then $(Z_j^s) \underset{i-1}{\cong} (Z_{j+1}^{\pm t})$ for some s, t depending on j such that

$1 \leqslant s < t$, so $Z_j^s, Z_{j+1}^{\pm t}$ are conjugate in G_{i-1}. Hence for some s, t $Z^s, T^{\pm t}$ are conjugate in G_{i-1} hence in G_i. Now $T \in C_i$ so $T^e \underset{i}{=} I$. Hence $Z^{es} = I$ in G_i hence in G_n. But $Z \in J_n$; this contradicts 5.12 below.

5.10. Definition. Let $k \geqslant 1$. Put $Y \in L_{n+1}^k$ if Y or Y^{-1} is a C.A. of X^{ke} for some $X \in C_{n+1}$. Put $L_{n+1} = L_{n+1}^1$ and $G_{n+1} = \Gamma(C_1, ..., C_{n+1})$.

Note that if $Y \in L_{n+1}^k$ then each C.A. of Y is in L_{n+1}^k. Because of this we shall occasionally consider the elements of L_{n+1}^k as cyclic words.

5.10A. Informal Remark. We are interested primarily in L_{n+1} and subelements of L_{n+1}; L_{n+1}^k and subpowerelements enter only incidentally, for technical reasons.

5.11. Proposition. *If $Y \in L_{n+1}^k$ then Y is (n)-bounded by str^+.*

Proof. Let $X \in C_{n+1}$. Then $X \in J_n$ so by Axioms 8a and 9 any power of X is (n)-bounded by str^+.

5.12. Corollary. *If $X \in J_n$ and $s \geqslant 1$ then $X^s \neq I$ in G_n; hence G_n contains elements of infinite order. $Y \in L_{n+1}^k$ implies $Y \neq I$ in G_n.*

Proof. Let $X \in J_n$. If $X^s \underset{n}{=} I$, $s \geqslant 1$, then since X^s is (n)-bounded by $str^+ \leqslant b_\infty$ we have by Axiom 10 that $X^s \equiv I$, a contradiction

(($str^+ \leqslant b_\infty$))

5.13. Corollary. *(i) If $W^e \neq I$ in G_n then W has infinite order in G_n. (ii) If e is odd then G_n has no elements of order 2.*

Proof. (i) By Axiom 11b some conjugate of W in G_n lies in J_n so by 5.12 has infinite order.

(ii) Let $X^2 \underset{n}{=} I$. By (i) $X^e \underset{n}{=} I$, so if e is odd $X \underset{n}{=} I$.

5.14. Proposition. *Let $X, Y \in J_n$. (i) If $X \rho_{n-1} Y$ then $X \rho_n Y$. (ii) If $X \sigma_{n-1} Y$ then $X \sigma_n Y$.*

Proof. By Axioms 7, 8.

5.15. Proposition. *Let* $A \in C_j$, $B \in C_i$ *and* $(A^a) \underset{j-1}{\cong} (B^b)$ *where* $a, b \neq 0$ *and* $1 \leqslant i \leqslant j \leqslant n+1$. *Then* $i = j$ *and* $|a| = |b|$.

Proof. Let $i < j$. Now $B^e = I$ in G_i hence in G_{j-1}. By Axiom 4 $A^{ae} = I$ in G_{j-1} contrary to $A \in J_{j-1}$.

Let $i = j$. Then $|a| = |b|$ otherwise $A\, E_m\, B$ or $B\, E_m\, A$ where $m = j - 1$.

5.16. Proposition. *Let* $A \in L_i$, $B \in L_j$, $1 \leqslant i \leqslant j \leqslant n+1$.

(i) *If* $i < j$ *then* $\beta(A,B) < q' L(B)$.

(ii) *If* $i = j$ *then* $A \cdot B \equiv I$ *or* $\beta(A,B) < qL(B)$.

Proof. We have $B \equiv Y^e \equiv UT$, $A^{-1} \equiv X^e \equiv US$ where $L(U) = \beta(A,B)$, and some C.A.I. X', Y' of X, Y belong to C_i, C_j respectively.

Let $L(U) \geqslant L(X) + L(Y)$. By 2.18A we have $X^a \equiv Y^b$ for some $a, b \geqslant 1$. Hence $(X'^{\pm a}) \equiv (Y'^{\pm b})$. By 5.15 $a = b$ and $i = j$. Hence $X \equiv Y$ and $A \cdot B \equiv I$.

Let $L(U) < L(X) + L(Y)$ and $i < j$. If $\beta \geqslant q'L(B) = q'eL(Y)$ then $L(X) > L(Y)$, so $\beta < 2L(X) \leqslant feL(X)$. $\quad$ (($fe \geqslant 2$, $q'e \geqslant 2$))
Hence $X\, \sigma_0\, Y$, so $X'\, \sigma_0\, Y'$ and $X'\, \sigma_{i-1}\, Y'$ contrary to 5.9 (ii).

Similarly when $i = j$, by 5.9 (i).

5.17. Proposition. *If* AB *is a subpowerelement of* L_n *then*

$$s^n(AB) \leqslant s^n(A) + s^n(B) + 2/e.$$

Proof. Immediate from Axiom 16.

5.18. Proposition. *Let* $AB \in \mathsf{L}_n^k$ *and* $s_{AB}(A) < h_0 - \epsilon_3$. *Then* $s^n(A) = \sup s_{AC}(A)$ *over all* C *such that* $AC \in \mathsf{L}_n$.

Proof. Let $AD \in \mathsf{L}_n^k$. If $s_{AD}(A) \geqslant h_0$ then $s^n(A) \geqslant h_0 > r_0$ so

$$h_0 \leqslant s^n(A) \leqslant s_{AB}(A) + \epsilon_3 < h_0. \qquad ((h_0 > r_0))$$

Hence $s_{AD}(A) < h_0$. By Axiom 18 there exists AQ in L_n with $s_{AQ}(A) = S_{AD}(A)$.

5.19. Proposition. Let A, B be subpowerelements of L_n, $s^n(A) > r_0$ and $A \subset B$. Then $s^n(A) \leqslant s^n(B) + \epsilon_3$. Hence if A, B are subpowerelements of L_n then $A \subset B$ implies $s^n(A) \leqslant \text{Max}(r_0, s^n(B) + \epsilon_3)$.

Proof. By Axioms 15, 19.

5.20. Proposition. *Let* $X \in L_n$. *Then* $1 - 2/e \leqslant s^n(X) \leqslant 1 + \epsilon_3$.

Proof. By Axiom 13, $1 - 2/e \leqslant s_X(X) \leqslant 1$. Now use Axiom 19.

5.21. Proposition. *If* $t < s$ *and* X *is a subelement of* L_t *and of* L_s *then* $s^s(X) \leqslant r'_0$.

Proof. Say $XA \in L_t$, $XB \in L_s$. By Axiom 20 $s_{XB}(X) < r'_0 < h_0 - \epsilon_3$. Let $XZ \in L_s$. Then $s_{XZ}(X) < r'_0$. By 5.18, $s^s(X) \leqslant r'_0$. $((r'_0 < h_0 - \epsilon_3))$

5.22. Proposition. *(i) If* A *is an n-urelement then* $s^n(A) > r_0$. *(ii) If* A *is an n-urelement,* B *is an n-urelement and* B *contains* A *then* B *is an n-urelement. (iii). If* $s^n(A) > r_0 + \epsilon_3$ *then* A *is an n-urelement.*

Proof. (iii) Let $B \subset A$. By 5.19 $s^n(A) \leqslant s^n(B) + \epsilon_3$ hence $s^n(B) > r_0$.

5.23. Proposition. *Let* AB *be an n-subpowerelement. Then* $s^n(A) + s^n(B) - s^n(AB)$ *is less than* $2\epsilon_3 + \epsilon_1^*$, $r_0 + 2\epsilon_3$ or $2r_0 + 2\epsilon_3$ *according as both, one or neither of* A, B *are n-urelements.*

Proof. Say $ABC \in L_n^k$. Then $s^n(A) \leqslant s_{ABC}(A) + \epsilon_3$ and $s^n(B) \leqslant s_{BCA}(B) + \epsilon_3$ in the first case and the result follows by Axiom 16a. In the second case say A is an urelement. Since $A \subset AB$ and $s^n(B)$ is less than $r_0 + \epsilon_3$ the result follows by 5.19. The third case is true since $-s^n(AB) \leqslant 0$.

5.24. Proposition. *(i) If A is an n-urelement and for some B, k* $AB \in \mathsf{L}_n^k$ *and* $s_{AB}(A) < h_0$ *then A is n-normal. (ii) If A is n-normal then* $s^n(A) \leqslant 1 + \epsilon_3$.

Proof. (i) By Axiom 18. (ii) holds since $s_{AB}(A) \leqslant 1$ and by Axiom 19.

5.25. Proposition. *Let A be n-normal. If AB and AC are in* L_n *then* $B = C$ *in* G_{n-1}.

Proof. We first show that there exists $AD \in \mathsf{L}_n$ with $s_{AD}(A) \geqslant r_0$. For if not, then $s_{AB}(A) < r_0 < h_0 - \epsilon_3$ so by 5.18 $r_0 < s^n(A) = \sup s_{AQ}(A)$ over $AQ \in \mathsf{L}_n$. Axiom 20 $B = D$ and $C = D$ in G_{n-1}.

5.26. Proposition. *Let* $r_0 + \epsilon_3 < b < h_0 - \epsilon_4$. *Then the following are equivalent*

1. *W is n-bounded by b.*
2. *If U is an n-urelement contained in W then* $s^n(U) < b$.
3. *If U is n-normal and contained in W then* $s^n(U) < b$.

Proof. Clearly 1 implies 2 and 2 implies 3. Assume 3 and suppose 1 is false. Then W contains X where $s^n(X) \geqslant b$. We show no left subword X^{L} of X satisfies $b \leqslant s^n(X^{\mathsf{L}}) < h_0$. For otherwise X^{L} is a subelement of L_n by Axiom 18; if X^{L} is an urelement then it is normal so by 3 we have a contradiction; if X^{L} is not an urelement then $s^n(X^{\mathsf{L}}) \leqslant r_0 + \epsilon_3 < b$.

Thus in particular $s^n(X) \geqslant h_0$. Take ϵ such that $b < b + \epsilon < h_0 - \epsilon_4$. Then Y exists such that $s_{XY}(X) > h_0 - \epsilon$. Thus $0 \leqslant b < h_0 - \epsilon < s_{XY}(X)$ and $h_0 - \epsilon - b > \epsilon_4$. By Axiom 23 $b \leqslant s_{XY}(X^{\mathsf{L}}) < h_0 - \epsilon < h_0$ and we have a contradiction.

5.27. Proposition. *Let W be n-bounded by* $b \leqslant h_0$. *Then a subword K of W is an n-urelement if and only if it is n-normal. Each n-normal subword of W is contained in a unique maximal n-normal subword.*

Let X be E.R. and let no C.A. of X be a subpowerelement of

L_n. Let $[X]$ be n-bounded by $b \leqslant h_0$. Then each n-normal subword of $[X]$ is contained in a unique maximal n-normal subword.

Proof. By Axioms 18 and 22.

5.28. Proposition. *Let X be E.R. and for all $s \geqslant 1$ let $[X^s]$ be n-bounded by h_0. Let $[X^t]$ contain R where $S^n(R) > 2(r_0 + \epsilon_3) + 2/e$. Then $L(R) < 2L(X)$.*

Proof. $s^n(R) < h_0$ hence R is a subelement of L_n. By 5.22, R is an urelement hence normal. We may suppose X^t begins with R: $X^t \equiv RT$. For any integer $N \geqslant 3$ consider X^{tN} and let the translates of R in it be $R, R^+, R^{++}, \ldots$. Assume $L(R) \geqslant 2L(X)$. Then R meets or touches R^{++}, so $R^+ \subset R \cup R^{++}$. Let $R \cap R^+ \equiv E$; then $R^+ \equiv EF$, $F \subset R^{++}$. At least one of E, F is normal otherwise.

$$s(R^+) \leqslant s(E) + s(F) + 2/e \leqslant 2(r_0 + \epsilon_3) + 2/e.$$

If F is normal then so is $R^+ \cap R^{++}$. Since $R^+ \cap R^{++} \equiv R \cap R^+$ we see that in any case $R \cap R^+$ is normal. Hence the unique maximal normal subwords containing R, R^+ coincide. Thus the subword of X^{tN} generated by all translates of R is normal. Hence so is $(RT)^{N-1}R$. By repeated use of 5.23.

$$(N-1)s(RT) + s(R) < s((RT)^{N-1}R) + (N-1)(2\epsilon_3 + \epsilon_1^*).$$

Since $s(RT) \geqslant s(R) - \epsilon_3$ we contradict the boundedness of X^{tN} by taking N sufficiently large.

5.28A. *Note.* Let $ABC \in \Pi$ where AB and BC are subpowerelements and B is an urelement. Then ABC is a subpowerelement (and an urelement.)

Proof. AB is an urelement by 5.22. By Axiom 22, $\overline{B} = \overline{AB}$. Similarly $\overline{B} = \overline{BC}$. Hence $\overline{B}$ contains AB and BC so contains ABC.

5.29. Proposition. *Let A be an n-subpowerelement and let P, Q be* *subwords of A where* $s^n(P), s^n(Q) > max(s^n(A) - a, r_0 + \epsilon_3)$. *Let* $s^n(A) \geqslant 2a + 4\epsilon_3 + \epsilon_1^*$. *Then P, Q meet and*

$$s^n(P \cap Q) > min(s^n(A) - a - r_0 - \epsilon_3 - 2/e, s^n(A) - 2a - 3\epsilon_3 - \epsilon_1^* - 2/e).$$

Hence if $r_0 \geqslant a + 2\epsilon_3 + \epsilon_1^*$ *then*

$$s^n(P \cap Q) > s^n(A) - a - r_0 - \epsilon_3 - 2/e.$$

Proof. *If P, Q are disjoint then* $A \equiv CPDQE$ *say. Then*

$$s(CPD) + s(QE) \leqslant s(A) + (2\epsilon_3 + .\epsilon_1^*)$$
$$s(A) - a < s(P) \leqslant s(CPD) + \epsilon_3$$
$$s(A) - a < s(Q) \leqslant s(QE) + \epsilon_3.$$

Adding, $2s(A) - 2a < s(A) + 4\epsilon_3 + \epsilon_1^*$, contrary to hypothesis.

The result is trivial if one of P, Q contains the other.

Now let P, Q overlap say $P \equiv RS$, $Q \equiv ST$. If T is not an urelement then $s(A) - a < s(Q) \leqslant s(S) + (r_0 + \epsilon_3) + 2/e$. If T is an urelement then

$$s(RS) + s(T) \leqslant s(RST) + (2\epsilon_3 + \epsilon_1^*)$$
$$s(A) - a \quad < s(RS)$$
$$s(A) - a \quad < s(ST) \leqslant s(S) + s(T) + 2/e$$
$$s(RST) \quad \leqslant s(A) + \epsilon_3$$

Adding, $s(A) - 2a < (2\epsilon_3 + \epsilon_1^*) + \epsilon_3 + s(S) + 2/e$.

5.30. Corollary. *In the hypothesis of* 5.29, *delete* $s^n(Q) > s^n(A) - a$ *and replace* $s^n(A) \geqslant 2a + 4\epsilon_3 + \epsilon_1^*$ *by* $s^n(Q) \geqslant a + 4\epsilon_3 + \epsilon_1^*$. *Then the conclusion of* 5.29 *holds if* $s^n(A)$ *is replaced by* $s^n(Q) + a$.

5.31. Proposition. *Let HUVWK be a subelement of* L_n *of size a. Then*

$$s(V) \geqslant s(UV) - r_0 - \epsilon_3 - 2/e$$
$$\text{or } s(V) \geqslant s(VW) - r_0 - \epsilon_3 - 2/e$$
$$\text{or } s(V) \geqslant s(UV) + s(VW) - a - 3\epsilon_3 - \epsilon_1^* - 2/e.$$

Proof. If U is not an urelement then

$$s(UV) \leqslant s(U) + s(V) + 2/e \leqslant r_0 + \epsilon_3 + s(V) + 2/e.$$

Similarly if W is not an urelement. Now let U, W be urelements. Then

$$s(UV) + s(W) \leqslant s(UVW) + (2\epsilon_3 + \epsilon_1^*)$$
$$s(VW) \leqslant s(V) + s(W) + 2/e$$
$$s(UVW) \leqslant a + \epsilon_3.$$

Hence $s(UV) + s(VW) \leqslant s(V) + a + 3\epsilon_3 + \epsilon_1^* + 2/e$ as required.

5.32. Remark. Let $1 \leqslant s \leqslant n$. If $W \equiv ABC$, where AB is s-bounded by a and BC is s-bounded by c, and if B is not a subelement of L_s (or more generally if no s-subelement contained in W properly contains B) then W is s-bounded by Max(a, c).

(For if N is any subelement contained in W then $N \subset AB$ or $N \subset BC$ or N properly contains B.)

§6

Informal Summary

In this section we show that G_n is a Generalized Tartakovskii Group and hence derive some properties of it. In particular if $A = B$ in G_n and S is a subword of A one can find, under suitable conditions, a corresponding subword of B (by making use of 8.1 of Chapter I).

6.1. Lemma. *(i)* $L_1, \ldots, L_n$ *satisfy the Generalizing Tartakovskii* *conditions. (ii) If $R = I$ in G_n, $R \not\equiv I$ then R contains a subword C, $L(R) \geqslant L(C) \geqslant 2$ and $(\exists v, D)$ $CD \in \mathsf{L}_v$, $1 \leqslant v \leqslant n$,*

$$s_{CD}(C) > \begin{cases} 1 - 3(r + \epsilon_5) - 2\epsilon_2 - 12/e \; (= k_1) & \text{if } v < n, \\ 1 - 3(r_0 + \epsilon_5) - 2\epsilon_2 - 12/e \; (= k_1^0) & \text{if } v = n. \end{cases}$$

Proof. (i) We prove that the Generalized Tartakovskii Conditions hold if we take $\epsilon_0 = \epsilon_1 = 2/e$, $1/p = r + \epsilon_5$, $1/p_0 = r_0 + \epsilon_5$, $1/p_0' = r_0' + \epsilon_5$, where ϵ_5 is a (new) parameter to be determined.

Let $AX_1 \in \mathsf{L}_t$, $X_2^{-1}B \in \mathsf{L}_s$ and $X_1 \cong X_2$, $AX_1 \equiv Ex$, $X_2^{-1}B \equiv yF$, $P \equiv xE \cdot Fy$ (cf. Chapter I, 6.2).

First let $t < s$ and $s_{X_1A}(X_1) \geqslant 1/p$. $((\epsilon_5 \geqslant 2\epsilon_2 + 4/e))$
Then since $1/p > \epsilon_2$ we have $L(X_1) \geqslant 2$. Hence $X_1 \equiv c_1Ud_1$, $X_2 \equiv c_2Ud_2$, $xE \equiv d_1Ac_1U \in \mathsf{L}_t$, $Fy \equiv U^{-1}c_2^{-1}Bd_2^{-1} \in \mathsf{L}_s$. Thus

$$r \leqslant r + \epsilon_5 - 2\epsilon_2 - 4/e \leqslant s_{X_1A}(X_1) - 2\epsilon_2 - 4/e \leqslant s_{Ud_1Ac_1}(U).$$

By Axiom 20 $P \equiv xE \cdot Fy = V \cdot Q \cdot V^{-1}$ in G_{t-1} where $Q \in \mathsf{L}_s$.

In a similar manner we deal with the case $t < s$ and $s_{X_2^{-1}B}(X_2^{-1}) \geqslant 1/p_0'$ and with the case $t = s$ and $s_{X_1A}(X_1) \geqslant 1/p_0$.
Finally we have $((r_0' + 2/e + \epsilon_2 \leqslant r_0 \leqslant r))$
$\epsilon_1 + \epsilon_2 + 1/p_0' \leqslant 1/p_0 \leqslant 1/p$, $1 \geqslant 5(1/p + \epsilon_1)$ $((r \leqslant 1/5 - \epsilon_5 - \epsilon_1))$

(ii) By Theorem 4 of Part I, R contains A, $A \cong A'$, $A'B \in \mathsf{L}_v$ for some v, $1 \leqslant v \leqslant n$ and $s_{A'B}(A') > 1 - 3/p - 4\epsilon_1$ (if $v = n$ replace p by p_0). Now

$$1 - 3/p - 4\epsilon_1 \geqslant \epsilon_2 + (2\epsilon_2 + 4/e) \quad ((1 - 3(r + \epsilon_5) - 4\epsilon_1 \geqslant 3\epsilon_2 + 4/e))$$

so in particular $L(A') \geqslant 2$. Thus $A' \equiv c'Cd'$, $A \equiv cCd$. Let $D \equiv d'Bc'$. Then $CD \in \mathsf{L}_v$. Now $s_{A'B}(A') \leqslant s_{CD}(C) + 2\epsilon_2 + 4/e$. Hence $s_{CD}(C) > 1 - 3/p - 4\epsilon_1 - (2\epsilon_2 + 4/e)$ as required.

6.2. Lemma. *Let X, Y be equal in G_n and not both I. Let each of X, Y be $((n-1))$-bounded by k_2 and n-bounded by k_2^0 where*

$$k_2 = 1 - 4(r + \epsilon_5) - 14/e - 2\epsilon_2$$

and k_2^0 arises from k_2 by replacing r by r_0. Then there are divisions $P_1, \ldots, P_m$ and $Q_1, \ldots, Q_m$ of X and Y respectively with the properties listed in 8.1 of Chapter I.

Proof. Let $ST \in L_i$ where $1 \leqslant i \leqslant n$ and let S be $*$-contained in X or Y. If $L(S) \leqslant 1$ then $s_{ST}(S) \leqslant \epsilon_2 < 1 - 4/p - 5\epsilon_1 \leqslant 1 - 4/p_0 - 5\epsilon_1$. If $L(S) \geqslant 2$ then $S \equiv cS'd$ and $S' \subset X$ or $S' \subset Y$. If $i < n$ then $s_{ST}(S) \leqslant 2\epsilon_2 + 4/e + s^i(S') < 2\epsilon_2 + 4/e + k_2 = 1 - 4/p - 5\epsilon_1$. Similarly if $i = n$.

6.3. Definitions. Let $A \underset{n}{=} B$, where not both of A, B are empty. Let A, B be $((n-1))$-bounded by k_2 and n-bounded by k_2^0. Let $P_1, \ldots, P_m$ and $Q_1, \ldots, Q_m$ be divisions as in 6.2. Consider any P_j of non-zero type. Let P_j' be the largest common subword of P_j and A and similarly define Q_j'. A point of A is called *left nice* if it is the left end point of A or of some P_j' or if it is an interior point of some P_i of zero type. We define left nice points of B similarly and note that they are in one to one correspondence with those of A. Dually we have the concept of *right nice.*

Call a non-empty subword S of A *nice* if its left end point is left nice and its right end point is right nice. Note that there will be a corresponding nice subword of B; denote it by S'.

The notation $S \underset{n}{\sim} S'$ means that there exist $A, B, P_1, \ldots, P_m$, $Q_1, \ldots, Q_m$ as above for which S, S' are corresponding nice words.

6.4. Note. If $A \underset{n}{=} B$ where not both of A, B are empty and each is $((n-1))$-bounded by k_2 and n-bounded by k_2^0 then $A \underset{n}{\sim} B$.

6.5. Note. In 6.3, note that the division $P_1, \ldots, P_m$ of A induces a division of S and similarly a division of S' is induced. We shall define *left* and *right nice* points and *nice* subwords of S with respect to these induced divisions exactly as in 6.3.

For any subword T of S, T is nice in S if and only if T is *nice* in A.

6.6. Note. If $S \underset{n}{\sim} S'$ and T, T' are corresponding nice words (of S, S' respectively) then $T \underset{n}{\sim} T'$.

6.6A. Definition. If $X \underset{n}{\sim} Y$ with divisions $P_1, \ldots, P_m, Q_1, \ldots, Q_m$ and $p \leqslant n$, we say that $X \underset{n}{\sim} Y$ induces $A \underset{p}{\sim} B$ if A, B are corresponding nice subwords (of X, Y respectively) and for $j = 1, \ldots, m$ if P_j has type $t_j \neq 0$ and $P'_j \subset A$ then $t_j \leqslant p$.

(Note that if P_j has type $t_j \neq 0$ then either P'_j is disjoint from A or $P'_j \subset A$.)

6.7. If P_j has type $t_j \neq 0$ then $P_j \equiv LP'_jM$, $Q_j \equiv NQ'_jO$ where L, M, N, O have length 0 or 1. Put $X'^{-1}_{j-1} \equiv L^{-1} \cdot X^{-1}_{j-1} \cdot N$, $X'_j \equiv 0 \cdot X_j \cdot M^{-1}$. Then $P'^{-1}_j X'^{-1}_{j-1} Q'_j X'_j \in \mathsf{L}_{t_j}$.

With the assumptions of 6.3 let S, S' be corresponding nice words. We shall define the *end words* E, K for S, S', with respect to the given divisions $P_1, \ldots, P_m$ and $Q_1, \ldots, Q_m$, as follows.

If the left end point of S is the left end point of P'_j put $E \equiv X'^{-1}_{j-1}$; otherwise put $E \equiv I$. If the right end point of S is the right end point of some P'_j put $K \equiv X'_j$; otherwise put $K \equiv I$.

Thus in all cases $S \underset{n}{=} ES'K$.

Moreover if $S \equiv AB$, $S' \equiv A'B'$ and A, A' are corresponding nice words (end words E_1, K_1 say) then B, B' are corresponding nice words (end words E_2, K_2 say) and

$$E_1 \equiv E, \quad K_1 \equiv E_2^{-1}, \quad K_2 \equiv K.$$

6.7A. Proposition. *With the assumption of 6.3 let P_i have type t_i $(i = 1, \ldots, m)$. Then for $i = 1, \ldots, m$ either $X'^{-1}_{i-1} \equiv I$ or*

$$s_C(X'^{-1}_{i-1}) - \epsilon_5 - 2\epsilon_2 - 4/e < r'_0, r_0 \text{ or } r \tag{$*$}$$

according as t_i is $>, =$ or $< t_{i-1}$; similarly $X'_i \equiv I$ or $s_D(X'_i)$ satisfies $(*)$ according as t_i is $>, =$ or $< t_{i+1}$; here

$$C \equiv X'^{-1}_{i-1} Q'_i X'_i P'^{-1}_i, \qquad D \equiv X'_i P'^{-1}_i X'^{-1}_{i-1} Q'_i.$$

If $t_i \neq 0$ then P'_i, Q'_i are t_i-normal and

$$s_E(P'_i) + s_F(Q'_i) > \begin{cases} k_3 & \text{if } t_i < n \\ k_3^0 & \text{if } t_i = n \end{cases}$$

where $E \equiv P_i' X_i'^{-1} Q_i'^{-1} X_{i-1}'$, $F \equiv Q_i' X_i' P_i'^{-1} X_{i-1}'^{-1}$ *and* X_i', X_{i-1}' *have* t_i*-size less than* r_6 *if* $t_i < n$, r_6^0 *if* $t_i = n$. *(Here* k_3, k_3^0, r_6, r_6^0 *are new parameters with values as in the proof below.)*

Finally if $j = t_i \neq 0$ *then*

$$s^j(P_i') > \begin{cases} k_3 - k_2 & \text{if } j < n \\ k_3^0 - k_2^0 & \text{if } j = n \end{cases}$$

and similarly for Q_i'.

Proof. If $t_i < t_{i+1}$ then $s_W(\hat{X}_i) < 1/p$ for an appropriate W in the notation of Chapter I, 8.1. Now $\hat{X}_i$ is a subword of X_i' and in fact $X_i' \equiv H\hat{X}_i K$ where H, K have length $\leqslant 1$. Hence $s_D(X_i') < 2\epsilon_2 + 1/p + 4/e$, that is, $s_D(X_i') - 2\epsilon_2 - \epsilon_5 - 4/e < r$. Similarly if $t_i = t_{i+1}$ or $t_i > t_{i+1}$. Next $1 - 2/e \leqslant s_E(P_i') + s_F(Q_i') + s_C(X_{i-1}'^{-1}) + s_D(X_i') + 6/e$, hence $s_E(P_i') + s_F(Q_i') > 1 - 8/e - 2(r + 2\epsilon_2 + \epsilon_5 + 4/e) = k_3$. Let k_3^0 arise from the expression for k_3 by replacing r by r_0. Then if $t_i = n$ we may replace k_3 by k_3^0.

If $s(X_i') > r_0$ then $s(X_i') \leqslant s_D(X_i') + \epsilon_3 < (r + 2\epsilon_2 + \epsilon_5 + 4/e) + \epsilon_3 = r_6$. This remains true if $s(X_i') \leqslant r_0$. (Let r_6^0 arise from the expression for r_6 by replacing r by r_0.)

Finally if $j = t_i < n$ then

$$s^j(P_i') + k_2 > s^j(P_i') + s^j(Q_i') > k_3$$

hence $s^j(P_i') > k_3 - k_2 > r_0 + \epsilon_3$. Thus P_i' is j-normal. $((k_3 - k_2 > r_0 + \epsilon_3))$

Similarly if $j = n$. $((k_3^0 - k_2^0 > r_0 + \epsilon_3))$

6.7B. If $P_1, \ldots, P_m$ is a division of S then, if $t_i \neq 0$, we denote the maximal t_i-normal subword of S determined by P_i' by $\overline{P_i'}$ or simply $\overline{P}_i$.

6.8. Proposition. *Let* $A \subset X \underset{p}{\sim} Y$, *where* $p < n$, A *is a subelement of* L_n *and* $s(A) > u_1 = 2(r_0' + \epsilon_2) + 8/e$. *Then* $A \equiv A'A^0A''$ *and* $I \equiv A^0 \underset{p}{\sim} Y^M$ *for some* Y^M $(Y^M \subset Y)$ *and*

$$s^n(A'), s^n(A'') \leqslant u_2 = r'_0 + \epsilon_2 + 2/e.$$

Hence $s^n(A^0) \geqslant s^n(A) - u_1$.

Proof. Say $X \equiv P_1, ..., P_m$, $Y \equiv Q_1, ..., Q_m$. Any point of A other than its right end point is either the left end point or an interior point of some P_i. By 5.21, any common subword J of A and a P_i of non-zero type satisfies $s^n(J) \leqslant r'_0$. A subword K of A of length 1 satisfies $s^n(K) \leqslant \epsilon_2$. Using 5.17 the result follows.

6.9. *Note.* We say A^0 arises from A by "deleting partial P's". We shall always assume that A^0 is chosen maximally.

More generally let $X \underset{n}{\sim} Y$ and let B be any subword of X. If $(\exists C)\, I \not\equiv C \subset B$, C is a nice subword of X then the union of all such C is nice in X; we denote it by B^0. The corresponding nice subword $B^{0'}$ of Y is called the *nice image* of B. If $A \subset B \subset X$ and A^0, B^0 exist then $A^0 \subset B^0$.

6.10. Proposition. *Let* $A \subset X \underset{p}{\sim} Y$, *where* $p < n$, *A is a subelement of* L_n *and* $s^n(A) > u_3 = 2(r'_0 + \epsilon_2) + 8/e + s_{25}$. *Let X, Y be (p)-bounded by* b_{25}. *Then* $A \supset A^0 \underset{p}{\sim} Y^M \supset B$ *where B is a subelement of* L_n *and* $s^n(B) > s^n(A) - u_4$ *where* $u_4 = u_1 + r_{25}$.

Any such B is called a weak image of A.

Proof. By 6.8 $A \supset A^0 \underset{p}{\sim} Y^M$. By Axiom 25

$$s^n(A^0) \geqslant s^n(A) - u_1 > s_{25}, \qquad s^n(B) > s^n(A^0) - r_{25}$$

so $s^n(B) > s^n(A) - u_1 - r_{25}$.

6.10A. *Note.* Let $X \underset{n-1}{\sim} Y$ where X, Y are $((n-1))$-bounded by b_{25}. Let Y be n-bounded by x where $u_3 < x < h_0 - \epsilon_2 - 2/e - u_4$. Then X is n-bounded by $x + u_4$.

Proof. If not then $A \subset X$, $s(A) \geqslant x + u_4$. Let A^l be the largest left subword of A such that $s(A^l) < x + u_4$. Define c and A^{L} by

$A^{\mathrm{L}} \equiv A^l c$. Then $u_3 < x + u_4 \leqslant s(A^{\mathrm{L}}) \leqslant s(A^l) + \epsilon_2 + 2/e < x + u_4 + \epsilon_2 + 2/e < h_0$. By Axiom 18, A^{L} is a subelement of L_n. By 6.10, Y contains an n-subelement of size $> s(A^{\mathrm{L}}) - u_4 \geqslant x$, which is a contradiction.

6.10B. *Note.* In 6.11, 6.12, 6.13 below we shall assume that X, Y are n-bounded by h_0 (cf. 5.27).

6.11. Proposition. *Let $A \subset X \underset{p}{\sim} Y$, where $p < n$, A is a subelement of L_n and $s^u(A) > u_5 \geqslant u_3$, where u_5 is a new parameter. Let X, Y be (p)-bounded by b_{25}. Thus by 6.10, $A \supset A^0 \underset{p}{\sim} Y^M \supset B$, where $s^n(B) > s^n(A) - u_4 > u_5 - u_4$. Let $J \subset Y^M$ and $s^n(J) > u_5 - u_4$. Then J, B are n-normal and determine the same maximal normal subword M of Y. In particular M is independent of the choice of B so it may be denoted by $Im(A)$. Also $s^n(Im(A)) > s^n(A) - u_4 - \epsilon_3$. Finally, A is normal.*

Proof. Since $s(B) > u_3$, $B \supset B^0 \underset{p}{\sim} X^M \supset C$ and $\quad ((u_5 \geqslant 2u_4 + u_3))$

$$s(C) > s(B) - u_4 > s(A) - 2u_4 > u_5 - 2u_4.$$

Also

$$J \supset J_0 \sim K_1 \supset K \supset K^0 \sim J_1 \supset J_2 \quad \text{and}$$

$$s(K) > s(J) - u_4 > u_5 - 2u_4,$$

$s(J_2) > s(K) - u_4$. Note that K and X^M are contained in A. Now $s(C) > r_0$ so $s(X^M) \geqslant s(C) - \epsilon_3 > s(A) - 2u_4 - \epsilon_3 = s(A) - a$, say. Use 5.30 taking X^M, K for P, Q respectively. $\quad ((u_5 - 2u_4 > r_0))$
We have

$$s(X^M), s(K) > r_0 + \epsilon_3, \qquad ((a = 2u_4 + \epsilon_3, u_5 - a > r_0 + \epsilon_3))$$

$s(K) \geqslant a + 4\epsilon_3 + \epsilon_1^*$. Hence,
$((u_5 - 2u_4 > r_0 + \epsilon_3, u_5 - 2u_4 \geqslant a + 4\epsilon_3 + \epsilon_1^*))$

$$s(K \cap X^M) > s(K) - a - r_0 - \epsilon_3 - 2/e \qquad ((r_0 \geqslant a + 2\epsilon_3 + \epsilon_1^*))$$

$((u_5 - 2u_4 - a - r_0 - \epsilon_3 - 2/e = \alpha \text{ say}, \alpha > u_3, \alpha - u_4 \geqslant r_0 + \epsilon_3))$

Now $K \cap X^M \supset K_0 \cap X^M \sim J_1 \cap B^0 \supset N$, where $s(N) > \alpha - u_4 \geqslant r_0 + \epsilon_3$. Thus $N \subset J_1 \subset J_0 \subset J$, $N \subset B^0 \subset B$. Since N is normal and contained in J and B, the result follows.

6.12. Proposition. *Let $A \subset X \underset{p}{\sim} Y$, where $p < n$ and $s^n(A) > u_6$. Let X, Y be (p)-bounded by b_{25}. Then $Im(Im\, A)$ exists and is $\bar{A}$, the maximal normal subword of X determined by A. Hence if A is maximal normal $|s(Im\, A) - s(A)| < u_4 + \epsilon_3$.*

Proof.

$$A \supset A^0 \sim Y^M \supset B,\ s(B) > s(A) - u_4 > u_6 - u_4, \qquad ((u_6 \geqslant u_5))$$

B is normal and $\bar{B} = \operatorname{Im} A$. $s(\bar{B}) \geqslant s(B) - \epsilon_3 > u_6 - u_4 - \epsilon_3 \geqslant u_5$ (parameter condition number 1). Hence $\bar{B} \supset \bar{B}^0 \sim X^M \supset C$, $\bar{C} = \operatorname{Im} \bar{B}$, $s(C) > s(\bar{B}) - u_4$ and $s(C) > u_6 - 2u_4 - \epsilon_3 \geqslant u_3$ (P.C. 2). By 6.10, $C \supset C^0 \sim Y^m \supset D$, $s(D) > s(C) - u_4 > u_6 - 3u_4 - \epsilon_3 > r_0$ (P.C. 3) and therefore $s(Y^m) > s(D) - \epsilon_3$. Thus we have $s(Y^m) > s(\bar{B}) - 2u_4 - \epsilon_3$. Apply 5.30 taking A, P, Q, a to be $\bar{B}$, Y^m, B, $2u_4 - \epsilon_3$. For this we need $u_6 - u_4 - \epsilon_3 - 2u_4 - \epsilon_3 > r_0 + \epsilon_3$ (P.C. 4), $u_6 > 3u_4 + 3\epsilon_3 + \epsilon_1^*$ (P.C. 5), $r_0 \geqslant \epsilon_3 + \epsilon_1^* + 2u_4$ (P.C. 6). Hence

$$\begin{aligned} s(Y^m \cap B) &> s(B) - r_0 - \epsilon_3 - 2/e \\ &> u_6 - u_4 - r_0 - \epsilon_3 - 2/e = u_7 \end{aligned}$$

Now $Y^m \cap Y^M \supset Y^m \cap B$ so if $u_7 > r_0$ (P.C. 7) then $s(Y^m \cap Y^M) \geqslant u_7 - \epsilon_3 > s_{25}$ (P.C. 8). By Axiom 25, $Y^m \cap Y^M \sim C^0 \cap A^0 \supset N$ where $s(N) > u_7 - \epsilon_3 - r_{25} > r_0 + \epsilon_3$ (P.C. 9). Thus N is normal. But $N \subset C^0 \subset C \subset \bar{C}$ and $N \subset A^0 \subset A$, so $\bar{A} = \bar{C} = \operatorname{Im}(\operatorname{Im} A)$.

6.13. Proposition. *Let $X \underset{p}{\sim} Y$ where $p < n$ and X, Y are (p)-bounded by b_{25}. Let A, B be distinct maximal n-normals of X of size greater than u_6 and A be left of B. Then $Im\, A$ is left of $Im\, B$.*

Proof. We have

$$A \supset A^0 \sim A^{0'} \supset S,\ \bar{S} = \operatorname{Im} A$$

$$B \supset B^0 \sim B^{0'} \supset T,\ \bar{T} = \operatorname{Im} B.$$

By 6.12, $\bar{S} \neq \bar{T}$; but $\bar{S}, \bar{T}$ are maximal so neither can contain the other. Also neither of S, T can contain the other. Now the left end points of A^0, B^0 satisfy L.E.P. $A^0 \leqslant$ L.E.P. B^0; also R.E.P. $A^0 \leqslant$ R.E.P. B^0. Also L.E.P. $S <$ L.E.P. $B^{0'}$ otherwise S is contained in $B^{0'}$ and $\bar{S} = \bar{T}$. Hence S is left of T. If $\bar{T}$ were left of $\bar{S}$ we would have $T \subset \bar{S} \cap \bar{T}$ and hence, since T is normal $\bar{S} = \bar{T}$. Thus $\bar{S}$ is left of $\bar{T}$.

6.14. Proposition. *Let $X \underset{n}{\sim} Y$ with decompositions $P_1, \ldots, P_k$ and $Q_1, \ldots, Q_k$. Let X, Y be $((n-1))$-bounded by k_4 and n-bounded by k_5. If $1 \leqslant j < m \leqslant k$ and $t_j = t_m = n$ then $\bar{P}_j \neq \bar{P}_m$ and $\bar{Q}_j \neq \bar{Q}_m$.*

Proof. It is sufficient to prove $\bar{Q}_j \neq \bar{Q}_m$. We shall do this by assuming $\bar{Q}_j = \bar{Q}_m$, deducing that $\bar{P}_j = \bar{P}_m$ and afterwards obtaining a contradiction from this. Thus the subword $(Q'_j, Q'_m) \equiv Q'_j E Q'_m$ say of Y is a subelement of L_n, say $Q'_j E Q'_m S \in L_n$. By 6.7A, 5.25 and since

$$P'^{-1}_j X'^{-1}_{j-1} Q'_j X'_j \quad \text{and} \quad P'^{-1}_m X'^{-1}_{m-1} Q'_m X'_m$$

belong to L_n we get

$$EQ'_m S = X'_j P'^{-1}_j X'^{-1}_{j-1} \quad \text{and} \quad SQ'_j E = X'_m P'^{-1}_m X'^{-1}_{m-1} \quad \text{in } G_{n-1}.$$

Let

$$P_j \equiv LP'_j M,\quad Q_j \equiv NQ'_j O,\quad P_m \equiv L'P'_m M',\quad Q_m \equiv N'Q'_m O'.$$

The subword (P'_j, P'_m) of X is equal in G_{n-1} to

$$(X'^{-1}_j EQ'_m SX'_{j-1})^{-1} MP_{j+1} \cdot \cdots \cdot P_{m-1} L' (X'^{-1}_m SQ'_j EX'_{m-1})^{-1}.$$

Now $P_{j+1} \cdot \cdots \cdot P_{m-1} = X_j^{-1} Q_{j+1} \cdot \cdots \cdot Q_{m-1} X_{m-1}$ and $E \equiv OQ_{j+1} \cdots Q_{m-1} N'$. By 6.7 we deduce that $(P'_j, P'_m) = X'^{-1}_{j-1} S^{-1} Q'^{-1}_m E^{-1} Q'^{-1}_0 S^{-1} X'_m$; thus

$$Q'^{-1}_j X'_{j-1} (P'_j, P'_m) X'^{-1}_m Q'^{-1}_m = Q'^{-1}_j S^{-1} Q'^{-1}_m E^{-1} Q'^{-1}_j S^{-1} Q'^{-1}_m$$

where the right side is a subpowerelement of L_n by 5.28A.

$Q'^{-1}_j X'_{j-1} P'_j$ is a subelement of L_n so by Axiom 24 it is equal in G_{n-1} to $(Q'^{-1}_j)^{\mathrm{L}} X'' P'^{\mathrm{R}}_j$, which is an n-subelement and is $((n-1))$-bounded by $k_4 + q_{24}$; the hypothesis of Axiom 24 is satisfied since $k_4 \leqslant b_{24}$ (P.C. 1), $k_3^0 < s(P'_j) + s(Q'_j) \leqslant k_5 + s(Q')$ implies $s(Q') > k_3^0 - k_5 \geqslant s_{24}$ (P.C. 2). Moreover, $s(P'^{\mathrm{R}}_j) > s(P'_j) - r_{24} > k_3^0 - k_5 - r_{24} > r_0 + \epsilon_3$ (P.C. 3) so P'^{R}_j is normal and therefore not a subelement of $L_1 \cup \cdots \cup L_{n-1}$ (cf. 5.21). Similarly for $P'_m X'^{-1}_m Q'^{-1}_m$, $Q'_m S Q'_j$; in particular they are equal in G_{n-1} to say $P'^{\mathrm{L}}_m Z Q'^{-1\mathrm{R}}_m$, $Q'^{\mathrm{L}}_m S' Q'^{\mathrm{R}}_j$ respectively. Thus

$$Q'^{-1\mathrm{L}}_j X'' \underline{(P'^{\mathrm{R}}_j, P'^{\mathrm{L}}_m)} Z Q'^{-1\mathrm{R}}_m = Q'^{\mathrm{R}-1}_j S'^{-1} \underline{Q'^{\mathrm{L}-1}_m E^{-1} Q'^{\mathrm{R}-1}_j} S'^{-1} Q'^{\mathrm{L}-1}_m$$

where the right side is an n-subpowerelement and each side is $((n-1))$-bounded by $k_4 + q_{24} \leqslant b_{25}$ (P.C. 4) since the two underlined words are subwords of X, Y respectively. Now let $A \equiv Q'^{-1\mathrm{L}}_j X'' P'^{\mathrm{R}}_j$ and $B \equiv P'^{\mathrm{L}}_m Z Q'^{-1\mathrm{R}}_m$. By Axiom 24 and 6.7A $s(A), s(B) \geqslant \frac{1}{2} > u_6$ (P.C. 5). With respect to the above equation, Im A, Im B both coincide with its right side, hence by 6.12 $\bar{A} \equiv \bar{B}$. Thus $\overline{(P'^{\mathrm{R}}_j, P'^{\mathrm{L}}_m)}$ is a subpowerelement. Now P'^{R}_j, P'^{L}_m are normal so $\overline{P'^{\mathrm{R}}_j} = \overline{P'^{\mathrm{L}}_m}$ and $\overline{P_j} = \overline{P_m}$.

Now where $(P'_j, P'_m) \equiv P'_j K P'_m$ we have $s(P'_j) + s(KP'_m) \leqslant k_5 + (2\epsilon_3 + \epsilon_1^*)$ and $s(P') \leqslant s(KP') + \epsilon_3$; hence $s(P'_j) + s(P'_m) \leqslant k_5 + (3\epsilon_3 + \epsilon_1^*)$. Similarly $s(Q'_j) + s(Q'_m) \leqslant k_5 + (3\epsilon_3 + \epsilon_1^*)$. By adding the last two inequalities and using 6.7A we obtain $2k_3^0 < 2k_5 + 2(3\epsilon_3 + \epsilon_1^*)$. But $k_3^0 > k_5 + 3\epsilon_3 + \epsilon_1^*$ (P.C. 6) so we have the desired contradiction.

6.14A. Proposition. *Let* $SX \underset{p}{\sim} TY$, *where* $p < n$, SX *and* TY *are* (p)*-bounded by* b_{25}, S, T *are maximal n-normal and*

$$s^n(S) \geqslant u_6 + u_4 + \epsilon_3, \qquad s^n(T) > u_6.$$

Then $\operatorname{Im} S = T$ *and* $\operatorname{Im} T = S$.

Proof. Since $s(S) > u_6$ we have $\operatorname{Im}(\operatorname{Im} S) = S$. Now if B is a weak image of S, then $s(B) > s(S) - u_4$ hence $\operatorname{Im} S = \overline{B}$ has n-size greater than $s(S) - u_4 - \epsilon_3 \geqslant u_6$. If $\operatorname{Im} S \neq T$ then T is left of $\operatorname{Im} S$, hence $\operatorname{Im} T$ is left of $\operatorname{Im}(\operatorname{Im} S)$, a contradiction. Therefore $\operatorname{Im} S = T$ and $S = \operatorname{Im}(\operatorname{Im} S) = \operatorname{Im} T$.

6.14B. Proposition. *Let* $A \underset{p}{\sim} B$, *where* $p < n$ *and* A, B *are* (p)-*bounded by* b_{25}. *Let* A, B *be subelements of* L_n. *Let* $s^n(A) > \operatorname{Max}(s_{25}, r_0 + r_{25})$. *Then* $|s^n(A) - s^n(B)| < r_{25} + \epsilon_3$.

Proof. By Axiom 25, B contains C such that $s(C) > s(A) - r_{25} > r_0$. Hence $s(C) \leqslant s(B) + \epsilon_3$ and $s(A) - r_{25} < s(B) + \epsilon_3$. If $s(B) \geqslant s(A)$ then we have symmetry and the result follows. Now let $s(B) < s(A)$. Then

$$0 < s(A) - s(B) < r_{25} + \epsilon_3$$

and the result follows.

6.15. Proposition. *Let* $X \underset{n}{\sim} Y$ *with divisions* $P_1, \ldots, P_m$ *and* $Q_1, \ldots, Q_m$. *Let* X, Y *be* $((n-1))$-*bounded by* k_6 *and* n-*bounded by* k_7. *Let* $X \underset{n}{\sim} Y$ *induce* $A \underset{n-1}{\sim} B$ *and let* B *be a subelement of* L_n *with* $s^n(B) > u_8$. *For some* p *let* Q_p *have type* n. *Then* B *is not contained in* Q_p.

Proof. Assume $B \subset \overline{Q}_p$. Since Q'_p is disjoint from B we may assume without loss of generality that it is right of B and that X, Y begin with A, B respectively. Now we have say $P'^{-1}_p X'^{-1}_{p-1} Q'_p X'_p \in \mathsf{L}_n$

$$P_p \equiv L P'_p M, \quad Q_p \equiv N Q'_p O, \quad X'^{-1}_{p-1} \equiv L^{-1} X^{-1}_{p-1} N, \quad X'_p \equiv O X_p M^{-1}.$$

Now (In Y, $Q'_p) \equiv Q_1 \cdot \cdots \cdot Q_{p-1} \cdot N Q'_p \equiv B U Q'_p$ for some U and

$BUQ'_pS \in \mathsf{L}_n$ for some S. Note that none of $Q_1, ..., Q_{p-1}$ have type n by 6.14. By 5.25, $SBU \underset{n-1}{=} X'_p P'^{-1}_p X'^{-1}_{p-1}$. Also in G_{n-1}

$$(\text{In}\, X, P_{p-1}) = X_0^{-1}(\text{In}\, Y, Q_{p-1})X_{p-1} = X_0^{-1}BU \cdot N^{-1}X_{p-1}$$

Hence $Q'_pSX_0(\text{In}\, X, P_{p-1})LP'_pX'^{-1}_pQ'^{-1}_p =$
$Q'_p(SBU)(N^{-1}X_{p-1} \cdot L)P'_pX'^{-1}_pQ'^{-1}_p = I$.

Now apply Axiom 24 to $P'_pX'^{-1}_pQ'^{-1}_p$; this is possible since $k_3^0 < s(P'_p) + s(Q'_p) \leqslant k_7 + s(Q'_p)$, so $s(Q'_p) > k_3^0 - k_7 \geqslant s_{24}$ (P.C. 1) and since P'_p, Q'_p are $((n-1))$-bounded by $k_6 \leqslant b_{24}$ (P.C. 2). Hence there exists $P'^{\mathrm{L}}_pX''Q'^{-1\mathrm{R}}$ equal to $P'_pX'^{-1}_pQ'^{-1}_p$ in G_{n-1} and $((n-1))$-bounded by $k_6 + q_{24}$ and $s(P'^{\mathrm{L}}_p) > s(P'_p) - r_{24} > k_3^0 - k_7 - r_{24} > r'_0$ (P.C. 3).

Let $P \equiv (\text{In}\, X, P_{p-1}) \cdot L$, $Q \equiv (\text{In}\, Y, Q_{p-1}) \cdot N$. Then $P \equiv AP'$, $Q \equiv BQ'$ say. Now T exists such that X_0, T are end words (cf. Axiom 31A) for A, B; hence $A = X_0^{-1} \cdot B \cdot T$. By Axiom 31B, since $s(B) > u_8 \geqslant s_{31}$ (P.C. 4) and $k_6 \leqslant b_{31}$ (P.C. 5) there exists $B^{\mathrm{L}}FA^{\mathrm{R}}$ $((n-1))$-bounded by $k_6 + q_{31}$ and equal to $X_0 \cdot A$ in G_{n-1}. A^{R} is not a subelement of $\mathsf{L}_1 \cup \cdots \cup \mathsf{L}_{n-1}$ and $s(B^{\mathrm{L}}) > s(B) - r_{31} > u_8 - r_{31} \geqslant s_{24}$ (P.C. 6).

Hence $I = Q'_pSX_0PP'^{\mathrm{L}}_pX''_pQ'^{-1\mathrm{R}}_p = Q'_pSB^{\mathrm{L}}FA^{\mathrm{R}}P'P'^{\mathrm{L}}_pX''_pQ'^{-1\mathrm{R}}_p$. By Axiom 24 there exists $Q'^{\mathrm{L}}_pS'B^{\mathrm{LR}}$ equal to Q'_pSB^{L} in G_{n-1} and $((n-1))$-bounded by $k_6 + q_{24}$; also

$$s(B^{\mathrm{LR}}) > s(B^{\mathrm{L}}) - r_{24} > u_8 - r_{31} - r_{24} > r'_0 \quad \text{(P.C. 7)}$$

Thus

$$I = \underline{Q'^{\mathrm{L}}_pS'B^{\mathrm{LR}}}\,\underline{FA^{\mathrm{R}}P'}\,\underline{P'^{\mathrm{L}}_pX''_pQ'^{-1\mathrm{R}}_p}$$

where the right side is $((n-1))$-bounded by $\text{Max}(k_6 + q_{24}, k_6 + q_{31}) \leqslant b_\infty$ (P.C. 8). This is a contradiction by Axiom 10.

6.16. Definition. Let $A \subset X \underset{n}{\sim} Y$ with divisions $P_1, ..., P_k$ and $Q_1, ..., Q_k$. Let X, Y be $((n-1))$-bounded by k_8 and n-bounded by k_9. Let A be maximal n-normal in X, $s^n(A) > u_9$. We define $\text{Im}'\,A$ to be a certain maximal n-normal of Y as follows.

Case (a). $(\exists i) P_i$ has type n, $A = \overline{P_i}$.

Then i is unique by 6.14 since $k_8 \leqslant k_4$ (P.C. 1) and $k_9 \leqslant k_5$ (P.C. 2). Put $\mathrm{Im}' A = \overline{Q_i}$.

Case (b). Not Case (a).

Suppose $J \subset A \cap P_j'$ for some j, where P_j' has type t_j. If $t_j < n$, then $s^n(J) \leqslant r_0' < r_0 + \epsilon_3$. If $t_j = n$ then $s^n(J) \leqslant r_0 + \epsilon_3$, otherwise J would be normal and $A = \overline{P_j}$. Similarly to 6.8 we find $A \equiv A' A^0 A''$ and $I \not\equiv A^0 \underset{n-1}{\sim} Y^M$, $s(A')$, $s(A'') \leqslant (r_0 + \epsilon_3) + \epsilon_2 + 2/e$. Hence $s(A) < s(A^0) + \alpha$ where

$$\alpha = 2(r_0 + \epsilon_3 + \epsilon_2 + 2/e) + 4/e \leqslant u_9 - u_5 - \epsilon_3 \quad \text{(P.C. 3)}.$$

Thus $s(A^0) > s(A) - \alpha > u_9 - \alpha \geqslant u_5 + \epsilon_3 \geqslant u_5$. Since $k_8 \leqslant b_{25}$ (P.C. 4) we have by 6.11 $A^0 \underset{n-1}{\sim} Y^M \supset H$ where H is n-normal. (Im A^0 is the maximal normal of Y^M determined by H.) Put $\mathrm{Im}' A = \overline{H}$, the maximal normal of Y determined by H. For later note that A^0 is normal.

6.17. Note. With the same hypothesis as 6.16 suppose $X \underset{n}{\sim} Y$ induces $Z \underset{n-1}{\sim} T$. Let Case (b) apply and let $A^0 \subset Z$. Then

$$\mathrm{Im}' A = \overline{\mathrm{Im}(Z \cap A)}.$$

Proof. $(A \cap Z)^0$ is A^0. Since $A \cap Z \supset (A \cap Z)^0 \equiv A^0 \sim Y^M \supset H$, we have by 6.11 that $\mathrm{Im}(A \cap Z)$ is the maximal normal of T determined by H. Hence $\overline{\mathrm{Im}(A \cap Z)} = \overline{H} = \mathrm{Im}' A$.

6.18. With the same hypothesis $\mathrm{Im}' \overline{P_i} = \overline{Q_i}$ if $t_i = n$.

Proof. By 6.7A, $k_3^0 < s(P_i') + s(Q_i') < k_9 + s(P_i')$ so $s(P_i') > k_3^0 - k_9 \geqslant r_0$ (P.C. 1). Hence $s(\overline{P_i}) > k_3^0 - k_9 - \epsilon_3 \geqslant u_9$ (P.C. 2).

6.19. Proposition. *Let $A \subset X \underset{n}{\sim} Y$ where X, Y are $((n-1))$-bounded by k_8' and n-bounded by k_9', A is maximal n-normal and $s^n(A) > u_{10}$; then Im′(Im′ A) exists and equals A.*

Proof. The hypothesis of 6.16 holds $((k_8' \leqslant k_8, k_9' \leqslant k_9, u_{10} \geqslant u_9))$.

If Case (a) applies, then $A = \overline{P}_i$ $\mathrm{Im}'A = \overline{Q}_i$. By 6.18 $\mathrm{Im}'\overline{Q}_i = \overline{P}_i$ so the result follows.

Now let Case (b) apply. Then $A \supset A^0 \underset{n-1}{\sim} Y^M \supset H, \overline{H} = \mathrm{Im}'A$. Now $\mathrm{Im}'\overline{H}$ exists since $s(\overline{H}) > s(H) - \epsilon_3 > s(A^0) - u_4 - \epsilon_3 > u_{10} - \alpha - u_4 - \epsilon_3 > u_9$ (P.C. 1). We show Case (a) cannot apply to $\overline{H}$. Assume $\overline{H} = Q_i$ and $t_i = n$. Since $s(H) > u_{10} - \alpha - u_4 > u_1$ we have by 6.8 $H \supset H^0 \sim A^{0M}$ $s(H^0) > s(H) - u_1 > u_8$. Now use 6.15 $((k_8' \leqslant k_6, k_9' \leqslant k_7))$; we obtain that H^0 is not contained in any $\overline{Q}_p, Q_p$ of type n. But $H^0 \subset \overline{H} = \overline{Q}_i$, contradiction.

Thus Case (b) applies and $\overline{H} \supset \overline{H}^0 \underset{n-1}{\sim} C \supset D, \overline{D} = \mathrm{Im}'\overline{H} = \mathrm{Im}'\mathrm{Im}'A$. Now $\overline{H}^0$ and Y^M are nice and contain $\overline{H}^0$, hence $Z \underset{n-1}{\sim} T$, where $T \equiv \overline{H}^0 \cup Y^M$, $Z \equiv C \cup A^0$. Since $A^0 \subset Z$ and $\overline{H}^0 \subset T$ we have $\mathrm{Im}'A = \overline{\mathrm{Im}(Z \cap A)}$, $\mathrm{Im}'\overline{H} = \overline{\mathrm{Im}(T \cap \overline{H})}$. Now $A^0 > A - \alpha > u_{10} - \alpha > u_6 + \epsilon_3$. Now A^0 is normal and $A^0 \subset Z \cap A$, so $Z \cap A$ has size greater than u_6 and $\mathrm{Im}\mathrm{Im}\, Z \cap A = Z \cap A$. Now $\overline{H}^0 \subset \overline{H} \cap T \subset \overline{H}$ so $\overline{\overline{H} \cap T} = \overline{H} = \overline{\mathrm{Im}\, Z \cap A}$. Now $\overline{H} \cap T$, $\mathrm{Im}\, Z \cap A$ are maximal normal in T so $\overline{H} \cap T = \mathrm{Im}\, Z \cap A$. Thus

$$\overline{\mathrm{Im}(\overline{H} \cap T)} = \overline{Z \cap A} = \mathrm{Im}'\overline{H} = A.$$

6.20. Corollary. *With the same hypothesis if A is left of A' then $\mathrm{Im}'A$ is left of $\mathrm{Im}'A'$.*

Proof. Similar to 6.13.

6.20A. Informal Note. $\mathrm{Im}\,A$ is defined only if A is n-normal. $\mathrm{Im}\,A$ is unique. $\mathrm{Im}'A$ is defined only if A is maximal n-normal. $\mathrm{Im}'A$ is unique.

In general if A has a weak image this is not unique. The definition of weak image does not require A to be n-normal.

If $X \underset{p}{\sim} Y$, $p < n$ then $X \underset{n}{\sim} Y$. Thus $\mathrm{Im}A$ and $\mathrm{Im}'A$ may both exist, but if so they coincide.

6.21. Definition. (Cyclic version of 6.16). Let (X), (Y) be cyclic words $((n-1))$-bounded by k_8 and n-bounded by k_9. Let no C.A. of X or of Y be a subelement of $L_1 \cup \cdots \cup L_n$. Let $A \subset (X) \underset{n}{-} (Y)$;

hence some cyclic arrangements X', Y' of X, Y respectively satisfy $X' \underset{n}{=} Y'$ with divisions $P_1, \dots, P_m, Q_1, \dots, Q_m$. Let A be maximal n-normal with $s^n(A) > u_9$. We define $\mathrm{Im}'A$ to be a certain maximal n-normal subword of (Y) as follows; the definition depends on the choice of $X', Y', P_1, \dots, P_m, Q_1, \dots, Q_m$.

X'^2 (and Y'^2) are $((n-1))$-bounded by k_8 and n-bounded by k_9; for if N is a subelement of $L_1 \cup \cdots \cup L_n$ and $N \subset X'^2$ then $L(N) < L(X')$ since no C.A. of X is a subelement of $L_1 \cup \cdots \cup L_n$, hence N is contained in a C.A. of X.

Now $X'^2 \underset{n}{=} Y'^2$ with divisions

$$P_1, \dots, P_m, P_1, \dots, P_m \qquad Q_1, \dots, Q_m, Q_1, \dots, Q_m$$

except that if both P_1 and P_m are of type 0 then the two terms P_m, P_1 have to be replaced by the single term $P_m \cdot P_1$ and similarly for Q_m, Q_1.

Also A is maximal normal in X'^2; let B be $\mathrm{Im}'A$ in the sense of 6.16. Thus B is maximal normal in Y'^2. As before $L(B) < L(Y')$ so $B \subset (Y)$. Define $\mathrm{Im}'A$ to be the maximal normal subword of (Y) determined by B.

6.21A. Note. We may now delete Axiom 26′ provided that, from now on, we interpret Im U of Axiom 26′ to be $\mathrm{Im}'U$ in the sense of 6.21. Thus we identify $b'_{26'}, b_{26'}, c_{26}$ with k_9, k_8, u_9 respectively.

6.22. Proposition. *Let $X \underset{n}{\sim} Y$ with divisions $P_1, \dots, P_m$ and $Q_1, \dots, Q_m$. Let X, Y be $((n-1))$-bounded by k_6 and n-bounded by k_7. Let some P_i be of type n; we may write $\overline{P}_i \equiv EP_i'F$. Then $s^n(E) \leqslant v_1$, where $v_1 = (r_0 + \epsilon_3) + 2\epsilon_2 + u_8 + 6/e$; similarly for $s^n(F)$.*

Proof. The result is true if $L(E) \leqslant 2$, so assume $L(E) > 2$.

Note first that if w is a subword of X of unit length and if R.E.P. w is left nice, then either L.E.P. w = L.E.P. X or one of the two end points of w is right nice.

Now R.E.P. E is left nice so $E \equiv E_1 J$ say, where $L(J) \leqslant 1$ and R.E.P. E_1 is right nice. If L.E.P. E_1 is right nice then $E_1 \equiv E_2 E_3$

say, where $L(E_2) \leqslant 1$ and L.E.P. E_3 is left nice and E_3 is nice; this conclusion is clearly also true when L.E.P. E_1 is left nice. By 6.15 we have $s(E_3) \leqslant u_8$, so $s(E) \leqslant 2\epsilon_2 + u_8 + 4/e$.

Finally suppose L.E.P. E_1 is neither left nice nor right nice. Then it is an interior point of some P'_k and clearly $k < i$. Let $A = P'_k \cap E$. Then $s^n(A) \leqslant r_0 + \epsilon_3$. Now $E \equiv AH$ say. The result follows if $L(H) \leqslant 1$ so assume $L(H) \geqslant 2$. R.E.P. A is right nice so $H \equiv H_1 H_2$, $L(H_1) \leqslant 1$ and L.E.P. H_2 is left nice. H_2 has the form $H_3 J$. Thus $E \equiv AH_1 H_3 J$ and H_3 is nice, so $s(H_3) \leqslant u_8$ and the result follows.

6.23. Proposition. *Let $A \subset X \underset{n}{\sim} Y$ with divisions $P_1, \ldots, P_k$ and $Q_1, \ldots, Q_k$. Let X, Y be $((n-1))$-bounded by k'_8 and n-bounded by k'_9. Let A be maximal n-normal in X with $s^n(A) > u_{10}$. Then $Im' A$ exists and in Case (a) of 6.16*

$$|s^n(A) + s^n(Im' A) - 1| \leqslant v_2,$$

where $v_2 = Max(4v_1 + 4\epsilon_3 + \epsilon_1^* + 8/e,\ 1 - k_3^0 + 2\epsilon_3)$; in Case (b)

$$|s^n(A) - s^n(Im' A)| < v_3,$$

where $v_3 = \epsilon_3 + u_4 + 2(r_0 + \epsilon_3 + \epsilon_2) + 8/e$.

Proof. By 6.19 $\mathrm{Im}' A$ exists and $\mathrm{Im}'(\mathrm{Im}' A) = A$.

First consider Case (a). Here $A = \bar{P}_i$, $\mathrm{Im}' A = \bar{Q}_i$. Write $\bar{P}_i = EP'_i F$. Then $s(\bar{P}_i) \leqslant 4/e + s(P'_i) + 2v_1$ since $u_{10} \geqslant u_8$. Similarly $s(\bar{Q}_i) \leqslant 4/e + s(Q'_i) + 2v_1$. We have $P'^{-1}_i X'^{-1}_{i-1} Q'_i X'_i \in L_n$. Now $s(P'_i) \leqslant s(P'^{-1}_i X'^{-1}_{i-1}) + \epsilon_3$ since by 6.7A $s(P'_i) > r_0 + \epsilon_3 > r_0$. Similarly, $s(Q'_i) \leqslant s(Q'_i X'_i) + \epsilon_3$. Also

$$s(P'^{-1}_i X'^{-1}_{i-1}) + s(Q'_i X'_i) \leqslant s(P'^{-1}_i X'^{-1}_{i-1} Q'_i X'_i) + 2\epsilon_3 + \epsilon_1^*$$
$$\leqslant 1 + 2\epsilon_3 + \epsilon_1^*.$$

Combining these inequalities we obtain

$$s(\bar{P}_i) + s(\bar{Q}_i) - 1 \leqslant 2(4/e + 2v_1) + 2\epsilon_3 + 2\epsilon_3 + \epsilon_1^*$$

Conversely $s(P_i') \leqslant s(\bar{P}_i) + \epsilon_3$, $s(Q_i') \leqslant s(\bar{Q}_i) + \epsilon_3$ and by 6.7A, since $k_8' \leqslant k_2$, $k_9' \leqslant k_2^0$, we have $k_3^0 < s(P_i') + s(Q_i')$. Hence

$$k_3^0 < s(\bar{P}_i) + s(\bar{Q}_i) + 2\epsilon_3$$

and the result follows.

Now consider Case (b). In the proof of 6.19 we have $\bar{H} = \mathrm{Im}' A$ and $s(\bar{H}) > u_{10} - \alpha - u_4 - \epsilon_3 = u_{10} - v_3$. Indeed we may replace u_{10} by $s(A)$ to obtain $s(\mathrm{Im}' A) > s(A) - v_3$. But since $\mathrm{Im}'(\mathrm{Im}' A) = A$ we have by symmetry $s(A) > s(\mathrm{Im}' A) - v_3$.

6.24. Proposition. *Let $AX \underset{n}{\sim} BY$, where both AX and BY are $((n-1))$-bounded by k_8' and n-bounded by k_9'. Let A and B be maximal n-normal of size $> u_{10}$. Then $Im'A = B$.*

Proof. $\mathrm{Im}'(\mathrm{Im}' A) = A) = A$ and $\mathrm{Im}'(\mathrm{Im}' B) = B$.

Case 1. $A = \bar{P}_i$ and $\mathrm{Im}' A = \bar{Q}_i$.

Here $k_3^0 < s(P_i') + s(Q_i') < k_9' + s(Q_i')$, so $s(Q_i') > k_3^0 - k_9' \geqslant u_{10} + \epsilon_3 > r_0$. Now $Q_i' \subset \bar{Q}_i$ so $s(\bar{Q}_i) \geqslant s(Q_i') - \epsilon_3 > u_{10}$. Obviously B is left of, or equal to, $\bar{Q}_i$ so by 6.20 $\mathrm{Im}' B$ is left of, or equal to, A. Hence $\mathrm{Im}' B = A$ and $B = \mathrm{Im}'(\mathrm{Im}' B) = \mathrm{Im}' A$.

Case 2. Case 1 does not hold.

Here $A \supset A^0 \underset{n-1}{\sim} A^{0\prime} \supset H$, $\bar{H} = \mathrm{Im}' A$. Since A is a left subword of AX, A^0 is a left subword of AX and $A^{0\prime}$ is a left subword of BY.

If $A^{0\prime} \subset B$ then $H \subset B$ so $\bar{H} = B$ as required. Now let $B \subset A^{0\prime}$. B does not have the form $\bar{Q}_j$ (Q_j of type n) otherwise $Q_j' \subset B \subset A^{0\prime}$ so $P_j' \subset A^0 \subset A$ and $\bar{P}_j = A$, a contradiction. Hence

$$B \supset B^0 \underset{n-1}{\sim} B^{0\prime} \supset K, \qquad \bar{K} = \mathrm{Im}' B.$$

But $B^0 \subset B \subset A^{0\prime}$ so $B^{0\prime} \subset A^0$. Thus $K \subset A^0 \subset A$. Hence $\bar{K} = A$, i.e., $\mathrm{Im}' B = A$. Thus $B = \mathrm{Im}'(\mathrm{Im}' B) = \mathrm{Im}' A$.

6.25. Proposition. *Let $R \subset Y \underset{n}{\sim} Z$, where Y, Z are $((n-1))$-bounded by k_8'' and n-bounded by k_9'' and $s(R) > u_{10}'$. Then $Im'(R) = S$ exists and, in case (b) of Definition 6.16, there exists R^M, S^M (subwords of R, S respectively) each of size $> s(R) - u_{10}''$ such that $R^M \underset{n-1}{\sim} S^M$.*

Proof. Put $k_8'' = \text{Min}(k_8', b_{25}, k_6)$, $k_9'' = \text{Min}(k_9', k_7)$.

By 6.19, $\text{Im}' R = S$ exists since $u_{10}' \geqslant u_{10}$. Let (b) of 6.16 hold. By 6.19 $R \supset R^0 \underset{n-1}{\sim} Z^M \supset M$, $\overline{M} = S$; also $s(R^0) > s(R) - \alpha$ and

$$s(M) > s(R^0) - u_4 > s(R) - \alpha - u_4 > u_{10}' - \alpha - u_4 > u_5$$

By 6.11

$$B \subset Y^M \underset{n-1}{\sim} M^0 \subset M$$

where $s(B) \supset s(M) - u_4 > s(R) - \alpha - 2u_4 > u_{10}' - \alpha - 2u_4 > r_0$.

Hence $s(Y^M) > s(R) - \alpha - 2u_4 - \epsilon_3$.

Now $Y^M \subset R$ and $M^0 \subset S$. The result follows since $u_{10}'' = \alpha + 2u_4 + \epsilon_3$.

6.25A. Note. Let $R \subset Y \underset{n}{\sim} Z$ and let $\text{Im}' R = S$ exist. Assume Case (a) of 6.16 applies. Put $R^M = P_i'$, $S^M = Q_i'$; then clearly $R^M \underset{n}{\sim} S^M$.

6.26. Proposition. *Let $p < n$. Let $H \subset A \underset{p}{\sim} B$ where A, B are (p)-bounded by $b \leqslant k_{52}$ and H is a subelement of L_{p+1} of size $> v_4$. Then a weak image J of H exists and there exist $U \subset H$, $V \subset J$ such that $U \underset{p}{\sim} V$ (end words E, K say) and there exists words $U^{\text{L}} E V^{\text{R}}$, $V^{\text{L}} F U^{\text{R}}$, (p)-bounded by $b + q_{31}$ such that $U \underset{p}{=} U^{\text{L}} E V^{\text{R}} \cdot K \underset{p}{=} E \cdot V^{\text{L}} F U^{\text{R}}$. Also U^{L}, U^{R}, V^{L}, V^{R} are not subelements of $\mathsf{L}_1 \cup \cdots \cup \mathsf{L}_p$.*

Proof. Since $v_4 > u_3 + u_4 > u_3$ and $k_{52} \leqslant b_{25}$ we have $H \supset H^0 \sim B^M \supset J$ and $s(J) > s(H) - u_4 > v_4 - u_4 > u_3$. Hence $L \subset U \sim V \subset J$ where V is J^0. Now $s(V) \geqslant s(J) - u_1 > v_4 - u_4 - u_1 > s_{31}$ and $k_{52} \leqslant b_{31}$ so there are words $U^{\text{L}} E V^{\text{R}}$, $V^{\text{L}} F U^{\text{R}}$ (p)-bounded by $b + q_{31}$, and $U^{\text{L}} E V^{\text{R}} = U \cdot K^{-1}$ (cf. Axiom 31B). V^{R} is not a subelement of $\mathsf{L}_1 \cup \cdots \cup \mathsf{L}_p$ and neither is U^{L}. Similarly for $V^{\text{L}} F U^{\text{R}}$.

6.27. Proposition. *Let $A \underset{p}{\sim} B$ where A, B are (p)-bounded by $b \leqslant k_{53}$ and $p \leqslant n$. Let S, T be distinct maximal p-normal subwords of A of size $> v_5$. Then*

1. $S' = Im' S$, $T' = Im' T$ exist

2. *$A \underset{p}{\sim} B$ induces $(S^{\mathrm{R}}, T^{\mathrm{L}}) \sim (S'^{\mathrm{R}}, T'^{\mathrm{L}})$ (end words E, K)*
3. *Words $S^{\mathrm{RL}}ET'^{\mathrm{LR}}$, $S'^{\mathrm{RL}}FT^{\mathrm{LR}}$ exist, (p)-bounded by $b + h_{31}$ and $(S^{\mathrm{R}}, T^{\mathrm{L}}) \underset{p}{=} S^{\mathrm{RL}}ET'^{\mathrm{LR}} \cdot K \underset{p}{=} E \cdot S'^{\mathrm{RL}}FT^{\mathrm{LR}}$*
4. *S^{RL}, T'^{LR}, $S!^{\mathrm{RL}}$, T^{LR} are p-normal.*

Proof. S' and T' exist since $k_{53} \leqslant k_8, k_9$ and $v_5 \geqslant u_9$. If S, S' satisfy (a) of 6.16 put $S^M \equiv P_i, S'^M \equiv Q_i$. Now $k_3^0 < s(P_i) + s(Q_i) < s(P_i) + k_{53}$, so $s(P_i) > k_3^0 - k_{53}$. Similarly $s(Q_i) > k_3^0 - k_{53}$.

If S, S' satisfy (b) then by 6.25, since $v_5 \geqslant u'_{10}$ and $k_{53} \leqslant k''_8, k''_9$, we have $S^M \sim S'^M$ where S^M, S'^M have size greater than $s(S) - u''_{10}$. Similarly $T^M \sim T'^M$. Hence $(S^{\mathrm{M}}, T^{\mathrm{M}}) \sim (S'^{\mathrm{M}}, T'^{\mathrm{M}})$ which we may also write $(S^{\mathrm{R}}, T^{\mathrm{L}}) \sim (S'^{\mathrm{R}}, T'^{\mathrm{L}})$; each of $S^{\mathrm{R}}, T^{\mathrm{L}}, S'^{\mathrm{R}}, T'^{\mathrm{L}}$ has size greater than $\mathrm{Min}(v_5 - u''_{10}, k_3^0 - k_{53}) - \epsilon_3 > g_{31}$. Axiom 31C is available since $k_{53} \leqslant f_{31}$ and the result follows.

6.28. Corollary. *In 6.26 or 6.27 let E', K' be end words for $A \underset{p}{\sim} B$. Then words $A \underset{p}{\sim} B$. Then words $A^{\mathrm{L}}EB^{\mathrm{R}}$, $B^{\mathrm{L}}FA^{\mathrm{R}}$ exist, (p)-bounded by*

$$b + q_{31} \quad (\textit{for } 6.26)$$

$$b + h_{31} \quad (\textit{for } 6.27),$$

and $A \underset{p}{=} A^{\mathrm{L}}EB^{\mathrm{R}} \cdot K' \underset{p}{=} E' \cdot B^{\mathrm{L}}FA^{\mathrm{R}}$.

Proof. Consider 6.27. Say $A \equiv A_1 S^{\mathrm{RL}} A_2, B \equiv B_1 T'^{\mathrm{LR}} B_2$. Put $A^{\mathrm{L}} \equiv A_1 S^{\mathrm{RL}}$, $B^{\mathrm{R}} \equiv T'^{\mathrm{LR}} B_2$. Then $A^{\mathrm{L}}EB^{\mathrm{R}}$ is $((p-1))$-bounded by $k_{53} + h_{31}$ since $S^{\mathrm{RL}}, T'^{\mathrm{LR}}$ are not subelements of $L_1 \cup \cdots \cup L_{p-1}$. If it were not p-bounded by $k_{53} + h_{31}$ it would contain a p-subelement N where N contains S^{RL} or T'^{LR} properly; say $N \supset S^{\mathrm{RL}}$. Now $S \equiv S_1 S^{\mathrm{RL}} S_2$ say. Let $S^{\mathrm{RL}}C$ be maximal normal in $S^{\mathrm{RL}}ET'^{\mathrm{LR}}$. Then $N \subset S_1 S^{\mathrm{RL}} C$, and

$$s(N) \leqslant s(S_1 S^{\mathrm{RL}} C) + \epsilon_3 \leqslant s(S_1) + s(S^{\mathrm{RL}}) + a_{31} + 2/e + \epsilon_3.$$

Now

$$s(S_1) + s(S^{\mathrm{RL}}) < s(S_1 S^{\mathrm{RL}}) + \mathrm{Max}(2\epsilon_3 + \epsilon_1^*, r_0 + 2\epsilon_3)$$

and

$$s(S_1 S^{\mathrm{RL}}) \leqslant \epsilon_3 + k_{53},$$

so

$$s(N) < a_{31} + 2/e + 2\epsilon_3 + k_{53} + \mathrm{Max}(2\epsilon_3 + \epsilon_1^*, r_0 + 2\epsilon_3) \leqslant k_{53} + h_{31}.$$

The required equations follow directly from the definition of end words (6.7). The argument for 6.26 is similar but easier.

6.29. Remark. If $A_1A_2A_3A_4A_5 \underset{p}{\sim} B_1B_2B_3B_4B_5$ induces $A_2 \underset{p}{\sim} B_2$ (end words E_2, K_2) and $A_4 \underset{p}{\sim} B_4$ (end words E_4, K_4) and if 6.28 applies to $A_2 \underset{p}{\sim} B_2$, $A_4 \underset{p}{\sim} B_4$ so that in particular $A_2 \underset{p}{\equiv} A_2^{\mathrm{L}}EB_2^{\mathrm{R}} \cdot K_2$, $A_4 \underset{p}{\equiv} E_4 \cdot B_4^{\mathrm{L}}F'A_4^{\mathrm{R}}$ then

$$A_1A_2A_3A_4A_5 \underset{p}{\equiv} A_1A_2^{\mathrm{L}}EB_2^{\mathrm{R}}B_3B_4^{\mathrm{L}}F'A_4^{\mathrm{R}}A_5.$$

Proof. Trivial (cf. 6.7).

§7

Informal Summary

We define L_{n+1}^k (cf. Axiom 12) for any positive integer k. We find that if $Y \in L_{n+1}^k$ then Y is conjugate in G_n to T^k for some $T \in L_{n+1}$ (cf. 5.10). We show any element T of L_{n+1} contains m subelements of L_n where $e \mid m$, $m > 0$. These determine km subwords of Y. If $Y \equiv AB$ then, roughly speaking, $s_{AB}(A)$ is defined to be a/m, where a is the number of these subwords contained in A (cf. Axiom 13).

7.1. Definition. By Axiom 11, $\underset{n}{\cong}$ is an equivalence relation on the set of cyclic words (n)-bounded by b_2. By 5.10 if $Z \in L_{n+1}$ then Z is a C.A.I of X^e for some X in C_{n+1}. By 5.11 Z is (n)-bounded by $str^+ \leqslant b_2$ (P.C. 1). Thus there is an induced equivalence relation $\underset{n}{\cong}$ on L_{n+1}. Hence the following is also an equivalence relation on L_{n+1}:

$$X \underset{n}{\cong} Y \quad \text{or} \quad X \underset{n}{\cong} Y^{-1}.$$

Choose a set S of representatives for this equivalence relation and define $Rep(L_{n+1})$ by

$$X \in Rep(L_{n+1}) \text{ if } X \text{ or } X^{-1} \in S$$

Thus $Rep(L_{n+1}) \subset L_{n+1}$. Clearly if $X \in Rep(L_{n+1})$ then $X^{-1} \in Rep(L_{n+1})$.

If $Z, Z' \in Rep(L_{n+1})$ and $Z \underset{n}{\cong} Z'$ we have $Z \equiv Z'^{\pm 1}$ (in § 13 we shall prove $Z \equiv Z'$); for if Z^{ϵ} and $Z'^{\epsilon'} \in S$ where $\epsilon, \epsilon' = \pm 1$ then $Z^{\epsilon} \cong (Z'^{\epsilon'})^{\epsilon' \epsilon}$ hence $Z^{\epsilon} \equiv Z'^{\epsilon'}$ so $Z \equiv Z'^{\epsilon' \epsilon}$. (If Z, Z' are viewed as linear words this conclusion reads: Z is a C.A.I. of Z'.)

Put Y in the set L^k_{n+1} if there exist $R, Z_0, Z_1, \ldots, Z_n$ such that $R \in Rep(L_{n+1})$, $k \geqslant 1$, $Z_n \equiv (R^k)$, $Z_0 \equiv (Y)$ and

$$Z_n \underset{n}{\rightarrow} Z_{n-1} \underset{n-1}{\rightarrow} \cdots \underset{1}{\rightarrow} Z_0.$$

Write L_{n+1} for L^1_{n+1}.

Note that $Z_i \in \mathsf{L}^k_{n+1}$ $(i = 0, 1, \ldots, n)$; this is trivial for $i = 0$ and if $i > 0$ we have $Z_n \underset{n}{\rightarrow} \cdots \underset{i+1}{\rightarrow} Z_i$ or $i = n$; now Z_i is (i)-bounded by $s_0 \leqslant \text{Min}(a_{28}, b_{28})$ (P.C.) and by Axiom 8 $(Z_i) \underset{i}{\cong} (Z_i)$; by Axiom 28 $(Z_i) \underset{i}{\rightarrow} \cdots \underset{1}{\rightarrow} (Z_i)$. Hence $Z_n \underset{n}{\rightarrow} \cdots \underset{i+1}{\rightarrow} Z_i \underset{i}{\rightarrow} \cdots \underset{1}{\rightarrow} Z_i$, as required.

7.2. Proposition. *L^k_{n+1} is contained in L^k_{n+1}, hence $L_{n+1} \subset \mathsf{L}_{n+1}$.*

Proof. Let $Y \in L^k_{n+1}$ so that $Y = T^k$, $T \in L_{n+1}$. Let R be the representative of T so that $(R) \underset{n}{\cong} (T)$. By Axiom 11, $(R^k) \underset{n}{\cong} (T^k)$. By Axiom 28 and since $str^+ \leqslant \text{Min}(a_{28}, b_{28})$ (P.C. 1), $(R^k) \rightarrow \cdots \rightarrow (T^k)$, hence $Y \in \mathsf{L}^k_{n+1}$.

7.3. Proposition. *Let $Z \in L_{n+1}$ where $n \geqslant 1$. Then either*

(i) (Z) contains ke distinct subwords, each maximal n-normal of size greater than r'' (where r'' is a new parameter) for some $k \geqslant 1$, or

(ii) (Z) contains e disjoint n-subelements each of size greater than u_{11}.

Proof. Suppose $(Z) \equiv (X^e)$ contains R and $s^n(R) \geqslant 2(r_0 + \epsilon_3) + 2/e$. Then by 5.28 since $str^+ \leqslant h_0$ (P.C. 1) we have $L(R) < 2L(X)$. Since R is n-normal, and so is any translate of R and since $e \geqslant 6$ (P.C. 2) we can find three disjoint translates; any C.A. of Z contains two so cannot be an n-subpowerelement. By Axiom 22, R is contained in a unique maximal normal.

Case 1. (Z) contains a maximal n-normal R where $s(R) > r''$.

Since $r'' > 2(r_0 + \epsilon_3) + 2/e$ (P.C. 3) we have $L(R) < 2L(X)$. Consider the set of all such R. Note that kernels exist since $r'' > 2(r_0 + \epsilon_3) + 4/e$ (P.C. 4). The set is closed to translation, hence the number of elements is a non-zero multiple of e.

Case 2. Not Case 1.

Some C.A.I. of X, say Y satisfies $Y \in C_{n+1} \subset J_n \subset J_{n-1}$. By 5.8 $Y >_{n-1} Z' \in C_n$. By Axiom 29 some $\delta_f(Y)$ contains a subelement of L_n of size greater than q_0; for if $Y^s \underset{n-1}{=} U^{-1}Z'^t U$ then $Y^{se} \underset{n-1}{=} U^{-1}Z'^{te} U \underset{n}{=} 1$, contrary to 5.12 since $Y \in J_n$. Thus for some C.A. X' of X, $X'^u \equiv JK$ where J is a subelement of L_n of size greater than q_0 and $u = \{fe\}$ (the least integer $\geqslant fe$). Thus $J \equiv X'^v X'^{\mathsf{L}}$ where $v \leqslant u - 1$. Hence $q_0 < vs(X') + s(X'^{\mathsf{L}}) + 2v/e$. Therefore $s(X')$ or $s(X'^{\mathsf{L}})$ is greater than $(q_0 - 2v/e)/(v+1) \geqslant (q_0 - 2v/e)/u \geqslant q_0/u - 2/e > q_0/(fe+1) - 2/e \geqslant u_{11}$ (P.C. 5). Since $q_0 > 2f$ we have $q_0 - 2v/e > 0$.

7.4. Proposition. *Let $Y_1, \ldots, Y_b$ be disjoint k-subelements contained in U and let each be of size greater than u_{12}. Let $k < n$ and let $U \underset{k-1}{\sim} V$, where U, V are $((k-1))$-bounded by k_{10}. For some i let $J \subset (Y_{i+1}, Y_{i+d})$ where J is a subelement of L_{k+1}, $s^{k+1}(J) > u_{13}$. Then weak images $Y'_1, \ldots, Y'_b$ exist and (Y'_{i-1}, Y'_{i+d+2}) contains J' where $s^{k+1}(J') > s(J) - u_4$, provided $1 \leqslant i-1 \leqslant i+d+2 \leqslant b$. If (Y'_r, Y'_s) contains K where $s^k(K) > u_{14}$ then (Y_r, Y_s) contains K' where $s^k(K') > s^k(K) - u_4$.*

Proof. Since $u_{12} \geqslant u_3$, $k_{10} \leqslant b_{25}$ (P.C. 1, 2) we have by 6.10 $Y_v \supset Y_v^0 \sim Y_v^{0\prime} \supset Y'_v$, $s(Y'_v) > s(Y_v) - u_4$. Now

$$J \subset (Y_{i+1}, Y_{i+d}) \subset (Y_i^0, Y_{i+d+1}^0) \underset{k-1}{\sim} (Y_i^{0\prime}, Y_{i+d+1}^{0\prime}) \equiv L, \text{ say}$$

By 6.10 since $u_{13} \geqslant u_3$ (P.C. 3), L contains J', $s(J') > s(J) - u_4$. But $J' \subset L \subset (Y'_{i-1}, Y'_{i+d+2})$. Lastly

$$K \subset (Y'_r, Y'_s) \subset (Y_r^{0\prime}, Y_s^{0\prime}) \underset{k-1}{\sim} (Y_r^0, Y_s^0) \subset (Y_r, Y_s);$$

using 6.10 K' exists since $u_{14} \geqslant u_3$ (P.C. 4).

7.5. Corollary. *Let $Y_1, \ldots, Y_b$ be disjoint k-subelements contained in U each of size greater than $x \geqslant u_{12}$. Let $k < n$ and let $U \underset{k}{\sim} V$ where U, V are ((k – 1))-bounded by k_{10} and k-bounded by k_{11}. For $i = 0, 1, \ldots, b - d$ let (Y_{i+1}, Y_{i+d}) contain a subelement J_i of L_{k+1}, $s^{k+1}(J_i) > y \geqslant u_{13}$. Let $W \equiv (Y_1, Y_b)$ not contain a k-subelement of size greater than $z \leqslant u_{15}$. Then weak images Y'_v of Y_v exist ($v = d + 2, d + 3, \ldots, b - d - 1$), $s(Y'_v) > x - u_4$, and any $d + 4$ consecutive Y'_v generate a word containing a subelement of L_{k+1} of size $> y - u_4$. Finally (Y'_{d+2}, Y'_{b-d-1}) does not contain a k-subelement of size greater than $z + u_4$, provided $z \geqslant u_{14} - u_4$.*

Proof. (Y_1, Y_d) is not a subelement of $\mathsf{L}_1 \cup \cdots \cup \mathsf{L}_k$ otherwise J_0 would be, contrary to $u_{13} \geqslant r'_0$ (P.C. 1). Clearly $L(Y_{d+1}) > 1$ since $u_{12} \geqslant \epsilon_2$ (P.C. 2). Hence $W^0 \supset (Y_{d+2}, Y_{b-d-1})$, $W^0 \underset{k}{\sim} V^M$. We next prove $W^0 \underset{k-1}{\sim} V^M$. If not, $P'_i \subset W^0$ where P_i has type k. Now V^M is k-bounded by k_{11}, so $k_3^0 < s(P'_i) + s\ Q'_i) < s\ P'_i) + k_{11}$. Thus $s(P'_i) > k_3^0 - k_{11} \geqslant u_{15} \geqslant z$ contrary to hypothesis. $((k_3^0 - k_{11} \geqslant u_{15}))$

We now have $(Y_{d+2}, Y_{b-d-1}) \subset W^0 \underset{k-1}{\sim} V^M$. By 7.4 Y'_v exists ($v = d + 2, \ldots, b - d - 1$) and any $d + 4$ consecutive Y'_v contain a $(k+1)$-subelement of size $> y - u_4$. If (Y'_{d+2}, Y'_{b-d-1}) contained a k-subelement of size greater than $z + u_4$ then W would contain a k-subelement of size $> z$. $((u_{14} - u_4 \leqslant u_{15}))$

7.5A. Note. See 11.5, 11.6 for an extension of 7.4, 7.5.

7.6. Remark. If $n(A, X, Y)$ temporarily denotes the nice image of $A \subset X$ in Y (c.f. 6.9) then if $n(A, X, Y) = A_1$, $n(A_1, X_1, Y_1) = A_2, \ldots$, $n(A_r, X_r, Y_r) = A_{r+1}$, we call A_{r+1} the *iterated nice image* of A in Y, sometimes we omit the word "iterated" if it is clear from the context that iterated is meant.

7.7. Proposition. *Let $A \subset X \underset{k}{\to} Y$ where $k < n$ and A is a subelement of L_{k+1}, $s^{k+1}(A) > u_{16}$. Then*

(i) A contains subelements $X_1, \ldots, X_a$ of L_k, and t exists, as in Axiom 30. In particular $|a/t - s^{k+1}(A)| < \epsilon_{30}$.

(ii) In Case (a) of Axiom 30, $X_4, \ldots, X_{a-3}$ have images in Y (in the sense of Axiom 26), each being maximal k-normal of size $> u_{17}$

In Case (b) $X_{j'+1}, \ldots, X_{a-j'}$ have images in Y for some j', these being disjoint k-subelements of size $> u_{18}$.

(iii) Let j be 3 in case (a) and the least value of j' in Case (b). Then $j/a < u_{19}$ and the iterated nice image of (X_1, X_a) hence of A in Y contains the images of $X_{j+1}, \ldots, X_{a-j}$.

Proof. (i) holds since $u_{16} \geqslant s_{30}$ (P.C. 1) and since X is (k)-bounded by $s_0 \leqslant b_{30}$ (P.C. 2).

Now assume (a) of Axiom 30 holds. Then $s^k(X_v) > r^* \geqslant w_{26}$ (P.C. 3). If $(Y) \equiv (X)$ there is nothing to do since $r^* \geqslant u_{17}$ (P.C. 4); thus we may assume that (Y) arises from (X) by simultaneous replacement of some maximal normals $M_1, \ldots, M_k$, each of size $> r'$. Since $s_0 \leqslant 1 - z_{26}$ (P.C. 5) $r' \geqslant z_{26}$ (P.C. 6), we have by Axiom 26 that there are cyclic words X_1', X_2' such that $(X) \underset{k-1}{\text{—}} X_1' \underset{k}{\text{—}} X_2' \underset{k}{\text{—}} (Y)$ where $X_1', X_2', (Y)$ are k-bounded by $1 - r' + t_{26}$ and $((k-1))$-bounded by $s_0 + s_{26}, s_0, s_0$ respectively. If X_v is an M_i, its image Y_v in the sense of Axiom 26 satisfies $s(Y_v) > 1 - s(X_v) - r_{26} > 1 - - s_0 - r_{26} \geqslant u_{17}$ (P.C. 7). If X_v is not an M_i its image Y_v satisfies

$$s(Y_v) > s(X_v) - r_{26} > r^* - r_{26} \geqslant u_{17} \quad \text{(P.C. 8)}.$$

Now assume (b) of Axiom 30 holds. Then $s^k(X_v) > q^*$, the X_v are disjoint and A contains no k-subelement of size $> c^*$. Again we may suppose that Y arises from X by simultaneous replacement of maximal normals $M_1, \ldots, M_k$ each of size $> r'$, since $q^* \geqslant u_{18}$ (P.C. 9).

The required j will be $3d + 15$, where d is the least integer $\geqslant t \cdot u_{20}$ where $u_{20} = u_{13} + 2u_4 + \epsilon_{30}'$ (P.C. 10). We show first that $a - 2j > 0$ and $d/t - \epsilon_{30}' \geqslant u_{13} + 2u_4$.

Since $t \geqslant e$ and $a > (s(A) - \epsilon_{30})t > (u_{16} - \epsilon_{30})t$ we have $j/a < (3(tu_{20} + 1) + 15)/(u_{16} - \epsilon_{30})t = (18/e + 3u_{20})/(u_{16} - \epsilon_{30}) < u_{19}$

(P.C. 11). But $u_{19} \leqslant \frac{1}{2}$ (P.C. 12) so $a - 2j > 0$. The remaining inequality is obvious.

We now want to apply 7.5 to $(X) \underset{k-1}{\longrightarrow} X_1'$, $X_1' \underset{k}{\longrightarrow} X_2'$ and $X_2' \underset{k}{\longrightarrow} (Y)$ in succession. The boundedness conditions hold since $s_0 + s_{26} \leqslant k_{10}$, $1 - r' + t_{26} \leqslant k_{11}$ (P.C. 13, 14). Next, $s(X_v) > q^* \geqslant u_{12} + 2U_4$ (P.C. 15) and $s^{k+1}(X_{i+1}, X_{i+d}) > d/t - \epsilon_{30}' \geqslant u_{13} + 2u_4$. Finally (X_1, X_a) does not contain a k-subelement of size $> c^* \leqslant u_{15} - 2u_4$ (P.C. 16) and also $c^* \geqslant u_{14} - u_4$.

Hence images Y_v of X_v in Y exist for $v = j+1, \ldots, a-j$ and $s(Y_v) > q^* - 3u_4 \geqslant u_{18}$ (P.C. 17).

7.7A. Corollary. $a - 2j > u_{20}' \geqslant 6$.

Proof. If $j = 3$ then $a > e(u_{16} - \epsilon_{30}) \geqslant 6 + u_{20}'$. If $j \neq 3$ then

$$\begin{aligned} a - 2j &> t(u_{16} - \epsilon_{30}) - 2(18 + 3tu_{20}) \\ &= t(u_{16} - 6u_{20} - \epsilon_{30}) - 36 \\ &\geqslant e(u_{16} - 6u_{20} - \epsilon_{30}) - 36 = u_{20}' \geqslant 6 \end{aligned}$$

7.7B. Corollary. $j/t < 3u_{20} + 18/e$.

Proof. If $j = 3$ then $j/t = 3/t \leqslant 3/e \leqslant 3u_{20} + 18/e$. If $j \neq 3$ then $j = 3d + 15 < 3(tu_{20} + 1) + 15$. Hence

$$j/t < 3u_{20} + 18/t \leqslant 3u_{20} + 18/e.$$

7.8. Let $W \in L_{n+1}^k$ where $n \geqslant 1$, so that

$$Z_n \underset{n}{\rightarrow} \cdots \underset{2}{\rightarrow} Z_1 \underset{1}{\rightarrow} Z_0 \equiv W$$

$Z_n \equiv (R^k)$, $R \in \mathrm{Rep}(L_{n+1})$. We shall describe certain subwords of $Z_n, \ldots, Z_2, Z_1$. First consider Z_n. By 7.3 Z_n contains m distinct n-subelements B_i $(i = 1, 2, \ldots, m)$ where $e \mid m$, say $m = he$ and either

1. the B_i are maximal normal of size $> r''$, or
2. the B_i are disjoint and of size $> u_{11}$.

Next we describe certain subwords A_i of Z_{n-1}.

If 2 holds Z_n does not contain a maximal normal of size $> r''$. Since $r' \geqslant r''$ (P.C. 1) we have $(Z_{n-1}) \equiv (Z_n)$; put $A_i = B_i$ $(i = 1, 2, \ldots, m)$. Now let 1 hold. If $(Z_{n-1}) \equiv (Z_n)$ again put $A_i = B_i$. If not then Z_{n-1} arises from Z_n by simultaneous replacement of some set of maximal normals of size $> r'$. Since $r'' \geqslant w_{26}$ (P.C. 2) B_i has image A_i in Z_{n-1}. Moreover $s(A_i) > u_{18}$ since $1 - s_0 - r_{26} \geqslant r'' - r_{26} \geqslant u_{18}$ (P.C. 3, 4).

Now let γ be an r-tuple of integers, $r \geqslant 1$, and assume A_γ is a subelement of L_{n-r+1} contained in Z_{n-r} and $s(A_\gamma) > u_{18}$. Then if $n - r > 0$ we have $A_\gamma \subset Z_{n-r} \underset{n-r}{\rightarrow} Z_{n-r-1}$. By 7.7 since $u_{18} \geqslant u_{16}$ (P.C. 5) A_γ contains subelements $B_{\gamma i}$ $(i = 1, 2, \ldots, a_\gamma)$ of L_{n-r}, and t_γ exists, with the following properties. Since $q^* \leqslant r^*$, $s(B_{\gamma i}) > q^*$; $B_{\gamma i}$ $(i = 1 + j_\gamma, \ldots, a_\gamma - j_\gamma)$ have images in Z_{n-r-1} say $A_{\gamma i}$; since $u_{17} \geqslant u_{18}$, $s(A_{\gamma i}) > u_{18}$; $j_\gamma / a_\gamma < u_{19}$
$|a_\gamma / t_\gamma - s(A_\gamma)| < \epsilon_{30}$. $\qquad ((q^* \leqslant r^*, u_{17} \geqslant u_{18}))$

We summarize the above in the following diagram.

$$
\begin{array}{cccccccccccc}
Z_n \underset{n}{\rightarrow} & Z_{n-1} \underset{n-1}{\rightarrow} & Z_{n-2} \underset{n-2}{\rightarrow} & Z_{n-3} \underset{n-3}{\rightarrow} & \cdots & Z_{n-r} \underset{n-r}{\rightarrow} & \cdots & Z_1 \underset{1}{\rightarrow} & Z_0 \\
B_i^n & A_i^n & & & & & & & \\
& B_{ij}^{n-1} & A_{ij}^{n-1} & & & & & & \\
& & B_{ijk}^{n-2} & A_{ijk}^{n-2} & & & & & \\
& & & & \cdots & & & & \\
& & & & & A_{(r)}^{n-r+1} & & & \\
& & & & & B_{(r+1)}^{n-r} & & & \\
& & & & & & \cdots & & \\
& & & & & & & A_{(n-1)}^{2} & \\
& & & & & & & B_{(n)}^{1} & A_{(n)}^{1}
\end{array}
$$

In this diagram, superscript i denotes subelement of L_i and (k) denotes a k-tuple.

Since $u_{11} \geqslant q^*$, $r'' \geqslant u_{11}$, $u_{11} \geqslant u_{18}$ (P.C. 6, 7, 8) all B's have size $> q^*$ and all A's have size $> u_{18}$.

Call $B_{\gamma i}$ *admissible* if $A_{\gamma i}$ exists, i.e., if $1 + j_\gamma \leqslant i \leqslant a_\gamma - j_\gamma$. Also call $B_1, \ldots, B_m$ admissible.

We refer to the above notation as the *A-B notation.*

7.9. Definition. By a *proper descendant* of any A_μ where μ is a t-tuple, $1 \leqslant t \leqslant n-1$, we mean any $A_{\mu\nu}$ or any admissible $B_{\mu\nu}$, where ν is an s-tuple for some s such that $1 \leqslant s$ and $s + t \leqslant n$.

By a *proper descendant* of any admissible B_μ we mean A_μ or any proper descendant of A_μ.

By a *descendant* of X we mean X or a proper descendant of X.

Let $AD_{n-r}(B_i)$ mean the subword of Z_{n-r} generated by all A-descendants of B_i in Z_{n-r}, where $n \geqslant r > 0$. $BD_{n-r}(B_i)$ is defined similarly if $n > r \geqslant 0$. The purpose of 7.9A–7.9C below is to prove:

Proposition. *If $n \geqslant r > 0$ then kernels exist for the collection of subwords $AD_{n-r}(B_i)$ $(i = 1, 2, \ldots, m)$ of Z_{n-r} and similarly kernels exist for $BD_{n-r}(B_i)$ $(i = 1, 2, \ldots, m)$ where $n > r \geqslant 0$.*

7.9A. Let $n > n - r > 0$ and consider the diagram

$$
\begin{array}{ll}
Z_{n-r} & \underset{n-r}{\rightarrow} Z_{n-r-1} \\
A_\gamma^{n-r+1} & \\
B_{\gamma i}^{n-r} & A_{\gamma i}^{n-r}
\end{array}
$$

where γ is a fixed r-tuple. Either $(Z_{n-r-1}) \equiv (Z_{n-r})$ or by Axiom 26 there are cyclic words X_1, X_2 such that

$$Z_{n-r} \underset{n-r-1}{\overline{\qquad}} X_1 \underset{n-r}{\overline{\qquad}} X_2 \underset{n-r}{\overline{\qquad}} Z_{n-r-1}$$

By 7.7 the (iterated) nice image of A_γ in Z_{n-r-1} contains all $A_{\gamma i}$ and A_γ^0 (taken with respect to $Z_{n-r} \underset{n-r-1}{\overline{\qquad}} X_1$) contains all admissible $B_{\gamma i}$.

Now let δ be another r-tuple, let A_γ, A_δ be disjoint and let $B_{\gamma i}$, $B_{\delta j}$ be admissible. Then it follows that $B_{\gamma i} \cap B_{\delta j}$ and $A_{\gamma i} \cap A_{\delta j}$ are empty. Thus all descendants of A_γ are disjoint from all descendants of A.

7.9B. Proposition. *There exist c_i, d_i such that (B_{ic_i}, B_{id_i}) is disjoint from all A_j, $j \neq i$ where $B_{ic_i}, \ldots, B_{id_i}$ are admissible and $d_i + 1 - c_i \geqslant 7$.*

Proof. This is trivial if the A_i are disjoint so we may assume all A_i are maximal normal of size $> r'' - r_{26}$ (cf. 7.8). By 7.7 and Axiom 30, A_i contains $B_{i1}, \ldots, B_{ia}$, and t exists such that $|a/t - s(A_i)| < \epsilon_{30}$ and if $1 \leqslant w \leqslant a$ then $s(B_{il}, B_{iw}) > w/t - \epsilon'_{30}$.

We show that w can be chosen so that $a - 2w \geqslant 7$ and $w/t - \epsilon'_{30} \geqslant r_0 + \epsilon_3$. This will imply that (B_{i1}, B_{iw}) is normal and $(B_{i\ w+1}, B_{i\ a-w})$ is therefore disjoint from all A_j, $j \neq i$; since $B_{i\ j+1}, \ldots, B_{i\ a-j}$ are admissible and $a - 2j \geqslant 7$ for some j the required result follows.

Put $w = \{\alpha t\}$, the least integer $\geqslant \alpha t$, where $\alpha = r_0 + \epsilon_3 + \epsilon'_{30}$. Then $a - 2(w-1) > t(s(A_i) - \epsilon_{30} - 2\alpha) \geqslant 8$, since

$$(r'' - r_{26}) - \epsilon_{30} - 2(r_0 + \epsilon_3 + \epsilon'_{30}) \geqslant 8/e \qquad \text{(P.C. 1)}.$$

Thus $a - 2w \geqslant 7$.

7.9C. Kernels exist for the $BD_n(B_i)$, i.e., for the B_i, since the B_i are either disjoint or maximal normal of size $> r'' \geqslant 2(r_0 + \epsilon_3) + 4/e$ (P.C. 2).

Kernels exist for the $AD_{n-1}(B_i) = A_i$ since the A_i are disjoint or maximal normal of size $> r'' - r_{26} \geqslant 2(r_0 + \epsilon_3) + 4/e$ (P.C. 3). This can also be seen from 7.9B.

Kernels exist for the $BD_{n-1}(B_i)$ since by 7.9B, (B_{ic_i}, B_{id_i}) is contained in $BD_{n-1}(B_i)$ and disjoint from A_j, $j \neq i$, hence from $BD_{n-1}(B_j)$.

Now consider the $AD_{n-2}(B_i)$. In 7.7 we showed that, in Case (a), the nice image of (X_1, X_a) in Y contains the images of $X_4, \ldots, X_{a-3}$. By the same argument we have in the notation of 7.9B that the nice image of (B_{ic_i}, B_{id_i}) in Z_{n-2} contains A_{iq} if $c_i + 3 \leqslant q \leqslant d_i - 3$; there is at least one such q. Hence if $j \neq i$, A_{iq} is disjoint from the nice image of A_j in Z_{n-2}, which contains $AD_{n-2}(B_j)$. But $A_{iq} \subset AD_{n-2}(B_i)$.

Finally kernels exist in all remaining cases since by 7.9A all descendants of A_{iq} are disjoint from all descendants of A_{js} where $j \neq i$ and s is arbitrary.

7.10. We next want to define $s_{XY}(X)$ where $XY \in \mathsf{L}^k_{n+1}$ but first we make a few preliminary remarks.

Let R_i mean $AD_0(B_i)$ $(i = 1, 2, ..., m)$. Thus the R_i are subwords of Z_0 for which kernels exist. Let H_i be the kernel of R_i. Then $Z_0 \equiv (L)$, $L \equiv H_1 E_1 H_2 E_2 \cdots H_m E_m$ for some words E_i. If $L \equiv RST$ for some R, S, T we can consider the number of R_i contained in S and the number of R_i contained in TR.

Let $(XY) \equiv Z_0$ for some X, Y. Then C, D exist such that $L \equiv CD$, $XY \equiv DC$, and we may consider the number x of R_i contained in X. x is determined once C, D are chosen, but x depends on this choice.

Say $R_{i+1}, ..., R_{i+x}$ are contained in X; then X contains the kernels of each of these, so $L(X) \geqslant x$.

7.11. Definition. If $XY \in \mathsf{L}^k_{n+1}$ we define $s_{XY}(X)$ and $s^{n+1}(X)$ as follows. There exist $R \in \mathrm{Rep}(L_{n+1})$, $Z_0, Z_1, ..., Z_n$ where $Z_n \equiv (R^k)$, $Z_0 \equiv (XY)$, $Z_n \underset{n}{\rightarrow} \cdots \underset{1}{\rightarrow} Z_0$. There exist A's and B's as in 7.8–7.9C. There exist R_i $(i = 1, 2, ..., m)$, C, D and x as in 7.10. Put $d = m/k$. Since ke divides m, d is an integer divisible by e.

Put $s_{XY}(X) = \sup x/d$.

Put $s^{n+1}(X) = \sup s_{XY}(X)$, taken over all Y such that $(\exists k)\, XY \in \mathsf{L}^k_{n+1}$.

By 7.10 we have $s_{XY}(X) \leqslant L(X)/e$, $s^{n+1}(X) \leqslant L(X)/e$.

7.12. Proposition. *Let* $XY \in \mathsf{L}^k_{n+1}$. *Then*

1. $0 \leqslant s_{XY}(X) \leqslant k$,
2. *If* X *is* I *then* $s_{XY}(X) = 0$,
3. *if* Y *is* I *then* $s_{XY}(X) \geqslant k - 2/e$,
4. $s_{XY}(X) = s_{X^{-1}Y^{-1}}(X^{-1})$,
5. $L(X) = 1$ *implies* $s_{XY}(X) = 0 \leqslant \epsilon_2$.

We also have

6. $s_{BCDA}(B) \leqslant s_{ABCD}(ABC)$,
7. $s_{ABC}(AB) \leqslant s_{ABC}(A) + s_{BCA}(B) + 2/e$.

Proof. These results follow from Definition 7.11 in a straightforward manner.

7.13. Proposition. (i) *If* $Y \in \mathsf{L}^k_{n+1}$ *then* Y *is conjugate in* G_n *to an element of* L^k_{n+1}

(ii) $[L_1 \cup \cdots \cup L_{n+1}] = [\mathsf{L}_1 \cup \cdots \cup \mathsf{L}_{n+1}]$ *where* $[\cdots]$ *denotes the normal subgroup of* Π *generated by* $\cdots$.

Proof. (i) If $X \underset{i}{—} Z$ then X is conjugate to Z in G_i. Hence if $X \underset{i}{\rightarrow} Z$, where $1 \leqslant i \leqslant n$ then any linearizations of X, Z are conjugate in G_i. Hence, using 7.1, if $Y \in \mathsf{L}^k_{n+1}$ then Y is conjugate in G_n to R^k for some $R \in L_{n+1}$.

(ii) This follows from 7.2 and the special case $k = 1$ of (i).

7.14. Proposition. *Let* $Y \in \mathsf{L}^k_r \cap \mathsf{L}^h_s$, *where* $1 \leqslant r \leqslant n + 1$ *and* $1 \leqslant s \leqslant n + 1$. *Then* $r = s$.

N.B. It will be proved later that $k = h$ also.

Proof. The result is true if $r \leqslant n$ and $s \leqslant n$ by Axiom 12, so assume $r = n + 1$. If $s \leqslant n$ then by 7.13 (i) we may assume that Y is conjugate in G_{s-1} to an element Z of L^h_s. Since $Z \underset{s}{=} I$ we have $Y = I$ hence $Y \underset{n}{=} I$. But Y is conjugate in G_n to an element T of $L^{k^s}_{n+1}$, so $T \underset{n}{=} I$, in contradiction to 5.12. Thus $s = n + 1 = r$.

7.15. Proposition. *Let* $s^{n+1}(X) \geqslant \frac{1}{2}$. *Then* X *contains a subword* UUU *where* U *is E.R.*

Proof. By 7.11 there exist Y, k such that $XY \in \mathsf{L}^k_{n+1}$, and there exists $R, Z_0, \ldots$ such that $x/d > \frac{1}{4}$. Hence $x \neq 0$, so X contains at least one R_i. Therefore X contains A_γ for some n-tuple γ. Now A_γ is a subelement of $\mathsf{L}_1 = L_1$ of size $> u_{18}$; say $A_\gamma U \equiv X^e \in L_1$ where some C.A.I. of X is in C_1, so X is E.R. Now

$$L(A_\gamma) > u_{18}\, eL(X) \geqslant 3L(X) \text{ since } u_{18} \geqslant 3/e \qquad ((u_{18} \geqslant 3/e))$$

7.16. Proposition. *In the notation of* 7.8 *consider* $A_{\gamma 1}, A_{\gamma t}$ *where* γ *is an* $(r-1)$*-tuple,* $1 \leqslant r \leqslant n - 1$ *and* $t \geqslant 2$. *Let* N *be an integer* *less than* u_{34}. *Then for* $i = 1, 2, ..., N$ $A_{\gamma 1i}$ *is left of and disjoint from all* $A_{\gamma tj}$.

Proof. This is trivial if $A_{\gamma 1}, A_{\gamma t}$ are disjoint so we may assume that $t = 2$ and $A_{\gamma 1}, A_{\gamma t}$ overlap.

If $r = 1$ we have to prove that A_{1i} is left of and disjoint from all A_{2j} $(i = 1, 2, ..., N)$. By 7.9B, 7.9C, if $c_1 + 3 \leqslant q \leqslant d_1 - 3$ then A_{1q} is disjoint from the nice image of A_2 in Z_{n-2} hence is disjoint from and left of all A_{2j}. Now by 7.9B, $(\exists a, t, w)$ $a/t > s(A_1) - \epsilon_{30}$, $d_1 = a - w$, $a - 2w \geqslant 7$, $t \geqslant e$. Hence

$$d_1 - 3 \geqslant \tfrac{1}{2}a + \tfrac{1}{2} > \tfrac{1}{2}(1 + e(r'' - r_{26} - \epsilon_{30})) \geqslant u_{34} - 1 > N - 1,$$

and the result follows.

If $r > 1$ then as part of the diagram in 7.8 we have

$$\begin{array}{ll} A_\gamma & \\ & \\ B_{\gamma i} & A_{\gamma i} \end{array}$$

and by 7.7 the $A_{\gamma i}$ have size $> u_{17}$, otherwise $A_{\gamma 1}, A_{\gamma 2}$ would be disjoint. The argument of 7.9B and the relevant part of 7.9C remains valid since $\frac{1}{2}(1 + e(u_{17} - \epsilon_{30})) \geqslant u_{34} - 1 > N - 1$ and (instead of P.C. 1 in 7.9B) $u_{17} - \epsilon_{30} - 2(r_0 + \epsilon_3 + \epsilon'_{30}) \geqslant 8/e$.

7.16A. Corollary. *Let* λ_r *be the r-tuple* $(1, 1, ..., 1)$. *Let* ρ *be an r-tuple,* $\rho \neq \lambda_r$. *Let* $1 \leqslant r \leqslant n - 1$. *Then for* $i = 1, 2, ..., N$ $A_{\lambda_r i}$ *is left of and disjoint from all* $A_{\rho j}$.

Proof. $\rho = \lambda_a s\sigma$ where $a \geqslant 0$, $s \geqslant 2$ and σ is an $(r - a - 1)$-tuple. Consider $A_{\lambda_a 1}, A_{\lambda_a s}$. If $1 \leqslant i \leqslant N$ then by 7.16 $A_{\lambda_a 1i}$ is left of and disjoint from all $A_{\lambda_a sj}$. If σ is not a 0-tuple $A_{\lambda_a 11}$ is left of and disjoint from $A_{\lambda_a sk}$, where k is the first term of σ; say $\sigma = k\sigma'$. For any $(r - a - 2)$-tuples α, β $A_{\lambda_a 11\alpha}$ is left of and disjoint from $A_{\lambda_a sk\beta}$. Taking $\alpha = A_{r-a-2}$, $\beta = \sigma'$ we have A_{λ_r} is left of and disjoint from A_ρ and the result follows (for arbitrary i).

If σ is a 0-tuple then $\rho = \lambda_a s$, $\lambda_r = \lambda_a 1$ and the result follows.

§8

Informal Summary

The relations $\overline{n}$ and $\underset{n}{\sim}$ are closely related to $\underset{n}{=}$ but they are not equivalence relations. The purpose of this section is to introduce the relation of n-almost equality, denoted by $\underset{n}{\hat{=}}$ which is an equivalence relation. If S, S' are n-normal and S is a left subword of S' then for any T, T' such that $ST \in L_n$, $S'T' \in L_n$ we have $ST = S'T'$ in G_{n-1} (cf. 5.25); because of this if XS and XS' are words (X arbitrary) then they are equivalent in some sense. Thus we consider words of the form XS where X is arbitrary and S is n-normal, calling these right n-objects, and we show how to define an equivalence relation for these, generalizing $\underset{n}{=}$. This idea extends to n-objects, which begin and end with an n-normal subword. It turns out that if $U \underset{n}{\sim} V$ and A, B are distinct maximal n-normal subwords of U then the n-objects (A, B) and $(\mathrm{Im}'A, \mathrm{Im}'B)$ are n-almost equal.

8.1. Definition. A *left n-preobject* is a word O together with subwords S, X, H, X' such that $O \equiv SX \equiv SHX'$ and

1. O is $((n-1))$-bounded by k_{12} and n-bounded by k_{13},
2. S is a subelement of $\mathrm{Rep}(L_n)$ and $s^n(S) > u_{21}$,$((u_{21} \geqslant r_0 + \epsilon_3))$
3. SH is maximal n-normal in O,
4. $s^n(H) < u_{22}$.

Note that S is normal.

8.2. Definition. Let $O_i \equiv S_iH_iX_i'$ $(i = 1, 2)$ be left n-preobjects. Temporarily, write $O_1 \subset' O_2$ if $X_1 \equiv X_2$ and $S_1 \equiv S_2^{\mathrm{R}}$; it follows that $H_1 \equiv H_2$, $X_1' \equiv X_2'$ (cf. 5.27). Call O_1 a *left n-adjustment* of O_2 if $O_1 \subset' O_2$ or $O_2 \subset' O_1$. $((k_{13} < h_0))$

8.2A. This is an equivalence relation; to prove transitivity, let O_1 be a left adjustment of O_2, and O_2 a left adjustment of O_3. The only non-trivial case is $O_2 \subset' O_1$ and $O_2 \subset' O_3$; then $S_1 \equiv AS_2$, $S_3 \equiv BS_2$ say. Now U, V exist such that $UAS_2, VBS_2 \in \mathrm{Rep}(L_n)$. By Axiom 33, since $s(S_2) > u_{21} \geqslant s_{33}$ (P.C. 1), $UA \equiv VB$. If say

$L(A) \leqslant L(B)$ then $B \equiv B_1 A$ so $S_1 \equiv S_3^R$ and $O_1 \subset' O_3$. Similarly if $L(A) \geqslant L(B)$.

8.3. Proposition. *Let* $AS_1X_1 \underset{n}{=} BS_2X_2$, *where each side is* $((n-1))$-*bounded by* k_{12} *and n-bounded by* k_{13} *and where* S_1X_1 *and* S_2X_2 *are left n-preobjects. Assume* A, B *do not contain an n-subelement of size* $> u_{23}$. *Then either*

1. $S_i \equiv S_i'N_iS_i''$, N_i *normal* $(i = 1, 2)$ *and* $N_1S_1''X_1 \underset{n}{=} N_2S_2''X_2$, $N_1 \equiv N_2$, or

2. $S_i \equiv S_i'N_iS_i''$, N_i *normal* $(i = 1, 2)$ *and J exists such that* $N_1S_1''X_1 \underset{n}{=} JN_2S_2''X_2$, *hence* $S_1'^{-1}A^{-1}\cdot BS_2' \underset{n}{=} J$, *and* $N_1^{-1}JN_2$ *is a subelement of* $Rep(L_n)$.

Note. A supplementary result is given in 8.5.

Proof. Only conditions 1, 2 of 8.1 will be used in this proof. By 6.2, AS_1X_1, BS_2X_2 have divisions $P_1, \ldots, P_m$ and $Q_1, \ldots, Q_m$ say. $((k_{12} \leqslant k_2,\ k_{13} \leqslant k_2^0))$

Case 1. For some P_i' of type n, S_1 and P_i' determine the same maximal normal subword of AS_1X_1: briefly $\bar{S}_1 = \bar{P}_i'$.

Thus $\bar{S}_1 \equiv EP_i'F$ say and $u_{21} < s(S_1) \leqslant s(\bar{S}_1) + \epsilon_3$.

If $\bar{Q}_i' \neq \bar{S}_2$ then $\bar{S}_2$ is left of $\bar{Q}_i'$, for if $\bar{Q}_i'$ were left of $\bar{S}_2$ then $s(\bar{Q}_i') \leqslant u_{23} + (r_0 + \epsilon_3) + 2/e \leqslant K_3^0 - k_{13} - \epsilon_3$ (P.C. 1), but $k_3^0 < s(P_i') + s(Q_i')$, so $s(Q_i') > k_3^0 - k_{13} > r_0$ (P.C. 2) and $s(\bar{Q}_i') > k_3^0 - k_{13} - \epsilon_3$, a contradiction. Now 6.18 and 6.20 are available since $k_{12} \leqslant k_8, k_8'$ and $k_{13} \leqslant k_9, k_9'$ (P.C. 3), $u_{21} - \epsilon_3 > u_{10}$ (P.C. 4), $k_3^0 - k_{13} - \epsilon_3 > u_{10}$ (P.C. 5); hence Im$'$ $\bar{S}_2$ is left of Im$'$ $\bar{Q}_i' = \bar{P}_i' = \bar{S}_1$. As above Im$'\,\bar{S}_2$ has size $\leqslant u_{23} + (r_0 + \epsilon_3) + 2/e = d$ say. But Im$'$ $\bar{S}_2$ has size greater than either $1 - s(\bar{S}_2) - v_2$ or $s(\bar{S}_2) - v_3$, hence in either case is greater than d because both $1 - k_{13} - v_2$, $u_{21} - \epsilon_3 - v_3 \geqslant u_{23} + (r_0 + \epsilon_3) + 2/e$ (P.C. 6, 7).

Thus $\bar{Q}_i' = \bar{S}_2$. Now consider $\bar{S}_1 \equiv EP_i'F$. Since $s(E) \leqslant v_1$ (cf. 6.22), $s_1 \cap E$ has size $\leqslant \text{Max}(v_1 + \epsilon_3, r_0)$, hence

$$s(S_1 \cap P_i') > \xi = u_{21} - 2\text{Max}(v_1 + \epsilon_3, r_0) - 4/e.$$

Similarly $s(S_2 \cap Q_i') > \xi \geqslant m_{33}$ (P.C. 8). Put $N_1 \equiv S_1 \cap P_i'$,

$N_2 \equiv S_2 \cap Q_i'$, $P_i' \equiv P_i'^{L} N_1 P_i'^{R}$, $Q_i' \equiv Q_i'^{L} N_2 Q_i'^{R}$. N_1, N_2 are normal since $\xi > r_0 + \epsilon_3$ (P.C. 9). Thus $N_1^{-1} P_i'^{L-1} X_{i-1}'^{-1} Q_i'^{L} N_2$ is a subelement of L_n. By Axiom 33″ there is a subelement $N_1^{-1} J N_2$ of $\mathrm{Rep}(L_n)$ equal to it in G_{n-1}. Put $S_i \equiv S_i' N_i S_i''$ $(i = 1, 2)$. Now from $P_i \cdot \cdots \cdot P_m = X_{i-1}^{-1} \cdot Q_i \cdot \cdots \cdot Q_m$ it follows easily that $N_1 S_1'' X_1 = J N_2 S_2'' X_2$.

Case 2. Not Case 1.

As in 6.16 we find $S_1 \equiv S_1' S_1^0 S_1''$, $s(S_1') \leqslant (r_0 + \epsilon_3) + \epsilon_2 + 2/e = \beta$ $s(S_1'') \leqslant \beta$, $s(S_1) \leqslant 4/e + 2\beta + s(S_1^0) = s(S_1^0) + \alpha$ and finally $S_1^0 \underset{n-1}{\sim} V \supset D$, where $s^n(D) > s(S_1^0) - u_4 > u_{21} - \alpha - u_4$. Similarly $S_2 \equiv S_2' S_2^0 S_2''$, $S_2^0 \underset{n-1}{\sim} W$. Put $N_1 \equiv S_1^0 \cap W$, $N_2 \equiv V \cap S_2^0$. We prove $I \not\equiv N_1 \underset{n-1}{\sim} N_2$ and either N_1 or N_2 has size $> s_{33}$; the result will follow by Axiom 33.

This is clear if $S_1^0 \subset W$ or $W \subset S_1^0$, since $s(S_1^0) > s(S_1) - \alpha > u_{21} - \alpha > s_{33}$ (P.C. 10).

There remains the case S_1^0 left of W, without loss of generality. Hence V is left of S_2^0 and $D \subset V \subset B S_2' S_2^0$. Now $s(D \cap B) \leqslant u_{23}$ and $s(D \cap S_2') \leqslant \mathrm{Max}(\beta + \epsilon_3, r_0) = \beta + \epsilon_3$. Thus $s(D) \leqslant s(D \cap S_2^0) + \eta$, where $\eta = u_{23} + (\beta + \epsilon_3) + 4/e$. Hence $s(D \cap S_2^0) > u_{21} - \alpha - u_4 - \eta > r_0$ (P.C.11) and $s(D \cap S_2^0) \leqslant s(V \cap S_2^0) + \epsilon_3$. Hence

$$s(V \cap S_2^0) > u_{21} - \alpha - u_4 - \eta - \epsilon_3 > s_{33} \quad \text{(P.C.12)}.$$

If $A \equiv B \equiv I$ note that S_i^0 is a left subword of S_i, hence N_i is a left subword of S_i $(i = 1, 2)$.

8.4. Given a left n-preobject SHX' we shall define a certain left adjustment of it.

There exists T such that $TS \in \mathrm{Rep}(L_n)$. Now $s(TS) \geqslant 1 - 2/e > r_0$ and TSH is an n-subpowerelement (cf. 5.28A) so $s(TSH) \geqslant 1 - 2/e - \epsilon_3$. Use Axiom 23 taking $\alpha = \frac{1}{2} - \epsilon_4$, $\beta = \frac{1}{2} + \epsilon_4$, $A = (TSH)^{-1}$. This is possible since $\epsilon_4 < \frac{1}{2} - 2/e$. Hence for some A^{L}, $\frac{1}{2} - \epsilon_4 \leqslant s(A^{L}) < \frac{1}{2} + \epsilon_4$. Hence for some $(TSH)^{R}$, $|s(TSH)^{R}) - \frac{1}{2}| \leqslant \epsilon_4$. $(TSH)^{R}$ does not have the form H^{R} otherwise $s(H) > \frac{1}{2} - \epsilon_4 - \epsilon_3 \geqslant u_{22}$ Thus $(TSH)^{R}$ is $(TS)^{R} H$ and

$$s((TS)^{R}) > \tfrac{1}{2} - \epsilon_4 - u_{22} - 2/e \geqslant u_{21}\,.$$

Thus $(TS)^R HX'$ is a left preobject with the size s of its leftmost maximal normal subword satisfying $|s - \frac{1}{2}| \leqslant \epsilon_4$. For each $O \equiv SHX'$ make a choice of $(TS)^R HX'$ and call it the *standard left adjustment* of O. Denote it by $st_L(O)$. Clearly we may assume it is independent of X' in the sense that if $O \equiv SHX'$ and $O_1 \equiv SHX'_1$ are left n-preobjects then for some U $st_L(O) \equiv UHX'$ and $st_L(O_1) \equiv UHX'_1$. If O_1 is an adjustment of O_2 then $st_L(O_1) = = st_L(O_2)$.

8.5. Remark. In 8.3, suppose $A \equiv B \equiv I$ and suppose alternative 1. holds. Then N_i is a left subword of S_i $(i = 1, 2)$ and one of S_1, S_2 is a left subword of the other.

Proof. By the remark at the end of Case 2. in the proof of 8.3 $S_i \equiv N_i S_i''$ $(i = 1, 2)$. Now N_1 or N_2 has size $> s_{33}$ and $N_1 \underset{n-1}{\sim} N_2$. Say $S_i T_i \in \operatorname{Rep} L_n$ $(i = 1, 2)$. By Axiom 33, $N_1 \equiv N_2$ and $S_1'' T_1 \equiv S_2'' T_2$. Hence $S_1 T_1 \equiv S_2 T_2$ so one of S_1, S_2 is a left subword of the other.

8.5′. Note. In 8.3 suppose $A \equiv B \equiv I$. Then

$$|s(\overline{S}_1) - s(\overline{S}_2)| < v_3 \text{ or } |s(\overline{S}_1) + s(\overline{S}_2) - 1| \leqslant v_2.$$

(By 6.24, 6.23 since $k_{12} \leqslant k_8'$, $k_{13} \leqslant k_9'$ and $s(\overline{S}_1) \geqslant s(S_1) - - \epsilon_3 > u_{21} - \epsilon_3 > u_{10}$.)

8.5A. Proposition. *Let* $AS_1X_1 \underset{n}{=} BS_2X_2$, *with divisions* $P_1, \ldots, P_m$ *and* $Q_1, \ldots, Q_m$ say, *where each side is* $((n-1))$*-bounded by* k_{12} *and* n*-bounded by* k_{13} *and* S_1X_1, S_2X_2 *are left* n*-preobjects* $(A, B$ *being arbitrary). Let* $S_i \equiv S_i' N_i S_i''$ $(i = 1, 2)$ *where either*

(1) S_1 *is not contained in any* P_j *of type* n,
S_2 *is not contained in any* Q_k *of type* n,
the equation induces $N_1 \equiv N_2$ (*hence* $N_1 S_1'' X_1 \underset{n}{=} N_2 S_2'' X_2$),
$s(N_i) > u_{28}$, *or*

(2) *for some* P_i' *of type* n, $N_1 \equiv S_1 \cap P_i'$, $N_2 \equiv S_2 \cap Q_i'$, *say* $P_i' \equiv P_{i1}' N_1 P_{i2}'$, $Q_i' \equiv Q_{i1}' N_2 Q_{i2}'$, *and* $s(N_i) > m_{33}$ (*hence* J *exists such that* $N_1^{-1} J N_2$ *is a subelement of* $Rep(L_n)$ *and is equal in* G_{n-1} *to* $N_1^{-1} P_{i1}'^{-1} X_{i-1}'^{-1} Q_{i1}' N_2$ *and* $N_1 S_1'' X_1 \underset{n}{=} J N_2 S_2'' X_2$).

Then if i = 1 or 2 is chosen,

$$S_1^* X_1 \underset{n}{=} S_2^* X_2$$

where the left side and right side are left adjustments of $S_1 X_1$, $S_2 X_2$ *respectively and* $S_i^* X_i$ *is* $st_L(S_i X_i)$. *Hence* $S_1 \cdot S_1^{*-1} \cdot S_2^* \cdot S_2^{-1} \underset{n}{=} A^{-1} \cdot B$.

Note. The reader may find it convenient to note now that if the hypothesis of 8.3 holds then the hypothesis of 8.5A holds (cf. the proof of 8.5B below).

Proof of 8.5A. First assume (1).

1^0. Since $u_{28} \geqslant s_{33}$, one of S_1', S_2' is a right subword of the other (cf. Axiom 33) so we may assume $S_1' \equiv S_2'^{\mathrm{R}}$. (Also one of S_1'', S_2'' is a left subword of the other.) Thus $S_1' N_1 S_1'' X_1 \underset{n}{=} S_1' N_2 S_2'' X_2$; briefly $S_1 X_1 = S_2^{\mathrm{R}} X_2$. As usual let $S_i H_i$ be maximal normal in $S_i X_i \equiv S_i H_i X_i'$ $(i = 1, 2)$.

2^0. We next show that, with respect to the equation $S_1 X_1 = S_2^{\mathrm{R}} X_2$, $\mathrm{Im}'(S_1 H_1) = S_2^{\mathrm{R}} H_2$. Both sides are $((n-1))$-bounded by $k_{12} \leqslant k_8'$ and n-bounded by $k_{13} \leqslant k_9'$; also $s(S_1 H_1) > u_{21} - \epsilon_3 \geqslant u_{10}$. By definition of Im′ (cf. 6.16), $S_1 H_1 \supset (S_1 H_1)^0 \underset{n-1}{\sim} (S_1 H_1)^{0'} \supset \bar{H}$, $\bar{H} \equiv \mathrm{Im}'(S_1 H_1)$. Now in the present case $(S_1 H_1)^0$ is a left subword of $S_1 X_1$ containing $S_1' N_1$, so $(S_1 H_1)^{0'}$ is a left subword of $S_2^{\mathrm{R}} X_2$ containing $S_1' N_2$. Since $s(N_2) > u_{28} \geqslant u_5 - u_4$, we have by 6.11 that $\bar{N}_2 = \bar{H}$, i.e., $S_2^{\mathrm{R}} H_2 = \mathrm{Im}'(S_1 H_1)$.

3^0. By 6.23, $|s(S_1 H_1) - s(S^{\mathrm{R}} H_2)| < v_3$. Now one of S_1'', S_2'' is a left subword of the other. We suppose $S_2'' \equiv S_1'' W$; the argument in the other case is similar. Thus $S_2^{\mathrm{R}} \equiv S_1' N_2 S_1'' W \equiv S_1 W$. We find an upper bound for $s(W)$.

If $s(W) > r_0 + \epsilon_3$ then

$$\begin{aligned} s(W) + s(S_1) &< s(S_1 W) + 2\epsilon_3 + \epsilon_1^* \\ &< s(S_1 W H_2) + 3\epsilon_3 + \epsilon_1^* \end{aligned}$$

$$s(S_1 W H_2) \quad < v_3 + s(S_1 H_1)$$

$$s(S_1 H_1) \quad < s(S_1) + u_{22} + 2/e$$

hence $s(W) < 3\epsilon_3 + \epsilon_1^* + v_3 + u_{22} + 2/e = a$, say.

Thus $s(W) < \mathrm{Max}(r_0 + \epsilon_3, a) = b$, say.

4^0. Let $KS_1 \in \mathrm{Rep}(L_n)$. Choose $(KS_1)^{\mathrm{R}}H_1$ to have size s satisfying $|s - \frac{1}{2}| \leqslant \epsilon_4$ (cf. 8.4). Since $S_1X_1 = S_1WX_2$ we have $(KS_1)^{\mathrm{R}}X_1 = (KS_1)^{\mathrm{R}}WX_2$, where the left side is $st_{\mathrm{L}}(S_1X_1)$. We prove that the right side is a left preobject, hence a left adjustment of S_2X_2.

$$s((KS_1)^{\mathrm{R}}WH_2) \leqslant (\tfrac{1}{2} + \epsilon_4 + \epsilon_3) + b + u_{22} + 4/e < k_{13}$$

Since $\frac{1}{2} - \epsilon_4 \leqslant s \leqslant s((KS_1^{\mathrm{R}})) + u_{22} + 2/e$ we have

$$s(KS_1^{\mathrm{R}}W) > s((KS_1^{\mathrm{R}})) - \epsilon_3 > \tfrac{1}{2} - \epsilon_4 - u_{22} - 2/e - \epsilon_3 \geqslant u_{21}$$

as required.

5^0. Alternatively choose $(KS_1W)^{\mathrm{R}}H_2$ to have size s, $|s - \frac{1}{2}| \leqslant \epsilon_4$. Now for any W^{R} we have

$$s(W^{\mathrm{R}}H_2) \leqslant s(W^{\mathrm{R}}) + u_{22} + 2/e \leqslant \mathrm{Max}(r_0, \epsilon_3 + b) + u_{22} + 2/e < \tfrac{1}{2} - \epsilon_4 \leqslant s.$$

Hence $(KS_1W)^{\mathrm{R}}H_2$ has the form $(KS_1)^{\mathrm{R}}WH_2$. Thus $(KS_1)^{\mathrm{R}}X_1 = (KS_1)^{\mathrm{R}}WX_2$ and the right side is $st_{\mathrm{L}}(S_2X_2)$; it remains to show that the left side is a left preobject. Its size is at most

$$s((KS_1)^{\mathrm{R}}) + u_{22} + 2/e \leqslant (\tfrac{1}{2} + \epsilon_4 + \epsilon_3) + u_{22} + 2/e < k_{13}.$$

Also since $s \leqslant s((KS_1)^{\mathrm{R}}) + s(W) + s(H_2) + 4/e$ we have

$$s((KS_1)^{\mathrm{R}}) \geqslant \tfrac{1}{2} - \epsilon_4 - b - u_{22} - 4/e > u_{21},$$

as required.

Now assume (2).

6^0. Put $W \equiv P_i'^{-1}X_{i-1}'^{-1}Q_i'$. Then $W \equiv P_{i2}'^{-1}N_1^{-1}J_1N_2Q_{i2}'$, where $J_1 \equiv P_{i1}'^{-1}X_{i-1}'^{-1}Q_{i1}'$. Hence $J_1 = J$ in G_{n-1}. Note that F, E exists such that $P_{i2}'F \equiv S_1''H_1$, $Q_{i2}'E \equiv S_2''H_2$. By 5.28A, $F^{-1}WE$ is a subpower-element of L_n since $s(Q_i') > k_3^0 - s(P_i') > k_3^0 - k_{13} \geqslant r_0 + \epsilon_3$ and similarly for P_i'. Let D arise from $F^{-1}WE$ by replacing J_1 by J;

thus $D \equiv H_1^{-1} S_1''^{-1} N_1^{-1} J N_2 S_2'' H_2$. Then by Axiom 33'',

$$|s(D) - s(F^{-1}WE)| < p_{33}.$$

By 5.20, $s(W) \leqslant 1 + \epsilon_3$ and $s(WX_i') \geqslant 1 - 2/e$. Now $s(WX_i') \leqslant s(W) + r_6^0 + 2/e$ (cf. 6.7A) and $s(F^{-1}WE) \leqslant s(W) + s(F) + s(E) + 4/e \leqslant s(W) + 2v_1 + 4/e$ (cf. 6.22). Thus

$$s(D) < p_{33} + (1 + \epsilon_3) + 2v_1 + 4/e = a, \text{ say}$$

and $s(D) > s(F^{-1}WE) - p_{33} > s(W) - \epsilon_3 - p_{33} > b$, where $b = 1 - 4/e - r_6^0 - \epsilon_3 - p_{33}$.

Since $s(D) > b > \frac{1}{2} + \epsilon_4$, some left subword D^{L} is such that $|s(D^{\mathrm{L}}) - \frac{1}{2}| \leqslant \epsilon_4$. We have $D \equiv D^{\mathrm{L}} D^{\mathrm{R}}$, say.

7^0. We prove that $D^{\mathrm{L}} \equiv H_1^{-1} M^{-1}$, $D^{\mathrm{R}} \equiv KH_2$ for some M, K.

If $D^{\mathrm{L}} \subset H_1^{-1}$ then we would have $\frac{1}{2} - \epsilon_4 \leqslant s(D^{\mathrm{L}}) \leqslant s(H_1) + \epsilon_3 < u_{22} + \epsilon_3 \leqslant \frac{1}{2} - \epsilon_4$, a contradiction. Thus $H_1^{-1} \subset D^{\mathrm{L}}$ as required.

Now $b < s(D) \leqslant s(D^{\mathrm{L}}) + s(D^{\mathrm{R}}) + 2/e \leqslant \frac{1}{2} + \epsilon_4 + s(D^{\mathrm{R}}) + 2/e$, so $s(D^{\mathrm{R}}) > b - \frac{1}{2} - \epsilon_4 - 2/e = c$, say. Also $c > r_0 + \epsilon_3$

If $D^{\mathrm{R}} \subset H_2$, then $s(D^{\mathrm{R}}) \leqslant s(H_2) + \epsilon_3 < u_{22} + \epsilon_3 \leqslant c$, a contradiction. Thus $H_2 \subset D^{\mathrm{R}}$ as required.

8^0. Since $N_1 S_1'' H_1 X_1' = JN_2 S_2'' H_2 X_2'$ we have $X_1' = DX_2'$, hence $MH_1 X_1' = KH_2 X_2'$. Note that $MH_1 X_1'$ is $st_{\mathrm{L}}(S_1 X_1)$. We show that $KH_2 X_2'$ is a left preobject; it will therefore be a left adjustment of $S_2 X_2$ as required.

Now

$$s(KH_2) = s(D^{\mathrm{R}}) < s(D) - s(D^{\mathrm{L}}) + 2\epsilon_3 + \epsilon_1^*$$

$$< a - (\tfrac{1}{2} - \epsilon_4) + 2\epsilon_3 + \epsilon_1^* \leqslant k_{13}$$

Also $c < s(D^{\mathrm{R}}) = s(KH_2) < s(K) + u_{22} + 2/e$, hence $s(K) > c - u_{22} - 2/e \geqslant u_{21}$ as required.

9^0. Similarly it can be shown that some left adjustment of $S_1 X_1$ is equal in G_n to $st_{\mathrm{L}}(S_2 X_2)$.

8.5B. Proposition. *Let* $O_i \equiv S_i X_i \equiv S_i H_i X_i'$ $(i = 1, 2)$ *be left n-pre-*

objects and let $O_1 \underset{n}{=} O_2$. *Then* $S_1'X_1 \underset{n}{=} S_2'X_2$, *where* $S_1'X_1$ *is* $st_L(O_1)$ *and* $S_2'X_2$ *is a left adjustment of* O_2. *Hence* $S_1'^{-1} \cdot S_2' \underset{n}{=} S_1^{-1} \cdot S_2$.

Proof. The hypothesis of 8.3 holds so the conditions of 8.3 hold. We deduce that the hypothesis of 8.5A holds (hence 8.5B follows). This is clear if Case (1) of the proof of 8.3 holds; we then have (2) of 8.5A.

Now let Case (2) hold.

Here $N_1 \;{}_{n-1}{\simeq}_1\; N_2$ where N_1 or N_2 has size $> u_{21} - \alpha - u_4 - \eta - \epsilon_3 = d$, say. By 6.14B, since $d > s_{25}$ and $d > r_0 + r_{25}$ we have for $i = 1, 2$ that $s(N_i) > d - r_{25} - \epsilon_3 \geqslant u_{28}$.

8.6. We have the dual notions of a *right n-preobject* $O \equiv X'HS$, a *right adjustment* and $st_R(O)$.

8.6A. Definition. An *n-preobject* is a word O which is both a left n-preobject SHX' and a right n-preobject $X_1'H_1S_1$ and also $X' \not\equiv I$, $X_1' \not\equiv I$.

8.6B. A left adjustment of a preobject is a preobject; so is a right adjustment. Let O_1, O_2 be preobjects. Call O_1 an *adjustment* of O_2 if it arises from O_2 by a left adjustment followed by a right adjustment (equivalently, a right adjustment followed by a left adjustment). This is an equivalence relation.

Let the standard adjustment of the preobject O mean $st_R(st_L(O))$ and be denoted by $st(O)$.

8.7. Definition. Let O_1, O_2 be n-preobjects. We say O_1, O_2 are *n-almost equal,* and write

$$O_1 \underset{n}{\hat{=}} O_2$$

if there are adjustments O_1', O_2' of O_1, O_2 respectively such that $O_1' \underset{n}{=} O_2'$.

8.7A. Proposition. *n-almost equality is an equivalence relation on the set of n-preobjects.*

Proof. Let $O_1 \underset{n}{\doteq} O_2$, $O_2 \underset{n}{\doteq} O_3$. For suitable adjustments O'_1 of O_1; O'_2, O''_2 of O_2; O''_3 of O_3 we have $O'_1 \underset{n}{=} O'_2$, $O''_2 \underset{n}{=} O''_3$. Therefore some adjustment of O'_1 is equal in G_n to $st(O'_2)$, which is the same as $st(O''_2)$, which is equal in G_n to some adjustment of O''_3. Thus some adjustment of O_1 is equal in G_n to some adjustment of O_3, as required.

8.8. Definition. An *n-object* is a word (S, T) which is $((n-1))$-bounded by k'_{12} and n-bounded by k'_{13}, where S, T are maximal n-normal of size $> u_{24}$ and S is left of T.

8.8A. With each n-object we shall associate certain n-preobjects, as follows.

If S, T overlap let $S \equiv S_1 A$, $T \equiv BT_1$ where $A \equiv B \equiv S \cap T$. (Thus S_1, T_1 are the kernels of S, T.) $S \cap T$ is not normal.

If S, T are disjoint let $S \equiv S_1 A$, $T \equiv BT_1$ where A, B are not normal (but are otherwise arbitrary).

In both cases $(S, T) \equiv S_1 U T_1$ for some U and

$$s(S_1) > u_{24} - (r_0 + \epsilon_3) - 2/e > t_{33}.$$

By Axiom 33′ there are words S_1^*, T_1^* of the form $S_1^{\mathrm{L}} EZFS_1^{\mathrm{R}}$, $T_1^{\mathrm{L}} E'Z'F'T_1^{\mathrm{R}}$ respectively where $S_1 \underset{n-1}{=} S_1^*$, $T_1 \underset{n-1}{=} T_1^*$, S_1^*, T_1^* are n-subelements and Z, Z' are subelements of $\mathrm{Rep}(L_n)$. Temporarily call S_1^*, T_1^* with their given factorizations *associates* of S, T respectively. Now (S, T) is equal in G_{n-1} to $S_1^* U T_1^*$ which contains $ZFS_1^{\mathrm{R}} UT_1^{\mathrm{L}} E'Z'$; we show that this last word is a preobject, and we call it an *associate* of the given n-object (S, T).

First, $s(Z) > s(S_1) - r_{33} > u_{24} - (r_0 + \epsilon_3) - 2/e - r_{33} > u_{21}$ Next, since $k'_{12} \leqslant b_{33}$ S_1^* is $((n-1))$-bounded by $k'_{12} + c_{33} \leqslant k_{12}$; since neither S_1^{R} nor T_1^{L} is a subelement of $L_1 \cup \cdots \cup L_{n-1}$ the word (Z, Z') under consideration is $((n-1))$-bounded by k_{12}; so is $S_1^* U T_1^*$ (cf. 5.32).

If (Z, Z') were not n-bounded by k_{13}, there would be an n-subelement A in it of size $\geqslant k_{13} \geqslant k'_{13} + u_4 > u_3$; by 6.10, since $k_{12} \leqslant b_{25}$, (S, T) would contain a subelement of size $> s(A) - - u_4 \geqslant k'_{13}$.

The maximal normal subword of (Z, Z') containing Z has the form $ZFS_1^{\mathrm{R}}H' \equiv ZH$ and it is required to prove that $s(H) < u_{22}$. Now

$$s(Z) > s(S_1) - r_{33} \geqslant s(S) - (r_0 + \epsilon_3) - 2/e - r_{33}.$$

S_1^*H' is maximal normal in $S_1^*UT_1^*$ and has size greater than

$$s(Z) - \epsilon_3 > u_{24} - (r_0 + \epsilon_3) - 2/e - r_{33} - \epsilon_3 > u_6 + u_4 + \epsilon_3$$

Also $s(S) > u_{24} > u_6 + u_4 + \epsilon_3$. By 6.14A, S and S_1^*H' are images of each other. Hence

$$s(S_1^*H') < s(S) + u_4 + \epsilon_3 < s(Z) + a,$$

where $a = (r_0 + \epsilon_3) + 2/e + r_{33} + u_4 + \epsilon_3$. Since $ZH \equiv ZFS_1^{\mathrm{R}}H' \subset S_1^*H'$, $s(ZH) \leqslant s(Z) + a + \epsilon_3$. If $s(H) > r_0 + \epsilon_3$ then $s(H) + s(Z) \leqslant s(ZH) + 2\epsilon_3 + \epsilon_1^*$. Since $a + \epsilon_3 + (2\epsilon_3 + \epsilon_1^*) < u_{22}$ it follows that $s(H) < u_{22}$. The result follows since $r_0 + \epsilon_3 < u_{22}$.

8.8B. Definition. A *left n-object* is a word SX which is $((n-1))$-bounded by k'_{12} and n-bounded by k'_{13}, where S is maximal n-normal of size $> u_{24}$.

8.8C. Note. The theory of left n-objects can be developed along very similar lines to the theory of n-objects (which occupies Sections 8 and 9) but it is easier and it may be left to the reader (in particular the ϵ, κ-theory in Section 9 is trivial for left n-objects). Thus with each left n-object one can associate a left n-preobject; two left n-objects O, O' are left n-almost equal if there exists associated left n-preobjects P, P' which are left n-almost equal (i.e., some left adjustment of P is equal in G_n to some left adjustment of P').

Similarly for right n-objects.

8.9. Proposition. *Let* $SAT \underset{n}{\sim} UBV$ *with divisions* $P_1, \ldots, P_m$ *and* $Q_1, \ldots, Q_m$, *so that* $SAT \underset{n}{=} X_0^{-1} \cdot UBV \cdot X_m$. *Let* SAT, UBV *be*

$((n-1))$-bounded by k_{14} and n-bounded by k_{15}. Let S, T, U, V be n-normal of size $> u_{25}$ where $\overline{S} \neq \overline{T}$ (i.e., the maximal normals determined by S, T are different) and $\overline{U} \neq \overline{V}$. Then

$$ES^{\mathrm{R}}AT^{\mathrm{L}}F \underset{n}{=} E'U^{\mathrm{R}}BV^{\mathrm{L}}F'$$

where each side is $((n-1))$-bounded by k'_{14} and n-bounded by k'_{15} and $s(S^{\mathrm{R}}) > s(S) - u_{26}$, $s(T^{\mathrm{R}}) > s(T) - u_{26}$, and similarly for U, V. Also

$$S^{\mathrm{R}-1}E^{-1}\cdot E'U^{\mathrm{R}} \underset{n-1}{=} S^{-1}\cdot X_0^{-1}\cdot U \text{ and}$$

$$V^{\mathrm{L}}F'\cdot F^{-1}T^{\mathrm{L}-1} \underset{n-1}{=} V\cdot X_m\cdot T^{-1}.$$

Finally each of E, F, E', F' is either an n-subelement of size $< u_{27}$ or is a possibly empty subelement of $\mathsf{L}_1 \cup \cdots \cup \mathsf{L}_{n-1}$.

Proof. We first modify SAT, UBV on the left to obtain an equation $ES^{\mathrm{R}}AT = E'U^{\mathrm{R}}BV\cdot X_m$.

Case 1. P_1 has type 0.

Then $X_0 \equiv I$; take $E \equiv E' \equiv I$, $S^{\mathrm{R}} \equiv S$, $U^{\mathrm{R}} \equiv U$.

Case 2. P_1 has type i, $0 < i < n$.

Then $S \equiv P'_1 S^{\mathrm{R}}$, $U \equiv Q'_1 U^{\mathrm{R}}$. Also $s(S) \leqslant r'_0 + 2/e + s(S^{\mathrm{R}})$ hence $s(S) < s(S^{\mathrm{R}}) + u_{26}$ since $r'_0 + 2/e < u_{26}$. Now $P_1'^{-1}X_0'^{-1}Q'_1X'_1 \in \mathsf{L}_i$ and $X'_0 \equiv X_0$. Also $k_3 < s(P'_1) + s(Q'_1) < s(P'_1) + k_{14}$ hence $s(P'_1) > k_3 - k_{14} > s_{24}$. By Axiom 24 since $k_{14} \leqslant b_{24}$ $P_1'^{-1}X_0'^{-1}Q'_1 \underset{i-1}{=} P_1'^{\mathrm{R}-1}\xi^{-1}Q_1'^{\mathrm{R}}$ where the right side is an i-subelement. Let $E \equiv \xi P_1'^{\mathrm{R}}$, $E' \equiv Q_1'^{\mathrm{R}}$. Then

$$(1) \qquad ES^{\mathrm{R}}AT = E'U^{\mathrm{R}}BV\cdot X_m.$$

The left side is $((i-1))$-bounded by $k_{14} + q_{24}$ since $P_1'^{\mathrm{R}}$ is not a subelement of $\mathsf{L}_1 \cup \cdots \cup \mathsf{L}_{i-1}$ (for $s(P_1'^{\mathrm{R}}) > s(P'_1) - r_{24} > k_3 - k_{14} - r_{24} > r'_0$). Now consider its i-bound. Since $\xi < s(X'_0) + d_{24}$ and $s(X'_0) < r_6$ (cf. 6.7A) the i-bound is $< 2/e + (r_6 + d_{24} + \epsilon_3) + k_{14} = d$, say. The j-bound ($j = 1+1, \ldots, n-1$) is $k_{14} + 2/e + r'_0$, and the n-bound is $k_{15} + 2/e + r'_0$. Note that

$k_{14} + q_{24}$, d and $k_{14} + 2/e + r'_0$ are $\leqslant k'_{14}$ and $k_{15} + 2/e + r'_0 \leqslant k'_{15}$.

Case 3. P_1 has type n.

Say $SAT \equiv P'_1 X$, $UBV \equiv Q'_1 Y$. Now $k_3^0 < s(P'_1) + s(Q'_1) < s(P'_1) + k_{15}$ hence $s(P'_1) > k_3^0 - k_{15} > s_{24}$. Hence $P_1'^{-1} X_0'^{-1} Q'_1 = P_1'^{R-1} \xi^{-1} Q_1'^{R}$ and $\xi P_1'^{R} X = Q_1'^{R} Y \cdot X_m$. Consider the left side of this last equation. It is $((n-1))$-bounded by $k_{14} + q_{24} \leqslant k'_{14}$. Now $s(\xi) < s(X'_0) + d_{24} < r_6^0 + d_{24}$ hence it is n-bounded by $2/e + (r_6^0 + d_{24} + \epsilon_3) + k_{15}$

Next, $s(Q_1'^{R}) > s(Q'_1) - r_{24} > k_3^0 - k_{15} - r_{24} > r_0 + \epsilon_3$

Now either $s(Q_1'^{L}) \leqslant r_0 + \epsilon_3$ or $s(Q_1'^{L}) + s(Q_1'^{R}) \leqslant s(Q'_1) + 2\epsilon_3 + \epsilon_1^*$. Hence $s(Q_1'^{L}) \leqslant \mathrm{Max}(r_0 + \epsilon_3, r_{24} + 2\epsilon_3 + \epsilon_1^*) = a$, say. If $U \subset Q_1'^{L}$ then $s(U) \leqslant s(Q_1'^{L}) + \epsilon_3 \leqslant a + \epsilon_3 \leqslant u_{25}$, a contradiction. Thus $U \equiv Q_1'^{L} U^{R}$ say and $s(U) \leqslant s(Q_1'^{L}) + s(U^{R}) + 2/e < a + s(U^{R}) + 2/e < s(U^{R}) + u_{26}$ since $a + 2/e < u_{26}$. Similarly $S \equiv P_1'^{L} S^{R}$.

Let $E \equiv \xi$, $E' \equiv I$. Then $s(E) < r_6^0 + d_{24} < u_{27}$. Also (1) holds and $S^{R} E^{-1} E' U^{R} = S^{-1} X_0^{-1} U$.

There is of course a dual argument where the right sides are modified. Since $s(S^{R}) > u_{25} - u_{26} > r_0 + \epsilon_3$, S^{R} is n-normal; similarly T^{L} is n-normal. Therefore $S^{R} A T^{L}$ is not a subelement of $L_1 \cup \cdots \cup L_n$. This shows that the left modifications and the right modifications are independent and the proposition follows.

8.9A. Corollary. *In 8.9 assume further that SAT, UBV are $((n-1))$-bounded by $x \leqslant k_{14}$ and n-bounded by $y \leqslant k_{15}$. Then $ES^{R}AT^{L}F$, $E'U^{R}BV^{L}F'$ are $((n-1))$-bounded by $x + u'_{14}$ and n-bounded by $y + u'_{15}$ where*

$$u'_{14} = Max(q_{24}, 2/e + r_6 + d_{24} + \epsilon_3, 2/e + r'_0)$$

$$u'_{15} = Max(2/e + r'_0, 2/e + r_6^0 + d_{24} + \epsilon_3).$$

Proof. Immediate from the proof of 8.9.

8.10. Proposition. *Let $(S, T) \equiv SX$, $(S', T') \equiv S'X'$ be n-objects and let $S' \equiv S^{R}$, $X \equiv X'$* (hence $T \equiv T'$). *Let S_1^*, T_1^* be associates of*

S, T and let P be the associated n-preobject of (S, T) determined by S_1^*, T_1^*. Let $S_1'^*$ be an associate of S'. Then T_1^* is an associate of T' and the associated n-preobject P' of (S', T') thereby determined is n-almost equal to P.

8.10A. Corollary. *Any two n-preobjects associated with a given n-object are n-almost equal.*

Proof of 8.10. We have $S \equiv S_1 A$, where A is not normal and, if S, T overlap then A is $S \cap T$. Similarly $S' \equiv S_1' A'$. Also

$$S_1^* \equiv S_1^{\mathrm{L}} EZFS_1^{\mathrm{R}}, \qquad S_1'^* \equiv S_1'^{\mathrm{L}} E'Z'F'S_1'^{\mathrm{R}}.$$

Now $S \equiv S^{\mathrm{L}} S^{\mathrm{R}} \equiv S^{\mathrm{L}} S'$, so $S_1^{\mathrm{L}} EZFS_1^{\mathrm{R}} A \underset{n-1}{=} S^{\mathrm{L}} S_1'^{\mathrm{L}} E'Z'F'S_1'^{\mathrm{R}} A'$, each side being an n-subelement.

Now $s(FS_1^{\mathrm{R}}) < q_{33}$ by Axiom 33′, so $s(FS_1^{\mathrm{R}} A) < q_{33} + r_0 + \epsilon_3 + 2/e$ and hence any subword of $FS_1^{\mathrm{R}} A$ has size $< q_{33} + r_0 + \epsilon_3 + 2/e + \epsilon_3 < u_{23}$. Similarly for $F'S_1'^{\mathrm{R}} A'$.

By the dual of 8.3 and by 8.5, $Z \equiv WN_1 Y$, $Z' \equiv W'N_2 Y'$, where N_1, N_2 are normal and $S_1^{\mathrm{L}} EWN_1 \underset{n}{=} S^{\mathrm{L}} S_1'^{\mathrm{L}} E'W'N_2$, $N_1 \equiv N_2$; note that alternative 2. of 8.3 cannot apply since we are considering equality in G_{n-1}. Hence

$$(1) \qquad N_1 YFS_1^{\mathrm{R}} AX = N_2 Y'F'S_1'^{\mathrm{R}} A'X.$$

Now $T \equiv BT_1$, where B is not normal ($B \equiv S \cap T$ if S, T overlap) and T_1^* is $T_1^{\mathrm{L}} E''Z''F''T_1^{\mathrm{R}}$, say. Thus $X \equiv DT_1$ for some D. Thus the associated preobjects under consideration are

$$ZFS_1^{\mathrm{R}} ADT_1^{\mathrm{L}} E''Z'', \qquad Z'F'S_1'^{\mathrm{R}} A'DT_1^{\mathrm{L}} E''Z''$$

respectively; briefly (Z, Z''), (Z', Z'').

Now $S_1^{\mathrm{L}} E(Z, Z'') = S^{\mathrm{L}} S_1'^{\mathrm{L}} E'(Z', Z'')$; applying 8.5A to this equation there are left adjustments (Z^*, Z'') of (Z, Z''), (Z'^*, Z'') of (Z', Z'') which are equal in G_n. Hence $(Z, Z'') \underset{n}{\hat{=}} (Z', Z'')$ as required.

8.10B. For later note that $Z \cdot Z^{*-1} \cdot Z'^{*} \cdot Z'^{-1} = E^{-1} S_1^{\mathrm{L}-1} S^{\mathrm{L}} S_1'^{\mathrm{L}} E'$.

8.10C. Proof of 8.10A. By the special case of 8.10 in which S' is S, we see that the associated preobject is independent, up to n-almost equality, of the choice of S_1^*. By symmetry it is also independent of the choice of T_1^*.

8.11. Definition. Let O_1, O_2 be n-objects. Then O_1, O_2 are *n-almost equal*, $O_1 \underset{n}{\triangleq} O_2$, if there exist associated n-preobjects P_1, P_2 of O_1, O_2 respectively such that $P_1 \underset{n}{\triangleq} P_2$.

8.12. Proposition. *n-almost equality is an equivalence relation on the set of n-objects.*

Proof. This is an easy consequence of 8.10A.

8.13. Proposition. *Again let $(S, T) \equiv SX$, $(S', T') \equiv S'X'$ be n-objects with $S' \equiv S^{\mathrm{R}}$, $X \equiv X'$, $T \equiv T'$. Then $(S, T) \underset{n}{\triangleq} (S', T')$.*

Briefly: $(S, T) \underset{n}{\triangleq} (S^{\mathrm{R}}, T)$.

Proof. Immediate from 8.10.

8.14. Proposition. *Let $(S, T) \underset{n}{\sim} (U, V)$, where each side is an n-object and is $((n-1))$-bounded by k_{16} and n-bounded by k_{17}. Let S, T, U, V have size greater than u_{28}. Then $(S, T) \underset{n}{\triangleq} (U, V)$.*

Proof. Let the kernels of S, T be S_1, T_1. Then $(S, T) \equiv S_1 A T_1$ and $u_{28} < s(S) \leqslant s(S_1) + (r_0 + \epsilon_3) + 2/e$ and similarly for T. In the same way we have $(U, V) \equiv U_1 B V_1$, say.

The hypothesis of 8.9 holds since $k_{16} \leqslant k_{14}$, $k_{17} \leqslant k_{15}$ and $s(S_1) > u_{28} - r_0 - \epsilon_3 - 2/e > u_{25}$. Hence by 8.9, 8.9A,

$$(1) \qquad E S_1^{\mathrm{R}} A T_1^{\mathrm{L}} F \underset{n}{=} E' U_1^{\mathrm{R}} B V_1^{\mathrm{L}} F'$$

where each side is $((n-1))$-bounded by $k_{16} + u'_{14} \leqslant k'_{12}$ and n-bounded by $k_{17} + u'_{15} \leqslant k'_{13}$; $s(S_1^{\mathrm{R}}) > s(S_1) - u_{26}$ (and similarly for $T_1^{\mathrm{L}}, U_1^{\mathrm{R}}, V_1^{\mathrm{L}}$); E is an n-subelement of size $< u_{27}$

or is a subelement of $L_1 \cup \cdots \cup L_{n-1}$ (and similarly for F, E', F').

$S_1^{\mathrm{R}} A T_1^{\mathrm{L}}$ and $U_1^{\mathrm{R}} B V_1^{\mathrm{L}}$ are n-objects since, e.g.,

$$s(\overline{S_1^{\mathrm{R}}}) \geqslant s(S_1^{\mathrm{R}}) - \epsilon_3 > u_{28} - u_{26} - \epsilon_3 \geqslant u_{24}.$$

Note that the kernels of $\overline{S_1^{\mathrm{R}}}, \overline{T_1^{\mathrm{L}}}$ are $S_1^{\mathrm{R}}, T_1^{\mathrm{L}}$.

We next consider associated preobjects $(Z_1, Z_2), (Z_3, Z_4)$ of these objects (cf. 8.8A): we have $S_1^{\mathrm{R}} \underset{n-1}{=} S_1^{\mathrm{RL}} E_1 Z_1 F_1 S_1^{\mathrm{RR}}$ with similar equations for $T_1^{\mathrm{L}}, U_1^{\mathrm{R}}, V_1^{\mathrm{L}}$, hence

$$(2) \qquad E S_1^{\mathrm{RL}} E_1 (Z_1, Z_2) F_2 T_1^{\mathrm{LR}} F \underset{n}{=} E' U_1^{\mathrm{RL}} E_3 (Z_3, Z_4) F_4 V_1^{\mathrm{LR}} F'$$

We want to apply 8.3, with Z_1, Z_3 for S_1, S_2. By Axiom 33′, each side of (2) is $((n-1))$-bounded by $k_{16} + u'_{14} + c_{33} \leqslant k_{12}$. Since the left side of (1) is n-bounded by $k_{17} + u'_{15}$ and is equal in G_{n-1} to the left side of (2) we have by 6.10 that the left side of (2) is n-bounded by $k_{17} + u'_{15} + u_4 \leqslant k_{13}$. Now $s(S_1^{\mathrm{RL}} E_1) < q_{33}$, so if E is an n-subelement then $s(E S_1^{\mathrm{RL}} E_1) < q_{33} + u_{27} + 2/e$, hence any subword of $E S_1^{\mathrm{RL}} E_1$ has size $\mathrm{Max}(r_0, q_{33} + u_{27} + 2/e + \epsilon_3) \leqslant u_{23}$; if E is not an n-subelement, then by 5.21 and 5.19 any n-subelement contained in $E S_1^{\mathrm{RL}} E_1$ has size $< r'_0 + \mathrm{Max}(r_0, q_{33} + \epsilon_3) + 2/e \leqslant u_{23}$

Thus 8.3 and 8.5A are available, hence a left adjustment $(Z'_1, Z_2 F_2 T_1^{\mathrm{LR}} F)$ of $(Z_1, Z_2 F_2 T_1^{\mathrm{LR}} F)$ is equal in G_n to a left adjustment $(Z'_3, Z_4 F_4 V_1^{\mathrm{LR}} F')$ of $(Z_3, Z_4 F_4 V_1^{\mathrm{LR}} F')$. By the duals of 8.3, 8.5A, a right adjustment (Z'_1, Z'_2) of (Z'_1, Z_2) is equal in G_n to a right adjustment (Z'_3, Z'_4) of (Z'_3, Z_4). Hence $(Z_1, Z_2) \triangleq (Z_3, Z_4)$, i.e., $S_1^{\mathrm{R}} A T_1^{\mathrm{L}} \triangleq U_1^{\mathrm{R}} B V_1^{\mathrm{L}}$. By 8.13, $S_1 A T_1 \triangleq U_1 B V_1$, i.e., $(S, T) \triangleq (U, V)$.

8.15. Corollary. *Let $U \underset{n}{\sim} V$, where U, V are $((n-1))$-bounded by k_{16} and n-bounded by k_{17}. Let A, B be distinct maximal n-normal subwords in U of size greater than u_{29}. Then $Im'A$, $Im'B$ exist and $(A, B) \underset{n}{\triangleq} (Im'A, Im'B)$. In particular (A, B), $(Im'A, Im'B)$ are n-objects.*

Proof. $C \equiv \mathrm{Im}'A$ and $D \equiv \mathrm{Im}'B$ exist since $k_{16} \leqslant k'_8$, $k_{17} \leqslant k'_9$,

$u_{29} \geqslant u_{10}$. We shall show that $(A^R, B^L) \underset{n}{\sim} (C^R, D^L)$ where each side is an n-object and we shall then deduce from 8.14 that $(A^R, B^L) \underset{n}{\hat{=}} (C^R, D^L)$; the result will then follow from 8.13.

In the notation of 6.16, first suppose $A = \overline{P_i}$ (P_i of type n). Then $P_i' \subset A$, $Q_i' \subset C$; say $A \equiv A_1 P_i' A_2$, $C \equiv C_1 Q_i' C_2$. Put $A^R \equiv P_i' A_2$, $C^R \equiv Q_i' C_2$. Since $k_3^0 < s(P_i') + s(Q_i') < s(P_i') + k_{17}$ we have $s(A^R) \geqslant s(P_i') - \epsilon_3 > k_3^0 - k_{17} - \epsilon_3 \geqslant u_{28}$. Similarly $s(C^R) > u_{28}$.

In the opposite case, we have that A has the form $A' A^0 A''$, where $s(A^0) > s(A) - b$, $b = 2(r_0 + \epsilon_3 + \epsilon_2 + 4/e)$ and

$$A^0 \underset{n-1}{\sim} A^{0\prime} \supset G, \quad \overline{G} = C.$$

Since by 6.19 $\mathrm{Im}' C = A$ we also have $C \equiv C' C^0 C''$, $C^0 \underset{n-1}{\sim} C^{0\prime} \supset H$, $\overline{H} = A$. Let p, q, p', q' be the left end points of A^0, C^0, $A^{0\prime}$, $C^{0\prime}$ respectively. Then we may assume q is left of p' or coincides with p'. Now $s(G) > s(A^0) - u_4 > u_{29} - b - u_4 > u_3$, so $G \supset G^0 \underset{n-1}{\sim} G^{0\prime}$. Since $G \subset C$ we have $G^0 \subset C^0 \cap A^{0\prime}$. Put $K' \equiv C^0 \cap A^{0\prime}$, $K \equiv C^{0\prime} \cap A^0$. Then $A^0 \equiv KL$, $C^0 \equiv MK'N$ say. Take A^R to be $A^0 A''$ and take C^R to be $K' N C''$. Note that $G \subset C^R$ since $G \subset C \cap A^{0\prime}$.

We show A^R, C^R have size $> u_{28}$. $s(A^R) \geqslant s(A^0) - \epsilon_3 > u_{29} - b - - \epsilon_3 \geqslant u_{28}$. $s(C^R) \geqslant s(G) - \epsilon_3 > u_{29} - b - u_4 - \epsilon_3 > u_{28}$.

We define B^L, D^L similarly. Thus $(A^R, B^L) \underset{n}{\sim} (C^R, D^L)$ where each side is an n-object. By 8.14 $(A^R, B^L) \underset{n}{\hat{=}} (C^R, D^L)$, as required.

8.16. Axiom. Let P, P' be n-preobjects. Let P be an adjustment of P'. Let $P \underset{n}{=} P'$. Then $P \equiv P'$.

§9

Informal Summary

In this section we give a deeper discussion of n-almost equality. The main results are (i) if (S, U, T) is a word where S, U, T are distinct maximal n-normal subwords and if (S, U, T) and its subword (S, U) are n-objects then they are not n-almost equal. (One could, e.g., use this to show that $\mathrm{Im}' A$ (cf. 6.16) is independent of the divisions $P_1, \ldots, P_k$ $Q_1, \ldots, Q_k$ chosen.) (ii) Proposition 9.14 which allows us to construct a new pair of n-almost equal objects from given pairs, under certain conditions.

First we consider n-preobjects.

9.1. Definition. Suppose given n-preobjects P_1, P_2 such that $P_i \underset{n}{\hat{=}} P_2$. Choose adjustments P_3, P_4 of P_1, P_2 respectively such that $P_3 \underset{n}{=} P_4$. For $i = 1, 2, 3, 4$ we have, by 8.6A, $P_i \equiv S_i H_i X_i' \equiv X_i'' H_i' T_i$ say; put $U_i \equiv S_i H_i$, $V_i \equiv H_i' T_i$ so that $P_i \equiv (U_i, V_i)$ and U_i, V_i are maximal n-normal in P_i.

Put $\epsilon \equiv U_1 \cdot U_3^{-1} \cdot U_4 \cdot U_2^{-1}$; then $\epsilon \equiv S_1 \cdot S_3^{-1} \cdot S_4 \cdot S_2^{-1}$ since $H_1 \equiv H_3$ and $H_2 \equiv H_4$.

Put $\kappa \equiv V_2^{-1} \cdot V_4 \cdot V_3^{-1} \cdot V_1 \equiv T_2^{-1} \cdot T_4 \cdot T_3^{-1} \cdot T_1$.

Then $P_1 \underset{n}{=} \epsilon \cdot P_2 \cdot \kappa$.

We say that P_1, P_2 *produce* ϵ and κ. Let $E(P_1, P_2)$ be the subset of Π consisting of all such ϵ as P_3, P_4 vary. Similarly $K(P_1, P_2)$ denotes the set of all κ.

9.2. Proposition. *Let $\epsilon, \epsilon' \in E(P_1, P_2)$. Then $\epsilon \underset{n}{=} \epsilon'$. Similarly if $\kappa, \kappa' \in K(P_1, P_2)$ then $\kappa \underset{n}{=} \kappa'$.*

Proof. Let P_j denote an adjustment of P_1 or of P_2 according as j is odd or even. Let $P_j \equiv (U_j, V_j)$.

Let $P_5 = P_6$ and $\epsilon' \equiv U_1 U_5^{-1} U_6 U_2^{-1}$; it is required to prove that $\epsilon \underset{n}{=} \epsilon'$, i.e.,

(1) $$U_3^{-1} U_4 \underset{n}{=} U_5^{-1} U_6.$$

Let P_{11} be $st(P_1)$; then also P_{11} is $st(P_j)$ if j is odd. By 8.5B, since $P_3 = P_4$, $st(P_3)$ equals some adjustment of P_4, i.e., $(\exists P_8) P_{11} = P_8$, and moreover

(2) $$U_3^{-1} U_4 = U_{11}^{-1} U_8$$

Since $P_5 = P_6$,

(3) $$(\exists P_{10})\, P_{11} = P_{10} \quad \text{and} \quad U_5^{-1} U_6 = U_{11}^{-1} U_{10}.$$

Hence $P_8 = P_{10}$. But P_8 is an adjustment of P_{10} hence by 8.16 we have P_8 is P_{10}; thus U_8 is U_{10}; thus (2) and (3) imply (1).

Similarly $\kappa \underset{n}{=} \kappa'$.

9.3. Proposition. *Let* $\epsilon_1 \in E(P_1, P_2)$, $\epsilon_2 \in E(P_2, P_3)$, $\epsilon_3 \in E(P_1, P_3)$. *Then* $\epsilon_1 \cdot \epsilon_2 \underset{n}{=} \epsilon_3$.

Proof. Let P_1', P_3' be adjustments of P_1, P_3. Let P_2', P_2'' be adjustments of P_2. Then we may suppose $P_1' = P_2', P_2'' = P_3'$, $\epsilon_1 \equiv U_1 U_1'^{-1} U_2' U_2^{-1}$, $\epsilon_2 \equiv U_2 U_2''^{-1} U_3' U_3^{-1}$ in an obvious notation. By the preceding proof (cf. eq. (2)) we may suppose P_2' is $st(P_2)$ and P_2'' is $st(P_2)$. Thus $P_2' \equiv P_2''$, $U_2' \equiv U_2''$ and $P_1' \underset{n}{=} P_3'$. Hence $\epsilon_1 \cdot \epsilon_2 \equiv U_1 U_1'^{-1} U_3' U_3^{-1} \in E(P_1, P_3)$ and the result follows by 9.2.

9.4. Proposition. *If* $\epsilon \in E(P_1, P_2)$ *then* $\epsilon^{-1} \in E(P_2, P_1)$. *If* $P_1 \hat{=} P_2$ *then* $P_1^{-1} \hat{=} P_2^{-1}$ *and if* $\epsilon \in E(P_1^{-1}, P_2^{-1})$ *then* $\epsilon^{-1} \in K(P_1, P_2)$.

Proof. This is trivial.

Now we consider **n**-objects.

9.5. Definition. Given an n-object (S, T). For a choice of associates S_1^*, T_1^* of S, T there is determined $S_1^{\mathrm{L}} EZFS_1^{\mathrm{R}} UT_1^{\mathrm{L}} E'Z'F'T_1^{\mathrm{R}} \equiv W$ say, where (Z, Z') is an associated preobject and $(S, T) \underset{n-1}{=} W$. Write $W \equiv W_1 W_2 W_3$ where $W_2 \equiv (Z, Z')$. Although W_2 depends on the choices of S_1^*, T_1^* note that W_1 depends on the choice of S_1^* only, and W_3 depends only on the choice of T_1^*. Temporarily call $W_1 W_2 W_3$ *a triplet* for (S, T).

Now suppose given two n-objects O_1, O_2 such that $O_1 \underset{n}{\hat{=}} O_2$. Choose triplets $W_1 W_2 W_3$, $V_1 V_2 V_3$ for O_1, O_2. Now $W_2 \hat{=} V_2$; choose ϵ_1, κ_1 such that W_2, V_2 produce ϵ_1, κ_1. Put

$$\epsilon \equiv W_1 \cdot \epsilon_1 \cdot V_1^{-1}, \qquad \kappa \equiv V_3^{-1} \cdot \kappa_1 \cdot W_3.$$

We say that O_1, O_2 *produce* ϵ, κ. Let $E(O_1, O_2)$ denote the subset of Π consisting of all ϵ (as the choices vary) and let $K(O_1, O_2)$ be the set of all κ.

Note that $O_1 \underset{n}{=} \epsilon \cdot O_2 \cdot \kappa$.

9.5A. Remark. Let (S, T) and (S^{R}, T) be objects. Let S_1^*, T_1^* be associates for (S, T) and let $S_1'^*, T_1'^*$ be associates for (S^{R}, T).

Moreover let $T_1^* \equiv T_1'^*$. Let $A_1 A_2 A_3$ and $B_1 B_2 B_3$ be the triples thereby determined. Then in the notation of 8.10 we have

$$A_2 \equiv (Z, Z''), \quad B_2 \equiv (Z', Z''), \quad A_1 \equiv S_1^{\mathrm{L}} E, \quad B_1 \equiv S_1'^{\mathrm{L}} E'.$$

Then 8.10B states that a member of $E(A_2, B_2)$ is equal in G_n to $A_1^{-1} \cdot S^{\mathrm{L}} \cdot B_1$ (recall that $S \equiv S^{\mathrm{L}} S^{\mathrm{R}}$). It is also clear that $I \in K(A_2, B_2)$. Thus S^{L} is equal in G_n to a member of $E((S, T), (S^{\mathrm{R}}, T))$, proving 9.7 below.

9.6. Proposition. *If* $\epsilon, \epsilon' \in E(O_1, O_2)$ *then* $\epsilon \underset{n}{=} \epsilon'$. *If* $\kappa, \kappa' \in K(O_1, O_2)$ *then* $\kappa \underset{n}{=} \kappa'$.

Proof. Let $W_1' W_2' W_3'$, $V_1' V_2' V_3'$ be another choice of triplets for O_1, O_2 and let W_2', V_2' produce ϵ_2, κ_2. Let $\epsilon' \equiv W_1' \epsilon_2 V_1'^{-1}$; it is required to show that $\epsilon \underset{n}{=} \epsilon'$. We note that $W_2 \triangleq W_2'$ by 8.10A. Suppose we prove:

(a) If $\epsilon_3 \in E(W_2, W_2')$ then $I \underset{n}{=} W_1 \cdot \epsilon_3 \cdot W_1'^{-1}$.

Then by symmetry if $\epsilon_4 \in E(V_2', V_2)$ then $I \underset{n}{=} V_1' \cdot \epsilon_4 \cdot V_1^{-1}$.

Now $\epsilon_1 \underset{n}{=} \epsilon_3 \cdot \epsilon_2 \cdot \epsilon_4$ by 9.3, hence

$$\epsilon \underset{n}{=} W_1 \cdot \epsilon_3 \cdot \epsilon_2 \cdot \epsilon_4 \cdot V_1^{-1} \underset{n}{=} W_1' \cdot \epsilon_2 \cdot V_1'^{-1} \equiv \epsilon'.$$

Thus for 9.6 it is sufficient to prove (a).

Say $W_1 W_2 W_3$ is determined by S_1^*, T_1^* and $W_1' W_2' W_3'$ is determined by S_1', T_1'. Then S_1^*, T_1' determines a triplet $W_1'' W_2'' W_3''$ say. Suppose:

(b) if $\epsilon_5 \in E(W_2, W_2'')$ then $I = W_1 \epsilon_5 W_1''^{-1}$; and

(c) if $\epsilon_6 \in E(W_2'', W_2')$ then $I = W_1'' \epsilon_6 W_1'^{-1}$.

Then (a) follows since $\epsilon_5 \cdot \epsilon_6 = \epsilon_3$.

Now consider the special case of 9.5A in which S^{L} is I; this gives (c).

Also by 9.5A with $S^{\mathrm{L}} \equiv I$ we have $I \in K(W_2'', W_2')$; as another instance of this same fact we have $I \in K(W_2^{-1}, W_2''^{-1})$. Hence by 9.4 $I \in E(W_2, W_2'')$. Therefore $\epsilon_5 \underset{n}{=} I$; but $W_1 \equiv W_1''$, so (b) follows.

9.7. Proposition. *Let* (S, T) *and* (S^{R}, T) *be objects* (*as in* 8.10).

Then where $S \equiv S^{\mathrm{L}} S^{\mathrm{R}}$ *we have* S^{L} *is equal in* G_n *to an element of* $E((S,T),(S^{\mathrm{R}},T))$.

Proof. This is immediate from 9.5A.

9.8 Proposition. *Let* $\epsilon_1 \in E(O_1, O_2)$, $\epsilon_2 \in E(O_2, O_3)$, $\epsilon_3 \in E(O_1, O_3)$. *Then* $\epsilon_1 \cdot \epsilon_2 \underset{n}{=} \epsilon_3$.

Proof. Say $W_1 W_2 W_3$, $V_1 V_2 V_3$ are triplets for O_1, O_2, $\epsilon' \in E(W_2, V_2)$ and $\epsilon_1 \equiv W_1 \epsilon' V_1^{-1}$. Say $V_1' V_2' V_3'$, $U_1 U_2 U_3$ are triplets for O_2, O_3, $\epsilon'' \in E(V_2', U_2)$ and $\epsilon_2 \equiv V_1' \epsilon'' U_1^{-1}$.

Let $\epsilon_4 \in E(V_2, U_2)$, $\epsilon_5 \equiv V_1 \epsilon_4 U_1^{-1}$. Then $\epsilon_5 \underset{n}{=} \epsilon_2$ by 9.6. Let $\bar{\epsilon} \in E(W_2, U_2)$ and $\epsilon_6 \equiv W_1 \bar{\epsilon} U_1^{-1}$; then $\epsilon_6 \in E(O_1, O_3)$. Thus $\epsilon_1 \cdot \epsilon_2 \underset{n}{=} \epsilon_1 \cdot \epsilon_5 \underset{n}{=} W_1 \epsilon' \epsilon_4 U_1^{-1} \underset{n}{=} W_1 \bar{\epsilon} U_1^{-1} \underset{n}{=} \epsilon_6 \underset{n}{=} \epsilon_3$.

9.8A. Note. Proposition 9.4 is valid for n-objects.

9.9. Definition. Let O_1, O_2 be n-objects with $O_1 \underset{n}{\hat{=}} O_2$. Then $E'(O_1, O_2)$ denotes the subset of Π given by

$$X \in E'(O_1, O_2) \quad \text{if } (\exists \epsilon)\, X \underset{n}{=} \epsilon \in E(O_1, O_2).$$

(Thus $E'(O_1, O_2)$ is an equivalence class for equality in G_n.)

Similarly we define $K'(O_1, O_2)$.

9.9A. Proposition. *Let* $O_1 \equiv (S,T)$, $O_2 \equiv (S^{\mathrm{R}}, T)$ *be n-objects* (*as in* 8.10). *Then* (*i*) $S^{\mathrm{L}} \in E'(O_1, O_2)$ *and* (*ii*) $I \in K'(O_1, O_2)$.

Proof. Let O_1, O_2 produce ϵ, κ. Then by 9.7 $S^{\mathrm{L}} \underset{n}{=} \epsilon$, proving (i). Now $S^{\mathrm{L}} O_2 \equiv O_1 \underset{n}{=} \epsilon \cdot O_2 \cdot \kappa$. Hence $\kappa \underset{n}{=} I$, proving (ii).

9.10. Proposition. *Let* $O_1 \equiv (S,T)$, $O_2 \equiv (U,V)$ *be n-objects* $((n-1))$-*bounded by* k_{16} *and n-bounded by* k_{17} *such that* S, T, U, V *have size* $> u_{28}$. *Then*

(*i*) *if* $O_1 \underset{n}{\sim} O_2$ *with divisions* $P_1, \ldots, P_m$ $Q_1, \ldots, Q_m$, *so that* $O_1 \underset{n}{=} X_0^{-1} \cdot O_2 \cdot X_m$ *and by* 8.14 $O_1 \underset{n}{\hat{=}} O_2$, *then* $X_0^{-1} \in E'(O_1, O_2)$ *and* $X_m \in K'(O_1, O_2)$. *In particular*

(ii) if $O_1 \underset{n}{=} O_2$ then $I \in E'(O_1, O_2), I \in K'(O_1, O_2)$.

Proof. Consider the proof of 8.14; using the same notation we have $O_1 \equiv S_1 A T_1$, $O_2 \equiv U_1 B V_1$. Let $O_1' \equiv S_1^R A T_1^L$, $O_2' \equiv U_1^R B V_1^L$. Also eq. (2) holds and

$$(Z_1', Z_2 F_2 T_1^{LR} F) \underset{n}{=} (Z_3', Z_4 F_4 V_1^{LR} F')$$

From these two equations we obtain that if K denotes $Z_1 \cdot Z_1'^{-1} \cdot Z_3' \cdot Z_3^{-1}$ then $K \underset{n}{=} (E S_1^{RL} E_1)^{-1} E' U_1^{RL} E_3$. But $K \in E((Z_1, Z_2), (Z_3, Z_4))$, so $S_1^{RL} E_1 K (U_1^{RL} E_3)^{-1} \in E(O_1', O_2')$. Thus $E^{-1} \cdot E' \in E'(O_1', O_2')$. Let $O_1'' \equiv S_1 A T_1^L$, $O_2'' \equiv U_1 B V_1^L$. Then O_1'', O_2'' are objects and where $S_1 \equiv S_1^L S_1^R$, $U_1 = U_1^L U_1^R$ we have by 9.7 that $S_1^L \in E'(O_1'', O_1')$. Similarly $U_1^L \in E'(O_2'', O_2')$, hence $U_1^{L-1} \in E'(O_2', O_2'')$ (cf. 9.8A). By 9.8, $L \equiv S_1^L E^{-1} E' U_1^{L-1} \in E'(O_1'', O_2'')$. By the dual of 9.9A (ii), $I \in E'(O_1, O_1'')$ and $I \in E'(O_2'', O_2)$. Hence by 9.8 $L \in E'(O_1, O_2)$. As in 8.9 we have $S_1^{R-1} E^{-1} E' U_1^R \underset{n}{=}$ $\underset{n}{=} S_1^{-1} X_1^{-1} U_1$, i.e., $L \underset{n}{=} X_0^{-1}$, hence $X_0^{-1} \in E'(O_1, O_2)$.

If O_1, O_2 produce ϵ, κ then $O_1 = \epsilon O_2 \kappa = X_0^{-1} O_2 \kappa$. Hence $X_m \underset{n}{=} \kappa$, i.e., $X_m \in K'(O_1, O_2)$.

9.11. Proposition. *Let $(S, T) \underset{n}{\hat{=}} (S', T')$, where each side is an n-object, ((n – 1))-bounded by k_{18} and n-bounded by k_{19}. Let U be maximal n-normal in (S, T), $U \neq S, T$. Let S, T, S', T' have size $> u_{30}$ and let $s(U) > u_{30}'$. Then there is a maximal n-normal subword U' of (S', T'), $U' \neq S', T'$, such that $(S, U) \underset{n}{\hat{=}} (S', U')$ and $s(U') > u_{30}$ Further,*

(i) $|s(U) + s(U') - 1| < v_2 + u_4 + \epsilon_3$, or
(ii) $|s(U) - s(U')| < v_3 + u_4 + \epsilon_3$.

Proof. 1°. Let S_1, U_1, T_1 be the kernels of S, U, T. Then $(S, T) \equiv S_1 A U_1 B T_1$ say. Also

$$(S, T) \underset{n-1}{=} S_1^L E Z S_1^R A U_1 B T_1^L E' Z' F' T_1^R \equiv W$$

say, where (Z, Z') is an associated preobject of (S, T). Take any associated preobject (Z'', Z^+) for (S', T'); say

$$(S', T') \equiv S_1'CT_1' \underset{n-1}{=}_1 S_1'^{\mathrm{L}} E''Z''F''S_1'^{\mathrm{R}} CT_1'^{\mathrm{L}} E^+Z^+F^+T_1'^{\mathrm{R}} \equiv W'.$$

Now (S, U) is an object since $k_{18} \leqslant k_{12}'$, $k_{19} \leqslant k_{13}'$ and $s(U) > u_{30}' \geqslant u_{24}$

2°. Next we show W is an object.

By the argument of 8.8A, W is $((n-1))$-bounded by $k_{18} + c_{33} \leqslant k_{12}'$ and n-bounded by $k_{19} + u_4 \leqslant k_{13}'$, also $s(Z) > s(S_1) - r_{33} > u_{30} - (R_0 + \epsilon_3) - 2/e - r_{33} \geqslant \epsilon_3 + {}$ $+ \operatorname{Max}(u_{24}, u_6, u_{29})$. Hence $s(Z) > \operatorname{Max}(u_{24}, u_6, u_{29}) \geqslant u_{24}$. Similarly W ends with a maximal normal $\overline{Z}'$ of size $> u_{24}$. Now $s(S) > u_{30} \geqslant u_6 + u_4 + \epsilon_3$ so by 6.14A Im $S = \overline{Z}$. Similarly Im $T = \overline{Z}'$. Hence $\overline{Z} \neq \overline{Z}'$.

3°. We have $(Z, Z') \underset{n}{\triangleq} (Z'', Z^+)$ so for suitable adjustments, one of which will be measured

$$(4) \qquad (Z_1, Z_1') \underset{n}{=} (Z_1'', Z_1^+).$$

Now $U \subset AU_1B \subset W$; moreover U is maximal normal in W. Hence U is maximal normal in (Z_1, Z_1'). With respect to (4) let U'' be $\operatorname{Im}'(U)$. We may apply 6.23 since $k_{18} + c_{33} \leqslant k_8'$, $k_{19} + u_4 \leqslant k_9'$, $s(U) > u_{30}' \geqslant u_{10}$. Hence either $|s(U) + s(U') - 1| \leqslant v_2$ or $|s(U) - s(U'')| < v_3$. In the first case $s(U'') \geqslant 1 - s(U) - v_2 > 1 -$ $- (k_{19} + u_4) - v_2 > u_{30} + u_4 + \epsilon_3$, while in the second case $s(U'') > s(U) - v_3 > u_{30}' - v_3 > u_{30} + u_4 + \epsilon_3$. Hence $s(U'') > u_{30} > u_6$. Now $U'' \subset (Z'', Z^+) \subset W'$. Let U' be the image of U'' in (S', T'); then $U' \neq S', T'$; (by 6.20 $U'' \neq \overline{Z}_1'', \overline{Z}_1^+$). By 6.12 $|s(U') - s(U'')| < u_4 + \epsilon_3$ since $k_{18} + c_{33} \leqslant b_{25}$. Hence $s(U') > s(U'') -$ $- u_4 - \epsilon_3 > u_{30}$. Also (i) or (ii) holds.

We now make use of 8.15, noting first that $k_{18} + c_{33} \leqslant k_{16}$, $k_{19} + u_4 \leqslant k_{17}$ and U, U'', U' have size $> u_{30} \geqslant u_{29}$. Then

$$(S, U) \triangleq (\operatorname{In} W, U) \triangleq (Z, U) \triangleq (Z_1, U) \triangleq (Z_1'', U'') \triangleq (Z'', U'')$$
$$\triangleq (\operatorname{In} W', U'') \triangleq (S', U').$$

9.12. Corollary. *Let $O \equiv (S, U, T)$ be an n-object, where S, U, T are (distinct) maximal normal-n subwords and $s(S), s(T) > u_{30}$ while*

$s(U) > u'_{30}$. Let O be $((n-1))$-bounded by k_{18} and n-bounded by k_{19}. Then $O' \equiv (S, U)$ is an n-object and O' is not n-almost equal to O.

Proof. Assume $O \underset{n}{\triangleq} O'$. Now Z exists such that Z is maximal normal in O, $s(Z) > u_{30}$, S is left of Z, Z is left of or coincides with U, $(S, U, T) \triangleq (S, Z)$; namely we could take Z to be U, recalling that $u'_{30} > u_{30}$. Assume now that Z is chosen leftmost.

By 9.11, $(S, Z) \equiv (S, Z', Z)$, $s(Z') > u_{30}$, $(S, U) \triangleq (S, Z')$, S is left of Z', Z' is left of Z, Z' is maximal normal in (S, Z) (hence in O). But $(S, U, T) \triangleq (S, U) \triangleq (S, Z')$, in contradiction to the choice of Z.

9.12A. The corresponding result for left n-objects (cf. 8.8B, 8.8C) is as follows. Let $O \equiv SX$ be a left n-object containing a maximal n-normal subword U different from the maximal n-normal subword S; thus $O \equiv YUZ$ and $Y \not\equiv I$. Let $s(S) > u_{30}$, $s(U) > u'_{30}$. Let O be $((n-1))$-bounded by k_{18} and n-bounded by k_{19}. Then UZ is a left n-object not left n-almost equal to SX.

9.13. Proposition. Let $O \equiv (S, U, T)$, $O' \equiv (S', U', T')$ be n-objects *$((n-1))$-bounded by k_{20} and n-bounded by k_{21} and $S, U, \ldots, T'$ are maximal normal subwords of size greater than u_{31}. Let O, O' be n-almost equal and produce ϵ_1, κ_1. Let (S, U), (S', U') be n-almost equal and produce ϵ_2, κ_2. Then $\epsilon_1 \underset{n}{=} \epsilon_2$.*

Proof. 1°. $O \equiv S_1 A U_1 B T_1$ say where S_1, U_1, T_1 are the kernels of S, U, T. Let subscripts $1, 2, \ldots, 6$ on the letters E, Z, F refer to $S, U, \ldots, T'$ respectively. Then

$$(S, U, T) \underset{n-1}{=} S_1^{\mathrm{L}} E_1 Z_1 F_1 S_1^{\mathrm{R}} A U_1 B T_1^{\mathrm{L}} E_3 Z_3 F_3 T_1^{\mathrm{R}} \equiv W$$

say where (Z_1, Z_3) is an associated preobject. Similarly

$$(S', U', T') \underset{n-1}{=} S_1'^{\mathrm{L}} E_4 Z_4 F_4 S_1'^{\mathrm{R}} A' U_1' B' T_1'^{\mathrm{L}} E_6 Z_6 F_6 T_1'^{\mathrm{R}} \equiv W'$$

say. Since $(Z_1, Z_3) \underset{n}{\triangleq} (Z_4, Z_6)$ we have for suitable adjustments, one of which will be taken standard,

(5) $$(Z_1^*, Z_3^*) \underset{n}{=} (Z_4^*, Z_6^*)$$

and $Z_1 \cdot Z_1^{*-1} \cdot Z_4^* \cdot Z_4^{-1} \in E((Z_1, Z_3), (Z_4, Z_6))$.

Now $U \subset AU_1B \subset (Z_1^*, Z_3^*)$; moreover U is maximal normal in (Z_1^*, Z_3^*). Similarly U' is maximal normal in (Z_4^*, Z_6^*). As in the proof of 9.11 we find that W is $((n-1))$-bounded by $k_{20} + c_{33} \leqslant k'_{12}, k'_8, k_{16}$ and W is n-bounded by $k_{21} + u_4 \leqslant k'_{13}, k'_9, k_{17}$. And further

$$s(Z_1) > u_{31} - r_0 - \epsilon_3 - 2/e - r_{33} \geqslant \epsilon_3 + \mathrm{Max}(u_{24}, u_{29})$$

$$s(U_1) > u_{31} - 2(R_0 + \epsilon_3) - 4/e \geqslant \epsilon_3 + \mathrm{Max}(u_{24}, u_{29})$$

$$s(Z_2) > s(U_1) - r_{33}.$$

2°. With respect to (5) let U'' be Im$'$ U; we note that $s(U) > u_{31} > \mathrm{Max}(u_{10}, u_{29})$. We prove that U'' is U'. Now by 8.15

(6) $$(Z_1^*, U) \doteq (Z_4^*, U'').$$

Now $(S, U) \doteq S_1AU_1 \doteq S_1^{\mathrm{L}}E_1Z_1F_1S_1^{\mathrm{R}}AU_1 \doteq (Z_1, U_1) \doteq (Z_1^*, U)$. Similarly $(S', U') \doteq (Z_4^*, U')$. Hence by (5) and (6) $(Z_4^*, U'') \doteq (Z_4^*, U')$. By 9.12, U'' is U'.

3°. From (1) and the fact that Im$'$ U is U' we obtain, as in the proof of 8.15, that

(7) $$(Z_1^*, U^{\mathrm{L}}) \underset{n}{\sim} (Z_4^*, U'^{\mathrm{L}})$$

where each side is an n-object. Clearly in the usual notation X_0^{-1} for (3) is I. Moreover by the proof of 8.15 U^{L} and U'^{L} have size $> u_{28}$. By 9.10, $(Z_1^*, U^{\mathrm{L}}) \doteq (Z_4^*, U'^{\mathrm{L}})$ and $I \in E'((Z_1^*, U^{\mathrm{L}}), (Z_4^*, U'^{\mathrm{L}}))$. Now

$$(Z_1^*, U^{\mathrm{L}}) \doteq (Z_1^*, U) \doteq (Z_1^*, U_1) \doteq (Z_1^*, U_1^{\mathrm{L}}E_2Z_2F_2U_1^{\mathrm{R}}) \doteq (Z_1^*, Z_2),$$

hence $I \in E'((Z_1^*, U^{\mathrm{L}}), (Z_1^*, Z_2))$. Similarly $I \in E'((Z_4^*, U'^{\mathrm{L}}), (Z_4^*, Z_5))$.

Hence $I \in E'(Z_1^*, Z_2), (Z_4^*, Z_5))$. Next

$$Z_1 \cdot Z_1^{*-1} \in E'((Z_1, Z_2), (Z_1^*, Z_2)),$$

$$Z_4^* \cdot Z_4^{-1} \in E'((Z_4^*, Z_5), (Z_4, Z_5)).$$

Hence $Z_1 \cdot Z_1^{*-1} Z_4^* \cdot Z_4^{-1}$ is an element of $E'((Z_1, Z_2), (Z_4, Z_5))$; by the above it is also an element of $E'((Z_1, Z_3), (Z_4, Z_6))$.

Now $(Z_1, Z_2), (Z_4, Z_5)$ are associated preobjects for $S_1 A U_1$, $S_1' A' U_1'$ respectively. Hence if $S_1 A U_1$, $S_1' A' U_1'$ produce ϵ_3, κ_3 then $\epsilon_1 = \epsilon_3$.

Since $I \in E'(S_1 A U_1, (S, U))$ and $I \in E'(S_1' A' U_1', (S', U'))$ we obtain $\epsilon_1 \underset{n}{=} \epsilon_2$, as required.

9.14. Proposition. (*i*) *Let O_1, O_2, O_3 be n-objects. Let $O_1 \equiv X O_2$, $X \not\equiv I$, and let $O_3 \equiv O_2 Y$, $Y \not\equiv I$. Thus we have, say*

$$O_1 \equiv (S, T, U), \quad O_2 \equiv (T, U), \; O_3 \equiv (T, U, V).$$

Then $X O_2 Y \equiv (S, T, U, V)$ is an n-object.

(*ii*) *Suppose, similarly, that*

$$O_1' \equiv (S', T', U'), \quad O_2' \equiv (T', U'), \quad O_3' \equiv (T', U', V')$$

are n-objects. For $i = 1, 2, 3$ let $O_i \underset{n}{\triangleq} O_i'$ and let O_i, O_i' be $((n-1)$-bounded by k_{20} and n-bounded by k_{21}. Let $S, T, \ldots, U', V'$ have size $> u_{31}$. Then

$$(S, T, U, V) \underset{n}{\triangleq} (S', T', U', V').$$

Proof. (i) is trivial since O_2 is not a subelement of $L_1 \cup \cdots \cup L_n$ (cf. 5.32).

(ii). Let O_i, O_i' produce ϵ_i, κ_i $(i = 1, 2, 3)$. Now by 9.13 $\epsilon_2 \underset{n}{=} \epsilon_3$ and dually $\kappa_1 \underset{n}{=} \kappa_2$. Let S_1, T_1, U_1, V_1 be the kernels of S, T, U, V; then $(S, T, U, V) \equiv S_1 A T_1 B U_1 C V_1$ say. Let subscripts $1, 2, \ldots, 8$ on the letters E, Z, F refer to $S, T, \ldots, V'$ respectively. Say $U \equiv L U_1 M$. Then

$$(S, T, U) \underset{n-1}{=} S_1^{\mathrm{L}} E_1 Z_1 F_1 S_1^{\mathrm{R}} A T_1 B U_1^{\mathrm{L}} E_3 Z_3 F_3 U_1^{\mathrm{R}} M \equiv W$$

say, where (Z_1, Z_3) is an associated preobject. Similarly

$$(S', T', U') \underset{n-1}{=} S_1'^{\mathrm{L}} E_5 Z_5 F_5 S_1'^{\mathrm{R}} A' T_1' B' U_1'^{\mathrm{L}} E_7 Z_7 F_7 U_1'^{\mathrm{R}} M' \equiv W'$$

say, where (Z_5, Z_7) is an associated preobject of (S', T', U'). Since $(Z_1, Z_3) \triangleq (Z_5, Z_7)$ we have for suitable adjustments $(Z_1^*, Z_3^*) = (Z_5^*, Z_7^*)$ and

$$\epsilon_1 \underset{n}{=} S_1^{\mathrm{L}} E_1 (Z_1 \cdot Z_1^{*-1} \cdot Z_5^* \cdot Z_5^{-1}) (S_1'^{\mathrm{L}} E_5)^{-1}.$$

Now $W = (S, T, U) = \epsilon_1 (S', T', U') \kappa_1 = \epsilon_1 W' \kappa_1$. On substituting for ϵ_1 and using $U = U_1^{\mathrm{L}} E_3 Z_3 F_3 U_1^{\mathrm{R}}$ we obtain

(8) $$Z_1^* F_1 S_1^{\mathrm{R}} A (T_1, U_1) M = Z_5^* F_5 S_1'^{\mathrm{R}} A' (T_1', U_1') M' \cdot \kappa_1.$$

Similarly by considering O_3, O_3' we obtain

(9) $$P(T_1, U_1) C V_1^{\mathrm{L}} E_4 Z_4^* = \epsilon_3 \cdot P'(T_1', U_1') C' V_1'^{\mathrm{L}} E_8 Z_8^*$$

where $T \equiv P T_1 Q$, $T' \equiv P' T_1' Q'$.

Now $P(T_1, U_1) M \equiv O_2 = \epsilon_2 O_2' \kappa_2 = \epsilon_3 O_2' \kappa_1 = \epsilon_3 P'(T_1', U_1') M' \kappa_1$ hence by (8) and (9) we have

$$Z_1^* F_1 S_1^{\mathrm{R}} A (T_1, U_1) C V_1^{\mathrm{L}} E_4 Z_4^* = Z_5^* F_5 S_1'^{\mathrm{R}} A' (T_1', U_1') C' V_1'^{\mathrm{L}} E_8 Z_8^*.$$

The left side is an adjustment of an associated preobject of (S, T, U, V). The right side is an adjustment of an associated preobject of (S', T', U', V'). Therefore $(S, T, U, V) \triangleq (S', T', U', V')$.

9.15. Proposition. *Let (S, T), (U, V) be n-objects producing ϵ, κ and let them be (n)-bounded by b where $k_{24}' \leqslant b \leqslant k_{24}$. Let S, T, U, V have n-size $> u_{39}$. Then there is a word $S^{\mathrm{L}} C V^{\mathrm{R}}$, (n)-bounded by $b + u_{40}$* and such *that $S^{\mathrm{L}} C V^{\mathrm{R}} \underset{n}{=} \epsilon \cdot (U, V) \underset{n}{=} (S, T) \cdot \kappa^{-1}$. Moreover S^{L}, V^{R} are n-normal and $s(S^{\mathrm{L}}) > s(S) - u_{40}'$, $s(V^{\mathrm{R}}) > s(V) - u_{40}'$, $s(S^{\mathrm{L}}) > s(\overline{S^{\mathrm{L}}}) - u_{40}''$ and $s(V^{\mathrm{R}}) > s(\overline{V^{\mathrm{R}}}) - u_{40}''$.*

Proof. 1°. By 8.8A we may write

$$S \equiv S'A, \quad T \equiv BT', \quad U \equiv U'C, \quad V \equiv DV'$$

(10) $$(S, T) \underset{n-1}{=} (S'^{\mathrm{L}} EZFS'^{\mathrm{R}} A, BT'^{\mathrm{L}} E_1 Z_1 F_1 T'^{\mathrm{R}})$$

(11) $$(U, V) \underset{n-1}{=} (U'^{\mathrm{L}} E_2 Z_2 F_2 U'^{\mathrm{R}} C, DV'^{\mathrm{L}} E_3 Z_3 F_3 V'^{\mathrm{R}})$$

where (Z, Z_1), (Z_2, Z_3) are associated preobjects, and adjustments of these are equal in G_n:

(12) $$(Z^* FS'^{\mathrm{R}} A, BT'^{\mathrm{L}} E_1 Z_1^*) \underset{n}{=} (Z_2^* F_2 U'^{\mathrm{R}} C, DV'^{\mathrm{L}} E_3 Z_3^*)$$

We may take the left side (Z^*, Z_1^*) to be the standard adjustment, so that

$$|s(\overline{Z^*}) - \tfrac{1}{2}| \leqslant \epsilon_4, \qquad |s(\overline{Z_1^*}) - \tfrac{1}{2}| \leqslant \epsilon_4.$$

As in 8.8A, Z, Z_1, Z_2, Z_3 have n-size greater than

$$u_{39} - (r_0 + \epsilon_3) - 2/e - r_{33} \quad (= \beta \text{ say})$$

and

(13) $$s(Z) > s(S) - (r_0 + \epsilon_3) - 2/e - r_{33}.$$

The right sides of (10), (11) are $((n-1))$-bounded by $k_{24} + c_{33} \leqslant b_{25}$ and n-bounded by $k_{24} + u_4$ since $u_3 < k_{24} < h_0 - \epsilon_2 - 2/e - u_4$ (cf. 6.10A).

By 5.11 and 5.32 both sides of (12) are $((n-1))$-bounded by $\mathrm{Max}(str^+, k_{24} + c_{33}) = k_{24} + c_{33} \leqslant k_8'$. (Z^*, Z_1^*) is n-bounded by $\mathrm{Max}(\frac{1}{2} + \epsilon_4 + \epsilon_3, k_{24} + u_4) = k_{24} + u_4 \leqslant k_9'$, since any subelement of L_n contained in it is in $\overline{Z^*}$ or in $\overline{Z_1^*}$ or in (Z, Z_1). Since (Z_2^*, Z_3^*) is a preobject it is n-bounded by $k_{13} \leqslant k_9'$, and $\overline{Z_2^*}, \overline{Z_3^*}$ have size $> u_{21} - \epsilon_3 > u_{10}$. By 6.24 $\mathrm{Im}'(\overline{Z^*}) = \overline{Z_2^*}$ and $\mathrm{Im}' \overline{Z_1^*} = \overline{Z_3^*}$ (since $\frac{1}{2} - \epsilon_4 > u_{10}$). By 6.23 $|s(\overline{Z_2^*}) - \frac{1}{2}| \leqslant \epsilon_4 + \mathrm{Max}(v_2, v_3) = \beta_1$ say and similarly for $\overline{Z_3^*}$. Hence each of Z^*, Z_1^*, Z_2^*, Z_3^* has n-size $> \frac{1}{2} - \beta_1 - u_{22} - 2/e = \alpha_0$ say (cf. 8.1).

We wish to apply Axiom 31C to (12). Each side is $((n-1)$-bounded by $k_{24}+c_{33} \leqslant f_{31}$ and n-bounded by α_1 where $\alpha_1 = \text{Max}(k_{24}+u_4, \frac{1}{2}+\beta_1) \leqslant f_{31}$. Each of $\overline{Z^*}, \overline{Z_1^*}, \overline{Z_2^*}, \overline{Z_3^*}$ has size $\geqslant \frac{1}{2} - \beta_1 > g_{31}$. By Axiom 31C there is a word $\overline{Z^{*\text{L}}}\, G\overline{Z_3^{*\text{R}}}$ (n)-bounded by $\alpha_2 + h_{31}$, where $\alpha_2 = \text{Max}(k_{24}+c_{33}, \alpha_1)$, $s(\overline{Z^{*\text{L}}}) > s(\overline{Z^*}) - j_{31}$ (similarly for $\overline{Z_3^{*\text{R}}}$) and

(14) $$\overline{Z^{*\text{L}}}\, G\overline{Z_3^{*\text{R}}} \underset{n}{=} (Z_2^* F_2 U'^{\text{R}} C, DV'^{\text{L}} E_3 Z_3^*)$$

Let $A_1 \equiv S'^{\text{L}} EZ \cdot Z^{*-1}$, $A_2 \equiv U'^{\text{L}} E_2 Z_2 \cdot Z_2^{*-1}$. Then $\epsilon \underset{n}{=} A_1 \cdot A_2^{-1}$, so by premultiplying (14) by A_1 and postmultiplying by $Z_3^{*-1} \cdot Z_3 F_3 V'^{\text{R}}$ we obtain

(15) $$J \underset{n}{=} \epsilon \cdot (U, V)$$

where $J \equiv S'^{\text{L}} EZ \cdot Z^{*-1} \cdot \overline{Z^{*\text{L}}}\, G\overline{Z_3^{*\text{R}}} \cdot Z_3^{*-1} \cdot Z_3 F_3 V'^{\text{R}}$.

2°. Let $Y \equiv Z \cdot Z^{*-1} \cdot \overline{Z^{*\text{L}}}$. We shall show that (i) one of $Y, \overline{Z^{*\text{L}}}$ is a right subword of the other, (ii) one of Y, Z is a left subword of the other, (iii) the intersection B of the subwords $Y \cap \overline{Z^{*\text{L}}}$, $Y \cap Z$ of Y is non-empty and satisfies $s^n(B) > \beta - \alpha_5 - 2/e > r_0$ where $\alpha_5 = j_{31} + \epsilon_3 + (2\epsilon_3 + \epsilon_1^*)$, (iv) if $Z \subset Y$, say $Y \equiv ZW$ then $\overline{Z^{*\text{L}}} = Z^* W$, (v) if $Y \subset Z$, say $Z \equiv YF$, then $Z^* \equiv \overline{Z^{*\text{L}}} F$, (vi) Y is a subpowerelement of L_n hence is $((n-1))$-bounded by str^+.

It will follow that $J' \equiv S'^{\text{L}} EYG\overline{Z_3^{*\text{R}}}$ is $((n-1))$-bounded by $\text{Max}(\alpha_2 + h_{31}, k_{24} + c_{33}) = \alpha_3$ say; to see this we use 5.32 and note that if N is a subpowerelement of $L_1 \cup \cdots \cup L_{n-1}$ contained in $S'^{\text{L}} EY$ but not in Y then $N \subset S'^{\text{L}} EZ$ because $s(N \cap Y) \leqslant r_0'$. In the same way if $Y_3 = \overline{Z_3^{*\text{R}}} \cdot Z_3^{*-1} \cdot Z_3$ we obtain

$$J \equiv S'^{\text{L}} EYGY_3 F_3 V'^{\text{R}}$$

is $((n-1))$-bounded by α_3 also. The n-bound of J will be considered later.

To prove (i)–(vi) we note first that $-Z \equiv -Z^*$ and $Z^*- \equiv \overline{Z^{*\text{L}}}-$. The case when one of $Z, Z^*, \overline{Z^{*\text{L}}}$ is a subword of both the other two is easy; B is in this case Z, Z^* or $\overline{Z^{*\text{L}}}$ so $s(B) > \text{Min}(\beta, \alpha_0, \frac{1}{2} - \beta_1 - j_{31}) = \beta$.

There remains the case $\overline{Z^*}^{L} \subset Z^*$ and $Z \subset Z^*$. Say $Z^* = \overline{Z^*}^{L} R$. If $s(R) > r_0 + \epsilon_3$ then $s(\overline{Z^*}^{L}) + s(R) \leqslant s(Z^*) + (2\epsilon_3 + \epsilon_1^*)$ and $s(Z^*) \leqslant s(\overline{Z^*}) + \epsilon_3 \leqslant s(\overline{Z^*}^{L}) + j_{31} + \epsilon_3$, hence $s(R) < \alpha_4$, where $\alpha_4 = j_{31} + \epsilon_3 + (2\epsilon_3 + \epsilon_1^*)$. Hence $s(R) < \mathrm{Max}(r_0 + \epsilon_3, \alpha_4) = \alpha_5$ say. Thus $R \subset Z$; otherwise $s(Z) \leqslant s(R) + \epsilon_3 < \alpha_5 + \epsilon_3 < \beta$. Thus $Z^* \equiv PQR$, $Z \equiv QR$, $\overline{Z^*}^{L} \equiv PQ$ and $Y \equiv Q$ for some P, Q, R. B above is just Q so $\beta < s(Z) \leqslant s(Q) + \alpha_5 + 2/e$; $\alpha_4 = \alpha_5$ since $j_{31} \geqslant r_0$.

Consider the n-bound of J'. Let $\overline{Z^*}^{L} H$, say, be maximal normal in $\overline{Z^*}^{L} G Z_3^{*R}$. Then $S'^{L} EYH$ is maximal normal in J'.

If $Y \subset Z$ then $s(S'^{L}EY) \leqslant s(S'^{L}EZ) + \epsilon_3 < k_{24} + u_4 + \epsilon_3$. Now in general $s(A) > s(AB) - x$, A normal, implies $s(B) < \mathrm{Max}(r_0 + \epsilon_3, x + (2\epsilon_3 + \epsilon_1^*)$. Hence by Axiom 31C, $s(H) < \mathrm{Max}(r_0 + \epsilon_3, a_{31} + (2\epsilon_3 + \epsilon_1^*))$ $(= \alpha_6$ say) and $s(S'^{L}EYH) \leqslant k_{24} + u_4 + \epsilon_3 + \alpha_6 + 2/e$.

If $Z \subset Y$ then $Y \equiv ZW$ and $\overline{Z^*}^{L} \equiv Z^*W$ say. Now

$$s(Z^*) > s(\overline{Z^*}) - u_{22} - 2/e > s(\overline{Z^*}^{L}) - \epsilon_3 - u_{22} - 2/e$$

so $s(W) < \mathrm{Max}(r_0 + \epsilon_3, \epsilon_3 + u_{22} + 2/e + (2\epsilon_3 + \epsilon_1^*)) = \alpha_7$ say. Now $s(S'^{L}EYH) \leqslant s(S'^{L}EZ) + s(W) + s(H) + 4/e < k_{24} + u_4 + \alpha_6 + \alpha_7 + {}$ $+ 4/e = \alpha_8$ say. Thus J' is n-bounded by $\mathrm{Max}(\alpha_2 + h_{31}, \alpha_8) = \alpha_9$ say.

Thus J also is n-bounded by α_9.

3°. $Y \cap Z$ may be denoted by Z^{L}; it contains B so $s(Z^{L}) \geqslant \beta - \alpha_5 - 2/e - \epsilon_3 = \alpha_{10}$ say. Thus

$$J \equiv S'^{L} E Z^{L} H Z_3^{R} F_3 V'^{R}$$

for some H, Z_3^{R}. Also $s(Z_3^{R}) \geqslant \alpha_{10}$.

Now $S'^{L}EZFS'^{R}A$ $(\equiv S^*$, say) $\underset{n-1}{=} S$ and $S^* \supset Z^{L}$. The weak image of Z^{L}, say Z', exists since $k_{24} + c_{33} \leqslant \mathrm{Min}(b_{25}, b_{31})$ and $s(Z^{L}) > \alpha_{10} > u_3$. Thus $Z^{L} \supset \xi \underset{n-1}{\simeq} \eta \supset Z'$ where $\xi = (Z^{L})^0$ and $s(Z') > s(Z^{L}) - u_4 > \alpha_{10} - u_4 > r_0$. Hence

$$(16) \qquad s(\eta) > s(Z^{L}) - u_4 - \epsilon_3 > \alpha_{10} - u_4 - \epsilon_3 > s_{31}$$

By Axiom 31B there is a word $\eta^{L} F \xi^{R}$, $((n-1))$-bounded by $k_{24} + c_{33} + q_{31} = \alpha_{11}$ and equal in G_{n-1} to E. $\xi \underset{n-1}{=} \eta \cdot K^{-1}$ where

E, K are end words for $\eta \sim \xi$. Also $\eta^{\mathrm{L}}, \xi^{\mathrm{R}}$ are not subelements of $L_1 \cup \cdots \cup L_{n-1}$ and

(17) $\quad s(\eta^{\mathrm{L}}) > s(\eta) - r_{31}$.

Where $S \equiv \theta_1 \eta \theta_2$, $S^* \equiv \theta_3 \xi \theta_4$, $Z^{\mathrm{L}} \equiv \theta_5 \xi \theta_6$ we have $S'^{\mathrm{L}} E Z^{\mathrm{L}} \equiv S'^{\mathrm{L}} E \theta_5 \xi \theta_6 \equiv \theta_3 \xi \theta_6 \underset{n-1}{=} \theta_1 \eta \cdot K^{-1} \cdot \theta_6 \underset{n-1}{=} \theta_1 \eta^{\mathrm{L}} F \xi^{\mathrm{R}} \theta_6$ and this last word is $((n-1))$-bounded by $\mathrm{Max}(k_{24}, \alpha_{11}, str^+)$ and may be written $S^{\mathrm{L}} F Z^{\mathrm{LR}}$; we note that Z^{LR} is not a subelement of $L_1 \cup \cdots \cup L_{n-1}$ since it contains ξ^{R}. Similarly $Z_3^{\mathrm{R}} F_3 V'^{\mathrm{R}} \underset{n-1}{=} Z_3^{\mathrm{RL}} F' V^{\mathrm{R}}$. Hence

(18) $\quad J \underset{n-1}{=} S^{\mathrm{L}} F Z^{\mathrm{LR}} H Z_3^{\mathrm{RL}} F' V^{\mathrm{R}} \equiv S^{\mathrm{L}} C V^{\mathrm{R}}$, say

Now $S^{\mathrm{L}} C V^{\mathrm{R}}$ is $((n-1))$-bounded by $\mathrm{Max}(k_{24}, \alpha_{11}, str^+, \alpha_3) \leqslant \mathrm{Min}(b_{25}, k_{24} + u_{40})$ and by 6.10A, $S^{\mathrm{L}} C V^{\mathrm{R}}$ is n-bounded by $\alpha_9 + u_4 \leqslant k_{24} + u_{40}$, because $u_3 < \alpha_9 < h_0 - \epsilon_2 - 2/e - u_4$.

4°. Consider $Z^{\mathrm{L}} \equiv Y \cap Z$.

(a) If $Y \subset Z$ then $Y \equiv Z^{\mathrm{L}}$, say $Z \equiv YF$, $Z^* \equiv \overline{Z^{*\mathrm{L}}} F$. Now $s(\overline{Z^{*\mathrm{L}}}) > s(\overline{Z^*}) - j_{31} > s(Z^*) - \epsilon_3 - j_{31}$ so $s(F) < \mathrm{Max}(r_0 + \epsilon_3, \epsilon_3 + j_{31} + (2\epsilon_3 + \epsilon_1^*) = \alpha_{12}$ say. Hence

(19) $\quad s(Z) \leqslant s(Z^{\mathrm{L}}) + \alpha_{12} + 2/e$

(b) If $Y \supset Z$ then $Z^{\mathrm{L}} \equiv Z$ and (19) remains true. Now $s(S^{\mathrm{L}}) \geqslant s(\eta^{\mathrm{L}}) - \epsilon_3$ hence from (13), (16), (17) and (19) we obtain

(20) $\quad s(S^{\mathrm{L}}) > s(S) - u'_{40}$

where $u'_{40} = (\alpha_{12} + 2/e) + \epsilon_3 + (u_4 + \epsilon_3) + r_{31} + (r_0 + \epsilon_3 + 2/e + r_{33})$. Now $s(S^{\mathrm{L}}) > u_{39} - u'_{40} > r_0 + \epsilon_3$ so S^{L} is normal.

5°. (a) Each side of the equation $S^* \underset{n-1}{=} S$ is $((n-1))$-bounded by b_{25} and $s(S) > u_{39} > \mathrm{Max}(s_{25}, r_0 + r_{25})$ so by 6.14B

(21) $\quad s(S^*) < s(S) + r_{25} + \epsilon_3$.

(b) Since Z is normal

(22) $s(S'^{\mathrm{L}}EZ) \leqslant s(S^*) + \epsilon_3$.

(c) From (2) we have

(23) $s(S'^{\mathrm{L}}EYH) < s(S'^{\mathrm{L}}EZ) + \alpha_6 + \alpha_7 + 4/e$

(d) Consider (18). $S^{\mathrm{L}}CV^{\mathrm{R}}$ begins with a maximal normal $\overline{S^{\mathrm{L}}}$ and J begins with a maximal normal, namely $S'^{\mathrm{L}}EYH$. We shall apply 6.14A.

$S^{\mathrm{L}}CV^{\mathrm{R}}$ is $((n-1))$-bounded by b_{25} and J is $((n-1))$-bounded by $\alpha_3 \leqslant b_{25}$. Now $s(S^{\mathrm{L}}) > u_{39} - u'_{40}$ hence $s(\overline{S^{\mathrm{L}}}) > u_{39} - u'_{40} - \epsilon_3 >$ $> u_6 + u_4 + \epsilon_3$. Since $Y \supset B$,

$$s(S'^{\mathrm{L}}EYH) \geqslant s(B) - \epsilon_3 > \beta - \alpha_5 - 2/e - \epsilon_3 > u_6.$$

Hence

(24) $s(S'^{\mathrm{L}}EYH) = s(\mathrm{Im}\,\overline{S^{\mathrm{L}}}) > s(\overline{S^{\mathrm{L}}}) - u_4 - \epsilon_3$.

By (20)–(24)

$$s(\overline{S^{\mathrm{L}}}) - s(S^{\mathrm{L}}) < (u_4 + \epsilon_3) + (\alpha_6 + \alpha_7 + 4/e) +$$

$$+ \epsilon_3 + r_{25} + \epsilon_3 + u'_{40} = u''_{40}.$$

6°. The parameter conditions involving k_{24} can be put in the form

$$k_{24} \leqslant \lambda_i \ (i = 1, ..., a), \qquad k_{24} \geqslant \mu_i \ (i = 1, ..., b)$$

where λ_i, μ_i are expressions involving parameters different from k_{24}. Of course it will be shown later (Section 21) that these P.C.'s are consistent. Now define

$$k'_{24} = \mathrm{Max}(\mu_1, ..., \mu_b)$$

and temporarily write m for $\mathrm{Min}(\lambda_1, ..., \lambda_a)$. Then the above proof remains valid if k_{24} is replaced by b where $k'_{24} \leqslant b \leqslant m$; in particular if b satisfies $k'_{24} \leqslant b \leqslant k_{24}$, (since $k'_{24} \leqslant k_{24} \leqslant m$).

§ 10

Informal Summary

In this section we discuss a problem of which the following is a special case. Suppose $W_1 \underset{n}{\sim} W_2 \underset{n}{\sim} \cdots \underset{n}{\sim} W_m$ and A_1 is a maximal n-normal subword of W_1. Let $A_2 \subset W_2$ be $\mathrm{Im}' A_1$, let $A_3 \subset W_3$ be $\mathrm{Im}' A_2$, and so on. Since, for any B, the size of $\mathrm{Im}' B$ may be less than the size of B, it is not obvious that all of $A_2, ..., A_m$ exist. The problem is to find sufficient conditions, independent of m, for all of $A_2, ..., A_m$ to exist.

10.1. We suppose given a sequence $W_1, W_2, ..., W_m$ where for each i W_i is either a linear word or a circular word and W_i is $((n-1))$-bounded by k_{22} and n-bounded by k_{23}. Let $1 \leqslant i \leqslant m-1$; if W_i is circular let there be given a linearization X_i, i.e., X_i is a specified C.A. of $\mathrm{lin}(W_i)$; if W_{i+1} is circular let there be given a linearization Y_{i+1}.

For $i = 1, 2, ..., m-1$ assume one of the following conditions is satisfied.

1a: $W_i^{-1} \equiv W_{i+1}$ where W_i, W_{i+1} are linear

or 1b: $X_i^{-1} \equiv Y_{i+1}$

or 2a: $W_i \underset{n}{\sim} W_{i+1}$ where W_i, W_{i+1} are linear, divisions of W_i, W_{i+1} being given

or 2b: $X_i \underset{n}{=} Y_{i+1}$, divisions being given

or 3a: a linear word J_i is given such that W_i, W_{i+1} are specified subwords of J_i (here W_i and W_{i+1} are linear)

or 3b: W_i is linear and is a given potential subword of Y_{i+1}

or 3c: W_{i+1} is linear and is a given potential subword of X_i.

10.1A. Assume that each circular W_i $(i = 1, 2, ..., m)$ has at least four maximal n-normal subwords of size $> u_{32}$; we remark that this condition is stronger than is necessary for the following theory in this section but it is satisfied in all subsequent applications.

10.2. A maximal n-normal subword A of W_i, where $1 \leqslant i \leqslant m-1$, determines in a natural way a maximal normal subword $f_i(A)$ of W_{i+1} (the notation is temporary) under the following conditions.

In Cases 1a, 1b A^{-1} is a subword of W_{i+1}; let $f_i(A)$ be this subword. (Thus there is no restriction on A in Cases 1a, 1b.)

In Case 2a, define $f_i(A)$ to be $\mathrm{Im}'A$ if $s(A) > u_{10}$; note that 6.16, 6.19 hold since $k_{22} \leqslant k_8, k_8'$ and $k_{23} \leqslant k_9, k_9'$; $(u_{10} \geqslant u_9)$.

In Case 2b, define $f_i(A)$ to be $\mathrm{Im}'A$, in the sense of 6.21, if $s(A) > u_{10}$; we note that no C.A. of W_i or W_{i+1} is a subelement of $L_1 \cup \cdots \cup L_n$ because of 10.1A.

Thus in Cases 2a, 2b, $f_i(A)$ is defined if and only if $s(A) > u_{10}$.

In Cases 3a, 3b, 3c, if A is properly contained in $W_i \cap W_{i+1}$ define $f_i(A)$ to be the subword A of W_{i+1}.

Under similar conditions, a maximal normal subword B of W_{i+1} determines a maximal normal subword $g_{i+1}(B)$ of W_i; again the notation is only temporary. Note that if $g_{i+1}(f_i(A))$ exists it is A, and if $f_i(g_{i+1}(B))$ exists it is B.

10.3. Now assume further that each linear W_i contains two given proper maximal normal subwords λ_i, μ_i of size $> u_{32}$, (where λ_i is left of μ_i); call these *extremal* for W_i. A subword of W_i is then called *central* if it is a subword of (λ_i, μ_i). Note that the extremal subwords have size $> u_{10}$ since $u_{32} \geqslant u_{10}$. Moreover we assume assume

In case 1a, $f_i(\lambda_i) = \mu_{i+1}$ and $f_i(\mu_i) = \lambda_{i+1}$.

In case 2a, $f_i(\lambda_i) = \lambda_{i+1}$ and $f_i(\mu_i) = \mu_{i+1}$.

In case 3a, at least one of $f_i(\lambda_i)$, $g_{i+1}(\lambda_{i+1})$ exists and is central and at least one of $f_i(\mu_i)$, $g_{i+1}(\mu_{i+1})$ exists and is central; (this statement could be weakened).

If W is circular, any subword is called *central.*

The notation $A \to B$, where A is a subword of W_i, means that A, B are central and either $f_i(A) = B$ or $g_i(A) = B$.

Note that if A, A' are subwords of W_i and $A \to B$, $B \to A'$ then A' is A.

10.4. Lemma. *Let the assumptions of* 10.1–10.3 *be made. Let* A_0 *be a maximal n-normal subword of one of* $W_1, W_2, \ldots, W_m$ *such that* $s(A_0) > u_{32}$. *Let* $A_0 \to A_1 \to \cdots \to A_k$. *Then* $s(A_k) > u_{33}$.

Proof. 1°. Say $A_j \subset W_{a_j}$; then a_{j+1} is $a_j \pm 1$. We may assume that the sequence $a_0, a_1, ..., a_k$ has the form $p, p+1, ..., p+k$ or the form $p, p-1, ..., p-k$; indeed without loss of generality we may assume it has the form $0, 1, ..., k$. Thus $A_j \subset W_j$.

We use induction on k. The lemma is true for $k = 0$ since $u_{32} \geqslant u_{33}$. For the induction step we can assume $k > 0$; $A_0, A_1, ..., A_{k-1}$ have size $> u_{33}$; none of $A_1, ..., A_k$ has size $> u_{32}$; the pair W_0, W_1 satisfies 2a or 2b (otherwise $u_{32} < s(A_0) = s(A_1)$).

Call a maximal normal subword E_1 of W_1 *admissible* if there exists q, $1 \leqslant q \leqslant k$, and a maximal normal subword E_q of W_q such that $E_q \to E_{q-1} \to \cdots \to E_1 (E_j \subset W_j)$ and $s(E_q) > u_{32}$; hence $E_q, E_{q-1}, ..., E_1$ have size $> u_{33}$.

If W_1 is linear then λ_1 is left of A_1 and A_1 is left of μ_1. If W_1 is circular there exists maximal normal subwords B', C' of size $> u_{32}$ such that (B', A_1, C') is a subword of W_1. Thus in either case we may consider the rightmost admissible subword B_1 left of A_1 and the leftmost admissible subword C_1 right of A_1, and (B_1, A_1, C_1) is a subword of W_1.

2°. We prove that for $i = 1, 2, ..., k$ there exist sequences

$$B_1 \to B_2 \to \cdots \to B_i, \qquad C_1 \to C_c \to \cdots \to C_i$$

where, for all j, B_j and C_j are subwords of W_j. It follows that B_j and C_j have size $> u_{33}$.

This is true for $i = 1$. Assume that it is true for some i, where $i < k$. Now A_i lies between B_i, C_i since the operations of 10.2 preserve "betweenness". We suppose $(C_i, A_i, B_i) \subset W_i$ (the argument is similar for the case $(B_i, A_i, C_i) \subset W_i$).

In case 1a, 1b B_{i+1} and C_{i+1} clearly exist. They also exist in cases 2a, 2b since $u_{33} \geqslant u_{10}$.

In case 3a C_{i+1} exists if $f_i(\lambda_i)$ exists and is central. In the opposite case, $D_i = g_{i+1}(\lambda_{i+1})$ exists and is central. If D_i is left of C_i or coincides with it then C_{i+1} exists. Now let C_i be left of D_i. Then, since D_i is left of A_i we have

$$\lambda_{i+1} \to D_i \to D_{i-1} \to \cdots \to D_1$$

so D_1 is admissible and between A_1, C_1, a contradiction.

Similarly B_{i+1} exists.

In Case 3b B_{i+1}, C_{i+1} clearly exist.

In Case 3c let $D_i = g_{i+1}(\lambda_{i+1})$. If D_i is left of C_i or coincides with it then C_{i+1} exists. If C_i is left of D_i we obtain a contradiction as in Case 3a. Similarly B_{i+1} exists.

3°. Let $1 \leqslant i \leqslant k$. We prove that if B_i is left of C_i then $(B_1, C_1) \triangleq (B_i, C_i)$ while if C_i is left of B_i then $(B_1, C_1) \triangleq (C_i, B_i)^{-1}$. This is clear if $i = 1$; note that (B_i, C_i) is an object since $k_{22} \leqslant k'_{12}$, $k_{23} \leqslant k'_{13}, u_{33} \geqslant u_{24}$. Assume the result is true for some $i < k$. If B_i is left of C_i, then

$$(B_i, C_i) \begin{cases} \equiv (C_{i+1}, B_{i+1})^{-1} & \text{in Cases 1a, 1b} \\ \triangleq (B_{i+1}, C_{i+1}) & \text{in Case 2a, 2b} \\ \equiv (B_{i+1}, C_{i+1}) & \text{in Cases 3a, 3b, 3c} \end{cases}$$

by 8.15, since $k_{22} \leqslant k_{16}, k_{23} \leqslant k_{17}, u_{33} \geqslant u_{29}$. Hence $(B_1, C_1) \triangleq (B_{i+1}, C_{i+1})$ or $(B_1, C_1) \triangleq (C_{i+1}, B_{i+1})^{-1}$. Similarly if C_i is left of B_i. Since the pair W_0, W_1 satisfies 2a or 2b and since B_1, C_1 have size $> u_{33} \geqslant u_{10}$, $B_0 = g_1(B_1)$, $C_0 = g_1(C_1)$ exist; indeed by 8.15 $(B_0, C_0) \triangleq (B_1, C_1)$. Similarly $(B_0, A_0) \triangleq (B_1, A_1)$. We may suppose that the pair W_{k-1}, W_k satisfies 2a or 2b, otherwise $s(A_k) = s(A_{k-1}) > u_{33}$ and the Lemma follows. By 6.23 either

$$|s(A_{k-1}) + s(A_k) - 1| \leqslant v_2 \quad \text{or} \quad |s(A_{k-1}) - s(A_k)| < v_3.$$

In the first case $s(A_k) \geqslant 1 - v_2 - k_{23} \geqslant u_{33}$ and the Lemma follows. Thus we may suppose $s(A_k) > u_{33} - v_3 \geqslant u'_{30} > u_{30}$. Since $u_{33} - v_3 > u_{24}, u_{29}$ we have as before (B_1, A_1) is n-almost equal to (B_k, A_k) or $(A_k, B_k)^{-1}$ according as B_k is left of or right of A_k. We shall suppose that B_k is left of A_k; the other case is similar.

Thus $(B_0, A_0) \triangleq (B_1, A_1) \triangleq (B_k, A_k)$. Also $A_0 \subset (B_0, C_0) \triangleq (B_k, C_k)$; we wish to apply 9.11. As for A_k, we find $s(B_0) > u_{33} - v_3 \geqslant u'_{30} > u_{30}$; similarly for C_0. Further, $k_{22} \leqslant k_{18}, k_{23} \leqslant k_{19}$, C_k and B_k have size $> u_{33} > u_{30}$ and $s(A_0) > u_{32} \geqslant u_{33} > u'_{30}$. By 9.11 (B_k, C_k) contains A'_0 such that $(B_0, A_0) \triangleq (B_k, A'_0)$. From the proof of 9.11 it is easy to deduce that either

$$s(A'_0) > 1 - (k_{23} + u_4) - v_2 - u_4 - \epsilon_3 \geqslant u_{33}$$

or

$$s(A'_0) > u_{32} - v_3 - u_4 - \epsilon_3 \geqslant u_{33}.$$

Now $(B_k, A_k) \triangleq (B_k, A'_0)$; we wish to apply 9.12 to yield that A_k is A'_0. Now 9.12 is available since $s(A_k) > u'_{30}$, $s(A'_0) > u_{33} > u'_{30}$ and $s(B_k) > u_{33} > u_{30}$. Therefore $s(A_k) > u_{33}$ and the Lemma is proved.

§11

Informal Summary

The purpose of this section is to prove 11.8. Roughly, this implies that if $Z_n \underset{n}{\rightarrow} \cdots \underset{1}{\rightarrow} DE$, $U_n \underset{n}{\rightarrow} \cdots \underset{1}{\rightarrow} DF$, $Z_n \in \operatorname{Rep} L_{n+1}$ and D is large enough in DE, say $s_{DE}(D) > c$, then $Z_n \equiv D'E'$, $U_n \equiv D''F'$, $D' \underset{n}{\sim} D''$, $s_{D'E'}(D') > c - \epsilon$ and $E'^{-1} \cdot F'$ is conjugate in G_n to $E^{-1} \cdot F$. Hence a cancellation hypothesis between two elements of L_{n+1} is reduced to a relation $\underset{n}{\sim}$ between subwords of two corresponding elements of Rep L_{n+1}.

11.1 Proposition. *Let*

(1) $\quad X \underset{k}{\sim} Y$ *(end words E, K)*

where $k \leqslant n$ and X, Y are $((n-1))$-bounded by k_{16} and n-bounded by k_{17}. Let $X \equiv X_0 X_1 X_2 X_3 X_4$, $Y \equiv Y_0 Y_1 Y_2 Y_3 Y_4$. Let (1) induce

(2) $\quad X_1 \underset{t}{\sim} Y_1,$

(3) $\quad X_3 \underset{s}{\sim} Y_3.$

(Thus $0 \leqslant t \leqslant k$, $0 \leqslant s \leqslant k$). Let $t \geqslant 1$, $s \geqslant 1$, and let

T_1, T_2 be maximal t-normal in X and contained in X_1,
T'_1, T'_1 be maximal t-normal in Y and contained in Y_1,
S_1, S_2 be maximal s-normal in X and contained in X_3,
S'_1, S'_2 be maximal s-normal in Y and contained in Y_3.

In (2) let $Im' T_i = T_i'$ $(i = 1, 2)$; *in (3) let* $Im' S_i = S_i'$ $(i = 1, 2)$. *Let* T_1, T_2 *have t-size* $> u_{29}$; *let* S_1, S_2 *have size-s* $> u_{29}$. *Thus by* 8.15, *since* $k_{16} \leqslant k_{17}$, $(T_1, T_2) \underset{t}{\hat{=}} (T_1', T_2')$; *let this pair of objects produce* ϵ, κ. *Similarly let* (S_1, S_2), (S_1', S_2') *produce* ϵ', κ'.

Then

(i) $(T_1, S_2) \underset{k}{=} \epsilon \cdot (T_1', S_2') \cdot \kappa'$ (cf. 2.8)

(ii) *if X, Y are E.R. and* $K \equiv E^{-1}$ *then where*

$$(X) \equiv ((T_1, S_2)A), \quad (Y) \equiv ((T_1', S_2')B)$$

we have $A \underset{k}{=} \kappa'^{-1} \cdot B \cdot \epsilon^{-1}$.

Proof. (i) By the proof of 8.15

$$(4) \qquad (T_1^{\mathrm{R}}, T_2^{\mathrm{L}}) \underset{t}{\sim} (T_1'^{\mathrm{R}}, T_2'^{\mathrm{L}})$$

where each side is a t-object and $T_1^{\mathrm{R}}, T_1'^{\mathrm{R}}$ have size $> u_{28}$. Similarly $T_2^{\mathrm{L}}, T_2'^{\mathrm{L}}$ have size $> u_{28}$. By 9.10 (i), if the pair of objects (4) produce ϵ_4, κ_4 then $\epsilon_4 \underset{t}{=} E_4$, $\kappa_4 \underset{t}{=} F_4$, where E_4, F_4 are the end words for (4). Similarly

$$(5) \qquad (S_1^{\mathrm{R}}, S_2^{\mathrm{L}}) \underset{s}{\sim} (S_1'^{\mathrm{R}}, S_2'^{\mathrm{L}})$$

and $\epsilon_5 \underset{s}{=} E_5$, $\kappa_5 \underset{s}{=} F_5$.

By 9.7, 9.10 (ii) and 9.8, $\epsilon \underset{t}{=} T_1^{\mathrm{L}} \cdot \epsilon_4 \cdot T'^{\mathrm{L}-1}$; hence $\epsilon \underset{t}{=} T_1^{\mathrm{L}} \cdot E_4 \cdot T_1'^{\mathrm{L}-1}$. Similarly $\kappa' \underset{s}{=} S_2'^{\mathrm{R}-1} \cdot F_5 \cdot S_2^{\mathrm{R}}$. Now $(T_1^{\mathrm{R}}, S_2^{\mathrm{L}}) \underset{k}{=} E_4 \cdot (T_1'^{\mathrm{R}}, S_2'^{\mathrm{L}}) \cdot F_5$ by definition of end words. Hence (i) follows.

(ii) Say $X \equiv U(T_1^{\mathrm{R}}, S_2^{\mathrm{L}})V$, $Y \equiv U'(T_1'^{\mathrm{R}}, S_2'^{\mathrm{L}})V'$. Then we have $V \underset{k}{=} F_5^{-1} V' \kappa$, $U \underset{k}{=} EU'E_4^{-1}$, $S_2^{\mathrm{R}} A T_1^{\mathrm{L}} \equiv VU$, $S_2'^{\mathrm{R}} B T_1'^{\mathrm{L}} \equiv V'U'$; hence (ii) follows.

11.1A. Note. In the same way we have that if X, Y are E.R. and $((n-1))$-bounded by k_{16}, n-bounded by k_{17} and $X \underset{n}{\sim} Y$ (end words E, E^{-1}) and if T_1, T_2 are maximal n-normal in X of size $> u_{29}$ and $\mathrm{Im}' T_i = T_i'$ $(i = 1, 2)$ then $A \underset{n}{=} \kappa^{-1} \cdot B \cdot \epsilon^{-1}$, where $(T_1, T_2), (T_1', T_2')$ produce ϵ, κ and $(X) \equiv ((T_1, T_2)A)$, $(Y) \equiv ((S_1, S_2)B)$.

11.2. Definition. Let $X_1, \ldots, X_{q-1}$ and $Y_2, \ldots, Y_q$ be given where for $i = 1, 2, \ldots, q-1$ either $X_i \underset{n}{\sim} Y_{i+1}$ or $X_i \underset{n}{-} Y_{i+1}$. A *k-path* where $0 \leqslant k \leqslant n$ is a sequence

(6) $$A_1 \underset{k}{\sim} B_2 \supset A_2 \underset{k}{\sim} B_3 \supset \cdots \supset A_{q-1} \underset{k}{\sim} B_q$$

where $X_i \underset{n}{\sim} Y_{i+1}$ (or $X_i \underset{n}{-} Y_{i+1}$) induces $A_i \underset{k}{\sim} B_{i+1}$ $(i = 1, \ldots, q-1)$.

11.2A. Let S be a k-subelement contained in A_1. We say that S has *images throughout* the k-path (1) if

1. S is maximal k-normal in X_1:
2. $\mathrm{Im}' S = T$ exists with respect to $A_1 \underset{k}{\sim} B_2$;
3. T is maximal k-normal in Y_2;
4. either $q = 2$ or $T \subset A_2$ and T has images throughout the k-path $A_2 \underset{k}{\sim} B_3 \supset A_3 \cdots \underset{k}{\sim} B_q$.

11.2B. Let S be a $(k+1)$-subelement contained in A_1, $k + 1 \leqslant n$. We say that S has *weak images throughout* (1) if

1. the weak image T of S in B_2 exists;
2. either $q = 2$ or $T \subset A_2$ and T has weak images throughout the k-path $A_2 \underset{k}{\sim} B_3 \supset A_3 \cdots \underset{k}{\sim} B_q$.

Sometimes we shall omit the word "weak" in using 11.2B.

11.3. Note. Let S, T, U be disjoint $(k+1)$-subelements having weak images throughout the k-path (6). Then the iterated nice image of T exists throughout the k-path; for the iterated nice image of T is disjoint from and between the weak images of S, U (cf. 2.6).

11.3A. The conclusion of 7.5 can now be expressed as follows. There is a $(k-1)$-path, with respect to $U \underset{k}{\sim} V$, beginning with a subword of (Y_1, Y_b) such that $Y_{d+2}, \ldots, Y_{b-d-1}$ have weak images throughout.

Recalling that in the proof of 7.7, in case (b), we used 7.5 three times in succession, the conclusion of 7.7 may be expressed as follows.

In Case (a) of Axiom 30 there is a k-path, with respect to

(*) $$(X) \underset{k-1}{-} X' \underset{k}{-} X' \underset{k}{-} (Y)$$

beginning with a subword of A in which each of $X_{j+1}, \ldots, X_{a-j}$ has images throughout.

In Case (b) there is a $(k-1)$-path with respect to $(*)$ beginning with a subword of A in which each of $X_{j+1}, \ldots, X_{a-j}$ has weak images throughout.

11.4. Remark. Suppose that $W_1 \underset{n}{-} W_2 \underset{n}{-} \cdots \underset{n}{-} W_q$. Put $X_i \equiv W_i$ $(i = 1, \ldots, q-1)$, $Y_i \equiv W_i$ $(i = 2, \ldots, q)$; thus $X_i \underset{n}{-} Y_{i+1}$ $(i = 1, \ldots, q-1)$. Let (6) be a corresponding k-path. Let C_1 be a subword of X_1 containing A_1. Then the iterated nice image of C_1 exists throughout, i.e.,

(7) $$C_1^0 \underset{n}{\sim} C_2 \supset C_2^0 \underset{n}{\sim} C_3 \supset \cdots \supset C_{q-1}^0 \underset{n}{\sim} C_q$$

Moreover the terms of (7) contain the terms of (6) respectively.

Proof. A_1 is nice and contained in C_1 so $A_1 \subset C_1^0$, say $C_1^0 \underset{n}{\sim} C_2$. Hence $B_2 \subset C_2$. Thus $A_2 \subset C_2$, $A_2 \subset C_2^0$ and so on.

11.5. We consider two conditions:

Condition A. If $1 \leqslant s \leqslant n$ and X is a subpowerelement of L_{n+1} and of L_s then $s^{n+1}(X) \leqslant r_0'$.

(cf. 5.21, noting that in 5.21 t and s are $\leqslant n$.)

Condition B. If $A \underset{p}{\sim} Y$, where A is an $(n+1)$-subelement, $p \leqslant n-1$, $s^{n+1}(A) > s_{25}$ and A, Y are (p)-bounded by b_{25} then Y contains a subelement of L_{n+1} of size greater than $s(A) - r_{25}$.

(Condition B is implied by, but does not imply, the statement that Axiom 25 holds for $n+1$.)

Thus in Conditions A, B if we replace n by $n-1$ we obtain true statements.

11.6. Remark. Assume Conditions A, B. Then 7.4, 7.5 remain valid if $k = n$.

Proof. Using Condition A it may be verified that 6.8 holds if A is a subelement of L_{n+1} and $p \leqslant n-1$. Hence using Condition B it may be seen that 6.10 holds for A a subelement of L_{n+1} and

$p \leqslant n-1$. The proofs of 7.4, 7.5 are now seen to remain valid for $k = n$; (in the first sentence of the proof of 7.5 Condition A is needed again).

11.7. Proposition. *Let*

$$Z_n \underset{n}{\rightarrow} \cdots \rightarrow Z_{n-r} \underset{n-r}{\rightarrow} Z_{n-r-1} \equiv (DE)$$

$$U_{n-r} \underset{n-r}{\rightarrow} U_{n-r-1} \underset{n-r-1}{=} (DF)$$

where

1. $Z_n \in Rep(L^k_{n+1})$
2. $0 \leqslant r < n$
3. (DF) *is* $((n-r-1))$*-bounded by* u_{35}.
4. *In the notation of 7.8 let the A-descendants of* $B_1, \ldots, B_g$ *in* Z_{n-r-1} *be contained in D except possibly for the five leftmost of* B_1 *and the five rightmost of* B_g, *where g is such that* $1 \leqslant g \leqslant m$ *and* $kg/m > u_{36}$.
5. *If* $r = 0$ *assume Conditions A, B.*

Then

$$Z_{n-r} \equiv (JE')$$

$$U_{n-r} \underset{n-r}{=} (JF')$$

where:

1. (JF') *is* $((n-r))$*-bounded by* u_{35}.
2. $E^{-1} \cdot F \underset{n-r}{=} T^{-1} \cdot E'^{-1} \cdot F' \cdot T$ *for some T.*
3. *If* $r > 0$, *J contains all A-descendants of* $B_1, \ldots, B_g$ *in* Z_{n-r} *except possibly for the five leftmost of* B_1 *and the five rightmost of* B_g. *If* $r = 0$, $J \supset (B_{i+s}, B_{g-s})$, *where* $k(g-2s)/m > kg/m - u_{37}$.
4. *In* $U_{n-r} \rightarrow U_{n-r-1} - (DF)$ *let the iterated nice image of D in* U_{n-r} *be* D^* *say and in* $U_{n-r} - (JF')$ *let the nice image of J in* U_{n-r} *be* J^* *say; then* J^* *is contained in* D^*. *In particular* J^*, D^* *exist.*

Note. If $r = 0$ we interpret 4 of the hypothesis to mean that the A-descendants of $B_6, \ldots, B_{g-5}$ in Z_{n-r-1} are contained in D (where g is such that $1 \leqslant g \leqslant m$ and $kg/m > u_{36}$).

We take a similar interpretation when $r = 1$ in 3 of the conclusion; also in 11.8, conclusion 3, when $p = n - 1$.

11.7A. Proposition. *Proposition* 11.7 *remains valid when modified as follows. Let* U_{n-r}, U_{n-r-1} *be linear words and let*

$$U_{n-r} \underset{n-r}{\rightarrow} U_{n-r-1} \underset{n-r-1}{=} GDH$$

where GDH *is* $((n-r-1))$*-bounded by* u_{35} (1, 2, 4, 5 *of the hypothesis being unchanged). Then* $U_{n-r} \underset{n-r}{=} G'JH'$ *is* $((n-r))$*-bounded by* u_{35}, $G \cdot E^{-1} \cdot H \underset{n-r}{=} G' \cdot E'^{-1} \cdot H'$, 3 *remains valid and* $J^* \subset D^*$ (*where* J^*, D^* *are defined by analogy with* 4).

Proof of 11.7. Let $m' = m/k$.

1°. Write Z, U, U', k for Z_{n-r}, U_{n-r}, U_{n-r-1}, $n-r$. Then $1 \leqslant k \leqslant n$ and

$$Z \underset{k}{\rightarrow} (DE)$$

$$U \underset{k}{\rightarrow} U' \underset{k-1}{=} (DF)$$

Let $r > 0$, i.e., $k < n$. Let the A-descendants of $B_1, \ldots, B_g$ in Z_{n-r} be V_i $(i \in I)$. (Thus I is in one-one correspondence with the set of r-tuples γ such that A_γ^{n-r+1} appears in the diagram of 7.8 and begins with one of $1, 2, \ldots, g$. If $i \neq j$ the subwords V_i, V_j of Z_{n-r} may coincide.) Each V_i is a $(k+1)$-subelement contained in Z so is (k-bounded) by s_0. By Axiom 30 since $s_0 \leqslant b_{30}$ and $s(V_i) > u_{18} \geqslant s_{30}$ we have, that V_i contains k-subelements V_{ij} of size $> q^*$. Since $q^* \geqslant s_{30}$ we have, if $k \geqslant 2$, that each V_{ij} contains subelements V_{ijk} of L_{k-1} of size $> q^*$; and so on. Summarizing:

$$V_i^{k+1} \supset V_{ij}^k \supset V_{ijk}^{k-1} \supset \cdots \supset V_{(k+1)}^1 \tag{8}$$

Let $r = 0$ $(k = n)$. Put $V_1 \equiv (B_1, B_g)$ and put $V_{1i} = B_i$ $(i = 1, \ldots, g)$. Then $V_1 \subset Z_n \in L_{n+1}$, V_{1i} is an n-subelement and $s(V_{1i}) > q^*$. Thus (8) also holds for $r = 0$ but i has the fixed value 1.

2°. By Axiom 26 cyclic words Y_1, Y_2, Y_3, Y_4 exist such that

(9) $$Z \underset{k-1}{\to} Y_1 \underset{\bar{k}}{\to} Y_2 \underset{\bar{k}}{\to} (DE)$$

(10) $$(DF) \underset{k-1}{\to} U' \underset{\bar{k}}{\to} Y_4 \underset{\bar{k}}{\to} Y_3 \underset{k-1}{\to} U$$

(Possibly $Z \equiv Y_1 \equiv Y_2 \equiv (DE)$ or $U \equiv Y_3 \equiv Y_4 \equiv U'$.). We say a subword A of Z *clears a t-path* if there is a t-path, with respect to (9) followed by (10), beginning with a subword of A.

3°. Consider a fixed V_i and assume the V_{ij} are maximal normal; more precisely if $r > 0$ assume the V_{ij} satisfy (a) of Axiom 30, while if $r = 0$ assume the V_{1j} $(= B_j)$ are maximal normal of size $> r''$ (cf. the beginning of 7.8).

If $r > 0$, V_i is an A-descendant of some B_j so has size $> u_{18}$; recall that $u_{18} \geqslant u_{16}$. By 7.7 V_{ij} $(j = 4, ..., a-3)$ have images V'_{ij} in (DE) of size $> u_{17} \geqslant u_6$ and they are contained in the iterated nice image of (V_{i1}, V_{ia}) in (DE). Now (cf. 7.8) these V'_{ij} are the A-descendants of V_i in $Z_{n-r-1} \equiv (DE)$ so by 4 of the hypothesis V'_{ij} $(j = 10, ..., a-9)$ are *properly* contained in D, so are maximal normal in (DF).

If $r = 0$ then by 7.8 the B_j $(= V_{1j})$ have images A_j $(= V'_{1j}$ say) in Z_{n-1} of size $> r'' - r_{26} \geqslant u_{17}$. Writing a instead of g, $A_4, ..., A_{a-3}$ are contained in the iterated nice image of $V_1 = (V_{11}, V_{1a})$ in (DE) and again $A_j = V'_{1j}$ $(j = 10, ..., a-9)$ are properly contained in D.

Since $s(V'_{ij}) > u_6$, V'_{ij} $(j = 11, ..., a-10)$ have images V''_{ij} in U' contained in the nice image of $(V'_{i\,10}, V'_{i\,a-9})$ in U' and by 6.11 they have size $> s(V'_{ij}) - u_4 - \epsilon_3 > u_{17} - u_4 - \epsilon_3 > v_{26}$. Now $U \to U'$, so by Axiom 26″ the V''_{ij} have images V'''_{ij} in U. Further, V'''_{ij} $(j = 14, ..., a-13)$ are contained in the iterated nice image of $(V''_{i\,11}, V''_{a-10})$ in U.

Summarizing, V_i clears a k-path and V_{ij} $(j = 14, ..., a-13)$ have images throughout. $a - 26 \geqslant 4$ since

$$a > e(V_i) - \epsilon_{30}) > e(u_{18} - \epsilon_{30}) \geqslant 29 \quad (r > 0)$$

$$a = g > e \cdot u_{36} \geqslant 29 \quad (r = 0).$$

4°. Assume $r > 0$ and suppose that $V_{i1}, ..., V_{ia}$ satisfy (b) of Axiom 30.

In the present case, by 7.7, and 11.3A there is a $(k-1)$-path, with respect to (2) beginning with a subword of (V_{i1}, V_{ia}) in which V_{ij} $(j = p+1, ..., a-p)$ have images throughout (p here is the j of 7.7 (iii)); let V'_{ij} be the image in (DE) of V_{ij}. The V'_{ij} are the A-descendants of V_i in (DE), so V'_{ij} $(j = p+6, ..., a-p-5)$ are contained in D.

We shall apply 7.5 four times in succession; it will follow that there is a $(k-1)$-path with respect to (3) beginning with a subword of $(V'_{i\,p+6}, V'_{i\,a-p-5})$ in which some of the V'_{ij} have images throughout. Thus V_i will clear a $(k-1)$-path in which some of the V_{ij} have images throughout.

The words $Z, Y_1, Y_2, (DE), U', Y_4, Y_3, U$ are $((k-1))$-bounded by $s_0 + s_{26}$ and k-bounded by $1 - r' + t_{26}$. (DF) is $((k-1))$-bounded by u_{35}. Since $U' \underset{k-1}{=} (DF)$ and since $\mathrm{Max}(s_0 + s_{26}, u_{35}) \leqslant b_{25}$ we have by 6.11 that (DF) is k-bounded by $1 - r' + t_{26} + u_4$. Since $\mathrm{Max}(s_0 + s_{26}, u_{35}) \leqslant k_{10}$ and $1 - r' + t_{26} + u_4 \leqslant k_{11}$, all the words are $((k-1))$-bounded by k_{10} and k-bounded by k_{11}.

Let $\alpha = \epsilon'_{30} + u_{13} + 6u_4$. Let d be the integer defined by $d - 1 < t\alpha \leqslant d$ (where t arises from Axiom 30). Let u_{38} be defined by $u_{38} = u_{18} - \epsilon_{30} - 14\alpha - 14/e$. We have $a/t > s(V_i) - \epsilon_{30} > u_{18} - \epsilon_{30}$.

We show $a - 14d > e \cdot u_{38}$. Indeed $a - 14(d-1) > t(u_{18} - \epsilon_{30} - 14\alpha) = t(14/e + u_{38}) \geqslant 14 + eu_{38}$.

Write Y_j for V_{ij}. Any d consecutive Y_j's generate a subelement of L_{k+1} of size $> d/t - \epsilon'_{30} \geqslant \alpha - \epsilon'_{30} = u_{13} + 6u_4$. We have $s(Y_j) > q^* > u_{12} + 6u_4$. Next, since (Y_1, Y_a) contains no k-subelement of size $> c^*$, it does not contain a k-subelement of size greater than $u_{15} - 6u_4$, since $c^* \leqslant u_{15} - 6u_4$

In general, if 7.5 is applied n times in succession the number of Y's having images throughout is $b_n = b - 2n(2n + d - 1)$; (we used $b_1 = b - 2(d+1)$ in 7.5 and $b_3 = b - 2(3d + 15)$ in 7.7). In the present case the number of V_{ij} having images throughout is $a - 14(13 + d) - 10 = a - 14d - 192$ (since ten are 'lost' in passing from (DE) to (DF)) $> e \cdot u_{38} - 192 > 3$.

5°. Let $r = 0$ and suppose that $V_{11}, ..., V_{1a}$ are small in the sense that they are disjoint and of size $> u_{11}$ (cf. the beginning of 7.8).

Since Conditions A, B hold, 11.6 is available. Recall that V_{1j} is B_j. Any s consecutive B_j's generate a subword of Z $(= Z_n)$ hence a subelement of L_{n+1} of size $\geqslant s/m'$ (cf. 7.11), so of size $> s/m' - \epsilon'_{30}$. (B_1, B_g) does not contain an n-subelement of size $> r''$ (cf. 7.3). Thus the argument of 4° remains valid for $r = 0$, where u_{11}, r'' take the place of q^*, c^* since $u_{11} > u_{12} + 6u_4$ and $r'' \leqslant u_{15} - 6u_4$.

6°. Let V_γ^{p+1} where γ is a $(k+1-p)$-tuple occur in (1) and let $p \geqslant 1$. We say V_γ *satisfies* (*R*) if all $V_{\gamma i}$ satisfy (b) of Axiom 30, V_γ clears a $(p-1)$-path and $V_{\gamma i}$ $(i \in I_\gamma)$ have images throughout; here I_γ is a set of integers of the form

$$\{s_\gamma + 1, s_\gamma + 2, \dots, s_\gamma + t_\gamma\}, \quad t_\gamma \geqslant 4.$$

We say V_γ *satisfies* (S) if all $V_{\gamma i}$ satisfy (a) of Axiom 30, V_γ clears a p-path and $V_{\gamma i}$ $(i \in I_\gamma)$ have images throughout.

We have already proved that each V_i^{k+1} satisfies (S) or (R), if we interpret 'satisfies (a) or (b) of Axiom 30' when $r = 0$ to mean 'satisfies 1 or 2 at the beginning of 7.8'.

Assume V_γ^{p+1} satisfies (R) and let $i_0 \in I'_\gamma = s_\gamma + 2, \dots, s_\gamma + t_\gamma - 1$ we shall prove that $V_{\gamma i_0}^p$ satisfies (R) or (S) (if $p \geqslant 2$).

V_γ clears a $(p-1)$-path and all $V_{\gamma i}$ $(i \in I_\gamma)$ have images throughout, so the iterated nice image of $V_{\gamma i_0}$ exists throughout the $(p-1)$-path by 11.3. Let $J = V_{\gamma i_0}$, $J_j = V_{\gamma i_0 j}$.

First assume the J_j $(j = 1, \dots, a)$ satisfy (a) of Axiom 30; we shall prove that J satisfies (S). Since $p \leqslant k$, all the words of the $(p-1)$-path are $((p-1))$-bounded by $\mathrm{Max}(s_0 + s_{26}, u_{35}) \leqslant$ $\leqslant \mathrm{Min}(k_{16}, k_{17}, k_{18}, k_{19}, k'_8, k'_9)$. The path begins say $Z^M \underset{p-1}{\sim} Y_1^M \supset Y_1^m \underset{p-1}{\sim} Y_2^M$, where $J \subset Z^M$ and $J^0 \underset{p-1}{\sim} J^{0'} \supset J^{0'0} \underset{p-1}{\sim} J^{0'0'}$. Clearly $J^{0'} \subset Y_1^M$ but by considering 11.3 we obtain $J^{0'} \subset Y_1^m$. Since $s(J_i) > r^* > \mathrm{Min}(u_{29}, u_{10})$, the J_i have images J'_i in Y_1^M and $(J'_2, J'_{a-1}) \subset J^{0'}$; moreover by 8.15 we have $(J_r, J_s) \underset{p-1}{\hat{=}} (J'_r, J'_s)$ for $r, s \in \{1, 2, \dots, a\}, r < s$.

Now $s(J'_i) > r^* - v_3 > r^* - v_3 - u_4 - \epsilon_3 > \mathrm{Min}(u_{29}, u_{10})$, so $J'_2, \dots, J'_{a-1}$ have images $J''_2, \dots, J''_{a-1}$ in Y_2^M and $(J''_3, J''_{a-2}) \subset J^{0'0'}$. Also $(J'_r, J'_s) \hat{=} (J''_r, J''_s)$ for $r, s \in \{2, 3, \dots, a-1\}, r < s$. Hence for this range of r, s $(J_r, J_s) \hat{=} (J''_r, J''_s)$.

Let $3 \leqslant x \leqslant a-2$; we prove $s(J''_x) > r^* - v_3 - u_4 - \epsilon_3$. We have $(J_{x-1}, J_{x+1}) \hat{=} (J''_{x-1}, J''_{x+1})$. Since $s(J_x) > r^* > \text{Max}(u_{30}, u'_{30})$ then by 9.11 there exists $\bar{J}_x$ such that $(J_{x-1}, J_x) \hat{=} (J''_{x-1}, \bar{J}_x)$ and

$$s(\bar{J}_x) > r^* - v_3 - u_4 - \epsilon_3 > \text{Max}(u_{30}, u'_{30})$$

By 9.12, $\bar{J}_x$ is J''_x and the result follows.

By repeating the above argument we find that images of $J_8, \ldots, J_{a-7}$ exist throughout the $(p-1)$-path.

Note that $a > e(s(J) - \epsilon_{30}) > e(q^* - \epsilon_{30}) > 17$.

Finally assume the J_j $(j = 1, \ldots, a)$ satisfy (b) of Axiom 30; we show that J satisfies (R). Consider the part of 4° from 'The words $Z, Y_1, Y_2, \ldots$' to the end; it applies in the present case since $s(J) > q^* \geqslant u_{18}$, and we obtain that the number of J_j's having images throughout is (at least) $a - 14(13 + d) > e \cdot u_{38} - 182 > 3$.

7°. For later, note that the part of the argument of 6° from "First assume the J_j ..." to "... $J_8, \ldots, J_{a-7}$ exist throughout the $(p-1)$-path" can be applied to the situation of 3°; with V_i, V_{ij}, k replacing $J, J_j, p-1$ respectively. To see this, note that all words of (2), (3) are $((k-1))$-bounded by $\text{Max}(s_0 + s_{26}, u_{35}) \leqslant$ $\leqslant \text{Min}(k_{16}, k_{18}, k'_8)$ and all words except (DF) are k-bounded by $1 - r' + t_{26}$; but since $U' \underset{k-1}{=} (DF)$ and $\text{Max}(s_0 + s_{26}, u_{35}) \leqslant b_{25}$ we see from 6.10A that (DF) is k-bounded by $1 - r' + t_{26} + u_4 \leqslant$ $\leqslant \text{Min}(k_{17}, k_{19}, k'_9)$.

The result is that (in the notation of 3°)

$$(V_{ir}, V_{is}) \underset{k}{\hat{=}} (V'_{ir}, V'_{is}) \underset{k}{\hat{=}} (V''''_{ir}, V''''_{is}) \text{ if } 14 \leqslant r < s \leqslant a - 13.$$

8°. Let the leftmost, rightmost V_i $(i \in I)$ be denoted by V_1, V_f say; these coincide if and only if $r = 0$. Choose $i_1 \in I$; then V_{i_1} satisfies (S) or (R). If it satisfies (R) and $k \geqslant 2$, choose $i_2 \in I'_{i_1}$. Then $V_{i_1 i_2}$ satisfies (S) or (R). If it satisfies (R) and $k \geqslant 3$, choose $i_3 \in I'_{i_1 i_2}$. Then $V_{i_1 i_2 i_3}$ satisfies (S) or (R). Eventually either $V_{i_1 i_2 \ldots i_w}$ satisfies (S) for some w, $1 \leqslant w \leqslant k$, or $V_{i_1 i_2 \ldots i_k}$ satisfies (R). We consider, in particular, two such sequences:

(i) $i_1 = 1$ and $i_2, i_3, \ldots$ chosen minimally. Denote this sequence by $\lambda = \{a_1, a_2, \ldots\}$ $(a_1 = 1)$.

(ii) $i_1 = f$ and $i_2, i_3, \ldots$ chosen maximally; call this sequence $\rho = \{b_1, b_2, \ldots\}$ $(b_1 = f)$.

We shall assume that both V_λ and V_ρ satisfy (S); if one or both satisfy (R) the subsequent argument becomes easier since 0-paths are easily dealt with.

Let V_λ, V_ρ be subelements say of L_{t+1}, L_{s+1} respectively. Then $1 \leqslant t \leqslant k$, $1 \leqslant s \leqslant k$, λ is a $(k+1-t)$-tuple and ρ is a $(k+1-s)$-tuple.

V_λ clears a t-path and $V_{\lambda i}$ $(i \in I_\lambda)$ has images throughout. The $V_{\lambda i}$ satisfy (a) of Axiom 30; the case when $r = 0$ and V_1 satisfies (S) is easy but exceptional and will be considered later; thus V_λ, V_ρ are disjoint. Let the two leftmost $V_{\lambda i}$ $(i \in I_\lambda)$ be say T_1, T_2. Let the two rightmost of the $V_{\rho i}$ $(i \in I_\rho)$ be S_1, S_2. Let their images in (DE) be T_1', T_2', S_1', S_2' and let their images in U be $T_1'', T_2'', S_1'', S_2''$.

Then $(T_1, T_2) \underset{t}{\widehat{=}} (T_1', T_2') \underset{t}{\widehat{=}} (T_1'', T_2'')$. We have, say, $Z \equiv ((T_1, S_2)\alpha)$ $DE \equiv (T_1', S_2')\, D^{\mathrm{R}} E D^{\mathrm{L}}$ and $U \equiv (T_1'', S_2'')\gamma$.

We now use 11.1 three times; this is possible since

$$\operatorname{Min}(r'', r^*) - v_3 - u_4 - \epsilon_3 > u_{29},$$

$$s_0 + s_{26} \leqslant \operatorname{Max}(s_0 + s_{26}, u_{35}) \leqslant k_{16},$$

$$1 - r' + t_{26} < 1 - r' + t_{26} + u_4 \leqslant k_{17}.$$

If $(T_1, T_2), (T_1', T_2')$ produce ϵ, κ and $(S_1, S_2), (S_1', S_2')$ produce ϵ', κ' then, making use of 9.8 we obtain $(T_1, S_2) \underset{k}{\equiv} \epsilon \cdot (T_1', S_2') \cdot \kappa'$ and $\alpha \underset{k}{\equiv} \kappa'^{-1} D^{\mathrm{R}} E D^{\mathrm{L}} \epsilon^{-1}$. Let $(T_1', T_2'), (T_1'', T_2'')$ produce ϵ_1, κ_1 and let $(S_1', S_2'), (S_1'', S_2'')$ produce ϵ_2, κ_2. Using 11.1 four times we obtain similarly $(T_1', S_2') \underset{k}{\equiv} \epsilon_1 (T_1'', S_2'') \kappa_2$ and $D^{\mathrm{R}} F D^{\mathrm{L}} = \kappa_2^{-1} \gamma \epsilon_1^{-1}$.

Let $(T_1, T_2), (T_1'', T_2'')$ produce ϵ_4, κ_4 and let $(S_1, S_2), (S_1'', S_2'')$ produce ϵ_5, κ_5. Then $\epsilon \cdot \epsilon_1 \underset{k}{\equiv} \epsilon_4$, $\kappa_2 \cdot \kappa' = \kappa_5$. Hence

$$(11) \qquad (T_1, S_2) \underset{k}{\equiv} \epsilon_4 \cdot (T_1'', S_2'') \cdot \kappa_5$$

Also $\epsilon D^{\mathrm{L}-1} E^{-1} F D^{\mathrm{L}} \epsilon^{-1} \underset{k}{\equiv} \alpha^{-1} \kappa'^{-1} D^{\mathrm{R}} F D^{\mathrm{L}} \epsilon^{-1} = \alpha^{-1} \kappa_5^{-1} \gamma \epsilon_4^{-1}$; thus the last word is conjugate to $E^{-1} \cdot F$ in G_k.

9°. We shall show that words $\theta_j, W_j, W_j'', \phi_j, E_0, F_0$ exist such that

$$(12) \qquad (T_1, S_2) \equiv \theta_1 W_1 \theta_2 W_2 \theta_3$$

$$(13) \qquad (T_1'', S_2'') \equiv \phi_1 W_1'' \phi_2 W_2'' \phi_3$$

$$(14) \qquad \epsilon_4^{-1} \cdot \theta_1 W_1 \underset{k}{\equiv} \phi_1 W_1''^{\mathrm{L}} E_0 W_1^{\mathrm{R}}$$

$$(15) \qquad W_2 \theta_3 \cdot \kappa_5^{-1} \underset{k}{\equiv} W_2^{\mathrm{L}} F_0 W_2''^{\mathrm{R}} \phi_3$$

where $W_1^{\mathrm{R}}, W_2^{\mathrm{L}}, W_1''^{\mathrm{L}}, W_2''^{\mathrm{R}}$ are non-empty. We shall then have the following.

$(T_1'', S_2'')\gamma \underset{k}{\equiv} \epsilon_4^{-1} \cdot (T_1, S_2) \cdot \kappa_5^{-1} \cdot \gamma = V'$, where

$$V' \equiv \phi_1 W_1''^{\mathrm{L}} E_0 W_1^{\mathrm{R}} \theta_2 W_2^{\mathrm{L}} F_0 W_2''^{\mathrm{R}} \phi_3 \gamma,$$

$U \underset{k}{=} V$, where $V \equiv (V')$. For the required J in the conclusion of Proposition 11.7 take $W_1^{\mathrm{R}} \theta_2 W_2^{\mathrm{L}}$ and take $E' \equiv W_2^{\mathrm{R}} \theta_3 \alpha \theta_1 W_1^{\mathrm{L}}$, $F' \equiv F_0 W_2''^{\mathrm{R}} \phi_3 \gamma \phi_1 W_1''^{\mathrm{L}} E_0$, where $W_i \equiv W_i^{\mathrm{L}} W_i^{\mathrm{R}}$ $(i = 1, 2)$. Part 2 of the conclusion will follow since by (14), (15)

$$E'^{-1} \cdot F' = W_1^{\mathrm{L}-1} \theta_1^{-1} \alpha^{-1} (\theta_3^{-1} W_2^{\mathrm{R}-1} F_0 W_2''^{\mathrm{R}} \phi_3) \gamma (\phi_1 W_1''^{\mathrm{L}} E_0)$$

$$= W_1^{\mathrm{L}-1} \theta_1^{-1} (\alpha^{-1} \kappa_5^{-1} \gamma \epsilon_4^{-1}) \theta_1 W_1^{\mathrm{L}}$$

so $E'^{-1} \cdot F'$ is conjugate in G_k to $E^{-1} \cdot F$ (by 8°).

Obviously the nice image J^* in U of the subword J of V is contained in (T_1'', S_2'').

10°. If a subword of Z clears a u-path (cf. 2°) whose last term is T, say, then it induces a u-path with respect to (3) whose first term is a subword D^M of D and whose last term is T. By 11.4 the iterated nice image D^* of D with respect to (3) contains T.

Hence $D^* \supset T_1''$, $D^* \supset S_2''$; hence $D^* \supset (T_1'', T_2'')$.

11°. Let $r > 0$. Each V_i $(i \in I)$ is, in the notation of 7.8 of the form A_γ^{n-r+1} where γ is an r-tuple.

If $r = 1$, the V_i $(i \in I)$ are $A_1, \ldots, A_g$ cf. 7.8 so we may take $I = \{1, 2, \ldots, g\}$, $V_i = A_i$, $f = g$. Thus kernels exist for $V_1, \ldots, V_f$.

If $r \geqslant 2$ consider the A-descendants of $B_1, \ldots, B_g$ in Z_{n-r+1}; temporarily denote the leftmost, rightmost by X, Y respectively and the remaining ones by W_j ($j \in J$). By 7.16A, if N is the largest integer $< u_{34}$ the first N A-descendants of X in Z_{n-r} are left of and disjoint from all A-descendants in Z_{n-r} of W_j ($j \in J$) and Y. (Dually for Y.) Thus there is no loss of generality in denoting these N A-descendants by $V_1, V_2, \ldots, V_N$ or in assuming (i) f is an integer (ii) $f \geqslant 2N$. The rightmost N A-descendants of Y in Z_{n-r} may then be denoted by $V_{f-N+1}, \ldots, V_f$. Kernels exist for $V_1, \ldots, V_N$ and for $V_{f-N+1}, \ldots, V_f$. Since $N \geqslant u_{34} - 1 > 5$ we have $N \geqslant 6$, $f \geqslant 12$. Also $g > eu_{36} \geqslant 12$ so $g \geqslant 13$.

12°. Let $r > 0$. Now $T_1 \subset V_\lambda \subset V_1, S_2 \subset V_\rho \subset V_f$.
Put $(T_1, S_2) \equiv T_1 A S_2$, $(T_1'', S_2'') \equiv T_1'' B S_2''$. Note that $A \supset (V_3, V_{f-2})$.

Recalling 8.8A, 8.8B we have

$$T_1 \underset{t-1}{=} T_1^{\mathrm{L}} EZFT_1^{\mathrm{R}}, \qquad T_1'' \underset{t-1}{=} T_1''^{\mathrm{L}} E'Z'F'T_1''^{\mathrm{R}}$$

where Z, Z' are subelements of $\mathrm{Rep}(L_t)$. As in 9.5

$$\epsilon_4 \underset{t}{=} T_1^{\mathrm{L}} EZZ^{*-1} Z'^{*} Z'^{-1} E'^{-1} T_1''^{\mathrm{L}-1}$$

for a suitable Z^* of $\mathrm{Rep}(L_t)$ such that one of Z, Z^* is a right subword of the other, and a suitable Z'^* with similar properties. Hence

$$\text{(16)} \qquad T_1^{-1} \epsilon_4 T_1'' \underset{t}{=} (Z^* F T_1^{\mathrm{R}})^{-1} (Z'^* F' T_1''^{\mathrm{R}}).$$

Similarly $S_2 \underset{s-1}{=} S_2^{\mathrm{L}} E_2 Z_2 F_2 S_2^{\mathrm{R}}$, $S_2'' \underset{s-1}{=} S_2''^{\mathrm{L}} E_3 Z_3 F_3 S_2''^{\mathrm{R}}$. By (11),

$$\text{(17)} \qquad Z^* F T_1^{\mathrm{R}} A S_2^{\mathrm{L}} E_2 Z_2^* \underset{k}{=} Z'^* F' T_1''^{\mathrm{R}} B S_2''^{\mathrm{L}} E_3 Z_3^*.$$

Next we show that each side of (17) is $((k))$-bounded by b_{25}. $T_1^{\mathrm{R}} A S_2^{\mathrm{L}}$ is not a subelement of $L_1 \cup \cdots \cup L_k$ since it contains V_3 and $s^{k+1}(V_3) > u_{18} > r_0'$ (cf. 5.21); hence we need only show that $H \equiv Z^* F T_1^{\mathrm{R}} A S_2^{\mathrm{L}}$ is $((k))$-bounded by b_{25}. Now $T_1^{\mathrm{R}} A S_2^{\mathrm{L}} \subset Z$, so is $(\boldsymbol{k})$-bounded by s_0. $Z^* F T_1^{\mathrm{R}}$ is a subword of a t-preobject so by 8.1 is $((t-1))$-bounded by k_{12} and t-bounded by k_{13}. Now T_1^{R} is not a subelement of $L_1 \cup \cdots \cup L_{t-1}$ so H is $((t-1))$-bounded by $\mathrm{Max}(k_{12}, s_0) \leqslant b_{25}$.

Next consider the t-bound of H. Where $T_2 \underset{t-1}{=} T_2^{\mathrm{L}} E_1 Z_1 F_1 T_2^{\mathrm{R}}$, $T_2'' \underset{t-1}{=} T_2''^{\mathrm{L}} E'' Z'' F'' T_2''^{\mathrm{R}}$ say, we have

$$(T_1, T_2) \qquad \underset{t-1}{=} (T_1^{\mathrm{L}} EZFT_1^{\mathrm{R}}, T_2^{\mathrm{L}} E_1 Z_1 F_1 T_2^{\mathrm{R}})$$

$$(T_1'', T_2'') \qquad \underset{t-1}{=} (T_1''^{\mathrm{L}} E'Z'F'T_1''^{\mathrm{R}}, T_2''^{\mathrm{L}} E''Z''F''T_2''^{\mathrm{R}})$$

$$(*) \qquad (Z^*FT_1^{\mathrm{R}}, T_2^{\mathrm{L}} E_1 Z_1^*) \underset{t}{\overline{=}} (Z'^*F'T_1''^{\mathrm{R}}, T_2''^{\mathrm{L}} E''Z''^*)$$

where each side of $(*)$ is a t-preobject and one side is standard.

We claim that $\overline{Z^*}$ has size $\leqslant \frac{1}{2} + \epsilon_4 + \mathrm{Max}(v_2, v_3) = \alpha$ say.

This is true if the left side of $(*)$ is standard. In the opposite case $|\overline{Z'^*} - \frac{1}{2}| \leqslant \epsilon_4$; but by 8.5' $|\overline{Z^*} - \overline{Z'^*}| < v_3$ or $|\overline{Z^*} + \overline{Z'^*} - 1| \leqslant v_2$ so the result is true in this case also.

Let N be a subelement of L_t contained in H. Note that $Z^*FT_1^{\mathrm{R}}$ is maximal normal in H. Let $N \equiv N_1 N_2$ where $N_2 = N \cap AS_2^{\mathrm{L}}$. If N_2 is non-empty then N_1 is not normal and $s(N_2) < s_0$, hence $s(N) < s_0 + (r_0 + \epsilon_3) + 2/e$. If N_2 is empty then $s(N) \leqslant \alpha + \epsilon_3$. Hence H is t-bounded by b_{25} since $\mathrm{Max}(s_0 + r_0 + \epsilon_3 + 2/e, \alpha + \epsilon_3) < b_{25}$.

Let $t < i \leqslant k$. If N is an i-subelement contained in H then $s(N \cap Z^*F) \leqslant r_0'$, so H is i-bounded by $s_0 + r_0' + 2/e \leqslant b_{25}$.

The left side of (10) contains A hence V_4 so applying 6.10 $V_4 \supset V_4^0 \underset{k}{\sim} V_4' \supset V_4''$ since $s^{k+1}(V_4) > u_{18} \geqslant u_3$. Here V_4'' has size $> s(V_4) - u_4 > r_0'$. Similarly for V_{f-3}. (17) has the form

$$(18) \qquad A_1 V_4^0 A_2 V_{f-3}^0 A_3 \underset{k}{\overline{=}} B_1 V_4' B_2 V_{f-3}' B_3$$

where $A_1 = Z^*FT_1^{\mathrm{R}} A^{\mathrm{L}}$, $A_3 = A^{\mathrm{R}} S_2^{\mathrm{L}} E_2 Z_3^*$. We prove $(V_4', V_{f-3}') \subset B$; we may then write $B_1 \equiv Z'^*F'T_1''^{\mathrm{R}} B^{\mathrm{L}}$, $B_3 \equiv B^{\mathrm{R}} S_2''^{\mathrm{L}} E_3 Z_3^*$. Let $W \equiv V_3 \backslash V_4$. Then $W \supset W^0 \sim W' \supset X$ where X is a $(k+1)$-subelement of size $> r_0'$. Now $Z'^*F'T_1''^{\mathrm{R}}$ is a subelement of L_t so does not contain W' by 5.21. Thus $V_4' \subset B$. (Since W is disjoint from V_4, W' is disjoint from V_4'.) Dually $V_{f-3}' \subset B$.

From (11) we have say

$$A_1 = B_1 X_1^{-1}, \quad V_4^0 = X_1 V_4' X_2, \quad A_2 = X_2^{-1} \cdot B_2 X_3^{-1}$$

$$V_{f-3}^0 = X_3 \cdot V_{f-3}' \cdot X_4, \qquad A_3 = X_4^{-1} \cdot B_3$$

Axiom 31B is applicable to $V_4^0 \sim V_4'$ since V_4^0, V_4' are (k)-bounded by $s_0 \leqslant b_{31}$ and by 6.8 $s(V_4^0) \geqslant s(V_4) - u_1$ (since $u_{18} > u_1) > u_{18} - u_1 > s_{31}$. Hence there is a word $V_4'^{\mathrm{L}} C V_4^{0\mathrm{R}}$ (k)-bounded by $s_0 + q_{31}$ and equal in G_k to $V_4' \cdot X_2$, where $V_4'^{\mathrm{L}}$ is not a subelement of $L_1 \cup \cdots \cup L_k$ and $s(V_4^{0\mathrm{R}}) > s(V_4^0) - r_{31} > u_{18} - u_1 - r_{31} > r_0'$. Similarly $V_{f-3}^{0\mathrm{L}} D V_{f-3}'^{\mathrm{R}} = X_3 V_{f-3}'$. Now

$$(12') \qquad (T_1, S_2) \equiv T_1 A^{\mathrm{L}} V_4^0 A_2 V_{f-3}^0 A^{\mathrm{R}} S_2$$

$$(13') \qquad (T_1'', S_2'') \equiv T_1'' B^{\mathrm{L}} V_4' B_2 V_{f-3}' B^{\mathrm{R}} S_2''.$$

We show

$$(14') \qquad \epsilon_4^{-1} \cdot T_1 A^{\mathrm{L}} V_4^0 \quad = T_1'' B^{\mathrm{L}} V_4'^{\mathrm{L}} C V_4^{0\mathrm{R}}$$

$$(15') \qquad V_{f-3}^0 A^{\mathrm{R}} S_2 \cdot \kappa_5^{-1} = V_{f-3}^{0\mathrm{L}} D V_{f-3}'^{\mathrm{R}} B^{\mathrm{R}} S_2.$$

$$\begin{aligned}
\text{The right side of } (7') &= T_1'' B^{\mathrm{L}} V_4' \cdot X_2 = T_1'' B^{\mathrm{L}} X_1^{-1} V_4^0 \\
&= \epsilon_4^{-1} T_1 (Z^* F T_1^{\mathrm{R}})^{-1} (Z'^* F' T_1''^{\mathrm{R}}) B^{\mathrm{L}} X_1^{-1} V_4^0 \text{ (by 16)} \\
&= \epsilon_4^{-1} T_1 (A_1 A^{\mathrm{L}-1})^{-1} (B_1 B^{\mathrm{L}-1}) B^{\mathrm{L}} X_1^{-1} V_4^0 \\
&= \epsilon_4^{-1} T_1 A^{\mathrm{L}} V_4^0.
\end{aligned}$$

Similarly (15′) is proved. Thus (12′), (13′), (14′), (15′) have the same form as (12), (13), (14), (15), taking W_1, W_2, W_1'', W_2'' to be $V_4^0, V_{f-3}^0, V_4', V_{f-3}'$.

Thus 2 of the conclusion of 11.7 is proved.

In the present case J is $V_4^{0\mathrm{R}} A_2 V_{f-3}^{0\mathrm{L}}$ and

$$JF' \equiv V_4^{0\mathrm{R}} A_2 V_{f-3}^{0\mathrm{L}} D V_{f-3}'^{\mathrm{R}} B^{\mathrm{R}} S_2'' \gamma T_1'' B^{\mathrm{L}} V_4'^{\mathrm{L}} C$$

which is (k)-bounded by $s_0 + q_{31} \leqslant u_{35}$; thus 1 follows.

By 9° the nice image J^* in U of the subword J of V is contained in (T_1'', S_2''). By 10° $D^* \supset (T_1'', S_2'') \supset J^*$. Thus 4 follows.

Finally $J \supset A_2 \supset (V_6, V_{f-5})$ so 3 follows.

13°. Let $r = 0$ and let V_1 satisfy (R) (cf. 6°). Thus the V_{1i} $(= B_i)$ are disjoint, V_1 clears an $(n-1)$-path, V_{1i} $(i \in I_1)$ have images throughout, $I_1 = \{s_1+1, ..., s_1+t_1\}$, $I_1' = \{s_1+2, ..., s_1+t_1-1\}$ (for some s_1, t_1), $T_1 \subset V_\lambda \subset V_{1\,s_1+2}$, $S_2 \subset V_\rho \subset V_{1\,s_1+t_1-1}$.

Since V_1 clears an $(n-1)$-path, (4) may be improved by replacing $k\ (=n)$ by $n-1$. Now $T_1 A S_2 \supset A \supset (V_{1\,s_1+3}, V_{1\,s_1+t_1-2})$. Put $a = s_1+4$, $b = s_1+t_1-3$. Then if $a \leqslant p \leqslant b$ we have from the analogue of (17), (18) $V_{1p} \supset V_{1p}^0 \sim V_{1p}' \subset B$. Hence

$$(T_1, S_2) \equiv T_1 A^{\mathrm{L}} V_{1a}^0 A_2 V_{1b}^0 A^{\mathrm{R}} S_2$$

$$(T_1'', S_2'') \equiv T_1'' B^{\mathrm{L}} V_{1a}' B_2 V_{1b}' B^{\mathrm{R}} S_2''.$$

As before

$$(T_1'', S_2'') \underset{n-1}{=} T_1'' B^{\mathrm{L}} V_{1a}'^{\mathrm{L}} C V_{1a}^{0\mathrm{R}} A_2 V_{1b}^{0\mathrm{L}} D V_{1b}'^{\mathrm{R}} B^{\mathrm{R}} S_2'' \equiv H$$

hence $U \underset{n-1}{=} (H\gamma)$ where $(H\gamma)$ is $((n-1))$-bounded by $s_0 + q_{31} \leqslant \leqslant \mathrm{Min}(b_{25}, u_{35})$. Now U is n-bounded by s_0 hence by 6.10 $(H\gamma)$ is n-bounded by $s_0 + u_4 \leqslant u_{35}$, proving 1. We have $U \underset{n}{=} (H\gamma)$.

2 and 4 follow as before.

J is here $V_{1a}^{0\mathrm{R}} A_2 V_{1b}^{0\mathrm{L}}$ so contains $(V_{1\,a+1}, V_{1\,b-1}) = (B_{a+1}, B_{b-1}) = = (B_{a+1}, B_{g-a})$, since $g = 2s_1 + t_1$ so $a + b = g + 1$. We must prove $(g-2a)/m' > g/m' - u_{37}$, i.e., $(2s_1+8)/m' < u_{37}$. In 4° we have $\frac{1}{2}(14d+192)/t < 7\alpha + 103/e$, so analogously from 5° we deduce $s_1/m' \leqslant 7\alpha + 103/e$, where $\alpha = \epsilon_{30}' + u_{13} + 6u_4$. Hence $(2s_1+8)/m' \leqslant 14\alpha + 214/e < u_{37}$; thus 3 holds.

14°. Let $r = 0$ and let V_1 satisfy (S). V_1 clears an n-path and V_{1j} $(=B_j)$ $(j = 14, ..., a-13)$ have images throughout (cf. 3°). Also $a = g > m'u_{36}$. Put $T_1 = B_{14}$, $T_2 = B_{15}$, $S_1 = B_{g-14}$, $S_2 = B_{g-13}$, and let their images in U be $T_1'', T_2'', S_1'', S_2''$. Say

$$Z \equiv (T_1, T_2) J (S_1, S_2) \alpha$$

$$U \equiv (T_1'', T_2'') P (S_1'', S_2'') \gamma.$$

Again we have (11) and the two equations immediately following (11).

We wish to apply 9.15 to $(T_1, T_2), (T_1'', T_2'')$. Each is (n)-bounded by s_0; also $k_{24}' \leqslant s_0 \leqslant k_{24}$. Now $s(T_1) > r'' > u_{39}$ and by 7°

$s(T_1'') > r'' - v_3 - u_4 - \epsilon_3 > u_{39}$. Hence by 9.15 there is a word $T_1''^{L} C T_2^{R}$, (n)-bounded by $s_0 + u_{40}$ and equal in G_n to $\epsilon_4^{-1} \cdot (T_1, T_2)$. Similarly there is a word $S_1^{L} D S_2''^{R} = (S_1, S_2) \cdot \kappa_5^{-1}$. Here $T_1''^{L}, T_2^{R}, S_1^{R}, S_2''^{L}$ are n-normal. Consider $T_1''^{L} C T_2^{R} J S_1^{L} D S_2''^{R} \gamma \equiv V'$. Then $U \underset{n}{=} (V')$, and (V') is (n)-bounded by $s_0 + u_{40} \leqslant u_{35}$ (we note that if N is a subpowerelement of L_n contained in (V') then N cannot properly contain T_2^{R} since T_2 is maximal normal in Z. Similarly N cannot properly contain $T_1''^{L}, S_1^{L} S_2''^{R}$).

Thus 1, 2, 4 follow as before.

J contains $B_{17}, \ldots, B_{g-16}$ so it remains to prove that $2s/m' < u_{37}$ where $s = 16$. But $2s/m' = 32/m' \leqslant 32/e < u_{37}$.

11.8. Lemma. *Let*

$$Z_n \underset{n}{\to} \cdots \underset{1}{\to} Z_0 \equiv (DE), \qquad U_k \underset{k}{\to} \cdots \underset{1}{\to} U_0 \equiv (DF)$$

where the following conditions hold.

1. $Z_n \in Rep(L_{n+1}^k)$.
2. $p = Min(n, k) \geqslant 1$.
3. *If* $p = n$ *then Conditions A, B hold.*
4. *In the notation of 7.8, all A-descendants of* $B_1,\!,\, B_g$ *in* Z_0 *are contained in D, where g is such that* $1 \leqslant g \leqslant m$, $kg/m > u_{36}$.

Then

$$Z_p \equiv (D_p E_p)$$

$$(*) \qquad U_p \underset{p}{=} (D_p F_p)$$

where:

1. $(D_p F_p)$ *is* (p)*-bounded by* u_{35}.
2. $E_p^{-1} \cdot F_p \underset{p}{=} T^{-1} \cdot E^{-1} \cdot F \cdot T$ for some T.
3. *If* $p < n$ *then* D_p *contains all A-descendants of* $B_1,\ldots, B_g$ *in* Z_p *except possibly for the five leftmost of* B_1 *and the five rightmost of* B_g. *If* $p = n$, $D_p \supset (B_{1+s}, B_{g-s})$ *where* $k(g-2s)/m > kg/m - u_{37}$.
4. *In* $U_p \underset{p}{\to} \cdots \underset{1}{\to} U_0 \equiv (DF)$ *let the iterated nice image of D in* U_p *be denoted by* I_p; *in* $(*)$ *let the nice image of* D_p *in* U_p *be* D_p^*; *then* $D_p^* \subset I_p$. *In particular* D_p^*, I_p *exist.*

Proof. Call the conclusion $P(p)$. Then $P(0)$ is true, taking D_0, E_0, F_0 to be D, E, F respectively. Assume $P(q)$ is true for some q, $0 \leqslant q < p$. Then $D_q^* \subset I_q$. We wish to apply 11.7 to

$$Z_{q+1} \to Z_q \equiv D_q E_q$$

(19) $$U_{q+1} \to U_q - D_q F_q$$

This is possible since if $r = 0$ then $n - 1 = q < p \leqslant n$, so $p = n$; hence Conditions A, B hold.

If $X \subset U_q$ let $f(X)$ denote the iterated nice image of X in U_{q+1}, with respect to $U_{q+1} \to U_q$.

By 11.7 we obtain

$$Z_{q+1} \equiv D_{q+1} E_{q+1}$$

(20) $$U_{q+1} - D_{q+1} F_{q+1}$$

and $J^* \subset D^*$, where J^* is the nice image of D_{q+1} in U_{q+1} under (20) and D^* is the iterated nice image of D_q in U_{q+1} under (19). Now $f(D_q^*) = D^*$, $I_{q+1} = f(I_q)$ and $D_{q+1}^* = J^*$. Since $D_q^* \subset I_q \subset U_q$ we have $f(D_q^*) \subset f(I_q)$ i.e., $D^* \subset I_{q+1}$. But $D_{q+1}^* = J^* \subset D^*$, hence $D_{q+1}^* \subset I_{q+1}$.

11.9. Lemma. (*Linear version of* 11.8 (*cf.* 11.7A)). *Let*

$$Z_n \to \cdots \to Z_0 = (DE), \qquad U_k \to \cdots \to U_0 \equiv GDH$$

where $U_k, \ldots, U_0$ *are linear words. Let* 1, 2, 3, 4 *of the hypothesis of* 11.8 *hold. Then*

$$Z_p \equiv (D_p E_p)$$

(*) $$U_p \mathrel{\overline{\overline{\scriptstyle p}}} G_p D_p H_p$$

where

1. $G_p D_p H_p$ *is* (p)*-bounded* by u_{35}.

2. $G \cdot E^{-1} \cdot H \underset{p}{=} G_p \cdot E_p^{-1} \cdot H_p$.

3. *Part* 3 *of the conclusion of* 11.8 *holds*

4. *In* $U_p \to \cdots \to U_0 \equiv GDH$ *let the iterated nice image of* D *in* U_p *be* I_p; *in* (∗) *let the nice image of* D_p *in* U_p *be* D_p^*; *then* $D_p^* \subset I_p$.

§12

Informal Summary

Axiom 20 may be regarded as the main axiom. In this section we prove (in 12.2, 12.18) that part (a) of Axiom 20 holds for $n + 1$. (The rather deeper result that part (b) holds is given in Section 13.)

We also prove that Axiom 25 holds for $n + 1$.

12.1. Consider the following condition.

Condition C. If $A \underset{n-1}{\sim} C$, $AB \in L_{n+1}$ and $CD \in L_n$, then $L(A)/L(AB) < u_{41}$.

12.2. Proposition. *Assume Condition C. If* $DE \in \mathsf{L}_{n+1}$, $DF \in \mathsf{L}_{k+1}$ $0 \leqslant k < n$ *then* $s_{DE}(D) < r_0'$.

Proof. Assume $s_{DE}(D) \geqslant r_0'$. Put $\epsilon = 1/e^2$. Then by 7.11 $Z_n \underset{n}{\to} \cdots \underset{1}{\to} Z_0 \equiv (DE)$, $Z_n \in \text{Rep}(L_{n+1})$. There exist B_i, denoted without loss of generality by $B_1, \ldots, B_g$ such that all A-descendants of them in Z_0 are contained in D, $1 \leqslant g \leqslant m$, $g/m > r_0' - \epsilon$. Also $Z_k' \underset{k}{\to} \cdots \underset{1}{\to} Z_0' \equiv (DF)$, $Z_k' \in \text{Rep}(L_{k+1})$.

Case 1. $k = 0$.

(1a) Let $n = 1$. If Z_0 is different from Z_1 then by 3.7 $A_1, \ldots, A_g$ are maximal 1-normal and contained in D. Now

$$g > m(r_0' - \epsilon) \geqslant e(r_0' - 1/e^2) > 1$$

so we have the contradiction that a subelement D of L_1 contains two distinct maximal 1-normal subwords. Hence $Z_0 \equiv Z_1$. Thus $DE \in L_2$, $DF \in L_1$. By 5.16 $L(D) < q'L(DE)$. But $|L(D)/L(DE) - g/m| < 3/e$ (cf. 2.16A) hence $g/m < 3/e + q' \leqslant r_0' - - 1/e^2$, a contradiction.

(1b) Let $n > 1$. Let A_γ be an A-descendant of one of $B_1, \ldots, B_g$

in Z_1. By 7.8, $Z_1 \supset A_\gamma \supset B_{\gamma i}$ and $A_{\gamma i} \subset Z_0$ $(i = 1 + j, \ldots, a - j)$. All $A_{\gamma i} \subset D$. If the $B_{\gamma i}$ are maximal 1-normal we have the contradiction that two distinct maximal 1-normal subwords are contained in D.

Now recall 7.5, 7.7. Any x consecutive $B_{\gamma i}$'s generate a subword of A_γ of size $> x/t - \epsilon'_{30}$. Hence any $x + 12$ consecutive $A_{\gamma i}$'s generate a word containing a 2-subelement of size $> x/t - \epsilon'_{30} - 3u_4$. In particular, taking $x + 12 = a - 2j$, D contains a 2-subelement of L_2 of size

$$> (a - 2j - 12)/t - \epsilon'_{30} - 3u_4$$

$$> (u_{18} - 6u_{20} - \epsilon_{30}) - 36/t - 12/t - \epsilon'_{30} - 3u_4 \geqslant \beta$$

where β temporarily denotes $u_{18} - 6u_{20} - \epsilon_{30} - 48/e - \epsilon'_{30} - 3u_4$. Now $\beta \geqslant r'_0 + u_4 > r'_0$ in contradiction to 5.21.

Case 2. $k > 0$.

By Lemma 11.8 since $g/m > r'_0 - 1/e^2 > u_{36}$

$$Z_k \equiv D_k E_k$$

$$(\dagger) \qquad Z'_k \;\overline{\scriptstyle k}\; (D_k F_k)$$

(2a) Let $n > k + 1$. Now if A_γ is any A-descendant in Z_{k+1} of B_2, $A_\gamma \supset B_{\gamma i}$, $A_{\gamma i} \subset D_k$. If the $A_{\gamma i}$ are maximal $(k + 1)$-normal then by $(\dagger)$ Z'_k contains two maximal $(k + 1)$-normal subwords (cf. 3° of 11.7A), a contradiction.

Hence we may assume as in Case 1 that D_k contains a $(k + 2)$-subelement of size $> \beta$. Now $k + 2 \leqslant n$ so by $(\dagger)$ Z'_k contains a $(k + 2)$-subelement of size $> \beta - u_4 \geqslant r'_0$, a contradiction.

(2b) Let $n = k + 1$. First suppose Z_{n-1} is the same as Z_n. Then D_k is a subelement of L_{n+1} hence of L_{n+1}. From (2a) if N is a subelement of $L_1 \cup \cdots \cup L_{n-1}$ contained in D_k then $s^{n+1}(N) \leqslant r'_0$. By considering 6.8 and $(\dagger)$ we have $D_k \supset D_k^0 \underset{k}{\sim} D'_k \subset Z'_k$ where $s^{n+1}(D_k^0) > s^{n+1}(D_k) - u_1$. But $s(D_k) \geqslant (g - 2)/m > r'_0 - 1/e^2 - 2/e$. Now $|L(D_k)/L(D_k E_k) - (g - 2)/m| < 3/e$ so

$$L(D_k)/L(D_k E_k) > -3/e + r'_0 - 1/e^2 - 2/e \geqslant u_{41}$$

contradicting Condition C.

Finally let Z_{n-1} be different from Z_n. Then $B_1, \ldots, B_g$ are maximal n-normal, D_k contains A_6, A_7 hence by (†) Z'_k contains two maximal n-normal subwords, a contradiction.

12.3. Corollary. *Condition C implies Condition A.*

12.4. Note. If A is a subelement of $\mathrm{Rep}(L_n)$ of size $> s_{33}$ then there is a unique B such that $AB \in \mathrm{Rep}(L_n)$; this is by Axiom 33 since $A \underset{n-1}{\sim} A$. Temporarily denote this B by A'.

12.5. Definition. Subelements A_1, A_2 of $\mathrm{Rep}(L_n)$ of size $> s_{33}$ *satisfy* (0) if $A_1 A'_1$ is a C.A. of $A_2 A'_2$; *satisfy* (1) if $(A_1 A'_1)^{-1}$ is a C.A. of $A_2 A'_2$.

12.5A. Let each of i, j, k be 0 or 1. If A_1, A_2 satisfy (i) and A_2, A_3 satisfy (j) then A_1, A_3 satisfy (k) where $i + j \equiv k \pmod 2$.

12.5B. A_1, A_2 cannot satisfy both (0), (1).

Proof. Assume not. Then $Z \equiv A_1 A'_1$ is a C.A. of Z^{-1}. Say $Z \equiv CD$, $Z^{-1} \equiv DC$. Then $CD \equiv C^{-1} D^{-1}$ so $C^2 \equiv I$, $D^2 \equiv I$. Since e is odd, $\Pi = G_0$ has no elements of order 2, hence $C \equiv I$, $D \equiv I$, $Z \equiv I$ which is a contradiction. ((e odd))

12.6. Proposition. *Let $EZF \underset{n-1}{\sim} AZ''B$ where each side is $((n-1))$-bounded by k_{25} and Z, Z'' are subelements of $\mathrm{Rep}(L_n)$ of size $> u_{42}$. Assume that if X is a subelement of L_n contained in E or F then $s(X) < u'_{43}$. Then Z, Z'' satisfy* (0).

Proof. Since $k_{25} \leqslant b_{25}$ and $u_{42} \geqslant u_3$ we have $Z'' \supset Z''^0 \underset{n-1}{\sim} W \supset A$ where $s(A) > s(Z'') - u_4$. Now $A \subset EZF$; consider $A \cap Z \equiv J$. Either J is A or J is Z or without loss of generality $A \subset EZ$ and $A \cap E$ has size $< u'_{43}$ so $s(J) > s(A) - 2/e - u'_{43}$. Thus in any case

$s(J) > u_{42} - u_4 - 2/e - u'_{43} = \alpha$ say. Now $A \supset J \supset J^0 \sim X$, where $X \subset Z''^0 \subset Z''$, $s(J^0) > \alpha - u_1 > s_{33}$. Hence by Axiom 33, $X \equiv J^0$. Thus Z'', Z satisfy (0).

12.7. Proposition. *Let X, Y be (n)-bounded by k_{26} and $X \underset{n}{\sim} Y$ with divisions $P_1, \ldots, P_k$ and $Q_1, \ldots, Q_k$. Let T, U be maximal n-normal in X, Y respectively of size $> u_{44}$ and images of each other in the sense of* 6.16. *As in Axiom* 33$'$ *let*

$$T \underset{n-1}{=} T^{\mathrm{L}} E_1 Z_1 F_1 T^{\mathrm{R}}, \qquad U \underset{n-1}{=} U^{\mathrm{L}} E_2 Z_2 F_2 U^{\mathrm{R}}.$$

Then

(i) *if $T = \bar{P}_i$, $U \equiv \bar{Q}_i$* (where *P_i, Q_i have type n*) *then Z_1, Z_2 satisfy* (1) *of* 12.5;

(ii) *in the opposite case, Z_1, Z_2 satisfy* (0) *of* 12.5.

Proof. (i) In the notation introduced in 6.7 we have $T \supset P'_i$, $U \supset Q'_i$, $P'^{-1}_i X'^{-1}_{i-1} Q'_i X'_i \in \mathsf{L}_n$. Now $s(P'_i) + s(Q'_i) > k_3^0$ hence $s(P'_i) > k_3^0 - k_{26} > t_{33}$ and $P'_i \equiv P'^{\mathrm{L}}_i E_3 Z_3 F_3 P'^{\mathrm{R}}_i$, $s(Z_3) > k_3^0 - k_{26} - r_{33} > s_{33}$ so $Z_3 Z'_3 \in \mathrm{Rep}(L_n)$. Similarly $Q'_i \equiv Q'^{\mathrm{L}}_i E_4 Z_4 F_4 Q'^{\mathrm{R}}_i$, $Z_4 Z'_4 \in \mathrm{Rep}(L_n)$. By Axiom 33$^+$ (using $k_3^0 - k_{26} > j_{33}$) $Z_3^{-1} Z'^{-1}_3$ is a C.A. of $Z_4 Z'_4$, i.e., Z_3, Z_4 satisfy (1). Now

$$U^{\mathrm{L}} E_2 Z_2 F_2 U^{\mathrm{R}} \underset{n-1}{=} U \equiv U^{\mathrm{L}} Q'_i U^{\mathrm{R}} \underset{n-1}{=} U^{\mathrm{L}} Q'^{\mathrm{L}}_i E_4 Z_4 F_4 Q'^{\mathrm{R}}_i U^{\mathrm{R}}$$

so by 12.6 Z_4, Z_2 satisfy (0) ($u_{44} - r_{33} > u_{42}$; $k_3^0 - k_{26} - r_{33} > u_{42}$; $q_{33} + \epsilon_3 < u'_{43}$; $r_0 < u'_{43}$; $k_{26} \leqslant b_{33}$; $k_{26} + c_{33} \leqslant k_{25}$.) Similarly Z_1, Z_3 satisfy (0). Hence by 12.5A Z_1, Z_2 satisfy (1).

(ii) 6.25 is available since $k_{26} \leqslant k''_8$, $k_{26} \leqslant k''_9$ and $u_{44} \geqslant u'_{10}$. Hence there are subwords T_1, U_1 of T, U each of size $> s(T) - u''_{10} > u_{44} - u''_{10}$ ($= \alpha_1$ say) and such that $T_1 \underset{n-1}{\sim} U_1$. As in the proof of 8.9 we deduce that

$$(*) \qquad H_1 T_1^M K_1 \underset{n-1}{=} H_2 U_1^M K_2$$

for some H_1, H_2, K_1, K_2 which are subelements of $\mathsf{L}_1 \cup \cdots \cup \mathsf{L}_{n-1}$; we require $k_{26} \leqslant k_{14}$, $\alpha_1 > u_{25}$. Moreover $s(T_1^M) > s(T_1) - 2u_{26}$,

$s(U_1^M) > s(U_1) - 2u_{26}$ and each side of (*) is $((n-1))$-bounded by

$$\alpha_2 = k_{26} + \mathrm{Max}(q_{24}, r_6 + d_{24} + 2/e + \epsilon_3, 2/e + r_0').$$

By Axiom 33′, since $\alpha_1 - 2u_{26} > t_{33}$ and $\alpha_2 \leqslant b_{33}$,

$$H_1 T^{\mathrm{ML}} E_3 Z_3 F_3 T_1^{\mathrm{MR}} K_1 \underset{n-1}{=} H_2 U_1^{\mathrm{ML}} E_4 Z_4 F_4 U_1^{\mathrm{MR}} K_2$$

where each side is $((n-1))$-bounded by $\alpha_2 + c_{33} \leqslant k_{25}$; also $s^n(T_1^{\mathrm{ML}} E_3) < q_{33}$ and $s(Z_3) > s(T_1^{\mathrm{M}}) - r_{33} > \alpha_1 - 2u_{26} - r_{33} > u_{42}$. If X is a subelement of L_n contained in $H_1 T_1^{\mathrm{ML}} E_3$ then $s(X) < r_0' + 2/e + \mathrm{Max}(r_0, q_{33} + \epsilon_3) \leqslant u_{43}'$. By 12.6 Z_3, Z_4 satisfy (0).

Let $T \equiv T' T_1 T''$, $U \equiv U' U_1 U''$, $T_1 \equiv T_1^{\mathrm{L}} T_1^{\mathrm{M}} T_1^{\mathrm{R}}$, $U_1 \equiv U_1^{\mathrm{L}} U_1^{\mathrm{M}} U_1^{\mathrm{R}}$. Now $T^{\mathrm{L}} E_1 Z_1 F_1 T^{\mathrm{R}} \underset{n-1}{=} T' T_1^{\mathrm{L}} T_1^{\mathrm{ML}} E_3 Z_3 F_3 T_1^{\mathrm{MR}} T_1^{\mathrm{R}} T''$. 12.6 is available since $k_{26} \leqslant b_{33}$, $k_{26} + c_{33} \leqslant k_{25}$, $s(Z_1) > s(T) - r_{33} > u_{44} - r_{33} > u_{42}$ (by above $s(Z_3) > u_{42}$). If X is an n-subelement in $T^{\mathrm{L}} E_1$ then $s(X) < \mathrm{Max}(q_{33} + \epsilon_3, r_0) < u_{43}'$. Hence Z_1, Z_3 satisfy (0). Similarly Z_4, Z_2 satisfy (0). Hence by 12.5A Z_1, Z_2 satisfy (0).

12.8. Proposition. *Let*

(1) $$Y \equiv Y_1 Y_2 Y_3 R Y_4 Y_5 Y_6 \underset{n}{\sim} X_1 X_2 X_3 S X_4 X_5 X_6$$

where each side is (n)*-bounded by* k_{27}, *induce* $Y_2 \underset{n}{\sim} X_2$ *(end words* E_2, K_2 *say) and* $Y_5 \underset{n}{\sim} X_5$ *(end words* E_5, K_5*). Let* R, S *be maximal* n*-normal of size* $> u_{45}$ *and images of each other in* (1). *Let*

$$Y_2 \cdot K_2^{-1} \underset{n}{=} Y_2^{\mathrm{L}} E X_2^{\mathrm{R}} \ (Y_2^{\mathrm{L}} \not\equiv I, X_2^{\mathrm{R}} \not\equiv I), \quad E_5^{-1} \cdot Y_5 \underset{n}{=} X_5^{\mathrm{L}} F Y_5^{\mathrm{R}},$$

$$(X_5^{\mathrm{L}} \not\equiv I,\ Y_5^{\mathrm{R}} \not\equiv I)$$

and hence

(2) $$Y \underset{n}{=} Y_1 Y_2^{\mathrm{L}} E X_2^{\mathrm{R}} X_3 S X_4 X_5^{\mathrm{L}} F Y_5^{\mathrm{R}} Y_6$$

where the three right sides are (n)-bounded by $k_{27} + u_{46}$. *For* (2) *let the image of S in Y be S'. Then S' is R.*

Proof. By 6.25 we have $R^{\mathrm{M}} \sim S^{\mathrm{M}}$ (end words E', K' say) since $k_{27} \leqslant k_8'', k_9''$ and $u_{45} \geqslant u_{10}'$. (If $R = \bar{P}_i$, $S = \bar{Q}_i$ let $R^{\mathrm{M}} = P_i'$, $S^{\mathrm{M}} = Q_i'$; then $s(R^{\mathrm{M}}) > k_3^0 - k_{27}$.) Say $R \equiv R^{\mathrm{L}} R^{\mathrm{M}} R^{\mathrm{R}}$, $S \equiv S^{\mathrm{L}} S^{\mathrm{M}} S^{\mathrm{R}}$. Thus

$$Y_1 Y_2 Y_3 R^{\mathrm{L}} R^{\mathrm{M}} \sim X_1 X_2 X_3 S^{\mathrm{L}} S^{\mathrm{M}}, \quad Y_1 Y_2 Y_3 R^{\mathrm{L}} R^{\mathrm{M}} = Y_1 Y_2^{\mathrm{L}} E X_2^{\mathrm{R}} X_3 S^{\mathrm{L}} S^{\mathrm{M}} \cdot K'.$$

We claim that

$$Y_1 Y_2 Y_3 R \triangleq Y_1 Y_2 Y_3 R^{\mathrm{L}} R^{\mathrm{M}} \triangleq Y_1 Y_2^{\mathrm{L}} E X_2^{\mathrm{R}} X_3 S^{\mathrm{L}} S^{\mathrm{M}} \triangleq$$

$$\triangleq Y_1 Y_2^{\mathrm{L}} E X_2^{\mathrm{R}} X_3 S \triangleq (\text{In } Y, S')$$

where the terms are right objects.

The first and second terms are right objects since $k_{24} \leqslant u_{46} + k_{27} \leqslant k_{12}', k_{13}'$; $u_{45} \geqslant u_{24}$; $s(R^{\mathrm{M}}) > \mathrm{Min}(u_{45} - u_{10}'', k_3^0 - k_{27}) > u_{24} + \epsilon_3$. They are almost equal by 8.13. The third term is a right object similarly. Apply 8.15 to (2); this is possible since $k_{27} + u_{46} \leqslant k_{16}, k_{17}$ and $u_{45} > u_{29}$. Hence the fourth and fifth terms are right objects and are almost equal. The third and fourth terms are almost equal by 8.13. Now 6.23 is applicable to (2) since $k_{27} + u_{46} \leqslant k_8', k_9'$ and $u_{45} > u_{10}$. Hence $s(S') > s(S) - v_3 > u_{45} - v_3 > u_{30}'$ or $s(S') > 1 - v_2 - k_{27} > u_{30}'$. By 9.12A, R is S' $(k_{27} \leqslant k_{18}, k_{19}; u_{45} > u_{30}')$.

12.9. Proposition. *Let* (*i*) $Y \underset{n}{=} Y^{\mathrm{L}} A Y^{\mathrm{R}}$ *and* (*ii*) $A \underset{n}{=} A^{\mathrm{L}} B A^{\mathrm{R}}$, *so that* (*iii*) $Y \underset{n}{=} Y^{\mathrm{L}} A^{\mathrm{L}} B A^{\mathrm{R}} Y^{\mathrm{R}}$, *where all these words are* (*n*)*-bounded by* k_{28}. *Let* $S \subset A$ *and let S be maximal n-normal in* $Y^{\mathrm{L}} A Y^{\mathrm{R}}$. *Let* $T \subset B$ *and let T be maximal n-normal in* $Y^{\mathrm{L}} A^{\mathrm{L}} B A^{\mathrm{R}} Y^{\mathrm{R}}$. *Let* $s(S)$, $s(T) > u_{47}$. *Let S' be maximal n-normal in Y. Let* A^{L}, A^{R} *be non-empty. Let S', S be images of each other in* (*i*) *and let S, T be images of each other in* (*ii*). *Then S', T are images of each other in* (*iii*).

Proof. With respect to (iii) let $T' = \text{Im}'\, T$; by 6.23 this exists since $k_{28} \leqslant k'_8, k'_9$ and $u_{47} \geqslant u_{10}$; also either $|s(T) + s(T') - 1| \leqslant v_2$ or $|s(T) - s(T')| < v_3$. There are two similar inequalities for S, S'.

Using right n-objects, $(\text{In } Y, T') \triangleq Y^{\text{L}} A^{\text{L}} (\text{In } B, T)$ (cf. 8.15, noting that $k_{28} \leqslant k_{16}, k_{17}$ and $u_{47} > u_{29}$). Similarly, $(\text{In } Y, S') \triangleq Y^{\text{L}} (\text{In } A, S)$ and $(\text{In } A, S) \triangleq A^{\text{L}} (\text{In } B, T)$. By premultiplying both sides of the last expression by Y^{L} and using the transitivity we obtain $(\text{In } Y, S') \triangleq (\text{In } Y, T')$. From this follows by the dual of 9.12A that S' is T' (we note that $k_{28} \leqslant k_{18}, k_{19}$ and each of S', T' has size at least $\text{Min}(1 - v_2 - k_{28}, u_{47} - v_3) > u'_{30} > u_{30}$.)

12.10. We introduce the following temporary notation. In the notation of 7.8, if X is a family of A-descendants in Z_{n-r}, where $1 \leqslant r \leqslant n-1$, let $\mathcal{D}(X)$ be the following family of A-descendants in Z_{n-r-1}: let Y arise from X by deleting the leftmost 11 and the rightmost 11 members, for each $A_\gamma \in Y$ put all $A_{\gamma i}$ in $\mathcal{D}(X)$. $\mathcal{D}^2(X)$ means $\mathcal{D}(\mathcal{D}(X))$.

12.11. Proposition. *Assume the following:*

$Z_n \underset{n}{\rightarrow} \cdots \underset{1}{\rightarrow} Z_0$ *where* $Z_n \in Rep(L_{n+1})$; $Z_p \equiv (CD)$ *for some* p, $1 \leqslant p \leqslant n$; *C contains:*

$B_1, \ldots, B_g$ *if* $p = n$

all members of $\mathcal{D}^{n-p-1}(A_9, \ldots, A_{g-8})$ *if* $p < n$,

where $g/m > u_{48}$; $Y \underset{p}{\sim} C$ *where Y is (p)-bounded by* k_{29}.

Then there exist Z'_{p-1}, C', D' *such that*

1. $Z_p \underset{p}{\rightarrow} Z'_{p-1} \equiv (C'D')$.
2. $Y^{\text{M}} \underset{p-1}{\sim} C'$ *for some subword* Y^{M} *of* Y.
3. *In* $Z_n \underset{n}{\rightarrow} \cdots \underset{p+1}{\rightarrow} Z_p \underset{p}{\rightarrow} Z'_{p-1}$, C' *contains all members of* $\mathcal{D}^{n-p}(A_9, \ldots, A_{g-8})$.

Proof. 1°. Say $P_1, \ldots, P_m$ and $Q_1, \ldots, Q_m$ are divisions for $Y \underset{p}{\sim} C$. If Q_i has type p denote by R_i the maximal p-normal subword of Z_p containing Q_i. Since $k_3^0 < s(P_i) + s(Q_i)$ we have $s(Q_i) > k_3^0 - k_{29} > r' + \epsilon_3$; hence $s(R_i) > r'$. Let Z'_{p-1} arise from Z_p by simultaneous replacement of all R_i; if there are no R_i let $Z'_{p-1} \equiv Z_p$. Then $Z_p \underset{p}{\rightarrow} Z'_{p-1}$ so by Axiom 26 $Z_p \underset{p-1}{=} Y_1 \underset{p}{=} Y_2 \underset{p}{=} Z'_{p-1}$, say.

Case 1. $p < n$.

Let the members of $\mathcal{D}^{n-p-1}(A_9, ..., A_{g-8})$ be α_i ($i \in I$); they are $(p+1)$-subelements. As in 11° in the proof of 11.7 denote the leftmost members by $\alpha_1, \alpha_2, ...$ and the rightmost members by $..., \alpha_{f-1}, \alpha_f$. Consider the A-descendants of any α_i in Z'_{p-1}. By 7.8 and 4° of 11.7 α_i contains p-subelements α_{is} ($s = 1, ..., a$ say) and either

(a) α_i clears a p-path, or

(b) α_i clears a $(p-1)$-path,

in which $\alpha_{i\,j+1}, ..., \alpha_{i\,a-j}$ have images throughout; denote the image of α_{is} in Y_1, Y_2, Z'_{p-1} by $\alpha'_{is}, \alpha''_{is}, \beta_{is}$ respectively, $s = j+1, ..., a-j$. If (a) holds these maximal p-normal. The β_{is}, $s = j+1, ..., a-j$, are the A-descendants of α_i in Z'_{p-1}.

If (b) holds we see from 7.4 and Axiom 30 part 4 that $(\alpha'_{i\,j+1}, \alpha'_{i\,a-j})$ in Y_1 and the three similarly defined subwords of Z_p, Y_2, Z'_{p-1} each contain a $(p+1)$-subelement of size $> ((a-2j)-4h)/t - hu_4 - \epsilon'_{30}$ where $h = 0, 1, 2, 3$ for Z_p, Y_1, Y_2, Z'_{p-1} respectively. Hence since $s(\alpha_i) > u_{18}$ each contains a $(p+1)$-subelement of size greater than σ where

$$\sigma = (u_{18} - 6u_{20} - \epsilon_{30}) - 36/e - 12/e - 3u_4 - \epsilon'_{30}$$

(cf. 7.7A). Since $\sigma > r'_0$ each is not a p-subelement. In this case, moreover, if $(\beta_{i\,j+1}, \beta_{i\,a-j})$ contains a p-subelement J then α_i contains a p-subelement of size $> s(J) - 3u_4$; hence by Axiom 30 $s(J) \leqslant c^* + 3u_4$ (note that $u_{14} + 2u_4 \leqslant c^* + 3u_4$).

2°. We prove that Z'_{p-1} is p-bounded by $\beta = \xi + \epsilon_3$ where

$$\xi = 1 - k_3^0 + k_{29} + \epsilon_3 + r_{26}.$$

Suppose M is maximal p-normal in Z'_{p-1} of size $\geqslant \xi$. By Axiom 26″, since $\xi > v_{26}$, there exists N, maximal normal in Z_p such that $|s(N) + s(M) - 1| < r_{26}$; for if $|s(N) - s(M)| < r_{26}$ then $\xi \leqslant s(M) < s(N) + r_{26} < s_0 + r_{26} \leqslant k_{29} + r_{26} < \xi$, a contradiction. Thus N is replaced in $Z_p \to Z'_{p-1}$, i.e., N is of the form R_i. Hence $s(M) < 1 + r_{26} - (k_3^0 - k_{29} - \epsilon_3) = \xi$, a contradiction.

Similarly Y_2 is p-bounded by β. Since $Z_p \underset{p-1}{\overset{}{=}} Y_1$ we have by 6.10A that Y_1 is p-bounded by $s_0 + u_4 \leqslant \beta$; $(s_0 + s_{26} \leqslant b_{25}$, $u_3 < s_0 < h_0 - \epsilon_2 - 2/e - u_4)$.

Since $s_0 + s_{26} \leqslant \beta$ and $k_{29} \leqslant \beta$, each of $Y, Z_p, Y_1, Y_2, Z'_{p-1}$ is (p)-bounded by β.

3°. We use the following notation for the p- or $(p-1)$-path cleared by α_i: $H_i \sim J_i \supset K_i \sim L_i \supset M_i \sim N_i$. Then $\alpha_i \supset H_i, J_i \subset Y_1$, $L_i \subset Y_2$ and $N_i \subset Z'_{p-1}$. Now $H_i \subset \alpha_i \subset C \underset{p}{\sim} Y$ so $H_i \supset C_i \underset{p}{\sim} D_i$ where C_i is H_i^0 (with respect to $C \underset{p}{\sim} Y$); that H_i^0 exists will appear below.

First suppose the α_{is} satisfy (a) of Axiom 30. Then they are maximal p-normal of size $> r^* > r^* - 3v_3 > u_9$ so by 6.16 their images in Y (with respect to $C \sim Y$) exist since $\beta \leqslant k_8, k_9$. In particular α_{is}^0 exists. Now $H_i \supset \alpha_{is}$ so $C_i \equiv H_i^0 \supset \alpha_{is}^0$; hence C_i contains properly two of the α_{is}.

Using 6.28, 6.27, which are available since $r^* > r^* - 3v_3 > v_5$ and $\beta \leqslant k_{53}$, and putting $\beta_1 = \beta + \mathrm{Max}(h_{31}, q_{31})$ we have the following.

(*) There are words $C_i^{\mathrm{L}} F_i D_i^{\mathrm{R}}, D_i^{\mathrm{L}} E_i C_i^{\mathrm{R}}$ (p)-bounded by β_1 such that $D_i \underset{p}{=} D_i^{\mathrm{L}} E_i C_i^{\mathrm{R}} \cdot K'_i \underset{p}{=} E'_i \cdot C_i^{\mathrm{L}} F_i D_i^{\mathrm{R}}$ where E'_i, K'_i are end words for $D_i \sim C_i$. None of $D_i^{\mathrm{L}}, C_i^{\mathrm{R}}, C_i^{\mathrm{L}}, D_i^{\mathrm{R}}$ is a subelement of $L_1 \cup \cdots \cup L_p$.

The last part of (*) holds because of the four words properly contains a maximal p-normal subword.

Now suppose the α_{is} satisfy (b). Then from 1° H_i contains a $(p+1)$-subelement W_i of size greater than σ. Now $\sigma > u_1$ so $C_i \equiv H_i^0 \supset W_i^0, s(W_i^0) > s(W_i) - u_1$. 6.28, 6.26 are available since $\beta \leqslant k_{52}, \sigma > v_4$ so we have (*) again.

Hence (cf. 6.7)

$$Y \underset{p}{=} (\mathrm{In}\, Y, D_1^{\mathrm{L}}) E (C_1^{\mathrm{R}}, C_f^{\mathrm{L}}) F (D_f^{\mathrm{R}}, \mathrm{Fin}\, Y) \tag{3}$$

where $E \equiv E_1$, $F \equiv F_f$; the right side is (p)-bounded by β_1.

4°. If the α_{is} satisfy (a) of Axiom 30 then all of $\alpha_{is}, \alpha'_{is}, \alpha''_{is}, \beta_{is}$ have size $> r^* - 3v_3$; this is by 6.23 since $\beta \leqslant k'_8, k'_9$ and $r^* - 2v_3 > u_{10}$ and $1 - \beta - v_2 > r^* - v_3$.

Note that $(C_1^{\mathrm{R}}, C_f^{\mathrm{L}})$ contains (H_4, H_{f-3}) properly.

Just as we obtained (1) we can find an equation

$$(H_4, H_{f-3}) \underset{p}{=} H_4^{\mathrm{L}} E' (J_4^{\mathrm{R}}, J_{f-3}^{\mathrm{L}}) F' H_{f-3}^{\mathrm{R}} \tag{4}$$

and similarly

(5) $(K_6, K_{f-5}) \underset{p}{=} K_6^{\mathrm{L}} E''(L_6^{\mathrm{R}}, L_{f-5}^{\mathrm{L}}) F'' K_{f-5}^{\mathrm{R}}$

(6) $(M_8, M_{f-7}) \underset{p}{=} M_8^{\mathrm{L}} E'''(N_8^{\mathrm{R}}, N_{f-7}^{\mathrm{L}}) F''' M_{f-7}^{\mathrm{R}}$.

Combining (1) to (4) we have for some H, K

(7) $Y \underset{p}{=} H(N_8^{\mathrm{R}}, N_{f-7}^{\mathrm{L}}) K$

and the right side of (7) is (p)-bounded by β_1.

5°. Let θ be a temporary parameter. Let V be maximal p-normal in Z'_{p-1} of size $> \theta \geqslant v_{26}$ and contained in (N_{10}, N_{f-9}). By Axiom 26″ it has images S, T, U say in Z_p, Y_1, Y_2 respectively. Also either $|s(S) - s(V)| < r_{26}$ or $s(S) > r'$. We have $U \subset (M_8, M_{f-7}) \subset (L_8, L_{f-7})$, $T \subset (K_6, K_{f-5}) \subset (J_6, J_{f-5})$ and S is contained in (H_4, H_{f-3}) which is properly contained in $(C_1^{\mathrm{R}}, C_f^{\mathrm{L}}) \subset C \underset{p}{\sim} Y$.

Let $\mathrm{Im}' S = R \subset Y$; this exists by 6.19, and $\mathrm{Im}' R = S$, since $\beta \leqslant \mathrm{Min}(k'_8, k'_9)$, $r' > u_{10}$ and $s(V) - r_{26} > \theta - r_{26} > u_{10}$. Hence by 6.23 $|s(R) + s(S) - 1| \leqslant v_2$ or $|s(R) - s(S)| < v_3$.

By 12.8 R, S are images of each other in (1); for $\beta \leqslant k_{27}$, $\mathrm{Min}(r', \theta - r_{26}) > u_{45} + v_3$, $1 - v_2 - s_0 > u_{45}$, $\beta_1 \leqslant u_{46} + \beta$.

Now $S \subset (H_4, H_{f-3}) \sim (J_4, J_{f-3}) \supset T$; in fact S, T are images of each other ($\mathrm{Im}' S = T$ but since $s(S) > u_{10}$ also $\mathrm{Im}' T = S$). Now

$$\mathrm{Min}(r' - 2v_3, 1 - v_2 - s_0 - v_3, 1 - \beta - v_2) \geqslant \theta - r_{26} - 2v_3 = \gamma \geqslant u_{10},$$

hence V, S, R, T, U have size $\geqslant \gamma$.

By 12.8 since $\gamma > u_{45}$,

S, T are images of each other in (2)

T, U are images of each other in (3)

U, V are images of each other in (4).

By 12.9 since $\beta_1 \leqslant k_{28}$ and $\gamma > u_{47}$ R, V are images of each other in (5).

6°. We first claim that $(N_{10}, N_{f-9}) \equiv A(N_{10}^{\mathrm{R}}, N_{f-9}^{\mathrm{L}})C \equiv ABC$ for some A, B, C where A, C are not subelements of $L_1 \cup \cdots \cup L_p$ and B is nice with respect to (5), hence $Y^M \underset{p}{\sim} B$ for some Y^M.

To see this, first let α_{10} satisfy (b) of 1°. Then N_{10} contains a subelement X of L_{p+1}. Let $N_{10} \equiv A_1 A_2$ where A_1 is a subelement of $L_1 \cup \cdots \cup L_p$ and A_1 is chosen maximally. Then $A_2 \not\equiv I$ by 1°; say $A_2 \equiv cA_3$. Let $A_3 \equiv A_4 A_5$ where A_4 is a maximal subelement of $L_1 \cup \cdots \cup L_p$. Then $A_5 \not\equiv I$ since otherwise $X \subset A_1 c A_4$ and $s(X) = s(A_1 c A_4 \cap X) \leqslant 4/e + 2r'_0 + \epsilon_2 < \sigma < s(X)$.
Thus the required A will have the form $A_1 c A_4^{\mathrm{L}}$. If α_{10} satisfies (a) then it is easy to select A.

Next we prove that in

$$(8) \qquad Y^M \underset{p}{\sim} B$$

p may be replace by $p-1$. Assume not. Then some P_i, Q_i for (8) have type p; let $R' = \overline{P}_i \subset Y^M$ and $V' = \overline{Q}_i \subset B$. Now $V' \subset Z'_{p-1}$. Let $V \supset V'$ be maximal normal in Z'_{p-1} and let $R \supset R'$ be maximal normal in Y. Then $V \subset (N_{10}, N_{f-9})$ by the choice of A, C. Now $k_3^0 < s(P'_i) + s(Q'_i) < \beta + (s(V) + \epsilon_3)$ so $s(V) > k_3^0 - \beta - \epsilon_3 > \theta$. Similarly $s(R) > k_3^0 - \beta - \epsilon_3 > \theta$.

Using the notation R, S, T, U, V of 5° (which is consistent with the present use of R, V) choose for each of R, S, T, U, V a fixed decomposition as in Axiom 33′ (e.g., $V \underset{p-1}{=} V^{\mathrm{L}} EZFV^{\mathrm{R}}$ say); this is possible since $\gamma > t_{33}$ and $\beta \leqslant b_{33}$. We say that a pair of them satisfy (0) or (1) according as the corresponding pair of Z's satisfy (0) or (1) of 12.5; note that each Z has size $\gamma - r_{33} > s_{33}$.

By 12.7, R, V satisfy (1) (since $\beta_1 \leqslant k_{26}, \beta \leqslant k_{26}, \gamma > u_{44}$).

By Remark 12.12 below, one row of the following partial table holds, where an entry 0 or 1 is read as the statement that the indicated pair satisfies (0) or (1):

R, S	S, T	T, U	U, V	R, V
	0	0	0	
	0	1	0	
	0	0	1.	

By the definition of Z'_{p-1} and by 12.7,

R, S satisfy (1) $\leftrightarrow$ S is replaced (i.e., S is an M_j in Axiom 26)

Thus the column R, S has entries 0, 1, 1 in rows 1, 2, 3 respectively. By 12.3A the column R, V has entries 0, 0, 0 respectively, i.e., in all cases R, V satisfy (0). By 12.5B we have a contradiction.

(6) now reads $Y^M \underset{p-1}{\sim} B$ and since $B \subset Z'_{p-1}$ 1, 2 of 12.11 hold. B contains (N_{12}, N_{f-11}) so contains all A-descendants in Z'_{p-1} of all α_i except $\alpha_1, \ldots, \alpha_{11}$ and $\alpha_{f-10}, \ldots, \alpha_f$. Hence 3 holds.

7°. *Case* 2. $p = n$

Recalling the definition of the B_i (cf. 7.8), first suppose the B_i are maximal normal of size $> r''$. C contains B_i $(i = 1, \ldots, n)$; let their images in Y be Y_i and let their images in Z'_{p-1} be A_i. As in 2°, Z'_{p-1} is p-bounded by β and as in 3°
$Y \underset{n}{=} (\text{In}\, Y, Y_1^{\mathrm{L}}) E(B_2^{\mathrm{R}}, B_{g-1}^{\mathrm{L}}) F(Y_g^{\mathrm{R}}, \text{Fin}\, Y)$ and $(B_4, B_{g-3}) \underset{n}{=}$
$\underset{n}{=} B_4^{\mathrm{L}} E'(A_5^{\mathrm{R}}, A_{g-4}^{\mathrm{L}}) F' B_{g-3}^{\mathrm{R}}$; $(r'' > r^*)$. Hence $Y \underset{n}{=} H(A_5^{\mathrm{R}}, A_{g-4}^{\mathrm{L}}) K$ and $Y^M \sim (A_7, A_{9-6})^0$, where the right side contains $A_9, \ldots, A_{g-8}$ as required.

Finally suppose the B_i are not as above. Then by 7.3 (Z_n) does not contain a maximal n-normal subword of size $> r''$. It follows that $Y \underset{n-1}{\sim} C$, otherwise by the beginning of 1° Z_n would contain a maximal n-normal subword of size $> r' > r''$. Moreover there are no replacements in $Z_n \to Z_{n-1}$. Thus $Z'_{p-1} \equiv Z'_{n-1}$ is just Z_{n-1}. Also $Z_n \equiv Z_{n-1}$. For the required Y^{M}, C' take Y, C.

12.12. Remark. When we prove, later, that Axiom 26 holds for $n + 1$ it will appear that, in the notation of Axiom 26, (i) if U is an M_j then exactly one of $\text{Im}\, U_1$, $\text{Im}\, U_2$ arises as in Case (a) of 6.16; (ii) if U is not an M_j then both $\text{Im}\, U_1$, $\text{Im}\, U_2$ arise as in Case (b) of 6.16. Further, (iii) $\text{Im}\, U$ always arises as in Case (b). We shall therefore add these properties to Axiom 26.

12.13. Corollary. *With the hypothesis of* 12.11, *Y contains an $(n + 1)$-subelement of size $\geqslant g/m - 40/e$.*

Proof. We have by 12.11 $Z_n \underset{n}{\to} \cdots \underset{p+1}{\to} Z_p \underset{p}{\to} Z'_{p-1} \equiv C'D'$. Now Z'_{p-1} is $((p-1))$-bounded by s_0 (by Axiom 26) and $s_0 \leqslant b_{28}, a_{28}$ so by Axiom 28

$$Z_n \underset{n}{\to} \cdots \underset{p+1}{\to} Z_p \underset{p}{\to} Z'_{p-1} \underset{p-1}{\to} \cdots \underset{1}{\to} Z'_{p-1}$$

and 12.11 is again available. Hence by induction

$$Z_n \to \cdots \to Z_p \to Z'_{p-1} \to H_{p-2} \to \cdots \to H_1 \to H_0$$

where $H_0 \equiv C''D''$ and $Y^m \underset{0}{\sim} C''$, i.e., $Y^m \equiv C''$ and C'' contains all members of $\mathcal{D}^{n-1}(A_9, A_{g-8})$. Hence C'' contains an A-descendant in H_0 of A_{20} and one of A_{g-19} (cf. 7.9). Hence C'' contains all A-descendants in H_0 of $A_{21}, \ldots, A_{g-20}$. Hence $s(C'') \geqslant (g-40)/m \geqslant g/m - 40/e$.

12.14. Proposition. *Condition B holds.*

Proof. Assume $p \geqslant 1$ since the result is trivial if $p = 0$. We have $Z_n \to \cdots \to Z_0 \equiv AB$, where $Z_n \in \operatorname{Rep} L_{n+1}$ and all A-descendants in Z_0 of $B_1, \ldots, B_g$ are contained in A and $g/m > s(A) - \epsilon > s_{25}$, for some preassigned ϵ satisfying $0 < \epsilon < s(A) - s_{25}$. Now since $b_{25} \leqslant b'_{25}$ we have by Axiom 28 say $A_p \underset{p}{\to} \cdots \underset{2}{\to} A_1 \underset{1}{\to} A$. Hence $A_p \underset{p}{=} A$ (cf. Axiom 26L part (ii)). By 11.9 since $p < n$ and $s_{25} \geqslant u_{36}$

$$Z_p \equiv A'B'$$

$$A \underset{p}{=} A_p \underset{p}{=} GA'H.$$

where A' contains all A-descendants of $B_1, \ldots, B_g$ in Z_p except for the five leftmost of B_1 and the five rightmost of B_g. They are $(p+1)$-subelements; denote them by α_i $(i \in I)$ and denote the leftmost four and the rightmost four by $\alpha_1, \ldots, \alpha_4$ and $\alpha_{f-3}, \ldots, \alpha_f$ respectively. Then $\alpha_i \supset \alpha_i^0 \underset{p}{\sim} \alpha'_i \subset A$. Also $(\alpha_1^0, \alpha_f^0) \underset{p}{\sim} (\alpha'_1, \alpha'_f) \subset A$ where each side is (p)-bounded by b_{25} (since $s_0 \leqslant b_{25}$). Let β_i be a weak image of α_i in (α'_1, α'_f) and let γ_i be a weak image of β_i in Y, with respect to $A \underset{p}{\sim} Y$, $(i \neq 1, f)$. Then $Y^{\mathrm{M}} \equiv (\gamma_2, \gamma_{f-1}) \underset{p}{=} \underset{p}{=} \gamma_2^{\mathrm{L}} E(\beta_2, \beta_{f-1}) F \gamma_{f-1}^{\mathrm{R}}$ (cf. 6.29) and $(\beta_4, \beta_{f-3}) = \beta_4^{\mathrm{L}} E'(\alpha_4^{\mathrm{R}}, \alpha_{f-3}) F' \beta_{f-3}^{\mathrm{R}}$ hence

$$Y^{\mathrm{M}} \underset{p}{=} H(\alpha_4^{\mathrm{R}}, \alpha_{f-3}^{\mathrm{L}}) K \equiv HA'^{\mathrm{M}}K$$

say. Hence $Z_p \supset (A'^{\mathrm{M}})^0 \underset{p}{\sim} Y^{\mathrm{MM}}$ and $(A'^{\mathrm{M}})^0$ contains (α_6, α_{f-5}).

We wish to apply 12.13. We have $b_{25} \leqslant k_{29}$.

If $p+1<n$ let the A-descendants of $B_1, \ldots, B_g$ in Z_{p+1} be $\delta_j (j \in J)$ of which the leftmost, rightmost are δ_1, δ_h. Then A' contains all A-descendants of these except for the first five of δ_1 and the last five of δ_h hence $(A'^{\mathrm{M}})^0$ contains all except for the first ten of δ_1 and the last ten of δ_h. Thus $(A'^{\mathrm{M}})^0$ contains all descendants of $B_2, \ldots, B_{g-1}$. The hypothesis of 12.13 holds if $(A'^{\mathrm{M}})^0$ contains all A-descendants of $g'-38$ consecutive B_i's for some $g' > m \cdot u_{48}$, that is if $(*)\ g-2 > m \cdot u_{48} - 38$. Put $g' = g+36$.

If $p = n-1$, A' contains $A_6, \ldots, A_{g-5}$ so $(A'^{\mathrm{M}})^0$ contains $A_{11}, \ldots, A_{g-10}$. The hypothesis of 12.13 holds if g' exists such that $g-20 \geqslant g'-16$, $g' > m \cdot u_{48}$, that is $(**)\ g-20 > m \cdot u_{48} - 16$. Put $g' = g-4$ in this case.

Now $(g-4)/m > s_{25} - 4/e > u_{48}$ so $(*)$ and $(**)$ hold.

By 12.13 Y contains an $(n+1)$-subelement of size greater than

$$g'/m - 40/e \geqslant g/m - 44/e > s(A) - \epsilon - 44/e \geqslant s(A) - r_{25}$$

since $r_{25} > 44/e$.

12.15. Proposition. *If $Z_n \underset{n}{\rightarrow} \cdots \underset{1}{\rightarrow} CD$, where $Z_n \in Rep(L_{n+1})$ and C is an n-subelement of size $> u_{49}$ then Z_{n-1} contains an n-subelement of size $> s(C) - u_{37} - u_4$.*

Proof. Say $Z'_{n-1} \underset{n-1}{\rightarrow} \cdots \underset{1}{\rightarrow} CE$, where $CE \in \mathsf{L}_n$, $Z'_{n-1} \in \mathrm{Rep}(L_n)$. Using the A–B notation for this sequence let all A-descendants of $B_1, \ldots, B_g$ be contained in C; we may assume $g/m > u_{49} > u_{36}$. By 11.8, 11.5

$$Z'_{n-1} \equiv C'E'$$

$$Z_{n-1} \underset{n-1}{=} C'D'$$

where $C' \supset (B_{1+s}, B_{g-s})$, $(g-2s)/m > g/m - u_{37} > u_{49} - u_{37} > u_3$. Hence $s^n(C') \geqslant (g-2s)/m > u_{49} - u_{37}$. Finally C' has weak image in Z_{n-1} of size $> s(C') - u_4$ so the result follows $(u_{35} \leqslant b_{25})$.

12.16. Proposition. *Assume Condition C. Then Axiom 25 holds for $n+1$. That is, if $A \underset{p}{\sim} Y$ where A is an $(n+1)$-subelement of size $> s_{25}$, $p < n+1$ and A, Y are (p)-bounded by b_{25}, then Y contains an $(n+1)$-subelement of size $> s(A) - r_{25}$.*

Proof. This is true if $p < n$ by 12.14 so we assume $p = n$ and $A \underset{n-1}{\not\sim} Y$. Now $Z_n \underset{n}{\rightarrow} \cdots \underset{1}{\rightarrow} Z_0 \equiv AB$ for some B, $Z_n \in \mathrm{Rep}(L_{n+1})$, where all A-descendants in Z_0 of $B_1, \ldots, B_g$ are contained in A and we may assume $g/m > s(A) > s_{25} > u_{36}$. Also $A_n \underset{n}{\rightarrow} \cdots \underset{1}{\rightarrow} A$, by Axiom 28.

Now 11.9 is applicable by 12.3, 12.4. Hence

$$Z_n \equiv A'B'$$

$$A \underset{n}{=} A_n \underset{n}{=} GA'H$$

and A' contains (B_{1+s}, B_{g-s}) where $(g-2s)/m > g/m - u_{37}$.

Case 1. The B_i are maximal normal of size $> r''$.

Let their images in A be C_i and let the images of the C_i in Y be D_i $(i = a, \ldots, b)$ where $a = 1 + s$, $b = g - s$. Then $Y^{\mathrm{M}} \equiv (D_a, D_b) = D_a^{\mathrm{L}} E(C_{a+1}^{\mathrm{R}}, C_{b-1}^{\mathrm{L}}) F D_b^{\mathrm{R}}$ and

$$(C_{a+1}^{\mathrm{R}}, C_{b-1}^{\mathrm{L}}) \supset (C_{a+3}, C_{b-3}) = C_{a+3}^{\mathrm{L}} E'(B_{a+4}^{\mathrm{R}}, B_{b-4}^{\mathrm{L}}) F' C_{b-3}^{\mathrm{R}}.$$

Hence $Y^{\mathrm{M}} \underset{n}{=} HA'^{\mathrm{M}} K$, where $A'^{\mathrm{M}} \supset (B_{a+6}, B_{b-6}) \equiv (B_{s+7}, B_{g-s-6})$. Also $Y^{\mathrm{MM}} \underset{n}{\sim} (B_{s+7}^0, B_{g-s-6}^0) \supset (B_{s+8}, B_{g-s-7}) \equiv A''$ say. Let $g' = g - 2(s+7)$. Then

$$g'/m = (g-2s)/m - 14/m > g/m - u_{37} - 14/m$$

$$> s(A) - u_{37} - 14/e > s_{25} - u_{37} - 14/e > u_{48}.$$

By 12.13 Y contains an $(n+1)$-subelement of size $> g'/m - 40/e > s(A) - u_{37} - 54/e \geqslant s(A) - r_{25}$, since $r_{25} \geqslant u_{37} + 54/e$ $(b_{25} \leqslant k_{29})$.

Case 2. Not Case 1.

Here $Z_n \equiv Z_{n-1}$. Since $A \underset{n-1}{\not\sim} Y$, A contains an n-subelement of size $> k_3^0 - b_{25} - \epsilon_3 > u_{49}$ (cf. 6.7A). By 12.15, $Z_n \equiv Z_{n-1}$ contains an n-subelement of size $> (k_3^0 - b_{25} - \epsilon_3) - u_{37} - u_4 \geqslant r''$, a contradiction.

12.17. Proposition. *Let $k < n$ and let $A_1 TA_2 \underset{k}{\sim} B_1 SB_2$ (end words E'', K'' say) where each side is (k)-bounded by k_{30}, induce $T \underset{k}{\sim} S$ (end words E, K say). Let $TT' \in Rep(L_{k+1})$, $SS' \in \mathsf{L}_{k+1}$ and let T, S be of size $> u_{50}$. Then*

$$A_1 \cdot T'^{-1} \cdot A_2 \underset{k}{=} E'' \cdot B_1 \cdot S'^{-1} \cdot B_2 \cdot K''.$$

Proof. We shall prove $(*)$ $T'^{-1} \underset{k}{=} E \cdot S'^{-1} \cdot K$. Now (cf. 6.7) $A_1 = E'' \cdot B_1 \cdot E^{-1}$, $T = E \cdot S \cdot K$, $A_2 = K^{-1} \cdot B_2 \cdot K''$ so from $(*)$ follows $A_1 \cdot T'^{-1} \cdot A_2 = E'' \cdot B_1 \cdot E^{-1} E \cdot S'^{-1} \cdot K \cdot K^{-1} \cdot B_2 K''$ as required.

We apply Axiom 31B to $T \underset{k}{\sim} S$. This is possible since $u_{50} \geqslant s_{31}$ and $k_{30} \leqslant b_{31}$. We obtain

$$(9) \qquad T^{\mathrm{L}} E' S^{\mathrm{R}} \underset{k}{=} T \cdot K^{-1} = E \cdot S$$

$$(10) \qquad S^{\mathrm{L}} F' T^{\mathrm{R}} = S \cdot K \quad = E^{-1} \cdot T$$

where S^{L}, S^{R} have size $> s(S) - r_{31} > u_{50} - r_{31} > r_0 + \epsilon_3$. If we had $S \equiv S^{\mathrm{L}} P S^{\mathrm{R}}$ then by 5.23 $s(S^{\mathrm{L}}) + s(PS^{\mathrm{R}}) < s(S) + 2\epsilon_3 + \epsilon_1^*$, $s(S^{\mathrm{R}}) < s(PS^{\mathrm{R}}) + \epsilon_3$. Hence $s(S) < 2r_{31} + 3\epsilon_3 + \epsilon_1^*$, a contradiction since $u_{50} > 2r_{31} + 3\epsilon_3 + \epsilon_1^*$. Hence $S \equiv PQR$, $S^{\mathrm{L}} \equiv PQ$, $S^{\mathrm{R}} \equiv QR$ for some P, Q, R. Hence by 5.31 $s(Q) > \mathrm{Min}(u_{50} - r_{31} - r_0 - \epsilon_3 - 2/e$, $u_{50} - 2r_{31} - 2(\epsilon_3 + \epsilon_1^*) - 4/e) = \alpha$ say. By (9) and (10) we have

$$T^{\mathrm{L}} E' Q F' T^{\mathrm{R}} \underset{k}{=} T, \quad T^{\mathrm{L}} E' \underset{k}{=} E \cdot P, \quad F' T^{\mathrm{R}} \underset{k}{=} R \cdot K.$$

Now $Z \underset{k}{\rightarrow} \cdots \underset{1}{\rightarrow} SS'$, $Z \in \mathrm{Rep}(L_{k+1})$ and by Axiom 28 since $k_{30} + q_{31} \leqslant a_{28}$, $str^+ \leqslant b_{28}$ $T \underset{k}{\rightarrow} \cdots \underset{1}{\rightarrow} T^{\mathrm{L}} E' Q F' T^{\mathrm{R}}$. By 11.9 since $\alpha > u_{36} + 1/e^2$

$$Z \equiv \sigma W, \quad s(\sigma) > \alpha - 1/e^2 - u_{37} > u_1 + s_{33}.$$

$$T \underset{k}{=} U \sigma V$$

and

(11) $U \cdot W^{-1} \cdot V \underset{k}{=} T^{\mathrm{L}} E'(RS'P)^{-1} F \cdot T^{\mathrm{R}}$.

Now σ, T are subelements of $\mathrm{Rep}(L_{k+1})$ and for some T^{M}

(12) $T^{\mathrm{M}} \underset{k}{\sim} \sigma^0$ (end words E_1, K_1 say)

Let $T \equiv T^{\mathrm{L}} T^{\mathrm{M}} T^{\mathrm{R}}$, $\sigma \equiv \sigma_1 \sigma^0 \sigma_2$. Then $T^{\mathrm{L}} \underset{k}{=} U\sigma_1 \cdot E_1^{-1}$, $T^{\mathrm{R}} = K_1^{-1} \cdot \sigma_2 V$. Now $s(\sigma^0) > s(\sigma) - u_1 > s_{33}$ so by (4) and Axiom 33 we have $T^{\mathrm{M}} \equiv \sigma^0$, $T^{\mathrm{R}} T' T^{\mathrm{L}} \equiv \sigma_2 W \sigma_1$. Further $E_1 \equiv K_1 \equiv I$. Thus by (11)

$$T^{\mathrm{L}} \sigma_1^{-1} W^{-1} \sigma_2^{-1} T^{\mathrm{R}} = T^{\mathrm{L}} E' P^{-1} S'^{-1} R^{-1} F' T^{\mathrm{R}}$$

$$T^{\mathrm{L}} T^{\mathrm{L}-1} T'^{-1} T^{\mathrm{R}-1} T^{\mathrm{R}} = E \cdot S'^{-1} \cdot K$$

and (*) follows.

12.18. Proposition. *Let $DE \in \mathsf{L}_{k+1}$, $DF \in \mathsf{L}_{n+1}$, $k < n$. Let $s_{DE}(D) \geqslant r$. Then $E^{-1} \cdot F$ is conjugate in G_k to an element of L_{n+1}.*

Proof. *Case* 1. $k \geqslant 1$.

1°. We have $Z'_k \underset{k}{\to} \cdots \underset{1}{\to} DE$, $Z'_k \in \mathrm{Rep}(L_{k+1})$

$$(Z_n \underset{n}{\to} \cdots \underset{k+1}{\to}) Z_k \underset{k}{\to} \cdots \underset{1}{\to} DF,\ Z_n \in \mathrm{Rep}(L_{n+1}).$$

By 11.8 since $r > u_{36}$

$$Z'_k \equiv D'E'$$

(13) $Z_k \underset{k}{=} D'F'$

$s(D') > r - u_{37} - 1/e^2$ and $E^{-1} \cdot F$ is conjugate in G_k to $E'^{-1} \cdot F'$.

From (13), $D' \supset D'^0 \underset{k}{\sim} Z_k^{\mathrm{M}} \supset S$ and $S \supset S^0 \underset{k}{\sim} T \supset T'$ where $T \subset D'^0$ and $s^{k+1}(S) > s(D') - u_4$. Since $S^0 \subset S$ we have $S^0 J \in \mathsf{L}_{k+1}$ for some J. Also $TT' \in \mathrm{Rep}\, L_{k+1}$ for some T''. Now

$$s(S^0) > s(S) - u_1 > r - u_{37} - 1/e^2 - u_4 - u_1$$

$$> \mathrm{Max}(u_{50}, r_0 + \epsilon_3, v_{26}).$$

We have $s(T') > s(S) - u_4 > r - u_{37} - 1/e^2 - 2u_4 > r_0$, hence $s(T') \leqslant s(T) + \epsilon_3$ and $s(T) > r - u_{37} - 1/e^2 - 2u_4 - \epsilon_3 > u_{50}$.

Let $Z_k \equiv (S^0\alpha)$, $D' \equiv D_1' T D_2'$ and let Y, Z be end words for $S^0 \sim T$. By (13) $\alpha \underset{k}{=} Z^{-1} D_2' F' D_1' Y^{-1}$, so $S^0\alpha = Y \cdot T D_2' F' D_1' Y^{-1}$.

By 12.17 $J^{-1} \cdot \alpha = Y T''^{-1} D_2' F' D_1' Y^{-1}$; but we may take T'' to be $D_2' E' D_1'$, so $J^{-1}\alpha = Y \cdot D_1'^{-1}(E'^{-1}F')D_1' Y^{-1}$; ($T$ and S^0 are (k)-bounded by $s_0 \leqslant k_{30}$).

2°. Let $\bar{S}$ be the maximal $(k+1)$-normal subword of Z_k determined by S^0; say $\bar{S} \equiv A S^0 B$, $(\bar{S}M) \equiv Z_k$. Then $\alpha \equiv BMA$. Say $\bar{S}U \in \mathsf{L}_{k+1}$. Now $Z_{k+1} \underset{k+1}{\rightarrow} Z_k$ so $\bar{S}$ determines a maximal $(k+1)$-normal subword of Z_{k+1}, say R', by Axiom 26″ (since $s(\bar{S}) \geqslant s(S^0) - \epsilon_3 > v_{26}$). Either $|s(R') + s(\bar{S}) - 1| < r_{26}$ and $s(R') > r'$ or $|s(R') - s(S)| < r_{26}$. Hence $s(R') > \mathrm{Min}(r', s(S^0) - \epsilon_3 - r_{26}) = r'$, since

$$-\epsilon_3 - r_{26} - u_1 - u_4 + r - u_{37} - 1/e^2 > r'.$$

By Axiom 26A, there exists H such that $Z_{k+1} \underset{k+1}{\rightarrow} H$ and any linearization of H is conjugate in G_k to $U^{-1} \cdot M$. Thus $H \in \mathsf{L}_{n+1}$. But $AS^0BU \in \mathsf{L}_{k+1}$ and $S^0 J \in \mathsf{L}_{k+1}$, so by 5.29 $J \underset{k}{=} BUA$.

In G_k, $E^{-1}F$ is conjugate to $E'^{-1}F'$ which is conjugate to $J^{-1}\alpha = A^{-1}U^{-1}B^{-1}BMA$. Thus $E^{-1}F$ is conjugate to $U^{-1}M$ hence to H.

3°. *Case* 2. $k = 0$. The argument of 2° remains valid if we take S, S^0 to be D, α to be F and J to be E ($E' \equiv E$, $F' \equiv F$).

§13

Informal Summary

In 13.11 it is proved that the remaining part (b) of Axiom 20 holds for $n + 1$. Afterwards some other axioms are proved for $n + 1$ including Axioms 19, 40, 33.

13.1 Definition. If X and Y are cyclic words then $X \underset{n}{\cong} Y$ means either $X \equiv Y$ or for some p, $1 \leqslant p \leqslant n$, X, Y properly contain maximal p-normal subwords S, T respectively, say $X \equiv (SA)$, $Y \equiv (TB)$ $(A \not\equiv I, B \not\equiv I)$, such that SAS, TBT are p-objects which are almost equal; let them produce ϵ, κ say; and further $SA \underset{p}{=} \epsilon \cdot TB \cdot \epsilon^{-1}$.

13.2. Note. Let X be E.R. and not a power and let $f \geqslant 2$ be an integer. For some u, v let X^f be n-bounded by u and $((n-1))$-bounded by v. Let S, T be maximal n-normal in X^f and let S^+ be T. Then X^g is n-bounded by u and $((n-1))$-bounded by v $(g = 1, 2, 3, \ldots)$.

Proof. Let N be a p-subelement in X^g $(p \leqslant n)$. Let the translates of S in $W \equiv XX^gX$ be S_i $(i = a, \ldots, b)$; these are maximal normal in W. If N is in some S_i then $N \subset S \subset X^f$. If N contains some S_i then $p = n$ since $r_0' < r_0 + \epsilon_3$ so $N \equiv S_i \equiv S \subset X^f$. Consider the remaining case. Now $N \subset (S_a, S_b)$ so $N \subset (S_i, S_{i+h})$ where i is maximal, h is minimal. If $p \geqslant 2$ then $N \supset (S_{i+1}, S_{i+h-1}) \supset S_{i+1}$, a contradiction. Hence $N \subset (S_i, S_{i+1}) \equiv (S, T) \subset X^f$.

13.3. Notation. If A, B are subwords of W then $(A, \widehat{B})$ means $(A, B) \cdot B^{-1}$; $(\widehat{A}, B)$ means $A^{-1} \cdot (A, B)$.

13.4. Proposition. *Assume*

1. X^f *is* $((n-1))$*-bounded by* k_{31} *and* n*-bounded by* k_{32} *where* X *is* E.R., *not a power, and* $f \geqslant 2$. $S_1, \ldots, S_p$ *are maximal* n*-normal subwords of* X^f *of size* $> u_{51}$ *and properly closed to translation; thus*

$$(\exists a)\, S_i^+ = S_{i+a} \quad (i = 1, \ldots, p-a) \quad \textit{and} \quad 1 \leqslant p - a.$$

2. *Similarly for* Y^g *and* $T_1, \ldots, T_p$*; say*

$$T_j^+ = T_{j+b} \quad (j = 1, \ldots, p-b),\ 1 \leqslant p - b.$$

3. *For some* $k \geqslant a + b - d + 2$, *where* $d = (a, b)$ *we have*

$$(S_i, S_j) \underset{n}{\triangleq} (T_i, T_j) \text{ whenever } 1 \leqslant i < j \leqslant k.$$

Then

(i) $(S_i, S_j) \underset{n}{\triangleq} (T_i, T_j)$ *whenever* $1 \leqslant i < j \leqslant p$.

(ii) *If* $(S_1, S_2), (T_1, T_2)$ *produce* ϵ_1, κ_2 *and if* w, *where* $3 \leqslant w < p$, *is divisible by a and by b then*

$$(S_1, \widehat{S}_{w+1}) \underset{n}{\equiv} \epsilon_1 \cdot (T_1, \widehat{T}_{w+1}) \cdot \epsilon_1^{-1} \quad \textit{and} \quad (X^{w/a}) \underset{n}{\cong} (Y^{w/b})$$

13.4A. Remark. Under the above hypothesis we may take p as large as we please, without loss of generality; for by 13.2 we may increase f and then form the set of all translates of $S_1, \ldots, S_p$ in the new word; similarly for Y^g.

Proof of 13.4. We have $k_{31} \leqslant \text{Min}(k'_{12}, k_{20})$, $k_{32} \leqslant \text{Min}(k'_{13}, k_{21})$, $u_{51} \geqslant \text{Max}(u_{24}, u_{31})$.

(i) It is sufficient to prove (i) when $p = k + 1$. Let $D_v \equiv (S_v, S_{v+2})$, $C_v \equiv (T_v, T_{v+2})$ $(v = 1, \ldots, k-2)$. Then $D_1, \ldots, D_{k-2}$ have period a in the sense that $D_t \triangleq D_{t+a}$ $(t = 1, \ldots, k-2-a)$: indeed $D_t \equiv D_{t+a}$. They also have period b since

$$D_t \triangleq C_t \equiv C_{t+b} \triangleq D_{t+b} \quad (t = 1, \ldots, k-2-b).$$

Since $k - 2 \geqslant a + b - d$ we have by 2.17 that they have period d: $D_t \triangleq D_{t+d}$ $(t = 1, \ldots, k-2-d)$. Since S_{k+1} exists we have $(S_{k-1}, S_{k+1}) \equiv D_{k-1-a}$. Also $(T_{k-1}, T_{k+1}) \equiv C_{k-1-b} \triangleq D_{k-1-a}$ since $k-1-b \equiv k-1-a \pmod d$. Thus $(S_{k-1}, S_{k+1}) \triangleq (T_{k-1}, T_{k+1})$. Similarly we may prove $(S_k, S_{k+1}) \triangleq (T_k, T_{k+1})$. But $(S_{k-1}, S_k) \triangleq (T_{k-1}, T_k)$ and if $w \leqslant k-2$ $(S_w, S_{k-1}, S_k) \triangleq$ $\triangleq (T_w, T_{k-1}, T_k)$, so by 9.14 $(S_w, S_{k+1}) \triangleq (T_w, T_{k+1})$.

(ii) By (i).

(1) $\qquad (S_1, S_{w+2}) \triangleq (T_1, T_{w+2})$.

Say $(S_r, S_{r+1}), (T_r, T_{r+1})$ produce ϵ_r, κ_{r+1} $(r = 1, \ldots, p-1)$; then by 9.13 and its dual the pair in (1) produce ϵ_1, κ_{w+2}. Since a divides w, $(S_1, \widehat{S}_{w+1})$ has the form $X_1^{w/a}$ where X_1 is a C.A. of X. Similarly $(T_1, \widehat{T}_{w+1}) \equiv Y_1^{w/b}$ for some C.A. Y_1 of Y. Since $X_1^{w/a}(S_{w+1}, S_{w+2}) \underset{n}{=} \epsilon_1 \cdot Y_1^{w/b}(T_{w+1}, T_{w+2}) \kappa_{w+2}$ we have $X_1^{w/a} \epsilon_{w+1} \underset{n}{=} \epsilon_1 \cdot Y_1^{w/b}$. Now $(S_{w+1}, S_{w+2}) \equiv (S_1, S_2)$ and $(T_{w+1}, T_{w+2}) \equiv (T_1, T_2)$ so clearly $\epsilon_1 \underset{n}{=} \epsilon_{w+1}$. Hence $X_1^{w/a} \underset{n}{=} \epsilon_1 Y_1^{w/b} \xi_1^{-1}$.

Since $w \geqslant 3$, $(T_1, T_2) \subset Y_1^{w/b}$, $(S_1, S_2) \subset X_1^{w/a}$ so $(X_1^{w/a}) \cong (Y_1^{w/b})$ and hence $(X^{w/a}) \cong (Y^{w/b})$.

13.5. Proposition. *Let X^f and $S_1, \ldots, S_p$ satisfy hypothesis 1 of 13.4. Let $(S_{c+1}, S_{c+m}) \underset{n}{\hat{=}} (S_{d+1}, S_{d+m})^{-1}$ and assume that whenever i, j satisfy $c+1 \leqslant i < j \leqslant c+m$ we have*

$$(2) \qquad (S_i, S_j) \underset{n}{\hat{=}} (S_{g(i)}^{-1}, S_{g(j)}^{-1})$$

where $g(i) = c + m + d + 1 - i$. Then $m \leqslant a + 1$.

Proof. Assume $m \geqslant a+2$. Without loss of generality we may suppose $c \leqslant d$, $1 \leqslant c+1 \leqslant a$, $1 \leqslant d+1 \leqslant a$ and $c = 0$. Thus

$$(3) \qquad (S_1, S_m) \hat{=} (S_{d+m}^{-1}, S_{d+1}^{-1})$$

By 13.4, since $m \geqslant a + a - a + 2$ we have

$$(4) \qquad (S_1, S_{m+d}) \hat{=} (S_{d+m}^{-1}, S_1^{-1})$$

and (2) holds if $1 \leqslant i < j \leqslant m+d$. Call the left side, right side of (4) say A, B. Let A, B produce ϵ, κ; let B, A produce ϵ'', κ''; let A^{-1}, B^{-1} produce ϵ', κ'. Then by 9.8A

$$\epsilon \underset{n}{=} \kappa'^{-1}, \quad \kappa \underset{n}{=} \epsilon'^{-1}, \quad \epsilon \underset{n}{=} \epsilon''^{-1}, \quad \kappa \underset{n}{=} \kappa''^{-1}$$

But A^{-1} is B, B^{-1} is A so we may take $\epsilon'' \underset{n}{=} \epsilon'$, $\kappa'' \underset{n}{=} \kappa'$ and we get $\epsilon \underset{n}{=} \kappa$. Now $A \underset{n}{=} \epsilon \cdot B \cdot \kappa$ so $I \underset{n}{=} B \cdot \epsilon \cdot B \cdot \epsilon$. By 5.13 we deduce $B \cdot \epsilon \underset{n}{=} I$ that is, $\epsilon \underset{n}{=} A$. Now (3) implies

$$(3') \qquad (S_1, S_m) \hat{=} (S_{d+m+a}^{-1}, S_{d+1+a}^{-1})$$

and in the same way that (4) follows from (3) we obtain

$$(4') \qquad A' \equiv (S_1, S_{m+d+a}) \doteq (S^{-1}_{m+d+a}, S^{-1}_1) \equiv B'.$$

As above if A', B' produce ϵ_0, κ_0 then $\epsilon_0 \underset{n}{=} A'$. Let (S_1, S_2), $(S^{-1}_{d+m+a}, S^{-1}_{d+m+a-1})$ produce ϵ_1, κ_1. Then by 9.13 $\epsilon_1 \underset{n}{=} \epsilon_0$. But

$$(S^{-1}_{d+m+a}, S^{-1}_{d+m+a-1}) \equiv (S^{-1}_{d+m}, S^{-1}_{d+m-1})$$

so comparing with (4) we have by 9.13 $\epsilon_1 \underset{n}{=} \epsilon$. Thus $\epsilon_0 \underset{n}{=} \epsilon$ and $A \underset{n}{=} A'$, i.e., $(\widehat{S}_{m+d}, S_{m+d+a}) \underset{n}{=} I$. This contradicts 6.1 since $k_{31} < k_1, k_{32} < k^0_1$.

13.6. We recall that $-A \equiv -B$ means that one of A, B is a right subword of the other (cf. 2.4). Although this relation is not transitive in general, it is transitive if we restrict $A, B, \ldots$ to be subwords of elements of Rep L_n of size $> s_{33}$ (cf. 8.2 or Axiom 33).

13.7. Proposition. *Let X^e and $S_1, \ldots, S_p$ satisfy hypothesis 1 of 13.4. Let $(S_1, S_m) \underset{n}{\doteq} (S_{c+1}, S_{c+m})$ where $1 \leqslant c+1 \leqslant a$ and assume that whenever i, j satisfy $1 \leqslant i < j \leqslant m$ then $(S_i, S_j) \underset{n}{\doteq} (S_{c+i}, S_{i+j})$. Let $m \geqslant a+2$. Then either $c = 0$ or T, Q exist such that*

$$(X^e) \underset{n-1}{=} (T^e) \underset{n}{=} (Q^{a'e})$$

where $a' = a/d$, $d = (a, c)$. Also $(Q^{a'e})$ is $((n-1))$-bounded by $k_{31} + c_{33} + c_{39}$ and n-bounded by k_{33}.

Proof. Assume $c \neq 0$. Since $m \geqslant a + a - a + 2$ we may take m as large as we please by 13.4. Let $1 \leqslant i < j$.

1°. We prove $(S_i, S_j) \doteq (S_{i+d}, S_{j+d})$. Put $D_v = (S_v, S_{v+j-i})$ $(v = 1, 2, \ldots)$. Then $D_v \doteq D_{v+a}$ because $D_v \equiv D_{v+a}$, and $D_v \doteq D_{c+v}$, so by 2.17 $D_v \doteq D_{v+d}$.

2°. We have $k_{31} \leqslant b_{33}, k_{31} + c_{33} \leqslant \mathrm{Min}(k'_{12}, k_{20}, b_{39}, k_{12})$ and $k_{32} + u_4 \leqslant \mathrm{Min}(k'_{13}, k_{21}, b'_{39}, k_{13})$.

Let the kernels of $S_1, \ldots, S_p$ be $A_1, K_2, K_3, \ldots, K_{p-1}, A_p$. Then $s(K_i) > u_{51} - 2(r_0 + \epsilon_3) - 4/e > \mathrm{Max}(t_{33}, u_4 + \epsilon_3)$. We may write

$A_1 \equiv A'_1 K_1, A_p \equiv K_p A'_p$ where $K_i \equiv K_{i+a}$ $(i = 1, 2, ..., p-a)$. We may also write $(S_1, S_p) \equiv A'_1 K_1 C_1 \cdots K_{p-1} C_{p-1} K_p A'_p$ and $C_i \equiv C_{i+a}$ $(i = 1, 2, ..., p-a-1)$. (K_1, K_{a+1}) is a C.A. of X, say X_1, and $X_1 \equiv K_1 C_1 K_2 C_2 \cdots K_a C_a$. Now as in Axiom 33′

$$K_i \underset{n-1}{=} K_i^{\mathrm{L}} E_i Z_i F_i K_i^{\mathrm{R}} \equiv K'_i, \text{ say,}$$

so $X_1 \underset{n-1}{=} Y \equiv K'_1 C_1 \cdots K'_a C_a$. Thus $(X^e) \underset{n-1}{=} (Y^e)$ and (Y^e) is $((n-1))$-bounded by $k_{31} + c_{33}$ and n-bounded by $k_{32} + u_4$ by 6.10A $(u_3 < k_{32})$.

Now $(S_i, S_j) \equiv S_i^{\mathrm{L}} (K_i, K_j) S_j^{\mathrm{R}} \doteqdot (K_i, K_j) \doteqdot (K'_i, K'_j) \doteqdot (Z_i, Z_j)$. Hence $(Z_i, Z_j) \doteqdot (Z_{i+d}, Z_{j+d})$. Note that if $Z_i B$ and DZ_j denote temporarily the maximal n-normal subwords of (Z_i, Z_j) containing Z_i, Z_j then since $s(F_i K_i^{\mathrm{R}}) < q_{33}$ and $B \equiv F_i K_i^{\mathrm{R}} H$ where H is not normal we have

$$s(B) < q_{33} + (r_0 + \epsilon_3) + 2/e = \alpha_2 \text{ (say)} < u_{22}.$$

Let (Z_i, Z_j), (Z_{i+d}, Z_{j+d}) produce $\epsilon_{ij}, \kappa_{ij}$. By 9.13, ϵ_{ij} is independent of j (mod G_n) so we may write ϵ_i instead of ϵ_{ij}; similarly $\kappa_{ij} = \kappa_j$.

Since $(Z_i, \hat{Z}_j) \equiv (Z_i, Z_{j+1}) \cdot (Z_j, Z_{j+1})^{-1}$ we have

$$(Z_i, \hat{Z}_j) = \epsilon_i \cdot (Z_{i+d}, \hat{Z}_{j+d}) \cdot \epsilon_j^{-1}. \tag{5}$$

Take $i = sd$, $j = (s+1)d$, $s \geqslant 1$. Put $H_s \equiv (Z_{sd}, \hat{Z}_{(s+1)d})$; then by (5), $\epsilon_{sd}^{-1} \cdot H_s \cdot \epsilon_{(s+1)d} = H_{s+1}$. Next

$$(Z_d, \hat{Z}_{a+d}) \equiv H_1 H_2 \cdots H_{a'}$$

$$= H_1 \cdot (\epsilon_d^{-1} H_1 \epsilon_{2d}) \cdot (\epsilon_{2d}^{-1} \epsilon_d^{-1} H_1 \epsilon_{2d} \epsilon_{3d}) \cdots$$

$$= (H_1 \epsilon_d^{-1})^{a'} \epsilon_d \epsilon_{2d} \cdots \epsilon_{a'd}$$

Now $\epsilon_{sd} \in E'((Z_{sd}, Z_{(s+1)d}), (Z_{(s+1)d}, Z_{(s+2)d}))$ so by 9.8

$$\epsilon_d \epsilon_{2d} \cdots \epsilon_{a'd} \in E'((Z_d, Z_{2d}), (Z_{(a'+1)d}, Z_{(a'+2)d}))$$

But $(a'+1)d \equiv d$ and $(a'+2)d \equiv 2d$ (mod a) so $\epsilon_d \epsilon_{2d} \cdots \epsilon_{a'd} \underset{n}{=} I$. Hence $(Z_d, \hat{Z}_{a+d}) \underset{n}{=} ((Z_d, \hat{Z}_{2d}) \cdot \epsilon_d^{-1})^{a'}$.

3°. Consider the word $A \equiv (Z_{2d-1}, \hat{Z}_{2d}) \cdot \epsilon_d^{-1} \cdot Z_d$. We shall prove that $A = Z_{2d-1}^{\mathrm{L}} H Z_d^{\mathrm{R}}$ for some H where $s(Z_{2d-1}^{\mathrm{L}}) > s(Z_{2d-1}) - d_{39}$ and $s(Z_d^{\mathrm{R}}) > u_{52}$.

Write Z, Z' for Z_d, Z_{2d} respectively. For any $j > d$, the pair of preobjects $(Z, Z_j), (Z', Z_{j+d})$ produce ϵ_d, κ_j. Adjustments are equal in G_n, say

(6) $$(Z^*, Z_j^*) = (Z'^*, Z_{j+d}^*) \text{ (left side standard).}$$

In particular, $-Z^* \equiv -Z$ and $-Z' \equiv -Z'^*$. Also $\epsilon_d = Z \cdot Z^{*-1} \cdot Z'^* \cdot Z'^{-1}$ by 9.1. Hence $A \equiv (Z_{2d-1}, Z') \cdot Z'^{-1} \cdot \epsilon_d^{-1} \cdot Z \equiv (Z_{2d-1}, Z') \cdot (Z'^*)^{-1} Z^*$.

By Note 8.5A and the proof of (2) of 8.5A we see that if (2) of 8.5A holds then (ii) $Z^{*-1} Z'^*$ is a subpowerelement of Rep L_n; if (1) of 8.5A holds then by 8.5 we have (i) $Z^* F_1 \equiv Z'^* F_2$ where F_1 or F_2 is I.

4°. Assume (i).

Here $A \equiv (Z_{2d-1}, Z') \cdot F_2 \cdot F_1^{-1}$. In (2) let Z^*B and Z'^*B' be maximal normal. If $F_2 \equiv I$ then by 8.5′ $|s(Z^*B) - s\ Z^*F_1B')| < v_3$. As in the proof of 8.5A we obtain

$$s(F_1) \leqslant \mathrm{Max}(r_0 + \epsilon_3, v_3 + 2/e + (2\epsilon_3 + \epsilon_1^*) + \epsilon_3 + \alpha_2) = \alpha_3 \text{ say.}$$

Similarly if $F_1 \equiv I$ then $s(F_2) \leqslant \alpha_3$.

Suppose F_1 is I. Then $A \equiv (Z_{2d-1}, Z') \cdot F_2$. Now $-Z'^* \equiv -Z'$ hence there is no cancellation or amalgamation in the product $Z' \cdot F_2$ and $-Z'^* F_2 \equiv -Z' F_2$. Now $-Z \equiv -Z^* \equiv -Z'^* F_2$. Thus A ends with $Z' F_2$ and $-Z' F_2 \equiv -Z$.

For any i, since $s(S_i) > u_{51}$ we have by Axiom 33′

$$s(Z_i) > s(K_i) - r_{33} > u_{51} - 2(r_0 + \epsilon_3) - 4/e - r_{33} = \beta \text{ say}$$

$$> \mathrm{Max}(u_{21}, u_{31}, s_{39}, u_{24} + \epsilon_3)$$

Hence $s(Z' F_2) \geqslant s(Z') - \epsilon_3 > \beta - \epsilon_3 \geqslant u_{52}$. Also $s(Z_i) < k_{32} + u_4$.

Suppose F_2 is I. Then $-Z' \equiv -Z'^* - Z^* F_1$. Now

$s(F_1) < \alpha_3 < \beta - \epsilon_3 < s(Z') - \epsilon_3$. Hence $Z' \equiv Z''F_1$ say. Thus $-Z''F_1 \equiv -Z^*F_1$ and $-Z'' \equiv -Z^* \equiv -Z$. A ends with Z''.

Next $s(Z') \leqslant s(Z'') + s(F_1) + 2/e < s(Z'') + \alpha_3 + 2/e$. Hence $s(Z'') > s(Z') - \alpha_3 - 2/e > \beta - \alpha_3 - 2/e \geqslant u_{52}$. Thus if (i) holds then A begins with Z_{2d-1}.

5°. Assume (ii).

Then $K_1 Z^{*-1} \cdot Z'^* \equiv K_2 R'$ where $R' \in L_n$ and K_1 or K_2 is I. Also $A \equiv (Z_{2d-1}, Z') \cdot R'^{-1} \cdot K_2^{-1} \cdot K_1$. Now $T^{-1} Z' \in L_n$ for some T and $-R' \equiv -Z'^* \equiv -Z'$; hence $T^{-1} Z' \equiv R'$.

By Axiom 39, $(Z_{2d-1}, Z') \cdot Z'^{-1} T \underset{n-1}{=} Z^{\mathrm{L}}_{2d-1} E T^{\mathrm{R}}$ say, so $A \underset{n-1}{=} Z^{\mathrm{L}}_{2d-1} E T^{\mathrm{R}} \cdot K_2^{-1} \cdot K_1$.

We prove $-T^{\mathrm{R}} K_2^{-1} \equiv -Z K_1^{-1}$. If $K_1 \equiv I$ we have $-T^{\mathrm{R}} \equiv -T \equiv -R'^{-1}$ hence $-T^{\mathrm{R}} K_2^{-1} \equiv -R'^{-1} K_2^{-1} \equiv -Z'^{*-1} Z^* \equiv -Z^* \equiv -Z$. If $K_2 \equiv I$ we have $-T^{\mathrm{R}} \equiv -R'^{-1} \equiv -Z'^{*-1} Z^* K_1^{-1} \equiv -Z^* K_1^{-1} \equiv -Z K_1^{-1}$.

Suppose K_1 is I. Then $A \underset{n-1}{=} Z^{\mathrm{L}}_{2d-1} E T^{\mathrm{R}} K_2^{-1}$ and $-T^{\mathrm{R}} K_2^{-1} \equiv -Z$. Also $s(T^{\mathrm{R}} K_2^{-1}) \geqslant s(T^{\mathrm{R}}) - \epsilon_3 > s(T) - d_{39} - \epsilon_3$ and $s(T) + k_{32} + u_4 > s(T) + s(Z') \geqslant 1 - 2/e$. Hence

$$s(T^{\mathrm{R}} K_2^{-1}) > -d_{39} - \epsilon_3 - k_{32} - u_4 + 1 - 2/e \geqslant u_{52}.$$

Suppose K_2 is I. Then $-T^{\mathrm{R}} \equiv -Z K_1^{-1}$ and $K_1 Z^{*-1} Z'^* \equiv R'$. From (2) and 8.5', $1 - v_2 \leqslant s(Z^*B) + s(Z'^*B') \leqslant s(Z^*) + s(B) + s(Z'^*) + s(B') + 4/e \leqslant s(Z^*) + s(Z'^*) + 2\alpha_2 + 4/e$. Since Z^* and Z'^* are normal we have by 5.23 $s(K_1) + s(Z^*) + s(Z'^*) \leqslant s(R') + (2\epsilon_3 + \epsilon_1^*) + (r_0 + 2\epsilon_3)$. Also $s(R') \leqslant 1$. Hence $s(K_1) \leqslant v_2 + 2\alpha_2 + 4/e + r_0 + \epsilon_1^* + 4\epsilon_3 = d'$ say. Now $s(T^{\mathrm{R}}) > s(T) - d_{39} > 1 - 2/e - k_{32} - u_4 - d_{39} = d''$ say. Since $d'' > d' + \epsilon_3$ we have $T^{\mathrm{R}} \equiv U K_1^{-1}$ for some U. Also $s(U) \geqslant s(T^{\mathrm{R}}) - s(K_1) - 2/e > d'' - d' - 2/e \geqslant u_{52} > r_0 + \epsilon_3$. In particular U is normal. Since $-U K_1^{-1} \equiv -Z K_1^{-1}$ we have $-U \equiv -Z$. Also $A \equiv Z^{\mathrm{L}}_{2d-1} E U$. Finally $s(Z^{\mathrm{L}}_{2d-1}) > s(Z_{2d-1}) - d_{39}$, which is also true when K_1 is I.

6°. Thus in all cases $A \underset{n-1}{=} Z^{\mathrm{L}}_{2d-1} U Z^{\mathrm{R}} \equiv A'$, say (for some U) where $s(Z^{\mathrm{L}}_{2d-1}) > s(Z_{2d-1}) - d_{39}$, $s(Z^{\mathrm{R}}_d) > u_{52}$.

If (i) holds the $((n-1))$-bound of A' is the same as that of (Z_{2d-1}, Z_{2d}) and the n-bound exceeds that of (Z_{2d-1}, Z_{2d}) by $2/e + \alpha_3 + \epsilon_3$.

If (ii) holds the $((n-1))$-bound of A' is the same as that of

$A'' \equiv Z^{\mathrm{L}}_{2d-1} ET^{\mathrm{R}}$ while the n-bound of A' exceeds that of A'' by $2/e + s(K_2) + \epsilon_3$. Now

$$s(Z^*) + s(Z'^*) \leqslant s(Z^*B) + s(Z'^*B') + 2\epsilon_3 \leqslant 1 + v_2 + 2\epsilon_3$$

$$s(K_2) + s(R') \leqslant s(K_2R') + (r_0 + 2\epsilon_3)$$

$$1 - 2/e \leqslant s(R')$$

$$s(K_2R') = s(Z^{*-1}Z'^*) \leqslant s(Z^*) + s(Z'^*) + 2/e$$

hence $s(K_2) \leqslant 2/e + v_2 + 2\epsilon_3 + r_0 + 2\epsilon_3 + 2/e = \gamma$, say.

Hence A' is $((n-1))$-bounded by $k_{31} + c_{33} + c_{39}$ and n-bounded by $\mathrm{Max}(\theta, \varphi)$, where $\theta = k_{32} + u_4 + 2/e + \alpha_3 + \epsilon_3$ and $\varphi = \mathrm{Max}(k_{32} + u_4, 1 - \beta + \epsilon_3) + g_{39} + 2/e + \gamma + \epsilon_3$.

7°. $(Z_d, Z_{2d-1}) \equiv Z_d V Z_{2d-1}$ say. Put $B \equiv (Z_d, \hat{Z}_{2d}) \cdot \epsilon_d^{-1} \cdot Z_d$. If $d = 1$ then $d = 2d - 1$ so $B \equiv A \equiv Z^{\mathrm{L}}_d U Z^{\mathrm{R}}_d$. If $d > 1$

$$B \equiv (Z_d, Z_{2d-1}) \cdot Z_{2d-1}^{-1} \cdot (Z_{2d-1}, \hat{Z}_{2d}) \epsilon_d^{-1} \cdot Z_d = Z_d V Z_{2d-1} U Z^{\mathrm{R}}_d$$

Hence in any case $B \equiv Z^{\mathrm{L}}_d W Z^{\mathrm{R}}_d$ where $s(Z^{\mathrm{L}}_d) > s(Z_d) - d_{39}$ and $s(Z^{\mathrm{R}}_d) > u_{52}$. Now $s(Z^{\mathrm{L}}_d) + s(Z^{\mathrm{R}}_d) - s(Z_d) > s(Z^{\mathrm{R}}_d) - d_{39} > u_{52} - d_{39} > 3\epsilon_3 + \epsilon_1^*$ so $Z^{\mathrm{R}}_d \cdot Z_d^{-1} \cdot Z^{\mathrm{L}}_d = Z^{\mathrm{M}}_d$.

Thus $(Z_d, \hat{Z}_{a+d}) \underset{n}{\hat{=}} ((Z_d, \hat{Z}_{2d}) \cdot \epsilon_d^{-1})^{a'} = (B \cdot Z_d^{-1})^{a'} = Z^{\mathrm{L}}_d \cdot (W Z^{\mathrm{M}}_d)^{a'} Z^{\mathrm{L}-1}_d$. Hence $Z^{\mathrm{L}-1}_d \cdot (Z_d, \hat{Z}_{a+d}) \cdot Z^{\mathrm{L}}_d = (W Z^{\mathrm{M}}_d)^{a'}$; but the left side is a linearization of Y, so $(Y^e) \underset{n}{=} ((W Z^{\mathrm{M}}_d)^{a'e})$.

By 5.31, where $Z_d \equiv PQR$, $Z^{\mathrm{L}}_d \equiv PQ$, $Z^{\mathrm{M}}_d \equiv Q$, $Z^{\mathrm{R}}_d \equiv QR$, either $s(Q) \geqslant \mathrm{Min}(\beta - d_{39}, u_{52}) - r_0 - \epsilon_3 - 2/e > r_0 + \epsilon_3$ or $s(Q) \geqslant u_{52} - d_{39} - 3\epsilon_3 - \epsilon_1^* - 2/e > r_0 + \epsilon_3$. Also $B \equiv PQWQR$.

We discuss the bounds of $(WQ)^{a'} \equiv K$, say; for this purpose we may regard a' as being as large as we please. Since Q is not a subelement of $L_1 \cup \cdots \cup L_{n-1}$ the $((n-1))$-bound of K is at most that of QWQ hence of B, i.e., $k_{31} + c_{33} + c_{39}$.

For the n-bound first note that A is not an n-subelement, hence neither is B since $B \supset A$.

In general if a word H begins with a subelement Z of Rep L_n and Z is normal and if ZB is maximal normal in H call B the *left excess* of H (in Z).

In Case (i) (cf. 4°) the left and right excess for A is the same for (Z_{2d-1}, Z') so each has size $< \alpha_2$ (cf. 2°). In Case (ii) (cf. 5°) the left and right excesses for A are the same as for $Z_{2d-1}^{\mathrm{L}} ET^{\mathrm{R}}$, so by Axiom 39 each has size $< \alpha_2'' = \mathrm{Max}(r_0 + \epsilon_3, f_{39} + (2\epsilon_3 + \epsilon_1^*))$. Put $\alpha_2' = \mathrm{Max}(\alpha_2, \alpha_2'')$. If $B \not\equiv A$ the left excess of B is the left excess of (Z_d, Z_{2d-1}) so has size $< \alpha_2'$; the right excess of B is the same as that of A. Let Y_1, Y_2 be the excesses for B. Then $PQWQR \equiv B \equiv PQY_1X_1 \equiv X_2Y_2QR$. The Q in QR is not contained in PQY_1, so $X_1 \equiv X_1'R, X_1' \not\equiv I$. Similarly $X_2 \equiv PX_2', X_2' \not\equiv I$. Hence $QWQ \equiv QY_1X_1' \equiv X_2'Y_2Q$. Now $K \supset QWQWQ \equiv X_2'Y_2QY_1X_1'$ so $N \equiv Y_2QY_1$ is maximal normal in K, and $s(N) < s(Y_2) + 2/e + k_B < \alpha_2' + 2/e + k_B$, where k_B denotes the n-bound for B given below. Any n-urelement in K is either in N or in QWQ so its size $\leqslant \mathrm{Max}(k_B, s(N) + \epsilon_3)$. Hence K is n-bounded by $k_B + \alpha_2' + 2/e + {} + \epsilon_3 \leqslant k_{33}$.

Since the right excess for $Z_d VZ_{2d-1}^{\mathrm{L}}$ is the same as that of $Z_d VZ_{2d-1}$ its size is $< \alpha_2$; hence $k_B = \mathrm{Max}(\sigma, \varphi) + d_2 + 2/e + \epsilon_3$.

13.8. Proposition. *Let X be E.R. and not a power and let (X^f), where $f \geqslant 2$, be $((n-1))$-bounded by k_{34} and n-bounded by k_{35} and be such that no linearization is a subpowerelement of L_n; similarly for Y and Y^g. Let $(X^f) \supset \xi \underset{n}{\sim} \eta \subset (Y^g)$ where $L(\xi)/L(X) > u_{53}$ and not $\xi \underset{n-1}{\sim} \eta$. Then there exist maximal n-normal subwords $S_1, \ldots, S_m$ of (X^f), which are contained in ξ and size $> u_{33}$, and maximal n-normal subwords $T_1, \ldots, T_m$ of (Y^g), contained in η and of size $> u_{33}$, satisfying the following conditions.*

1. *$L((S_1, S_m))/L(X) > u_{53} - 4$.*
2. *$Im'(S_i) = T_i$ and $Im'(T_i) = S_i$ $(i = 1, \ldots, m)$.*
3. $S_i^+ = S_{i+a}$ $(i = 1, \ldots, m-a)$ for some a where $m - a \geqslant 1$ (cf. 2.16).
4. *If $T_1^+ \subset (T_1, T_m)$ then $T_i^+ = T_{i+b}$ $(i = 1, \ldots, m-b)$ for some b, $m - b \geqslant 1$.*
5. *If $(X) \equiv (Y)$ then denoting the translates of the S_i in (X^f) by $S_1, \ldots, S_m, S_{m+1}, \ldots, S_p$, we have for some c, $1 \leqslant c+1 \leqslant a$, that $(T_i, T_j) \equiv (S_{c+i}, S_{c+j})$ whenever $1 \leqslant i \leqslant j \leqslant m$; in particular $T_i \equiv S_{c+i}$.*
6. *If $(X) \equiv (Y^{-1})$ then for some c, $1 \leqslant c+1 \leqslant a$, $(T_i, T_j) \equiv (S_{c+i}, S_{c+j})^{-1}$ whenever $1 \leqslant i \leqslant j \leqslant m$.*

Note: $(Y) \not\equiv (Y^{-1})$ *since* Π *has no elements of order* 2 (*since* e *is odd*).

Proof. 1°. Let ξ and η have divisions $P_1, P_2, \ldots$ and $Q_1, Q_2, \ldots$. Since ξ contains at least one P_i' of type n we may consider the n-normal subword A of ξ of maximal size. Since $k_3^0 < s(P_i') + s(Q_i') < s(P_i') + k_{35}$ we have $s(A) > k_3^0 - k_{35}$. By an argument similar to 5.28, $L(A) < 2L(X)$. Let the translates of A that are contained in ξ be $A_1, A_2, \ldots, A_u$. Kernels exist since $k_3^0 - k_{35} > 2(r_0 + \epsilon_3) + 2/e$. By 2.16A, $|L(\xi)/L(X) - u| < 3$, so $u_{53} < u + 3$. Also $L(\xi) < L((A_2, A_{u-1})) + 4L(X)$. Let $\bar{A}_i$ be the maximal normal subword of ξ determined by A_i ($i = 1, \ldots, u$). These are distinct and of size $> k_3^0 - k_{35} - \epsilon_3$. Let the image of $\bar{A}_i$ in η be B_i ($i = 1, \ldots, u$). For any i, either B_i has the form $\bar{Q}_j$, Q_j of type n, or by 6.23 $|s(B_i) - s(\bar{A}_i)| < v_3$. ($k_{34} \leqslant k_8'$, $k_{35} \leqslant k_9'$, $k_3^0 - k_{35} - \epsilon_3 > u_{10}$). In either case $s(B_i) > k_3^0 - k_{35} - \epsilon_3 - v_3$.

$\bar{A}_2, \ldots, \bar{A}_{u-1}$ are maximal normal in (X^f); $B_2, \ldots, B_{u-1}$ are maximal normal in (Y^g). Call $\bar{A}_2, \bar{A}_{u-1}, B_2, B_{u-1}$ *extremal.* As in Section 10 call a subword of ξ *central* if it is contained in $(\bar{A}_2, \bar{A}_{u-1})$; similarly for η. Note that $L((\bar{A}_2, \bar{A}_{u-1})) > L(\xi) - 4L(X) > (u_{53} - 4)\, L(X)$.

2°. We consider the following operations.

(1) $U \to V$ where U is central and maximal normal in ξ (or η) of size $> u_{10}$ and V is $\mathrm{Im}'(U)$; thus V is central. Call this operation v.

(2a) $U \to U^+$ where U and U^+ are central in ξ or in η.

(2b) $U \to U^-$ similarly. Call these translation operations t_1, t_2.

(3) The following operations g_ξ, g_η are defined only if $(X) \equiv (Y)$. Let the distinct maximal normal subwords of (X^f) of size $> u_{54}$ be $U_1, \ldots, U_r$. Define U_j for any integer j to be U_k where $j \equiv k \pmod r$ and $1 \leqslant k \leqslant r$. Then for all i, $U_i^+ = U_{i+h}$ where $h = r/f$. Since $u_{54} \leqslant k_3^0 - k_{35} - \epsilon_3 - v_3$, $(\bar{A}_2, \bar{A}_{u-1})$ is (U_{i_1}, U_{j_1}) and (B_2, B_{u-1}) is (U_{i_2}, U_{j_2}) say. Since $u - 1 > u_{53} - 4 \geqslant 2$ we have $\bar{A}_3 = U_{i_1+h}$ so $i_1 + h \leqslant j_1$. We say U_{i_2} corresponds to U_{i_1+s} where $i_2 \equiv i_1 + s \pmod h$ and $0 \leqslant s < h$; further say U_{i_2+v} corresponds to U_{i_1+s+v} for $v = 0, 1, \ldots, z$ where $z = \mathrm{Min}(j_2 - i_2, j_1 - i_1 - s)$. Put $g_\eta(U_{i_2+v}) = U_{i_1+s+v}$, $g_\xi(U_{i_1+s+v}) = U_{i_2+v}$. Then $g_\eta(B_2)$ exists.

If $z = j_2 - i_i$ then $g_\eta(B_{u-1})$ exists; if $z = j_1 - i_1 - s$ then $g_\xi(\bar{A}_{u-1})$ exists.

(4) If $(X) \equiv (Y^{-1})$ we define f_ξ, f_η as follows. $(\bar{A}_2, \bar{A}_{u-1})$ is (U_{i_1}, U_{j_1}) and (B_2, B_{u-1}) is $(U_{j_2}^{-1}, U_{i_2}^{-1})$ say. For the same range of v as before put $f_\eta(U_{i_2+v}^{-1}) = U_{i_1+s+v}$, $f_\xi(U_{i_1+s+v}) = U_{i_2+v}^{-1}$. $f_\eta(B_{u-1})$ exists and one of $f_\eta(B_2), f_\xi(\bar{A}_{u-1})$ exists.

3°. Call a central subword of ξ (or η) *initial special* if it is either extremal or the maximal normal subword of ξ (or η) containing a P_i' (or Q_i') of type n; its size is $> k_3^0 - k_{35} - \epsilon_3 - v_3 > u_{32}$.

Call a central subword of ξ or η *special* if it arises from an initial special word by a sequence of the above operations. Clearly any special word is maximal normal.

We claim that each special word has size $> u_{33}$. This is by Section 10; for we can associate with any sequence of operations a sequence $W_1, W_2, \ldots$ where each W_i is one of ξ, η if the operations f_ξ, f_η do not occur, or one of $\xi, \xi^{-1}, \eta, \eta^{-1}$ if these operations do occur; thus operation (1) above corresponds to the operation f_i of 10.2 Case 2a and $W_i \equiv W_{i+1} \equiv \xi$ (or η); (2a), (2b), (3) above correspond to the operation f_i of 10.2 Case 3a. For example consider (2a) with U, U^+ central in ξ. Then $\xi = AUB = CU^+D$ say where $L(C) = L(A) + L(X)$. ξ is a subword of some power X^N of X: $X^N \equiv E\xi F \equiv GUH$ where $G \equiv EA, H \equiv BF$. Now identify CU^+D with that subword of X^N defined by the requirement that the U^+ of CU^+D be identified with the U of GUH. Here $W_i \equiv AUB$, $W_{i+1} \equiv CU^+D$, $J_i \equiv X^N$. Operation $f_\xi\colon \xi \to \eta$ in (4) above can be factorized into an operation $\xi \to \eta^{-1}$ of the form of 3a of 10.2 followed by an operation $\eta^{-1} \to \eta$ of the form 1a. Note that $k_{34} \leqslant k_{22}$ and $k_{35} \leqslant k_{23}$.

4°. Let the special words in ξ be $S_1, \ldots, S_m$; then S_1 is $\bar{A}_2$, S_m is $\bar{A}_{u-1}$. Since $s(S_i) > u_{33} \geqslant u_{10}$, $\mathrm{Im}'(S_i) = T_i$ exists; hence $v(S_i) = T_i$, so T_i is special in η. Thus the special words in η are precisely $T_1, \ldots, T_m$. If $S_i^+ \subset (S_1, S_m)$ then $t_1(S_i) = S_i^+$ so S_i^+ is special. Thus for some a $S_i^+ = S_{i+a}$ $(i = 1, 2, \ldots, m-a)$ and $1 \leqslant m - a$. If $T_1^+ \subset (T_1, T_m)$ define b by $T_1^+ = T_b$; then $T_i^+ = T_{i+b}$ $(i = 1, \ldots, m-b)$ and $1 \leqslant m - b$.

Finally we note that 5 and 6 hold.

13.9. Corollary. *With the hypothesis of* 13.8 *either* $\xi \underset{n}{\sim} \eta$ *induces* $\xi' \underset{n}{\sim} \eta'$ *where* $L(\xi')/L(X) > u_{53} - 6$ *and* $L(\eta')/L(Y) < 5$ *or (ii) of 13.4 holds.*

Proof. Since $\mathrm{Im}' S_i = T_i$, there are words $\xi' \equiv (S_1^{\mathrm{R}}, S_m^{\mathrm{L}})$ and $\eta' \equiv (T_1^{\mathrm{R}}, T_m^{\mathrm{L}})$ such that $(S_2, S_{m-1}) \subset \xi' \underset{n}{\sim} \eta' \subset (T_1, T_m)$. Now $(S_1, S_m) \equiv U(S_2, S_{m-1})V$ for some U, V; also $L(U) \leqslant L(X)$, $L(V) \leqslant L(X)$. Hence

$$L(\xi')/L(X) \geqslant L((S_2, S_{m-1}))/L(X) \geqslant L((S_1, S_m))/L(X) - 2$$

$$> u_{53} - 6.$$

If $T_1^+ \not\subset (T_1, T_m)$ then $L((T_1, T_m)) < L(Y) + L(T_1^+) < 3L(Y)$.

If $T_1^+ \subset (T_1, T_m)$ and if $m < a + b - d + 2$, where $d = (a, b)$ then $L((T_1, T_m))/L(Y) < 3 + m/b$ and $m/a > -3 + L((S_1, S_m))/L(X) > u_{53} - 7$. Since $d \geqslant 1$ we have

$$m \leqslant a + b, \qquad a(u_{53} - 7) < a + b$$

$$a < b/(u_{53} - 8) \qquad m/b \qquad < 1 + 1/(u_{53} - 8) \leqslant 2$$

since $u_{53} \geqslant 9$. Hence $L(\eta')/L(Y) \leqslant L((T_1, T_m))/L(Y) < 5$.

There remains the case $T_1^+ \subset (T_1, T_m)$ and $m \geqslant a + b - d + 2$. By 13.4, 8.15 the desired result follows, since $k_{34} \leqslant \mathrm{Min}(k_{16}, k_{31})$, $k_{35} \leqslant \mathrm{Min}(k_{17}, k_{32})$.

13.10. Proposition. *Condition C holds.*

Proof. Assume not; then $A \underset{n-1}{\sim} C$, $AB \in L_{n+1}$, $CD \in L_n$ and $L(A)/L(AB) \geqslant u_{41}$. Let p be minimal such that $A \underset{p}{\sim} C$; then $p \leqslant n - 1$.

If $p = 0$ then $A \equiv C$ and $\beta(D^{-1}C^{-1}, AB) \geqslant L(A)$. By 5.16 $\beta < q'L(AB)$ hence $L(A)/L(AB) < q' \leqslant u_{41}$, a contradiction.

Hence $p > 0$. Since $str^+ \leqslant k_{34}, k_{35}$ (cf. 5.11) and $u_{41} > u_{53}/e$ the hypothesis of 13.8 holds. Say $AB \equiv X^e$, $CD \equiv Y^e$. By 13.9 (also cf. 3.1) either (i) $A' \underset{p}{\sim} C'$ (hence $A' \underset{n-1}{\sim} C'$), $L(A')/L(AB) > (u_{53} - 6)/e \geqslant q'$, $L(C')/L(CD) < 5/e \leqslant f$, or (ii) for some integers $a, b \neq 0$, $(X^{eb}) \underset{p}{\cong} (Y^{ea})$.

If (ii) holds then $X^{eb} \underset{p}{=} T^{-1} \cdot Y^{ea} \cdot T$ for some T. Now $Y^e \equiv CD \underset{n}{=} I$ hence $(AB)^b \underset{n}{=} I$, contrary to 5.12.

Thus (i) holds. Now X is a C.A.I. of X_0, where $X_0 \in C_{n+1} \subset J_n \subset J_p$ and Y is a C.A.I. of $Y_0 \in C_n \subset J_{n-1} \subset J_p$. By 2.11, $A' \underset{n-1}{\sim} C'$ has the form $\epsilon_{q'}(X_0) \underset{n-1}{\sim} \delta_f(Y_0)$, i.e., $Y_0 \sigma_{n-1} X_0$ contrary to 5.9 (ii).

13.10A. Note. We have now proved that conditions A, B, C hold (cf. 12.3, 12.4).

13.11. Proposition. *Let $DE, DF \in \mathsf{L}_{n+1}$ and let $s_{DE}(D) \geqslant r_0$. Then $DE \underset{n}{=} DF$.*

Hence by 13.10, 12.2 *and* 12.18, *Axiom* 20 *is true for* $n + 1$.

Proof. We have

(1) $Z_n \underset{n}{\rightarrow} \cdots \underset{1}{\rightarrow} DE$

(2) $U_n \underset{n}{\rightarrow} \cdots \underset{1}{\rightarrow} DF$

where $Z_n, U_n \in \mathrm{Rep}(L_{n+1})$; say $Z_n \equiv (X^e)$, $U_n \equiv (Y^e)$. Using the $A - B$ notation for (1) let all A-descendants of $B_1, \ldots, B_g$ in DE be contained in D. If $\epsilon > 0$, in particular if $\epsilon = 1/e^2$, we may assume that $g/m > s_{DE}(D) - \epsilon$. By 11.8, since $r_0 - 1/e^2 > u_{36}$

$$Z_n \equiv D'E'$$

$$U_n \underset{n}{=} D'F'$$

$$E'^{-1} \cdot F' \underset{n}{=} T^{-1} \cdot E^{-1} \cdot F \cdot T$$

$$D' \supset (B_{1+s}, B_{g-s}),$$

$$(g - 2s)/m > g/m - u_{37} > r_0 - 1/e^2 - u_{37} > u_1$$

Now $D' \supset D'^0 \underset{n}{\sim} U_n^{\mathrm{M}}$; we may write Z_n^{M} for D'^0, so $Z_n^{\mathrm{M}} \underset{n}{\sim} U_n^{\mathrm{M}}$.

By 12.2 if A is any subelement of L_i and of L_{n+1}, where $i \leqslant n$, then $s^{n+1}(A) \leqslant r_0'$; thus 6.8 holds for $n + 1$.

Now $D' \equiv \alpha_1 D'^0 \beta_1$ say; as in 6.8 α_1 and β_1 have $(n + 1)$-size $\leqslant r_0' + \epsilon_2 + 2/e$. Let w be the least integer $> m(r_0' + \epsilon_2 + 2/e)$. If $\alpha_1 \supset (B_{1+s}, B_{w+s})$ then $s(\alpha_1) \geqslant w/m > r_0' + \epsilon_2 + 2/e$, a contradiction; hence $(B_{w+s+2}, B_{g-s-w-1}) \subset D'^0 \equiv Z_n^{\mathrm{M}}$. Now by 2.16A

$$|(g - 2(s + w + 1))/m - L(B_{w+s+2}, B_{g-s-w-1})/L(X^e)| < 3/e$$

hence

$$L(Z_n^{\mathrm{M}})/L(X^e) > -3/e + (g - 2(s + w + 1)/m$$

$$> -3/e + r_0 - 1/e^2 - u_{37} - 2(r_0' + \epsilon_2 + 4/e = \alpha \text{ sa}$$

Thus $L(Z_n^{\mathrm{M}})/L(X) > \alpha e \geqslant u_{53}$.

Case 1. $Z_n^{\mathrm{M}} \not\equiv U_n^{\mathrm{M}}$.

Let $Z_n^{\mathrm{M}} \underset{p}{\sim} U_n^{\mathrm{M}}$ where p is minimal; thus $0 < p \leqslant n$. Since $str^+ \leqslant \mathrm{Min}(k_{34}, k_{35})$ we may apply 13.8 and 13.9. By either (i) $Z_n^{\mathrm{MM}} \underset{p}{\sim} U_n^{\mathrm{MM}}$ (hence $\underset{n}{\sim}$), $L(U_n^{\mathrm{MM}})/L(Y) < 5$ and $L(Z_n^{\mathrm{MM}})/L(X) > [\alpha e] - 6$, or (ii) the conclusion of 13.4 holds; in particular a, b exist such that if $w \geqslant 3$, $a \mid w$, $b \mid w$ then $(X^{w/a}) \underset{p}{\cong} (Y^{w/b})$ (hence $\underset{n}{\cong}$).

Assume (i) holds. Now X is a C.A.I. of X_0 and Y is a C.A.I. of Y_0, for some X_0, Y_0 in C_{n+1}. Since $[\alpha e] - 6 \geqslant qe$ and $5 \leqslant fe$ we have $Z_n^{\mathrm{MM}} \equiv \epsilon_q(X_0)$, $U_n^{\mathrm{MM}} \equiv \delta_f(Y_0)$ and $Y_0 \, \rho_n \, X_0$ contrary to 5.9(i).

Hence (ii) holds, and $a = b$, otherwise $X_0 \rho_n Y_0$ or $Y_0 \rho_n X_0$. Take $w = ae$. From 13.8, $(S_1, S_m) \subset Z_n^{\mathrm{M}}$ and $(T_1, T_m) \subset U_n^{\mathrm{M}}$. Also (cf. (ii) of 13.4) $(S_1, S_2) \underset{p}{\triangleq} (T_1, T_2)$ and

(7) $$(S_1, S_2) \underset{p}{=} \epsilon_1 \cdot (T_1, T_2) \cdot \kappa_2$$

(8) $$Z_n' \underset{p}{=} \epsilon_1 \cdot U_n' \cdot \epsilon_1^{-1}$$

where $Z_n' \equiv (S_1, \widehat{S}_{ae+1})$ and $U_n' \equiv (T_1, \widehat{T}_{ae+1})$.

Z_n' is a C.A. of X^e, i.e., of Z_n and U_n' is a C.A. of U_n. We have say $Z_n' \equiv (S_1, S_2)Z'$, $U_n' \equiv (T_1, T_2)U'$. By (7), (8) we have

(9) $$\kappa_2 \cdot Z' \underset{p}{=} U' \cdot \epsilon_1^{-1}$$

Since $Z_n^{\mathrm{M}} \subset D'$ we may write $D' \equiv D_1(S_1, S_2)D_2$. Now $Z_n \equiv D'E'$ so

(10) $$Z' \equiv D_2 E' D_1$$

Since $U_n \underset{n}{-} D'F'$ induces $U_n^{\mathrm{M}} \underset{p}{\sim} Z_n^{\mathrm{M}}$, which induces $(T_1, T_2) \underset{p}{\triangleq}$ $\underset{p}{\triangleq} (S_1, S_2)$, then by 11.1A,

(11) $D_2 F' D_1 \underset{n}{=} \kappa_2^{-1} \cdot U' \cdot \epsilon_1^{-1}$,

since $u_{33} \geqslant u_{29}$ and $\mathrm{Max}(u_{35}, str^+) \leqslant \mathrm{Min}(k_{16}, k_{17})$.

By (9), (10), (11) $F' \underset{n}{=} E'$; therefore $E \underset{n}{=} F$ and $DE \underset{n}{=} DF$.

Case 2. $Z_n^{\mathrm{M}} \equiv U_n^{\mathrm{M}}$.

Some C.A. of Z_n has the form $Z_n^{\mathrm{M}} Z'$ and some C.A. of U_n has the form $U_n^{\mathrm{M}} U'$. Now $L(Z_n^{\mathrm{M}})/L(X^e) > \alpha \geqslant q$, so by 5.16 (ii) we have $Z' \equiv U'$. Now $Z_n^{\mathrm{M}} \subset D'$; say $D' \equiv D_1 Z_n^{\mathrm{M}} D_2$; but $Z_n \equiv D'E'$, so $Z' \equiv D_2 E' D_1$. Since $U_n \underset{n}{-} D'F'$ induces $U_n^{\mathrm{M}} \equiv Z_n^{\mathrm{M}}$, $U' \underset{n}{=} D_2 F' D_1$. Hence $E' \underset{n}{=} F'$, $E \underset{n}{=} F$, $DE \underset{n}{=} DF$.

13.12. Proposition. *Let the hypothesis of* 13.7 *hold and let* $X^e \in Rep(L_{n+1})$. *Then* $c = 0$.

Proof. Assume $c \neq 0$. Then $(X^e) \underset{n-1}{-} (T^e) \underset{n}{-} (Q^{a'e})$ where $a' \geqslant 2$. We may suppose that $X \in C_{n+1}$, hence $X \in J_n$. Now (X^e), (T^e), $(Q^{a'e})$ are (n)-bounded by $\mathrm{Max}(k_{31} + c_{33} + c_{39}, k_{32} + u_4, k_{33}) \leqslant \mathrm{Min}(c_{28}, b_2, d_{28})$, hence by Axioms 28A, 8, 11 we have $(X^e) \underset{n}{\cong} (Q^{a'e})$. By Axiom 28B $(Q^{a'e}) \underset{n}{\cong} (P^{a'e})$ where the right side is (n)-bounded by $str \leqslant b_2$; hence

$$(*) \qquad (X^e) \underset{n}{\cong} (P^{a'e})$$

If $P^{a'} \underset{n}{=} I$, then $X^e \underset{n}{=} I$ contrary to 5.12. Hence $P \in J_n$ by Axiom 9. By (*) $X E_n P$ hence $X \rho_n P$; this contradicts 5.9 (i) since $X \in C_{n+1}$.

13.13. Proposition. *Let* $AA', BB' \in Rep(L_{n+1})$. *Let* $A \underset{n}{\sim} B$, *with divisions* $P_1, \ldots, P_r$ $Q_1, \ldots, Q_r$ *say and let* $L(A)/L(AA') > u_{55}$. *Then* $A \equiv B$, $A' \equiv B'$, $r = 1$, *and* P_1 *has type zero.*

Proof. It is sufficient to prove that no P_i has type > 0; for then $r = 1$ and P_1 has type 0, so $A \equiv B$. By 5.16 (ii) since $u_{55} \geqslant q$ we have $AA' \equiv BB'$.

Let p be the maximal type occurring and assume $p \neq 0$. Thus $A \underset{p}{\sim} B$. Now $u_{55} \geqslant u_{53}/e$. As in Case 1 of the proof of 13.11, since $\lceil u_{55} e \rceil - 6 \geqslant qe$, we have $(X^e) \underset{p}{\cong} (Y^e)$, i.e., $(AA') \underset{p}{\cong} (BB')$ (hence $\underset{n}{\cong}$).

Since AA' and $BB' \in \text{Rep}(L_{n+1})$, AA' is a C.A.I. of BB' (cf. 7.1). By 13.8 parts 5 and 6 we have

$$(12) \qquad (T_i, T_j) \equiv (S_{c+i}, S_{c+j})^{\pm 1}, \qquad 1 \leqslant i \leqslant j \leqslant m$$

and $1 \leqslant c+1 \leqslant a$. Now $m > 2a$ since $m/a > L(S_1, S_m)/L(X) - 3 > u_{53} - 7 \geqslant 2$ so $m > a+1$ and $m \geqslant a+2$ (cf. 2.16A).

By 13.5 we may replace the ± 1 in (12) by $+1$, since $u_{33} \geqslant u_{51}$, $str^+ \leqslant k_{31}, k_{32}$.

By 13.7 either $c = 0$ or T, Q exist such that

$$(X^e) \underset{p-1}{-} (T^e) \underset{p}{-} (Q^{a'e}) \text{ where } a' > 1.$$

By 13.12 we have $c = 0$. Some P_i has type p. For some w we have $P'_i \subset S_w$ (cf. the beginning of 3° in the proof of 13.8); hence $Q'_i \subset T_w \equiv S_{c+w} \equiv S_w$. We shall prove $(*)$ $P'_i \cap Q'_i$ has size $> s_{40}$. Assuming $(*)$, let $J \equiv P'_i \cap Q'_i$; since $P'^{-1}_i X'^{-1}_{i-1} Q'_i X'_i \in \mathsf{L}_p$, some element of L_p has the form $J^{-1}AJB$, in contradiction to Axiom 40, since $str^+ \leqslant b_{40}$.

It remains to prove $(*)$. Now $k_3^0 < s(P'_i) + s(Q'_i)$ and by 5.11 $s(S_w) < str^+$. If $Q'_i \subset P'_i$ then $J \equiv Q'_i$ and $s(J) > k_3^0 - str^+ \geqslant s_{40}$. Similarly if $P'_i \subset Q'_i$.

If P'_i, Q'_i were disjoint (say P'_i is left of Q'_i) then abbreviating $L(X)$ to X and using 5.23

$$\begin{aligned}
P'_i K Q'_i &\leqslant S_w + \epsilon_3 \\
P'_i K + Q'_i &\leqslant P'_i K Q'_i + (2\epsilon_3 + \epsilon_1^*) \\
P'_i + K &\leqslant P'_i K + (r_0 + 2\epsilon_3)
\end{aligned}$$

Adding,

$$P'_i + Q'_i + K \leqslant S_w + \epsilon_3 + 2\epsilon_3 + \epsilon_1^* + r_0 + 2\epsilon_3$$

which is a contradiction since $k_3^0 > str^+ + 5\epsilon_3 + \epsilon_1^* + r_0$.

If P'_i, Q'_i overlap, say $P'_i \equiv DE$, $Q'_i \equiv EF$; then

$$\begin{aligned} DEF &\leqslant str^{+} + \epsilon_3 \\ DE + F &\leqslant DEF + r_0 + 2\epsilon_3 \\ EF &\leqslant E + F + 2/e \\ k_3^0 &< DE + EF \end{aligned}$$

Hence

$$s(E) > k_3^0 - (str^{+} + \epsilon_3 + r_0 + 2\epsilon_3 + 2/e) \geqslant s_{40}.$$

13.14. Proposition. *No element of* L_{n+1} has the form $J^{-1}AJB$, *where* $s_W(J) \geqslant r_0$, $W \equiv JBJ^{-1}A$.

Proof. We have (i) $Z_n \to \cdots \to JBJ^{-1}A$ where $Z_n \in \mathrm{Rep}(L_{n+1})$. Taking the inverse of every term we have
(ii) $Z_n^{-1} \to \cdots \to JB^{-1}J^{-1}A^{-1}$.
Denote (ii) also by $U_n \to \cdots \to JV$ where $V \equiv B^{-1}J^{-1}A^{-1}$. As in the proof of 13.11, $Z_n^{\mathrm{M}} \underset{n}{\sim} U_n^{\mathrm{M}}$, $L(Z_n^{\mathrm{M}})/L(X^e) > \alpha \geqslant u_{55}$ where α is as in proof of 13.11. By 13.13 $Z_n^{\mathrm{M}} \equiv U_n^{\mathrm{M}}$ and Z_n is a C.A. of U_n. Thus Z_n is a C.A. of Z_n^{-1}. This is impossible since Π has no elements of order 2, as e is odd.

13.15. Proposition. *Let* $DE, DF \in \mathsf{L}_{n+1}$ *and* $s_{DE}(D) > r_0$. *Then*

$$s_{DF}(D) \geqslant s_{DE}(D) - \epsilon_3.$$

Proof. As in the proof of 13.11, $Z_n \equiv D'E'$ and $U_n \underset{n}{-} D'F'$. Further $U_n^{\mathrm{M}} \underset{n}{\sim} D'^0$ ($\equiv Z_n^{\mathrm{M}}$). As in 13.14 $U_n^{\mathrm{M}} \equiv D'^0$ and Z_n is a C.A. of U_n. Now

$$(B_{w+s+2}, B_{g-s-w-1}) \subset D'^0 \equiv U_n^{\mathrm{M}}.$$

Let I' be the iterated nice image of D in U_n with respect to $U_n \to \cdots \to DF$. Let D^* be the nice image of D' in U_n with respect to $U_n - D'F'$. By 11.8, $D^* \subset I'$. But D^* is U_n^{M}, so $I' \supset D^* \supset (B_{w+s+2}, B_{g-s-w-1})$. Hence in $U_n \to \cdots \to DF$, all A-descendants of $B_{w+s+2}, \ldots, B_{g-s-w-1}$ are contained in D. Therefore

$$s_{DF}(D) \geqslant (g - 2(s + w + 1))/m$$

$$> s_{DE}(D) - \epsilon_3$$

since $\epsilon_3 = 1/e^2 + u_{37} + 2(r_0' + \epsilon_2 + 4/e)$.

13.16. Corollary. *Axiom* 19 *is true for* $n + 1$.

Proof. The above proof remains valid if $DE \in \mathsf{L}^l_{n+1}$, $DF \in \mathsf{L}^k_{n+1}$.

13.17. Proposition. *Let* $XY \in \mathsf{L}_{n+1}$. *As in Definition* 7.11 *let a sequence* $Z_n \to \cdots \to Z_0 \equiv (XY)$ *and* A*'s,* B*'s,* C, D *be chosen. Let* σ *denote this set of choices, and in the notation of* 7.11 *denote* x/d *(here* $k = 1$ *so* $d = m$*) by* $s_{XY,\sigma}(X)$. *Let* τ *be another set of choices. Then*

(i) $|s_{XY,\sigma}(X) - s_{XY,\tau}(X)| < \epsilon_3 + 4/e$

(ii) $s_{XY}(X) \geqslant s_{XY,\sigma}(X) \geqslant s_{XY}(X) - \epsilon_3 - 4/e$.

Proof. Consider σ, which may be called a set of choices for the pair X, Y. σ determines in an obvious way a set σ' of choices for Y, X. Let the number of R_i contained in Y be y. Then $x + y \leqslant m \leqslant x + y + 4$ and $y/m = s_{YX,\sigma'}(Y)$. Hence if we put

$$(13) \qquad t_\sigma = 1 - s_{XY,\sigma}(X) - s_{YX,\sigma'}(Y)$$

then $0 \leqslant t_\sigma \leqslant 4/e$ (since $e|m$).

Case 1. $s_{XY,\sigma}(X) > r_0 + \epsilon_3$.

Since $s_{XY,\sigma}(X) > r_0$, the proof of 13.15 without change implies that $s_{XY,\tau}(X) > s_{XY,\sigma}(X) - \epsilon_3$. Hence $s_{XY,\tau}(X) > r_0$ so again $s_{XY,\sigma}(X) > s_{XY,\tau}(X) - \epsilon_3$. Hence

$$(14) \qquad |s_{XY,\tau}(X) - s_{XY,\sigma}(X)| < \epsilon_3$$

so (i) holds in this case.

Case 2. Not Case 1.

By (1), $s_{YX,\sigma'}(Y) = 1 - t_\sigma - s_{XY,\sigma}(X) \geqslant 1 - 4/e - r_0 - \epsilon_3 > r_0 + \epsilon_3$ so by Case 1

(15) $\quad |s_{YX,\sigma'}(Y) - s_{YX,\tau'}(Y)| < \epsilon_3$

By (15), (13) and the equation obtained from (13) by replacing σ, σ' by τ, τ' respectively

$$|1 - s_{XY,\sigma}(X) - t_\sigma - (1 - s_{XY,\tau}(X) - t_\tau)| < \epsilon_3.$$

Now $|t_\tau - t_\sigma| \leqslant 4/e$ hence (i) holds in case 2 also.

To prove (ii) we may, given $\epsilon > 0$, choose τ such that

$$s_{XY,\tau}(X) > s_{XY}(X) - \epsilon.$$

The result (ii) now follows from (i).

13.18. Corollary. *Let $ABC \in \mathsf{L}_{n+1}$. Then*

$$s_{ABC}(A) + s_{BCA}(B) \leqslant s_{ABC}(AB) + \epsilon_1^*.$$

Proof. Given $\epsilon > 0$ choose σ such that

$$s_{ABC}(A) - \epsilon < s_{ABC,\sigma}(A).$$

For σ let a, b, x be the number of R_i (in the notation of 7.10) contained in A, B, AB respectively; then $a + b \leqslant x$ so for choices σ', σ'' determined by σ in the obvious way

$$s_{ABC,\sigma}(A) + s_{BCA,\sigma'}(B) \leqslant s_{ABC,\sigma''}(AB).$$

Now $s_{ABC,\sigma''}(AB) \leqslant s_{ABC}(AB)$ and

$$s_{BCA}(B) \leqslant \epsilon_3 + 4/e + s_{BCA,\sigma'}(B).$$

By adding these four inequalities the result follows since $\epsilon_1^* = \epsilon_3 + 4/e$.

13.19. Note. Recalling 7.12 we now have that the following "size axioms" are true for $n+1$: Axioms 13, 14, 15, 16, 16a, 17, 19.

13.20. Proposition. *Axiom* 40 *holds for* $n+1$.

Proof. Suppose $J^{-1}AJB \in L_{n+1}, s^{n+1}(J) > s_{40}$. Let $W \equiv JBJ^{-1}A$. Since $s_{40} \geqslant r_0$,

$$s_W(J) \geqslant s^{n+1}(J) - \epsilon_3 > s_{40} - \epsilon_3 = r_0.$$

This contradicts 13.14.

13.21. Proposition. *If* $CD \in Rep(L_{n+1})$ *then*

$$|s_{CD}(C) - L(C)/L(CD)| < \epsilon_1^* + 3/e.$$

Proof. Take a choice σ based on the trivial sequence $CD \underset{n}{\rightarrow} CD \underset{n-1}{\rightarrow} \cdots \underset{1}{\rightarrow} CD$. Let c be the number of B_i contained in C. Then $s_{CD,\sigma}(C) = c/m$. Since by 7.3 the translate of any B_i is a B_i we have by 2.16A

$$|c/m - L(C)/L(CD)| < 3/e.$$

By 13.17, $|s_{CD,\sigma}(C) - s_{CD}(C)| < \epsilon_3 + 4/e = \epsilon_1^*$. The result follows.

13.22. Proposition. *Axiom* 33 *holds for* $n+1$.

Proof. Let $AA', BB' \in Rep(L_{n+1}), A \underset{n}{\sim} B, s^{n+1}(A) > s_{33}$. Since $s_{33} \geqslant r_0, s_{AA'}(A) \geqslant s^{n+1}(A) - \epsilon_3 > s_{33} - \epsilon_3$. Now

$$L(A)/L(AA') > s_{AA'}(A) - \epsilon_1^* - 3/e > s_{33} - \epsilon_3 - \epsilon_1^* - 3/e \geqslant u_{55}.$$

By 13.13, $A \equiv B$ and $A' \equiv B'$.

13.23. Note. In the proof of 13.22 we proved slightly more than was necessary, namely, if $P_1, \ldots, P_r\ Q_1, \ldots, Q_r$ denote divisions for $A \underset{n}{\sim} B$ then $r = 1$ and P_1 has type 0 (cf. the beginning of 13.13). Thus from now on we may interpret Axiom 33 in this strengthened form.

§ 14

Informal Summary

Another key axiom, Axiom 30, is shown to hold for $n + 1$.

14.1. Proposition. *Axiom 23 holds for $n + 1$, i.e., if $AB \in \mathsf{L}^k_{n+1}$ and $0 \leqslant \alpha < \beta < s_{AB}(A)$ and $\beta - \alpha > \epsilon_4$ then for some left subword A^{L} of A, $\alpha \leqslant s_{AB}(A^{\mathrm{L}}) < \beta$.*

Proof. For some σ, we have $\beta < s_{AB,\sigma}(A)$. Now $s_{AB,\sigma}(A) = x/d$ where x, d are integers and e divides d. For any A^{L}

(1) $$s_{AB,\sigma}(A^{\mathrm{L}}) = y/d$$

where y is an integer, $0 \leqslant y \leqslant x$. Also if y is an integer, $0 \leqslant y \leqslant x$ then for some A^{L} we have (1). Now by 13.17 (ii)

$$s_{AB,\sigma}(A^{\mathrm{L}}) \leqslant s_{AB}(A^{\mathrm{L}}) \leqslant s_{AB,\sigma}(A^{\mathrm{L}}) + \epsilon_3 + 4/e$$

so if (i) $\alpha \leqslant y/d < \beta - \epsilon_3 - 4/e$ then $\alpha \leqslant s_{AB}(A^{\mathrm{L}}) < \beta$ as required.

Thus it is sufficient to show there is an integer y such that

$$d\alpha \leqslant y < \mathrm{Min}(d(\beta - \epsilon_3 - 4/e), x).$$

This is so if (ii) $d(\beta - \alpha - \epsilon_3 - 4/e) \geqslant 1$ and (iii) $x - d\alpha \geqslant 1$. Now (ii) holds since $\beta - \alpha > \epsilon_4 = 5/e + \epsilon_3 \geqslant 1/d + 4/e + \epsilon_3$, and (iii) holds since $x/d - \alpha > \beta - \alpha > 1/e \geqslant 1/d$.

14.2. Proposition. *Let $Y \in \mathsf{L}^h_{n+1} \cap \mathsf{L}^k_{n+1}$. Then $h = k$. Hence by 7.14, Axiom 12 holds for $n + 1$.*

Proof. Suppose not; then $h < k$ say. We have

$$Z^h \to \cdots \to Y \equiv YA$$

$$Z'^k \to \cdots \to Y \equiv YB \qquad (Z, Z' \in \mathrm{Rep}(L_{n+1}))$$

where $A \equiv B \equiv I$. By 11.8,

$$Z^h \equiv Y'A'$$

$$(*) \qquad Z'^k \underset{n}{=} Y'B'$$

and $A' \underset{n}{=} B'$. Now $s(Y) \geqslant h - 2/e \geqslant 1 - 2/e$ so $s(Y') \geqslant 1 - 2/e - - u_{37} > u_1 + s_{33}$ so from (*) (also cf. 6.8) $Y' \supset Y'^0 \equiv C \underset{n}{\sim} D$ say and $Z'^k \equiv DD'$, $Y' \equiv Y_1 C Y_2$ and $s(C) > s_{33}$. We claim that $C^{\mathrm{L}}, D^{\mathrm{L}}$ exist such that $C^{\mathrm{L}} \underset{n}{\sim} D^{\mathrm{L}}$ and one of them has size $> s_{33}$ and $L(C^{\mathrm{L}}) \leqslant L(Z)$, $L(D^{\mathrm{L}}) \leqslant L(Z')$.

$CY_2A'Y_1 \equiv Z_0^h$ where Z_0 is a C.A. of Z. Let C' be the left subword of C such that $C' \equiv C \cap Z_0$. Now $s(Z_0) \geqslant 1 - 2/e > s_{33} + u_1$ so $C'^0 \sim D_1$ where D_1 is a left subword of D. If $D_1 \subset Z'$ the result follows. If $Z' \subset D_1$ then $s(Z') > s_{33} + u_1$ so $Z'^0 \sim C^{\mathrm{L}}$.

By Axiom 33 for $n + 1$ $C^{\mathrm{L}} \equiv D^{\mathrm{L}}$ and $Z' \equiv Z_0$. Now (*) induces $C^{\mathrm{L}} \sim D^{\mathrm{L}}$ so (cf. 13.23) $DD' = CY_2B'Y_1 = CY_2A'Y_1$. But

$$DD' \equiv Z'^k \equiv Z_0^h Z'^{k-h} \equiv CY_2A'Y_1Z'^{k-h} \quad \text{hence } Z'^{k-h} = I.$$

Thus $S^{e(k-h)} \underset{n}{=} I$ for some $S \in C_{n+1}$. Now $S \in C_{n+1} \subset J_n$, contradicting 5.12.

14.3. Proposition. *Axiom* 30 *holds for* $n + 1$.

We first need a lemma.

14.4. Lemma. *Let* $1 \leqslant w \leqslant n$. *Let* $S_1, \ldots, S_k$ *be subelements of* L_w *and contained in a word* X. *Let* $T_1, \ldots, T_k$ *be subelements of* L_w *contained in* Y. *Let* X, Y *be* (w)*-bounded by* η *where* $\eta \leqslant k_{52}, k_{53}$. *Also assume either* (*A*) $X \underset{w-1}{\sim} Y$, *the* S_i *are disjoint of size* $> v_4$, T_i *is a weak image of* S_i, *there is a positive integer* N *such that* $k > 2N + 5$ *and neither* (S_{i+1}, S_{i+N}) nor (T_{i+1}, T_{i+N}) *is a subelement of* $\mathsf{L}_1 \cup \cdots \cup \mathsf{L}_w$ $(i = 0, 1, \ldots, k - N)$ *or* (*B*) $X \underset{w}{\sim} Y$, S_i *is maximal normal of size* $> v_5$ *and* $Im' S_i = T_i$ $(i = 1, \ldots, k)$, $N = 5$ and $k > 2N + 5$.

Then (*i*) $(T_1, T_k) \supset M \underset{w}{\sim} M'$ (*end words* E, K); $M \cdot K^{-1} \underset{w}{=} M^{\mathrm{L}} F M'^{\mathrm{R}}$ *where the right side is* (w)*-bounded by* $\eta^+ = \eta + Max(u_4, q_{31}, h_{31})$; $M'^{\mathrm{R}} \supset (S_{N+3}, S_{k-3})$ *and* $M^{\mathrm{L}}, M'^{\mathrm{R}}$ *are not subelements of* $\mathsf{L}_1 \cup \cdots \cup \mathsf{L}_w$ *and* (*ii*) $(T_1, T_k) \supset M \underset{w}{\sim} M'$; $M \underset{w}{=} M^{\mathrm{L}} E M'^{\mathrm{M}} F M^{\mathrm{R}}$

where the right side is (w)-bounded by η^+ *and* $M^{\mathrm{L}}, M^{\mathrm{R}}$ *are not subelements of* $L_1 \cup \cdots \cup L_w; M'^{\mathrm{M}} \supset (S_{N+3}, S_{k-N-2})$.

Proof of 14.4. (A) For any S_i we have by 6.26, $S_i \supset U_i \underset{w-1}{\sim} V_i \subset T_i$ (end words E_i, K_i) and there exist words $U_i^{\mathrm{L}} D_i V_i^{\mathrm{R}}$, $V_i^{\mathrm{L}} F_i U_i^{\mathrm{R}}$ ((w − 1))-bounded by $\eta + q_{31}$ where $U_i^{\mathrm{L}}, V_i^{\mathrm{R}}, V_i^{\mathrm{L}}, U_i^{\mathrm{R}}$ are not subelements of $L_1 \cup \cdots \cup L_{w-1}$. If $1 < a < b < k$ for some a, b then

$$(\dagger) \qquad (V_1, V_k) \underset{\omega-1}{=} (V_1, V_a^{\mathrm{L}}) F_a (U_a^{\mathrm{R}}, U_b^{\mathrm{L}}) D_b (V_b^{\mathrm{R}}, V_k) \equiv J' \text{ say}$$

and the right side is ((w − 1))-bounded by $\eta + q_{31} \leqslant \eta^+$. Since the left side is w-bounded by η the right side is w-bounded by $\eta + u_4 \leqslant \eta^+$.

If $1 < a < c < b < k$ then

$$(V_1, V_c) = (V_1, V_a^{\mathrm{L}}) F_a (U_a^{\mathrm{R}}, U_c) \cdot K_c^{-1} \equiv J \cdot K_c^{-1} \text{ say}$$

and since $J \subset J'$, J is (w)-bounded by η^+. Moreover $(V_1, V_a^{\mathrm{L}}) \supset (T_2, T_{a-}$ which is not a subelement of $L_1 \cup \cdots \cup L_w$ if $a - 2 \geqslant N$ and $(U_a^{\mathrm{R}}, U_c) \supset (S_{a+1}, S_{c-1})$ is not such a subelement if $c - a - 1 \geqslant N$. Now $(T_1, T_k) \supset (V_1, V_c)$ so (i) follows if we take $a = N + 2$, $b = k - 1$, $c = k - 2$.

In (†) if $a - 2 = N$ and $k - b - 1 = N$ then (V_1, V_a^{L}) and (V_b^{R}, V_k) are not subelements of $L_1 \cup \cdots \cup L_w$. Also $(U_a^{\mathrm{R}}, U_b^{\mathrm{L}}) \supset (S_{N+3}, S_{k-N-2})$

(B) By 6.27 if $1 \leqslant i < k$ then $(S_i^{\mathrm{R}}, S_{i+1}^{\mathrm{L}}) \sim (T_i^{\mathrm{R}}, T_{i+1}^{\mathrm{L}})$ (end words E_i, K_{i+1}) and there exist words $S_i^{\mathrm{RL}} D_i T_{i+1}^{\mathrm{LR}}$, $T_i^{\mathrm{RL}} F_i S_{i+1}^{\mathrm{LR}}$ (w)-bounded by $\eta + h_{31}$. Also by the proof of 6.27 $S_i \supset U_i \sim V_i \subset T_i$ $(i = 1, ..., k)$ and $(U_i, U_{i+1}) \equiv (S_i^{\mathrm{R}}, S_{i+1}^{\mathrm{L}})$, $(V_i, V_{i+1}) \equiv (T_i^{\mathrm{R}}, T_{i+1}^{\mathrm{L}})$. Thus $U_i \sim V_i$ has end words E_i, K_i. By 6.28, $(V_1, V_k) = (V_1, T_5^{\mathrm{RL}}) F_5 (S_6^{\mathrm{LR}}, U_k)$, $K_k^{-1} \equiv J \cdot K_k^{-1}$ where J is (w)-bounded by $\eta + h_{31} \leqslant \eta^+$. Now (V_1, T_5^{RL}) properly contains T_3 which is maximal normal so (V_1, T_5^{RL}) is not a subelement of $L_1 \cup \cdots \cup L_w$. Similarly for (S_6^{LR}, U_k). Now $(T_1, T_k) \supset (V_1, V_k)$ and $(S_6^{\mathrm{LR}}, U_k) \supset (S_8, S_{k-3})$. Also

$$(V_1, V_k) = (V_1, T_5^{\mathrm{RL}}) F_5 (S_6^{\mathrm{LR}}, S_{k-5}^{\mathrm{RL}}) D_{k-5} (T_{k-4}^{\mathrm{LR}}, V_k)$$

and $(S_6^{\mathrm{LR}}, S_{k-5}^{\mathrm{RL}}) \supset (S_8, S_{k-7})$.

Proof of 14.3.

1°. We have $Z_n \underset{n}{\to} \cdots \underset{1}{\to} PQ$ where $Z_n \in \operatorname{Rep} L_{n+1}$ and we may assume $s_{PQ,\sigma}(P) > s(P) - 1/e^2 > s_{30} - 1/e^2$. P is (n)-bounded by b_{30}. Using the A-B notation, Z_n contains subelements B_i of L_n $(i = 1, \ldots, m)$; let $R_i = AD_0(B_i)$. Then $R_1, \ldots, R_a$ say are contained in P and $s_{PQ,\sigma}(P) = a/m$. We shall prove that each R_i in P contains a subelement of L_n. This is trivial if $n = 1$ so let $n \geqslant 2$. Consider the diagram

$$\begin{array}{lll} Z_{n-r} & \underset{n-r}{\to} & Z_{n-r-1} \\[2ex] A_\gamma^{n-r+1} & & \\[2ex] B_{\gamma i}^{n-r} & & A_{\gamma i}^{n-r} \end{array}$$

where γ is an r-tuple; here $n > r > 0$. Say the range of i is $1, 2, \ldots, f$ for $B_{\gamma i}$ to exist; $1 + j, \ldots, f - j$ for $A_{\gamma i}$ to exist. Recalling 7.7 we have say $Z_{n-r} \underset{n-r-1}{\overline{\quad}} X_1 \underset{n-r}{\overline{\quad}} X_2 \underset{n-r}{\overline{\quad}} Z_{n-r-1}$.

2°. If the $B_{\gamma i}$ are disjoint of size $> q^*$ then by 11.3A we have say

$$Z_{n-r} \supset N_1 \underset{w-1}{\simeq} N_2 \supset N_3 \underset{w-1}{\simeq} N_4 \supset N_5 \underset{w-1}{\simeq} N_6, \quad w = n - r,$$

where $N_2 \subset X_1, N_4 \subset X_2, N_6 \subset Z_{n-r-1}$ and for $i = j + 1, \ldots, f - j$ there exist $C_{\gamma i}$ and $D_{\gamma i}$ such that $B_{\gamma i} \subset N_1, C_{\gamma i} \subset N_3, D_{\gamma i} \subset N_5$, $A_{\gamma i} \subset N_6$ and the weak images of $B_{\gamma i}, C_{\gamma i}, D_{\gamma i}$ are $C_{\gamma i}, D_{\gamma i}, A_{\gamma i}$ respectively. Also each of $B_{\gamma i}, C_{\gamma i}, D_{\gamma i}, A_{\gamma i}$ has size $> q^* - 3u_4 \geqslant u_{85}$. Also $j = 3d + 15$, $d \geqslant t \cdot u_{20} > d - 1$. By 7.8, $s(A_\gamma) > u_{18}$ so $f/t > s(A_\gamma) - \epsilon_{30} > u_{18} - \epsilon_{30}$ by Axiom 30. Let $N = d + 12$. Since $s^{w+1}(B_{\gamma\, v+1}, B_{\gamma\, v+d}) > d/t - \epsilon'_{30} \geqslant u_{20} - \epsilon'_{30} = u_{13} + 2u_4$ we have, where X ranges through the symbols A, B, C, D, $s(X_{\gamma\, v+1}, X_{\gamma\, v+N}) > u_{13} - u_4 > r'_0$ $(v = j, \ldots, f - j - N)$; hence $(X_{\gamma\, v+1}, X_{\gamma\, v+N})$ is not a subelement of $\mathsf{L}_1 \cup \cdots \cup \mathsf{L}_w$. The N_i are $((w - 1))$-bounded by $s_0 + s_{26} \leqslant k_{56}$. Since N_1 is w-bounded by s_0 we have N_2, N_4, N_6 are w-bounded respectively by $s_0 + u_4, s_0 + 2u_4, s_0 + 3u_4 \leqslant k_{56}$; thus all the N_i are (w)-bounded by k_{56}.

3°. If the $B_{\gamma i}$ are maximal normal of size $> r^*$ then for $i = 4, \ldots, f - 3$ there exist $C_{\gamma i}, D_{\gamma i}$ in X_1, X_2 respectively such that

$$\mathrm{Im}'\, B_{\gamma i} = C_{\gamma i}, \qquad \mathrm{Im}'\, C_{\gamma i} = D_{\gamma i}, \qquad \mathrm{Im}'\, D_{\gamma i} = A_{\gamma i}$$

Z_{n-r} is w-bounded by s_0; X_1 is w-bounded by $s_0 + u_4$. We now make the assumption that $(A_{\gamma 1}, A_{\gamma f})$ is w-bounded by k_{54}.

We prove that $(D_{\gamma 1}, D_{\gamma f})$ is w-bounded by k_{55}. Suppose not. Then it contains a w-subelement S of size $\geqslant k_{55} > r_0 + \epsilon_3$. Let $\bar{S}$ be maximal normal in X_2. Then $(D_{\gamma 1}, D_{\gamma f}) \supset \bar{S}$ and $s(\bar{S}) \geqslant s(S) - \epsilon_3 \geqslant k_{55} - \epsilon_3 > v_{26}$. By Axiom 26″, $\bar{S}$ has the form U_2 in Axiom 26. By 6.21A and 6.23, using $1 - r' + t_{26} \leqslant k'_9$ and $s_0 + s_{26} \leqslant k'_8$ and $k_{55} - \epsilon_3 > u_{10}$, we have either 1a:

$$|s(U_1) - s(U_2)| < v_3 \text{ or } 1\text{b: } |s(U_1) + s(U_2) - 1| \leqslant v_2$$

and either

$$2\text{a:} |s(U_2) - s(U_3)| < v_3 \text{ or } 2\text{b: } |s(U_2) + s(U_3) - 1| \leqslant v_2.$$

Hence $s(U_1) > u_{10}$ since $k_{55} - \epsilon_3 - v_3 > u_{10}$ and $1 - (1 - r' + t_{26}) - v_2 > u_{10}$. Hence

$$3\text{: } |s(U) - s(U_1)| < v_3.$$

By 12.12, 1b and 2b do not both hold. If 1a holds then $k_{55} - \epsilon_3 < s(U) + 2v_3 < s_0 + 2v_3 \leqslant k_{55} - \epsilon_3$, a contradiction. If 1a does not hold then 2a holds and $s(U_2) < k_{54} + v_3 \leqslant k_{55} - \epsilon_3$, a contradiction.

Thus $(X_{\gamma 1}, X_{\gamma f})$ is (w)-bounded by

$$\mathrm{Max}(s_0 + s_{26}, s_0 + u_4, k_{54}, k_{55}) \leqslant k_{56}$$

($X = A, B, C, D$). Also each $X_{\gamma i}$ has size $> r^* - 3v_3 \geqslant u_{86}$.

4°. We show that the number of $B_{\gamma i}$, $f - 2j$, satisfies

$$(2) \qquad (f - 2j) - 4 - 4(N + 2) > 2N + 5$$

where N is defined to be 5 if the $B_{\gamma i}$ are maximal normal as in 3°; ($N = d + 12$ in 2°). Thus in the situation of 2° we must prove that

$f - 2(3d + 15) - 4 > 6(d + 12) + 13$, i.e., $f > 12d + 119$. Now

$$(f - 12d - 119)/t > u_{18} - \epsilon_{30} - 12(u_{20} + 1/t) - 119/t$$

$$\geqslant u_{18} - \epsilon_{30} - 12u_{20} - 131/e.$$

Denote the last expression by u_{87}. Note that the inequalities remain true if u_{18} is replaced by $s(A_\gamma)$. (2) follows since $u_{87} > 0$. If $N = 5$ we must show that $f - 2j > 47$. But $f - 2j > u'_{20} > 47$ (cf. 7.7A).

5°. Since $k_{56} \leqslant \text{Min}(k_{52}, k_{53})$, $u_{85} \geqslant v_4$ and $u_{86} \geqslant v_5$ we may apply 14.4 three times to obtain

$$(A_{\gamma\, j+3}, A_{\gamma\, f-j-2}) \underset{w}{=} X(B_{\gamma\, j+3+3(N+2)}, B_{\gamma\, f-j-3(N+2)-2})\, Y \equiv$$

$$\equiv XA_\gamma^{\mathrm{M}}\, Y$$

say where the right side is (w)-bounded by $k_{57} = k_{56} +$ $+ \text{Max}(u_4, q_{31}, h_{31})$. Also X, Y are not subelements of $L_1 \cup \cdots \cup L_w$ and

$$s^{w+1}(A_\gamma^{\mathrm{M}}) > (f - 2j - 6(N + 2) - 4)/t - \epsilon'_{30} > u_{87} - \epsilon'_{30}.$$

The left side of the equation is the subword of Z_{w-1} generated by all A-descendants of A_γ except for the first two and the last two.

If $A_\delta \subset Z_{n-r}$ and A_γ, A_δ are disjoint then similarly

$$(A_{\gamma\, j+3}, A_{\delta\, f'-j'-2}) \underset{w}{=} X(A_\gamma^{\mathrm{M}}, A_\delta^{\mathrm{M}})\, Y'$$

say where the right side is (w)-bounded by k_{57} and X, Y' are not subelements of $L_1 \cup \cdots \cup L_w$. The left side is the subword of Z_{w-1} generated by all the A-descendants of A_γ, A_δ except the first two of A_γ and the last two of A_δ.

6°. Take any fixed A_i $(1 \leqslant i \leqslant a)$. We claim that if $0 \leqslant s \leqslant n - 2$ and $A_{\mathrm{L}}, A_{\mathrm{R}}$ are the leftmost, rightmost A-descendants of A_i in Z_s then

$$\text{(3)} \qquad R_i \equiv AD_0(A_i) \underset{s}{=} P(A_{\mathrm{L}}^{\mathrm{M}}, A_{\mathrm{R}}^{\mathrm{M}})\, Q$$

for some P, Q and some $A_{\mathrm{L}}^{\mathrm{M}} \subset A_{\mathrm{L}}, A_{\mathrm{R}}^{\mathrm{M}} \subset A_{\mathrm{R}}$, where the right side is (s)-bounded by k_{58}. This is true for $s = 0$ since $b_{30} \leqslant k_{58}$. Assume it is true for some s, $0 \leqslant s < n-2$. Let A_λ, A_ρ be the leftmost, rightmost A-descendants of A_i in Z_{s+1}; then $A_{\lambda\,1+j} \equiv A_{\mathrm{L}}$ and $A_{\rho\, f'-j'} \equiv A_{\mathrm{R}}$. The left side of (2) is (n)-bounded by b_{30} so the right side is $(s+1)$-bounded by $b_{30} + u_4 \leqslant k_{54}$ and has the form $PP'(A_{\lambda\,3+j}, A_{\rho\, f'-j'-2})Q'Q$. By 5°

$$X(A_\lambda^{\mathrm{M}}, A_\rho^{\mathrm{M}})Y \underset{s+1}{=} (A_{\lambda\,3+j}, A_{\rho\, f'-j'-2})$$

where the left side is $((s+1))$-bounded by k_{57}. Hence

$$R_i \underset{s+1}{=} PP'X(A_\lambda^{\mathrm{M}}, A_\rho^{\mathrm{M}})YQ'Q$$

where the right side is $(s+1)$-bounded by $\mathrm{Max}(b_{30} + u_4, k_{57}) = k_{58}$ and (s)-bounded by $\mathrm{Max}(k_{58}, k_{57}) = k_{58}$, thus is $((s+1))$-bounded by k_{58}. The A-descendants of A_i in Z_{n-2} are say $A_{i\,1+j}, \ldots, A_{i\, f-j}$ hence $R_i \underset{n-2}{=} P''(A_{i\,1+j}^{\mathrm{M}}, A_{i\, f-j}^{\mathrm{M}})Q''$. Now $X'A^{\mathrm{M}}Y' \underset{n-1}{=} (A_{i\,3+j}, A_{i\, f-j-2})$ and $s^n(A_i^{\mathrm{M}}) > u_{87} - \epsilon_{30}' > u_3$, hence $R_i \underset{n-1}{=} H_i A_i^{\mathrm{M}} K_i$ say where the right side is $((n-1))$-bounded by $k_{58} \leqslant b_{25}$. R_i is (n)-bounded by $b_{30} \leqslant b_{25}$ hence by 6.10 R_i contains a subelement S_i of L_n of size $> u_{87} - \epsilon_{30}' - u_4 > q^*$ namely a weak image of A_i^{M}.

7°. If Case 2 of 7.8 holds then the A_i are disjoint; hence the R_i are disjoint (cf. 7.7 (iii)). For the $X_1, \ldots, X_a$ of Axiom 30 take $S_1, \ldots, S_a$. Now Z_n does not contain a maximal normal subword of size $> r''$ and $(Z_{n-1}) \equiv (Z_n)$. Suppose that P contains an n-subelement of size $> c^*$. Since $c^* \geqslant u_{49}$, Z_{n-1} contains an n-subelement T of size $> c^* - u_{37} - u_4 > r'' + \epsilon_3$ (cf. 12.15) so $s(\overline{T}) \geqslant s(T) - \epsilon_3 > r''$, a contradiction.

8°. Now assume that Case 1 of 7.8 holds. Then the A_i are maximal normal and kernels exist but the A_i are not necessarily disjoint. Let $b = a$ or $a-1$, whichever is odd. Then $R_1, R_3, R_5, \ldots, R_b$ are disjoint and

$$(R_1, R_a) \equiv R_1 X_2 R_3 X_4 \cdots \underset{n-1}{=} H_1 A_1^{\mathrm{M}} K_1 X_2 H_3 A_3^{\mathrm{M}} K_3 X_4 \cdots \tag{4}$$

where the right side is $((n-1))$-bounded by k_{58}. Let i range over

$1, 2, 3, ..., a$ and let j range over $1, 3, 5, ..., b$. A weak image of A_j^{M} in (3) is still S_j. Now by 4°, 5° $s(A_i^{\mathrm{M}}) > (u_{87} - u_{18} + s(A_i)) - \epsilon'_{30}$. But by 7.8 $s(A_i) > r'' - r_{26}$; hence $s(\overline{A_i^{\mathrm{M}}}) > u_{87} - u_{18} + r'' - r_{26} - - \epsilon'_{30} = u_{88} > u_5$. Therefore $\mathrm{Im}(A_j^{\mathrm{M}}) = S_j$. Next

$$(5) \qquad H_1 A_1^{\mathrm{M}} K_1 X_2 H_3 \cdots {}_{n-1}{=}_1\, H_1 (A_1^{\mathrm{M}}, A_b^{\mathrm{M}}) Z$$

for some Z where the right side is $((n-1))$-bounded by k_{58}. Write (5) briefly as $A = B$. In (5) a weak image of $A_j^{\mathrm{M}} \subset B$ is $\overline{A_j^{\mathrm{M}}} \subset A$ since $(\mathrm{In}\, A, A_j^{\mathrm{M}}) \;{}_{n-1}{=}_1\; (\mathrm{In}\, B, A_j^{\mathrm{M}})$. Hence $\mathrm{Im}_5(A_j^{\mathrm{M}}(\subset B)) = \overline{A_j^{\mathrm{M}}} \subset A$ where the subscript 4 refers to eq. (5).

By 6.11 if $\mathrm{Im}\, X$ exists then $\mathrm{Im}(\overline{X}) = \mathrm{Im}\, X$; so $\mathrm{Im}_5(\overline{A_j^{\mathrm{M}}}(\subset B)) = \overline{A_j^{\mathrm{M}}}(\subset A))$. We wish to apply 6.13. Now $s(\overline{A_i^{\mathrm{M}}}) > u_{88} - \epsilon_3 > u_6$ $\mathrm{Im}_5(A_i^{\mathrm{M}}) > s(\overline{A_i^{\mathrm{M}}}) - u_4 - \epsilon_3 > u_{88} - u_4 - \epsilon_3 > u_6$. With respect to (4) let $\mathrm{Im}(\mathrm{Im}_5 \overline{A_i^{\mathrm{M}}}) = C_i$. Then by 6.13 C_i is left of C_{i+1} $(i = 1, 2, ..., a-1)$; the C_i are maximal normal in (R_1, R_a) of size $> u_{88} - 2u_4 - 2\epsilon_3 > r^*$. For the required $X_1, ..., X_a$ of Axiom 30 take $C_2, C_3, ..., C_{a-1}$; they are maximal n-normal in PQ. Also $C_j = \mathrm{Im}(\overline{A_j^{\mathrm{M}}}) = \overline{S_j} \supset S_j \subset R_j$. Let $P' \subset P$ and let the number of $C_2, ..., C_{a-1}$ in P' be γ and the number of $R_1, ..., R_a$ in P' be ρ. We prove that $|\rho - \gamma| \leqslant 6$.

Say $R_{i+1}, ..., R_{i+\rho} \subset P'$; then $R_{2u+1}, R_{2u+2}, ..., R_{2t+1} \subset P'$ where $2t + 1 - 2u \geqslant \rho - 2$ and S_{2u+1}, S_{2t+1} are in P'; hence so are C_{2u+2}, C_{2t}. Thus $\gamma \geqslant 2t - 2u - 1 \geqslant \rho - 4$. Now say $C_{i+1}, ..., C_{i+\gamma} \subset P'$; then C_{2u+1}, C_{2t+1} are in P' where $2t + 1 - 2u \geqslant \gamma - 2$ so R_{2u+3}, R_{2t-1} are in P' and $\rho \geqslant 2t - 2u - 3 \geqslant \gamma - 6$.

In Case 2 of 7.8 if γ is the number of $S_1, ..., S_a$ in P' then $\gamma \geqslant \rho \geqslant \gamma - 2$ so again $|\rho - \gamma| \leqslant 6$.

Now $s_{PQ\sigma}(P') = \rho/m$ so by 13.17

$$|\rho/m - s_{PQ}(P')| \leqslant \epsilon_3 + 4/e, \qquad |\gamma/m - \rho/m| \leqslant 6/m \leqslant 6/e$$

hence $|\gamma/m - s_{PQ}(P')| \leqslant \epsilon_3 + 10/e < \epsilon'_{30}$. If $P' = P$ then $\rho = a$ and since $|s_{PQ\sigma}(P) - s(P)| < 1/e^2$ we have

$$|\gamma/m - s^{n+1}(P)| < 6/e + 1/e^2 < \epsilon_{30}.$$

14.5. Proposition. *Axiom* 10 *holds for* $n+1$.

Proof. Let X be $((n+1))$-bounded by b_∞. Let $X \underset{n+1}{=} I$. Since Axiom 20 holds for $n+1$ (cf. 13.11), we have that 6.1 holds for $n+1$; hence if $X \not\equiv I$ then X contains a subelement C of L_v $(1 \leqslant v \leqslant n+1)$ of size greater than k_1 or k_1^0. But $s(C) < b_\infty \leqslant \mathrm{Min}(k_1^0, k_1)$. Hence $X \equiv I$.

§ 15

Informal Summary

Propositions 15.3, 15.3A below correspond to the informal results 4.4A and 4.4B. Using these propositions we can discuss simultaneous $(n+1)$-replacements and prove that Axiom 26 holds for $n+1$ (cf. 4.4C).

15.1. The statement that Axioms 18, 22 and 33′ hold for $n+1$ will be called *Condition* D_{n+1} or simply *Condition D*.

15.1A. It is convenient to proceed as follows. Assuming Condition D_{n+1} we shall develop in this section a theory of $(n+1)$-replacements, say T_{n+1}. Since D_n is true, so is T_n. Then, using T_n, we prove D_{n+1}.

15.2. Let $W \equiv XAY$ where A is $(n+1)$-normal. Call W' a *weak replacement* of A in W if $W' \underset{n}{=} X \cdot B^{-1} \cdot Y$ and $AB \in \mathsf{L}_{n+1}$ for some B.

By 5.25 for $n+1$ any two weak replacements are equal in G_n.

15.3. Proposition. *Let the (linear) word* $HZ \equiv BDZ$ *be* $((n-1))$-*bounded by* k_{36} *and n-bounded by* k_{37}. *Let* $ZT^{-1} \in Rep(L_{n+1})$ *and let DZ be maximal* $(n+1)$-*normal in BDZ where* $s(Z) > s(DZ) - u_{57}$ *and* $u_{58} > s(Z) > u_{59}$.

Then $BDZ \cdot Z^{-1}T \underset{n}{=} C'T^{\mathrm{R}}$ *where* $C'T^{\mathrm{R}}$ *is* (n)-*bounded by* k_{50}, $s(T^{\mathrm{R}}) > s(T) - u_{77}$ *and*

$$|s(DZ) + s(\overline{T^{\mathrm{R}}}) - 1| < u_{78}$$

and as usual $\overline{T^{\mathrm{R}}}$ *denotes the maximal normal word containing* T^{R}.

Proof. 1°. By Axiom 11a, $BD \cdot T \underset{n}{=} C^{-1}$, where C is (n)-bounded by $b_{\min}$. Put $H \equiv BD$. Then $H \cdot T \cdot C \underset{n}{=} I$. To apply 8.2 of Chapter I we require that if $SS' \in \mathsf{L}_i$, $i \leqslant n$, and S is $*$-contained in H, T or C then $s_{SS'}(S) \leqslant 1 - 5/p - 5\epsilon_1 - \epsilon_0 = 1 - 5(r + \epsilon_3) - 12/e = \beta$ say (if $i = n$, replace r by r_0 and β by β_0.). As in 6.2, this is so if H, T, C are $((n-1)$-bounded by $\beta' = \beta - 2\epsilon_2 - 4/e$ and n-bounded by $\beta_0' = \beta_0 - 2\epsilon_2 - 4/e$. Now Z is (n)-bounded by $str^+ \leqslant \beta' \leqslant \beta_0'$; C is (n)-bounded by $b_{\min} \leqslant \beta' \leqslant \beta_0'$; H is $((n-1))$-bounded by $k_{36} \leqslant \beta'$ and n-bounded by $k_{37} \leqslant \beta_0'$.

Hence there are divisions H_1, H_2, H_3 of H; T_1, T_2, T_3 of T; C_1, C_2, C_3 of C and

$$H_3 \underset{n}{\sim} T_1^{-1} \quad \text{(end words } U, I)$$

$$T_3 \underset{n}{\sim} C_1^{-1} \quad \text{(end words } V, I)$$

$$C_3 \underset{n}{\sim} H_1^{-1} \quad \text{(end words } W, I)$$

for some U, V, W. Let H_3' be the largest common subword of H, H_3 (possibly H_3' is I). Then $L(H_3') \geqslant L(H_3) - 1$. Similarly define H_1'. Now define H_2' by $H \equiv H_1' H_2' H_3'$. Similarly $T \equiv T_1' T_2' T_3'$ and $C \equiv C_1' C_2' C_3'$. By 5.19 if H_1' has size $> r_0$ then $s(H_1') - \epsilon_3 \leqslant s(H_1) \leqslant \epsilon_2 + s(H_1') + 2/e$.

Choose H_{32}' of maximal length such that $H_3' \equiv H_{31}' H_{32}'$ and $H_{32}' Z$ is a subelement of L_{n+1}. Then $H_{32}' Z \subset HZ \equiv BDZ$, so $H_{32}' Z \subset DZ$. Since $s(Z) > u_{59} \geqslant r_0 + \epsilon_3 > r_0$ we have

$$s(Z) \leqslant s(H_{32}' Z) + \epsilon_3$$

$$s(H_{32}' Z) \leqslant s(DZ) + \epsilon_3 < s(Z) + u_{57} + \epsilon_3$$

$$(1) \qquad -\epsilon_3 \leqslant s\ H_{32}' Z) - s(Z) < u_{57} + \epsilon_3$$

Now $H_3 Z \cdot Z^{-1} T_1 \underset{n}{=} U$ and $H_3 Z \underset{n}{\sim} T_1^{-1} Z$. Let $\bar{Z}$ denote the maximal normal subword of $H_3 Z$ containing Z. We shall prove that

$$(2) \qquad |s(T_1^{-1} Z) - s(\bar{Z})| < u_4 + \epsilon_3$$

By 12.16 and 13.10, Axiom 25 holds for $n+1$, hence so does 6.10. Thus $(\bar{Z})^0 \sim (\bar{Z})^{0'} \supset M$ where M is a subelement of L_{n+1} and $s(M) > s(\bar{Z}) - u_4$. Also $s(M) \leqslant s(T_1^{-1}Z) + \epsilon_3$ hence $s(\bar{Z}) < s(T_1^{-1}Z) + {} + \epsilon_3 + u_4$. Again, taking a weak image of $T_1^{-1}Z$, say N, $H_3 Z \supset N$, $s(N) > s(T_1^{-1}Z) - u_4$. By 6.11, since $s(T_1^{-1}Z) \geqslant s(Z) - \epsilon_3 > u_{59} - \epsilon_3 > u_5$ and $\mathrm{Max}(str^+, k_{36}, k_{37}) \leqslant b_{25}$ we have $\mathrm{Im}(T_1^{-1}Z) = \bar{N}$. Now $Z \subset H_3 Z$ and $s(Z) > u_{59} > u_5 - u_4$ hence $\bar{Z} = \bar{N}$ (again by 6.11). Hence $N \subset \bar{Z}$ and $s(N) \leqslant s(\bar{Z}) + \epsilon_3$ and $s(T_1^{-1}Z) < s(\bar{Z}) + \epsilon_3 + u_4$, proving (2).

Now $H_3 Z$ is $H_3' Z$ or $cH_3' Z$ ($L(c) = 1$) so either (i) $\bar{Z} \subset H_3' Z$ and $\bar{Z} \equiv H_{32}' Z$, or (ii) $\bar{Z} \equiv cH_3' Z$ and $H_3' \equiv H_{32}'$. In either case,

(3) $$-\epsilon_3 \leqslant s(\bar{Z}) - s(H_{32}' Z) \leqslant \epsilon_2 + 2/e$$

Now

(4) $$-\epsilon_2 - 2/e \leqslant s(T_1'^{-1} Z) - s(T_1^{-1} Z) \leqslant \epsilon_3.$$

Adding (1), (2), (3), (4) we find

$$-\alpha < s(T_1'^{-1} Z) - s(Z) < u_{57} + \alpha$$

where $\alpha = 3\epsilon_3 + u_4 + \epsilon_2 + 2/e$.

Next we show that $s(T_2') < 2\epsilon_2 + 4/e + r_0' = \beta_1$ say.

If $L(T_2) > 2$ then by 8.2 of Chapter I $T_2 \equiv cAd$ where A is a subelement of $L_1 \cup \cdots \cup L_n$. Now $T_2' \equiv c'Ad'$, $c \sim c'$, $d \sim d'$ hence $s^{n+1}(T_2') \leqslant 2\epsilon_2 + 4/e + s^{n+1}(A) \leqslant 2\epsilon_2 + 4/e + r_0'$. If $L(T_2) = 1$ or 2 then $L(T_2') = L(T_2) \leqslant 2\epsilon_2 + 2/e$. If $T_2 = I$ then $L(T_2') \leqslant 1$ so $s(T_2') \leqslant \epsilon_2$.

Thus

$$1 - 2/e \leqslant s(Z^{-1} T_1' T_2' T_3') \leqslant 4/e + s(Z^{-1} T_1') + \beta_1 + s(T_3')$$

$$\leqslant 4/e + s(Z) + u_{57} + \alpha + \beta_1 + s(T_3')$$

(I) $$s(T_3') + s(Z) - 1 > -(6/e + u_{57} + \alpha + \beta_1)$$

Hence T_3' is normal ($1 - u_{58} - (6/e + u_{57} + \alpha + \beta_1) > r_0 + \epsilon_3$).

$$s(Z^{-1}T_1') + s(T_2'T_3') < s(Z^{-1}T_1'T_2'T_3') + (2\epsilon_3 + \epsilon_1^*)$$

$$\leqslant 1 + (2\epsilon_3 + \epsilon_1^*).$$

Also $-\epsilon_3 + s(T_3') \leqslant s(T_2'T_3')$ and $s(Z) < s(Z^{-1}T_1') + \alpha$ so by adding,

(II) $$s(T_3') + s(Z) - 1 < \epsilon_3 + (2\epsilon_3 + \epsilon_1^*) + \alpha.$$

2°. We have $C_1^{-1} \underset{n}{\sim} T_3$ with end words V^{-1}, I.

Case 1. $C_1^{-1} \underset{n-1}{\not\sim} T_3$.

T_3 contains say P_i' where $k_3^0 < s(P_i') + Q_i' < s(P_i') + b_{\min}$. Hence T_3' contains a normal subelement of L_n of size $> k_3^0 - b_{\min} - \epsilon_2 - - 2/e - \epsilon_3 > u_{11}$. By 7.3, $T_1'T_2'T_3'Z^{-1}$ contains m distinct maximal normal subwords of size $> r''$ which as in 7.8 are denoted by $B_1, \ldots, B_m$. Let u of them be contained in T_3'. By 13.17 (ii) and 13.16

(III) $$u/m \geqslant (s(T_3') - \epsilon_3) - \epsilon_3 - 4/e.$$

Now $T_3 \supset T_3' \supset (B_{a+1}, \ldots, B_{a+u})$. 6.27 is available since $\mathrm{Max}(str^+, b_{\min}) \leqslant k_{53}$ and $r'' > v_5$. Let B_i' be the image of B_i in C_1^{-1}. For $i = a+2$, $(B_i'^{\mathrm{R}}, B_{i+1}'^{\mathrm{L}}) \underset{n}{\sim} (B_i^{\mathrm{R}}, B_{i+1}^{\mathrm{L}})$ which we abbreviates to $B' \underset{n}{\sim} B$ (end words E', K' say); thus $C_1^{-1} \equiv (C_1^{-1})^l B' (C_1^{-1})^r$ and $T_3 \equiv T_3^l B T_3^r$. Also there is a word $B_i'^{\mathrm{RL}} A B_{i+1}^{\mathrm{LR}} \underset{n}{=} B' \cdot K'^{-1}$. Let

$$D_0 \equiv C_3^{-1} C_2^{-1} (C_1^{-1})^l B_i'^{\mathrm{RL}} A B_{i+1}^{\mathrm{LR}} T_3^r.$$

Then $D_0 = C_3^{-1} C_2^{-1} (C_1^{-1})^l B' K'^{-1} T_3^r \underset{n}{=} C^{-1}$. Now $B_i'^{\mathrm{RL}} A B_{i+1}^{\mathrm{LR}}$ is (n)-bounded by $\mathrm{Max}(str^+, b_{\min}) + h_{31} = \lambda$ say. $B_i'^{\mathrm{RL}}$ and B_{i+1}^{LR} are not subelements of $L_1 \cup \cdots \cup L_{n-1}$ hence D_0 is $((n-1))$-bounded by λ (cf. 5.32). Since $D_0 \underset{n}{=} C^{-1}$, D_0 is n-bounded by $b_{\min} + u_4$ (cf. 6.10A). Hence D_0 is (n)-bounded by $\lambda + u_4 \leqslant k_{50}$. D_0 has the form $C_3^{-1} C_2^{-1} (C_1^{-1})^{\mathrm{L}} A T^{\mathrm{R}}$. T_3^r contains $B_{i+3}, \ldots, B_{a+u}$; hence

$$s(T^{\mathrm{R}}) \geqslant (u-4)/m \geqslant s(T_3') - 2\epsilon_3 - 4/e - 4/e$$

$$> (1 - s(Z)) - (6/e + u_{57} + \alpha + \beta_1) - 2\epsilon_3 - 8/e$$

$$= 1 - s(Z) - \rho, \text{ say.}$$

Now

(*) $s(Z)+s(T) \leqslant s(Z^{-1}T) + (2\epsilon_3 + \epsilon_1^*) \leqslant 1 + (2\epsilon_3 + \epsilon_1^*)$

Hence

(**) $$s(T^{\mathrm{R}}) > s(T) - \rho - (2\epsilon_3 + \epsilon_1^*)$$

Case 2. $C_1^{-1} \underset{n-1}{\sim} T_3$.

If (i) of 7.3 holds we may proceed as in Case 1. Thus we may assume that $T_1' T_2' T_3' Z^{-1}$ contains $B_1, ..., B_m$, disjoint n-subelements each of size $> u_{11}$. Again (III) holds. 6.26 is available since $\mathrm{Max}(str^+, b_{\min}) \leqslant k_{52}$ and $u_{11} > v_4$. Hence for $i = a + 2$ (in particular) a weak image B_i' of B_i exists and $B_i'^{\mathrm{M}} \sim B_i^{\mathrm{M}}$ (end words E', K' say) and there is a word $B_i'^{\mathrm{ML}} A B_i^{\mathrm{MR}} = B_i'^{\mathrm{M}} \cdot K'^{-1}$, $((n-1))$-bounded by $\mathrm{Max}(str^+, b_{\min}) + q_{31} = \lambda'$ say. As before there is a word $D \equiv C_3^{-1} C_2^{-1} (C_1^{-1})^{\mathrm{L}} A T^{\mathrm{R}}$, (n)-bounded by $\lambda' + u_4 \leqslant k_{50}$ and again (**) holds.

Thus in both cases $s(T^{\mathrm{R}}) - s(T) > -\rho - (2\epsilon_3 + \epsilon_1^*) \geqslant -u_{77}$. T^{R} is normal since $1 - u_{58} - \rho > r_0 + \epsilon_3$.

We now have

$$s(\overline{T^{\mathrm{R}}}) \geqslant s(T^{\mathrm{R}}) - \epsilon_3 > 1 - s(Z) - \rho - \epsilon_3$$

$$\geqslant 1 - (s(DZ) + \epsilon_3) - \rho - \epsilon_3$$

hence

(IV) $$s(DZ) + s(\overline{T^{\mathrm{R}}}) - 1 > -\rho - 2\epsilon_3 \geqslant -u_{78}$$

3°. Write M for $\overline{T^{\mathrm{R}}}$; then $D_0 \equiv B'M \equiv B'M^l T^{\mathrm{R}}$ say for some B'. By Axiom 33′ for $n+1$, $B'M^l T^{\mathrm{R}} \underset{n}{=} B'M^{\mathrm{L}} E_1 Z_1 F_1 M^{\mathrm{R}}$ (since $s(M) > 1 - u_{58} - \rho - \epsilon_3 > \mathrm{Max}(t_{33}, r_{33} + u_1)$ and $\mathrm{Max}(\lambda, \lambda') + u_4 \leqslant b_{33}$); also the right side is (n)-bounded by $\lambda'' = \mathrm{Max}(\lambda, \lambda') + U_4 + c_{33}$. Now $s(Z_1) > s(M) - r_{33} > u_1$ and $s(T^{\mathrm{R}}) > 1 - u_{58} - \rho > u_1$ so Z_1^0, $(T^{\mathrm{R}})^0$ exist, and $Z_1 \supset Z_1^0 \sim Z_1^{0\prime}$. Define T_1, T_2 by $(T^{\mathrm{R}})^0 \equiv T_1 T$ $T_1 = Z_1^{0\prime} \cap (T^{\mathrm{R}})^0$. If T_2 has size $> \alpha'$ (defined below) then since $\alpha' > u_3$ and $\lambda'' \leqslant b_{25}$ we have YF_1M^{R} contains a weak image J of T_2 where $Z_1 \equiv XZ_1^0 Y$; also $s(J) > \alpha' - u_4$. Let $J \equiv J_1 J_2$ where $J_1 \equiv Y \cap J$.

Either $s(J_1) \leqslant r_0$ or $s(J_1) \leqslant s(Y) + \epsilon_3 \leqslant u_2 + \epsilon_3$ (cf. 6.8). We have

$$s(J) \quad \leqslant s(J_1) + s(J_2) + 2/e$$

$$\alpha' - u_4 < \mathrm{Max}(r_0, u_2 + \epsilon_3) + 2/e + s(J_2)$$

$$s(J_2) \quad > \alpha' - u_4 - \mathrm{Max}(r_0, u_2 + \epsilon_3) - 2/e > r_0$$

$$s(J_2) \quad \leqslant s(F_1 M^{\mathrm{R}}) + \epsilon_3 < q_{33} + \epsilon_3$$

so defining α' by $q_{33} + \epsilon_3 = \alpha' - u_4 - \mathrm{Max}(r_0, u_2 + \epsilon_3) - 2/e$ we have a contradiction. It follows that

$$s((T^{\mathrm{R}})^0) \leqslant s(T_1) + s(T_2) + 2/e \leqslant s(T_1) + \alpha' + 2/e.$$

By 6.8, since $U(T^{\mathrm{R}})^0 \equiv T^{\mathrm{R}}$ for some U, $s(U) < u_2$; hence $s(T^{\mathrm{R}}) < u_2 + s((T^{\mathrm{R}})^0) + 2/e < s(T_1) + \alpha' + 4/e + u_2$. Hence

$$s(T_1) > 1 - u_{58} - \rho - \alpha' - 4/e - u_2 > \mathrm{Max}(s_{33}, r_0 + \epsilon_3).$$

By definition of T_1, $T^{\mathrm{R}} \supset T_1 \underset{n}{\sim} S_1 \equiv Z_1^0 \cap (T^{\mathrm{R}})^{0\prime} \subset Z_1$. Hence $T_1 \equiv S_1$ since Axiom 33 is true for $n+1$. Where $Z_1 \equiv Z_{11} S_1 Z_{12}$, $T^{\mathrm{R}} \equiv T^{\mathrm{RL}} T_1 T^{\mathrm{RR}}$ then (cf. 13.23) $T_1 T^{\mathrm{RR}} \underset{n}{=} S_1 Z_{12} F_1 M^{\mathrm{R}}$. Thus $D_0 \underset{n}{=} B' M^{\mathrm{L}} E_1 Z_{11} T_1 T^{\mathrm{RR}}$ which is (n)-bounded by λ'' (cf. 5.32). Let $D' \equiv M^{\mathrm{L}} E_1$, and $Z' \equiv Z_{11} T_1 T^{\mathrm{RR}}$. Now $M^{\mathrm{L}} E_1 Z_{11} T_1$ is a subelement of L_{n+1} since it is contained in $M^{\mathrm{L}} E_1 Z_1$, $T_1 T^{\mathrm{RR}}$ is a subelement of L_{n+1} and T_1 is normal. Hence $D'Z' \equiv M^{\mathrm{L}} E_1 Z_{11} T_1 T^{\mathrm{RR}}$ is a subelement of L_{n+1} (cf. 5.28A). Note that $-Z' \equiv -T$. Now $D'Z'$ is $(n+1)$-normal, $D'Z' \underset{n}{=} B'^{-1} \cdot D_0 \equiv M \equiv \overline{T^{\mathrm{R}}}$. Hence

$$\text{(i) } s(D'Z') > s(\overline{T^{\mathrm{R}}}) - u_4 - \epsilon_3$$

$$\text{(ii) } s(D'Z') < q_{33} + s(Z') + 2/e.$$

Case 1$'$. Either $Z' \subset T$ or $Z' \equiv VT$ where $s^{n+1}(V) \leqslant r_0'$.

Then $Z' \leqslant s(T) + \epsilon_3$ or $s(Z') \leqslant r_0' + s(T) + 2/e$ so in either case $s(Z') \leqslant r_0' + s(T) + 2/e + \epsilon_3$, hence by (i), (ii) and (*)

$$s(\overline{T^{\mathrm{R}}}) + s(Z) - 1 < u_4 + \epsilon_3 + q_{33} + 4/e + r_0' + \epsilon_3 +$$

$$+ (2\epsilon_3 + \epsilon_1^*) = \Delta$$

say. Since $s(DZ) - u_{57} < s(Z)$ we deduce

(V) $\qquad s(\overline{T^{\mathrm{R}}}) + s(DZ) - 1 < u_{57} + \Delta \leqslant u_{78}.$

Case 2′. Not Case 1′.

We shall obtain a contradiction. Since $-Z' \equiv -T$ we have $Z' \equiv VT, s^{n+1}(V) > r_0'$. In particular $V \not\equiv I$. Now

$$BD \cdot T \underset{n}{=} C^{-1} \underset{n}{=} D_0 \underset{n}{=} B'D'Z'$$

so

(†) $\qquad BD \cdot V^{-1} D'^{-1} B'^{-1} \underset{n}{=} I$

Now $-Z^{-1} T \equiv -T \equiv -Z' \equiv -VT$ hence $-Z^{-1} \equiv -V$ and $Z- \equiv V^{-1}-$. Therefore $BDZ- \equiv BDV^{-1}-$. Hence the dot can be removed from (†). Next we show DBV^{-1} is (n)-bounded by k_{37}. If $V^{-1} \equiv ZH$ for some H then $BDV^{-1} \equiv BDZH$. Now Z is not a subelement of $L_1 \cup \cdots \cup L_n$, BDZ is (n)-bounded by k_{37} and ZH is n-bounded by $str^+ \leqslant k_{37}$. If $V^{-1} \not\equiv ZH$ then $Z \equiv V^{-1} K$ for some K and $BDV^{-1} \subset BDZ$ which is (n)-bounded by k_{37}. $V^{-1} D'^{-1} B'^{-1} = (B'D'V)^{-1} \subset (B'D'Z')^{-1}$ which by above is (n)-bounded by λ''; V^{-1} is not a subelement of $L_1 \cup \cdots \cup L_n$. Hence the left side if (†), with dot removed, is (n)-bounded by $\mathrm{Max}(\lambda'', k_{37}) \leqslant b_\infty$ in contradiction to Axiom 10.

15.3A. Corollary. *Assume further that BDZ is* $(n+1)$*-bounded by* b *where* $u_{79} - u_{83} < b < 1 - u_{83}$. *Then*

(i) $C'T^{\mathrm{R}}$ *is* $(n+1)$*-bounded by* $\mathrm{Max}(b + u_{83}', 1 - s(DZ) + u_{78}')$

(ii) *If* Q *is maximal* $(n+1)$*-normal in* $C'T^{\mathrm{R}}$ *where* Q *is not* $\overline{T^{\mathrm{R}}}$ *and* $s(Q) > u_{79}'$ *then there is a maximal* $(n+1)$*-normal subword* Q^* *of BDZ, not DZ, and*

$$|s(Q) - s(Q^*)| < u_{84}'$$

(iii) *If P is maximal* $(n+1)$*-normal in BDZ, not DZ,* $s(P) > u'_{79}$ *then there is a maximal* $(n+1)$*-normal subword* P^* *of* $C'T^{\mathrm{R}}$*, not* $\overline{T^{\mathrm{R}}}$*, where* $|s(P) - s(P^*)| < u'_{84}$.

(iv) *If* Q^{**} *exists it is* Q*; if* P^{**} *exists it is* P.

(v) *If* $B \equiv Z'C$ *where* Z' *is a subelement of Rep* L_{n+1} *of size* $> u_{80}$ *we may take* $C' \equiv Z'^{\mathrm{L}}K$ *(for some* K*) in* (i)–(iv) *where* $s(Z'^{\mathrm{L}}) > s(Z') - u_{81}$.

Proof. 1°. First assume only that Q is $(n+1)$-normal in $C'T^{\mathrm{R}}$, $Q \not\subset \overline{T^{\mathrm{R}}}$ and $s(Q) > u_{79}$. Now $C'T^{\mathrm{R}} \equiv D_0 \equiv C_3^{-1}C_2^{-1}(C_1^{-1})^{\mathrm{L}}AT^{\mathrm{R}} \underset{n}{=} C^{-1} \equiv$
$\equiv C_3^{-1}C_2^{-1}C_1^{-1}$ so

(5) $\qquad C_1^{-1} \underset{n}{=} C_1^{-1\mathrm{L}}AT^{\mathrm{R}} \equiv D_1$ say

Also

(6) $\qquad T_3 \underset{n}{\sim} C_1^{-1}$ (end words V, I)

Write $Q_1 = Q \cap C_3^{-1}$, $Q_2 = Q \cap C_2^{-1}$, $Q_3 = Q \cap C_1^{-1\mathrm{L}}AT^{\mathrm{R}}$.

2°. We show that D_1 is $(n+1)$-bounded by α_2 where

$$\alpha_2 = 1 - u_{59} + (2\epsilon_3 + \epsilon_1^*) + 2u_4 + 2\epsilon_3.$$

Suppose not. Then it contains an $(n+1)$-subelement S of size $\geqslant \alpha_2 > u_3 + u_4$. Let S' be a weak image of S in C_1^{-1} (under (5)) (the right side of (5) is (n)-bounded by λ''). Then $s(S') > s(S) - u_4 > u_3$. Let S'' be a weak image of S' under (6). Then $s(S'') > \alpha_2 - 2u_4 > r_0 + \epsilon_3$ and $s(T_3) \geqslant s(S'') - \epsilon_3 > r_0$. Thus we have

$$u_{59} + s(T) - 1 < s(Z) + s(T) - 1 \leqslant (2\epsilon_3 + \epsilon_1^*)$$

$s(T_3) \leqslant s(T) + \epsilon_3$ and $s(S) - 2u_4 \leqslant s(S'') \leqslant s(T_3) + \epsilon_3$. By addition we obtain $s(S) < \alpha_2$, a contradiction.

3°. We show that $s(Q_3) \leqslant \mathrm{Max}(r_0 + \epsilon_3, u_6 + u_4 + 2\epsilon_3) = \beta_2$ say.

Assume not. Then Q_3 is normal and of size $> u_6 + u_4 + 2\epsilon_3$. Let $\widetilde{Q}_3$, $\widetilde{T^{\mathrm{R}}}$ be maximal normal in D_1 and containing Q_3, T^{R}. Now $D_1 \subset D_0$ so $\widetilde{T^{\mathrm{R}}} \subset \overline{T^{\mathrm{R}}}$. If $\widetilde{Q}_3 = \widetilde{T^{\mathrm{R}}}$ then $Q_3 \subset \overline{T^{\mathrm{R}}} \cap Q$; hence the subword $\overline{T^{\mathrm{R}}} \cup Q$ of D_0 is an $(n+1)$-subelement containing $\overline{T^{\mathrm{R}}}$ so is $\overline{T^R}$; thus $Q \subset \overline{T^R}$ which contradicts the hypothesis. Thus $\widetilde{Q}_3 \neq \widetilde{T}^R$. Now

$$s(\widetilde{T^R}) \geqslant s(T^R) - \epsilon_3 > 1 - s(Z) - \rho - \epsilon_3 > 1 - u_{58} - \rho - \epsilon_3$$

$$> u_6 + u_4 + \epsilon_3$$

and $s(\widetilde{Q_3}) \geqslant s(Q_3) - \epsilon_3 > u_6 + u_4 + \epsilon_3$. By applying 6.13 to (5), (6) we see that Im Im $\widetilde{T}^R$, Im Im $\widetilde{Q}_3$ are different maximal normal subwords of T_3; this is impossible.

4°. Thus $s(Q) - s(Q_1) \leqslant s(Q_2) + s(Q_3) + 4/e \leqslant r_0' + \beta_2 + 4/e = \beta_3$ say and $s(Q_1) > u_{79} - \beta_3 = \beta_4$ say $> u_3$. Now $Q_1 \subset C_3^{-1} \sim H_1$. Let Q' be a weak image of Q_1 in H_1. Then

$$s(Q') > s(Q_1) - u_4 > \beta_4 - u_4$$

$$> \mathrm{Max}(r_0 + \epsilon_3, \epsilon_3 + u_{57} + 2\epsilon_3 + \epsilon_1^*).$$

In $H_1 H_2 H_3 Z$, clearly Q' and Z are disjoint. Let Q^* be the maximal normal subword of HZ containing Q'. Then where $u_{84} = \beta_3 + u_4 + \epsilon_3$ we have $s(Q^*) \geqslant s(Q') - \epsilon_3 > s(Q_1) - u_4 - \epsilon_3 > s(Q) - u_{84}$.

We prove that $Q^* \neq DZ$. If $Q^* = DZ$ then $Q' \subset DZ$ so $Q' \subset D$. Also

$$s(Q') - \epsilon_3 + s(Z) \leqslant s(D) + s(Z) \leqslant s(DZ) + 2\epsilon_3 + \epsilon_1^*$$

$$< s(Z) + u_{57} + 2\epsilon_3 + \epsilon_1^*;$$

hence $s(Q') < \epsilon_3 + u_{57} + 2\epsilon_3 + \epsilon_1^*$, which is a contradiction.

We now have $s(Q_1) < s(Q') + u_4 < b + u_4$ so

$$s(Q) \leqslant s(Q_1) + \beta_3 < b + u_4 + \beta_3 = b + u_{83} < b + u_{83}'.$$

Now suppose Q_0 is some $(n+1)$-normal subword of $\overline{T^R}$. Then

$$s(Q_0) \leqslant s(\overline{T^R}) + \epsilon_3 < u_{78} + 1 - s(DZ) + \epsilon_3.$$

Since $u_{78} + \epsilon_3 \leqslant u'_{78}$ we have proved (i); indeed $C'T^{\mathrm{R}}$ is $(n+1)$-bounded by

$$\mathrm{Max}(b + u_{83}, 1 - s(DZ) + u_{78} + \epsilon_3) = \gamma \text{ say}$$

5°. If Q is now maximal normal then since $s(Q_1) > \beta_4 > u_6$ we have by 6.12 that $Q \cap C_3^{-1}$ and $Q^* \cap H_1$ are images of each other, with respect to $C_3^{-1} \sim H_1$, in the sense of 6.11.

6°. Now assume only that P is $(n+1)$-normal in BDZ, $P \not\subset DZ$ and $s(P) > u_{79}$. We argue similarly, as follows. Let $P_1 \equiv P \cap H_1$, $P_2 = P \cap H_2$, $P_3 = P \cap H_3 Z$. First, $s(P_3) \leqslant \beta_2$. (Assume not; then let $\tilde{Z}$, $\tilde{P}_3$ be maximal normal in $H_3 Z$. Thus $\tilde{Z} \subset DZ$. Also $\tilde{P}_3 \neq \tilde{Z}$. Now $H_3 \sim T_1^{-1}$ (end words U, I) so $H_3 Z \sim T_1^{-1} Z$. Since $u_{59} - \epsilon_3 > u_6$ we may use 6.13 and obtain a contradiction.)

Next $s(P) - s(P_1) \leqslant \beta_3$ and $s(P_1) > \beta_4 > u_6 \geqslant u_3$. Now $P_1 \subset H_1 \sim C_3^{-1}$; let P' be a weak image of P_1 in C_3^{-1}. Then

$$s(P') > s(P_1) - u_4 > \beta_4 - u_4 > \mathrm{Max}(r_0 + \epsilon_3, \Delta + \rho + 3\epsilon_3 + \epsilon_1^*).$$

In $C_3^{-1} C_2^{-1} D_1 \equiv D_0$, P' and T^{R} are disjoint; let P^* be maximal normal in D_0 containing P'. Now

$$s(P^*) \geqslant s(P') - \epsilon_3 > s(P_1) - u_4 - \epsilon_3 \geqslant s(P) - \beta_3 - u_4 - \epsilon_3 =$$

$$= s(P) - u_{84}.$$

If $P^* = \overline{T^{\mathrm{R}}}$ then $P' \subset \overline{T^{\mathrm{R}}}$ so where $\overline{T^{\mathrm{R}}} \equiv UT^{\mathrm{R}}$ we have $P' \subset \dot{U}$ and

$$s(P') - \epsilon_3 + s(T^{\mathrm{R}}) \leqslant s(U) + s(T^{\mathrm{R}}) \leqslant s(\overline{T^{\mathrm{R}}}) + 2\epsilon_3 + \epsilon_1^*.$$

By Case 1′ of 3° of 15.3 and (**) $s(Z') \leqslant s(T^{\mathrm{R}}) + r'_0 + 2/e + \epsilon_3 + \rho + 2\epsilon_3 + \epsilon_1^*$ and $s(\overline{T^{\mathrm{R}}}) < u_4 + \epsilon_3 + q_{33} + 2/e + s(Z')$ so $s(\overline{T^{\mathrm{R}}}) < s(T^{\mathrm{R}}) + \Delta + \rho$. Thus $s(P') \leqslant \Delta + \rho + 3\epsilon_3 + \epsilon_1^*$, a contradiction.

Thus $P^* \neq \overline{T^{\mathrm{R}}}$. Also if P is maximal normal then $P \cap H_1$ and $P^* \cap C_3^{-1}$ are images of each other with respect to $H_1 \sim C_3^{-1}$.

7°. If Q is maximal normal and of size $> u_{79} + u_{84}$ then $s(Q^*) > u_{79}$; write P for Q^*. Then by 5° $\mathrm{Im}(P \cap H_1) = \mathrm{Im}(Q^* \cap H_1)$

so $P^* \cap C_3^{-1} = Q \cap C_3^{-1}$. Since a normal subword is contained in a unique maximal normal subword, $P^* = Q$. Thus $Q^{**} = Q$ and $|s(Q) - s(Q^*)| < u_{84}$. In particular since $u'_{79} \geqslant u_{79} + u_{84}$ and $u_{84} \leqslant u'_{84}$ we have (ii).

A similar argument with the P of 6° yields (iii).

8°. Finally consider (v). Here $BDZ \equiv Z'CDZ$. Since $s(Z') > u_{80} \geqslant u_{79}$ we may write P for Z' and then $P_1 \equiv Z' \cap H_1$ has size $\geqslant s(P) - \beta_3 > \beta_4 > u_1$. With respect to $H_1 \sim C_3^{-1}$ (which has end words I, W) we have $V \equiv P_1^0 \underset{n}{\sim} N$ say and $s(V) > s(P_1) - u_1$. V is a left subword of Z'; N is a left subword of C_3^{-1}. Thus $V \sim N$ is analogous to the situation of 2° in 15.3, where $C_1^{-1} \underset{n}{\sim} T_3$. Thus $N \underset{n}{=} V^{\mathrm{L}} E N^{\mathrm{R}}$ where the right side is (n)-bounded by $\mathrm{Max}(\lambda, \lambda') + u_4$ and $s(V^{\mathrm{L}}) > s(V) - 2\epsilon_3 - 8/e$; also N^{R} is not a subelement of $\mathsf{L}_1 \cup \cdots \cup \mathsf{L}_{n-1}$. Thus $s(V^{\mathrm{L}}) > s(P) - u_{81}$ where $u_{81} = \beta_3 + u_1 + 2\epsilon_3 + + 8/e$. Where $C_3^{-1} \equiv NM$, we have $BD \cdot T \underset{n}{=} C^{-1} \underset{n}{=} D_0 \equiv C'T^{\mathrm{R}}$ and $D_0 \equiv NMC_2^{-1}(C_1^{-1})^{\mathrm{L}} A T^{\mathrm{R}} \underset{n}{=} V^{\mathrm{L}} E N^{\mathrm{R}} M C_2^{-1} (C_1^{-1})^{\mathrm{L}} A T^{\mathrm{R}} \equiv D'_0$ say. Thus D'_0 is $((n-1))$-bounded by $\mathrm{Max}(\lambda, \lambda') + u_4$; by 6.10A it is n-bounded by $\mathrm{Max}(\lambda, \lambda') + 2u_4 \leqslant b_{25}$ and $(n+1)$-bounded by $\gamma + u_4$. Thus (i) still holds since $u_{83} + u_4 \leqslant u'_{83}$ and $u_{78} + \epsilon_3 + u_4 \leqslant u'_{78}$.

If Q is maximal $(n+1)$-normal in D'_0 of size $> u'_{79}$ then since $u'_{79} > u_6$ we have, with respect to $D_0 = D'_0$, that $\mathrm{Im}\,\mathrm{Im}\,Q = Q$ and $|s(Q) - s(\mathrm{Im}\,Q)| < u_4 + \epsilon_3$. Hence $s\,(\mathrm{Im} Q) > u'_{79} - u_4 - \epsilon_3 \geqslant u_{79} + u_{84}$, so $|s((\mathrm{Im}\,Q)^*) - s(\mathrm{Im}\,Q)| < u_{84}$. Since $u_{84} + u_4 + \epsilon_3 \leqslant u'_{84}$, we see (ii) still holds. Also (iii) still holds since $u'_{79} - u_{84} > u_6$.

15.3B. Corollary. *With the hypothesis of* 15.3 *and the further hypothesis of* 15.3A, *but with n replaced by* $n-1$ *throughout (so that, e.g.,* $ZT^{-1} \in \mathit{Rep}\ \mathsf{L}_n$*) let* $BDZ \equiv JX$ *for some* X, *where* J *is a subelement of* $\mathit{Rep}\ \mathsf{L}_{n+1}$ *of size* $> u_{96} \geqslant s_{30}$; *thus* J *contains subelements* $C_1, \ldots, C_a$ *of* L_n *as in Axiom* 30. *Then we may take* $C'T^{\mathrm{R}} \equiv J^{\mathrm{L}} K \overline{T}^{\mathrm{R}}$ *for some* K, *where* J^{L} *contains* $C_1, \ldots, C_{v+1}, \ldots, C_{a-v}$ *where* $(a-v)/t > (a-2v)/t > s(J) - u_{97}$ *and* t *is as in Axiom* 30.

Proof. $BDZ \underset{n}{=} BD \cdot T \underset{n-1}{=} C'T^{\mathrm{R}}$. Let $P_1, P_2, \ldots\ Q_1, Q_2, \ldots$ be divisions for $JX \equiv \overline{BDZ} \underset{n}{=} \overline{C'T^{\mathrm{R}}}$. There is a P_i of type n, otherwise $ZT^{-1} \underset{n-1}{=} I$, and $\overline{P}_i \equiv \overline{Z}$, $\overline{Q}_i \equiv \overline{T}^{\mathrm{R}}$. Also $P_1, \ldots, P_{i-1}$ have type at most $n-1$. Let $J \equiv J_1 J_2$ and let $C_{a-u}, \ldots, C_a$ be the C's in J_2. Then

$s(J_2) \geqslant (u+1)/t - \epsilon'_{30}$ and $|s(J) - a/t| < \epsilon_{30}$. We shall have $s(J_2) > u_{98} = \mathrm{Max}(u_3, u_4 + r'_0, u_1 + r'_0)$ if $(u+1)/t > \epsilon'_{30} + u_{98}$, which holds if we define the integer u by $u + 1 > t(\epsilon'_{30} + u_{98}) \geqslant u$; we note that $\epsilon'_{30} + u_{98} < u_{96} - \epsilon_{30} < a/t$ so $1 \leqslant a - u \leqslant a$.

Hence $J_2 \supset J_2^0 \sim J_2^{0\prime} \supset F$ where $s(F) > u_{98} - u_4 > r'_0$ and $s(J_2^0) > u_{98} - u_1 > r'_0$ hence neither $J_2^{0\prime}$ nor J_2^0 is a subelement of $L_1 \cup \cdots \cup L_n$. Hence $J_2^0, J_2^{0\prime} \not\subset \overline{Z}, \overline{T^{\mathrm{R}}}$ respectively. Let $J \equiv J_3 J_2^0 J_4$ and $C'T^{\mathrm{R}} \equiv LJ_2^{0\prime}M$. Then $J_3 \underset{n-1}{\simeq} L$. Now $J_3 \supset J_1 \supset C_1, \ldots, C_{a-u-2}$. Let $v = u + 4$. Then $(a-u-2)/t > (a-2v)/t > s(J) - \epsilon_{30} - 2(\epsilon'_{30} + u_{98} + 4/e) = s(J) - u_{97}$. Now $C_{a-u-2} \subset J_3 \underset{n-1}{\simeq} L$; let C'' be a weak image in L. Then by 6.26 with $H = C_{a-u-2}$ we have $C'T^{\mathrm{R}} = (\mathrm{In}\, J, C^{\mathrm{L}}_{a-u-2})E(C''^{\mathrm{R}}, \mathrm{Fin}\, T^{\mathrm{R}})$. The right side has the form $J^{\mathrm{L}}EP$ where $P \supset J_2^{0\prime}M \supset \overline{T^{\mathrm{R}}}$ and $J^{\mathrm{L}} \supset C_1, \ldots, C_{a-v}$.

15.3C. Note. Let JAJ' be $((n-1))$-bounded by $\mathrm{Min}(b_{25}, b_{31})$ where J, J' are subelements of Rep L_{n+1} of size $> u_{96}$. Then $JAJ' \underset{n-1}{=} J^{\mathrm{L}}YJ'^{\mathrm{R}}$ where the right side is $((n-1))$-bounded by $\mathrm{Max}(str^+, b_{\min}) + q_{31}$, and J^{L} contains $C_1, \ldots, C_{a-v}$ as in 15.3B, and similarly for J'^{R}.

Proof. Consider $JAJ' \underset{n-1}{=} M$, where M is $((n-1))$-bounded by $b_{\min}$ and consider J_1, J_2, J_3 as in the proof of 15.3B.

15.4. We now discuss simultaneous replacements in cyclic words; we leave the easy modification of the subsequent argument necessary to discuss linear words to the reader.

15.5. Definition. Let W be a cyclic word (n)-bounded by s_0 and $(n+1)$-bounded by $1 - z_{26}$. Let the max. $(n+1)$-normal subwords of size $> z_{26}$ be $N_1, \ldots, N_h$ $(h \geqslant 0)$. If $h = 1$ assume $L(N_1) < L(W)$. Let σ be a sequence $\sigma_1, \ldots, \sigma_h$ where each σ_i is 0 or 1 and $\sigma_i = 1$ implies $s(N_i) \geqslant z$ (z is as in Axiom 26; the set of N_i such that $\sigma_i = 1$ corresponds to the set $M_1, \ldots, M_k$ of Axiom 26, but we do not assume here that $k \geqslant 1$.). If $h = 0$ let σ be the empty sequence. Any cyclic word W_3 arising from the following construction is said to arise from W by *simultaneous replacement* of those N_i such that $\sigma_i = 1$. (W_3 is not unique.)

1°. Consider N_i; it may be written ABC where A, C are of maximal length such that $s(A), s(C) \leqslant r_0 + \epsilon_3$; B is called the *formal kernel* of N_i and we have $s(B) \geqslant s(N_i) - a_1 > z_{26} - a_1 =$
$= a_2 \geqslant \mathrm{Max}(t_{33}, r_0 + \epsilon_3, s_{33} + r_{33})$, where $a_1 = 2(r_0 + \epsilon_3) + 4/e$.

2°. Let $h \neq 0$. Let $K_1, ..., K_h$ be the formal kernels of $N_1, ..., N_h$; then $K_1, ..., K_h$ are disjoint. (If $C_1, ..., C_h$ are the kernels of $N_1, ..., N_h$ then $K_i \subset C_i$ $(i = 1, ..., h)$. Choose a word $K_i' \equiv K_i^{\mathrm{L}} E_i Z_i F_i K_i^{\mathrm{R}}$ as in Axiom 33′. Then $K_i \underset{n}{=} K_i'$ and $s(Z_i) > s(K_i) - r_{33} > s_{33}$, so there is a unique T_i such that $Z_i T_i^{-1} \in \mathrm{Rep}\, L_{n+1}$. Let W_i' arise from W by replacing each K_i by K_i'. Then W_1' is (n)-bounded by $s_0 + c_{33} \leqslant s_0 + s_{26}$ and $W \underset{n}{=} W_1'$. Let W_1 arise from W by replacing each K_i by K_i' and each A_i such that $L(A_i) \geqslant 2$ by a word A_i' of the form $A_i^{\mathrm{L}} D A_i^{\mathrm{R}}$ $(A_i^{\mathrm{L}} \not\equiv I, A_i^{\mathrm{R}} \not\equiv I)$ which equals A_i in G_n in such a way that W_1 is (n)-bounded by $s_0 + s_{26}$ (for example, each A_i' could be A_i).

W_1 has the form $(Z_1 B_1 \cdots Z_h B_h)$ and $W \underset{n}{=} W_1$.

Define Q_i to be Z_i if $\sigma_i = 0$, T_i if $\sigma_1 = 1$. Consider the linear word $Q_1 \cdot B_1 \cdot \cdots \cdot Q_h \cdot B_h$.

First let $h \geqslant 2$, and consider $Q_1 \cdot B_1 \cdot Q_2$. We prove this is equal in G_n to a word $U \equiv Q_1^{\mathrm{L}} Y_1 Q_2^{\mathrm{R}}$ where U is (n)-bounded by k_{51} and $s(Q_1^{\mathrm{L}}) > s(Q_1) - u_{82}$, $s(Q_2^{\mathrm{R}}) > s(Q_2) - u_{82}$. There are three cases (i) $Z_1 B_1 \cdot T_2$ (and its dual), (ii) $T_1 \cdot B_1 \cdot T_2$, (iii) $Z_1 B_1 Z_2$.

Case (i). Write $Z_1 B_1 Z_2 \equiv BDZ$ where $Z \equiv Z_2$, $DZ \equiv \overline{Z}_2$. Then $B \equiv Z_1' C$ where $Z_1' \equiv Z_1 \backslash \overline{Z}_2$. We wish to use 15.3 and 15.3A (v). Since W is $(n+1)$-bounded by $s_0 + u_4$ the hypothesis except (†) $s(Z) > s(DZ) - u_{57}$ are satisfied ($s_0 + s_{26} \leqslant k_{36}, k_{37}$; $a_2 - r_{33} \geqslant u_{59}$; $s_0 + u_4 \leqslant \beta_0$ where $\beta_0 = \mathrm{Max}(s_0 + u_4, u_{79} - u_{83} + 1/e^2) < 1 - u_{83}$; $s_0 + u_4 \leqslant u_{58}$; $a_2 - r_{33} - 2/e - r_0 - \epsilon_3 > u_{80}$.) To verify (†) note that $W \underset{n}{=} W_1$ induces $K_2 = K_2^{\mathrm{L}} E_2 Z_2 F_2 K_2^{\mathrm{R}}$; considering $\mathrm{Im}\, Z_2$: $Z_2 \supset Z_2^{0'} \sim Z_2^{0'} \supset B$, $\overline{B} = \mathrm{Im}\, Z_2$. Now $Z_2^{0'} \subset K_2$ so $B \subset K_2$ and $\mathrm{Im}\, Z_2 = \mathrm{Im}\, \overline{Z}_2 = \overline{K}_2 = N_2$. Hence $s(N_1) > s(\overline{Z}_2) - u_4 - \epsilon_3$ and $s(Z) > s(N) - a_1 - r_{33} > s(\overline{Z}) - u_{57}$ since $u_{57} \geqslant u_4 + \epsilon_3 + a_1 + r_{33}$.

Thus $Z_1 B_1 \cdot T_2 \underset{n}{=} Z_1'^{\mathrm{L}} K T_2^{\mathrm{R}}$ and the result follows since $u_{82} \geqslant u_{81}$, $u_{82} \geqslant u_{77}$ and $k_{50} \leqslant k_{51}$.

Case (ii). Write $Z_1'^{\mathrm{L}} K T_2^{\mathrm{R}}$ in the form ZDB where $Z_1'^{\mathrm{L}} \equiv Z$, $ZD \equiv \overline{Z}$ and let $Z'' \equiv T_2^{\mathrm{R}} \backslash \overline{Z}$.

We check the same hypotheses as in Case (i). $k_{50} < k_{36}, k_{37}$.

$$s(\bar{Z}_2) > s(Z_2) - \epsilon_3 > z - a_1 - r_{33} - \epsilon_3 = \beta_1.$$

$$\text{Max}(\beta_0 + u'_{83}, 1 + u'_{78} - \beta_1, u_{79} - u_{83} + 1/e^2) < 1 - u_{83}.$$

$$s(Z_1'^{\text{L}} > s(Z_1') - u_{81} > a_2 - r_{33} - 2/e - r_0 - \epsilon_3 - u_{81} = a_2 - \xi, \text{ say}$$

$$> \text{Max}(u_{59}, r_0 + \epsilon_3, s_{33}).$$

$s_0 + u_4 + \epsilon_3 \leqslant u_{58}$. $s(Z_1'^{\text{L}}) > (a_2 - z_{26} + s(N_1) - \xi$; but from $Z_1 B_1 Z_2 \underset{n+1}{=} Z_1'^{\text{L}} K T_2^{\text{R}}$ we have $s(\bar{Z}_1'^{\text{L}}) < v_3 + s(\bar{Z}_1) < v_3 + u_4 + \epsilon_3 + s(N_1)$; hence $s(\bar{Z}_1'^{\text{L}}) > s(Z_1'^{\text{L}}) - \eta$ where $\eta = v_3 + u_4 + \epsilon_3 + z_{26} + \xi - a_2$. Now $\eta \leqslant u_{57}$.$s(T) \geqslant 1 - 4/e - (s_0 + c_{33})$.

$$s(T_2^{\text{R}} \backslash \bar{Z}_1'^{\text{L}}) > s(T_2^{\text{R}}) - 2/e - r_0 - \epsilon_3 > s(T) - u_{77} - 2/e - r_0 - \epsilon_3.$$

$$1 - s_0 - u_{77} - 6/e - c_{33} - r_0 - \epsilon_3 \geqslant u_{80}.$$

Where $Z_1'^{\text{L}} \tau^{-1} \in \text{Rep}\, L_{n+1}$ we have $Z_1'^{\text{L}} \tau^{-1} \equiv Z_1 T_1^{-1}$ hence $T_1 \cdot B_1 \cdot T_2 = \tau^{\text{L}} K Z''^{\text{R}}$ for some K, where $s(Z''^{\text{R}}) > s(Z'') - u_{81}$ and $s(\tau^{\text{L}}) > s(\tau) - u_{77}$. If T_1 contains τ^{L}, then τ^{L} has the form T_1^{L}; now since $s(Z_1'^{\text{L}}) \leqslant s(Z_1) + \epsilon_3$ we have $s(\tau) - u_{77} > s(T_1) - u_{82}$ since $u_{82} \geqslant 4/e + 3\epsilon_3 + \epsilon_1^* + u_{77}$. If τ^{L} contains T_1 the corresponding result is true trivially. Finally Z''^{R} has the form T_2^{R} and has size

$$> s(T_2) - u_{81} - u_{77} - 2/e - r_0 - \epsilon_3 \geqslant s(T_2) - u_{82}$$

Case (iii) is easy and is left to the reader.

Thus $Q_i \cdot B_i \cdot Q_{i+1} \underset{n}{=} Q_i^{\text{L}} Y_i Q_{i+1}^{\text{R}}$ $(i = 1, ..., h)$ where Q_{h+1} means Q_1. Also $s(Q_i^{\text{L}}), s(Q_i^{\text{R}}) > s(Q_i) - u_{82}$ and $s(Q_i) > v_6$. By 5.29, with $a = u_{82}$, since $v_6 - u_{82} > r_0 + \epsilon_3$ and $v_6 \geqslant 2u_{82} + 4\epsilon_3 + \epsilon_1^*$ we have that $Q_i^{\text{L}}, Q_i^{\text{R}}$ intersect; say $Q_i \equiv F_i G_i H_i$, $Q_i^{\text{L}} \equiv F_i G_i$, $Q_i^{\text{R}} \equiv G_i H_i$. Write Q_i^{M} for G_i. Also $s(Q_i^{\text{M}}) > s(Q_i) - v_7$ where $v_7 = \text{Max}(u_{82} + r_0 + \epsilon_3 + 2/e, 2u_{82} + 3\epsilon_3 + \epsilon_1^* + 2/e)$.

Let $W_3 \equiv (Q_1^{\text{M}} Y_1 \cdots Q_h^{\text{M}} Y_h)$; this is the required W_3 in the case under consideration. If $\sigma_i = 0$ for some i then $Q_i^{\text{M}} \equiv Z_i^{\text{M}} \subset W_1$, so $W_1 \,\overline{\overline{n+1}}\, W_3$; define W_2 to be W_3 in this case. If $\sigma_i = 1$ for all i let

W_2 arise from W_1 by using the sequence 0, 1, 1, ..., 1. Then $W_1 \overline{\overline{n+1}} W_2$ and $W_2 \overline{\overline{n+1}} W_3$. Thus in both cases

(7) $$W \overline{\overline{n}} W_1 \overline{\overline{n+1}} W_2 \overline{\overline{n+1}} W_3$$

Now let $h = 1$. Here $W \equiv (N_1 C) \equiv (K_1 A_1)$ for some C and $W_1 \equiv (Z_1 B_1)$. Also $Q_1 \cdot B_1 \equiv Q_1 \cdot B_1 \cdot Q_1 \cdot Q_1^{-1} = Q_1^{\mathrm{L}} Y_1 Q_1^{\mathrm{R}} \cdot Q_1^{-1}$ and $Q_1^{\mathrm{R}} \cdot Q_1^{-1} \cdot Q_1^{\mathrm{L}} \equiv Q_1^{\mathrm{M}}$ so as before take $(Q_1^{\mathrm{M}} Y_1)$ for the required W_3. (7) follows if $\sigma_1 = 0$, but without further conditions on C (7) does not follow if $\sigma_1 = 1$. As an example of a condition on C sufficient for (7) see 15.6C.

Finally consider $h = 0$. For the required W_3 take any cyclic word (n)-bounded by s_0 such that there exists a cyclic word W_1 such that $W \overline{\overline{n}} W_1 \overline{\overline{n}} W_3$ and W_1 is (n)-bounded by $s_0 + s_{26}$ (e.g., $W_1 \equiv W_3 \equiv W$).

15.5A. Note. If W is a kth power in Π, i.e., $W \equiv P^k$ for some P, and if $\sigma_i = 1$ if and only if $s(N_i) > \frac{1}{2}$ then clearly at least one W_3 exists which is a kth power in Π.

15.6. We introduce the concept of a simultaneous t-replacement modulo a set of $(t+1)$-subelements; $t < n$.

Let W be a cyclic word (t)-bounded by s_0. Let $S_1, \ldots, S_u$ be disjoint subwords of W, where each S_i is a $(t+1)$-subelement of size $> u_{99}$ $(u \geqslant 0)$. Let the maximal t-normal subwords of size $> z_{26}$ in W be $N_1, \ldots, N_h$ $(h \geqslant 0)$. Let $\mathrm{Max}(u, h) \geqslant 2$. If $h \neq 0$ let σ be the sequence $\sigma_1, \ldots, \sigma_h$ where $\sigma_i = 0$ or 1 and if $\sigma_i = 1$ then $s(N_i) > r'$. If $h = 0$ let σ be the empty sequence. We shall define a cyclic word W^σ called the result of simultaneously replacing the N_i with $\sigma_i = 1$ modulo $S_1, \ldots, S_u$.

Call S_j *preadmissible* if it does not properly contain an N_i. (If $h = 0$ all S_j are preadmissible.) In this case we can write $S_j \equiv A S_j' B$ where A is the left subword of maximal length such that $s(A) \leqslant r_0'$ and dually for B. Call S_j' *admissible*; thus $s(S_j') \geqslant s(S_j) - 2r_0' - 4/e$. If $h \geqslant 1$ let the formal kernel of N_i be K_i and call $K_1, \ldots, K_h$ *admissible.* Let there be k admissible words; they are disjoint and $k \geqslant \mathrm{Max}(u, h) \geqslant 2$.

If A is a subelement of L_m where $s(A) > t_{33}$ make a choice, arbitrary but henceforth fixed, $f_m(A)$ of a word of the form $A^{\mathrm{L}}EZFA^{\mathrm{R}}$ as in Axiom 33′. Let W_1 arise from W by replacing each admissible S_j' by $f_{t+1}(S_j')$, each admissible K_i by $f_t(K_i)$. Now $S_j' \underset{t}{=} f_{t+1}(S_j')$ and S_j' is (t)-bounded by $s_0 < b_{33}$ so the right side is (t)-bounded by $s_0 + c_{33}$. Any P_i of type t would have size $> k_3^0 - s_0 - c_{33} > z_{26} + \epsilon_3 > r_0 + \epsilon_3$, hence $\bar{P}_i$ has the form N_g and S_j' would meet K_g. Thus $S_j' \underset{t-1}{=} f_{t+1}(S_j')$ and hence $W \underset{t-1}{=} W_1$. Write $f_{t+1}(S_j') \equiv S_j'^{\mathrm{L}} E_j' J_j F_j' S_j'^{\mathrm{R}}$, $f_t(K_i) \equiv K_i^{\mathrm{L}} E_i Z_i F_i K_i^{\mathrm{R}}$. As before let Q_i be Z_i or T_i according as σ_i is 0 or 1. Corresponding to $Q_1 \cdot B_1 \cdot Q_2$ in 15.5 we now have two cases

1. $J_1 F_1 S_1'^{\mathrm{R}} A K_2^{\mathrm{L}} E_2 \cdot Q_2$ (and its dual)
2. $J_1 F_1 S_1'^{\mathrm{R}} A S_2'^{\mathrm{L}} E_2' J_2$

(the case 3. $Q_1 \cdot F_1 K_1^{\mathrm{R}} A K_2^{\mathrm{L}} E_2 \cdot Q_2$ has already been discussed in 15.5). Since $s(J_j) > s(S_j') - r_{33} > u_{99} - 2r_0' - 4/e - r_{33} > u_{96}$ we have by 15.3B, 15.3C that the word 1 or 2 is equal in G_t to (1′) $J_1^{\mathrm{L}} Y_1 Q_2^{\mathrm{R}}$ or (2′) $J_1^{\mathrm{L}} Y_1 J_2^{\mathrm{R}}$. We recall that 3 equals (3′) $Q_1^{\mathrm{L}} Y_1 Q_2^{\mathrm{R}}$. Although Y_1 is not unique we now make any choice of Y_1 to be fixed henceforth.

In 15.5 we had that the subwords Q_i^{L}, Q_i^{R} of Q_i intersect, the intersection being Q_i^{M}. We now observe that J_i^{L}, J_i^{R} intersect, in J_i^{M} say, where $s(J_i^{\mathrm{M}}) > s(J_i) - u_{97} - \epsilon_{30}'$.

Put $W_3 \equiv (\xi_1^{\mathrm{M}} Y_1 \cdots \xi_k^{\mathrm{M}} Y_k)$ where ξ_i^{M} is Q_i^{M} or J_i^{M}. Then $W_1 \underset{t}{=} W_3$ unless $k = h$ and all σ_i are 1 when, as before, $W_1 \underset{t}{=} W_2 \underset{t}{=} W_3$. Thus in any case $W \underset{t-1}{=} W_1 \underset{t}{=} W_2 \underset{t}{=} W_3$. W_3 is the required W^{σ}.

15.6A. The word (1′) is a function of $S_1 A K_2$ and σ_2; the word (2′) is a function of $S_1 A S_2$ only; the word (3′) is a function of $K_1 A K_2$, σ_1 and σ_2. Indeed, writing (1′), (2′) or (3′) as $\xi_1^{\mathrm{L}} Y \xi_2^{\mathrm{R}}$, we may say that ξ_1^{L} is a function of S_1 or a function of K_1 and σ_1; ξ_2^{R} is a function of S_2 or a function of K_2, σ_2. Hence ξ_i^{M} is a function of S_i or of K_i and σ_i.

15.6B. Suppose $u \geqslant 3$ and consider thesubword $V \equiv (S_{i-1}, S_i, S_{i+1})$ of W. Let σ and σ' be such that W^{σ}, $W^{\sigma'}$ exist. Assume that if $N_v \subset V$ properly then $\sigma_v = \sigma_v'$ $(v = 1, \ldots, h)$. We show that W^{σ}, $W^{\sigma'}$ have a common subword.

Each S_j contains at least one admissible word (S_j' or K_i); let A_l, A_m, A_n be admissible words contained in S_{i-1}, S_i, S_{i+1} respectively; thus $1 \leqslant l < m < n \leqslant k$ without loss of generality. Then the subword $(\xi_l^M, \xi_m^M, \xi_n^M)$ is the same in W^σ and $W^{\sigma'}$.

Since S_{i-1}, S_i, S_{i+1} are disjoint, their iterated nice images with respect to $W - W_1 - W_2 - W_3$, say $S_{i-1}^*, S_i^*, S_{i+1}^*$, are disjoint, so $S_i^* \subset (\xi_l^M, \xi_m^M, \xi_n^M)$; hence S_i^* for σ is the same as S_i^* for σ'.

15.6C. Note. If $h = 1$, $W \equiv (N_1 C)$ and C contains a $(t + 1)$-subelement S of size $> u_{99}$ we can consider a replacement of N_1 modulo S. (cf. the case $h = 1$ of 15.5.).

15.6D. Clearly a simultaneous t-replacement modulo a set of $(t + 1)$-subelements is a simultaneous replacement (cf. 15.5).

15.6E. Informal remark. In 15.5, a simultaneous replacement in a cyclic word W is discussed by subdividing the word into linear pieces. Unfortunately for some applications this subdivision is not fine enough because there may be too few maximal normal subwords in W. This is the reason for the construction in 15.6.

15.7. We consider 7.8 and specify certain A-descendants to be called *preferred* A-descendants.

Take any A-descendants A_γ in Z_r where $n > r > 0$. This contains $B_{\gamma i}$ $(i = 1, ..., a)$ as in Axiom 30.

We define the *centre* $C(A_\gamma)$ of A_γ to be the formal kernel of A_γ if $s(A_\gamma) > \mathrm{Min}(1 - s_0 - r_{26}, r^* - r_{26}) = r^* - r_{26}$; otherwise $C(A_\gamma) = A_\gamma$.

The admissible $B_{\gamma i}$ (cf. 7.8) are $B_{\gamma\, 1+j}, ..., B_{\gamma\, a-j}$ where $j/t < 3u_{20} + 18/e$ (7.7B). If $C(A_\gamma) \not\equiv A_\gamma$ we can choose w such that $s(B_{\gamma\, 1}, B_{\gamma\, w}) > r_0 + 2\epsilon_3$, namely (cf. 7.9B) w is the least integer $\geqslant t(r_0 + 2\epsilon_3 + \epsilon'_{30})$. Then $(B_{\gamma\, w+2}, B_{\gamma\, a-w-1}) \subset C(A_\gamma)$ and $(a - 2w - 2)/t > r^* - r_{26} - \epsilon_{30} - 4/e - 2(r_0 + 2\epsilon_3 + \epsilon'_{30}) = b$ say. Since $3u_{20} + 18/e < (r_0 + 2\epsilon_3 + \epsilon'_{30}) + 1/e$ we have $j \leqslant w + 1$ hence $B_{\gamma\, w+2}, ..., B_{\gamma\, a-w-1}$ are admissible. Let $p = w + 1$ in this case; if $C(A_\gamma) = A_\gamma$ put $p = j$. Then $B_{\gamma\, 1+p}, ..., B_{\gamma\, a-p}$ are admissible and contained in $C(A_\gamma)$; also $(a - 2p)/t > \mathrm{Min}(b, u'_{20}/e)$.

Recalling 7.7, the iterated nice image of $(B_{\gamma\, 1+p}, B_{\gamma\, a-p})$ contains $(A_{\gamma\, 1+p+j}, A_{\gamma\, a-p-j})$. Call $A_{\gamma\, 1+p+j}, ..., A_{\gamma\, a-p-j}$ the *preferred*

$A_{\gamma i}$ or the *preferred first* A-*descendants* of A_γ. Note that $(a - 2p - 2j)/t > \text{Min}(b, u'_{20}/e) - 6u_{20} - 36/e$.

Thus the iterated nice image of $C(A_\gamma)$ contains all preferred $A_{\gamma\, i}$.

A *preferred A*-descendant of A_γ is defined by

1. a preferred first A-descendant of A_γ is a preferred A-descendant;

2. if X is a preferred A-descendant of A_γ then any preferred first A-descendant of X is a preferred A-descendant of A_γ. A *preferred A*-descendant of B_i is A_i or any preferred A-descendant of A_i. A *preferred A*-descendant is a preferred A-descendant of some B_i.

Two A-descendants of the form $A_{\gamma\, i}, A_{\gamma\, i+1}$ are called consecutive.

15.7A. Proposition. *The centres of all preferred A-descendants in* Z_r *are disjoint* $(r = 0, 1, ..., n-1)$.

Proof. This is true for $r = n - 1$. Assume it true for some r, $n - 1 \geqslant r > 0$. Let P, Q be preferred A-descendants in Z_{r-1} and different subwords of Z_{r-1}. If P is $A_{\gamma\, i}$ and Q is $A_{\gamma\, j}$ for some γ (then $i \neq j$) then clearly $C(P)$ is not $C(Q)$. Now let P be $A_{\gamma\, i}$ and Q be $A_{\delta\, j}$ where $\gamma \neq \delta$. By the induction hypothesis $C(A_\gamma)$, $C(A_\delta)$, and hence their iterated nice images, are disjoint. Hence P, Q are disjoint and so are their centres.

15.8. We specialize the definition of L_{n+1} as follows, by fixing certain choices which so far have been arbitrary.

Let $R \in \text{Rep}\, L_{n+1}$ be fixed. Put $R_n \equiv R$. Take a set $B_1, ..., B_m$ as in 7.3; these are subelements of L_n ($> r''$ or $> u_{11}$); if a choice is involved make an arbitrary but henceforth fixed choice, subject to (*) below.

If R does not contain a maximal n-normal subword $> z_{26}$ put $R_{n-1} \equiv R$ and $A_i \equiv B_i$ $(i = 1, ..., m)$.

Now let $N_1, ..., N_h$ be the maximal n-normal subwords of R of size $> z_{26}$ and let $h \geqslant 1$. Choose a sequence σ such that $\sigma_i = 1$ implies $s(N_i) > r'$. Modulo the empty set of $(n+1)$-subelements define R_{n-1} to be R^σ. We next describe a set S of disjoint n-subelements in R_{n-1}. Consider A-descendants A_i $(i = 1, ..., m)$. If the B_i

have size $> r''$ then since $r'' > z_{26}$ A_i has the form $\overline{Q_j^{\mathrm{M}}}$. For S take the formal kernels of $A_1, \ldots, A_m$.

If the B_i do not have size $> r''$ then $\sigma_i = 0$ for all i since $r' > r''$, so $R_n \underset{n-1}{-} K \underset{n-1}{-} R^\sigma$ for some K. Since the formal kernels K_i have size $> r_0 > u_{11}$ we may in this case assume $(*)$ each B_i is a K_j. Let $(Z_{i-1}^{\mathrm{M}}, Z_{i+1}^{\mathrm{M}}) \equiv Z_{i-1}^{\mathrm{M}} E Z_{i+1}^{\mathrm{M}}$. Then the iterated nice image of B_i in R^σ is contained in E hence so is A_i. Since the B_i are disjoint so are the A_i. For S take the A_i's, making a fixed choice.

Now let R_{n-2} arise by simultaneous $(n-1)$-replacement in R_{n-1} modulo S. Also write S_{n-1} for S.

If R_r has been defined for some r, $n - 2 \geqslant r > 0$, let S_r be the set of centres of all preferred A-descendants in R_r. (The members of S_r are disjoint.) Let R_{r-1} arise from R_r by a simultaneous r-replacement modulo S_r. Finally L_{n+1} is the set of all R_0 obtained by this construction.

§16

Informal Summary

The main difficulty in this section is the proof of Axiom 22 for $(n + 1)$. We also deal with Axioms 33′, 31B, 31C and several minor axioms.

N.B. In some sections of Section 16 the discussion is informal and details are left to the reader.

16.1. The reader may now verify that Axioms 26, 26″, 26A, 26L, 26^- and 39 hold for $(n + 1)$.

16.2. Denote by (X, Y, p) that part of Definition 13.1 after "$1 \leqslant p \leqslant n$". Thus if X, Y are cyclic words then $X \underset{n+1}{\cong} Y$ means $X \equiv Y$ or for some p, $1 \leqslant p \leqslant n + 1$, (X, Y, p) holds.

16.3. Note. Axioms 4, 8 and 28A holds for $(n + 1)$.

(If $X \underset{n+1}{-} Y$ choose p to be minimal such that $X \underset{p}{-} Y$. Then $X \equiv Y$ or, by considering the normal subwords $\overline{P_i}$, $\overline{Q_i}$ where P_i has type p, (X, Y, p) holds.)

16.3A. Proposition. *If* $X \underset{n+1}{\cong} Y$ *then a maximal* $(n+1)$*-normal subword* S *of one of* X, Y, *where* $s(S) > u_{10}^*$, *determines a maximal* $(n+1)$*-normal subword* $Im^* S = S'$ *in the other such that either*

$$|s(S) - s(S')| < v_3^* \quad or \quad |s(S) + s(S') - 1| \leqslant v_2^*.$$

Proof. Cf. 9.11, taking $v_2^* = v_2 + u_4 + \epsilon_3$, $v_3^* = v_3 + u_4 + \epsilon_3$ and $u_{10}^* = u'_{30}$.

16.3B. Proposition. *Let* $(S, T) \underset{p}{\hat{=}} (S', T')$ *where each side is a p-object,* $p < n$, *and is* (p)*-bounded by* k_{59}. *Let* S, T, S', T' *have size* $> u_{30}$. *Let* R *be an n-subelement of size* $> u_{91}$, *contained in* (S, T). *Then* (S', T') *contains an n-subelement of size* $> s(R) - u_{92}$.

Proof. We have $k_{59} \leqslant k_{18}, k_{19}$. Recall the proof of 9.11. Let S_1, T_1 be the kernels of S, T; then $(S, T) \equiv S_1 A T_1$ say. An associated preobject (Z, Z') of (S, T) contains A. Similarly $(S', T') \equiv S'_1 C T'_1$ and some associated preobject (Z'', Z^+) contains C. Suitable adjustments of these preobjects are equal:

(1) $$(Z_1, Z'_1) \underset{p}{=} (Z''_1, Z^+_1).$$

Some subword R^M of R is disjoint from S, T and is such that $s(R^M) > u_{91} - 2r'_0 - 4/e > u_3$. Thus $R^M \subset A \subset (Z_1, Z'_1)$. Since $k_{59} + c_{33} \leqslant b_{25}$ and $k_{59} + u_4 \leqslant b_{25}$, the right side of (1) contains a weak image V of R^M and $s(V) > s(R^M) - u_4$. Some subword R' of V is contained in C and is such that $s(R') > s(V) - 2r'_0 - 4/e$. Thus $R' \subset (S', T')$ and $s(R') > s(R) - u_{92}$ since $u_{92} = 4r'_0 + 8/e + u_4$.

16.3C. Corollary. *Assume further that* $s(R) > u_{93} \geqslant u_{91}$. *Then* (S', T') *contains a maximal normal subword* R'' *of size* $\geqslant s(R) - u_{94}$. *If* R_1, R_2 *are distinct maximal normal subwords each of size* $> u_{93}$ *then* $R''_1 \neq R''_2$.

Proof. $s(R^M) > u_{93} - 2r'_0 - 4/e > u_6$, *hence Im* R^M is maximal normal in the right side of (1) and of size $> u_{93} - 2r'_0 - 4/e - u_4 - \epsilon_3 = t$, say. Also $s(C \cap (\text{Im } R^M)) \geqslant t - 2r'_0 - 4/e > r_0 + \epsilon_3$ and the maximal normal subword R'' of (S', T') thereby determined has size

$\geqslant t - 2r_0' - 4/e - \epsilon_3$. Thus $s(R'') > y - u_{94}$ since $u_{94} = 4r_0' + 8/e + u_4 + 2\epsilon_3$ and the first part follows. The second part follows from 6.13.

16.3D. Note. Let $X \underset{n}{\cong} Y$ where X, Y are (n)-bounded by k_{59} and X is $(n+1)$-bounded by x where $\frac{1}{2} \leqslant x < h_0 - \epsilon_2 - 2/e - u_4$. Then Y is $(n+1)$-bounded by $x + u_4$.

Proof. Recall 6.10A and use 16.3B.

16.3E. Proposition. *Axiom* 11 *holds for* $(n+1)$.

Proof. Let $X \underset{n+1}{\cong} Y$ and $Y \underset{n+1}{\cong} Z$. We may assume (X, Y, p) and (Y, Z, q) otherwise it is trivial to deduce that $X \underset{n+1}{\cong} Z$. Say $q \geqslant p$. Then Y contains a maximal q-normal subword and by 16.3C it determines a maximal q-normal subword of X. Hence (X, Z, q), so $X \underset{n+1}{\cong} Z$.

16.3F. Proposition. *Let B arise from A by a simultaneous $(n+1)$-replacement, where A, B are $((n+1))$-bounded by k_{60}. Then $A \underset{n+1}{\cong} B$.*

Proof. We have $A \underset{n}{-} W_1 \underset{n+1}{-} W_2 \underset{n+1}{-} B$. By Axioms 28A, 11 for $n+1$ we have $A \underset{n+1}{\cong} B$.

16.3G. Proposition. *Axiom* 28B *holds for* $n+1$.

Proof. We are given a cyclic word X, $((n+1))$-bounded by d_{28}. X is 1-bounded by $d_{28} \leqslant 1 - z_{26}$ so by simultaneous replacement of all maximal 1-normal subwords of size $> \frac{1}{2}$ we obtain say X_1, 1-bounded by $\frac{1}{2} + r_{26} \leqslant str$ (cf. Axiom 26). Now $d_{28} \leqslant k_{60}$ and $\frac{1}{2} + r_{26} \leqslant k_{60}$ so $X \underset{1}{\cong} X_1$. Consider the following hypothesis $P(t)$.

There is a sequence $X, X_1, \ldots, X_t$ where X_i arises from X_{i-1} by simultaneous replacement of all maximal i-normals of size $> \frac{1}{2}$. (Write $X_{i-1}(\frac{1}{2})Y$). $(i = 1, \ldots, t)$; $X \underset{t}{\cong} X_t$; X_t is t-bounded by $\frac{1}{2} + r_{26}$ and $((t-1))$-bounded by s_0; if X is a kth power in Π then so is X_t.

Then $P(1)$ is true. Assume $P(t)$ holds for some t where $1 \leqslant t \leqslant n$. Since $\frac{1}{2} + r_{26} \leqslant s_0 \leqslant k_{59}$ and $d_{28} \leqslant k_{59}$, X_t is $(t+1)$-bounded by $d_{28} + u_4 \leqslant 1 - z_{26}$. Hence $X_t(\frac{1}{2})X_{t+1}$ say, where X_{t+1} is $(t+1)$-bounded by $\frac{1}{2} + r_{26}$ and (t)-bounded by s_0. By 16.3F since $d_{28}, s_0 \leqslant k_{60}$, $X_t \underset{t+1}{\cong} X_{t+1}$. Hence $X \underset{t+1}{\cong} X_t \underset{t+1}{\cong} X_{t+1}$. Hence $P(t+1)$ holds (cf. 15.5A).

Hence by $P(n+1)$ $X \underset{n+1}{\cong} X_{n+1}$ where X_{n+1} is $(n+1)$-bounded by $\frac{1}{2} + r_{26}$ and (n)-bounded by $s_0 \leqslant d_{28}$; if X is a kth power then so is X_{n+1}.

By Axiom 28B (for n) $X_{n+1} \underset{n}{\cong} X'_{n+1}$ where X'_{n+1} is (n)-bounded by $\frac{1}{2} + r_{26} + u_4 \leqslant k_{59}$. Now $s_0 \leqslant k_{59}$ so X'_{n+1} is $(n+1)$-bounded by $\frac{1}{2} + r_{26} + u_4 \leqslant str$. Also $X \underset{n+1}{\cong} X_{n+1} \underset{n+1}{\cong} X'_{n+1}$.

16.3H. Proposition. *Let $X \underset{n}{\cong} Y$ where X, Y are (n)-bounded by k_{61}. Let X contain at least four maximal n-normal subwords of size $> v_8$. Then W exists such that $X \underset{n}{-} W \underset{n}{-} Y$ and W is (n)-bounded by $Max(k_{61}, k'_{24}) + u_{40}$.*

Proof. SAS has the form (S, U_1, U_2, U_3, S) where the U_i are maximal normal with images V_i in TBT, so that $TBT \equiv (T, V_1, V_2, V_3, T)$. Now by 9.15 $SAS = S^{\mathrm{L}}E(V_1^{\mathrm{R}}, V_3^{\mathrm{L}})FS^{\mathrm{R}}$ where the right side is (n)-bounded by $\mathrm{Max}(k_{61}, k'_{24}) + u_{40}$. The subwords S^{L}, S^{R} of S overlap, say $S \equiv LMN$, $S^{\mathrm{L}} \equiv LM$, $S^{\mathrm{R}} \equiv MN$. Hence $(SA) \equiv (MNAL)$ and

$$MNAL = L^{-1} \cdot SAS \cdot S^{-1} \cdot L = L^{-1} \cdot LME(V_1^{\mathrm{R}}, V_3^{\mathrm{L}})F = ME(V_1^{\mathrm{R}}, V_3^{\mathrm{L}})F.$$

Thus $(SA) \underset{n}{-} (ME(V_1^{\mathrm{R}}, V_3^{\mathrm{L}})F)$. We now show that $(ME(V_1^{\mathrm{R}}, V_3^{\mathrm{L}})F) \underset{n}{-} (TB)$. Recalling 13.1 we have $S = \epsilon \cdot T \cdot \kappa$. Now $S^{\mathrm{L}}EV_1^{\mathrm{R}} = \epsilon \cdot (T, V_1)$ and $V_3^{\mathrm{L}}FS^{\mathrm{R}} = (V_3, T) \cdot \kappa$. Hence

$$(V_1^{\mathrm{R}}, V_3^{\mathrm{L}})FME = (V_1^{\mathrm{R}}, V_3^{\mathrm{L}}) \cdot V_3^{\mathrm{L}-1} \cdot (V_3, T) \cdot T^{-1} \cdot (T, V_1) \cdot V_1^{\mathrm{R}-1}$$

which is a linearization of TB.

16.3J. Proposition. *Axiom* 28 *holds for* $n + 1$.

Proof. 1°. Let $X \underset{n+1}{=} Y$, where the linear words X, Y are $((n+1))$-bounded by b_{28}, a_{28} respectively. Now $b_{28} = s_0$. Consider the set H, possibly empty, of $\bar{P}_i$, where P_i has type $n + 1$. Then by simultaneous replacement of all such $\bar{P}_i$ we have $X \underset{n+1}{\to} Z$ say. Clearly $Z \underset{n}{=} Y$ and since Z is (n)-bounded by s_0 we have by Axiom 28 (for n) $Z \underset{n}{\to} \cdots \underset{1}{\to} Y$. Hence $X \underset{n+1}{\to} Z \underset{n}{\to} \cdots \underset{1}{\to} Y$.

2°. Now let $X \underset{n+1}{\cong} Y$, where the cyclic words X and Y are $((n + 1))$-bounded by $b_{28} = s_0$ and a_{28} respectively.

Case 1. Not $(X, Y, n + 1)$

Then $X \underset{n}{\cong} Y$. Let $X \underset{n+1}{\to} X'$ by the zero simultaneous $(n + 1)$-replacement (i.e., either $\sigma_i = 0$ for all i or σ is empty). Then $X \underset{n}{-} W_1 \underset{n}{-} W_2 \underset{n}{-} X'$ where X' is (n)-bounded by s_0. Hence by Axiom 28A $X \underset{n}{\cong} X'$, so $X' \underset{n}{\cong} Y$. By Axiom 28 (for n)

(2) $$X' \underset{n}{\to} \cdots \underset{1}{\to} Y$$

so

(3) $$X \underset{n+1}{\to} X' \underset{n}{\to} \cdots \underset{1}{\to} Y$$

Case 2. $(X, Y, n + 1)$ holds.

Here $SAS \underset{n+1}{\hat{=}} TBT$ where $X \equiv (SA)$, $Y \equiv (TB)$. Adjustments of corresponding preobjects are equal: $S'AS'' \underset{n+1}{=} T'BT''$. Again consider the set H defined as in 1°, but for the new equation. By Axiom 40. $S' \in H$ if and only if $S'' \in H$. Now any maximal normal subword N except S of X determines a maximal normal N' in $S'AS''$; let $X \underset{n+1}{\to} X'$ by simultaneous replacement of all N such that $N' \in H$ together with S if $S' \in H$. Then X' is (n)-bounded by s_0 and $X' \underset{n+1}{\cong} X$, hence $X' \underset{n+1}{\cong} Y$; indeed $X' \underset{n}{\cong} Y$. Hence (1) holds and since $X \underset{n+1}{\to} X'$ also (2) holds.

3°. Given a cyclic word W $((n + 1))$-bounded by $b'_{25} \leqslant d_{28}$ there exists by 16.3G a word Y, $((n + 1))$-bounded by *str*, such that $W \underset{n+1}{\cong} Y$. By 2° $Y \underset{n+1}{\to} \cdots \underset{1}{\to} W$.

16.4. Definition. In Axiom 30 let b be the smallest integer $\leqslant t(r'_0 + \epsilon'_{30})$; then if $b < a$, (X_1, X_{b+1}) and (X_{a-b}, X_a) each have size $> (b + 1)/t - \epsilon'_{30} > r'_0$. Call X_i *central* for P if $3 + b \leqslant i \leqslant a - b - 2$.

Note: If $s(P) > q^*$ then central X_i exist since $(a - 2b - 4)/t$

$> q^* - \epsilon_{30} - 2(r'_0 + \epsilon'_{30}) - 4/e > 0$. If $P \subset A \sim B$ then P^0 (cf. 6.8) exists and contains all central X_i.

Recall that the word "central" has been used in a different sense in Section 10.

16.4A. Definition. Let V be a $(t+1)$-subelement of size $> q^*$, where $1 \leqslant t \leqslant n$. V contains t-subelements V_i as in Axiom 30; if they satisfy (b) denote by $p_{t+1}V$ any central V_i. If the V_i satisfy (a) there are no words $p_{t+1}V$; in this case call (V_i, V_{i+1}) where V_i and V_{i+1} are central a *protected t-object* of V. If some $p_{t+1}V$ exists consider any $J \equiv p_u \cdots p_t p_{t+1} V$ where $p_{u-1}J$ does not exist. Then $u \geqslant 2$ and J is a $(u-1)$-subelement. If $u \geqslant 3$ then the $u-2$-subelements J_i contained in J by Axiom 30 satisfy (a); call (J_i, J_{i+1}) where both terms are central a *protected* $(u-2)$*-object* of V. If $u = 2$ J is a 1-subelement; we say that V *contains a 0-object* (even though we shall not define "0-object".)

16.4B. Note. If V is as in 16.4A and $V \subset B \underset{t}{\sim} C$, it is easy to deduce that there is induced $V^{\mathrm{M}} \underset{u}{\sim} C^{\mathrm{M}}$ where either $u \neq 0$ and V^{M} contains a u-object O, where O is a protected u-object of V, and $O \underset{u}{\hat{=}} O' \subset C^{\mathrm{M}}$ for some u-object O', or $u = 0$ (hence $V^{\mathrm{M}} \equiv C^{\mathrm{M}}$) and V contains a 0-object.

16.4C. Proposition. *Let A be a word containing two q-objects $(S, T), (S', T')$, where $q \leqslant n$, such that they produce ϵ, κ. Let*

$$(4) \qquad (In\, A, (S, T)) \underset{n}{=} (In\, A, (S', T')) \cdot \kappa$$

Then the subwords $(S, T), (S', T')$ of A coincide.

Proof. It is sufficient to show T, T' coincide so assume say T is left of T'. By cancelling $(\mathrm{In}\, A, \hat{T})$ from (4) we obtain

$$(5) \qquad T \underset{n}{=} (T, T') \cdot \kappa.$$

Adjustments of associated preobjects are equal:

(6) $$(Z_1^* F_1 S^{\mathrm{R}}, T^{\mathrm{L}} EZ^*) \underset{q}{=} (Z_2^* F_2 S'^{\mathrm{R}}, T'^{\mathrm{L}} E'Z'^*)$$

where

(7) $$T' \cdot \kappa \cdot T^{-1} \underset{q}{=} (T'^{\mathrm{L}} E'Z'^*) \cdot (T^{\mathrm{L}} EZ^*)^{-1}$$

By (5) $I \underset{n}{=} (T, \hat{T}') \cdot T' \cdot \kappa \cdot T^{-1}$ hence by (7)

(8) $$T^{\mathrm{L}} EZ^* \underset{n}{=} (T, T'^{\mathrm{L}} E'Z'^*)$$

The left side of (8) is a q-subelement. Taking divisions for (8), any P_i of non-zero type has size $> r_0'$ so by 5.21 no P_i has type $> q$. Hence in (8), $\underset{n}{=}$ may be replaced by $\underset{q}{=}$. (8) now has the form $M \underset{q}{=} (M_1, M_2)$ where M_1, M_2 are distinct maximal q-normal subwords of size $> u_{10}$ and M is q-normal. By 6.19, $\mathrm{Im}' M_1 = M = \mathrm{Im}' M_2$ hence $M_1 = M_2$, a contradiction.

16.5. Proposition. *If AD, D and DC are $(n+1)$-urelements then so is ADC. Hence Axiom 22 holds for $n+1$.*

Proof. 1°. We may assume that D is normal. For let $D \equiv D'X$ where D' is $(n+1)$-normal and put $A' \equiv A$, $C' \equiv XC$. Then $AD'X \supset AD' \supset D'$ so AD' is an urelement. Then $AD \equiv A'D'X \supset A'D' \supset D'$ and $D'C \equiv DC$ so $A'D'$ and $D'C'$ are urelements. Thus if $A'D'C'$ is an urelement then so is ADC.

2°. We have

$$Z_n \underset{n}{\rightarrow} \cdots \underset{1}{\rightarrow} Z_0 \equiv DCX \equiv DH$$

and

$$U_n \underset{n}{\rightarrow} \cdots \underset{1}{\rightarrow} U_0 \equiv DYA \equiv DK$$

for some X, Y, H, K, where $Z_n \equiv (B^k)$, $U_n \equiv (B'^l)$ and $B, B' \in \mathrm{Rep}\, L_{n+1}$. Since $s(D) > r_0$, the situation is similar to that of 13.11. As in 13.11 we use the A–B notation for the Z-sequence; say $B_1, \ldots, B_m \subset Z_n$ and all descendants of say $B_1, \ldots, B_g$ in Z_0 are contained in D. Then we may assume $kg/m > r_0$. As in 13.11 we have by 11.8

$$Z_n \equiv D_n H_n, \qquad U_n \underset{n}{-} D_n K_n$$

and there is induced $U_n^{\mathrm{M}} \underset{n}{\sim} Z_n^{\mathrm{M}}$. Moreover $L(Z_n^{\mathrm{M}})/L(B) > \alpha$, where α is as defined in 13.11. Since $\alpha > u_{55}$ (P.C.) we have by 13.13 $Z_n^{\mathrm{M}} \equiv U_n^{\mathrm{M}}$ and $B \equiv B'$. Let the B's for U_n be $B'_1, \ldots, B'_{m'}$. Then for some $a \geqslant 1$ $B'_{i+a} \equiv B'_i$ for all i and $m' = lea$. Similarly $B_{i+a} \equiv B_i$ for all i and $m = kea$. But we may take $B_j \equiv B'_j$ for all j.

Write R_n for $(B_{w+s+2}, B_{g-s-w-1}) \subset Z_n^{\mathrm{M}}$ and R'_n for the corresponding subword of U_n^{M}; then $R'_n \equiv (B'_{w+s+1}, B'_{g-s-w-1})$. $U_n - D_n K_n$ induces $U_n^{\mathrm{M}} \equiv Z_n^{\mathrm{M}}$ and hence induces $R'_n \equiv R_n$. We have

$$Z_n \equiv R_n S_n, \qquad U_n \equiv R'_n T_n, \qquad \text{for some } S_n, T_n.$$

Also

$$Z_r \equiv D_r H_r, \qquad U_r \underset{r}{-} D_r K_r \qquad (r = 0, 1, \ldots, n)$$

where $D_0, H_0, K_0 \equiv D, H, K$ respectively.

3°. Assume inductively the following.

1. $Z_r \equiv R_r S_r$, $U_r \equiv R'_r T_r$, $R_r \equiv R'_r$, $R_r \subset D_r$. If U is any subword of R_r let $f_r(U)$ be the corresponding subword of R'_r.

2. $R_n \equiv (B_{1+h}, B_{g-h})$, $h = w + s + 1$; $R_{n-1} \equiv (A_{h+2}, A_{g-h-1})$; if $r \leqslant n-2$, R_r has the form (P_1, P_2, Q_1, Q_2) where P_1, P_2 are two consecutive preferred A-descendants of B_{3+h} and Q_1, Q_2 are two consecutive preferred A-descendants of B_{g-h-2}.

3. $U_r \underset{r}{-} D_r K_r$ induces $R'_r \equiv R_r$.

4. If $r < n$ the images under f_r of the preferred A-descendants (for the Z-sequence) contained in R_r are the preferred A-descendants (for the U-sequence) contained in R'_r.

5. $(R_n S_n R'_n T_n) \underset{n}{\rightarrow} \cdots \underset{r+1}{\rightarrow} (R_r S_r R'_r T_r)$

Call this induction hypothesis $P(r)$. Note that $P(n)$ is true.

4°. If $P(0)$ is true then the proposition follows. To see this note that $R_0 \subset D_0 \equiv D$, say $D \equiv X R_0 Y$; then $S_0 \equiv YHX$. By 3 $R'_0 T_0 \underset{0}{-} DK$ induces $R'_0 \equiv R_0$, hence $T_0 \equiv YKX$. Now by 5 $R_0 S_0 R'_0 T_0 \in L_{n+1}^{k+l}$ and $(R_0 S_0 R'_0 T_0) \equiv (R_0 YHXR'_0 YKX) \equiv (HDKD) \equiv (DKDH) \equiv$ $\equiv (DYADCX)$. Hence ADC is a subpowerelement of L_{n+1}; since it contains D it is an urelement.

Thus it remains to prove that if $P(r)$ holds for some r where $n \geqslant r > 0$ then $P(r-1)$ holds.

5°. This subsection, 5°, is mainly devoted to recalling some arguments from the proof of 11.7. Consider

$$\begin{aligned} &R_r S_r \equiv Z_r \underset{r}{\rightarrow} Z_{r-1} \equiv D_{r-1} H_{r-1} \\ (9) \qquad &R'_r T_r \equiv U_r \underset{r}{\rightarrow} U_{r-1} \underset{r-1}{=} D_{r-1} K_{r-1} \end{aligned}$$

In 11.7 we obtained from this an equation

$$(10) \qquad U_r^{\mathrm{M}} \underset{r}{=} U_r^{\mathrm{ML}} E_r Z_r^{\mathrm{M}} F_r U_r^{\mathrm{MR}}.$$

Having obtained (10) we defined D_r, H_r, K_r by $D_r \equiv Z_r^{\mathrm{M}}$, $H_r \equiv Z'$, $K_r \equiv F_r U_r^{\mathrm{MR}} U' U_r^{\mathrm{ML}} E_r$, where $Z_r \equiv (Z_r^{\mathrm{M}} Z')$, $U_r \equiv (U_r^{\mathrm{M}} U')$. Since $R_r \subset D_r$ and $U_r \underset{r}{=} D_r K_r$ induces $R'_r \equiv R_r$ we see that (10) induces $R'_r \equiv R_r$.

We recall how (10) was obtained. Z_r contains a t-object (T_1, T_2) and an s-object (S_1, S_2); U_r contains objects (T'_1, T'_2), (S'_1, S'_2) where $(T_1, T_2) \underset{t}{\hat{=}} (T'_1, T'_2)$ (producing ϵ_1, κ_1 say) and $(S_1, S_2) \underset{s}{\hat{=}} (S'_1, S'_2)$ (producing ϵ_3, κ_3 say).

(i) If $r < n$ then $(T_1, T_2) \subset V_1$, $(S_1, S_2) \subset V_f$ where V_1, V_f are the leftmost, rightmost A-descendants of $B_1, \ldots, B_g$ in Z_r.

(ii) If $r = n$ and $B_1, B_2, \ldots$ are disjoint and of size $> u_{11}$ (cf. 7.8) then for some u, v $(T_1, T_2) \subset B_u$, $(S_1, S_2) \subset B_v$.

(iii) If $r = n$ and $B_1, B_2, \ldots$ are maximal n-normal of size $> r''$ then for some u, v $(T_1, T_2) \equiv (B_u, B_{u+1})$, $(S_1, S_2) \equiv (B_v, B_{v+1})$.

By a *D-word* we mean in these three cases respectively an A-descendant of $B_1, \ldots, B_g$ in Z_r; a B_i; a subword (B_i, B_{i+1}). Write D_T for V_1, B_u or (B_u, B_{u+1}) and define D_S similarly. Note that (T_1, T_2) is a protected object of D_T in cases (i), (ii), while in case (iii) $(T_1, T_2) \equiv D_T$.

Let D_2, D_3, D_4 be D-words such that D_T, D_2, D_3, D_4, D_S are disjoint (each term being left of the next) and $D_3 \subset R_r$. D_3 contains an object, say a u-object (U_1, U_2) and in the language of 11.7 D_3, which we also denote by D_U, clears a u-path with respect to (9) in which (U_1, U_2) has images throughout; in particular there is an object (U'_1, U'_2) in U_r such that $(U_1, U_2) \underset{u}{\hat{=}} (U'_1, U'_2)$ (producing ϵ_2, κ_2 say).

Say $(T_1, S_2) \equiv T_1 A S_2$, $(T'_1, S'_2) \equiv T'_1 B S'_2$; we have $(T_1, S_2) = \epsilon_1 \cdot (T'_1, S'_2) \cdot \kappa_3$ and

(11) $$C \equiv T_1^* A S_2^* \underset{r}{=} T_1'^* B S_2'^* \equiv D$$

(where C, D are defined by (11)). In (11), T_1^*, $T_1'^*$ are such that

(12) $$T_1^{-1} \cdot \epsilon_1 \cdot T'_1 \underset{t}{=} (T_1^*)^{-1} \cdot T_1'^*$$

Thus $A \equiv (D_T)^{\mathrm{R}} A_1 D_2 A_2 D_3 A_3 D_4 A_4 (D_s)^{\mathrm{L}}$ for some A_i $(i = 1, \ldots, 4)$. Still recalling 11.7 we have $D_i \supset D_i^0 \sim D'_i$ $(i = 2, 3, 4)$, these being induced by (11); let $(T_1^* D_T^{\mathrm{R}})^0 \sim V$ so that V is a left subword of D. Then V is disjoint from D'_2 so $T_1'^* \subset D'_1$. Now $D \equiv D'_1 D'_2 B' D'_3 B'' D'_4 D'_5$ say and

(13) $$D \underset{r}{=} D'_1 D_2'^{\mathrm{L}} E((D_2^0)^{\mathrm{R}}, (D_4^0)^{\mathrm{L}}) F D_4'^{\mathrm{R}} D'_5$$

Since $D \equiv T_1'^* B S_2'^*$, $T_1'^* \subset D'_1$ and dually $S_2'^* \subset D'_5$ we have from (13)

(14) $$B = B^{\mathrm{L}} E(D_2^{0\mathrm{R}}, D_4^{0\mathrm{L}}) F B^{\mathrm{R}}$$

for some B^{L}, B^{R}. (14) is the required equation (10) ($B \subset (T'_1, S'_2) \subset U_r$, $(D_2^{0\mathrm{R}}, D_4^{0\mathrm{L}}) \subset A \subset Z_r$).

6°. Now $(U_1, U_2) \subset D_U \equiv D_3 \subset (D_2^{0\mathrm{R}}, D_4^{0\mathrm{L}})$ so by (13) since either $(U_1, U_2) \equiv D_U$ or (U_1, U_2) is a protected object of D_U there is a u-object $(\ddot{U}_1, \ddot{U}_2) \subset D$ and using 9.10 (i) if $(\ddot{U}_1, \ddot{U}_2)$, (U_1, U_2) produce ϵ_5, κ_5 then

(15) $$(\text{In } D, \ddot{U}_2) = D'_1 D_2'^{\mathrm{L}} E(D_2^{0\mathrm{R}}, U_2) \cdot \kappa_5$$

Let $(T_1, U_2) \equiv T_1 A_0 U_2$, $(T'_1, U'_2) \equiv T'_1 B_0 U_2$. Just as we obtained (11), so we may obtain

(16) $$T_1^* A_0 U_2^* = T_1'^* B_0 U_2'^*$$

where (12) holds and hence $U'_2 \cdot \kappa_2 \cdot U_2^{-1} = U_2'^* \cdot U_2^{*-1}$. Hence

(17) $T_1^* A_0 U_2 = T_1'^* B_0 U_2' \cdot \kappa_2.$

(U_1, U_2) has image $(\dot{U}_1, \dot{U}_2)$ in D with respect to (11). Moreover $(\dot{U}_1, \dot{U}_2) \subset D_u'$. Let $(U_1, U_2), (\dot{U}_1, \dot{U}_2)$ produce ϵ_4, κ_4. Then

(18) $(\text{In } C, U_2) = (\text{In } D, \dot{U}_2) \cdot \kappa_4$

and

(19) $(\text{In } C, U_2^{\text{L}}) \sim (\text{In } D, \dot{U}_2^{\text{L}})$ (end words I, K say)

is induced by (11) where $K = \dot{U}_2^{\text{R}} \cdot \kappa_4 \cdot U_2^{\text{R}-1}$ and $U_2 \equiv U_2^{\text{L}} U_2^{\text{R}}$, $\dot{U}_2 \equiv \dot{U}_2^{\text{L}} \dot{U}_2^{\text{R}}$. Since $(U_1, U_2) \subset \dot{D}_U$ and $(\dot{U}_1, \dot{U}_2) \subset D_U', \dot{D}_2^0$ is disjoint from (U_1, U_2) and D_2' is disjoint from $(\dot{U}_1, \dot{U}_2)$; thus in (19), D_2^0 is contained in the left side and D_2' is contained in the right side. Hence

$$(\text{In } D, \dot{U}_2^{\text{L}}) \quad = (\text{In } D, D_2'^{\text{L}}) E(D_2^{0\text{R}}, U_2^{\text{L}}) \cdot K^{-1}$$

and

(20) $(\text{In } D, \dot{U}_2) \cdot \kappa_4 = (\text{In } D, D_2'^{\text{L}}) E(D_2^{0\text{R}}, U_2)$

From (17), (18) $(\text{In } D, U_2') \cdot \kappa_2 = (\text{In } D, \dot{U}_2) \cdot \kappa_4$; but (U_1', U_2'), $(\dot{U}_1, \dot{U}_2)$ produce $\epsilon_2^{-1} \cdot \epsilon_4, \kappa_4 \cdot \kappa_2^{-1}$. By 16.4C, $\dot{U}_2$ is U_2'. It follows that $(\dot{U}_1, \dot{U}_2) \equiv (U_1', U_2')$ so $\kappa_2 = \kappa_4$. Thus the left side of (20) can be replaced by $(\text{In } D, U_2') \cdot \kappa_2$ and hence by (15) we have $(\text{In } D, \ddot{U}_2) \cdot \kappa_5^{-1} = (\text{In } D, U_2') \cdot \kappa_2$. Since $(\ddot{U}_1, \ddot{U}_2), (U_1', U_2')$ produce $\epsilon_5 \cdot \epsilon_2, \kappa_2 \cdot \kappa_5$ we have by 16.4C that $\ddot{U}_2$ is U_2'.

Since (14), i.e., (10), induces $R_r' \equiv R_r$ and hence (13) induces $R_r' \equiv R_r$ and since $D_U \equiv D_3 \subset R_r$ we have $f_r((U_1, U_2))$ is $(\ddot{U}_1, \ddot{U}_2)$ hence is (U_1', U_2'); in particular $(U_1, U_2) \equiv (U_1', U_2')$.

7°. Consider (1) and, recalling 11.7, let (U_1'', U_2'') say be the image of (U_1, U_2) in D_{r-1}; let (U_1''', U_2''') be the image of (U_1'', U_2'') in U_{r-1}. Then (U_1', U_2') is the image of (U_1''', U_2''') in U_r.

Thus in the Z-sequence, U_2 has image U_2'' in Z_{r-1} and in the U-sequence $U_2' = f_r(U_2)$ has image U_2''' in U_{r-1}.

8°. We prove that if J is maximal r-normal in R_r, and properly contained in it, and if $s(J) > r'$ then J is replaced in $Z_r \to Z_{r-1}$ if and only if $f_r(J)$ is replaced in $U_r \to U_{r-1}$.

$R_r \subset D_r \equiv Z_r^{\mathrm{M}} \equiv (D_2^{0\mathrm{R}}, D_4^{0\mathrm{L}})$. Assume $r < n$. Where $(V_1, V_f) \equiv$ $\equiv V_1 C_0 V_f$ we have $C_0 \supset R_r \supset J$. Now V_1, V_f are $(r+1)$-subelements and by 11.7 they clear paths with respect to (9). Thus J has image say J'' in D_{r-1}, J'' has image say J''' in U_{r-1} and J''' has image say J' in U_r. Also $J' \subset (T_1', S_2')$. Since $J \subset (D_2^{0\mathrm{R}}, D_4^{0\mathrm{L}})$, J has images $\ddot{J}$ in D by (13) and $\dot{J}$ in D by (11). Making use of right r-objects we can follow the argument in 6° with J in the rôle of U_2. We conclude that $f_r(J)$ is J'.

If $r = n$ then since $s(J) > r'$, all B_i are maximal normal of size $> r''$ and we obtain the same conclusion by a simpler direct argument.

Recall 12.5 and if J is replaced in $Z_r \to Z_{r-1}$ put $\theta = 1$, otherwise put $\theta = 0$. If J' is replaced in $U_r \to U_{r-1}$ put $\varphi = 1$, otherwise put $\varphi = 0$. Since $\mathrm{Im}'(J'') = J'''$ with respect to $U_{r-1} \underset{r-1}{=} D_{r-1} K_{r-1}$ so by 12.7 J'', J''' satisfy (0). Since $J \equiv J'$ we have $\theta + \varphi \equiv 0 \pmod 2$ and the result follows.

9°. Let the preferred A-descendants in R_r be P_i $(i \in I)$. Their cetres C_i $(i \in I)$ which we denote also by $C_a, C_{a+1}, \ldots, C_b$ are disjoint. The set $\underline{S}_r$ for Z_r contains $C_a, \ldots, C_b$ hence the iterated nice image of C_v $(a < v < b)$ in Z_{r-1} is equal in Π to the iterated nice image of $f_r(C_v)$ in U_{r-1}. Define R_{r-1} as the word generated by the preferred first A-descendants of P_v $(v = a+1, \ldots, b-1)$ and let R'_{r-1} be generated by the preferred first A-descendants of $f_r(P_v)$ $(v = a+1, \ldots, b-1)$. Then $R_{r-1} \equiv R'_{r-1}$. Next, since the iterated nice image of R_r in Z_{r-1}, and hence that of C_v $(a < v < b)$, is contained in $D_{r-1} \subset$ we have $R_{r-1} \subset D_{r-1}$.

Recall that $(U_1, U_2) \subset D_3 \subset R_r \subset (D_2^{0\mathrm{R}}, D_4^{0\mathrm{L}})$. In Case (i) D_3 is an A-descendant; we now restrict D_3 to be a preferred A-descendant, not the two leftmost or the two rightmost, in R_r. Making use of the available choice of (U_1, U_2) within D_3 we may say that $f_{r-1}(U_2'')$ is U_2'''; indeed $f_{r-1}((U_1'', U_2''))$ is (U_1''', U_2''') so $(U_1'', U_2'') \equiv (U_1''', U_2''')$. Since (U_1''', U_2'''), (U_1'', U_2'') correspond in $U_{r-1} - D_{r-1} K_{r-1}$ the following lemma implies that $U_{r-1} - D_{r-1} K_{r-1}$ induces $(U_1''', U_2''') \equiv (U_1'', U_2'')$. It follows that $R'_{r-1} \equiv R_{r-1}$ is induced since if X is any non-empty subword of R_{r-1} such that $f_{r-1}(X) \equiv X$ is induced, then $R'_{r-1} \equiv R_{r-1}$ is induced.

16.5A. Lemma. *Let*

(20) $$A_1A_2(U_1, U_2)A_4A_5 \underset{n}{=} B_1B_2(V_1, V_2)B_4B_5$$

induce

(21) $$A_2(U_1, U_2)A_4 \underset{u}{\sim} B_2(V_1, V_2)B_4$$

where $(U_1, U_2), (V_1, V_2)$ *are u-objects and with respect to* (21) $Im' \, U_i = V_i$ $(i = 1, 2)$. *Also let* $(U_1, U_2) \equiv (V_1, V_2)$. *Then* (1) *induces* $(U_1, U_2) \equiv (V_1, V_2)$.

Proof. If $X \subset (U_1, U_2)$ let $f(X)$ be the corresponding subword of (V_1, V_2); thus $X \equiv f(X)$. Now the two u-objects produce I, I so we have

(22) $$A_1A_2(U_1, U_2) \underset{n}{=} B_1B_2(V_1, V_2)\cdot I$$

Since Im′ $U_2 = V_2$ and $U_2 \equiv V_2$ we see that Case (b) of 6.16 applies hence by 6.25 $U_2^{\mathrm{M}} \underset{u-1}{\sim} V_2^{\mathrm{M}}$ (cf. Axiom 40). Let j be minimal such that $U_2^{\mathrm{M}} \underset{j}{\sim} V_2^{\mathrm{M}}$. If $j = 0$ then $(A_1, U_2^{\mathrm{M}}) \underset{n}{=} (B_1, V_2^{\mathrm{M}})$; also $U_2^{\mathrm{M}} \equiv V_2^{\mathrm{M}}$, so where $U_2 \equiv U_2^{\mathrm{L}} U_2^{\mathrm{M}} U_2^{\mathrm{R}}$ and $V_2 \equiv V_2^{\mathrm{L}} V_2^{\mathrm{M}} V_2^{\mathrm{R}}$ we have by (22) $U_2^{\mathrm{R}} \underset{n}{=} V_2^{\mathrm{R}}$. Since $U_2 \equiv V_2$, this implies that $U_2^{\mathrm{R}} \equiv V_2^{\mathrm{R}}$, otherwise a subword of U would equal I in G_n, contrary to Axiom 10. Thus $f(U_2^{\mathrm{M}})$ is V_2^{M} and the required result follows since (20) induces $U_2^{\mathrm{M}} \equiv V_2^{\mathrm{M}}$.

Now let $j \neq 0$; we shall obtain a contradiction. Take a P_i of type j; then $P_i'^{-1} X_{i-1}'^{-1} Q_i' X_i' \in \mathsf{L}_j$ (cf. 6.7). Let $\bar{P}_i, \bar{Q}_i$ be the maximal normals of U_2, V_2 respectively determined by P_i, Q_i. Now $f(\bar{P}_i) \subset V_2$ and $f(\bar{P}_i) \neq \bar{Q}_i$ by Axiom 40; say $f(\bar{P}_i)$ is left of $\bar{Q}_i$. Thus $V_2 \equiv V_2'(f(P_i'), Q_i') V_2''$ and $U_2 \equiv U_2' P_i' U_2''$ say. Now $(B_1, Q_i') V_2'' \equiv$ $\equiv (B_1, V_2) = (A_1, P_i') U_2'' = (B_1, Q_i') X_i' U_2''$ hence $V_2'' = X_i' \cdot U_2''$. But $(f(P_i'), Q_i') V_2'' \equiv P_i' U_2''$ so $(f(P_i'), Q_i') \cdot X_i' \underset{n}{=} P_i'$. By Axiom 24 $Q_i' X_i' P_i'^{-1} \equiv Q_i'^{\mathrm{L}} W (P_i'^{\mathrm{L}})^{-1} \equiv H$ say. Write $H \equiv KL$ where K is of least length such that $s(K) > \frac{1}{2} s(H)$; hence $K \cap Q_i'^{\mathrm{L}} \equiv N$ is normal and $(f(P_i'), K) = L^{-1}$. This has the form $(M_1, M_2) = M$, which leads to a contradiction as in 16.4C; for $f(P_i')$ and K are j-normal, but if $(f(P_i'), K)$ were a j-subpowerelement, then $(f(P_i'), N)$ and hence $(f(P_i'), Q_i')$ would be subpowerelements, contrary to $f(\bar{P}_i) \neq \bar{Q}_i$.

16.6. Let $Z \in \mathrm{Rep}(L_{n+1})$ and let Z contain maximal n-normal subwords B_i $(i = 1, \ldots, m)$ of size $> r'$. Let $Z \underset{n}{\rightarrow} C'$, say B_i replaced or not according as $\beta_i = 1$ or 0 and let $Z \underset{n}{\rightarrow} D'$ using $\beta_i' = 0$ or 1. Let $Z \underset{n}{\rightarrow} C' \underset{n-1}{\rightarrow} \cdots \underset{1}{\rightarrow} C$ and $Z \underset{n}{\rightarrow} D' \underset{n-1}{\rightarrow} \cdots \underset{1}{\rightarrow} D$. Now let

$$\alpha_i = \beta_i \quad (i = 1, \ldots, q)$$

$$\alpha_i = \beta_i' \quad (i = q + 1, \ldots, m)$$

Then $Z \underset{n}{\rightarrow} B'$ say by replacing B_i if $\alpha_i = 1$. Making use of the standard choices (15.5A) one can show

$$Z \underset{n}{\rightarrow} B' \underset{n-1}{\rightarrow} \cdots \underset{1}{\rightarrow} C^{\mathrm{M}} E D^{\mathrm{M}} F$$

16.6A. In particular if D is Z (cf. 16.3A) we have constructed an element of L_{n+1} of the form $C^{\mathrm{M}} E Z^{\mathrm{M}} F$. Let $C \equiv C^{\mathrm{M}} X$ say. Then $X \underset{n}{=} E Z^{\mathrm{M}} F$. Let $C^{\mathrm{M}} \equiv C_1 C_2 C_3$; then $C_3 E Z^{\mathrm{M}} F C_1$ is a subelement of L_{n+1} equal in G_n to the subword $C_3 X C_1$ of C. In this way Axiom 33′ may be shown to hold for $n + 1$.

16.7. Axiom 33″ is easily proved to hold for $n + 1$.

16.8. Consider Axiom 33⁺. Say $Z_1 \underset{n}{\rightarrow} \cdots \underset{1}{\rightarrow} ABCD$. Now if the equation $A \underset{n}{=} A^{\mathrm{L}} EZFA^{\mathrm{R}}$ arises from $Z'' \underset{n}{\rightarrow} \cdots \underset{1}{\rightarrow} AQ$ say (cf. 16.6A) then $Z'' \equiv ZT$ for some T. Since A is normal $Z_1 \equiv Z''$ i.e., ZT is a C.A. of Z_1. Similarly $Z'T'$ is a C.A. of Z_1.

16.9. Consider Axiom $\overline{33}$. We have $Z_n \rightarrow \cdots \rightarrow Z_0 \equiv WV \in \mathsf{L}_{n+1}$ and $W \underset{n}{=} W^{\mathrm{L}} D W^{\mathrm{R}}$ with divisions say $P_1, P_2, \ldots$ and $Q_1, Q_2, \ldots$. Let the P_i of type n be $H_1, H_2, \ldots, H_s$; Any Q_i of non-zero type is contained in D. By 12.15 Z_{n-1} contains maximal n-normal subwords $K_1, \ldots, K_s$. The image L_i of K_i in Z_n in the present case is $> r'$.

Let the maximal normal subwords of size $> r'$ in Z_n be B_i $(i = 1, \ldots, m)$ and in $Z_n \rightarrow Z_{n-1}$ let $\beta_i = 1$ if B_i is replaced, 0 otherwise. Now $L_i = B_{f(i)}$ $(i = 1, \ldots, s)$.

Let $Z_n \rightarrow Z'$ by replacing B_j if (i) $\beta_j = 1$ and j does not have the form $f(i)$, or (ii) $\beta_j = 0$ and $j = f(i)$. One then shows

$$Z_n \to Z' \to \cdots \to PW^{RR}\,VW^{LL}$$

$$W^L DW^R V \underset{n-1}{=} W^{LL} PW^{RR} V$$

Hence $W^{LL} PW^{RR} \equiv W'$ is a subelement of L_{n+1} equal in G_{n-1} to $W^L DW^R$ and $W^L DW^R$ may be written $W'^L D' W'^R$. By induction $W^L DW^R$ is a subelement of L_{n+1}.

16.10. Axioms 31B and 31C for $n+1$ read briefly:

31B. If $T \underset{n}{\sim} Y$ where T is an $(n+1)$-subelement then there is a word $T^L FY^R$.

31C. If $(S, T) \underset{n+1}{\sim} (S', T')$ where S, T, S', T' are maximal $(n+1)$-normal there is a word $S^L AT'^R$.

To prove 31B, T contains subelements U_i of L_n by Axiom 30 for $n+1$. If these are maximal normal let two of them be U, V. Let $U' = \mathrm{Im}'\,U$ $V' = \mathrm{Im}'\,V$;then $(U^R, V^L) \underset{n}{\sim} (U'^R, V'^L)$ and by Axiom 31C we obtain the desired result. If the U_i are not maximal normal then $U_1 \subset T \underset{n-1}{\sim} Y$, hence $U_1^0 \underset{n-1}{\sim} Y^M$ and we may use Axiom 31B.

Now consider 31C. Let the divisions be $P_1, P_2, \ldots$ and $Q_1, Q_2, \ldots$: If S does not have the form $\bar{P}_i$, P_i of type $n+1$ then $S^L \underset{n}{\sim} S'^L$ and we may use 31B. Similarly for T. There remains the case when $S = \bar{P}_i$ $T = \bar{P}_j$; P_i and P_j of type $n+1$.

Now $(P_i, P_j) \underset{n+1}{\sim} (Q_i, Q_j)$ and it is sufficient to show that a word $P_i^L GQ_j^R$ exists which $\underset{n+1}{=} X_{i-1}^{-1}, (Q_i, Q_j)$.

By Axiom 33′ for $n+1$

$$P_i = P_i^L E_1 Z_1 F_1 P_i^R \equiv P_i^*, \qquad Q_i = Q_i^L E_3 Z_3 F_3 Q_i^R \equiv Q_i^*$$

$$P_j = P_j^L E_2 Z_2 F_2 P_j^R \equiv P_j^*, \qquad Q_j = Q_j^L E_4 Z_4 F_4 Q_j^R \equiv Q_j^*.$$

Now $P_i^{*-1} X_{i-1}^{-1} Q_i^* X_i \in L_{n+1}$ and contains (Z_1^{-1}, Z_3) hence J, K exist such that $KZ_1^{-1} JZ_2 \in \mathrm{Rep}(L_{n+1})$ and $J = E_1^{-1} P_i^{L-1} X_{i-1}^{-1} Q_i^L E_3$. Similarly $Q_j^* X_j P_j^{*-1} \supset (Z_4; Z_2^{-1})$ and $K' Z_4 J' Z_2^{-1} \in \mathrm{Rep}(L_{n+1})$. Now

$$(Z_1 F_1 P_i^R, P_j^L E_2 Z_2) J'^{-1} \underset{n+1}{=} J(Z_3 F_3 Q_i^R, Q_j^L E_4 Z_4)$$

In the left side replace $Z_2J'^{-1}$ by $K'Z_4$. By 15.3 the result is equal in G_n to $Z_1^{\mathrm{L}}WZ_4^{\mathrm{R}}$. Hence

$$Z_1^{\mathrm{L}}WZ_4^{\mathrm{R}} \underset{n+1}{=} J(Z_3F_3Q_i^{\mathrm{R}}, Q_j^{\mathrm{L}}E_4Z_4)$$

$$P_i^{\mathrm{L}}E_1Z_1^{\mathrm{L}}WZ_4^{\mathrm{R}}F_4Q_j^{\mathrm{R}} \underset{n+1}{=} X_{i-1}^{-1}(Q_i^{\mathrm{L}}E_3Z_3F_3Q_i^{\mathrm{R}}, Q_j^{\mathrm{L}}E_4Z_4F_4Q_j^{\mathrm{R}})$$

The left side is $P_i^{*\mathrm{L}}WQ_j^{*\mathrm{R}}$. Now $P_i \underset{n}{=} P_i^*$ so $P_i^{*\mathrm{L}} = P_i^{\mathrm{L}}EP_i^{*\mathrm{LR}}$.

Similarly $Q_j^{*\mathrm{R}} = Q_j^{*\mathrm{RL}}FQ_j^{\mathrm{R}}$ and $P_i^{\mathrm{L}}EP_i^{*\mathrm{LR}}WQ^{*\mathrm{RL}}FQ_j^{\mathrm{R}}$ is the required word.

16.11. Axiom 24 is easily shown to hold for $n+1$.

16.12. Consider Axiom 18. First note that if $Z \in \mathrm{Rep}(L_{n+1})$ then one can find a sequence $Z_n \to \cdots \to Z_0$ from Z to Z, i.e., $Z_n \equiv Z_0 \equiv Z$ such that (i) every Z_i has the form X_i^e and (ii) for $k = 1, 2, 3, \ldots$

$$Z_n^k \underset{n}{\to} \cdots \underset{i+1}{\to} Z_i^k \underset{i}{\to} \cdots \underset{1}{\to} Z_0^k$$

is a sequence from Z^k to Z^k.

Now suppose $AB \in \mathsf{L}_{n+1}^k$, $s_{AB}(A) < h_0$. Then $Z^k \underset{n}{\to} \cdots \underset{1}{\to} AB$, $Z \in \mathrm{Rep}(L_{n+1})$. By Axiom 33′ for $n+1$,

$$B \underset{n}{=} B^{\mathrm{L}}EZ_1FB^{\mathrm{R}}, \qquad Z_1Z_2 \equiv Z^k.$$

Now $Z_1 \equiv Z^{k-1}Z^{\mathrm{L}}$, $Z \equiv Z^{\mathrm{L}}Z^{\mathrm{R}}$, $Z^{\mathrm{R}} \equiv Z_2$, and

$$Z^k \to \cdots \to Z^{\mathrm{LR}}FB^{\mathrm{R}}AB^{\mathrm{L}}EZ^{k-1}Z^{\mathrm{LL}}.$$

Hence $Z \to \cdots \to Z^{\mathrm{LR}}FB^{\mathrm{R}}AB^{\mathrm{L}}EZ^{\mathrm{LL}}$; thus A is a subelement of L_{n+1}.

16.13. The reader will easily verify that the following Axioms hold for $n+1$: 2, 3, 6, 7, 21, 31A.

16.14. Note that Condition D now holds.

§ 17

Informal Summary

This section is mainly concerned with the proof of Axioms 11a, 11b for $n + 1$. In the case of 11a, one would like simply to take W' to be a word of smallest length such that $W' = W$. The author does not know whether or not such a W' is $((n + 1))$-bounded by any parameter or in particular by 1, so in the proof given an alternative to "smallest length" has been used.

17.1. Proposition. *Let $A \subset X \underset{n}{\sim} Y$ where A is a subelement of L_{n+1} and $s(A) > u_{60}$. Let $A \supset A^0 \underset{n}{\sim} A^{0'} \supset B$ where B is a weak image of A. Let A, B be (n)-bounded by $b \leqslant Min(b_{25}, b_{31})$. Then there is a word $A^{\mathrm{L}} EB^{\mathrm{R}}$, (n)-bounded by $b + q_{31}$ where neither A^{L} nor B^{R} is a subelement of $\mathsf{L}_1 \cup \cdots \cup \mathsf{L}_n$.*

Proof. See the proof of 6.26 and take $u_{60} = v_4$.

17.1A. Informal note. We shall be interested in applying 17.1, 17.2 when the (n)-bound of X, Y is greater than the (n)-bound of A, B.

17.2. Proposition. *Let $A \subset X \underset{n}{\sim} Y$ where A is a subelement of L_{n+1} and $s(A) > u_{61}$. Let $A \supset A^0 \sim A^{0'} \supset B$ where B is a weak image of A. Let A, B be (n)-bounded by $b \leqslant b_{25}, b_{31}$. Then there is a word $A^{\mathrm{L}} EB^{\mathrm{R}}$, (n)-bounded by $b + q_{31}$ where $s(A^{\mathrm{L}}) > s(A) - u_{62}$ and B^{R} is not a subelement of $\mathsf{L}_1 \cup \cdots \cup \mathsf{L}_n$. Also*

$$(*) \qquad s(A) > s(B) - u_4 - \epsilon_3.$$

Proof. $s(B) > s(A) - u_4 > u_{61} - u_4 > u_3$. Hence by 6.10 $C \subset B^{0'} \sim \sim B^0 \subset B$, $s(C) > s(B) - u_4$. Also $s(C) > u_{61} - 2u_4 > r_0$. Now $C \subset B^{0'} \subset A^0 \subset A$, so $s(C) \leqslant s(A) + \epsilon_3$ and $(*)$ follows.

Now $s(C) \leqslant s(B^{0'}) + \epsilon_3$ hence $s(B^{0'}) \geqslant u_{61} - 2u_4 - \epsilon_3 > s_{31}$, $r_0 + r_{31}$ and there is a word $B^{0'\mathrm{L}} EB^{0\mathrm{R}}$, (n)-bounded by $b + q_{31}$ and equal in G_n to $B^{0'} \cdot K^{-1} = E \cdot B^0$ for end words E, K. $B^{0\mathrm{R}}$ is not a subelement of $\mathsf{L}_1 \cup \cdots \cup \mathsf{L}_n$ and $s(B^{0'\mathrm{L}}) > s(B^{0'}) - r_{31} > r_0$. Now $A \equiv ZB^{0'}T$, $B \equiv X'B^0Y'$say. The required word $A^{\mathrm{L}} EB^{\mathrm{R}}$ is $ZB^{0'\mathrm{L}} EB^{0\mathrm{R}'} Y'$

We have

$$s(A^{\mathrm{L}}) \geqslant s(B^{0'\mathrm{L}}) - \epsilon_3 > s(B^{0'}) - r_{31} - \epsilon_3 \geqslant S(A) - u_{62}$$

since $u_{62} = 2u_4 + 2\epsilon_3 + r_{31}$.

For later we require $u_{61} > u_{62} + r_0 + \epsilon_3$.

17.3. Notation. We use the notation α_i ($i = 1, 2, 3, \ldots$) for certain expressions which will be defined explicitly in the present section (Section 17).

17.4. Definition. Let U be a linear word (n)-bounded by k_{39}. Call a sequence of subwords $A_1, \ldots, A_m$ ($m \geqslant 1$) of U *linearly admissible* if

1. Each A_i is a subpowerelement of $\mathrm{Rep}(L_{n+1})$ of size $> u_{63}$.
2. $A_1, \ldots, A_m$ are pairwise disjoint.
3. (A_i, A_{i+1}) is not a subpowerelement of L_{n+1} ($i = 1, \ldots, m-1$).

17.4A. Definition. Let U be (n)-bounded by k_{39}. If U has no linearly admissible subwords put $D'_{n+1}(U) = 0$. Otherwise, put

$$D'_{n+1}(U) = \mathrm{Sup} \sum_{i=1}^{m} s(A_i)$$

(taken over all linearly admissible sequences).

17.4B. Note. $D'_{n+1}(U)$ is finite.

Proof. Take any linearly admissible sequence $A_1, \ldots, A_m$. By 7.11, $s(A_i) \leqslant L(A_i)/e$, hence $\Sigma s(A_i) \leqslant \Sigma L(A_i)/e \leqslant L(U)/e$.

17.4C. Note. If $D'_{n+1}(U) > 0$ then $D'_{n+1}(U) > u_{63}$.

17.5. Definition. Let $W \in \Pi$. By Axiom 11a, since $k_{39} \geqslant b_{\min}$,

$$(\exists W')\; W \underset{n}{=} W',\; W' \text{ is } (n)\text{-bounded by } k_{39}.$$

Hence $D'_{n+1}(W')$ exists; put

$$D_{n+1}(W) = \operatorname{Sup} D'_{n+1}(W').$$

17.5A. $D_{n+1}(W)$ is 0 or $> u_{63}$.

17.5B. If $W_1 \underset{n}{=} W_2$ then $D_{n+1}(W_1) = D_{n+1}(W_2)$.

17.5C. Proposition. *$D_{n+1}(W)$ is finite.*

Proof. It is sufficient to show that there are only finitely many W' as in 17.5. This follows from the next proposition since $b_{\min} \leqslant s_0$.

17.5D. Proposition. *If $W_1 \in \Pi$ and W_1 is (n)-bounded by s_0, then there are only finitely many W_2 such that W_2 is (n)-bounded by k_{39} and $W_1 \underset{n}{=} W_2$.*

Proof. Consider any W_2. Then by Axiom 28 ($s_0 \leqslant b_{28}$; $k_{39} \leqslant a_{28}$)

$$W_1 \equiv U_n \underset{n}{\rightarrow} \cdots \underset{1}{\rightarrow} U_0 \equiv W_2.$$

Now if $W_1 \underset{n}{\rightarrow} X$ and $W_1 \underset{n}{\rightarrow} Y$ by simultaneous replacement of the same set of maximal n-normal subwords of W_1 then $X \underset{n-1}{=} Y$; hence there is a finite set $X_1, \ldots, X_N$ such that $W_1 \underset{n}{\rightarrow} X_i$ $(i = 1, \ldots, N)$ and if $W_1 \underset{n}{\rightarrow} Y$ then $(\exists j)$ $Y \underset{n-1}{=} X_j$, $1 \leqslant j \leqslant N$. Say $U_{n-1} \underset{n-1}{=} X_j$. Now $U_{n-1} \underset{n-1}{=} W_2$, so $X_j \underset{n-1}{=} W_2$. Now X_j is $((n-1))$-bounded by s_0, W_2 is $((n-1))$-bounded by k_{39} so by an induction hypothesis there are only finitely many choices for W_2.

17.6. Proposition. *Let $R', C' \in \Pi$. Let U be E.R. and let the cyclic word (U) be (n)-bounded by k_{39}. Let $R' \cdot U \cdot R'^{-1} \underset{n}{=} C'$. Let $A_1, \ldots, A_s$ be disjoint subwords of (U) where each A_i is a subelement of Rep(L_{n+1}) of size $> u_{63}$ and (A_i, A_{i+1}) is not a subelement of L_{n+1} $(i = 1, \ldots, s)$ (where A_{s+1} means A_1). Let $\Sigma\, s(A_i) > \alpha_1$. Then*

$C' \underset{n}{=} XVY$,

XVY is (n)-bounded by k_{39},

V is a subword of (U),

$\Sigma s'(A_i \cap V) > \Sigma s(A_i) - \alpha_1$,

where $s'(X)$ means $s(X)$ if $s(X) > u_{63}$ and 0 otherwise.

17.6A. Note. $s'(X) \leqslant s(X) \leqslant s'(X) + u_{63}$. Also if AB is a subelement of L_{n+1} then $s'(AB) \leqslant s'(A) + s'(B) + 2u_{63} + 2/e$.

Before proving 17.6 we need a lemma.

17.6B. Lemma. *Let $A_1, \ldots, A_r, B_1, \ldots, B_s$ be disjoint subwords of X where A_r is left of B_1 (and 2.6 applies). Let X be (n)-bounded by k_{39}. Let $A_1, \ldots, A_r$ and also $B_1, \ldots, B_s$ be linearly admissible. Let*

$$a = \sum_1^r s(A_i), \qquad b = \sum_1^s s(B_j).$$

Then there is a word X', (n)-bounded by k_{39} and equal to X in G_n, and there is a linearly admissible sequence $C_1, \ldots, C_t$ for X' such that

$$c = \sum_1^t s(C_k) \geqslant a + b - u_{89} \quad (u_{89} = Max(r_0 + \epsilon_3, m_{33}, 3\epsilon_3 + \epsilon_1^*))$$

Hence $D_{n+1}(X) \geqslant a + b - u_{89}$.

Proof. Assume $(A_r, B_1) \equiv A_r E B_1$ is an $(n+1)$-subpowerelement, since otherwise the result is trivial.

If $s(A_r) \leqslant u_{89}$ then $A_1, \ldots, A_{r-1}, B_1, \ldots, B_s$ is admissible and $c = a + b - s(A_r) \geqslant a + b - u_{89}$. Similarly if $s(B_1) \leqslant u_{89}$.

Now let $s(A_r), s(B_1) > u_{89}$. Now $u_{89} > r_0'$. By Axiom 33″ for $n+1$ there is a subelement $A_r D B_1 \equiv J$ of Rep L_{n+1} equal in G_n to $A_r E B_1$. Where $X \equiv X_1 A_r E B_1 X_2$ put $X' \equiv X_1 A_r D B_1 X_2$. By 5.11, 5.32 X' is (n)-bounded by k_{39}. Also $X \underset{n}{=} X'$. Now $s(A_r) + s(DB_1) \leqslant s(A_r D B_1) + (2\epsilon_3 + \epsilon_1^*)$ and $s(B_1) \leqslant s(DB_1) + \epsilon_3$. Hence $s(A_r D B_1) \geqslant s(A_r) + s(B_1) - u_{89} > s(A_r) > u_{63}$. Thus $A_1, \ldots, A_{r-1}, J, B_2, \ldots, B_s$ is admissible and

$$c = a - s(A_r) + s(J) + b - s(B_1) \geqslant a + b - u_{89}.$$

Proof of 17.6. First note that $u_{63} \geqslant u_3, u_{60}; b_{\min}, k_{39} \leqslant b_{25}$ and $\mathrm{Max}(b_{\min}, str^+) + q_{31} \leqslant b_{25}, b_{31}$.

Put $N = \Sigma s(A_i)$; then $N > \alpha_1$.

1°. Take C such that $C' \underset{n}{=} C$ and C is (n)-bounded by $b_{\min}$ (Axiom 11a). There is at least one R such that

(∗) R is (n)-bounded by $b_{\min}$ and U' exists such that

U' is a C.A. of U and $R \cdot U' \cdot R^{-1} \underset{n}{=} C$.
Namely, take R to be equal to R' in G_n and (n)-bounded by $b_{\min}$. Of the R satisfying (∗) choose one such that $D_{n+1}(R) < \operatorname{Inf}_R D_{n+1}(R) + 1/e^3$.

Let $R \cdot U' \underset{n}{=} D$ where D is (n)-bounded by $b_{\min}$. We can apply 8.2 of Chapter I since $b_{\min}, k_{39} \leqslant 1 - 5(r + \epsilon_3) - 12/e$ (cf. the beginning of the proof of 15.3). Hence there are divisions $R_1 R_2 R_3$ of R, $U_1 U_2 U_3$ of U', $D_1 D_2 D_3$ of D where
$R_3 \underset{n}{\sim} U_1^{-1}$ (end words θ, I)
$U_3 \underset{n}{\sim} D_3$ (end words φ, I)
$R_1 \underset{n}{\sim} D_1$ (end words I, ψ).
As in 15.3 write $R \equiv R_1' R_2' R_3'$ and similarly for U' and D.

2°. We have $U_1' \subset (U)$. We shall obtain a contradiction from

$$\Sigma s'(A_i \cap U_1') > \alpha_2.$$

Let $P_i = A_i \cap U_1'$ and let the P_i of size $> u_{63}$ be $P_p, \ldots, P_q$ (since $\alpha_2 \geqslant 0$ there is at least one). Then $\Sigma_p^q\, s(P_i) = \Sigma\, s'(P_i) > \alpha_2$. Now $P_i \subset U_1' \subset U_1$ so we may consider a weak image P_i' of P_i in R_3^{-1} $(i = p, \ldots, q)$. Say $R_3^{-1} \equiv X' P_q' Y'$, $U_1 \equiv X P_q Y$. By 17.1 there is a word $P_q^{\mathrm{L}} E P_q'^{\mathrm{R}}$, (n)-bounded by $\operatorname{Max}(b_{\min}, str^+) + q_{31}$. Also, if $s(P_q) > u_{61}$ then by 17.2 $s(P_q^{\mathrm{L}}) > s(P_q) - u_{62}$. Put $\bar{R}_3 \equiv X P_q^{\mathrm{L}} E P_q'^{\mathrm{R}} Y'$; then $\bar{R}_3 \underset{n}{=} R_3^{-1}$. Let $R^* \equiv R_1 \cdot R_2 \cdot Y'^{-1} P_q'^{\mathrm{R}-1} E^{-1} P_q^{\mathrm{L}-1}$. Then

$$C \underset{n}{=} R^* \cdot X P_q Y \cdot U_2 \cdot U_3 \cdot X P_q^{\mathrm{L}} E P_q'^{\mathrm{R}} Y' \cdot R_2^{-1} \cdot R_1^{-1} \equiv R'' \cdot U'' \cdot R''^{-1}$$

where we let $P_q \equiv P_q^{\mathrm{L}} P_q^{\mathrm{R}}$, $R'' \equiv R_1 \cdot R_2 \cdot Y'^{-1} P_q'^{\mathrm{R}-1} E^{-1}$ and $U'' \equiv P_q^{\mathrm{R}} Y \cdot U_2 \cdot U_3 \cdot X P_q^{\mathrm{L}}$.

3°. We show that $D_{n+1}(R'') < D_{n+1}(R) - 1/e^3$.

We have $\bar{R}_3 \supset X P_q^{\mathrm{L}} \supset P_p, \ldots, P_{q-1}, P_q^{\mathrm{L}}$. Now $R \underset{n}{=} R^*$ and R^* is (n)-bounded by $\operatorname{Max}(\operatorname{Max}(b_{\min}, str^+) + q_{31}, b_{\min}, k_{39}) = \beta_1$ say.

Now $\beta_1 = k_{39}$. R'' is (n)-bounded by $\mathrm{Max}(b_{\min}, str^+) + q_{31}$. Hence (cf. 5.28A, 5.22)

$$D_{n+1}(R) \geqslant \begin{cases} \sum_p^{q-1} s(P_i) + s(P_q^{\mathrm{L}}), \text{ if } s(P_q^{\mathrm{L}}) > r_0 + \epsilon_3 \quad (r_0 + \epsilon_3 > u_{63}) \\ \sum_p^{q-1} s(P_i), \text{ otherwise.} \end{cases}$$

If $s(P_q) \leqslant u_{61}$ then $\Sigma_p^{q-1} s(P_i) > \alpha_2 - u_{61}$ and $D_{n+1}(R) > \alpha_2 - u_{61} > 1/e^3$.

If $s(P_q) > u_{61}$ then $s(P_q^{\mathrm{L}}) > s(P_q) - u_{62} > r_0 + \epsilon_3$; if also $s(P_q^{\mathrm{R}}) > r_0 + \epsilon_3$ then $s(P_q) - u_{62} + s(P_q^{\mathrm{R}}) < s(P_q^{\mathrm{L}}) + s(P_q^{\mathrm{R}}) < s(P_q) + {} + 2\epsilon_3 + \epsilon_1^*$. Thus $s(P_q^{\mathrm{R}}) < \mathrm{Max}(r_0 + \epsilon_3, u_{62} + 2\epsilon_3 + \epsilon_1^*) = \alpha_3$.

Now $2/e + s(P_q^{\mathrm{L}}) + s(P_q^{\mathrm{R}}) + \Sigma_p^{q-1} s(P_i) > \alpha_2$ hence

$$D_{n+1}(R) > \alpha_2 - 2/e - s(P_q^{\mathrm{R}}) > \alpha_2 - 2/e - \alpha_3 > 1/e^3;$$

thus the result is true if $D_{n+1}(R'') = 0$.

Now let $D_{n+1}(R'') > 0$. Then S' exists, (n)-bounded by β_1 and $R'' \underset{n}{=} S'$, and there is an admissible sequence $B_1, \ldots, B_k$ for S' such that

$$\sum s(B_i) > D_{n+1}(R'') - \epsilon, \; \epsilon = 1/e^2.$$

Let C_i be a weak image of B_i in R''. There is a word $B_k^{\mathrm{L}} E C_k^{\mathrm{R}}$ (n)-bounded by $str^+ + 2q_{31}$, since $b_{\min} \leqslant str^+$. Also if $s(B_k) > u_{61}$ then $s(B_k^{\mathrm{L}}) > s(B_k) - u_{62}$. Hence the word R^{**} where

$$R^{**} \equiv (\mathrm{In}\, S', B_k^{\mathrm{L}}) E (C_k^{\mathrm{R}}, \mathrm{Fin}\, R'')$$

is (n)-bounded by β_1 since $str^+ + 2q_{31} \leqslant k_{39}$. Also $R^{**} \underset{n}{=} R''$. Now $R^* = R'' P_q^{\mathrm{L}-1} X^{-1}$ so

$$R^{**} P_q^{\mathrm{L}-1} X^{-1} \underset{n}{=} R^* \underset{n}{=} R$$

where there is no dot on the left side and it is (n)-bounded by β_1. Consider the sequence $B_1, \ldots, B_{k-1} B_k^{0\mathrm{L}}, P_q^{0\mathrm{L}-1}, P_{q-1}^{-1}, \ldots, P_p^{-1}$ where

X^0 means X is omitted if $s(X) \leqslant r_0 + \epsilon_3$. By 17.6B,

$$D_{n+1}(R) \geqslant \overset{0}{\sum_{1}^{k-1}} s(B_i) + \overset{0}{s(B_k^{\mathrm{L}})} + \overset{0}{s(P_k^{\mathrm{L}})} + \sum_{p}^{q-1} s(P_i) - \alpha_4$$

where $\alpha_4 = u_{89}$.

From the first part of 3° follows

$$\overset{0}{s(P_q^{\mathrm{L}})} + \sum_{p}^{q-1} s(P_i) > \alpha_2 - \mathrm{Max}(u_{61}, 2/e + \alpha_3)$$

Now $(\Sigma_1^{k-1} s(B_i) + s(B_k^0)^{\mathrm{L}}) - \Sigma_1^k s(B_i) = s(B_k^{0\mathrm{L}}) - s(B_k)$; if $s(B_k) \leqslant u_{61}$ this expression is $\geqslant -u_{61}$, while if $s(B_k) > u_{61}$ then as before $s(B_k^{\mathrm{R}}) < \alpha_3$ and the expression is

$$s(B_k^{\mathrm{L}}) - s(B_k) \geqslant -s(B_k^{\mathrm{R}}) - 2/e > -\alpha_3 - 2/e.$$

Hence $D_{n+1}(R) \geqslant -\alpha_4 + \alpha_2 - 2\mathrm{Max}(u_{61}, 2/e + \alpha_3) + \Sigma_1^k s(B_i) = = \alpha_5 + \Sigma_1^k s(B_i)$ say. Now $D_{n+1}(R) > \alpha_5 + D_{n+1}(R'') - \epsilon \geqslant D_{n+1}(R'') + + 1/e^3$ since we shall define α_2 by the equation $\alpha_5 = \epsilon + 1/e^3$.

This proves that $D_{n+1}(R'') < D_{n+1}(R) - 1/e^3$.

4°. Let $R'' \underset{n}{=} T$ where T is (n)-bounded by $b_{\min}$. Then $C \underset{n}{=} T \cdot U'' \cdot T^-$ and $D_{n+1}(R) > D_{n+1}(R'') = D_{n+1}(T)$, a contradiction.

5°. Thus we have $\Sigma\, s'(A_i \cap U_1') \leqslant \alpha_2$. Now $N \leqslant \Sigma\, s(A_i) = \Sigma\, s'(A_i)$ and $\Sigma\, s'(A_i) \leqslant 3(2u_{63} + 2/e) + \Sigma\, s'(A_i \cap U_1') + \Sigma\, s'(A_i \cap U_2') + + \Sigma\, s'(A_i \cap U_3')$. But $s'(A_i \cap U_2') = 0$, since $2\epsilon_2 + 4/e + r_0' \leqslant u_{63}$ (cf. 1° of 15.3)

$$\Sigma s'(A_i \cap U_3') \geqslant N - 3(2u_{63} + 2/e) - \alpha_2 = N_1, \text{ say.}$$

Let $Q_i = A_i \cap U_3'$ and say $Q_r, \ldots, Q_s$ have size $> u_{63}$ (there is at least one since $\Sigma_r^s\, s(Q_i) \geqslant N_1 > \alpha_1 - 3(2u_{63} + 2/e) - \alpha_2 > 0$. Now $Q_i \subset U_3 \sim D_3$ so we may consider a weak image Q_i' of Q_i in D_3 and there exists a word $Q_r'^{\mathrm{L}} E Q_r^{\mathrm{R}}$, (n)-bounded by $str^+ + q_{31}$ and

$$D_3 \underset{n}{=} (\mathrm{In}\, D_3, Q_r'^{\mathrm{L}}) E(Q_r^{\mathrm{R}}, \mathrm{Fin}\, U_3) \equiv D'', \text{ say}$$

and $D \underset{n}{=} D_1 \cdot D_2 \cdot D'' \equiv D^*$, say. Hence $C \underset{n}{=} R \cdot U' \cdot R^{-1} \underset{n}{=} D \cdot R^{-1} \underset{n}{=}$
$\underset{n}{=} D^* \cdot R^{-1}$. Put $H \equiv R^{-1}$; then there are divisions $D_1^* D_2^* D_3^*$ of D^*, $H_1 H_2 H_3$ of H and $C_1 C_2 C_3$ of C and

$D_3^* \sim H_1^{-1}$ (end words θ, I)
$H_3 \sim C_3$ (end words φ, I)
$C_1 \sim D_1^*$ (end words I, ψ)

Now $-D^* \equiv -D_3^* \equiv -(Q_r^{\mathrm{R}}, \mathrm{Fin}\, U_3)$; let $V \equiv D_3^* \cap (Q_r^{\mathrm{R}}, \mathrm{Fin}\, U_3)$.
Then J, D_3^{**} exists such that $(Q_r^{\mathrm{R}}, \mathrm{Fin}\, U_3) \equiv JV$, $D_3^* \equiv D_3^{**} V$.

6°. Assume that $\Sigma_r^s\, s'(V \cap Q_i) > \alpha_2$; we shall obtain a contradiction.

Put $S_i = V \cap Q_i$ and say $S_q, \ldots, S_s$ have size $> u_{63}$; then $\Sigma_q^s\, s(S_i) > \alpha_2$. Now $S_i \subset V \subset D_3^*$; let S_i' be a weak image in H_1^{-1} of S_i. There is a word $S_q'^{\mathrm{L}} E S_q^{\mathrm{R}}$, (n)-bounded by $str^+ + q_{31}$ and

$$H_1^{-1} \underset{n}{=} (\mathrm{In}\, H_1^{-1}, S_q'^{\mathrm{L}})\, E(S_q^{\mathrm{R}}, \mathrm{Fin}\, D_3^*)$$

where the right side is (n)-bounded by β_1. Now $U' \equiv J'JV$ say, and $V \equiv V_1 V_0$ for some V_1, where $V_0 \equiv (S_q^{\mathrm{R}}, \mathrm{Fin}\, D_3^*)$. Thus $H_1^{-1} \underset{n}{=} TEV_0 \equiv \bar{R}_0$, say, where $T \equiv (\mathrm{In}\, H_1^{-1}, S_q'^{\mathrm{L}})$. Let $R_0'' \equiv H_3^{-1} H_2^{-1} TE$ and $R_0^* \equiv R_0'' V_0$; thus $R_0^* \supset \bar{R}_0$. Also $R \underset{n}{=} H^{-1} = H_3^{-1} H_2^{-1} H_1^{-1} = R_0^*$. Now $C \underset{n}{=} R \cdot U' \cdot R^{-1} = R_0'' V_0 \cdot J'JV_1 V_0 \cdot V_0^{-1} R_0''^{-1} = R_0'' \cdot V_0 J'JV_1 \cdot R_0''^{-1}$, where $U_0'' \equiv V_0 J'JV_1$ is a C.A. of U' hence of U.

We shall show $D_{n+1}(R_0'') < D_{n+1}(R) - 1/e^3$. The method is the same as before but it may help the reader to give a sketch.

$R_0^* \supset \bar{R}_0 \supset V_0 \supset S_q^{\mathrm{R}}, S_{q+1}, \ldots, S_s$. Next it can be shown that R_0^* is (n)-bounded by k_{39} and hence as before

$$D_{n+1}(R) > \min(\alpha_2 - u_{61},\, \alpha_2 - 2/e - \alpha_3) > 1/e^2,$$

so the result is true if $D_{n+1}(R_0'') = 0$. Now let $D_{n+1}(R_0'') > 0$. Then S_0' exists, (n)-bounded by k_{39} and $R_0'' \underset{n}{=} S_0'$ and there is an admissible sequence $B_1, \ldots, B_k$ for S_0' such that $\Sigma\, s(B_v) > D_{n+1}(R_0'') - 1/e^2$. Let C_i be a weak image of B_i in R_0''. Then there is a word $R_0^{**} \equiv (\mathrm{In}\, S_0', B_k^{\mathrm{L}})\, E(C_k^{\mathrm{R}}, \mathrm{Fin}\, R_0'')$ and $R_0^{**} \underset{n}{=} R_0''$ so $R_0^{**} V_0 \underset{n}{=} R_0^* \underset{n}{=} R$. $R_0^{**} V_0$ contains subwords $B_1, \ldots, B_{k-1}, B_k^{\mathrm{L}}, S_q^{\mathrm{R}}, S_{q+1}, \ldots, S_s$ and proceeding as before we again obtain the result.

As in 4° the result just obtained leads to a contradiction.

Hence we have $\Sigma_r^s s'(V \cap Q_i) \leqslant \alpha_2$.

7°. Now

$$s'(Q_r^{\mathrm{R}}) + \sum_{r+1}^{s} s'(Q_i) = \sum_{r}^{s} s'(Q_i) - s'(Q_r) + s'(Q_r^{\mathrm{R}})$$

$$\geqslant N_1 - s'(Q_r) + s'(Q_r^{\mathrm{R}})$$

and JV contains $Q_r^{\mathrm{R}}, Q_{r+1}, \dots; Q_s$ hence

$$s'(Q_r^{\mathrm{R}}) + \sum_{r+1}^{s} s'(Q_i) \leqslant \sum s'(J \cap Q_i) + \sum s'(V \cap Q_i) + 2u_{63} + 2/e.$$

Therefore $\Sigma\, s'(J \cap Q_i) \geqslant N_1 - s'(Q_r) + s'(Q_r^{\mathrm{R}}) - \alpha_2 - 2u_{63} - 2/e > 0$. Hence $J \not\equiv I$ so $(Q_r^{\mathrm{R}}, \mathrm{Fin}\, U_3) \equiv JD_3^*$. Now $D_1^* D_2^* D_3^* \equiv D^* \equiv$ $\equiv J''(Q_r^{\mathrm{R}}, \mathrm{Fin}\, U_3) \equiv J''JD_3^*$ so $D_1^* D_2^* \equiv J''J$. Write $J \equiv J_1 J_2$ where $J_2 \equiv J \cap D_2^*$. Then

$$\sum s'(J \cap Q_i) \leqslant \sum s'(J_1 \cap Q_i) + \sum s'(J_2 \cap Q_i) + 2u_{63} + 2/e.$$

But $s'(J_2 \cap Q_i) = 0$ so

$$\sum s'(J_1 \cap Q_i) \geqslant N_1 - s'(Q_r) + s'(Q_r^{\mathrm{R}}) - \alpha_2 - 4u_{63} - 4/e = N_2 \text{ (say)} > 0$$

Hence $J_1 \not\equiv I$ and J_1 has the form $(D_1^*)^{\mathrm{R}} \subset D_1^* \sim C_1$. Let $T_i \equiv J_1 \cap Q_i \equiv J_1 \cap A_i \cap U_3'$ and let the T_i of size $> u_{63}$ be $T_a, \dots, T_b$ (there is at least one). Then $\Sigma_a^b\, s(T_i) \geqslant N_2$. Let T_i' be a weak image of T_i in C_1. Then

$$C_1 C_2 C_3 \underset{n}{=} (\mathrm{In}\, D_1^*, T_b^{\mathrm{L}})\, E(T_b'^{\mathrm{R}}, \mathrm{Fin}\, C_1)\, C_2 C_3$$

where the right side is (n)-bounded by k_{39} and contains $(T_a, \dots, T_{b-1}, T_b^{\mathrm{L}}) = V$ say; it is the required XVY. Also $V \subset U_3' \subset (U)$.

Now

$$\sum s'(A_i \cap V) = \sum_{a}^{b-1} s'(T_i) + s'(T_b^{\mathrm{L}})$$

$$\geqslant N_2 - s'(T_b) + s'(T_b^{\mathrm{L}}).$$

Now if $s(T_b) > u_{61}$ then $s(T_b^{\mathrm{L}}) > s(T_b) - u_{62} > u_{61} - u_{62} > u_{63}$, hence $s(T_b^{\mathrm{L}}) = s'(T_b^{\mathrm{L}})$ and $s(T_b) = s'(T_b)$. Therefore $s'(T_b^{\mathrm{L}}) > s'(T_b) - - u_{62}$. If $s(T_b) \leqslant u_{61}$ then $s'(T_b) \leqslant u_{61}$. Hence $s'(T_b^{\mathrm{L}}) - s'(T_b)$ is greater than $-u_{62}$ or $0 - u_{61}$ so in either case it is greater than $-u_{61}$. Similarly, $s'(Q_r^{\mathrm{R}}) - s'(Q_r) > -u_{61}$. Therefore

$$\sum s'(A_i \cap V) \geqslant -2u_{61} + N_1 - \alpha_2 - 4u_{63} - 4/e$$

$$= N - \alpha_1$$

if we define α_1 to be $2\alpha_2 + 5(2u_{63} + 2/e) + 2u_{61}$.

17.7. Corollary. $D_{n+1}(C') \geqslant \Sigma\, s(A_i) - \alpha_1 - 2(r_0 + \epsilon_3)$.

Proof. This is trivial if the right side is negative so let $\Sigma\, s(A_i) > \alpha_1 + 2(r_0 + \epsilon_3)$. Let $P_i \equiv A_i \cap V$. Those P_i of size $> u_{63}$ have the form (i) $P_a, \ldots, P_b$. Also

$$\sum_a^b s(P_i) = \sum s'(P_i) > \sum s(A_i) - \alpha_1 > 2(r_0 + \epsilon_3)$$

From the sequence (i) delete P_a if it is not normal to obtain (ii) $P_{a+1}, \ldots, P_b$, where

$$\sum_{a+1}^{k} s(P_i) = \sum_a^b s(P_i) - s(P_a) \geqslant \sum_a^b s(P_i) - (r_0 + \epsilon_3) > r_0 + \epsilon_3.$$

From the sequence (i) or (ii) delete P_b if it is not normal. For the new sequence $\Sigma\, s(P_i) \geqslant \Sigma_a^b\, s(P_i) - 2(r_0 + \epsilon_3) > 0$ so it is non-empty; it is also admissible, so

$$D_{n+1}(C') = D_{n+1}(XVY) \geqslant \sum s(A_i) - \alpha_1 - 2(r_0 + \epsilon_3).$$

17.8. Definition. Put $W \in J_n(x)$ if W is E.R. and (W) is (n)-bounded by x, and $W^e \neq I$ in G_n.

17.8A. Note. By Axiom 11b, if $W \in \Pi$, $W^e \neq I$ in G_n then some conjugate U of W in G_n is in $J_n(k_{39})$, because $cb_{\min} \leqslant k_{39}$.

17.9. Definition. If $U \in J_n(k_{39})$ we define $B'_{n+1}(U)$ as follows. Call $A_1, \ldots, A_q$ *admissible* for U if

1. $A_1, \ldots, A_q$ are disjoint subwords of (U).
2. Each A_i is a subelement of $\mathrm{Rep}(L_{n+1})$ of size $> u_{63}$.
3. (A_i, A_{i+1}) is not a subelement of L_{n+1} $(i = 1, \ldots, q)$ $A_{q+1} = A_1$.

Put $B'_{n+1}(U) = \mathrm{Sup}\, \Sigma_{i=1}^{q}\, s(A_i)$ (or 0 if no admissible sequence exists).

17.9A. $B'_{n+1}(U)$ is finite.

17.10. Definition. Let $W \in \Pi$, $W^e \neq I$ in G_{n+1}. Define $B_{n+1}(W)$ to be $\mathrm{Sup}\, B'_{n+1}(U)$ where U is conjugate to W in G_n and $U \in J_n(k_{39})$.

17.10A. Proposition. *$B_{n+1}(W)$ is finite.*

Proof. Choose U_0 where U_0 is conjugate to W in G_n and $U_0 \in J_n(k_{39})$. Consider any $U \in J_n(k_{39})$ conjugate to W in G_n. Then $R' \cdot U \cdot R'^{-1} \underset{n}{=} U_0$ for some R'. Let $A_1, \ldots, A_s$ be admissible for U. Then by 17.6, 17.7 either $\Sigma\, s(A_i) \leqslant \alpha_1$ or

$$B_{n+1}(U_0) \geqslant \sum s(A_i) - \alpha_1 - 2(r_0 + \epsilon_3).$$

Thus $\Sigma\, s(A_i)$ is bounded above, as required.

17.10B. Note. If $W_1^e \neq I$ in G_{n+1} and W_1, W_2 are conjugate in G_n then $B_{n+1}(W_1) = B_{n+1}(W_2)$.

17.11. Proposition. *If $W^e \neq I$ in G_{n+1} then there exists K such that (i) K is conjugate to W in G_{n+1}. (ii) If X is conjugate to W in G_{n+1} then $B_{n+1}(K) \leqslant B_{n+1}(X) + 1/e^3$. (iii) $K \in J_n(k_{39})$.*

Moreover, for any such K the cyclic word (K) is $(n+1)$-bounded by k_{40}.

Proof. Let L be conjugate to W in G_{n+1}. Then $L^e \neq I$ in G_{n+1} and

$B_{n+1}(L)$ exists. Of all such L choose L' such that $B_{n+1}(L') < \text{Inf } B_{n+1}(L) + 1/e^3$. Thus L' is conjugate to W in G_{n+1}. Now $L'^e \neq I$ in G_n so some conjugate K of L' in G_n is in $J_n(k_{39})$ and $B_{n+1}(K) = B_{n+1}(L')$.

It remains to show that (K) is $(n+1)$-bounded by k_{40}. Assume not; then (K) contains S where S is an $(n+1)$-subelement of size $\geqslant k_{40}$, so no linearization of (K) is a subelement of $L_1 \cup \cdots \cup L_n$. By Axiom 28B, since $k_{39} \leqslant d_{28}$, $(K) \underset{n}{\cong} (K')$ where (K') is (n)-bounded by $str = k_{41}$. Also (K') contains an image S' of S where $s(S') > s(S) - u_{64}$. Say $(K') \equiv (S'T')$. Since $k_{40} - u_{64} > t_{33}$ and $str \leqslant b_{33}$ we have $S' \underset{n}{=} S'^{\mathrm{L}} EZFS'^{\mathrm{R}} \equiv S''$, where S'' is (n)-bounded by $k_{41} + c_{33} \leqslant k_{39}$ and by 5.32, so is $(S''T')$. Also $s(Z) > s(S') - r_{33} > -r_{33} - u_{64} + k_{40} = \alpha_6$ and $\alpha_6 > \text{Max}(u_{63}, \alpha_1 + 1/e^2 + 1/e^3)$. We have $(S''T') \equiv (AZ)$ for some A, and $ZT^{-1} \in \text{Rep}(L_{n+1})$ for some T. Thus $Z \underset{n+1}{=} T$. Let $A \cdot T \underset{n}{=} C$ where C is (n)-bounded by $b_{\min}$ (cf. Axiom 11a). We may write $C \equiv Q^{-1}C'Q$ where Q is chosen maximally. Now write $X \text{ conj}_n Y$ if X, Y are conjugate in G_n; then

$$W \text{ conj}_{n+1} K \text{ conj}_n K' \equiv S'T' \underset{n}{=} S''T' \text{ conj}_\pi AZ \underset{n+1}{=} A \cdot T \underset{n}{=} C \text{ conj}_\pi C'.$$

Thus $W \text{ conj}_{n+1} C'$. Hence $B_{n+1}(K) < B_{n+1}(C') + 1/e^3$.

Now $K \text{ conj}_n AZ \in J_n(k_{39})$ and Z is admissible so $B_{n+1}(K) \geqslant s(Z)$. Therefore $B_{n+1}(C') > 0$. Hence $B_{n+1}(C') = \text{Sup } B'_{n+1}(U)$ over U such that $U \text{ conj}_n C'$ and $U \in J_n(k_{39})$. Thus U exists with an admissible sequence for U, say $\{A_i\}$, such that

$$\sum s(A_i) > B_{n+1}(C') - 1/e^2 \geqslant B_{n+1}(K) - 1/e^3 - 1/e^2$$

$$\geqslant s(Z) - 1/e^3 - 1/e^2 \geqslant \alpha_1.$$

Since $C' \text{ conj}_n U$ we have by 17.6 that $C' \underset{n}{=} XVY$, (n)-bounded by k_{39}, where $V \subset (U)$ and $\Sigma\, s'(A_i \cap V) > \Sigma\, s(A_i) - \alpha_1$. Put $P_i \equiv A_i \cap V$; then $P_j, \ldots, P_k$, say, have size $> u_{63}$ and $\Sigma_j^k\, s(P_i) > \Sigma\, s(A_i) - \alpha_1$. C' is (n)-bounded by $b_{\min}$; let P'_i be a weak image in C' of P_i. We have say $C' \equiv C'_1(P'_j, P'_k)\, C'_2$. If $j < k$

$$C' \underset{n}{=} C'_p P'^{\mathrm{L}}_j E(P^{\mathrm{R}}_j, P^{\mathrm{L}}_k) F P'^{\mathrm{R}}_k C'_2 \equiv C''.$$

This is true also when $j = k$ if we interpret $(P^{\mathrm{R}}_j, P^{\mathrm{L}}_k)$ to be P^{M}_j. C'' is (n)-bounded by k_{39}. Now $A \cdot T \underset{n}{=} C \equiv Q^{-1} C' Q \underset{n}{=} Q^{-1} C'' Q$ (no dots and $D \equiv Q^{-1} C'' Q$ is (n)-bounded by k_{39}. There are divisions $A_1 A_2 A_3$ of A, $T_1 T_2 T_3$ of T and $D_1 D_2 D_3$ of D and $A_3 \sim T_1^{-1}$, $T_3 \sim D_3$, $A_1 \sim D_1$. Let $\pi_j, ..., \pi_k$ mean $P^{\mathrm{R}}_j, P_{j+1}, ..., P_{k-1}, P^{\mathrm{L}}_k$ and let $Q_i \equiv \pi_i \cap D_3$.

Assume $\Sigma\, s'(Q_i) > s(T) + \alpha_7 > 0$; we shall obtain a contradiction.

Let the Q_i of size $> u_{63}$ be say $Q_l, ..., Q_m$. Then $\Sigma_l^m\, s(Q_i) > s(T) + \alpha_7$. Let Q'_i be a weak image of Q_i in T_3. Then

(1) $T_3 \underset{n}{=} (\mathrm{In}\ T_3, Q'^{\mathrm{L}}_l) E(Q^{\mathrm{R}}_l, Q^{\mathrm{L}}_m) F(Q'^{\mathrm{R}}_m, \mathrm{Fin}\ T_3) \equiv T''_3$, say

(if $l = m$ interpret $(Q^{\mathrm{R}}_l, Q^{\mathrm{L}}_m)$ as Q^{M}_l). T''_3 is (n)-bounded by k_{39}. Now

$$Z^{-1} T \equiv Z^{-1} T_1 T_2 T_3 \underset{n}{=} Z^{-1} T_1 T_2 T''_3$$

and by 5.32 the last word is (n)-bounded by k_{39}. By Axiom 28 since $str^+ \leqslant b_{28}$ and $k_{39} \leqslant a_{28}$ we have $Z^{-1} T \underset{n}{\rightarrow} \cdots \underset{1}{\rightarrow} Z^{-1} T_1 T_2 T''_3$ hence $Z^{-1} T_1 T_2 T''_3 \in \mathsf{L}_{n+1}$. Hence $(Q^{\mathrm{R}}_l, Q^{\mathrm{L}}_m)$ is a subelement of L_{n+1}. If $l + 2 < m$ then $(Q_{l+1}, Q_{l+2}) \equiv (A_{l+1}, A_{l+2})$ would be a subelement of L_{n+1}; hence $l + 2 \geqslant m$. Let

(2) $$\lambda = s'(Q^{\mathrm{R}}_l) + \sum_{l+1}^{m-1} s'(Q_i) + s'(Q^{\mathrm{L}}_m)$$

If $m - l = 2$ then either Q^{R}_l nor Q^{L}_m is normal. If $m - l = 1$ then not both of Q^{R}_l, Q^{L}_m are normal. Hence one of the summands H of (2) satisfies

$$\lambda \leqslant 2(r_0 + \epsilon_3) + s'(H).$$

Now

$$\lambda > s'(Q^{\mathrm{R}}_l) + s'(Q^{\mathrm{L}}_m) + s(T) + \alpha_7 - s'(Q_l) - s'(Q_m)$$

$$> s(T) + \alpha_7 - 2u_{61}$$

so

$$s'(H) - s(T) > \alpha_7 - 2u_{61} - 2(r_0 + \epsilon_3)$$

$$> (r_0 + \epsilon_3) + u_4.$$

Now $s(T) \geqslant 0$, so in particular $s'(H) = s(H)$. From (1), T_3 contains a weak image H' of H and $s(H') > s(H) - u_4 > r_0 + \epsilon_3$. Hence

$$s(T) \geqslant s(H') - \epsilon_3 > s(H) - u_4 - \epsilon_3 = s'(H) - u_4 - \epsilon_3.$$

Hence, defining α_7 to be $2u_{61} + 3(r_0 + \epsilon_3) + u_4 + \epsilon_3$ we obtain a contradiction. Hence $\Sigma\, s'(Q_i) \leqslant s(T) + \alpha_7$. We have

$$\sum s'(\pi_i) \leqslant \sum s'(\pi_i \cap D_1) + \sum s'(\pi_i \cap D_2) + \sum s'(\pi \cap D_3) + 2(2u_{63} + 2/\epsilon$$

$$\leqslant \sum s'(\pi_i \cap D_1) + s(T) + \alpha_7 + 2(2u_{63} + 2/e).$$

Now

$$\sum s'(\pi_i) > s'(P_j^{\mathrm{R}}) + s'(P_k^{\mathrm{L}}) + \sum s(A_i) - \alpha_1 - s(P_j) - s(P_k)$$

$$> \sum s(A_i) - \alpha_1 - 2u_{61}.$$

Hence

$$\sum s'(\pi_i \cap D_1) > -s(T) - \alpha_7 - 2(2u_{63} + 2/e) - \alpha_1 - 2u_{61} + \sum s(A_i)$$

$$= \sum s(A_i) - s(T) - \alpha_8 \ (= \mu \text{ say})$$

where

$$\alpha_8 = 2(2u_{63} + 2/e) + \alpha_1 + 2u_{61} + 2u_{61} + 3(r_0 + \epsilon_3) + u_4 + \epsilon_3.$$

Now $\Sigma\, s(A_i) > s(Z) - e^{-2} - e^{-3}$ and $s(Z) > \alpha_6$

$$s(T) + s(Z) \leqslant s(ZT^{-1}) + r_0 + 2\epsilon_3 \leqslant 1 + r_0 + 2\epsilon_3$$

so

$$\sum s(A_i) - s(T) > -1/e^2 - 1/e^3 + s(Z) - s(T)$$

$$> -1/e^2 + 2s(Z) - 1 - (r_0 + 2\epsilon_3) - 1/e^3$$

$$> 2\alpha_6 - 1 - (r_0 + 2\epsilon_3) - 1/e^2 - 1/e^3 = \alpha_9 > \alpha_8.$$

Thus $\mu > 0$. Let $R_i = \pi_i \cap D_1$. Let the R_i of size $> u_{63}$ be, say, $R_u, ..., R_v$; there is at least one: $\Sigma_u^v\, s(R_i) > \mu$. R_i has weak image R_i' in A_1; say $A \equiv X_1(R_u', R_v')X_2$. Then

$$AZ \underset{n}{=} X_1 R_u'^{\mathrm{L}}\, E(R_u^{\mathrm{R}}, R_v^{\mathrm{L}})\, F R_v'^{\mathrm{R}} X_2 Z \equiv G, \text{ say.}$$

G is E.R; and (G) is (n)-bounded by k_{39}, so $G \in J_n(k_{39})$. Now K conj$_n$ $AZ \underset{n}{=} G$. Consider the sequence

(3) $$R_u^{\mathrm{R}}, R_{u+1}, ..., R_{v-1}, R_v^{\mathrm{L}}, Z$$

of subwords of (G) and put $\mathrm{E} \equiv s'(R_u^{\mathrm{R}}) + \Sigma_{u+1}^{v-1}\, s'(R_i) + s'(R_v^{\mathrm{L}}) + s'(Z)$. Then

$$\mathrm{E} > s'(R_u^{\mathrm{R}}) + s'(R_v^{\mathrm{L}}) + s'(Z) + \mu - s'(R_u) - s'(R_v)$$

$$> s(Z) + \mu - 2u_{61} \equiv s(Z) + \Sigma\, s(A_i) - s(T) - \alpha_8 - 2u_{61}.$$

Now $s(Z) - s(T) > 1/e^2 + \alpha_9 + 1/e^3$ so the last expression is greater than $\Sigma\, s(A_i) + 1/e^2 - 2u_{61} + \alpha_9 - \alpha_8 + 1/e^3$. We can obtain an admissible sequence $B_1, B_2, ...$ from (3) as follows.

Delete R_u^{R} if not normal; delete R_v^{L} if not normal; then apply 17.6B to obtain a linearly admissible sequence $S_1, ..., S_p$ say, where $\Sigma\, s(S_i) \geq \mathrm{E} - 2(r_0 + \epsilon_3) - u_{89}$. If $p = 1$ the sequence is already admissible. If $p \geq 2$, apply 17.6B to $S_2, ..., S_p, S_1$ to obtain the required $B_1, ..., B_q$ where $\Sigma\, s(B_i) \geq \Sigma\, s(S_i) - u_{89}$. Hence $B_{n+1}(K) = B_{n+1}(AZ) \geq \Sigma\, s(B_i) > \mathrm{E} - u_{66}$ where $u_{66} = 2(r_0 + \epsilon_3 + u_{89})$. Hence

$$B_{n+1}(K) > \sum s(A_i) + 1/e^2 + 1/e^3 + \alpha_9 - 2u_{61} - \alpha_8 - u_{66}$$

$$> \sum s(A_i) + 1/e^2 + 1/e^3$$

since $\alpha_9 > 2u_{61} + \alpha_8 + u_{66}$, i.e., $2\alpha_6 - 1 - (r_0 + 2\epsilon_3) - 1/e^2 - 1/e^3 > 2u_{61} + \alpha_8 + u_{66}$. But we proved above that $\Sigma\, s(A_i) > B_{n+1}(K) - - 1/e^2 - 1/e^3$ so we have a contradiction.

17.12. Proposition. *Axiom* 11b *holds for* $n + 1$.

Proof. Let $W^e \neq I$ in G_{n+1}. Of the L such that L conj_{n+1} W choose L' such that $B_{n+1}(L') < \text{Inf}\, B_{n+1}(L) + 1/e^3$. By Axiom 11b, since $L'^e \neq I$, there exists K_0 such that K_0 is E.R., K_0 conj_n L' and (K_0) is (n)-bounded by $cb_{\min} \leqslant k_{39}$. Now $B_{n+1}(K_0) = B_{n+1}(L')$ and $K_0 \in J_n(k_{39})$. By 17.11 (K_0) is $(n + 1)$-bounded by k_{40}. Now $k_{40} = str$. Thus (K_0) is (n)-bounded by $cb_{\min}$ and $(n + 1)$-bounded by str. Since $cb_{\min} \leqslant str < str^+ + q_{31} \leqslant s_0$, (K_0) is $((n + 1))$-bounded by s_0.

Make a simultaneous $(n + 1)$-replacement in (K_0) by replacing those maximal normal subwords, if any, of size $> \frac{1}{2}$. Say $(K_0) \to (K_1)$. Then by Axioms 26, 26″ no maximal normal subword of (K_1) has size $\geqslant \frac{1}{2} + r_{26}$, since $\frac{1}{2} + r_{26} > v_{26}$; thus (K_1) is $(n + 1)$-bounded by $\frac{1}{2} + r_{26} + \epsilon_3 = k_{42}$. It is (n)-bounded by s_0. Similarly (K_2) exists such that $(K_1) \to (K_2)$ and (K_2) is n-bounded by k_{42} and $((n - 1))$-bounded by s_0; and so on to obtain finally (K_{n+1}), 1-bounded by k_{42}.

Now let S be a $(p + 1)$-subelement $(1 \leqslant p \leqslant n)$ in (K_{n+1}) of size $> u_{49}$. By 12.15 it determines a subelement S' of L_{p+1} in (K_{n+1-p}) and $s(S') > s(S) - u_{65}$ where $u_{65} = u_{37} + u_4$. Hence $s(S) < s(S') + + u_{65} < k_{42} + u_{65} = cb_{\min}$. Thus (K_{n+1}) is $((n + 1))$-bounded by $cb_{\min}$. Finally K_{n+1} conj_{n+1} K_0 conj_n L' conj_{n+1} W.

17.13. Corollary. *If* $W^e \neq I$ *in* G_{n+1} *then* K *exists such that* K $\textit{conj}_{n+1}$ W; X $\textit{conj}_{n+1}$ W *implies* $B_{n+1}(K) < B_{n+1}(X) + 1/e^3$, *and* $K \in J_{n+1}$, *where* J_{n+1} *is defined as follows.*

17.13A. Definition. Put $Y \in J_{n+1}$ if $Y^e \neq I$ in G_{n+1}, Y is E.R. and (Y) is $((n + 1))$-bounded by str (cf. Axiom 9).

Proof of 17.13. For the required K, take the K_0 of 17.12.

17.14. Note. Axiom 11a is just a linear version of Axiom 11b. The proof that Axiom 11a holds for $(n+1)$ is left to the reader.

17.15. Proposition. *Let T be E.R. and let (T) be (n)-bounded by k_{45}. Let $T^e \neq I$ in G_{n+1}. Let $B_{n+1}(T) \geqslant u_{71}$. Then $B_n(T) > 2B_{n+1}(T)$.*

Proof. As a preliminary, note that $k_{45} \leqslant b_{25}, b_{30}, b_{33}; u_{63} - u_4 > s_{30}$; $\mathrm{Min}(q^*, r^* - 2(r_0 + \epsilon_3) - 4/e) > t_{33}$ and $\frac{1}{2}u_{63} - \epsilon_{30} - 1/e^2 > r_0' + \epsilon_{30}' + 2/e$. U exists such that $U \operatorname{conj}_n T$, $U \in J_n(k_{39})$, and there is an admissible sequence $A_1, \ldots, A_q$ for U such that

$$\sum s(A_i) > B_{n+1}(T) - 1/e^2 \geqslant u_{71} - 1/e^2 > \alpha_1;$$

say, $R \cdot U \cdot R^{-1} \underset{n}{=} T$. By 17.6 $T \underset{n}{=} XVY$ where XVY is (n)-bounded by k_{39}, $V \subset (U)$ and $\Sigma\, s'(A_i \cap V) > \Sigma\, s(A_i) - \alpha_1$. Let $P_i = A_i \cap V$, and let those P_i of size $> u_{63}$ be $P_f, \ldots, P_g$, say. Then $\Sigma_f^g s(P_i) > \Sigma\, s(A_i) - \alpha_1 > B_{n+1}(T) - 1/e^2 - \alpha_1 > \frac{1}{2} B_{n+1}(T)$ since $u_{71} > 2(1/e^2 + \alpha_1)$.

For $i = f, \ldots, g$ let P_i' be a weak image in T of P_i; then $s(P_i') > s(P_i) - u_4 > \frac{1}{2} s(P_i)$ since $u_{63} > 2u_4$.

By Axiom 30 for $n+1$, P_i' contains subelements U_{ij} $(j = 1, \ldots, a_i)$ of L_n. Let C_{ij} be the kernels of the U_{ij} (i fixed); thus if the U_{ij} are disjoint then $C_{ij} \equiv U_{ij}$. Now (Axiom 33′)

$$C_{ij} \underset{n-1}{=} C_{ij}^{\mathrm{L}} E_{ij} Z_{ij} F_{ij} C_{ij}^{\mathrm{R}} \equiv C_{ij}''$$

say where $C_{ij}^{\mathrm{L}}, C_{ij}^{\mathrm{R}}$ are not subelements of $\mathsf{L}_1 \cup \cdots \cup \mathsf{L}_{n-1}$ and C_{ij}'' is $((n-1))$-bounded by $k_{45} + c_{33}$. Also $s(Z_{ij}) > q^* - r_{33} > u_{63}$. Let P_i'' arise from P_i' by replacing C_{ij} by C_{ij}'' for all j, and let T'' arise from T by replacing each P_i' by P_i''. Then $T \underset{n-1}{=} T''$ and T'' is $((n-1))$-bounded by $k_{45} + c_{33} \leqslant k_{39}$ (cf. 5.32).

For fixed i suppose the subword $(Z_{i\,a-1}, Z_{i\,b+1})$ of P_i'' is a subelement of L_n and $a < b$. Then $C_{ia}^{\mathrm{L}} E_{ia}(Z_{ia}, Z_{ib}) F_{ib} C_{ib}^{\mathrm{R}} \equiv U$ is a subelement of L_n say $UV \in \mathsf{L}_n$. Now $U \underset{n-1}{=} (C_{ia}, C_{ib}) \subset P_i'$ so $UV \underset{n-1}{=} (C_{ia}, C_{ib})V$. Since (C_{ia}, C_{ib}) is $((n-1))$-bounded by k_{45} $(C_{ia}, C_{ib})V \in \mathsf{L}_n$.

Since $a < b$, this implies that the U_{ij} are not maximal normal (cf. 5.28A), i.e., $C_{ij} \equiv U_{ij}$. Thus (U_{ia}, U_{ib}) is a subelement of L_n.

By Axiom 30, $s(U_{ia}, U_{ib}) > (b+1-a)/t_i - \epsilon'_{30}$ and $a_i/t_i > s(P'_i) - \epsilon_{30} > \frac{1}{2}s(P_i) - \epsilon_{30}$.

Next we show there are positive integers c_i, k_i such that

(i) $(c_i - 1)/t_i - \epsilon'_{30} > r'_0$

(ii) $k_i c_i + 1 \leqslant a_i$.

Let c_i be the least integer $> (r'_0 + \epsilon'_{30})\,t_i + 1 = \delta$ say, then (i) holds. Now $k_i c_i + 1 \leqslant k_i(\delta + 1) + 1 \leqslant a_i$ if $k_i \leqslant (a_i - 1)/(\delta + 1)$. Next,

$$(a_i/t_i - 1/t_i)/(\delta/t_i + 1/t_i) > (\tfrac{1}{2}s(P_i) - \epsilon_{30} - 1/e)/(r'_0 + \epsilon'_{30} + 2/e) =$$

$$= E_i \text{ say}$$

and $E_i > 1$. Thus for k_i we may take $[E_i]$. Hence $k_i > E_i - 1$.

Put $A_{is} = Z_{ij}$ where $j = 1 + sc_i$ $(s = 0, \ldots, k_i)$. Then we have shown that $(A_{is}, A_{i\,s+1})$ is not a subelement of L_n (cf. 5.21).

Thus the sequence $\{A_{is}\}$, $i = f, \ldots, g$, $s = 1, 2, \ldots, k_i$, is admissible for T'' (since, e.g., $(A_{i-1\,s'}, A_{is}) \supset (A_{i0}, A_{is})$). Hence

$$B_n(T) \geqslant \sum_{i=f}^{g} \sum_{s=1}^{k_i} s(A_{is}) > \sum_{i=f}^{g} u_{63}(E_i - 1).$$

We show that $u_{63}(E_i - 1) > 4s(P_i)$ and it will follow that $B_n(T) > 4\Sigma s(P_i) > 2B_{n+1}(T)$ as required. Let $\alpha = \epsilon_{30} + 1/e$, $\beta = r'_0 + \epsilon'_{30} + 2/e$. Then $E_i - 1 = (\frac{1}{2}s(P_i) - \alpha - \beta)/\beta > 4s(P_i)/u_{63}$ provided that

$$s(P_i)(\tfrac{1}{2}u_{63} - 4\beta) > u_{63}(\alpha + \beta). \tag{4}$$

But $s(P_i) > u_{63}$ and $\frac{1}{2}u_{63} > \alpha + 5\beta$ so (4) follows.

17.16. Proposition. *If (U) is (n)-bounded by k_{46} and $B_{n+1}(U) < u_{71}$ then (U) is $((n+1))$-bounded by k_{46}.*

Proof. If not then $(U) \equiv (U'S)$ where S is a subelement of L_{n+1} and $s(S) \geqslant k_{46} \geqslant \frac{1}{2}, t_{33}$. Now $S \underset{n}{=} S^{\mathrm{L}} EZFS^{\mathrm{R}} \equiv S'$ where S' is (n)-

bounded by $k_{46} + c_{33} \leqslant k_{39}$, and so is $U'S'$. Also $s(Z) > \frac{1}{2} - r_{33} > \frac{1}{4}$. Thus

$$u_{71} > B_{n+1}(U) \geqslant s(Z) > \tfrac{1}{4} > u_{71}$$

which is a contradiction ($k_{46} \leqslant b_{33}$).

17.17. Proposition. *$B_d^e = \Gamma(C_1, \ldots, C_{n+1}, J_{n+1})$. Also $C_1, \ldots, C_{n+1}$, J_{n+1} are pairwise disjoint. Hence Axiom 1 holds for $n+1$.*

Proof. Call the right side H say. Let $W \in \Pi$. If $W^e = I$ in G_{n+1} then $W^e = I$ in H. If $W^e \neq I$ in G_{n+1} then by 17.13 there exists $K \in J_{n+1}$ conjugate to W in G_{n+1}. Hence in H we have $I = K^e = T^{-1} \cdot W^e \cdot T$ for some T. So again $W^e = I$ in H.

$C_1, \ldots, C_{n+1}$ are disjoint since $C_{n+1} \subset J_n$. If $X \in J_{n+1}$ then $X^e \neq I$ in G_{n+1} while if $X \in C_i$ then $X^e = I$ in G_i hence in G_{n+1}.

17.18. Proposition. *If $TCTD \equiv P^e \in L_{n+1}$ and $L(T) \geqslant L(P)$ then $TC \equiv P^k$ for some k hence $(TC)^e \underset{n+1}{=} I$.*

Proof. This is trivial (cf. 2.18B).

17.19. Proposition. *Let $YAYB \in L_{n+1}$, $s(Y) > u_{72}$. Then if Y is (n)-bounded by k_{47} we have $(YA)^e = I$ in G_{n+1}.*

Proof. We have $Z \to \cdots \to YAYB$, $Z \equiv P^e \in \operatorname{Rep} L_{n+1}$. The first $Y \underset{n}{=} Y^{\mathrm{L}} E Z_1 F Y^{\mathrm{R}}$ and the second $Y \underset{n}{=} Y^l E' Z_2 F' Y^r$ ($u_{72} \geqslant t_{33}$; $k_{47} \leqslant b_{33}$). Also $(Z_1 C Z_2 D) \equiv Z$ and $Z_1 C Z_2 \underset{n}{=} Z_1 F Y^{\mathrm{R}} A Y^l E' Z_2$ by Section 16. Consider $Y^{\mathrm{L}} E Z_1 F Y^{\mathrm{R}} \underset{n}{=} Y^l E' Z_2 F' Y^r$. Let J be a weak image of Z_1 in the right side ($k_{47} + c_{33} \leqslant b_{25}$; $u_{72} - r_{33} > u_3$). Then $s(J \cap Z_2) \geqslant (u_{72} - r_{33} - u_4) - 4/e - 2\operatorname{Max}(r_0, q_{33} + \epsilon_3) = \delta$ say. Now $\delta > u_1$ and, where $Z_2^{\mathrm{M}} \equiv (J \cap Z_2)^0$, $s(Z_2^{\mathrm{M}}) > \delta - u_1 > s_{33}$. Say $Z_2^{\mathrm{M}} \sim Q$; then Q has the form Z_1^{M}. By Axiom 33 for $n+1$ $Z_2^{\mathrm{M}} \equiv Z_1^{\mathrm{M}}$. We may write $Z_i \equiv Z_i^{\mathrm{L}} Z_i^{\mathrm{M}} Z_i^{\mathrm{R}}$ ($i = 1, 2$). By 13.23 $Y^{\mathrm{L}} E Z_1^{\mathrm{L}} \underset{n}{=} Y^l E' Z_2^{\mathrm{L}}$. By 13.21 and Axiom 19, since $s_{33} \geqslant r_0$ we have $L(Z^{\mathrm{M}})/L(Z) > s_{33} - (\epsilon_1^* + 3/e) - \epsilon_3 > 1/e$. Now

$$Z \equiv (Z_1^{\mathrm{L}} Z_1^{\mathrm{M}} Z_1^{\mathrm{R}} C Z_2^{l} Z_2^{\mathrm{M}} Z_2^{\mathrm{R}} D) \text{ and } Z_1^{\mathrm{M}} \equiv Z_2^{\mathrm{M}}$$

hence $(Z_1^{\mathrm{M}} Z_1^{\mathrm{R}} C Z_2^{\mathrm{L}})^e \underset{n+1}{=} I$, by 17.18.

Now

$$YA = Y^{\mathrm{L}} E Z_1 F Y^{\mathrm{R}} A (Y^l E' Z_2 \cdot Z_2^{-1} E'^{-1} Y^{l-1})$$

$$= Y^{\mathrm{L}} E Z_1 C Z_2 \cdot Z_2^{-1} E'^{-1} Y^{l-1}$$

$$= Y^l E' (Z_2^{\mathrm{L}} Z_1^{\mathrm{M}} Z_1^{\mathrm{R}} C) E'^{-1} Y^{l-1}$$

Hence $(YA)^e \underset{n+1}{=} I$.

17.20. Proposition. *Axiom* 8a *holds for* $n+1$; *i.e., if* $X^e \neq I$ *in* G_{n+1} *and* (X) *is* $((n+1))$*-bounded by* $z \geqslant \frac{1}{2}$ *then* (X^t) *is* $((n+1))$*-bounded by* $z + a_8$.

Proof. If not we may assume that $(X^t) \supset S$ where S is a subelement of L_{n+1} of size $\geqslant z + a_8 = y$ say. Now $S \equiv Y^k Y^{\mathrm{L}}$ where Y is a C.A. of X is a C.A. of X and k is a non-negative integer. Since $Y^{\mathrm{L}} \subset Y \subset (X)$ we have $k \geqslant 1$. Clearly $Y^e \neq I$ in G_{n+1}.

If $k = 1$, $S \equiv Y^{\mathrm{L}} Y^{\mathrm{R}} Y^{\mathrm{L}}$. Since $s(S) \geqslant y$ and $s(Y) < z$ we have $s(Y^{\mathrm{L}}) > y - z - 2/e = a_8 - 2/e > u_{72}$. By 17.19 $(Y^{\mathrm{L}} Y^{\mathrm{R}})^e \underset{n+1}{=} I$, a contradiction ($z \leqslant k_{47}$).

If $k = 2$, $S \equiv YYY^{\mathrm{L}}$. One of Y, Y, Y^{L} has size $\geqslant \frac{1}{3}(y - 4/e) > r_0$ hence $s(Y) > \frac{1}{3}(y - 4/e) - \epsilon_3$ (5.19). Now $\frac{1}{3}(\frac{1}{2} - 4/e) - \epsilon_3 > u_{72}$. Now apply 17.19 to the subelement YY of L_{n+1}; we obtain $Y^e \underset{n+1}{=} I$.

If $k = 3$ we proceed similarly using $\frac{1}{4}(y - 6/e) > \mathrm{Max}(r_0, \epsilon_3 + u_{72})$.

If $k > 3$ then $k = 2s + \theta$ where $\theta = 0$ or 1. $2s + 1 \geqslant k > 3$ hence $s \geqslant 2$. One of Y^s, Y^s, $Y^\theta Y^{\mathrm{L}}$ has size $\geqslant \frac{1}{3}(y - 4/e)$ hence $s(Y^s) > u_{72}$ as above. Thus (i) $Y^{se} \underset{n+1}{=} I$. One of Y^{s-1}, Y has size $> \frac{1}{2}(\frac{1}{3}(4 - 4/e) - \epsilon_3 - 2/e) = \alpha$ (say) $> r_0$ so $s(Y^{s-1}) > \alpha - \epsilon_3$ hence (ii) $Y^{(s-1)e} \underset{n+1}{=} I$. By (i) and (ii) $Y^e \underset{n+1}{=} I$, contradiction.

§ 18

Summary

In this section we prove Axiom 8.16.

18.1. Notation. We shall write vA instead of $\mathrm{Im}'A$ (cf. 6.16). We shall also use the notation pA as follows. Suppose B is a subword of X and also a subword of another word Y; then any subword A of X such that $A \subset B$ is also a subword, which will be denoted by pA, of Y. Of course $A \equiv pA$. Thus p maps certain subwords of X to subwords of Y.

18.2. Proposition. *Let $AXC \underset{j}{\sim} BXD$, where each side is (j)-bounded by k_{48}, $j \leqslant n$ and A or B is I and C or D is I. Let M be maximal j-normal in AXC and properly contained in X. Let $s(M) > u_{73}$. Assume vM exists and let vM, pM be different subwords of BXD. Then there exists $k \geqslant 1$ such that for $s = 1, \ldots, k$*

(i) $(vp)^s M$ exists and has size $> u_{74}$; $(vp)^k M$ meets C, or

(ii) $(vp)^s M'$ exists and has size $> u_{74}$; $(vp)^k M'$ meets D, where M' is pM.

Proof. 1°. If J is Max j-normal in AXC or BXD and $s(J) > u_{10}$ then by 6.23 since $k_{48} \leqslant k_8', k_9'$, vJ exists and either $s(vJ) > s(J) - v_3$ or $s(vJ) \geqslant 1 - s(J) - v_2 > 1 - v_2 - k_{48} = \alpha$ say.

2°. pM is maximal normal in BXD. Both vM, vpM exist since $u_{73} > u_{10}$. Now vpM is different from M otherwise $vM = vvpM = pM$ by 6.19. Now $s(vM) > \mathrm{Min}(u_{73} - v_3, \alpha) > u_{10}$ so if vpM is left of M then by 6.20 $vvpM$ is left of vM, i.e., pM is left of $vppM$, i.e., M' is left of vpM'. Thus without loss of generality M is left of vpM; we shall prove (i).

We use the notation $K_1 = pM$, $M_1 = vK_1$, $pM_i = K_{i+1}$, $M_i = vK_i$.

$vpM = M_1$ has size $\geqslant \mathrm{Min}(\alpha, u_{73} - v_3) > u_{74} + v_3 > u_{74}$. It is sufficient to prove for $s = 1, 2, 3, \ldots$.

(∗) M_s exists $\Rightarrow s(M_s) > u_{74}$ and M_{s-1} is left of M_s (where M_0 means M). For if k is the largest integer s such that M_s exists then pM_k cannot exist, otherwise $M_{k+1} = vpM_k$ would exist because $u_{74} > u_{10}$; hence $M_k \not\subset X$, i.e., M_k meets C.

We may assume M_2 exists otherwise (∗) holds. Also

$s(M_2) > \mathrm{Min}(\alpha, s(M_1) - v_3) > u_{74}$. Now consider the following statement $P(t)$, $t \geqslant 2$: M_t exists and has size $> u_{74}$, $(M, M_2) \underset{j}{\hat{=}} (M_{t-2}, M_t)$ and $(M, M_1) \underset{j}{\hat{=}} (M_{t-2}, M_{t-1})$. $P(2)$ is true since $u_{74} > u_{24}$ and $k_{48} \leqslant k'_{12}, k'_{13}$ (cf. 8.8). Assume $P(2), \ldots, P(t)$ true. Then we may assume M_{t+1} exists; otherwise $(*)$ holds. Hence M_i is left of M_{i+1} $(i = 0, 1, \ldots, t)$. By 8.15 since $k_{48} \leqslant k_{16}, k_{17}$; $u_{74} > u_{29}$;

$$(M_{t-1}, M_{t+1}) \hat{=} (K_{t-1}, K_{t+1}) \equiv (M_{t-2}, M_t) \hat{=} (M, M_2)$$

and similarly $(M_{t-1}, M_t) \hat{=} (M, M_1)$. Apply 9.11 taking S, U, T, S', T' to be $M, M_1, M_2, M_{t-1}, M_{t+1}$; this is possible since $s(M_{t+1}) > \mathrm{Min}(\alpha, u_{74} - v_3) > u_{30}$, $\mathrm{Min}(\alpha, u_{73} - v_3) > u'_{30}$ and $k_{48} \leqslant k_{18}, k_{19}$. We conclude that U' exists where $(M_{t-1}, U') \hat{=} (M, M_1)$ and $s(U') > \mathrm{Min}(\alpha - v_3, u_{73} - 2v_3) - u_4 - \epsilon_3 = \beta > u'_{30}$. Thus $(M_{t-1}, U') \hat{=} (M_{t-1}, M_t)$. By 9.12 since $u_{74} > u'_{30} > u_{30}$ we have U' is M_t. Now $s(M_{t+1}) > \mathrm{Min}(\alpha, s(M_t) - v_3) \geqslant \mathrm{Min}(\alpha, \beta - v_3) > u_{74}$. Thus $P(t + 1)$ is true and $(*)$ is proved.

18.2A. Corollary. *The proposition remains true if u_{73}, u_{74} are replaced by $u_{73} + u_{90}, u_{74} + u_{90}$ respectively.*

Proof. Examine the P.C.'s in the proof of 18.2 and use $\alpha - u_{90} > 2v_3 + u_4 + \epsilon_3 + u_{74}$ and $u_{74} = u_{73} - u_{90}$.

18.3. Proposition. *Axiom* 8.16 *holds for $n + 1$*, i.e., *if O, O' are $(n + 1)$-preobjects and O is an adjustment of O' and also $O \underset{n+1}{=} O'$ then $O \equiv O'$.*

Proof. O is (S, T) where S, T are subelements of $\mathrm{Rep}(L_{n+1})$ of size $> u_{21}$, hence $(n + 1)$-normal, and S, T determine distinct maximal $(n + 1)$-normal subwords $\overline{S}, \overline{T}$ and where $\overline{S} \equiv SH$ then $s(H) < u_{22}$. Similarly for T. O is (n)-bounded by k_{12} and $(n + 1)$-bounded by k_{13}.

Similarly $O' \equiv (S', T')$.

Now $S \equiv AS_0$, $S' \equiv BS_0$ where A or B is I. Similarly $T \equiv T_0 C$, $T' \equiv T_0 D$, C or D being I.

1°. Let $(S_i, T_0) \equiv S_0 X'$; then $O \equiv AS_0 X'B$, $O' \equiv CS_0 X'D$. Let H, K be the kernels of S_0, T_0. Then $(S_0, T_0) \equiv HAK$ for some A. D exists such that D is (n)-bounded by $b_{\min}$ and $HAK \underset{n}{=} D$. Now $s(H) > s(S_0) - r_0 - \epsilon_3 - 2/e > u_{21} - r_0 - \epsilon_3 - 2/e > u_1$, so $H \supset H^0 \underset{n}{\sim} H^{0'}$ and $s(H^0) \geqslant s(H) - u_1$. H^0 has the form H^{L} and $H^{0'}$ the form D^{L}, so $H^{\mathrm{L}} \underset{n}{\sim} D^{\mathrm{L}}$. Similarly $K^0 \equiv K^{\mathrm{R}} \underset{n}{\sim} D^{\mathrm{R}}$. Also $D \equiv D^{\mathrm{L}} D^{\mathrm{M}} D^{\mathrm{R}}$ for some D^{M}. Since $s(H^{\mathrm{L}}) > u_{21} - r_0 - \epsilon_3 - 2/e - - u_1 = \theta$ say and since H^{L} is a subelement of Rep L_{n+1} we may apply Axiom 31B ($\theta > s_{31}$; $b_{\min}, str^+ \leqslant b_{31}$). We obtain a word $H^{\mathrm{LL}} E D^{\mathrm{LR}}$, (n)-bounded by $\mathrm{Max}(b_{\min}, str^+) + q_{31} = \varphi$ say. Now $\varphi \leqslant k_{12}$ and $s(H^{\mathrm{LL}}) > s(H^{\mathrm{L}}) - r_{31} > \theta - r_{31} > r_0 + \epsilon_3$. Similarly there is a word $D^{\mathrm{RL}} F K^{\mathrm{RR}}$ where $s(K^{\mathrm{RR}}) > \theta - r_{31}$. Moreover

$$HAK \underset{n}{=} H^{\mathrm{LL}} E D^{\mathrm{LR}} D^{\mathrm{M}} D^{\mathrm{RL}} F K^{\mathrm{RR}} \equiv J, \text{ say,}$$

where J is (n)-bounded by φ. Let $S_0^* \equiv H^{\mathrm{LL}}$, $S^* \equiv AS_0^*$, $T_0^* \equiv K^{\mathrm{RR}}$, $T^* \equiv T_0^* C$. Then $(S, T) \equiv A(S_0, T_0) C \underset{n}{=} AJC \equiv (S^*, T^*)$ and (S^*, T^*) is (n)-bounded by φ and $(n+1)$-bounded by $k_{13} + u_4$. Next

$$\begin{aligned} s(AS_0) &\leqslant s(A) + s(S_0) + 2/e \\ &\leqslant s(A) + 2/e + s(H^{\mathrm{L}}) + u_1 + r_0 + \epsilon_3 + 2/e \\ &\leqslant s(A) + 2/e + s(H^{\mathrm{LL}}) + r_{31} + u_1 + r_0 + \epsilon_3 + 2/e \leqslant s(S^*) + \psi, \end{aligned}$$

where $\psi = (R_0 + 2\epsilon_3) + 2/e + r_{31} + u_1 + r_0 + \epsilon_3 + 2/e$. Thus $s(S^*) > u_{21} - \psi = u_{21}^*$. By 6.14A, since $k_{12} \leqslant b_{25}$, $u_{21} - \epsilon_3 \geqslant u_6 + u_4 + \epsilon$ and $u_{21} - \psi - \epsilon_3 > u_6$ we have $\mathrm{Im}\,\overline{S^*} = \overline{S}$. Hence

$$\begin{aligned} s(\overline{S^*}) &< s(\overline{S}) + u_4 + \epsilon_3 < s(S) + u_{22} + 2/e + u_4 + \epsilon_3 \\ &< s(S^*) + u_{22}^* + 2/e \end{aligned}$$

where $u_{22}^* = \psi + u_{22} + u_4 + \epsilon_3$.

Similarly $BJD \equiv (S'^*, T'^*)$ where if X temporarily denotes S'^* or T'^* then $s(X) > u_{21}^*$ and $s(\overline{X}) < s(X) + u_{22}^* + 2/e$. Write $O^* \equiv (S^*, T^*)$,

$O'^* \equiv (S'^*, T'^*)$. Although these are not preobjects in general, $(S^*, T^*) \underset{n+1}{=} (S'^*, T'^*)$, $S'^* \equiv BS_0^*$, $T'^* \equiv S_0^* D$.

2°. We claim that it is sufficient to prove $O^* \equiv O'^*$. Let $O^* \equiv O'^*$. Then $AJC \equiv BJD$. Among the left subwords of J that are subelements of Rep L_{n+1} let U have maximal length. Then $H^{\mathrm{LL}} \subset U$. If $U \equiv J$ we would have $\overline{S^*} \equiv \overline{T^*}$ so Im $\overline{S^*}$ = Im $\overline{T^*}$, i.e., $\overline{S} = \overline{T}$. Hence $U \not\equiv J$. Say $J \equiv UV$. Thus $AUVC \equiv BUVD$. AU is the largest left subword of $AUVC$ which is a subelement of Rep L_{n+1}. Hence $AU \equiv BU$. Hence $C \equiv D$ and $O \equiv O'$.

3°. *Case 1.* $O^* \underset{n}{=} O'^*$

Write k_{12}^* for φ, k_{13}^* for $k_{13} + u_4$.

Take j minimal such that $O^* \underset{j}{=} O'^*$. If $j = 0$ there is nothing to do so let $j > 0$. There are divisions $P_1, P_2, \ldots$ and $Q_1, Q_2, \ldots$ where some P_i has type j. We have $O^* \equiv AJC$, $O'^* \equiv BJD$, $(S_0^*, T_0^*) \equiv J \equiv S_0^* X$ say. We first show

(†) S^* contains a maximal j-normal subword of size $> u_{75}$; by translation it is sufficient to show AS_0^* or BS_0^* does.

Since $k_3^0 < s(P_i') + s(Q_i') < s(P_i') + k_{12}^* \leqslant s(\overline{P}_i) + \epsilon_3 + k_{12}^*$ we have $s(\overline{P}_i) > k_3^0 - \epsilon_3 - k_{12}^* > u_{75}$; thus we may assume $\overline{P}_i \not\subset AS_0^*$, hence $P_i \subset S_0^* XC$ and similarly $\overline{Q}_i \subset S_0^* XD$. Now C or D is I; say C is I. Then $\overline{P}_i \subset S_0^* X$ so the subword $p\overline{P}_i$ of O'^* exists. We prove $p\overline{P}_i$ is not $\overline{Q}_i$. If $p\overline{P}_i \equiv \overline{Q}_i$ then by 6.22 since $k_{12}^* \leqslant k_6, k_7$ we have $Q_i' \subset p\overline{P}_i \equiv \overline{P}_i \equiv EP_i'F$ where $s(E), s(F) \leqslant v_1$. Hence $s(Q_i' \cap E) \leqslant \mathrm{Max}(r_0, v_1 + \epsilon_3)$ and

$$s(Q_i' \cap P_i') \geqslant s(Q_i') - 4/e - 2\mathrm{Max}(r_0, v_1 + \epsilon_3)$$

$$> k_3^0 - k_{12}^* - 4/e - 2\mathrm{Max}(r_0, v_1 + \epsilon_3) > s_{40}.$$

This contradicts Axiom 40 since $k_{12}^* \leqslant b_{40}$ (and $P_i'^{-1} X_{i-1}'^{-1} Q_i' X_i' \in \mathsf{L}_j$). By the dual of 18.2A (with $M = \overline{P}_i$), since

$$s(\overline{P}_i) > k_3^0 - \epsilon_3 - k_{12}^* > u_{73} + u_{90} \text{ and } k_{12}^* \leqslant k_{48},$$

a maximal j-normal subword of size $> u_{74} + u_{90} = u_{73}$ meets A or B and hence is contained in AS_0^* or BS_0^*. (Since S^* or S'^* is S_0^* we have $s(S_0^*) > u_{21}^* > r_0'$.) Similarly if D is I. (†) follows since $u_{73} = u_{75}$.

4°. Without loss of generality assume $B \equiv I$. If $A \equiv I$ then $C \underset{j}{=} D$, i.e., $C \underset{j}{=} I$ or $I \underset{j}{=} D$; since $k_{12}^* \leqslant b_\infty$ we have by Axiom 10 that $C \equiv D \equiv I$ and $O^* \equiv O'^*$ as required. Thus we may suppose that $A \not\equiv I$.

For the next part of the argument we shall assume only that

$$(1) \qquad AS_0^*XC \underset{j}{\sim} S_0^*XD$$

(instead of $\underset{j}{=}$). Let N_1 be the nice subword of the form $(AS_0^*)^{\mathrm{L}}$ having maximal length. Since $s(AS_0^*) > u_{21}^*$ we have as in 6.8 $s(N_1) > u_{21}^* - u_2 - 2/e = \gamma$ say. Then $N_1 \sim N_1' \subset S_0^*XD$. Let N_2 be the nice subword of S_0^*XD of the form $S_0^{*\mathrm{L}}$ having maximal length. Then $s(N_2) > \gamma > s_{33}$ and $N_2 \sim N_2' \subset AS_0^*XC$. Let $N_1 \cap N_2'$. Then $N \sim N' = N_1' \cap N_2$. Either N is N_1 or N' is N_2 so by Axiom 33 for $n+1$ $N \equiv N'$. Hence $s(N) > \gamma > r_0$. Each of $\overline{AS_0^*}$, $\overline{S_0^*}$ has size $> u_{21}^* - \epsilon_3 > u_6 + u_4 + \epsilon_3$ and $k_{12}^* \leqslant b_{25}$ hence (cf. 6.14A) $\operatorname{Im} \overline{AS_0^*} = \overline{S_0^*}$. Hence

$$s(\overline{S_0^*}) > s(AS_0^*) - u_4 - \epsilon_3$$

$$> s(A) + s(S_0^*) - (r_0 + 2\epsilon_3) - u_4 - \epsilon_3;$$

also $s(S_0^*) + u_{22}^* + 2/e > s(\overline{S_0^*})$. Thus $s(A) \leqslant u_{22}^* + 2/e + (r_0 + 2\epsilon_3) + u_4 + \epsilon_3 = a$, say. Next $A \subset N$ otherwise $s(N) \leqslant s(A) + \epsilon_3 \leqslant a + \epsilon_3 + y$, a contradiction. Thus $N \equiv AS_0^{*l}$ say and $s(S_0^{*l}) > \gamma - a - 2/e = b$, say. Now N' is a left subword of S_0^*, $N' \equiv S_0^{*\mathrm{L}}$ say and has size $> \gamma$. C, W exist such that $AS_0^*W \equiv C^e \in \operatorname{Rep} L_{n+1}$. By Axiom 19 and 13.21, $L(S_0^{*l})/L(C^e) > b - \epsilon_3 - (\epsilon_1^* + 3/e) = b'$ say and $b' > 5(a + \epsilon_1^* + 3/e) + 3/e > 5(L(A)/L(C^e)) + 3/e$ so $L(S_0^*)/L(C) > 5L(A)/L(C) + 3$. But $L(S_0^{*l})/L(C) > b'e > 1$ and since $AS_0^{*l} \equiv S_0^{*\mathrm{L}}$we see that $A \equiv C^\kappa$ for some $k > 0$.

By (†) AS_0^* contains a maximal j-normal subword F', which may be assumed proper, of size $> u_{75} = u_{73}$. Let F be the leftmost translate in AS_0^* of F'. Where $AS_0^* \equiv A'FB$ we have $L(A') \leqslant L(C)$. For any integer $N \geqslant 1$, $F \subset C^e \subset C^{eN}$ and we can consider the translates $F, F^1, F^2, \ldots$ of F in C^{eN}; here e.g., F^2 means F^{++}.

Now $L((F, F^{6k})) \leqslant 6k \cdot L(C) + 2L(C)$ so

$$L(A'(F, F^{6k})) < (6k+3)\,L(C) < kL(C) + L(S_0^{*l}) = L(AS_0^{*l}).$$

Thus $A'(F, F^{6k})$ is a left subword of AS_0^{*l}.

With respect to (1) let $vF = T_1, pT_1 = F_2, ..., vF_i = T_i, pT_i = F_{i+1}, ...$. F_1 means F. Since $u_{73} > u_{10}$ and $k_{12}^* \leqslant k_8', k_9'$, T_1 exists and $T_1 \equiv F$ so $s(T_1) > u_{73}$. Since (F, F^k) has length $kL(C) = L(A)$, $F^k \subset S_0^l$, so vpF^k exists; in fact it coincides with F. Thus $F_2 = pvF$ exists and coincides with F^k. Similarly F_{i+1} exists and coincides with F^{ik} $(i = 1, 2, ..., 6)$. Thus $T_1, T_2, ..., T_7$ exist. By 8.15, which is available since $k_{12}^* \leqslant k_{16}, k_{17}$ and $u_{73} > u_{29}$, $(T_1, T_3) \triangleq (F_1, F_3) \equiv (F, F^{2k})$; (since $AS_0^{*l} \equiv S_0^{*\mathrm{L}}$ we have even $(F_1, F_3) \equiv (T_1, T_3)$). Let p be maximal such that T_p exists; then $p \geqslant 7$. By 18.2 T_p meets D, hence $(\mathrm{In}\, O'^*, T_p) \supset S_0^* X \equiv (S_0^*, T_0^*)$, and $T_1, ..., T_p$ have size $> u_{74}$. Using that $pT_i = F_{i+1}$ we have $F_1, F_2, ..., F_p$ have size $> u_{74}$.

We prove by induction on s that if $1 \leqslant i < j \leqslant s$

$$(T_i, T_j) \triangleq (F^{(i-1)k}, F^{(j-1)k}) \quad (s = 3, 4, ..., p).$$

This is true for $s = 3$. Assume true for some $s, 3 \leqslant s < p$. Then T_{s+1} exists and by 8.15, since $u_{74} > u_{29}$, if $2 \leqslant u < v \leqslant s+1$

$$(T_u, T_v) \triangleq (F_u, F_v) \equiv (T_{u-1}, T_{v-1}) \triangleq (F^{(u-2)k}, F^{(v-2)k}) \equiv$$

$$\equiv (F^{(u-1)k}, F^{(v-1)k}).$$

In particular this holds when $u = s-1, v = s+1$ and when $u = s-1, v = s$. By induction hypothesis $(T_1, T_s) \triangleq (F, F^{(s-1)k})$. By 9.14, since $k_{12}^* \leqslant k_{20}, k_{21}$ and $u_{74} > u_{31}$, $(T_1, T_{s+1}) \triangleq (F, F^{sk})$. Thus the result is true for $s+1$.

We now have

$$(2) \qquad (T_1, T_p) \mathrel{\underset{j}{\triangleq}} (F, F^{(p-1)k}).$$

$S_0^* XD \equiv A'T_1R$ for some R. Since $L(A')/L(C^e) \leqslant 1/e$ we have either $s(A') \leqslant r_0$ or by Axiom 19 and 13.21 $s(A') \leqslant \epsilon_3 + 1/e + \epsilon_1^* + 3/e = \zeta'$ say; hence $s(A') \leqslant \mathrm{Max}(r_0, \zeta') = \zeta$ say. T_p meets D, so where $S_1 = S_0^* \cap (T_1, T_p)$ we have $s(S_1) \geqslant u_{21}^* - 2/e - \zeta > \mathrm{Max}(r_0 + \epsilon_3, u_{95})$.

5°. The word (T_1, T_p) contains T_0^* which has size $> u_{21}^*$. The maximal normal subwords of (T_1, T_p) determined by S_1, T_0^* are distinct and of size $> u_{95}$; we shall show later that, in view of (2), this is a contradictory situation.

6°. *Case* 2. Not Case 1.

Here $(S^*, T^*) \underset{n+1}{\equiv} (S'^*, T'^*)$. Using the same notation for the divisions, some P_i has type $n+1$.

We claim that $\bar{P}_i \neq \overline{S^*}$. For if $\bar{P}_i = \overline{S^*}$ then by 6.24, since $k_{12}^* \leqslant k_8'$; $k_{13}^* \leqslant k_9'$; $u_{21}^* - \epsilon_3 > u_{10}$, $\bar{Q}_i = \overline{S'^*}$. Let $\xi = \mathrm{Max}(r_0, v_1 + \epsilon_3)$. We have $u_{21}^* - 4/e - 2(\xi + \epsilon_2 + 2/e) > s_{40}$. Since $s(S_0^*) > \xi$, we can write $S_0^* \equiv AB'$ where A is of minimal length such that $s(A) > \xi$. Then where $A \equiv A_1 a$ $(L(a) = 1)$, $s(A) \leqslant \xi + \epsilon_2 + 2/e$ so $s(B') \geqslant s(S_0^*) - 2/e - (\xi + \epsilon_2 + 2/e) > \xi$. Thus $S_0^* \equiv ABC$ where $s(C) > \xi$, $s(B) > s_{40}$. Now $S_0^* \subset \bar{P}_i \equiv EP_i'F$ where E, F have size $\leqslant v_1$. Then $A \not\subset E$ and $C \not\subset F$, so $B \subset P_i'$. Similarly $B \subset Q_i'$. B is (n)-bounded by $k_{12}^* \leqslant b_{40}$; this is a contradiction by Axiom 40.

Similarly $\bar{P}_i \neq \overline{T^*}$. Hence $\bar{P}_i \subset (S_0^*, T_0^*)$ and $p\bar{P}_i$ exists. Now $p\bar{P}_i \neq \bar{Q}_i$ otherwise we could obtain a contradiction by Axiom 40. Since $S_0^* < \bar{P}_i < T_0^*$ we have $pS_0^* < p\bar{P}_i < pT_0^*$ so $p\bar{P} \neq \overline{S'^*}, \overline{T'^*}$.

Dually for $\bar{Q}_i$; in particular $p\bar{Q}_i$ exists. If $p\bar{P}_i < \bar{Q}_i$ then $\bar{P}_i < p\bar{Q}_i$; thus without loss of generality we may suppose $\bar{Q}_i < p\bar{P}_i$.

Now $\overline{S'^*} < \bar{Q}_i$; thus $\overline{S'^*}$ and $p\bar{P}_i$ are disjoint. Now $\overline{S^*} < p\bar{Q}_i < \bar{P}_i$ so $\overline{S^*}, \bar{P}_i$ are disjoint. Now $k_{12}^* \leqslant k_8'$, $k_{13}^* \leqslant k_9'$, $k_3^0 - k_{13}^* - \epsilon_3 > u_{10}$ and $u_{21}^* - \epsilon_3 > u_{10}$; thus we have $\overline{S'^*} < U < \bar{Q}_i$ where U is Im′ $p\bar{Q}_i$; it follows that $\overline{S'^*}$ and $\bar{Q}_i$ are disjoint. Also $s(U) > k_3^0 - k_{13}^* - \epsilon_3 - u_4$.

Take the leftmost $\bar{P}_i$ of type $n+1$. Then $P_1 \cdot P_2 \cdot \cdots \cdot P_{i-1}$ has the form S^*J for some J. S^*J meets or touches $\bar{P}_i$, so $S^{*\prime}J$ meets or touches $p\bar{P}_i$. $Q_i \cdot \cdots \cdot Q_{i-1}$ is disjoint from $p\bar{P}_i$ and contains S'^*, hence it has the form $S^{*\prime}J^{\mathrm{L}}$. Let U_1 be the part of U outside $\bar{Q}_i$. Then $S^{*\prime}J^{\mathrm{L}}$ contains S'^* and U_1 and

$$s(U_1) \geqslant s(U) - 2/e - (r_0 + \epsilon_3)$$

$$\geqslant k_3^0 - k_{13}^* - \epsilon_3 - u_4 - 2/e - (r_0 + \epsilon_3) = \eta \text{ say.}$$

Now $S^*J \underset{j}{\sim} S^{*\prime}J^{\mathrm{L}}$, i.e.,

(3) $AS_0^* J^{\mathrm{L}} J^{\mathrm{R}} \underset{j}{\sim} BS_0^* J^{\mathrm{L}}$

where $J \equiv J^{\mathrm{L}} J^{\mathrm{R}}$ and j is the maximum type of $P_1, \ldots, P_{i-1}$; thus $j \leqslant n$.

First suppose $j = 0$ or $A \equiv B$. Since S^*, S'^* are maximum subelements of Rep L_{n+1}, $j = 0$ implies $A \equiv B$. Hence $A \equiv B$, so $C \underset{n+1}{\equiv} D$. Since C or D is I we may use Axiom 10 to conclude that $C \equiv D$ and $O \equiv O'$.

Now let $j > 0$ and $A \not\equiv B$. We show that S_0^* contains maximal j-normal subwords of size $> u_{73}$. Consider a $\bar{Q}_i$ of type j. If $\bar{Q}_i \subset BS_0^*$ then, by considering translates, S_0^* contains a maximal normal subword as required. If $\bar{Q}_i \not\subset BS_0^*$ then $\bar{Q}_i \subset S_0^* J^{\mathrm{L}}$ and $p\bar{Q}_i$ exists and is different from $\bar{P}_i$ (by Axiom 40) so 18.2A is available ($k_{12}^* \leqslant k_{48}$; $k_3^0 - k_{12}^* - \epsilon_3 > u_{73} + u_{90}$); hence if $\bar{P}_i < p\bar{Q}_i$ then a maximal normal subword of size $> u_{73}$ meets A, but if $p\bar{Q}_i < \bar{P}_i$ then a maximal normal subword of size $> u_{73}$ meets B. Now (3) has the form

(3′) $AS_0^* J^{\mathrm{L}} J^{\mathrm{R}} \underset{j}{\sim} S_0^* J^{\mathrm{L}}$

or

(3″) $BS_0^* J^{\mathrm{L}} \underset{j}{\sim} S_0^* J^{\mathrm{L}} J^{\mathrm{R}}$

according as B or A is I; (3′) and (3″) are of the same form as (1). Hence the argument of 4° is available; in particular we have "T_p meets D" which shows that (3′) is impossible, hence (3″) holds. Hence T_p meets J^{R}, so (In $S_0^*, T_p) \supset S_0^* J^{\mathrm{L}} \supset (S_0^*, U_1)$. Hence (T_1, T_p) contains S_1 and U_1. We have seen that $s(S_1) \geqslant u_{95}$. Now note that $s(U_1) > \eta \geqslant u_{95}$, and S_1, U_1 are distinct maximal normal subwords of (T_1, T_p).

7°. Thus in both Case 1, Case 2 we have (2), where $p \geqslant 7$ and the right side is a subpowerelement of Rep L_{n+1} so is (j)-bounded by $str^+ \leqslant k_{59}$. The left side is (j)-bounded by $k_{12}^* \leqslant k_{59}$. Moreover (T_1, T_p) contains two distinct maximal normal subwords, each of size $> u_{95} \geqslant u_{93}$. By 14, since $u_{73} > u_{74} > u_{30}$, the right side of (2) contains two distinct maximal $(n + 1)$-normal subwords. This is a contradiction.

§19

19.1. Note. All axioms have now been proved true for $n+1$ except Axioms 5 and 29 which we shall discuss in this section. All previous results proved for n and not based on Axioms 5 or 29 are true for $n+1$.

19.2. We shall consider a sequence

$$Y_1, Y_2, \ldots, Y_k,\ k \geqslant 2,\ \ Y_i\, \sigma_{n+1}\, Y_{i+1}\ (i=1, \ldots, k-1).$$

Thus for $i=1, \ldots, k-1$ we have $Y_i, Y_{i+1} \in J_{n+1}$ and either

$$(Y_i^{s_i}) \underset{n+1}{\cong} (Y_{i+1}^{\pm t_i}),\ \ 1 \leqslant s_i < t_i$$

or some $\delta_f(Y_i) \underset{n+1}{\sim}$ some $\epsilon_{q'}(Y_{i+1})$.

19.2A. We refer to 19.2 as *Situation* A. A special case of Situation A is when $Y_1 \underset{n+1}{>} Y_k$: call this *Situation* A$^+$.

Situation B is as follows: $Y_1, Y_2, \ldots, Y_k \in J_{n+1}$ are given and for $i=1, \ldots, k-1$ $(Y_i^{s_i}) \underset{n+1}{\cong} (Y_{i+1}^{\pm t_i})$, $s_i \geqslant 1$, $t_i \geqslant 1$.

19.2B. We can find $\epsilon_i = \pm 1$ $(i=1, \ldots, k)$ such that $X_i \equiv Y_i^{\epsilon_i}$ and for $i=1, \ldots, k-1$ we have in Situation A

1. $(X_i^{s_i}) \cong (X_{i+1}^{t_i})$, $1 \leqslant s_i < t_i$, or
2. some $\delta_f(X_i) \sim$ some $\epsilon_{q'}(X_{i+1})$

and moreover $\delta_f(X_i)$ is a subword of X_i^m for some positive m (of length $\leqslant feL(X_i)$) and $\epsilon_{q'}(X_{i+1})$ is a subword of X_{i+1}^m for some positive m (of length $> q'eL(X_{i+1})$). Similarly in Situation B

3. $(X_i^{s_i}) \cong (X_{i+1}^{t_i})$, $s_i \geqslant 1$, $t_i \geqslant 1$.

19.3. Choose an integer $N > 0$ to be kept fixed. We show that if, for some i, s_i and t_i exist then they may be taken $\geqslant N$ and that if t_i and s_{i+1} both exist they may be taken to be equal; we shall therefore write s_{i+1} instead of t_i in 1 and 3.

We need only remark that $(A) \cong (B)$ implies $(A^r) \cong (B^r)$ for any integer $r \neq 0$, by Axiom 11.

Denote the left side of 1, 2, 3 by A_i and the right side by B_{i+1}. By 5.11, A_i and B_{i+1} are $((n+1))$-bounded by str^+.

19.4. Note. If 1 or 3 holds, then $A_i \cong B_{i+1}$, a maximal $(n+1)$-normal subword S of one of A_i, B_{i+1}, where $s(S) > u_{10}^*$, determines by 16.3A a maximal normal $\mathrm{Im}^* S = S'$ in the other such that either $|s(S) - s(S')| < v_3^*$ or $|s(S) + s(S') - 1| \leqslant v_2^*$. Also $v_2 \leqslant v_2^*$. $v_3 \leqslant v_3^*$ and $u_{10}^* \geqslant u_{10}$.

Moreover a cyclic word D_i exist such that $A_i \,\overline{\overline{\,{}_{n+1}}}\, D_i \,\overline{\overline{\,{}_{n+1}}}\, B_{i+1}$, $\mathrm{Im}'(\mathrm{Im}' S)$ is S' (where Im' is in the sense of 6.21) and D_i is $((n+1))$-bounded by $\mathrm{Max}(str^+, k_{24}') + u_{40} \leqslant k_{22}, k_{23}, k_{16}, k_{17}$.

19.5. Proposition. *In Situation A or B, let B_k contain a maximal $(n+1)$-normal subword S of size $> u_{67}$. Then A_1 contains a maximal $(n+1)$-normal subword of size $> u_{68}$.*

Proof. Since $u_{67} \geqslant u_{10}^*$ A_{k-1} contains a maximal $(n+1)$-normal subword of size $> u_{68} \geqslant u_{10}^*$ for either $s(\mathrm{Im}' S) \geqslant 1 - str^+ - v_2^* > u_{67} > u_{68}$ or $s(\mathrm{Im}' S) > u_{67} - v_3^* > u_{68}$.

Assume A_j contains a maximal $(n+1)$-normal J of size $> u_{68}$ for some j, $2 \leqslant j \leqslant k-1$. Then we show that A_{j-1} does.

We have the following cases

	A_{j-1}	B_j	A_j	B_{j+1}	
I	$X_{j-1}^{s_j-1}$	$\cong X_j^{s_j}$	$X_j^{s_j}$	$\cong X_{j+1}^{s_j+1}$	$B_j \equiv A_j$
II	$X_{j-1}^{s_j-1}$	$\cong X_j^{s_j}$	$\delta_f(X_j) \sim$	$\epsilon_{q'}(X_{j+1})$	$B_j \supset A_j$
III	$\delta_f(X_{j-1}) \sim$	$\epsilon_{q'}(X_j)$	$X_j^{s_j}$	$\cong X_{j+1}^{s_j+1}$	$B_j \subset A_j$
IV	$\delta_f(X_{j-1}) \sim$	$\epsilon_{q'}(X_j)$	$\delta_f(X_j) \sim$	$\epsilon_{q'}(X_{j+1})$	$B_j \supset A_j$

where we have identified one of B_j, A_j with a subword of the other.

Since J is a subword of a power of X_j we have $L(J) < 2L(X_j)$ so in Case III some translate of J lies in B_j. In the other case $J \subset A_j \subset B_j$. Thus B_j contains a maximal $(n+1)$-normal subword of size $> u_{68}$ hence, by translation, at least $s_j \geqslant N$ if B_j is cyclic and $> -3 + q'e$ if B_j is linear. If B_j is linear call the two outermost of these maximal normals *extremal*, we shall eventually identify this concept with the "extremal" in 10.3. Consider the sequence

$$(1) \qquad A_{j-1}, B_j, \ldots, A_{k-1}, B_k$$

For any linear B_v in (1) except B_k we define its extremal subwords similarly. For B_k take the outermost maximal normals of size $> u_{67}$ as its extremal subwords. For any linear A_i in (1) (then B_{i+1} is linear) define its extremal subwords as the images (cf. 6.16) with respect to $A_i \sim B_{i+1}$ of the extremal subwords of B_{i+1}; they have size $> u_{68} - v_3^* = u_{69}$.

Each term of (1) except A_{j-1} contains a maximal normal subword of size $> u_{68}$. A_{j-1} contains a maximal normal subword of size $> u_{69}$. Hence each cyclic term contains at least $s_i \geqslant N > 3$ maximal normals of size $> u_{69}$.

In Case IV we have for some A, B, C

$$\epsilon_{q'}(X_j) \equiv ABC, \qquad \delta_f(X_j) \equiv B$$

$$L(A) \geqslant 3L(X_j), \quad L(B) \geqslant 3L(X_j)$$

since $L(\epsilon_{q'}(X_j)) - L(\delta_f(X_j)) > L(X_j)(q'e - fe) > 8L(X_j)$ since $q' > f + 8/e$. Thus without loss of generality we may suppose that the subword of B_j with which A_j was identified is of the form of B above. Hence B is *central* in $\epsilon_{q'}(X_j)$, i.e., B lies between the extremals of $\epsilon_{q'}(X_j)$. For a cyclic word all subwords are *central.*

Call a subword of one of (1) C.N. if it is central and maximal normal. Say a C.N. word is *initial special* if its size is $> u_{69}$. Certain C.N. words are called *special* as follows:

1. Any initial special word is special.

2. If X is special in one of A_i, B_{i+1} and its image Y (Im′ or Im*) in the other exists (hence Y is maximal normal and central) then Y is special.

3. If X is special in one of B_i, A_i and is central in the other then it is special in the other.

4. If X is special and X' is a translate of X and X' is central then X' is special.

We claim that any special word U has size $> u_{33}$. There is a sequence $C_1, C_2, \ldots, C_k$ where $C_k \equiv U$, C_1 is initial special and C_{v+1} arises from C_v as in 2, 3 or 4 ($v = 1, \ldots, k-1$). Thus C_v is a subword of one of the words (1) say W_v; thus W_v is $A_{f(v)}$ or $B_{f(v)}$ say ($v = 1, \ldots, k$).

If C_{v+1} is Im* C_v then $W_v \cong W_{v+1}$; let D_v be chosen as in 19.4 such that $W_v - D_v - W_{v+1}$. The assumptions of Section 10 apply to the sequence obtained from $W_1, W_2, \ldots, W_k$ by inserting D_v between W_v, W_{v+1} whenever D_v exists ($str^+ \leqslant k_{22}, k_{23}; u_{69} \geqslant u_{32}$; $\mathrm{Min}(u_{69} - v_3, 1 - str^+ - v_2) \geqslant u_{32}$).

If C_{v+1} arises from C_v as in 2, then if $C_{v+1} = \mathrm{Im}'\, C_v$ we may regard $C_v \to C_{v+1}$ as a transformation 2a of 10.2; if $C_{v+1} = \mathrm{Im}^*\, C_v$ we may regard $C_v \to C_{v+1}$ as a pair of transformations of type 2b.

If C_{v+1} arises as in 3 we may regard $C_v \to C_{v+1}$ as a transformation 3a, 3b, 3c or the trivial case of 2b (when $\underset{n}{=}$ is $\equiv$).

Now let C_{v+1} arise from C_v as in 4. We consider two examples. (i) C_v is C.N. in $A_j \equiv \delta_f(X_j)$ and $C_{v+1} = C_v^+$ is central and $A_j \equiv PC_vQC_{v+1}R$. Then $W_v \equiv A_j$, $W_{v+1} \equiv A_j$. We may regard $C_v \to C_{v+1}$ as a transformation of the form 3a of Section 10 where for J_v we take the word constructed from two copies of A_j by identifying the C_v of the first copy with the C_{v+1} of the second copy; thus J_v is a subword of a power of X_j. (ii) $C_v \subset A_j \equiv X_j^{s_j}$. Here we may regard $C_v \to C_{v+1}$ as a transformation of the form 2b; the X_v of 2b is the linearization of A_j starting with C_v, the Y_{v+1} of 2b is the linearization of A_j starting with C_{v+1} (and $X_v \equiv Y_{v+1}$).

By 10.4, since $s(C_1) > u_{69} \geqslant u_{32}$, we see that $C_k \equiv U$ has size $> u_{33}$, as required.

Consider any term W of (1) and let its special subwords be $S_1, \ldots, S_r$ (then $r \geqslant 2$). If W is cyclic the translate S_1^+ is S_{p+1} say and $S_i^+ = S_{p+i}$ for all i (if $p + i > r$ let S_{p+i} mean S_q where $q \equiv p + i$ (mod r)). If W is linear and S_1^+ is central then $S_1^+ = S_{1+p}$ say and

(2) $$S_i^+ = S_{i+p} \quad (i = 1, \ldots, r-p).$$

If however S_1^+ is not central (then W is of the form A_v) choose any $p \geqslant r$ and then (2) holds vacuously. Call p the *period* of W.

2°. We prove that without loss of generality the periods of B_i, A_i are the same.

Case I is trivial. In the other three cases let the special subwords of W, where W is that one of B_i, A_i which is a subword of the other, be $S_1, \ldots, S_a$. These are central, hence special in the other word W' by the remark about Case IV so the special subwords of W' are $\ldots, S_1, S_2, \ldots, S_a, \ldots$. If $S_1^+ \subset (S_1, S_a)$ then $S_1^+ = S_{1+p}$ and both periods are p. In the opposite case the period of W is arbitrary $\geqslant a$ while the period p of W' is determined but $\geqslant a$. Hence the period of W can be taken to be p.

3°. Let π_i denote the period of B_i and A_i. For the remainder of 3°. We consider Situation A only. We prove

$$\pi_i > \pi_{i+1} \quad (i = j-1, \ldots, k-1).$$

If 1 of 19.2B holds then the number of special words in each side is the same; for any special word S has size $> u_{33} \geqslant u_{10}^*$ so $\mathrm{Im}^* S$ exists, hence $\mathrm{Im}^* S$ is special. Hence $\pi_i s_i = \pi_{i+1} s_{i+1}$. But $s_i < s_{i+1}$ hence $\pi_i > \pi_{i+1}$. Now let 2 hold. Let m be the number of special subwords in $\delta_f(X_i)$ and $\epsilon_{q'}(X_{i+1})$. Then by 2.16A $|m/\pi_i - l_i/L(X_i)| < 3$, $|m/\pi_{i+1} - l_{i+1}/L(X_{i+1})| < 3$ where l_i is the length of the subword generated by the two extremal subwords of A_i; similarly for l_{i+1}. Thus

$$l_i \leqslant feL(X_i), \quad l_{i+1} > q'eL(X_{i+1}) - 2L(X_{i+1})$$

Hence $m/\pi_i < 3 + fe \leqslant q'e - 5 < m/\pi_{i+1}$ so that $\pi_i > \pi_{i+1}$.

Note: Since $\pi_k \geqslant 1$ we have $\pi_j \geqslant k + 1 - j$. Also note that the number of special subwords in any B_i is $> \pi_i(q'e - 5)$ since $\pi_i N > \pi_i(q'e - 5)$ (take $N > q'e - 5$).

4°. Next we show that, in Situation A or B, there exist three special subwords A, B, C of B_k each of size $> u_{67}$ and special subwords A', B', C' of A_{j-1} such that

$$(A, B, C) \underset{n+1}{\hat{=}} (A', B', C') \quad \text{and} \quad (A, B) \underset{n+1}{\hat{=}} (A', B').$$

If follows that A_{j-1} contains a maximal normal subword of size $> u_{68}$ as required; to see this use 9.11 ($str^+ \leqslant k_{18}, k_{19}$; $u_{67} \geqslant u_{33} \geqslant u_{30}$; $u_{67} \geqslant u'_{30}$) and we have that $B'' \subset (A', B', C')$ exists where $(A, B) \triangleq (A', B'')$ and

$$s(B'') > \mathrm{Min}(1 - str^+ - v_2 - u_4 - \epsilon_3, u_{67} - v_3 - u_4 - \epsilon_3) > u'_{30}, u_{68}.$$

Hence $(A', B') \triangleq (A', B'')$. By 9.12 ($u_{33} \geqslant u'_{30}$) B'' is B'.

In Situation B, if we take r consecutive special subwords in B_{i+1} there are r corresponding special subwords in A_i and, since I applies, for any r consecutive specials in A_i there are r corresponding specials in B_i. Thus if we take r consecutive specials in B_k including three A, B, C of size $> u_{67}$ there will be r corresponding specials in A_{j-1} of which A', B', C' say correspond to A, B, C.

Note that, in any case, if S, T are distinct specials in some term of (1) then (S, T) is an $(n+1)$-object, and moreover if S', T' are specials in some term of (1) and $S' = \mathrm{Im}' S$, $T' = \mathrm{Im}' T$ then $(S, T) \triangleq (S', T')$; this is by 8.15 since $str^+ \leqslant k_{16}, k_{17}$ and $u_{33} \geqslant u_{29}$. This remains true if Im′ is replaced by Im* because, as we have seen, any D_i has $((n+1))$-bound $\leqslant \mathrm{Min}(k_{16}, k_{17})$.

Now consider Situation A. We prove that to any m consecutive specials of B_k where $m \leqslant \pi_k (q'e - 6)$, say $S_{d+1}, \ldots, S_{d+m}$, there exist m consecutive specials $U_{d+1}, \ldots, U_{d+m}$ of A_{j-1} such that $(S_{d+i}, S_{d+j}) \triangleq (U_{d+i}, U_{d+j})$ for all i, j such that $1 \leqslant i < j \leqslant m$.

Any r consecutive specials in B_{i+1} determine by 2 above r consecutive specials in A_i. Take r consecutive specials $\alpha_{h+1}, \ldots, \alpha_{h+r}$ in A_i. In Cases I, II, IV there are determined r consecutive specials of B_i. In Case III denote the specials of A_i by $\alpha_1, \ldots, \alpha_t$ without loss of generality and denote those of B_i by $\alpha_x, \ldots, \alpha_y$ where $1 \leqslant x < y \leqslant t$ and $1 \leqslant h+1 < h+r \leqslant t$. One of $\alpha_x, \alpha_{x+1}, \ldots, \alpha_{x+p-1}$ ($p = \pi_i$) is a translate of α_{h+1} say α_{c+1}. For $\alpha_x, \ldots, \alpha_y$ to contain $\alpha_{c+1}, \ldots, \alpha_{c+r}$ we require $c + r \leqslant y$ which is so if $y + 1 - x \geqslant p + r - 1$. Now $y + 1 - x > \pi_i(q'e - 5) \geqslant \pi_i + r > p + r - 1$ if $r \leqslant \pi_i(q'e - 6)$. Thus if we take r consecutive specials of B_k where $r \leqslant \pi_k(q'e - 6)$ then since $\pi_k \leqslant \pi_i$ ($i = j-1, \ldots, k$) there are r corresponding specials in A_{j-1} as required.

Let $T_1, T_2, \ldots, T_b$ be the specials in B_k. Then $b > \pi_k(q'e - 5) \geqslant 3\pi_k$

and at least three of $T_1, T_2, \ldots, T_{3\pi_k}$ have size $> u_{67}$. Since $3\pi_k \leqslant \pi_k(q'e - 4)$, $T_1, T_2, \ldots, T_{3\pi_k}$ have images $U_1, U_2, \ldots, U_{3\pi_k}$ in A_{j-1}.

19.6. Corollary. *In Situation A, if B_k contains properly a maximal $(n+1)$-normal subword of size $> u_{67}$ then k is at most equal to the number of maximal $(n+1)$-normal subwords of size $> u_{33}$ in X_1^e.*

Proof. Let there be r special subwords in A_1. Without loss of generality the period π_1 is at most r. But $\pi_1 \geqslant k$, hence $k \leqslant r$. Any special subword is maximal normal of size $> u_{33}$. If A_1 is $\delta_f(X_1)$ then $A_1 \subset X_1^e$. If A_1 is $X_1^{s_1}$ then A_1 contains X_1^4 and the number of specials, hence maximal normal subwords, in X_1^4 is $> \pi_1$, hence the number of maximal normals in X_1^e is $> \pi_1 \geqslant k$.

19.6A. For any word M let $\mu_i(M) = L(M)/L(X_i^{e_i})$ $(i = 1, \ldots, k)$

19.6B. In Situation A there are two cases: *Case* (a) j exists such that 2 of 19.2B holds (with i replaced by j); *Case* (b) not Case (a).

In Case (a) let i denote the largest such j. Then $A_i \sim B_{i+1}$ and B_{i+1} has the form $\epsilon_{q'}(X_{i+1})$. It may happen that $L(B_{i+1}) > qeL(X_{i+1})$. i.e., B_{i+1} also has the form $\epsilon_q(X_{i+1})$; if so define q'' to be q; if not define q'' to be q'. Thus $\mu_{i+1}(B_{i+1}) > q''$.

19.7. Proposition. *In Situation* A, *let B_k contain a maximal $(n+1)$-normal subword of size $> u_{67}$. Then in Case* (*a*) *there are subwords A of A_1, B of B_k such that $\mu_1(A) < f + 6/e$, $\mu_k(B) > q'' - 16/e$ and* (†) *there is a word $A^{\mathrm{L}}CBDA^{\mathrm{R}}$, $((n+1))$-bounded by k_{43} and equal to A in G_{n+1}. In Case* (*b*) *we have*

(*b*1) *if a subword B of B_k is given and $4/e \leqslant \mu_k(B) \leqslant \mu_k(B_k) - 12/e$ then there is a subword A of A_1 such that $\mu_1(A) < \mu_k(B) + 14/e$ and again* (†) *holds. Also $N/e \leqslant \mu_k(B_k)$.*

(*b*2) *if a subword A of A_1 is given and $\mu_1(A) > 9/e$ then there is a subword B of B_k such that $\mu_k(B) > \mu_1(A) - 12/e$ and again* (†) *holds.*

Proof. (b1) Let b be the number of specials of B_k in B. Then $|b/\pi_k - L(B)/L(X_k)| < 3$. Now $4L(X_k) \leqslant L(B) \leqslant L(B_k) - 12L(X_k)$ so $b > \pi_k(L(B)/L(X_k) - 3) \geqslant \pi_k \geqslant 1$, hence $b \geqslant 2$. Now $B_k \supset XBY$ where $L(X) = L(Y) = 6L(X_k)$. The number of specials in X is at least four since $\pi_k(6-3) \geqslant 3$. Thus there are specials $V_1, \ldots, V_m$ of B_k such that those in B are $V_5, \ldots, V_{m-4}$; $b = m - 8$. By 4° above there are specials $U_1, \ldots, U_m$ of A_1 such that $A \equiv (U_1, U_m) \triangleq (V_1, V_m)$. By 9.15, since $u_{33} \geqslant u_{39}$ and $str^+ \leqslant k_{24}, A \underset{n+1}{\equiv} U_1^{\mathrm{L}} E(V_2^{\mathrm{R}}, V_{m-1}^{\mathrm{L}}) F U_m^{\mathrm{R}}$, where the right side is $((n+1))$-bounded by $\mathrm{Max}(str^+, k'_{24}) + u_{40} = k_{43}$. We have $(V_2^{\mathrm{R}}, V_{m-1}^{\mathrm{L}}) \supset (V_4, V_{m-3}) \supset B$. Now $|L(A)/L(X_1) - m/\pi_1| < 3$ so

$$\mu_1(A) < m/e\pi_1 + 3/e < (b+8)/e\pi_k + 3/e$$

$$\leqslant b/e\pi_k + 11/e < \mu_k(B) + 14/e$$

where the right side contains $(V_4, V_{m-3}) \supset B$. Now $|L(A)/L(X_1) - m/\pi_1| < 3$ so

$$\mu(A) < m/e\pi_1 + 3/e < (b+8)/e\pi_k + 3/e$$

$$\leqslant b/e\pi_k + 11/e < \mu(B) + 14/e.$$

(b2). Let $\theta = \mu_1(A)$. Let the specials of A_1 in A be say $U_1, \ldots, U_m$. Then $m/\pi_1 > L(A)/L(X_1) - 3 = \theta e - 3 > 6$, so $m \geqslant 7$. These m specials of A_1 determine m specials $V_1, \ldots, V_m$ of B_k. Let $A' \equiv (U_1, U_m)$ $B \equiv (V_4, V_{m-3})$. Then

$$L(B)/L(X_k) > (m-6)/\pi_k - 3 \geqslant m/\pi_k - 9 \geqslant m/\pi_1 - 9 > \theta e - 12$$

$$A \supset (U_1, U_m) = U_1^{\mathrm{L}} E(V_2^{\mathrm{R}}, V_{m-1}^{\mathrm{L}}) F U_m^{\mathrm{R}}$$

$$(V_2^{\mathrm{R}}, V_{m-1}^{\mathrm{L}}) \supset B.$$

(a) In the proof of 19.5 we saw that r consecutive specials of A_i determine r of B_i, hence r of A_{i-1}, if $r \leqslant \pi_i(q'e - 6)$. Since $\pi_i < \pi_{i-1} < \cdots < \pi_1$ it follows that if $r \leqslant \pi_i(q'e - 6)$ then r consecutive specials of A_i determine r of A_1.

In the present case the total number t of specials in $A_i = \delta_f(X_i)$ satisfies $t/\pi_i < L(\delta_f(X_i))/L(X_i) + 3 \leqslant ef + 3 \leqslant q'e - 6$ so they determine specials $U_1, ..., U_t$ of A_1. The number of specials in B_{i+1} is also t, and

$$t/\pi_{i+1} > -3 + (L(\epsilon_{q''}(X_{i+1})) - 2L(X_{i+1}))/L(X_{i+1}) > q''e - 5.$$

If $i < k-1$, $B_{i+1} \subset X_{i+1}^{s_{i+1}} \cong \cdots \cong X_k^{s_k}$, so the specials of B_{i+1} determine t specials of B_k say $V_1, ..., V_t$. This is also true if $i = k-1$. Finally

$$L(V_4, V_{t-3})/L(X_k) > (t-6)/\pi_k - 3 \geqslant t/\pi_{i+1} - 9 > q''e - 14$$

$$L(U_1, U_t)/L(X_1) \quad < t/\pi_1 + 3 \leqslant t/\pi_i + 3 \leqslant ef + 6$$

$$(U_1, U_t) = U_1^{\mathrm{L}} E(V_2^{\mathrm{R}}, V_{t-1}^{\mathrm{L}}) F U_t^{\mathrm{R}}$$

$$(V_2^{\mathrm{R}}, V_{t-1}^{\mathrm{L}}) \supset (V_4, V_{f-3}).$$

19.8. Consider Situation A. For $i = 1, ..., k-1$ let $r(i)$ be the least integer q such that $A_i \underset{q}{\cong} B_{i+1}$ or $A_i \underset{q}{\sim} B_{i+1}$. Then $0 \leqslant r(i) \leqslant n+1$. If $r(i) > 0$ then A_i and B_{i+1} contain maximal $r(i)$-normals of size $> 1 - str^+ - v_2^* > u_{70}$.

Call $k-1$ *critical*; and if $1 \leqslant j \leqslant k-2$ call j *critical* if $r(j) > \mathrm{Max}(r(j+1), ..., r(k-1))$. Let the critical integers be $\lambda_1, \lambda_2, ..., \lambda_t$ in ascending order, where λ_t is $k-1$. Put $\lambda_0 = 0$.

Let $0 \leqslant i < j \leqslant t$, put $a = \lambda_i$, $b = \lambda_j$, and consider

$$A_{a+1}, B_{a+2}, A_{a+2}, ..., B_b, A_b, B_{b+1}$$

calling this sequence ξ_{ij}.

We shall consider two cases:

(1$^+$) $A_v \underset{r(v)}{\cong} B_{v+1} \quad (v = a+1, ..., b)$
i.e., 1 holds throughout ξ_{ij}, and moreover we do not have that 1 holds throughout $\xi_{i-1\,j}$ or that 1 holds throughout $\xi_{i\,j+1}$.

(2$^+$) $j = i+1$ and 1 does not hold throughout ξ_{ij}.

First consider (1$^+$): If $i > 0$ then a is critical; $i < j \leqslant t$ so

$a = \lambda_i < \lambda_t = k - 1$, hence $r(a) > \mathrm{Max}(r(a+1), \ldots, r(k-1)) \geqslant 0$, so B_{a+1} contains maximal $r(a)$-normal subwords of size $> u_{70}$ hence so does A_{a+1} since A_{a+1} is cyclic. Now Situation B holds for ξ_{ij} and Situation B is left-right symmetrical. By examination of the P.C.'s used, 19.5 holds if u_{67}, u_{68} are replaced by $u'_{67} = u_{70}, u'_{68} = u_{67}$ respectively, since $u_{70} > u_{67} > u_{68}$; $1 - str^+ - v_2^* > u'_{67}$; $1 - str^+ - v_2 - u_4 - \epsilon_3 > u'_{68}$; $u'_{67} > U'_{68} + v_3^*$; $u'_{67} > u'_{68} + v_3 + u_4 + \epsilon_3$. Hence B_{b+1} contains an $r(a)$-normal subword of size $> u'_{68} = u_{67}$. By 19.7 we have

(∗) if $B \subset B_{b+1}$ and $4/e \leqslant \mu(B) \leqslant (N-12)/e$ then there is a subword A of A_{a+1} such that $A = A^{\mathrm{L}} CBDA^{\mathrm{R}}$, which we abbreviate to $A = -B-$, and $\mu_{a+1}(A) < \mu_{b+1}(B) + 14/e$.

If $i = 0$ and $1 < t$ then $a = 0$, $1 \leqslant j$ so $\lambda_1 \leqslant \lambda_j = b$. Also $\lambda_1 < \lambda_t = k - 1$ and $r(\lambda_1) > \mathrm{Max}\ r(v)$ over $v > \lambda_1$ which is $\geqslant 0$. If $1 \leqslant \sigma < \lambda_1$ then $r(\sigma) \leqslant r(\lambda_1)$ since λ_1 is the smallest critical integer. B_{λ_1+1} contains an $r(\lambda_1)$-normal subword of size $> u_{70}$. We prove that B_{b+1} contains an $r(\lambda_1)$-normal subword of size $> u_{67}$. If $\lambda_1 = b$ this is trivial since $u_{70} \geqslant u_{67}$. If $\lambda_1 < b$ then $a + 1 = 1 \leqslant \lambda_1 + 1 \leqslant b$; hence A_{λ_1+1} is in ξ_{ij} so is cyclic, and contains an $r(\lambda_1)$-normal subword of size $> u_{70}$. Arguing as before, B_{b+1} contains an $r(\lambda_1)$-normal subword of size $> u_{67}$. Thus (∗) holds.

If $i = 0$ and $t = 1$ then $a = 0$, $b = k - 1$. If $r(b) \neq 0$ then B_{b+1} contains an $r(b)$-normal subword of size $> u_{70}$ and (∗) holds. If $r(b) = 0$ then $A_{a+1} \equiv B_{b+1}$; also $L(X_{a+1}) \geqslant L(X_{b+1})$ since $s_{a+1} \leqslant s_{b+1}$. Thus (∗) holds taking $A \equiv B$ (then $\mu_{a+1}(A) \leqslant \mu_{b+1}(B)$).

Consider (2^+): None of $r(a+1), \ldots, r(b-1)$ is $> r(b)$. If $r(b) > 0$ then B_{b+1} contains an $r(b)$-normal subword of size $> u_{70} > u_{67}$, hence by 19.7 there are subwords A of A_{a+1}, B of B_{b+1} where

$$\mu_{a+1}(A) < f + 6/e, \quad \mu_{b+1}(B) > q'' - 16/e, \quad \text{and} \quad A = -B-.$$

This is also true if $r(b) = 0$.

19.9. Proposition. *In Situation A^+ either*

(i) $X_i^{s_i} \underset{n+1}{\cong} X_{i+1}^{s_{i+1}}$ $(i = 1, \ldots, k-1)$, or

(ii) A_1 *contains a subword* A' *and* B_k *contains a subword* B' *where* $\mu_k(B') > q - 28/e$, $\mu_1(A') < f + 20/e$ *and* $A' \underset{n+1}{=} A'^{\mathrm{L}} CB'DA'^{\mathrm{R}}$ *where the right side is* $((n+1))$*-bounded by* k_{44}.

Proof. Assume (i) is false. Then some ξ_{ij} satisfies (2^+). Assume j chosen maximally. Thus either $j = t$ or ξ_{jt} satisfies (1^+). Also $j = i + 1$. Moreover, for the largest v such that $A_v \sim B_{v+1}$ we have

$\delta_f(X_v) \sim \epsilon_q(X_{v+1})$ (where we have q instead of q').

There is a subset $k_0, k_1, ..., k_p$ of $0, 1, ..., j$ such that $k_0 = 0$, $k_p = j$, $k_{p-1} = i$ and $\xi_{k_i k_{i+1}}$ satisfies (1^+) or (2^+) ($i = 0, ..., p-1$).

The following is true for $v = p - 1$ and we prove by induction that it is true for $v = 0, 1, ..., p - 1$.

(**) For $\xi_{k_v k_p}$ the first term contains A' and the last term contains B' such that $A' = -B'-$, $\mu(B') > q - 16/e$ and either (i) $\mu(A') < f + 20/e$ and $\xi_{k_v k_{v+1}}$ satisfies (1^+), or (ii) $\mu(A') < f + 6/e$.

Assume that (**) is true for some v, $0 < v \leqslant p - 1$. First suppose that (i) holds. Then $\xi_{k_{v-1} k_v}$ satisfies (2^+). Hence its first term contains A'' and its last term contains B'' where $A'' = -B''-$ and $\mu(B'') > q' - 16/e$, $\mu(A'') < f + 6/e$. Since $q' > f + 37/e$ we have $\mu(B'') > \mu(A') + 1/e$, so $L(B'') > L(A') + L(X_{d+1})$ where $d = \lambda_{k_v}$, hence some translate of A' is contained in B'', since $A' \subset A_{d+1}$, $B'' \subset B_{d+1}$. Hence $A'' = -B''- = -B'-$.

Now suppose that (ii) holds. If $\xi_{k_{v-1} k_v}$ satisfies (1^+) take the B of (*) to be A' if $\mu(A') \geqslant 4/e$ while if $\mu(A') < 4/e$ take B to satisfy $\mu(B) = 5/e$ (then B contains a translate of A'); this is possible since without loss of generality $\mu(B) < f + 6/e \leqslant (N - 12)/e$. Then $\mu(A) < \mu(B) + 14/e < f + 20/e$.

If $\xi_{k_{v-1} k_v}$ satisfies (2^+) some translate of A' is contained in B'' since $\mu(A') + 1/e < f + 7/e < q' - 16/e < \mu(B'')$.

Thus by induction, (**) is true for $v = 0$. Thus ξ_{ij} has first term containing A' and last term containing B' where $\mu(B') > q - 16/e$ and $\mu(A') < f + 20/e$. If $j = t$ we have finished since $k_{44} = k_{43}$. If $j \neq t$ apply (b2) with A defined to be B'; $(q - 16/e > 9/e)$.

19.10. Proposition. *There is no infinite sequence $Y_1, Y_2, Y_3, ...$ such that $Y_i \; \sigma_{n+1} \; Y_{i+1}$ ($i = 1, 2, 3, ...$), i.e., Axiom 5 holds for $n + 1$.*

Proof. Assume not. Again let $r(i)$ be the least q such that

$Y_i \sigma_q Y_{i+1}$. Then $0 \leqslant r(i) \leqslant n+1$. Let $\xi = \text{Max } r(i)$ for $i = 1, 2, 3, \ldots$. Then $\xi \leqslant n+1$. If $\xi < n+1$ then $Y_i \sigma_n Y_{i+1}$ $(i = 1, 2, \ldots)$ contrary to Axiom 5. Hence $\xi = n+1$. Consider $Y_p \sigma_{r(p)} Y_{p+1}$, where $r(p) = \xi$, and consider $Y_1, \ldots, Y_{p+1}$. Now Y_{p+1} contains a maximal ξ-normal subword so by 19.6 some N exists such that $p+1 \leqslant N$ then $r(i) < \xi \leqslant n+1$, so $Y_i \sigma_n Y_{i+1}$ $(i = N, N+1, \ldots)$, a contradiction.

19.11. Proposition. *Axiom 29 holds for $n+1$; i.e., if $X \in J_n$ $Y \in C_{n+1}$ and $X >_n Y$ then either*

(†) *some $\delta_f(X)$ contains a subelement of L_{n+1} of size $> q_0$, or*

(††) *$X^s \underset{n}{=} U^{-1} Y^t \cdot U$ for some $U \in \Pi$ and some s, t such that* $1 \leqslant s < |t|$.

Proof. 19.9 is available (with n instead of $n+1$), with $X_1 \equiv X^{\pm 1}$, $X_k \equiv Y^{\pm 1}$. If (i) holds then (††) holds. Now let (ii) hold. A subword A' of a positive power of X_1 and a subword B' of a positive power of X_k exist such that $\mu_1(A') < f + 20/e$, $\mu_k(B') > q - 28/e$, and $A' \underset{n}{=} A'^{\text{L}} CB'DA'^{\text{R}}$. Taking a subword of B' if necessary we may without loss of generality suppose B' is a subword of $X_k^e \equiv Y^{\pm e} \in L_{n+1}$. Now $s^{n+1}(B') > L(B')/L(X_k^e) - 3/e - \epsilon_1^* > q - 31/e - \epsilon_1^* > u_3$. Since 6.10 holds for $n+1$, A' contains a subelement of L_{n+1} of size $> q - 31/e - u_4 - \epsilon_1^*$ $(k_{44} \leqslant b_{25})$.

Let $A' \equiv A_1 A_2$ where $L(A_1) = [\frac{1}{2} L(A')]$. Then A_1 or A_2 contains a subelement of L_{n+1} of size $> \frac{1}{2}(q - 31/e - \epsilon_1^* - u_4) - 1/e = q_0$. Now $L(A_1) \leqslant L(A_2) \leqslant \frac{1}{2}(1 + L(A'))$.

$$1 + L(A') < L(X_1) + L(X_1)(ef + 20)$$

$$= L(X_1)(ef + 21) < 2ef\, L(X_1).$$

Hence A_1 and A_2 each have the form $\delta_f(X_1) = \delta_f(X)$.

§20

20.1. We now know that all the axioms hold for $n+1$ so it remains to complete the proof of the infiniteness of B_d^e as described in the introduction to Chapter II.

20.2. Proposition. *If for some $W \in \Pi$ we have*

$$W^e \neq I \quad in \quad \Gamma\left(\dot{\bigcup_{i=1}^{\infty}} C_i\right)$$

then $\bigcap_{n=0}^{\infty} J_n$ is not empty.

Proof. We use the results of Section 17. For $n = 1, 2, 3, \ldots$, choose W_n such that $W \operatorname{conj}_n W_n \in J_n$ and $X \operatorname{conj}_n W \Rightarrow B_n(W_n) < B_n(X) + \epsilon$ where $\epsilon = 1/e^3$. Since $W \operatorname{conj}_{n+1} W$, $B_{n+1}(W_{n+1}) < B_{n+1}(W) + \epsilon$. Since $W \operatorname{conj}_n W_n$, $B_{n+1}(W) = B_{n+1}(W_n)$. Hence

$$B_{n+1}(W_{n+1}) < B_{n+1}(W_n) + \epsilon.$$

Since $W_n \in J_n$, (W_n) is (n)-bounded by $str \leqslant k_{45}$. By 17.15 $B_{n+1}(W_n) < u_{71}$ or $B_n(W_n) > 2B_{n+1}(W_n)$.

Case 1. $B_n(W_n) > 2B_{n+1}(W_n)$ $(n = 1, 2, 3, \ldots)$

Then

$$B_1(W_1) > 2B_2(W_1) > 2(B_2(W_2) - \epsilon) > 2(2B_3(W_2) - \epsilon)$$
$$> 2(2(B_3(W_3) - \epsilon) - \epsilon) > \cdots$$

Let s be the least integer such that $B_1(W_1)/u_{63} \leqslant 2^s$ and $s \geqslant 1$.

Let $t > s$; we prove (i) $B_{t+1}(W_t) = 0$, (ii) $B_{t+1}(W_{s+1}) = 0$.

Assume $B_{t+1}(W_t) > 0$; then $B_{t+1}(W_t) > u_{63}$ and

$$B_1(W_1) > 2^t B_{t+1}(W_t) - \epsilon(2^{t-1} + 2^{t-2} + \cdots + 2) > 2^t u_{63} - \epsilon(2^t - 2)$$
$$> 2^t u_{63}(1 - \epsilon/u_{63}) > 2^{t-1} u_{63}$$

since $u_{63} > 2/e^3$. Hence $2^s \leqslant 2^{t-1} < B_1(W_1)/u_{63}$. This proves (i). Now W_t conj$_t$ W conj$_{s+1}$ W_{s+1} hence W_t conj$_t$ W_{s+1} and $0 = B_{t+1}(W_t) = B_{t+1}(W_{s+1})$ proving (ii).

Now we prove $W_{s+1} \in \cap J_n$. We have $W_{s+1} \in J_{s+1} \subset J_s \subset \cdots \subset J_0$. Assume inductively that $W_{s+1} \in J_t$ for some $t > s$. Then $B_{t+1}(W_{s+1}) = 0 < u_{71}$ and (W_{s+1}) is (t)-bounded by $str = k_{46}$. Hence by 17.16 (W_{s+1}) is $((t+1))$-bounded by str; thus $W_{s+1} \in J_{t+1}$.

Case 2. Not Case 1. For some $s = 1, 2, 3, \ldots$ $B_{s+1}(W_s) < u_{71}$. We prove $B_{t+1}(W_t) < u_{71}$ $(t = s, s+1, \ldots)$. This is true for $t = s$; assume inductively that $B_{t+1}(W_t) < u_{71}$ for some $t \geqslant s$. Then $B_{t+1}(W_{t+1}) < B_{t+1}(W_t) + \epsilon < u_{71} + \epsilon$. If $B_{t+2}(W_{t+1}) \geqslant u_{71}$ then by 17.15 $u_{71} > B_{t+1}(W_{t+1}) - \epsilon > 2B_{t+2}(W_{t+1}) - \epsilon \geqslant 2u_{71} - \epsilon$. Hence $\epsilon > u_{71} \geqslant 1/e^3$, a contradiction. Hence $B_{t+2}(W_{t+1}) < u_{71}$.

Now for $t \geqslant s$ W_s conj$_s$ W conj$_t$ W_t so W_s conj$_t$ W_t. Hence $B_{t+1}(W_s) = B_{t+1}(W_t) < u_{71}$. Now $W_s \in J_s \subset J_{s-1} \subset \cdots \subset J_0$. Assume $W_s \in J_t$ for some $t \geqslant s$. (W_s) is $((t))$-bounded by str and $B_{t+1}(W_s) < u_{71}$ hence (W_s) is $((t+1))$-bounded by str. Hence $W_s \in J_{t+1}$ and $W_s \in \cap J_n$ by induction.

20.3. Proposition. $\bigcap_{n=0}^{\infty} J_n$ *is empty. Hence by* 20.2 *we have*

$$B_d^e = \Gamma(C_1, C_2, C_3, \ldots).$$

Proof. Suppose not and let $U \in \cap J_n$. Let F be the least integer $> 1 + fe$. Now U^{F} has only a finite number of subwords $X_1, \ldots, X_r$. If X_i is a subelement of some L_j of size $> q_0$ then j is unique. Hence p exists such that U^{F} contains no subelement of $L_{p+1} \cup L_{p+2} \cup \cdots$ of size $> q_0$.

Let $m \geqslant 1$. Then $U \in J_m$ so $U \notin C_m$ and $U \in J_{m-1}$. By 5.7

$$(\exists Y)\, Y \in C_m,\ \ U >_{m-1} Y.$$

By Axiom 29 either

(a_m) some $\delta_f(U)$ contains a subelement of L_m of size $> q_0$, or

(b_m) $U^s \underset{m-1}{=} T^{-1} \cdot Y^t \cdot T$, $1 \leqslant s < |t|$.

Let $1 \leqslant m' < m$; we prove that not both (b_m), $(b_{m'})$ hold. Otherwise $U^{s'} \underset{m'-1}{=} T'^{-1} \cdot Y'^{t'} \cdot T'$, $Y' \in C_{m'}$ $Y'^e \in L_{m'}$ $Y'^e = I$ in

$G_{m'}$ hence in G_{m-1}. Hence in G_{m-1} we have $Y^{ts'e} = T \cdot U^{ss'e} \cdot T^{-1} = TT'^{-1} Y'^{t'se} T'T^{-1} = I$. Now $Y \in C_m \subset J_{m-1}$ so by 5.12 $Y^k \neq I$ in G_{m-1} $(k = 1, 2, \ldots)$, a contradiction.

Hence (a_m) holds for infinitely many m. Since $U^F \supset \delta_f(U)^{\pm 1}$ we have the desired contradiction.

Apart from the consistency of the P.C.'s, this completes the proof of the infiniteness of B_d^e for all sufficiently large odd e $(d \geqslant 2)$, and we note that B_d^e is a Generalized Tartakovskii Group.

§21. Consistency

21.1. Preliminary remarks. In the text, beginning with Section 5, the parameter conditions (P.C.), i.e., equations or inequalities involving parameters only, are emphasized only at first, since the reader will soon perceive that the P.C. in a given section of the text are easily identifiable, an understanding of the text being unnecessary for this purpose.

The parameters not in the axioms but arising in the text are mainly denoted by the letters u, k with numerical subscripts. Occasionally Greek letters are used for temporary abbreviations for expressions involving parameters.

21.2. The main parameters may be taken to have values as in the following table, where $\epsilon = 10^{-2}$.

Parameter	f	r_0'	u_{11}	r_0	r''	r'	r
Value	$e^{-\frac{7}{8}-\epsilon}$	$e^{-\frac{3}{4}}$	$e^{-\frac{3}{4}+\frac{\epsilon}{2}}$	$e^{-\frac{5}{8}}$	$e^{-\frac{5}{8}+\epsilon}$	$e^{-\frac{1}{2}}$	$e^{-\frac{3}{8}}$

Moreover

$$q' = r_0' - O(1/e) \quad q^* = u_{11} - O(r_0') \quad q_0 = q - O(r_0')$$

$$q = r_0 - O(r_0') \quad r^* = r'' - O(r_0) \quad c^* = r'' + O(r_0)$$

Here u_{11} and r'' do not occur in the axioms; they first occur in 7.3.

21.3. The parameters in the axioms not already mentioned have orders as follows.

$O(1/e)$: $s_{30}, \epsilon_2, \epsilon_{24}, \epsilon_{30}$

$O(r'_0)$: $a_{33}, d_{33}, h_{33}, r_{24}, r_{25}, r_{31}, r_{33}, s_{24}, s_{25}, s_{31}, \epsilon_1^*, \epsilon_3, \epsilon_4, \epsilon'_{30}$

$O(r_0)$: $a_8, a_{31}, c_{26}, c_{33}, c_{39}, d_{24}, d_{39}, d'_{39}, f_{39}, g_{31}, g_{39}, h_{31}, j_{31}, j_{33}, m_{33}, p_{33}, q_{24}, q_{31}, q_{33}, r_{26}, s_{26}, s_{33}, s_{39}, s_{40}, t_{26}, t_{33}, v_{26}, w_{26}, w_{33}, z_{26}, \epsilon_{33}$

$\frac{1}{2} + O(r_0)$: $b_{28}, b_{\min}, cb_{\min}, s_0, str, str^+$

$1 - O(r)$: $a_{28}, b_\infty, b_2, b_{24}, b_{25}, b'_{25}, b'_{26'}, b_{26'}, b_{30}, b_{31}, b_{33}, b_{39}, b'_{39}, b_{40}, c_{28}, d_{28}, f_{31}, f_{33}, g_{33}$

$1 - O(r_0)$: h_0.

21.4. The parameters in the main text may be taken to have orders as follows.

$O(r'_0)$: v_4

u_i for i = 1, 2, 3, 4, 8, 12, 13, 14, 16, 20, 37, 41, 48, 49, 60, 62, 63, 64, 65, 81, 91, 92, 94, 96, 97, 98, 99

$O(r_0)$: v_j for j = 1, 2, 3, 5, 8

u_i for i = 5, 6, 7, 9, 10, 21, 22, 23, 24, 25, 26, 27, 28, 29, 30, 31, 32, 33, 39, 40, 42, 43, 44, 45, 46, 47, 50, 51, 52, 54, 56, 57, 59, 61, 66, 67, 68, 69, 70, 71, 72, 73, 74, 75, 77, 78, 79, 80, 82, 83, 84, 89, 90, 93, 95

u'_j for j = 10, 15, 28, 30, 40, 43, 83, 78, 79, 84, 67, 68

$r_6^0, u''_{10}, u''_{40}, u^*_{10}, v_2^*, v_3^*$

$1 - O(r)$: k_i for i = 1, 2, 3, 4, 6, 8, 10, 12, 14, 16, 18, 20, 22, 24, 25, 26, 27, 28, 29, 30, 31, 34, 36, 43, 44, 47, 48, 52, 53, 54, 55, 56, 57, 58, 59, 60, 61

$k'_8, k''_8, k'_{12}, k'_{14}$

$1 - O(r_0)$: k_i for i = 5, 7, 9, 13, 15, 17, 19, 21, 23, 32, 33, 35, 37

$k_1^0, k_2^0, k_3^0, k'_9, k''_9, k'_{13}, k'_{15}, u_{58}$

$\frac{1}{2} + O(r_0)$: k_i for i = 39, 40, 41, 42, 45, 46, 50, 51

u_{35}, k'_{24}

$O(1/e)$: $\epsilon_1, \epsilon_5, u_{36}$

$O(r)$: r_6

$O(r'')$: u_{34}/e

Further asymptotic values: $k_{11} = 1 - c^*$, $u'_{14} = r$, $u_{15} = c^*$, $u_{17} = r^*$, $u_{18} = q^*$, $u_{19} = \frac{1}{2}$, $u_{38} = q^* - O(r'_0)$, $u_{53}/e = q'$, $u_{55} = q$, $u_{85} = q^*$, $u_{86} = r^*$, $u_{87} = q^*$, $u_{88} = r''$.

This completes the proof that for all sufficiently large odd e $(d \geqslant 2)$ B^e_d is infinite and is a Generalized Tartakovskii Group.

References

[1] M. Greendlinger, On Dehn's algorithm for the word problem, Comm. Pure App. Math. 13 (1960) 67–83.

[2] J. Leech, A problem on strings of beads, Math. Gaz. 41, No. 338 (1957) 277–278.

[3] R.C. Lyndon, On Dehn's algorithm, Math. Ann. 166 (1966) 208–228.

[4] M. Morse and C.A. Hedlund, Unending chess, symbolic dynamics and a problem in semi-groups, Duks Math. J. 11 (1944) 1–7.

[5] P.S. Novikov and S.I. Adjan, On infinite periodic groups, Izv. Akad. Nauk SSSR, Ser. Mat. 32, 212–214, 251–524, 709–731 (Russian).

[6] V.A. Tartakovskii, Solution of the word problem for groups with a k-reduced basis for k > 6, Izv. Akad. Nauk SSSR, Ser. Mat. 13 (1949), 483–494 (Russian).

THE ALGEBRAIC INVARIANCE OF THE WORD PROBLEM IN GROUPS

F.B. CANNONITO*

University of California, Irvine

This paper will be a sequel to my paper [2]; accordingly, I will lean very heavily on it for notation and concepts. Let K_1 and K_2 be two subclasses of the class of recursive functions such that $K_1 \subset K_2$ and $K_1 \neq K_2$. Later, candidates will be proposed for K_1 and K_2; for the present the intuition is this: functions in $K_2 \setminus K_1$ are more costly to compute than those in K_1. I will use K_1 and K_2 in connection with the word problem (w.p. henceforth) in finitely generated (f.g.) groups, in order to discover sufficient conditions on K_1 and K_2 which will insure that the "cost" of the solution of the w.p. is an *algebraic invariant* of the group. That is, if K_1 and K_2 are properly chosen, then either all f.g. presentations of the group have w.p. solvable by a function in K_1 or they all have w.p. solvable only by a function from $K_2 \setminus K_1$. When K_1 and K_2 are properly chosen I will call them "properly chosen."

The restriction to f.g. presentations is not artificial since groups can be presented on infinitely many generators so that the w.p. for the infinite system is unsolvable even though this is not so for f.g. presentations. On the other hand, it makes sense to ask if the cost is an algebraic invariant of f.g. presentations since Rabin has shown that when one f.g. presentation has a solvable w.p. so must presentation has a solvable w.p. so must any other, although conceivably the cost might vary. Of course, some reasonable measures of cost may not turn out to be algebraic invariants, neither must two distinct measures be compatible – that is, assign the

* This work was supported by AFOSR Grant No. 1321-67.

same *relative* cost on different groups. At any rate, in this paper I will only consider conditions on K_1 and K_2 which are sufficient to insure the invariance of cost and leave the other questions to another work.

Henceforth, for any positive integer n, let the expression $(*n)$ appearing in the margin indicate a condition on K_1 (and K_2), the integer to serve for bookkeeping purposes. Thus, assume a fixed Gödel numbering γ of the words on the alphabet $a_1, a_2, a_3 \ldots$ such that the free groups of rank $n < \infty$, denoted F_n and presented as $\langle a_1, \ldots, a_n | \rangle$, have the property

(*1) F_n has a K_1-admissible index γ consisting of the set $\gamma(F_n)$ of integers (indices), the (induced) multiplication $m : \gamma(F_n)^2 \to \gamma(F_n)$, and the (induced) inverse function $\mathrm{IN} : \gamma(F_n) \to \gamma(F_n)$.

Note. In [2] the admissible indices were specifically $\mathcal{E}^\alpha$-admissible, where $\mathcal{E}^\alpha$ is a member of the Grzegorczyk hierarchy. Here, K_1 (or K_2) will replace the $\mathcal{E}^\alpha$, mutatis mutandis. (Sometimes I will write K_i or K for either K_1 and K_2.)

It is possible to proceed as in [2] and define *arbitrary* K_i-admissible indices ($i = 1,2$) for f.g. groups. But with certain exceptions, discussed below, the most natural way to index a group is from the index γ of F_n, given a presentation Π of the group in the form $\langle a_1, \ldots, a_n | R, S, T, \ldots \rangle$. Such indices are called *standard indices* in [2]. Intuitively considered, a standard index for a group with presentation Π is obtained ny assigning each coset in F_n/N, (N = the normal closure of $R, S, T, \ldots$ in F_n) its minimal index. From the proof of Theorem 5.1 of [2] we see when the K_i satisfy

(*2) K_1 contains the constant functions and K_1 is closed under the functions $+$, $\cdot$, $\dot{-}$, substitution, bounded sums: $\sum_{u<v}$, and bounded products: $\prod_{u<v}$; (likewise for K_2),

then the propositional connectives, and the bounded quantifiers do not lead out of the class of K_1-decidable relations, and K_1 is closed under the bounded μ-operator. Identical properties hold for K_2.

Note. From (*2) it follows that K_1 (hence K_2) contains the Kalmar elementary functions; cf. Kleene [10] p. 285, for [a/b] is definable from the functions and operators given in (*2) given the K_1 definability of "$\leqslant$" and "$<$". But these follow from $\dot{-}$.

In particular, when (*1) and (*2) hold, the condition $K_1 \subset K_2$ implies that for a group presented as F_n/N, the solvability of its w.p. by a K_i function, ($i = 1,2$), implies that F_n/N has a K_i-admissible standard index, for this is precisely the content of Theorem 5.1 of [2].

Note. In order to define the relation $E(r, s)$ on the indices which holds when and only when the element with index r, denoted by g_r (likewise g_s) determines the same coset as g_s, it is necessary to assume the logical conjunction of K_1-decidable (resp. K_2-decidable) relations is K_1-decidable (resp. K_2-decidable), (cf. p. 380, line 11, of [2]). If the characteristic function C_R of a relation R is 0 on the extension of R and 1 otherwise, then $C_{R\&S} = C_R + C_S \dot{-} C_R \cdot C_S$. On the other hand, $C_{\neg R} = 1 \dot{-} C_R$ and so assuming (*2) gives the closure of the K_i-relations under all propositional connectives. It remains to be seen if under narrower assumptions on K_1 than those given, K_1 is sufficiently expressive so that (*1) is satisfied. In connection with this David Thompson, in a private communication, has shown that the F_n have $\mathcal{E}^2$-admissible standard indices, but not $\mathcal{E}^1$-standard indices.

Conversely, when F_n/N has a K_i-admissible (standard) index i, we want to conclude that the w.p. is K_i-decidable. For this to be so, we must be able to deduce the K_i-decidability of $\gamma(N)$. From Theorem 5.2 of [2], this will be the case when the natural homomorphism $\varphi : F_n \to F_n/N$ is K_i-computable with respect to γ and i.

At this point let me recall the definition of the nth iterate of a function f. That is, given f, a number-theoretic function, we define $g(n)$ "the nth iterate of f" by the recursion

$$\begin{cases} g(0) = f(0) , \\ g(n+1) = f(g(n)). \end{cases}$$

Since the operation in the set of indices corresponding to exponentiation of an element of the group is iteration of the induced multiplication function, and since the K-normal form property (cf. Definition 5.3 of [2]) involves such iteration, we obtain the next condition on K_1 and K_2, namely

(*3) the nth iterate of a K_1 (resp. K_2) function is a K_1 (resp. K_2) function.

Note. Condition (*3) taken with (*2) implies K_1 contains the primitive recursive functions. Cf. Péter [15] Section 7.

When all conditions thus far assumed on K_i are satisfied we conclude, via Theorem 5.2 and Theorem 5.3 of [2] that the K_i solvability of the w.p. in F_n/N is a consequence of the existence of a K_i-admissible standard index for F_n/N.

All that remains is to show the analog of Corollary 7.1 of [2] holds. That is, the K_i decidability of the w.p. is an algebraic invariant. Thus, assume F_n/N and F_m/M are two f.g. presentations of the same group and ψ is an isomorphism $F_n/N \approx F_m/M$. Further, assume F_n/N has a K_i-admissible standard index for fixed $i = 1,2$. I want to show F_m/M has a K_i-admissible standard index. For this purpose it is sufficient to show M is K_i-decidable in F_m since F_m has a K_i-admissible index by (*1). First, observe that if a group G has any K_i-admissible index i, then all its standard indices are K_i-admissible. For, suppose G is presented as F_m/M. Since F_n has a K_i-admissible index for which it has the K_1-normal form property and (*3) holds, the natural homomorphism $F_m \to F_m/M$ is K_i computable, whence the assertion (cf. Lemma 7.1 of [2]). To complete the argument, take F_n/N as G and invoke the equivalence of the K_i-decidability of the w.p. and the existence of a K_i-admissible standard index (cf. Theorem 7.1 of [2]). Thus, I have shown that (*1), (*2) and (*3) are sufficient to insure the algebraic invariance of the cost of the w.p. with respect to f.g. presentations. This may be summarized in the

Result A. Let K_1 and K_2 be as above. In order that K_1 and K_2 be properly chosen, it is sufficient that K_1 (resp. K_2) be closed under the primitive recursive functions and/or relations and the free groups F_n have K_1-admissible standard indices.

To some extent the program of discovering how to insure that K_1 and K_2 are properly chosen is incomplete, for I ought to show that there are groups whose w.p. is restricted to either K_1 or $K_2 \setminus K_1$. Succinctly put: that there is adequate pathology for K_1 and K_2? As it turns out, for the choices I will propose for K_1 and K_2 the "pathological" groups exist. There is, however, some doubt in my mind concerning whether or not one ought always to strive for *finitely presented* pathology. Of course, before it is known that finitely presented pathological groups exist – and with the hindsight we presently have regarding how difficult it was, initially at least, to obtain such groups – it probably is very reasonable to insist on producing finitely presented groups before agreeing that a (more or less) profound concept has been apprehended. However, with the exception, possibly, of the residually finite groups for which it is clear that finite presentability is fundamental (because the w.p. for f.g. residually finite (r.f.) groups may be unsolvable in general while the finitely presented r.f. groups have solvable w.p., cf. V.H. Dyson [4], or A.W. Mostowski [14]) it seems clear that for the purposes of algebra the more important concept is "finitely generated". Hence, if one produces f.g. pathological groups one ought to feel that the significance of the K_1 and K_2 is reasonably well established, especially if we can produce f.g. *recursively presented* pathological groups, for there is every reason to believe that having done so, the finitely presented pathological groups will be forthcoming through the Higman embedding technique (such reasons to become apparent below). Indeed, for the choices of K_1 and K_2 to be proposed, the Higman embedding does furnish the finitely presented pathological groups. Thus, I also want the condition

(*4) there exist f.g. recursively presented groups with w.p. solvable by K_1 functions (resp. by $K_2 \setminus K_1$ but not K_1 functions).

Note. The requirement that the group be both recursively presented and have a solvable w.p. is redundant since a f.g. group, which has a solvable w.p. with respect to a finite system of generators, can be recursively presented with respect to these generators because the multiplication table of the group is then r.e. Of course, the converse does not hold. That is, not every f.g. recursively presented group has a solvable w.p.

For the first candidates for K_1 and K_2, I will take the primitive recursive (p.r.) functions and the recursive functions, respectively. The F_n then have admissible p.r. indexings (in fact, elementary indexings) by Lemma 4.1 of [2]. Clearly (*n) for $n = 2, 3$ are immediate and (*4) follows from Theorem 7.2 of [2], choosing, e.g. $\alpha = 4$ and $\alpha = \omega + 1$. My student R.W. Gatterdam has shown in his doctoral dissertation [7] that the Higman embedding technique preserves p.r. solvability of the word problem. To fix ideas, let G be recursively presented and let the Higman process, denoted by H, take G into finitely presented G_H; in diagrams $G \overset{H}{\to} G_H$. Then, Gatterdam has shown that H is a p.r. process and the set of indices of the image of G in G_H is p.r. decidable in the solution to the w.p. of G_H. Hence, for G possessing K_1 (= p.r. functions)-admissible index, we obtain G_H with K_1-admissible index; that is, a finitely presentable group with K_1-decidable w.p. However for G possessing $K_2 \setminus K_1$ but not K_1-admissible index we obtain f.p. G_H with $K_2 \setminus K_1$ decidable w.p. but not K_1-decidable w.p. Because, if G_H has K_1 solvable w.p. given, say, by function f, then the function

$$g(x) = \begin{cases} f(x), & x \in H(G) \\ * & \text{, otherwise,} \end{cases}$$

solves the w.p. in G and is K_1 contrary to assumption; ("*" denotes any convenient value outside of $H(G)$). Hence, we have

Result B. Primitive recursiveness of the solution of the w.p. is an algebraic invariant of finitely generated groups.

Note. Boone has announced in his paper [1], footnote 4, p. 50, the existence of a finite presentation with recursive but not p.r. solvable w.p. Applying the analysis above, I am able to assert, for this group, the algebraic invariance of its word problem. But, in any event combining Result B, the case $\alpha = \omega + 1$ of [2], Theorem 7.2, or Theorems 7 and 8 of [3], and the construction of Gatterdam [7] gives

Result C. We can construct a finitely presented group G with algebraically invariant recursive but not p.r. solution to its w.p.

A somewhat more provocative choice for K_1 and K_2 devolves as follows: If $T(a, b, c)$ is the (intuitive) Kleene T-predicate, let $\mathcal{T}(x, y, z)$ be a formula of the normal number theory N either as given in Mendelson [13] or in Kino [9] or Kleene [10,11], which expresses T. That is, if ω is the standard model of N, let

$$\vdash_N \mathcal{T}(\mathbf{a}, \mathbf{b}, \mathbf{c}) \text{ when } \omega \models T(a, b, c),$$

and

$$\vdash_N \neg\, \mathcal{T}(\mathbf{a}, \mathbf{b}, \mathbf{c}) \text{ when not } \omega \models T(a, b, c).$$

We say a (number theoretic) function $f : \omega \to \omega$ is *provably recursive in N* if f has an index e such that

$$\vdash_N \forall x\, \exists y\; \mathcal{T}(\mathbf{e}, x, y).$$

It is easy to see not every recursive f is probably recursive in N (cf. Fischer [6], Theorem 5.1). In fact, as Enderton has shown [5], no recursive function need be provably recursive in N. I assume the formula $\mathcal{T}$ is chosen so that, at least, all p.r. functions are provably recursive; this is always possible as Kino has shown in [9].

Thus, let K_1 be the functions provably recursive in N and take K_2 to be all recursive functions. Then (*1) holds because of the assumption on $\mathcal{T}$ and my Lemma 4.1 in [2]. Likewise (*2) and (*3) hold because of the assumption on $\mathcal{T}$, and because if g is a function p.r. in provably recursive functions $f_1, f_2, ..., f_m$, then g is provably recursive (cf. Property 4.1in [5]). Condition (*4) is also immediate, cf. the discussion for the preceding choice of K_1 and K_2, mutatis mutandis, and the Theorem 8.3 of Fischer [6] which shows there is a recursive but not provably recursive set of integers. (Alternately, the existence of such a set follows from the existence of a recursive but not provably recursive enumeration of the provably recursive functions, the construction given in my Lemma 3.1 of [2], and Property 4.1 of Fischer [6]).

Note. From (*4) and the fact (alluded to above) that the Higman embedding process preserves provable recursiveness we deduce the amusing consequence, given in Result D below: there exists a finitely presentable group with intuitively solvable w.p. such that it is not possible to prove in formal number theory that the solution is a recursive function. Consider this result in the light of objections given by some mathematicians (e.g. Wang [17], p. 89, and Péter [15]) to the converse of Church's thesis: every recursive function is effectively computable. That is, a function is recursive if it has an index e such that

(+) $$\omega \models \forall x \exists y \ T(e, x, y).$$

However, nothing is said, in general, about how y is determined by x. Alternately, for the machine interpretation of $T(a, b, c)$ given by Kleene in [10], p. 243, nothing is said regarding how the length c of the computation depends, a priori, on the input b. It is not altogether convincing that functions for which property (+) holds are effectively when y is not known to depend on x in some satisfactory way. The argument from experience which goes – all functions known on intuitive grounds to be effectively computable are, in fact, recursive – really does not insure that some proper subclass of the recursive functions might not *already contain all such algorithms*. Yet, if one chooses to regard the class of functions provably recursive in (say) N (with respect to a "realistic" formula $\mathcal{T}$) as the true class of effectively computable functions we can deduce an anomalous version (Result D) of the Novikov-Boone theorem in which the "bad" group has, on intuitive grounds, a "solvable" w.p. despite the fact that on formal grounds its w.p. is unsolvable! (However, cf. Kleene [10], paragraph 3 of footnote 171 on p. 241 for a discussion of the converse to Church's thesis in which the converse is framed so that, in Kleene's view, it is not contestible. Kreisel has written me, as well as published elsewhere, concerning his view on Church's thesis. He argues that it is necessary to make a distinction between the concepts "mechanical" and "constructive" and, if one identifies these, then it is hard, if not impossible, to doubt Church's thesis. Presumably this means the converse is not in doubt whatever the view on mechanical vs. constructive.) In summary, then, we have by means of Gatterdam's analysis [7],

Result D. We can construct a finitely presented group with algebraically invariant recursive but not provably recursive solution to its w.p.

Note. The "effective" construction of the groups of Results C, D and E follows from the effectivity of Gatterdam's construction and from the effective construction of sets of integers with characteristic functions of suitable subrecursive degree of solvability. Here, by subrecursive degree, I mean solvable by a K_1 or K_2 function for properly chosen K_1 or K_2. The existence of suitable characteristic functions is not difficult to obtain. For example, refer to Lemma 3.1 of [2]. For the recursive vs. provably recursive case use the recursive but not provably recursive function which enumerates the provably recursive functions of N. The "effective" existence of such a function follows from the fact that N is an axiomatic formal theory, (cf. Mendelson [13], p. 29 (3)). Similarly one can show the Peter function is actually provably recursive and obtain Result E, mutatis mutandis.

Another choice for K_1 and K_2 are the p.r. functions and the provably recursive functions. The discussion above carries over immediately, it only being necessary to note that Rosza Péter's recursive but not p.r. function φ enumerating the p.r. functions can be shown, using Kino's characterization of provably recursive functions, to be provably recursive since φ is given by a transfinite induction up to ω^2; (cf. Péter [16], p. 125 f.) From this we obtain the existence of a set of integers which is provably recursive but not p.r. Such a set can be used to satisfy (*4) in the standard manner (cf. my Theorem 7.2 of [2] for the details). I state for the record the

Result E. We can construct a finitely presented group with algebraically invariant provably recursive but not p.r. solution to its w.p.

At this point it is tempting to try to carry out the above arguments for the finer degrees of the Grzegorczyk hierarchy. That is, take K_1 as $\mathcal{E}^{\alpha}$ and K_2 as $\mathcal{E}^{\alpha+k}$ for fixed $\alpha \geqslant 3$ and $k \geqslant 1$ (here $\alpha \geqslant 3$ since K_1 needs the elementary functions and probably a bit more). While to a great extent this can be carried out, certain imperfections, which threaten to be of a permanent nature, persist at the present state of science. Since this represents a departure from the point of view in my paper [2] I will go into some detail. The proofs, which are more or less automatic once the difficulties are perceived, will, as usual, be omitted or else published elsewhere.

The chief difficulty is encountered in the attempt to obtain an $\mathcal{E}^{\alpha}$ analog of Theorem 5.3 of [2] which holds for suitably chosen K_i (but which does not hold precisely as stated in [2]). This theorem provides the link whereby if we know that one of the quotients G/H of a K_i group G has a K_i admissible index then H is K_i decidable in G. Thus, the equivalence between the word problem being solvable by K_i functions and the existence of a K_i index for G is established by this theorem (in conjunction with the K_i analogs of Theorems 5.1 and 5.2 of [2]). However, for the choice say, of $\mathcal{E}^{\alpha}$ for K_1 and $\mathcal{E}^{\alpha+1}$ for K_2 the result of Theorem 5.3 does not go precisely over (despite the incorrect claim in [2] that it does). The difficulty lies in showing that the natural homomorphism $\varphi : G \to G/H$ is $\mathcal{E}^{\alpha}$-computable, even under the assumption G has the $\mathcal{E}^{\alpha}$ normal form property. For, if we denote by i and j the $\mathcal{E}^{\alpha}$ indices in G and G/H respectively, then the index normal form decomposition of an element x in $i(G)$ as $x = \tau_1^{\alpha_1} \circ \cdots \circ \tau_r^{\alpha_r}$ goes over under $\overline{\varphi}$ to $\overline{\varphi}(x) = k_{\tau_1}^{\alpha_1} \circ \cdots \circ k_{\tau_r}^{\alpha_r}$ in $j(G/H)$ and there is no way in general to provide an a priori $\mathcal{E}^\alpha$ bound on the iterated multiplication in $j(G/H)$ by which we obtain each $k_{\tau_i}^{\alpha_i}$. It is for this reason that I required in (*3) that K_1 be closed under iteration. Thus, the best that can be said for the case under consideration is that $\overline{\varphi}(x)$ is $\mathcal{E}^{\alpha+1}$ computable. Now, we usually apply this reault to the special case G is a free group of rank $n < \infty$ and i is a standard index, so the iterated multiplications in $i(G)$ which give $\tau_i^{\alpha_i}$ cause no trouble since an $\mathcal{E}^3$ bound is available in terms of the Kleene concatenation function $x * y$. But there seems to be no way to treat the iterations induced in $j(G/H)$ by the index normal form owing to the wholly arbitrary nature of the index j. If, in fact, j is a standard index then there is no difficulty (but this would be tantamount to assuming G/H had an $\mathcal{E}^\alpha$ solvable w.p., the very property we hoped to deduce from Theorem 5.3). As a consequence of this imperfection in the theory the algebraic invariance of the w.p. with respect to the $\mathcal{E}^\alpha$ heirarchy applies only to *standard* $\mathcal{E}^\alpha$ indices. It should be noted, however, that it is completely unreasonable to consider the w.p. with respect to any other kind of index because only the standard indices effectively relate words in the free group with their corresponding index through the technique of Gödelization. In fact, no way is presently known to construct an example of a non standard index despite the fact that allowance for the existence of these indices must be made. In summary then, all I can say is that $\varphi(x)$ is at most $\mathcal{E}^{\alpha+1}$ computable and accordingly, G/H has an $\mathcal{E}^{\alpha+1}$ solvable w.p. It is of minor comfort to know that this is the worst that can prevail; for, given that G is an $\mathcal{E}^\alpha$ computable group, we know that G has at most an $\mathcal{E}^{\alpha+1}$ standard index. Thus it is conceivable that a group G may have its word problem solvable precisely at level $\mathcal{E}^{\alpha+1}$ and not lower for any f.g. presentation and yet have an index for which it is an $\mathcal{E}^\alpha$ group. It can not have an index at levels less than α for this would, via the equivalence of w.p. and standard index, give rise to a standard index at $\mathcal{E}^\alpha$ and contradict the algebraic invariance at level $\alpha + 1$. For the sake of completeness I will list the easily demonstrated properties of $\mathcal{E}^\alpha$ indices for f.g. groups, $\alpha \geqslant 3$ (GH standing for "Grzegorczyk hierarchy"):

GH 1. f.g. subgroups of $\mathcal{E}^\alpha$ groups are $\mathcal{E}^{\alpha+1}$-standard.
GH 2. G is an $\mathcal{E}^\alpha$ group implies G is $\mathcal{E}^{\alpha+1}$-standard.
GH 3. G is $\mathcal{E}^\alpha$-standard implies all standard indices are at most $\mathcal{E}^\alpha$.
GH 4. G is $\mathcal{E}^\alpha$-standard *but not lower* for standard indices implies G is not an $\mathcal{E}^\beta$ group for any $\beta < \alpha - 1$.
GH 5. $\mathcal{E}^\alpha$ solvable w.p. is equivalent to $\mathcal{E}^\alpha$-standard. This property is algebraically invariant for f.g. presentations.

A second difficulty occurs in building $\mathcal{E}^\alpha$ indices for free products with amalgamation of $\mathcal{E}^\alpha$ groups. Again iteration is the culprit. For, if G_1 and G_2 are $\mathcal{E}^\alpha$ groups with isomorphic subgroups $H_1 < G_1$ and $H_2 < G_2$ and with the isomorphism $\varphi : H_1 \leftrightarrow H_2$ such that both H_1 and H_2 are $\mathcal{E}^\alpha$ decidable and φ and φ^{-1} are $\mathcal{E}^\alpha$ computable, then the best we can say is $G_1 *_\varphi G_2$ is at most $\mathcal{E}^{\alpha+1}$. This is because it is natural to build an index for $G_1 *_\varphi G_2$ from the indices of G_1 and G_2. However the reduction of the product of two normal forms $(hx_1 \ldots x_m)(ky_1 \ldots y_m) = h'z_1 \ldots z_p$ involves repeated iteration of the process which "slips k past x_m". In this process the index of the element in the amalgamated subgroup can grow arbitrarily large and hence there is no a priori $\mathcal{E}^\alpha$ bound for the entire iteration. Hence $G_1 *_\varphi G_2$ is at most $\mathcal{E}^{\alpha+1}$ is the best that can be said. The easily demonstrated revisions of [2] Section 6 are listed below (AFP stands for "amalgamated free product"):

AFP 1. If G_i is $\mathcal{E}^\alpha$-standard and $H_i < G_i$ is $\mathcal{E}^\alpha$-decidable for $i = 1, 2$ and φ is an isomorphism of H_1 and H_2 such that both φ and φ^{-1} are $\mathcal{E}^\alpha$ computable then $G_1 *_\varphi G_2$ is $\mathcal{E}^{\alpha+1}$-standard.
AFP 2. $G_1 * G_2$ is $\mathcal{E}^\alpha$-standard when G_1 and G_2 are.
AFP 3. If $G_1 = G_2$ and $H_1 = H_2$ and $\varphi = 1_G | H$ then $G_1 *_\varphi G_2$ is $\mathcal{E}^\alpha$ standard (under the conditions in AFP 1).

As a consequence of AFP 3 we can show the groups $G_{\alpha+1}$ presented in [2] as

$$\langle a, b, c, d; (a^n b^n)^n = (c^n d^n)^n,\ n \in U_{\alpha+1} \rangle$$

for certain $\mathcal{E}^{\alpha+1}_*$-decidable subsets $U_{\alpha+1}$ of ω, are, in fact, $\mathcal{E}^{\alpha+1}_*$-

standard and thus, have an $\mathcal{E}^{\alpha+1}$ algebraically invariant w.p. (although conceivably they could have anomalous $\mathcal{E}^{\alpha}$ indices).

The next flaw of the $\mathcal{E}^{\alpha}$ heirarchy turns up when we try to follow Gatterdam's analysis of the Higman embedding. The difficulty is this: at a certain point in the construction it is necessary to pass from a group G and an isomorphism $\varphi : H \to H' < G$ for some $H < G$ to a new group G_{φ} obtained from G by adjoining an infinite cycle t satisfying $tht^{-1} = \varphi(h)$ for all h in H. It can be seen that G_{φ} is embedded in a certain free product with amalgamation. Consequently this construction preserves K_1 cost ([7] Theorem 3.1). However, for the $\mathcal{E}^{\alpha}$ hierarchy Gatterdam shows the passage from G to G_{φ} possibly jumps two levels from $\mathcal{E}^{\alpha}$ to $\mathcal{E}^{\alpha+2}$ and the Higman technique needs (at this state of science) a countable number of applications building a tower T, viz. $G, G_{\varphi_1}, (G_{\varphi_1})_{\varphi_2}, \ldots$. Hence for the present it appears that the full class of p.r. functions is the minimum class which can accommodate the Higman construction. It remains to be seen if either the jump from $\mathcal{E}^{\alpha}$ to $\mathcal{E}^{\alpha+2}$ in passing from G to G_{φ} is necessary, or if (more likely) countably many applications do not force a p.r. index for the union of the tower T but rather, only a finite jump from $\mathcal{E}^{\alpha}$ to $\mathcal{E}^{\alpha+k}$ for some k suffices. A successful solution to either problem will suffice to show the Higman embedding preserves the $\mathcal{E}^{\alpha}$ hierarchy (albeit possibly entailing a finite shift from $\mathcal{E}^{\alpha}$ to $\mathcal{E}^{\alpha+k}$)*. Then because the groups $G_{\alpha+1}$ are recursively presented, we would conclude there is adequate pathology for the $\mathcal{E}^{\alpha}$ hierarchy. For the present all that we can say is for any $\beta \geqslant 3$ there is a finitely presented (henceforth f.p.) group whose word problem can not be solved at level lower than β. This is a consequence of Gatterdam's proof that the Higman embedding preserves p.r. solution to the w.p. Since the groups $G_{\alpha+1}$ are recursively presented and p.r. they are embeddable in a p.r. group which a fortiori resides at some (unknown) level γ of the $\mathcal{E}^{\alpha}$ hierarchy. Then the word problem is solvable at level $\gamma + 1$. Since the $G_{\alpha+1}$ have word problems at arbitrarily high levels we

* Cfr. this volume, "The computability of Group Constructions, Part I," by the author and R.W. Gatterdam for a solution to this problem showing the "strong Britton extension" only jumps two levels at most (Theorem 5.3).

get the assertion, for the set of γ's obtained by the embedding is necessarily cofinal in the natural numbers.

Note. It may be possible through the use of techniques employed by Boone in [1] to go directly from the sets $U_{\alpha+1}$ to f.p. groups with w.p. solvable at level $\mathcal{E}^{\alpha+k}$ but not lower (for $k < \infty$ depending uniformly on α). I do not know whether or not this program is feasible, but D.J. Collins and C.F. Miller III have expressed belief in the potential fruitfulness of such an approach to finding adequate pathology for the $\mathcal{E}^{\alpha}$ hierarchy.

Finally, with respect to the $\mathcal{E}^{\alpha}$ hierarchy let me observe that there is no p.r. group which contains a copy of every f.p. $\mathcal{E}^{\alpha}$ group. For suppose such a group G exists. Then being p.r. G is $\mathcal{E}^{\alpha}$ for some α. But all f.g. subgroups of $\mathcal{E}^{\alpha}$ groups are at most $\mathcal{E}^{\alpha+1}$ and we know there are f.p. groups at arbitrarily high levels of the $\mathcal{E}^{\alpha}$ hierarchy. Thus the assumption on G is impossible. (I am indebted to Gatterdam for pointing this out to me.)

Next, I want to comment on the cost of the w.p. for small cancellation groups such as the "less than 1/6" groups studied by Greendlinger [8] and Lipschutz [12]. If such a group, f.g. of course, has a presentation for which the relators may be enumerated by a K_1 function in order of increasing magnitude of, say, length or Gödel number, then its w.p. must be K_1 solvable. This is because in small cancellation groups the algorithm for solving the w.p. is a monotonic reduction process consisting of examining finitely many subwords of a word W which is suspected to define the identity in order to discover more than half of a symmetric relator. Because of the enumerating condition on the relators there are only finitely many candidates for comparison, all of which are obtainable by a K_1 function. Thus, if a fragment of a relator is found, it may be replaced in W by the inverse of its remaining part, which, being shorter than the part replaced, reduces (even before cancellations) the overall length of the original word. Thus, through iteration, which preserves K_1 by (*3), a word W defining the identity may be brought to the empty word by means of a K_1 process. For example, finitely presentable small cancellation groups must have an algebraically invariant K_1 w.p. On the other hand it is possible to construct an infinitely related recusively presented small cancellation group G which has a recursive but no p.r. solution to its w.p. (A typical construction for a recursively presented small cancellation group G with recursive but not p.r. word problem might go like this: let F_7 be the free group on $a_1, a_2, ..., a_7$ and let $H < F_7$ be the subgroup generated by all $a_1{}^k a_2{}^k ... a_7{}^k$, for $k \in U$, U a recursively decidable but not p.r. decidable set of integers. Take two copies of F_7 and amalgamate the subgroups H. Such a group is easily seen to be a "less than 1/6 group" and can be shown to have the required properties on its w.p. by imitating the technique in Cannonito [2]. No claim is made for elegance or economy in this example.) G can be embedded in a finitely presented group which, by Gatterdam's analysis, must have a $K_2 \setminus K_1$ decidable w.p. and which, accordingly, can not be a small cancellation group. The latter property could also be seen directly from the constructions required for the Higman embedding process.) I summarize this discussion as

Result F. A small cancellation group which has its relators enumerable in order of increasing length by a K_1 function, has a K_1 decidable w.p. In particular, finitely presented small cancellation groups have K_1 decidable w.p. (for suitable chosen K_i).

As a final comment, let me note that direct products, free products and free products with amalgamation (satisfying suitable conditions on the amalgam such as those given in [2] Corollary 6.3), applied finitely many times, preserves the algebraic invariance of the w.p.

I wish to acknowledge the many pleasurable and profitable discussions I have had on this work with my student R.W. Gatterdam.

References

[1] W.W. Boone, Word problems and recursively enumerable degrees of unsolvability. A sequel on finitely presented groups. Ann. of Math. 84 (1966) 49-84.

[2] F.B. Cannonito, Hierarchies of computable groups and the word problem, Jour. Symbolic Logic 31, No. 3 (1966) 376-392.

[3] F.B. Cannonito, On finitely generated groups with primitive recursive word problem, TAO technical report 63-65, Hughes Aircraft Company, 1963, 21-pp.

[4] V.H. Dyson, Notices Amer. Math. Soc. 11 (1964), p. 743, Abstract 616-617.

[5] H.B. Enderton, On provable recursive functions, Notre Dame Journal of Formal Logic IX, No. 1 (1968) 86-88.

[6] P.C. Fischer, Theory of provable recursive functions, Trans. Amer. Math. Soc. May (1965) 494-520.

[7] R.W. Gatterdam, Embeddings of primitive recursive computable groups, doctoral dissertation, University of California, Irvine, 1969.

[8] M.D. Greendlinger, Dehn's algorithm for the word problem, Comm. Pure and App. Math. 13 (1960) 67-83.

[9] A. Kino. On provably recursive functions and ordinal recursive functions. Jour. Math. Soc. Japan, 20 (1968) 456–476.

[10] S.C. Kleene, Introduction to Metamathematics (D. Van Nostrand, Princeton, N.J., 1952).

[11] S.C. Kleene, Mathematical Logic (John Wiley & Sons, 1967).

[12] S. Lipschutz, On square roots in eight *h*-groups, Comm. Pure and App. Math. 15 (1962) 39-43.

[13] E. Mendelson, Introduction to Mathematical Logic (D. Van Nostrand, Princeton, N.J., 1964).

[14] A.W. Mostowski, On the decidability of some problems in special classes of groups Fundamenta Mathematicae 59 (1966) 123-135.

[15] R. Péter, Rekursivität und Konstruktivität, Constructivity in Mathematics, ed. A. Heyting (North-Holland, Amsterdam, 1957) 226-233.

[16] R. Péter, Recursive Functions, third revised edition (Academic Press, New York, 1967).

[17] H. Wang, A survey of Mathematical Logic (North Holland, Amsterdam, 1963).

THE COMPUTABILITY OF GROUP CONSTRUCTIONS, PART I

F.B. CANNONITO* and R.W. GATTERDAM**
University of California, Irvine, University of Wisconsin, Parkside

The beginnings of computable algebra, especially as this applies to group theory, can be traced for all practical purposes to the fundamental paper of M.O. Rabin [9]. In this work Rabin showed the very natural equivalence between the existence of a "recursive realization" of the group in the natural numbers and the existence of a recursive solution to the word problem with respect to any finitely generated presentation of the group. Of course, the property of having word problem solvable with respect to a given presentation applies, in general, only to finitely generated (henceforth f.g.) presentations; there is no problem in presenting even f.g. groups on infinitely many generators so that the word problem with respect to such a presentation is unsolvable.

Subsequently Cannonito sharpened the concept of a computable group in [1] by superimposing the Grzegorczyk hierarchy, denoted "$(\mathcal{E}^{\alpha})_{\alpha \in \omega}$", and by showing that f.g. $\mathcal{E}^{\alpha}$-computable groups existed in profusion for infinitely many consecutive levels of the hierarchy. Although incorrectly stated in [1], it was shown in Cannonito [1,2] that with respect to *standard indices* of $\mathcal{E}^{\alpha}$ groups, the word problem is an *algebraic invariant*; that is, every f.g. presentation of an $\mathcal{E}^{\alpha}$-computable group has, with respect to a standard index, a word problem solvable by an $\mathcal{E}^{\alpha}$ function if *one* f.g. presentation of the group possesses this property.

However, certain inelegant aspects of the theory of $\mathcal{E}^{\alpha}$ comput-

* AFOSR-1321-67 (Grant).
** AFOSR-70-1870 (Grant).

able groups came to light as a consequence of the discovery that the proof in [1] of the equivalence of $\mathcal{E}^\alpha$ computable realizations and $\mathcal{E}^\alpha$ solvable word problems *only applies to standard indices.* These aspects are discussed, among other parts of the theory, in Cannonito [2]. Typical among the inelegancies is the inability, in general, to say more than this: if a f.g. group has an $\mathcal{E}^\alpha$ realization then its word problem is *not higher than* $\mathcal{E}^{\alpha+1}$. Furthermore, while the assertion in [1] claiming the groups G_α presented as

$$\langle a, b, c, d; (a^n b^n)^n = (c^n d^n)^n, n \in U_\alpha \rangle$$

where U_α is a set of natural numbers decidable only at $\mathcal{E}^\alpha$ or higher, had word problem solvable at level $\mathcal{E}^\alpha$ but not lower *was true*, the proof given was flawed in an essential detail; namely, contrary to the assertion in [1], Theorem 6.1, the best that can be said for the free product with amalgamation of $\mathcal{E}^\alpha$ groups (under suitable $\mathcal{E}^\alpha$ conditions on subgroups and isomorphisms, etc.) is that its level of computability is no higher than $\mathcal{E}^{\alpha+1}$. Thus applying the analysis in [1], to the groups G_α, we see the word problem is solvable no higher than $\mathcal{E}^{\alpha+1}$ *and* no lower than $\mathcal{E}^\alpha$. However, in this paper we show the word problem of the G_α actually resides no higher than $\mathcal{E}^\alpha$, owing to the fact that the G_α arise from a rather restricted type of free product with amalgamation in which both factors are copies and both subgroups to be amalgamated are also copies and the isomorphisms are the identities. This situation and its consequences are examined in detail in this paper with respect to a "relativized Grzegorczyk hierarchy" which seems to be appropriate for all countable groups whether finitely generated or not and whether possessed of a solvable word problem for f.g. presentations or not. Owing to the possible independent interest in this relative Grzegorczyk hierarchy we begin this paper with a discussion and proofs of this concept, which resembles the notion of relative computability as exposed in Davis [3], the essential idea being to close a level $\mathcal{E}^\alpha$ under the usual operations after first adding the characteristic function of some subset of the natural numbers.

In the next phase of the theory of $\mathcal{E}^\alpha$ groups the Higman embedding [6] of recursively presented groups into finitely presented (f.p.) groups was studied in the doctoral dissertation of Gatterdam [4]. It was shown there that the Higman embedding took f.g. groups with primitive recursive (p.r.) realization into f.p. groups which have a p.r. realization. Moreover, it was shown by Gatterdam that the actual embedding function was elementary (or, equivalently, $\mathcal{E}^3$) in the solution to the word problem of the receiver group. Hence, since the groups G_α are all recursively presented, they can be effectively embedded in f.p. p.r. groups and so there is an infinite spread of complexity of f.p. groups with respect to the $\mathcal{E}^\alpha$ hierarchy. But ideally one wants to show the Higman embedding actually preserves the relative Grzegorczyk hierarchy because by so doing one by-product would be the possibility to locate precisely on the hierarchy the receiver groups of the G_α. According to Gatterdam's analysis all that could be said is that the receiver groups must slip arbitrarily high up the hierarchy, but conceivably there could be great gaps and no method was at hand to locate the actual level of the hierarchy at which any particular receiver group resided. The reason for this dilemma can be found in a particular construction needed for the Higman proof which has come to be known as the strong Britton extension. The situation is this: we start with a f.g. group G and an isomorphism $H < G$ into G itself. If G has presentation, say,

$$\langle a_1, \ldots, a_n; R_1, \ldots \rangle$$

then we obtain a new group G_φ which embeds G with presentation

$$\langle a_1, \ldots, a_n, t; R_1, \ldots, tht^{-1} = \varphi(h) \text{ for all } h \in H \rangle .$$

Thus, G_φ is obtained by adjoining a (distinct) infinite cycle t which gives the isomorphism φ by conjugation. Since G_φ is a f.g. subgroup of a particular free product with amalgamation the computability of G_φ can be shown to lie no higher $\mathcal{E}^{\alpha+1}$ when G lies at level $\mathcal{E}^\alpha$. But the Higman embedding uses a countable number of such extensions resulting in a manyfold extension $G_{\varphi_1\varphi_2\varphi_3 \ldots}$ or, more simply, G_∞. Thus the level of computability can slip extremely high since it seems to require jumping possibly one level for each new infinite cycle adjoined. However, in this work we show (with respect to the relativized hierarchy) that when G is $\mathcal{E}^{\alpha}$ computable then G_{∞} is at most at level $\mathcal{E}^{\alpha+1}$. Thus, at this state of science the Higman embedding seems to actually preserve the relativized hierarchy from some point on for the remaining constructions do not appear to be troublesome at all. This will be the subject of Part II of this paper, to appear later, and will be based on the analysis we give in the present paper.

Additional constructions studied in this paper form the main stock in trade of infinite group theory. Among these are the "HNN extension" due to Higman, Neumann and Neumann. Also we show there can be no universal primitive recursive group containing a copy, even, of all f.p. p.r. groups.

For the convenience of the reader we give a flow chart showing the relationship the several sections bear to each other. The expressions "D2.3, L3.5, P2.6, and T2.7" refer respectively to Definition 2.3, Lemma 3.5, Proposition 2.6 and Theorem 2.7.

We will observe the following conventions: "N" denotes the natural numbers 0,1,2, ... and unmodified "integer" usually means natural number, exceptions to be explicitly mentioned in the context. We omit parentheses as much as possible, particularly in forming the composition of functions; thus, "$fg(x)$" rather than "$f(g(x))$". The characteristic function c_A of a subset A of N is 0 on A and 1 off A. The notations "$x \mapsto f(x)$" or "$x \overset{f}{\mapsto} y$" give the actual value assigned to x under the mapping f. We usually omit the f when this will cause no confusion. Finally, the n-tuple $(x_1, ..., x_n)$ is abbreviated to $x^{(n)}$ when it is not necessary to explicitly give the coordinates. The universal and existential quantifiers are respectively denoted by "$\forall$" and "$\exists$". All other notation will be defined *in situ.*

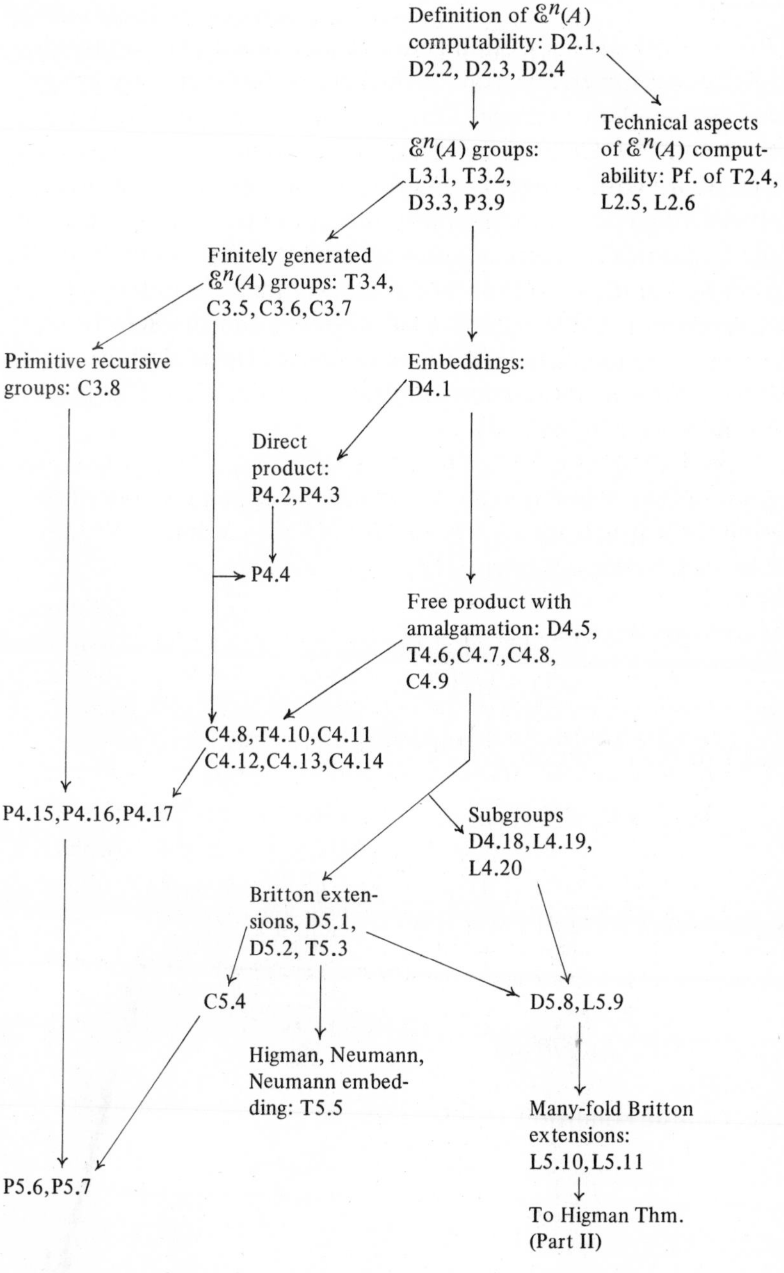
Definition of $\mathcal{E}^n(A)$ computability: D2.1, D2.2, D2.3, D2.4
Technical aspects of $\mathcal{E}^n(A)$ computability: Pf. of T2.4, L2.5, L2.6
$\mathcal{E}^n(A)$ groups: L3.1, T3.2, D3.3, P3.9
Finitely generated $\mathcal{E}^n(A)$ groups: T3.4, C3.5, C3.6, C3.7
Primitive recursive groups: C3.8
Embeddings: D4.1
Direct product: P4.2, P4.3
P4.4
Free product with amalgamation: D4.5, T4.6, C4.7, C4.8, C4.9
C4.8, T4.10, C4.11 C4.12, C4.13, C4.14
P4.15, P4.16, P4.17
Subgroups D4.18, L4.19, L4.20
Britton extensions, D5.1, D5.2, T5.3
C5.4
D5.8, L5.9
Higman, Neumann, Neumann embedding: T5.5
Many-fold Britton extensions: L5.10, L5.11
To Higman Thm. (Part II)
P5.6, P5.7

Fig. 1.

§2. Relative Grzegorczyk hierarchy

The purpose of this section is to define a relative Grzegorczyk hierarchy and to verify certain properties of this hierarchy are retained in the relative version. Our point of departure is the hierarchy defined by Grzegorczyk [5], but we modify the definition and relativize with respect to an arbitrary subset of the natural numbers $A \subset N$. The modifications are to use the characterization of the Grzegorczyk hierarchy due to Ritchie [10] and replace recursion by iteration plus additional initial functions in the manner of Robinson [11]. We relativize by adjoining the characteristic function c_A to the initial functions similar to Davis' definition of relative recursion and relative primitive recursion, [3]. Essential use is made of Ritchie [10].

Definition 2.1. A function $f : N^{k+1} \to N$ is defined by *limited recursion* from functions $g : N^k \to N$, $h : N^{k+2} \to N$ and $j : N^{k+1} \to N$ if it is given by the schema

$$f(x^{(k)}, 0) = g(x^{(k)})$$

$$f(x^{(k)}, y + 1) = h(x^{(k)}, y, f(x^{(k)}, y))$$

subject to the condition

$$f(x^{(k)}, y) \leqslant j(x^{(k)}, y).$$

The function $f : N \to N$ is defined from $h : N \to N$ and $j : N \to N$ by *limited iteration* if it is defined by the schema:

$$f(0) = c$$

$$f(x + 1) = hf(x)$$

subject to the condition

$$f(x) \leqslant j(x).$$

Of course the word "limited" in the definition above refers to the bounding functions j. Omitting the bounds j one has the definition of primitive recursion and iteration.

Following Ritchie, we use the "pairing" functions

$$J(x, y) = ((x + y)^2 + x)^2 + y$$

$$K(z) = E([z^{1/2}]) = [z^{1/2}] \dot{-} [[z^{1/2}]^{1/2}]^2$$

$$L(z) = E(z) = z \dot{-} [z^{1/2}]^2$$

where $[z^{1/2}]$ = largest integer whose square is less than z. The important properties of these functions are $KJ(x, y) = x$ and $LJ(x, y) = y$.

Inductively for $n \geqslant 2$ and $J = J^{(2)}$, $M_1{}^{(2)} = K$, $M_2{}^{(2)} = L$ we define

$$J^{(n+1)}(x^{(n+1)}) = J(J^{(n)}(x^{(n)}), x_{n+1})$$

$$M_i{}^{(n+1)}(z) = M_i{}^{(n)}K(z) \text{ for } 1 \leqslant i \leqslant n$$

$$M_{n+1}{}^{(n+1)}(z) = L(z).$$

Then the relation between recursion and iteration is given by Theorem 3.2 of [10]:

Theorem. *Let $f(x^{(n)}, y)$ be defined by primitive recursion*

$$f(x^{(n)}, 0) = g(x^{(n)}), \quad f(x^{(n)}, y + 1) = h(x^{(n)}, y, f(x^{(n)}, y))$$

for $n \geqslant 0$. Then

$$f(x^{(n)}, y) = \begin{cases} g(0^{(n)}) & \text{if } J^{(n+1)}(x^{(n)}, y) = 0 \\ Lf'J^{(n+1)}(x^{(n)}, y) & \text{otherwise} \end{cases}$$

where f' is defined by the iteration

$$f'(0) = 0$$

$$f'(z+1) = H''f'(z)$$

for

$$H''(w) = J(K(w)+1, H'(K(w), L(w)))$$

$$H''(z, w) = \begin{cases} H(0, g(0^{(n)})) & \textit{if } z = 0 \\ H(z, w) & \textit{otherwise} \end{cases}$$

$$H(z, w) = \begin{cases} g(M_1{}^{(n+1)}(z+1), \ldots, M_n{}^{(n+1)}(z+1) & \textit{if } M_{n+1}{}^{(n+1)}(z) = 0 \\ h(M_1{}^{(n+1)}(z), \ldots, M_{n+1}{}^{(n+1)}(z), w) & \textit{otherwise.} \end{cases}$$

The importance of this theorem is that in a class of functions containing the pairing functions $J, K, L, J^{(n)}, M_i^{(n)}$ and closed under substitution, closure under iteration implies closure under recursion. Moreover, J being monotone increasing $f(x^{(n)}, y) \leqslant j(x^{(n)}, y)$ implies $f'(z) \leqslant J(z, j(M_1^{(n+1)}(z), \ldots, M_{n+1}^{(n+1)}(z)))$ so closure under limited iteration implies closure under limited recursion.

We now define the relative Grzegorczyk hierarchy using the functions $f_n : N^2 \to N$ of Ritchie.

Definition 2.2.

$$f_0(x, y) = x + 1$$

$$f_1(x, y) = x + y$$

$$f_2(x, y) = xy$$

$$\left.\begin{aligned} &f_{n+1}(x, 0) = 1 \\ &f_{n+1}(x, y+1) = f_n(x, f_{n+1}(x, y)) \end{aligned}\right\} \text{ for all } n \geqslant 2.$$

Definition 2.3. Let $A \subset N$. Then the class of functions $\mathcal{E}^n(A)$ for $n \geqslant 2$ (with domain N^k for arbitrary k and Range N) is the smallest class of functions containing the initial functions

$$Z(x) = 0$$

$$U_m{}^n(x_1, ..., x_n) = x_m \qquad \text{for } 1 \leqslant m \leqslant n$$

$$f_0(x, y) = x + 1$$

$$f_1(x, y) = x + y$$

$$f_n(x, y)$$

$$E(x) = x \dot{-} [x^{1/2}]^2$$

$$c_A(x) = \begin{cases} 0 \text{ if } x \in A \\ 1 \text{ if } x \notin A \end{cases}$$

and closed under substitution and limited iteration.

Observe that $E(x) \in \mathcal{E}^n(A)$ implies the pairing functions are in $\mathcal{E}^n(A)$ so by the theorem stated above, $\mathcal{E}^n(A)$ is closed under limited recursion. Ritchie's main result is that the usual Grzegorczyk hierarchy is $\mathcal{E}^n = \mathcal{E}^n(N)$. Clearly $\mathcal{E}^n \subset \mathcal{E}^n(A)$ for every A. Also, $f_n(x, y) \in \mathcal{E}^n \subset \mathcal{E}^{n+1} \subset \mathcal{E}^{n+1}(A)$ implies $\mathcal{E}^n(A) \subset \mathcal{E}^{n+1}(A)$.

The following theorem is useful.

Theorem 2.4. *If $f : N^k \to N$ is defined by primitive recursion from functions in $\mathcal{E}^n(A)$ for $n \geqslant 2$ then $f \in \mathcal{E}^{n+1}(A)$.*

Proof. By the previous theorem it suffices to consider $f : N \to N$ given by iteration, $f(0) = c = h^0(c)$ and $f(x + 1) = hf(x) = h^{x+1}(c)$ (i.e., the composition $h \circ h \circ ... \circ h(c)$ $x + 1$ times). We must show there exists $j(x) \in \mathcal{E}^{n+1}(A)$ such that $f(x) \leqslant j(x)$.

The proof given is a modification of the techniques used in the proofs of Theorems 2.2 and 2.3 of [10]. The modifications allow the application of these methods directly to the class $\mathcal{E}^n(A)$ for $n \geqslant 2$. First we need some facts about $f_n(x, y)$. The proofs which are not included may be found in [10] as indicated or may be supplied by the reader using straightforward induction arguments.

(1) For $n \geqslant 2$, $f_n(x, 1) = x$. (pf: inductively $f_{n+1}(x, 1) = f_n(x, f_{n+1}(x, 0)) = f_n(x, 1)$ and $f_2(x, 1) = x$).

(2) For $n \geqslant 1$, $x \geqslant 2$, $f_n(x, y)$ is a strictly monotonic increasing function of y. (Lemma 1.1 of [10]).

(3) For $y \geqslant 1$ and all n, $f_n(x, y)$ is a strictly monotonic increasing function of x. (Lemma 1.2 of [10]).

(4) For $x \geqslant 2$, $y \geqslant 0$ and $n \geqslant 2$, $f_n(x, y) \leqslant f_{n+1}(x, y)$. (Lemma 1.3 of [10].

(5) For $p \geqslant 2$, $n \geqslant 3$, $x \geqslant 2$, and all q, $f_n(f_n(x, p), q) \leqslant f_n(x, f_{n-1}(p, q))$. (Theorem 1.1 of [10]).

(6) For $m > n \geqslant 2$, $x \geqslant 2$ and all y, z, $f_n(f_m(x, y), f_m(x, z)) \leqslant f_m(x, y+z)$. (Theorem 1.1 of [10]).

(7) For any $f(x^{(p)}) \in \mathcal{E}^n(A)$ with $n \geqslant 2$, there exists $k > 0$, such that when $x_i \geqslant 2$ for all $1 \leqslant i \leqslant p$, $f(x^{(p)}) < f_{n+1}(J^{(p)}(x^{(p)}), k)$.

Proof of (7). We show that the desired property holds for the initial functions and is preserved under substitution and limited iteration. For $x, y, x_i \geqslant 2$ we have:

$$Z(x) = 0 < f_{n+1}(x, 1) = x$$

$$U_m{}^p(x^{(p)}) = x_m < J^{(p)}(x^{(p)}) = f_{n+1}(J^{(p)}(x^{(p)}), 1) \quad \text{for } 0 \leqslant m \leqslant p$$

$$f_0(x, y) = x + 1 < 2x = f_2(x, 2) \leqslant f_{n+1}(x, 2) \leqslant f_{n+1}(J^{(2)}(x, y), 2)$$

$$f_1(x, y) = x + y < 2J^{(2)}(x, y) = f_2(J^{(2)}(x, y), 2) \leqslant f_{n+1}(J^{(2)}(x, y), 2)$$

$$f_n(x, y) \leqslant f_n(J^{(2)}(x, y), J^{(2)}(x, y)) = f_n(J^{(2)}(x, y), f_{n+1}(J^{(2)}(x, y), 1))$$
$$= f_{n+1}(J^{(2)}(x, y), 2) < f_{n+1}(J^{(2)}(x, y)$$

$$E(x) < x = f_{n+1}(x, 1)$$

$$c_A(x) < 2 \leqslant x = f_{n+1}(x, 1).$$

It is easy to verify by induction that $f_3(x, y) = x^y$ and that there exists $r, s > 0$ such that for all $x \geqslant 0$, $J^{(q)}(x, ..., x) \leqslant (rx)^s$ (the r and s depend on q). Then $J^{(q)}(x, ..., x) \leqslant (rx)^s = f_3(rx, s) = f_3(f_2(x, r), s) \leqslant f_3(f_3(x, r), s) \leqslant f_3(x, f_2(r, s))$. Thus for each q there exists k' such that $J^{(q)}(x, ..., x) < f_{n+1}(x, k')$.

Let $f(x^{(p)}) = h(g_1(x^{(p)}), ..., g_q(x^{(p)}))$ where $h(x^{(q)}) < f_{n+1}(J^{(q)}(x^{(q)}), k)$ and $g_i(x^{(p)}) < f_{n+1}(J^{(p)}(x^{(p)}), k_i)$ and let $k_0 = \max\{k_1, ..., k_q\}$. Then

$$
\begin{aligned}
f(x^{(p)}) &< f_{n+1}(J^{(q)}(g_1(x^{(p)}), ..., g_q(x^{(p)})), k) \\
&\leqslant f_{n+1}(J^{(q)}(f_{n+1}(J^{(p)}(x^{(p)}), k_1), ..., f_{n+1}(J^{(p)}(x^{(p)}), k_q)), k) \\
&\leqslant f_{n+1}(J^{(q)}(f_{n+1}(J^{(p)}(x^{(p)}), k_0), ..., f_{n+1}(J^{(p)}(x^{(p)}), k_0)), k) \\
&\leqslant f_{n+1}(f_{n+1}(f_{n+1}(J^{(p)}(x^{(p)}), k_0), k'), k) \\
&\leqslant f_{n+1}(f_{n+1}(J^{(p)}(x^{(p)}), f_n(k_0, k')), k) \\
&\leqslant f_{n+1}(J^{(p)}(x^{(p)}), f_n(f_n(k_0, k'), k)).
\end{aligned}
$$

Thus the desired property is preserved under substitution.

If $f(x)$ is defined by limited iteration from $h(x)$ and $j(x)$, then $f(x) \leqslant j(x) < f_{n+1}(x, k)$ for some k so limited iteration preserves the desired property. Thus we have verified (7).

Returning now to the proof of the theorem let f be defined by $f(0) = c = h^0(c)$ and $f(x+1) = hf(x) = h^{x+1}(c)$ for $h(x) \in \mathcal{E}^n(A)$ and $n \geqslant 2$. Then by (7), there exists k such that $h(y) < f_{n+1}(y, k)$ for $y \geqslant 2$. In particular $h^0(y) = y \leqslant \max(2, y) = f_{n+1}(\max(2, y), 1) = f_{n+1}(\max(2, y), f_{n+1}(k, 0))$ for all y. Inductively assume $h^x(y) \leqslant f_{n+1}(\max(2, y), f_{n+1}(k, x))$ for all y. Then

$$
\begin{aligned}
h^{x+1}(y) = hh^x(y) &\leqslant f_{n+1}(\max(2, h^x(y)), k) \\
&\leqslant f_{n+1}(f_{n+1}(\max(2, y), f_{n+1}(k, x)), k) \\
&\leqslant f_{n+1}(\max(2, y), f_n(f_{n+1}(k, x), k)) \\
&\leqslant f_{n+1}(\max(2, y), f_n(f_{n+1}(k, x), f_{n+1}(k, 1))) \\
&\leqslant f_{n+1}(\max(2, y), f_{n+1}(k, x+1))
\end{aligned}
$$

where the first line follows from (7), the second line from the inductive hypothesis and $f_{n+1}(\max(2, y), f_{n+1}(k, x)) \geqslant 2$, the third line from (5), the fourth line from (1) and the fifth line from (6).

Therefore $f(x) = h^x(c) \leqslant f_{n+1}(\max(2, c), f_{n+1}(k, x)) \in \mathcal{E}^{n+1}(A)$ as a function of x. □

Since by the theorem unlimited recursion leads from the class $\mathcal{E}^n(A)$ to the class $\mathcal{E}^{n+1}(A)$, the class $\bigcup_{n=2}^{\infty} \mathcal{E}^n(A)$ is closed under iteration and hence primitive recursion. Thus we have

Corollary 2.5. *The class $\bigcup_{n=2}^{\infty} \mathcal{E}^n(A)$ is the class of A-primitive recursive functions, and for all A and each $n \geqslant 2$, $\mathcal{E}^n(A)$ is properly contained in $\mathcal{E}^{n+1}(A)$.*

The special case $c(A) \in \mathcal{E}^m$ for some m, (i.e., A is $\mathcal{E}^m$ decidable) is of particular interest. Here $\mathcal{E}^n(A) \subset \mathcal{E}^p$ for $p = \max\{n, m\}$. However, suppose $n < m$ so $\mathcal{E}^n(A) \subset \mathcal{E}^m$. Then by the usual estimates an unbounded recursion leads from $\mathcal{E}^n(A)$ to $\mathcal{E}^{m+1}$. However we see from Theorem 4 that such an unbounded recursion leads from $\mathcal{E}^n(A)$ to $\mathcal{E}^{n+1}(A) \subset \mathcal{E}^m$. It should be noted that $c(A) \in \mathcal{E}^m$, $c(B) \in \mathcal{E}^m$ does not in general imply $\mathcal{E}^n(A) = \mathcal{E}^n(B)$.

The following lemma is useful in bounding certain recursive processes which appear later. It is of considerable use in Part II.

Lemma 2.6. *Let $f(x) = g(x, h(x))$ for g defined recursively by $g(x, 0) = x$ and $g(x, y+1) = k(x, y, g(x, y))$. Assume $g(x, y+1) \leqslant b(g(x, y))$ and $h(x)$, $b(z)$ are $\mathcal{E}^n$ computable for $n \geqslant 2$. Then $f(x)$ is bounded by an $\mathcal{E}^{n+1}$ computable function of x.*

Proof. Define

$$q(x, 0) = x$$

$$q(x, 1) = b(x)$$

$$q(x, y+1) = \sum_{z=0}^{q(x,y)} b(z)$$

so b being $\mathcal{E}^n$ computable, q is $\mathcal{E}^{n+1}$ computable. Clearly $g(x, 0) \leqslant q(x, 0)$ and $g(x, 1) \leqslant q(x, 1)$. Inductively $g(x, y+1) \leqslant b(g(x, y)) \leqslant \sum_{z=0}^{q(x,y)} b(z) = q(x, y+1)$ since $g(x, y) \leqslant q(x, y)$. Thus $f(x) \leqslant q(x, h(x))$ which is $\mathcal{E}^{n+1}$ computable.

Notice that no assumption is made above on the computability of k. A special case of note is $n = 3$ and k $\mathcal{E}^n(A)$ computable for $n \geqslant 4$. Then f is $\mathcal{E}^n(A)$ computable.

§3. $\mathcal{E}^n(A)$ groups and the word problem

Following the definition given in [1] we say a countable group G is an "$\mathcal{E}^n(A)$ group" (or is "$\mathcal{E}^n(A)$ computable") if it has an "index" (i, m, j) for i an injection of G onto an $\mathcal{E}^n(A)$ decidable subset of N, m an $\mathcal{E}^n(A)$ computable function $m : i(G) \times i(G) \to i(G)$ where m is given by $(i(g_1), i(g_2)) \overset{m}{\mapsto} i(g_1 g_2)$, and likewise for $j : i(G) \to i(G)$ given by $i(g) \overset{j}{\mapsto} i(g^{-1})$. For G_1 and G_2 $\mathcal{E}^n(A)$ computable groups with indices (i_1, m_1, j_1) and (i_2, m_2, j_2) respectively, we say a homomorphism $f : G_1 \to G_2$ is "$\mathcal{E}^n(A)$ computable" if $\hat{f} : i_1(G_1) \to i_2(G_2)$ by $\hat{f} i_1(g) \mapsto i_2 f(G)$ is an $\mathcal{E}^n(A)$ computable function. Note that the computability of f depends on the indices for G_1 and G_2. When not obvious from context, we will say "f is $\mathcal{E}^n(A)$ computable relative to indices (i_1, m_1, j_1) and (i_2, m_2, j_2)".

We freely use the results of §2, the $\mathcal{E}^3$ pairing functions, the $\mathcal{E}^3$ computable functions ## 1–21 of Kleene [7] p. 222f (note in particular ## 16 through ## 21, p. 230), the statements # A through # F of Kleene [7] p. 222f modified by replacing "primitive recursive" by "$\mathcal{E}^3$" and the concept of a group given by generators $a_1, a_2, \ldots$ and relations $R_1, R_2, \ldots$ (see [8]) which will be denoted $G = \langle a_1, \ldots; R_1, \ldots \rangle$.

We begin with a lemma stated as Lemma 4.1 of [1]. In the proof we use a slightly different index which is more convenient to use later.

Lemma 3.1. *A free group* $F = \langle a_1, \ldots; \rangle$ *on finitely or countably many generators is* $\mathcal{E}^3$.

Proof. F consists of freely reduced words of the form $w = a_{i_0}^{\alpha_0} \ldots a_{i_r}^{\alpha_r}$ for $\alpha_0, \ldots, \alpha_r$ positive or negative (but non-zero) integers. Write $\bar{\alpha} = 2\alpha$ if $\alpha > 0$ and $\bar{\alpha} = -2\alpha - 1$ if $\alpha < 0$. Then using the pairing functions of §2 we write for $w \neq \Lambda$ (the empty word)

$$i(w) = \prod_{k=0}^{r} p_k \exp J(i_k, \bar{\alpha}_k)$$

$$i(\Lambda) = 1$$

for p_k the kth prime starting with $p_0 = 2, p_1 = 3, \dots$. Denoting $GN(x) \leftrightarrow \forall k < 1hx((x)_k \neq 0)$ we see that $x \in i(F_n)$ for $F_n = \langle a_0, \dots, a_{n-1};\rangle$, finitely generated, iff $x = 1 \vee [GN(x) \wedge \forall k < 1hx((K((x)_k) < n) \wedge (L((x)_k) \neq 0))]$ and $x \in i(F_\infty)$ for $F_\infty = \langle a_0, \dots;\rangle$, countably generated iff $x = 1 \vee [GN(x) \wedge \forall k < 1hx(L((x)_k) \neq 0)]$.

To compute $m(x, y)$ one must "decode" x and y as words, freely reduce the concatenation of these words and then encode the result. It is clear that such a process can be interpreted by a recursion defined on $\mathcal{E}^3$ functions. Moreover since $m(x, y) \leqslant x * y$ (a relation we use later), the recursion is limited and m is an $\mathcal{E}^3$ function. Similarly it is clear that inversion, j, can be performed by a recursion defined on $\mathcal{E}^3$ functions and limited by $j(x) \leqslant p_{1hx} \exp((1hx)(\max_{0 \leqslant i < 1hx} (x)_i + 1))$ so j is $\mathcal{E}^3$ computable. □

Using the above lemma we can show that the study of $\mathcal{E}^n(A)$ groups is non-empty.

Theorem 3.2. *For every countable group G there exists $A \subset N$ such that G is an $\mathcal{E}^3(A)$ group.*

Proof. G being countable (or finite) there exists a presentation, $1 \to K \to F \to G \to 1$, as an exact sequence, for F free and at most countably generated hence $\mathcal{E}^3$. Let $A = i(K) \subset i(F) \subset N$. Following [1], Theorem 5.1 we define the $\mathcal{E}^3(A)$ predicate

$$E(x, y) \leftrightarrow x \in i(F) \wedge y \in i(F) \wedge m(j(x), y) \in i(K).$$

Thus the predicate E says "x and y are in the same coset modulo K". We define a unique index for each such coset, and hence for G, by

$$r(x) = \mu y \leqslant x(E(x, y)).$$

Now $i_G(G) = ri(F)$ and $x \in i_G(G) \leftrightarrow x \in i(F) \wedge r(x) = x$. We define $m_G(x, y) = rm(x, y)$ and note $m_G(x, y) \leqslant x * y$. Similarly $j_G(x) = rj(x)$. □

Theorem 3.2 suggests that if G is finitely generated (f.g.), then A is the encoded version of the word problem. The relation between $\mathcal{E}^n(A)$ groups and the word problem is made precise in the following definition, which, for the sake of the discussion further below, will be given for countably generated groups. However, note that the relationship between $\mathcal{E}^n(A)$ groups and the word problem depends on finite systems of generators, only.

Definition 3.3. A group G is said to be $\mathcal{E}^n(A)$ *standard* relative to an index (i, m, j) if i is defined by minimalization from a presentation $1 \to K \to F \to G \to 1$ for F free on at most countably many generators, $n \geqslant 3$, and $A \subset N$, the minimalization to be performed on $E(x, y)$ of the proof of Theorem 3.2 for $E(x, y)$ $\mathcal{E}^n(A)$ decidable.

In Definition 3.3 we do not require that $A = i(K)$ but merely that $i(K)$ be $\mathcal{E}^n(A)$ decidable so that $E(x, y)$ is $\mathcal{E}^n(A)$ decidable. Clearly the index given in Theorem 3.2 is a standard index. Intuitively, a finitely generated G is given a standard index by solving the word problem for a presentation of G by an $\mathcal{E}^n(A)$ process. Theorem 3.2 says that the word problem for a countable group can be solved for any given presentation and some A. Our next theorem shows that for finitely generated groups the level of computability of the word problem is independent of the f.g. presentation.

Theorem 3.4. *If G is f.g. and $\mathcal{E}^n(A)$ standard for $n \geqslant 3$ then any standard index of G is $\mathcal{E}^n(A)$.*

Proof. Let $1 \to K \to F \xrightarrow{\sigma} G \to 1$ be a presentation of G for F finitely generated. We show K is $\mathcal{E}^n(A)$ decidable by showing σ is $\mathcal{E}^n(A)$ computable relative to the index (i, m, j) on F and (i', m', j'), the given $\mathcal{E}^n(A)$ standard index of G. Then $x \in i(K) \leftrightarrow x \in i(F) \wedge \hat{\sigma}(x) = i'(1)$.

For $w \in F$, $\sigma(w)$ is computed as the product of images of generators of F corresponding to the spelling of w. Thus since F is finitely generated, $\hat{\sigma}(x)$ can be interpreted by a recursion on $\mathrm{lh}(x)$ involving m' where $\hat{\sigma}$ is $\mathcal{E}^n(A)$ computable if the recursion is bounded. Here $m'(y, z) \leqslant y * z$ since (i', m', j') is a standard index so $\hat{\sigma}(x)$ is bounded by $y_1 * y_2 * \ldots * y_r$ where each y_i as well as r can be computed from x by an $\mathcal{E}^3$ function. But $*_{i=1}^r y_i = y_1 * \ldots * y_r \leqslant$
$p_k \exp\left(k \sum_{i=1}^{r} \sum_{n=0}^{\mathrm{lh} y_i \dot{-} 1} (y_i)_n\right)$ for $k = \sum_{i=1}^{r} \mathrm{lh} y_i$. Thus the recursion is bounded and $\hat{\sigma}$ is $\mathcal{E}^n(A)$ computable. □

Corollary 3.5. *If G is f.g. and $\mathcal{E}^n(A)$ computable for $n \geqslant 3$ then G is $\mathcal{E}^{n+1}(A)$ standard.*

Proof. In the proof of Theorem 3.4 if G is $\mathcal{E}^n(A)$ but not $\mathcal{E}^n(A)$ standard, the recursion defining $\hat{\sigma}$ need not be bounded and therefore $\hat{\sigma}$ and hence $i(K)$ may be $\mathcal{E}^{n+1}(A)$ rather than $\mathcal{E}^n(A)$ by Theorem 2.4. □

Corollary 3.6. *If G is f.g. and $\mathcal{E}^n(A)$ computable for $n \geqslant 3$ then G is $\mathcal{E}^3(B)$ standard for B $\mathcal{E}^{n+1}(A)$ decidable.*

Proof. In the case of Corollary 3.5, set $B = i(K)$. Then B is $\mathcal{E}^{n+1}(A)$ decidable and G is $\mathcal{E}^3(B)$ standard by the usual construction of Theorem 3.2. □

Corollary 3.7. *If G is an $\mathcal{E}^n(A)$ computable group for $n \geqslant 3$ and $H < G$ is a finitely generated subgroup then H is $\mathcal{E}^{n+1}(A)$ standard. If G is $\mathcal{E}^n(A)$ standard then H is $\mathcal{E}^n(A)$ standard.*

Proof. In the proof of Theorem 3.4 and Corollary 3.5 we did not need σ surjective. Corollary 3.7 is then a restatement for σ not surjective and $H \cong F/K$. □

Corollary 3.8. *If G is f.g. and $\mathcal{E}^n$ standard but not $\mathcal{E}^{n-1}$ standard for $n \geq 4$, then G is not $\mathcal{E}^m$ for $m < n-1$.*

Proof. This is a restatement of Corollary 3.5 for $A = N$. □

Theorem 3.4 above is a mild generalization of a result due to Rabin, [9], that the computability of the word problem depends on G and not any of G's finitely generated presentations. Corollaries 3.5, 3.6, and 3.8 show the relationship between standard and non-standard indices, Corollary 3.8 is a special case for later use. Corollary 3.7 shows that the property of being $\mathcal{E}^n(A)$ standard is inherited by finitely generated subgroups although in general a finitely generated subgroup of an $\mathcal{E}^n(A)$ group may be $\mathcal{E}^{n+1}(A)$. From the proof of Corollary 3.7 we also see that the embedding $H \to G$ is $\mathcal{E}^{n+1}(A)$ computable since it can be computed by regarding an element in the index of H to be in $i(F)$ and applying the $\mathcal{E}^{n+1}(A)$ computable $\hat{\sigma}$. As a companion to the herditary result of Corollary 3.7 we see that under suitable conditions, quotient groups of $\mathcal{E}^n(A)$ groups are $\mathcal{E}^n(A)$.

Proposition 3.9. *If G is an $\mathcal{E}^n(A)$ group for $n \geq 3$ and $K < G$ is an $\mathcal{E}^n(A)$ decidable normal subgroup then G/K is $\mathcal{E}^n(A)$ computable.*

Proof. In the proof of Theorem 3.2 replace F by G and G by G/K and let (i, m, j) be the original index for G. The same definitions of E, c, m and j work as does the decidability criteria. □

It is of course not true that all quotient groups of $\mathcal{E}^n(A)$ are $\mathcal{E}^n(A)$ for if that were the case all groups would be $\mathcal{E}^3$ and, in particular, have a solvable word problem. Note however that if $B = i(K) \subset i(G)$ in Proposition 3.9 then by replacing A by

$$A \text{ join } B = \{x : 2 \mid x \Rightarrow [x/2] \in A \;\&\; 2\nmid x \Rightarrow [(x \dot{-} 1)/2] \in B$$

we see G/K is $\mathcal{E}^n(A \text{ join } B)$ computable.

§4. Free products with amalgamation

The free product with amalgamation is a useful construction when dealing with decision problems in groups since intuitively the normal form theorem yields decision procedures for such products modulo the decision procedures for the groups and the amalgamated subgroups. In the following we will study the free product with amalgamation for $\mathcal{E}^n(A)$ groups, $n \geqslant 3$, where the subgroups amalgamated are themselves $\mathcal{E}^n(A)$ as is the isomorphism relating them. In this context recall that if one decision problem has an $\mathcal{E}^p(B)$ solution and another an $\mathcal{E}^q(C)$ solution then they both have $\mathcal{E}^n(A)$ solutions for $n = \max(p, q)$ and $A = B$ join $C = \{x : 2 \mid x \Rightarrow [x/2] \in A \ \& \ 2 : x \Rightarrow [(x \dot{-} 1)/2] \in B\}$. In view of this our hypothesis will always involve a single computability class, $\mathcal{E}^n(A)$.

In general we require not only the index of the product but also the manner in which the original factors are embedded. The following definition is used to make such statements precise.

Definition 4.1. Let G and L be $\mathcal{E}^n(A)$ groups with indices (i, m, j) and (i', m', j') respectively. Suppose $\kappa : G \to L$ is an embedding. Then we say κ is an $\mathcal{E}^n(A)$ *embedding of* G *into* L (or simply G is $\mathcal{E}^n(A)$ *embedded* into L) if:

(i) $\hat{\kappa} : i(G) \to i'(L)$ is $\mathcal{E}^n(A)$ computable with respect to (i, m, j) and (i', m', j'), (i.e. κ is $\mathcal{E}^n(A)$ computable).

(ii) $\hat{\kappa} i(G) \subset i'(L)$ is $\mathcal{E}^n(A)$ decidable, (i.e., G is an $\mathcal{E}^n(A)$ decidable subgroup of L).

(iii) $\hat{\kappa}^{-1} : \hat{\kappa} i(G) \to i(G)$ is $\mathcal{E}^n(A)$ computable with respect to (i', m', j') and (i, m, j), (i.e., κ^{-1} is $\mathcal{E}^n(A)$ computable).

Notice that Definition 4.1 involves the particular choice of indices being used. In the following, the particular choice of indices is specified only when not obvious from context. Also observe that in the case where G is f.g. and (i, m, j) and (i', m', j') are standard indices, conditions (i) and (iii) are superfulous.

We demonstrate the type of result we desire for free products with amalgamation by first considering the more obvious situation for direct product, denoted $G_1 \times G_2$.

Proposition 4.2. *Let G_a be $\mathcal{E}^n(A)$ groups for $n \geqslant 3$ and $a = 1, 2$. Then $G = G_1 \times G_2$ is an $\mathcal{E}^n(A)$ group and the embeddings $G_a \to G$ are $\mathcal{E}^n(A)$ embeddings.*

Proof. Let G_a have indices (i_a, m_a, j_a) and give G an index by $i(g_1, g_2) = J(i_1(g_1), i_2(g_2))$. It is clear this index has all of the precribed properties. □

We may ask about the universal properties of the direct product. Clearly the projections are $\mathcal{E}^n(A)$ computable.

Proposition 4.3. *Under the assumptions of Proposition 4.2 assume also K is an $\mathcal{E}^n(A)$ group and $\alpha_a : K \to G_a$ are $\mathcal{E}^n(A)$ computable homomorphisms. Let $\pi_a : G \to G_a$ be the projections and $\alpha : K \to G$ the unique homomorphism such that $\pi_a \alpha = \alpha_a$. Then α is $\mathcal{E}^n(A)$ computable.*

Proof: $\hat{\alpha}(x) = J(\hat{\alpha}_1(x), \hat{\alpha}_2(x))$. □

As we saw the index given $G = G_1 \times G_2$ by Proposition 4.2 was the natural index with regard to the universal property. However, if G_1, G_2 (and hence G) are f.g. then it is not the standard index. Next we relate this index to the standard index.

Proposition 4.4. *Under the assumptions of Proposition 4.2 assume also that G_a are f.g. and $\mathcal{E}^n(A)$ standard. Then G is $\mathcal{E}^n(A)$ standard and the identity isomorphism on G is $\mathcal{E}^n(A)$ computable from G with standard index to G with index (i, m, j) of Proposition 4.2 and from G with index (i, m, j) to G with standard index.*

Proof. Let G_1 be generated by $a_1, ..., a_r$ and G_2 by $a_{r+1}, ..., a_s$. Consider $F = \langle a_1, ..., a_{r+s}; \rangle$ and $\alpha_1 : F \to G_1$ by $a_i \mapsto a_i$ for $i \leqslant r$, $a_i \mapsto 1$ for $i > r$ and $\alpha_2 : F \to G_2$ by $a_i \mapsto 1$ for $i \leqslant r$, $a_i \mapsto a_i$ for $i > r$. Then G_a being $\mathcal{E}^n(A)$ standard the α_a are $\mathcal{E}^n(A)$ computable and induce $\alpha : F \to G$ which is $\mathcal{E}^n(A)$ computable by Proposition 4.3. Thus $\ker\alpha$ is $\mathcal{E}^n(A)$ decidable so G is $\mathcal{E}^n(A)$ standard. This argument also shows that the identity isomorphism on G is $\mathcal{E}^n(A)$ computable from the standard index to the index of Proposition 4.2. Conversely since the G_a have standard indices it is an $\mathcal{E}^n(A)$ process to write $(g_1, g_2) = (w_1(a_i), w_2(a_i))$ as words on the generators (w_1 involves only $a_1, ..., a_r$ and w_2 only $a_{r+1}, ..., a_{r+s}$) and so associate (g_1, g_2) with $w = w_1(a_i)w_2(a_i) \in F$. Reducing w modulo $\ker\alpha$, the identity isomorphism on G is $\mathcal{E}^n(A)$ computable from the index of Proposition 4.2 to the standard index. □

We will proceed along the same lines for free product with amalgamation. That is, first we will develop an index which is natural, then verify the universal property and finally, restricting our attention to f.g. groups show the relationship to the standard index. In this case the natural index will reflect the normal form representation of elements in the free product. Informally this index will be called the normal form index. Since the normal form requires coset representatives in the factors modulo the amalgamated subgroup (amalgam for short) we use the following definition (compare [4] Definition 2.2).

Definition 4.5. Let G be an $\mathcal{E}^n(A)$ group and $H < G$ an $\mathcal{E}^n(A)$ decidable subgroup for $n \geqslant 3$. An $\mathcal{E}^n(A)$ *right coset representative system* for G mod H is an $\mathcal{E}^n(A)$ computable function $k : i(G) \to i(G)$ satisfying:

(i) $\quad x \in i(G) \quad y \in i(G) \quad m(x, j(y)) \in i(H) \leftrightarrow k(x) = k(y)$

(ii) $\quad m(x, jk(x)) \in i(H)$

(iii) $\quad x \in i(H) \to k(x) = i(1)$.

Intuitively k is a method of choosing right coset representatives for elements of G by an $\mathcal{E}^n(A)$ process. There always exists $\mathcal{E}^n(A)$

right coset representative systems. For example for $x \in i(G)$ define $k(x) = \mu y < x(y \in i(G) \wedge m(x, j(y)) \in i(H))$ if $x \notin i(H)$ and $k(x) = i(1)$ if $x \in i(H)$. Notice that any k as in the definition decomposes $x \in i(H)$ as $x = m(h(x), k(x))$ for $h(x) = m(x, jk(x))$, i.e. allows us to write $g = hg'$ for $h \in H$ and g' a particular coset representative by an $\mathcal{E}^n(A)$ process. Our use of such representative systems is seen in the following.

Theorem 4.6. *Let G_a be $\mathcal{E}^n(A)$ groups for $n \geqslant 3$ and $a = 1, 2$. Assume $H_a < G_a$ are $\mathcal{E}^n(A)$ decidable subgroups and $\varphi : H_1 \to H_2$ is an isomorphism such that φ and φ^{-1} are $\mathcal{E}^n(A)$ computable. Then for each choice of $\mathcal{E}^n(A)$ right coset representative systems $k_a : i_a(G_a) \to i_a(G_a)$ there is an $\mathcal{E}^{n+1}(A)$ index on $G_1 *_\varphi G_2$ (the free product of G_1 and G_2 amalgamating H_1 and H_2 by φ). The natural embeddings $G_a \to G$ are $\mathcal{E}^{n+1}(A)$ embeddings.*

Proof. This is a relativized version of Theorem 2.1 of [4] replacing the k_a of that proof by the given k_a. □

Corollary 4.7. *Let G_a be $\mathcal{E}^n(A)$ groups for $n \geqslant 3$ and $a = 1, 2$. Then $G_1 * G_2$ is an $\mathcal{E}^n(A)$ group and the natural embeddings $G_a \to G_1 * G_2$ are $\mathcal{E}^n(A)$ embeddings.*

Proof. Relativize Corollary 2.1.1 of [4]. □

We now have the universal property.

Corollary 4.8. *Under the assumptions of Theorem 4.6 assume also that K is an $\mathcal{E}^n(A)$ or an $\mathcal{E}^{n+1}(A)$ standard group and $\tau_a : G_a \to K$ are $\mathcal{E}^n(A)$ computable homomorphisms agreeing on the amalgam (i.e., if $h \in H_1$ then $\tau_1(h) = \tau_2\varphi(h)$). Then the unique homomorphism $\tau : G_1 *_\varphi G_2 \to K$ extending τ_1 and τ_2 (i.e., such that $\tau \mid G_1 = \tau_1$ and $\tau \mid G_2 = \tau_2$) is $\mathcal{E}^{n+1}(A)$ computable.*

Proof. Relativize Corollary 2.1.2 of [4] for K $\mathcal{E}^n(A)$. For K $\mathcal{E}^{n+1}(A)$ standard the bound on multiplication in K yields a bound on the recursion for $\overline{\tau}$. □

We now can relate indices of $G_1 *_\varphi G_2$ which arise from different choices of the coset representative systems (compare [4] Proposition 2.2).

Corollary 4.9. *Under the assumptions of Theorem 4.6 let $G_1 *_\varphi G_2$ have index (i, m, j) with respect to $\mathcal{E}^n(A)$ coset representative systems k_a and (i', m', j') with respect to $\mathcal{E}^n(A)$ coset representative systems k'_a. Then the identity isomorphism on $G_1 *_\varphi G_2$ is $\mathcal{E}^{n+1}(A)$ computable relative to (i, m, j) and (i', m', j').*

Proof. The embeddings $G_a \to G_1 *_\varphi G_2$ are $\mathcal{E}^n(A)$ computable with respect to (i_a, m_a, j_a) and (i', m', j') so extend to the identity isomorphism on $G_1 *_\varphi G_2$, $\mathcal{E}^{n+1}(A)$ computable with respect to (i, m, j) and (i', m', j') by Corollary 4.8 with $K = G_1 *_\varphi G_2$, $\mathcal{E}^{n+1}(A)$ computable using Lemma 2.6 to bound the recursion for $\overline{\tau}$. □

We now restrict our attention to f.g. groups and consider standard indices.

Theorem 4.10. *Let G_a be f.g. $\mathcal{E}^n(A)$ standard groups for $n \geqslant 3$ and $a = 1, 2$. Assume $H_a < G_a$ are $\mathcal{E}^n(A)$ decidable subgroups and $\varphi : H_1 \to H_2$ is an isomorphism such that φ and φ^{-1} are $\mathcal{E}^n(A)$ computable. Then $G_1 *_\varphi G_2$ is $\mathcal{E}^{n+1}(A)$ standard.*

Proof. We show that the word problem is $\mathcal{E}^{n+1}(A)$ decidable for a particular f.g. presentation of $G_1 *_\varphi G_2$. The argument will be a "spelling" argument but it should be clear that it can be encoded as a recursion defined on $\mathcal{E}^n(A)$ functions and hence $G_1 *_\varphi G_2$ is $\mathcal{E}^{n+1}(A)$ standard.

Let G_1 be generated by $a_1, \ldots, a_r$ and G_2 by $a_{r+1}, \ldots, a_{r+s}$. Since G_a are $\mathcal{E}^n(A)$ standard there is an $\mathcal{E}^n(A)$ process for recognizing if a word on the $a_1, \ldots, a_r$ is in H_1 and if so computing a word on $a_{r+1}, \ldots, a_{r+s}$ corresponding to its image under φ. Similarly one can compute φ^{-1} of a word in H_2. The statement that these processes are $\mathcal{E}^n(A)$ requires that the original indices of the G_a be standard.

Let w be any freely reduced word on the a_i. If the first symbol in w is a power of a_k for $1 \leqslant k \leqslant r$, write $w = w_1 \ldots w_p$ for each

$w_j \neq \emptyset$ and so that j odd implies w_j involves only symbols a_k for $1 \leqslant k \leqslant r$ and j even implies w_j involves only symbols a_k for $r+1 \leqslant k \leqslant r+s$. Similarly if the first symbol is a power of a_k for $r+1 \leqslant k \leqslant r+s$ write $w = w_1 .. w_p$ as above interchanging the roles of even and odd. In [8] such w_j are called syllables and p is called the syllable length of w. Clearly the decomposition of w into syllables is an $\mathcal{E}^3$ process.

We proceed by induction on the syllable length p of w. If $p = 1$ then $w = 1$ in $G_1 *_\varphi G_2$ iff $w = 1$ in G_1 or $w = 1$ in G_2, an $\mathcal{E}^n(A)$ decision. If $p > 1$ let $1 \leqslant q \leqslant p$ be the smallest integer such that either $w_q \in H_1$ or $w_q \in H_2$. If there is no such q, $w \neq 1$ in $G_1 *_\varphi G_2$. The search for q can be interpreted as bounded minimalization defined by the $\mathcal{E}^3$ function which decomposes w and the $\mathcal{E}^n(A)$ decision functions for H_a. If there is a q and $w_q \in H_1$, apply φ to w_q as described above and replace w by w' such that $w = w'$ in $G_1 *_\varphi G_2$ but w' has shorter syllable length than w. Similarly if $w_q \in H_2$ apply φ^{-1} to get w'. By induction we conclude that the word problem for $G_1 *_\varphi G_2$ is $\mathcal{E}^{n+1}(A)$ solvable.

Thus if $B \subset N$ is the encoded image of the kernel in the presentation of $G_1 *_\varphi G_2$ on $a_1, ..., a_{r+s}$, then $G_1 *_\varphi G_2$ is $\mathcal{E}^3(B)$ computable by Theorem 3.2 and $\mathcal{E}^{n+1}(A)$ computable since B is $\mathcal{E}^{n+1}(A)$ decidable by the above. □

Corollary 4.11. *Let G_a be f.g. $\mathcal{E}^n(A)$ standard groups for $n \geqslant 3$ and $a = 1, 2$. Then $G_1 * G_2$ is $\mathcal{E}^n(A)$ standard.*

Proof. In the proof of Theorem 4.10, the questions $w_k \in H_a$ are replaced by $w_k = 1$ in G_1 or G_2. Since the index of w' (and hence every succeeding reduction in the induction) is less than the index of w (there is no φ or φ^{-1} to possibly increase the index), the recursion used in an encoded version of the proof of Theorem 4.10 is bounded and the word problem $\mathcal{E}^n(A)$ decidable. □

We now study the relationship between the "normal form" index of Theorem 4.6 and the standard index of Theorem 4.10.

Corollary 4.12. *Under the assumptions of Theorem 4.10 let* (i', m', j') *be the* $\mathcal{E}^{n+1}(A)$ *standard index and* (i, m, j) *be the* $\mathcal{E}^{n+1}(A)$ *(normal form) index given by Theorem 4.6 arising from some* $\mathcal{E}^n(A)$ *right coset representative system. Then the identity isomorphism on* $G_1 *_\varphi G_2$ *is* $\mathcal{E}^{n+1}(A)$ *computable from the index* (i, m, j) *to the index* (i', m', j'). *It is* $\mathcal{E}^{n+1}(A)$ *computable from the index* (i', m', j') *to the index* (i, m, j).

Proof. Since G_a are f.g. and $\mathcal{E}^n(A)$ standard, there are $\mathcal{E}^n(A)$ quotient maps $\sigma_a : F^{(a)} \to G_a$ for $F^{(a)}$ f.g. free groups. Moreover for $x \in i_a(G_a)$, $x \in i_a(F^{(a)})$ with $\hat{\sigma}_a(x) = x$. Let $\sigma : F \to G_1 *_\varphi G_2$ be the $\mathcal{E}^{n+1}(A)$ quotient map for $F = F^{(1)} * F^{(2)}$ and $\tau_a : F^{(a)} \to F$ the $\mathcal{E}^3$ embeddings. Then the embedding $G_a \to G_1 *_\varphi G_2$ relative to (i_a, m_a, j_a) and (i', m', j') is given by $x \mapsto \hat{\sigma}\hat{\tau}(x)$ and so is $\mathcal{E}^{n+1}(A)$ computable. By Corollary 4.8 the extension of the embeddings, which is obviously the identity, is $\mathcal{E}^{n+1}(A)$ computable, the K of Corollary 4.8 being $G_1 *_\varphi G_2$ and having a standard index. The second statement of the Corollary is immediate from the fact that the quotient $\sigma : F \to G_1 *_\varphi G_2$ is $\mathcal{E}^{n+1}(A)$ computable relative to the index (i, m, j) (since F is f.g.) using Lemma 2.6 to bound the recursion. □

Corollary 4.13. *Under the assumptions of Corollary 4.11 let* (i', m', j') *be the* $\mathcal{E}^n(A)$ *standard index and let* (i, m, j) *be the* $\mathcal{E}^n(A)$ *(normal form) index given by Corollary 4.7. Then the identity isomorphism on* $G_1 * G_2$ *is* $\mathcal{E}^n(A)$ *computable from the index* (i, m, j) *to* (i', m', j') *and also from the index* (i', m', j') to the index (i, m, j).

Proof. In the proof of Corollary 4.12 the quotient map σ is $\mathcal{E}^n(A)$ computable when the amalgam is trivial and so by the remainder of the argument the identity is $\mathcal{E}^n(A)$ from the index (i, m, j) to (i', m', j'). The technique used to bound the index of a product in the proof of Corollary 4.7 ([4] Corollary 2.1.1) can be used to bound the recursion used in the computability of $\sigma : F \to G_1 * G_2$ relative to the index (i, m, j). Then σ is $\mathcal{E}^n(A)$ computable and hence so is the identity. □

By a standard technique Theorem 4.10 can be used to construct f.g. $\mathcal{E}^3(A)$ groups for any $A \subset N$ in the following sense.

Corollary 4.14. *For any $A \subset N$ there exists a f.g. $\mathcal{E}^3(A)$ standard group G_A such that if G_A is $\mathcal{E}^n(B)$ standard for $n \geqslant 3$ then A is $\mathcal{E}^n(B)$ decidable.*

Proof. Set $F' = \langle a, b; \rangle$, $F'' = \langle c, d; \rangle$, $H' < F'$ the subgroup generated by $\{ a^x ba^{-x} ; x \in A \}$ and $H'' < F''$ the subgroup generated by $\{ c^x dc^{-x} ; x \in A \}$. Then H' is freely generated by the $a^x ba^{-x}$ for $x \in A$ and so is $\mathcal{E}^3(A)$ decidable. Similarly for H''. Let $\varphi : H' \to H''$ be the $\mathcal{E}^3(A)$ computable isomorphism given by $a^x ba^{-x} \mapsto c^x dc^{-x}$ for $x \in A$. We consider $G_A = F' *_\varphi F''$. By theorem 4.10 G_A is $\mathcal{E}^4(A)$ computable but since $\hat{\varphi}$ and $\hat{\varphi}^{-1}$ are restrictions of the identity on N, the index of each successive w' (in the proof of Theorem 4.10) is less than or equal to $p_{1hx} \exp (J (4, \max_{0 \leqslant q < 1hx} L((x)_q)) \cdot 1hx)$

for x the index of w and hence the recursion used in solving the word problem is bounded so G_A is $\mathcal{E}^3(A)$ standard.

Suppose G_A is $\mathcal{E}^n(B)$ standard for $n \geqslant 3$. Then for any $x \in N$, to decide if $x \in A$, compute the index of $a^x ba^{-x} c^x d^{-1} c^{-x}$ in G_A by an $\mathcal{E}^n(B)$ process. This process is a decision procedure for A since $x \in A$ iff $a^x ba^{-x} c^x d^{-1} c^{-x} = 1$ in G_A. □

The above constructions for f.g. $\mathcal{E}^n(A)$ groups yield interesting corollaries for $\mathcal{E}^n = \mathcal{E}^n(N)$ groups. In particular we consider $\mathcal{E}^3(A)$ groups for A $\mathcal{E}^n$ decidable with $n > 3$.

Proposition 4.15. *Under the assumptions of Theorem 4.10 let $n \geqslant 3$ and A be $\mathcal{E}^m$ decidable for $m \geqslant n + 1$. Then $G_1 *_\varphi G_2$ has an $\mathcal{E}^m$ index for each $\mathcal{E}^n(A)$ coset representative system. Moreover $G_1 *_\varphi G_2$ is $\mathcal{E}^m$ standard and for $m \geqslant n + 1$ the identity isomorphism is $\mathcal{E}^m$ computable with respect to any of the above indices.*

Proof. Since $m \geqslant n + 1$ and A is $\mathcal{E}^m$ decidable $\mathcal{E}^{n+1}(A) \subset \mathcal{E}^m$ so the first part of the statement is immediate from Theorems 4.6 and 4.10. Similarly $\mathcal{E}^{n+1}(A) \subset \mathcal{E}^m$ for $m \geqslant n + 1$ so the second part follows from Corollaries 4.9 and 4.12. □

Of course we could specialize other statements to the case of $\mathcal{E}^n$ groups, e.g. Corollary 4.8. Of special interest is the following specialization of Corollary 4.14.

Proposition 4.16. *For $n \geqslant 4$ there exist $\mathcal{E}^n$ standard groups which are not $\mathcal{E}^{n-1}$ standard. For $n \geqslant 5$ there exist $\mathcal{E}^n$ standard groups which are not $\mathcal{E}^{n-2}$.*

Proof. In [1] it is shown that for $n \geqslant 4$ there exists sets $\mathcal{E}^n$ decidable but not $\mathcal{E}^{n-1}$ decidable. If A is such a set then the group G_A of Corollary 4.14 is $\mathcal{E}^3(A) \subset \mathcal{E}^n$ standard but not $\mathcal{E}^{n-1}$ standard since that would imply A $\mathcal{E}^{n-1}$ decidable. If $n \geqslant 5$ then G_A is not $\mathcal{E}^{n-2}$ since by Corollary 3.8 that would imply G_A is $\mathcal{E}^{n-1}$ standard. □

We say a group is primitive recursive (p.r.) if it is $\mathcal{E}^n$ for some n. We say a group is finitely presented (f.p.) if it has a f.g. presentation involving only finitely many relations. C.F. Miller III raised the question of whether there could exist a p.r., f.p. group containing as subgroups a copy of every p.r., f.p. group. As reported in [2] the answer is strongly negative.

Proposition 4.17. *There does not exist a p.r. group containing every f.g., p.r. group as a subgroup. There does not exist a p.r. group containing every f.p., p.r. group as a subgroup.*

Proof. By Proposition 4.16 there exist f.g. $\mathcal{E}^n$ groups not $\mathcal{E}^{n-2}$ for $n \geqslant 5$. Applying Theorem 5.1 of [4] there exist f.p. $\mathcal{E}^m$ groups for $m \geqslant n - 2$ and every $n \geqslant 5$. Now suppose there was a p.r. group G containing a copy of every p.r., f.g. (respectively f.p.) group as a subgroup. Then G is $\mathcal{E}^p$ for some p so by Corollary 3.7 each of its f.g. subgroups is $\mathcal{E}^{p+1}$ which is a contradiction since we have just seen there exist f.g. (respectively f.p.) $\mathcal{E}^m$ groups for arbitrarily high m. □

The remainder of this section is devoted to a technical devise for finding the computability of certain special subgroups of a free product with amalgamation. The computability of these subgroups is of importance for some later applications so the statements are included here for completeness. The casual reader may choose to ignore this material. We present a generalization of Definition 2.3, Proposition 2.3 and Lemma 2.1 of [4].

Definition 4.18. Let G be an $\mathcal{E}^n(A)$ group for $n \geqslant 3$ and $H < G$ an $\mathcal{E}^n(A)$ decidable subgroup. A subgroup $K < G$ is said to be H, $\mathcal{E}^n(A)$ *compatible* if it is $\mathcal{E}^n(A)$ decidable and there exists an $\mathcal{E}^n(A)$ right coset representative system for G mod H satisfying in addition to conditions (i) through (iii) of Definition 4.5:

(iv) $\qquad x \in i(K) \to k(x) \in i(K).$

What we require in Definition 4.18 is the existence of an $\mathcal{E}^n(A)$ right coset representative system for G mod H so that whenever a coset H_g satisfies $H_g \cap K \neq \emptyset$ then the representative of Hg is in K. Notice condition (iv) is trivially satisfied when $H < K < G$, all $\mathcal{E}^n(A)$ decidable, for then $g \in K$, $\hat{g} \in G$, $g\hat{g}^{-1} \in H$ implies $g\hat{g}^{-1} \in K$ so $\hat{g} \in K$ and any coset representative of Hg is in K. The following lemma characterizes compatibility and is more convenient than the definition for applications.

Lemma 4.19. *Let G be $\mathcal{E}^n(A)$ with index (i, m, j) such that $0 \notin i(G)$. Let $H < G$, $K < G$ be $\mathcal{E}^n(A)$ decidable subgroups. The following are equivalent:*

(1) *there exists an $\mathcal{E}^n(A)$ computable function $d : i(G) \to i(K) \cup \{0\}$ such that $d(x) = 0 \leftrightarrow \neg \exists y (y \in i(K) \wedge m(x, j(y)) \in i(H))$ and $d(x) \neq 0 \leftrightarrow m(x, jd(x)) \in i(H)$*

(2) *K is H, $\mathcal{E}^n(A)$ compatible*

(3) *H is K, $\mathcal{E}^n(A)$ compatible.*

Proof. Replace every occurrance of "*p.r.*" in the proof of Proposition 2.3 of [4] by "$\mathcal{E}^n(A)$". □

Observe that the condition $0 \notin i(G)$ in Lemma 4.19 is not critical since if $0 \in i(G)$, $i(g)$ can be replaced by $i(g) + 1$. Assume $H < G$, $K < G$ all $\mathcal{E}^n(A)$ and also that $\{H, K\} < G$, the subgroup of G generated by H and K, is $\mathcal{E}^n(A)$ decidable. Then we may take $d(x) = 0$ for $x \notin i(\{H, K\})$ since if for $g \in G$ there exists $\hat{g} \in K$ such that $g\hat{g}^{-1} = h \in H$ then $g = h\hat{g} \in \{H, K\}$. Thus to conclude K is H, $\mathcal{E}^n(A)$ compatible in G it suffices to show K is H, $\mathcal{E}^n(A)$ compatible in $\{H, K\}$. The use of the notion of compatibility is found in the following lemma.

Lemma 4.20. *Under the conditions of Theorem 4.6 let* $K_a < G_a$ *be* $\mathcal{E}^n(A)$ *decidable subgroups satisfying*

(i) K_a *is* H_a, $\mathcal{E}^n(A)$ *compatible for* $a = 1, 2$

(ii) $\varphi(K_1 \cap H_1) = K_2 \cap H_2$.

Then $K = \{K_1, K_2\} < G = G_1 *_\varphi G_2$ *satisfies*

(1) K *is* $\mathcal{E}^{n+1}(A)$ *decidable in* G *with respect to some normal form index on* G *hence* $\mathcal{E}^{n+1}(A)$ *decidable in* G *with respect to any normal form index or any standard index on* G *(i.e. the embedding* $K < G$ *is an* $\mathcal{E}^{n+1}(A)$ *embedding)*

(2) *for* $\varphi' = \varphi \mid K_1 \cap H_1$, $K = K_1 *_{\varphi'} K_2$

(3) $G_a \cap K = K_a$ *for* $a = 1, 2$.

Proof. Since by Corollaries 4.9 and 4.12 all normal form indices and all standard indices are related by an $\mathcal{E}^{n+1}(A)$ computable identity isomorphism on G it suffices to show K is $\mathcal{E}^{n+1}(A)$ decidable relative to the particular $\mathcal{E}^n(A)$ right coset representative systems with respect to which the K_a are H_a $\mathcal{E}^n(A)$ compatible. To show this as well as (2) and (3) replace every occurrance of "p.r." in the proof of Lemma 2.1 of [4] by "$\mathcal{E}^{n+1}(A)$" except that "p.r. right coset representative system" should be replaced by "$\mathcal{E}^n(A)$ right coset representative system". □

§5. Strong Britton extensions

In this section we consider a construction closely related to the free product with amalgamation, the so called strong Britton extension. We will refer to presentations of groups on generators and relations, viz. $G = \langle a_1, \ldots; R_1, \ldots \rangle$ and, when the meaning is clear, interpret this notation with liberty. For example, we may write

$\langle G, r;\rangle$ to mean $\langle a_1, ..., r; R_1, ...\rangle = G * \langle r;\rangle$; i.e. when the set of generators has some implied relations, we assume they are included in the relations of the presentation. We also use the notation $\{B\} < G$ for $B \subset G$, $\{B\}$ the subgroup of G generated by B. The following definitions are relativized versions of Definitions 3.1 and 3.2 of [4].

Definition 5.1. Let G be an $\mathcal{E}^n(A)$ group. An $\mathcal{E}^n(A)$ *isomorphism in* G is an isomorphism $\varphi : H \to H'$ such that H, H' are $\mathcal{E}^n(A)$ decidable subgroups of G and φ, φ^{-1} are $\mathcal{E}^n(A)$ computable (relative to the inherited index on H, H').

Definition 5.2. Let G be an $\mathcal{E}^n(A)$ group and φ an $\mathcal{E}^n(A)$ isomorphism in G. The *strong Britton extension of G by φ*, denoted by G_φ, is the group with presentation $\langle G, t; tht^{-1} = \varphi(h)\ \forall h \in \text{domain}\ \varphi\rangle$.

The $\mathcal{E}^n(A)$ computability of G and φ induces a computability structure in G_φ according to the following theorem and corollary.

Theorem 5.3. *Let G be an $\mathcal{E}^n(A)$ group and φ an $\mathcal{E}^n(A)$ isomorphism in G, for $n \geq 3$. Then G_φ is an $\mathcal{E}^{n+1}(A)$ group with index (i', m', j') inherited as a subgroup of a (particular) free product with amalgamation and its associated normal form index. The embedding $G \to G_\varphi$ is an $\mathcal{E}^{n+1}(A)$ embedding relative to (i', m', j').*

Proof. This is a relativized version of Theorem 3.1 of [4]. The proof is identical replacing $\mathcal{E}^n$ by $\mathcal{E}^n(A)$ using Lemma 2.6 to bound the recursion defining τ. □

Corollary 5.4. *If G is f.g. and $\mathcal{E}^n(A)$ standard in the hypothesis of Theorem 5.3, then G_φ is $\mathcal{E}^{n+1}(A)$ with index (i', m', j') and $G \to G_\varphi$ is an $\mathcal{E}^{n+1}(A)$ embedding. In addition, G_φ is $\mathcal{E}^{n+1}(A)$ standard with index $(\hat{\imath}, \hat{m}, \hat{\jmath})$. The identity isomorphism on G_φ is $\mathcal{E}^{n+1}(A)$ computable from index (i', m', j') to index $(\hat{\imath}, \hat{m}, \hat{\jmath})$ and $\mathcal{E}^{n+1}(A)$ computable from index $(\hat{\imath}, \hat{m}, \hat{\jmath})$ to index (i', m', j'). With respect to $(\hat{\imath}, \hat{m}, \hat{\jmath})$ the embedding $G \to G_\varphi$ is $\mathcal{E}^{n+1}(A)$ computable and an $\mathcal{E}^{n+1}(A)$ embedding.*

Proof. We modify the proof of Theorem 3.1 of [4]. Since G is f.g. and $\mathcal{E}^n(A)$ standard, L has an $\mathcal{E}^{n+1}(A)$ standard index, say $(\tilde{i}, \tilde{m}, \tilde{j})$, by Theorem 4.10. We compute $\tau : L \to L$ from index (i', m', j') to the index $(\tilde{i}, \tilde{m}, \tilde{j})$, where the latter index is an $\mathcal{E}^{n+1}(A)$ standard index. Corollary 4.8 shows τ_1 induced by τ_{11} and τ_{12} is automatically $\mathcal{E}^{n+1}(A)$ computable as is τ_2 induced by τ_{21} and τ_{22}. Similarly τ induced by τ_1 and τ_2 is $\mathcal{E}^{n+1}(A)$ computable. Thus $G_\varphi < L$ is $\mathcal{E}^{n+1}(A)$ decidable. Since the embedding $G \to L$ is an $\mathcal{E}^{n+1}(A)$ embedding and $G < G_\varphi$, this shows $G \to G_\varphi$ is an $\mathcal{E}^{n+1}(A)$ embedding.

To see G_φ is $\mathcal{E}^{n+1}(A)$ standard, observe that it is a f.g. subgroup of the $\mathcal{E}^{n+1}(A)$ standard group L and so Corollary 3.7 applies.

To see that the identity isomorphism on G_φ is $\mathcal{E}^{n+1}(A)$ computable from index (i', m', j') to index $(\hat{i}, \hat{m}, \hat{j})$ observe that τ can be defined from L to G_φ, the former with index (i', m', j') and the latter with index $(\hat{i}, \hat{m}, \hat{j})$, and that relative to these indices τ is $\mathcal{E}^{n+1}(A)$ computable by applications of Corollary 4.8 as above. The identity isomorphism in question is a restriction of τ to $G_\varphi < L$ so is $\mathcal{E}^{n+1}(A)$ computable. The identity isomorphism from index $(\hat{i}, \hat{m}, \hat{j})$ to (i', m', j') is $\mathcal{E}^{n+1}(A)$ computable since if $G_\varphi = F/K$ for F f.g. free, the quotient map $F \to G_\varphi$, where G_φ has index (i', m', j'), is $\mathcal{E}^{n+1}(A)$ computable using Lemma 2.6. Then $x \in \hat{\imath}(G_\varphi)$ can be viewed as an element in F and so the identity is $\mathcal{E}^{n+1}(A)$ computable. This also shows that $G \to G_\varphi$, with respect to $(\hat{i}, \hat{m}, \hat{j})$, is an $\mathcal{E}^{n+1}(A)$ embedding since with respect to (i', m', j') it is an $\mathcal{E}^{n+1}(A)$ embedding which may be carried over to the new index by the $\mathcal{E}^{n+1}(A)$ computable identity isomorphism. □

Using the above we obtain a version of the Higman, Neumann, Neumann Theorem.

Theorem 5.5. *Let G be an $\mathcal{E}^n(A)$ group for $n \geqslant 3$. Then G can be embedded into an $\mathcal{E}^{n+1}(A)$ standard group having two generators. The embedding is an $\mathcal{E}^{n+1}(A)$ embedding.*

Proof. By the relativized version of Theorem 3.2 of [4] G can be embedded into an $\mathcal{E}^{n+1}(A)$ group on three generators. Following the construction given in the proof of Corollary 3.2.1 of [4], G'

can be embedded in a two generator group of the form G'_ψ for ψ an $\mathcal{E}^{n+1}(A)$ isomorphism in G'. By Theorem 5.3, G'_ψ is $\mathcal{E}^{n+2}(A)$ but a simple application of Lemma 2.6 shows the recursion involved can be bounded and so G'_ψ is $\mathcal{E}^{n+1}(A)$. By Corollary 5.4, G'_ψ is $\mathcal{E}^{n+1}$ standard and the embedding $G \to G' \to G'_\psi$ is an $\mathcal{E}^{n+1}(A)$ embedding.

As before, the specialization of the above results to the case of $\mathcal{E}^3(A)$ groups for A $\mathcal{E}^n$ decidable is of interest.

Proposition 5.6. *If G is $\mathcal{E}^3(A)$ for A $\mathcal{E}^n$ decidable, $n \geq 4$, and φ is an $\mathcal{E}^{n-1}(A)$ isomorphism in G, G_φ is $\mathcal{E}^n$. If G is f.g. and $\mathcal{E}^3(A)$ standard for A $\mathcal{E}^n$ decidable, $n \geq 4$, and φ is an $\mathcal{E}^{n-1}(A)$ isomorphism in G, G_φ is $\mathcal{E}^n$ standard.*

Proof. Immediate from Theorem 5.3 and Corollary 5.4. □

Proposition 5.7. *If G is an $\mathcal{E}^3(A)$ group for A $\mathcal{E}^n$ decidable, $n \geq 4$, then G can be embedded into an $\mathcal{E}^n$ standard group on two generators. The embedding is an $\mathcal{E}^n$ embedding.*

Proof. Immediate from Theorem 5.5. □

In the remainder of this section we develop technical ideas for later application. Again, the casual reader may not wish to read further. First we obtain an analog of Lemma 4.20 for the strong Britton extensions (see [4] Definition 3.3 and Lemma 3.1).

Definition 5.8. Let G be an $\mathcal{E}^n(A)$ group, φ an $\mathcal{E}^n(A)$ isomorphism in G with domain H, and $K < G$ an $\mathcal{E}^n(A)$ decidable subgroup of G for $n \geq 3$. We say K is $\mathcal{E}^n(A)$ *invariant under* φ if

(i) K is H, $\mathcal{E}^n(A)$ compatible

(ii) K is $\varphi(H)$, $\mathcal{E}^n(A)$ compatible

(iii) $\varphi(H \cap K) = \varphi(H) \cap K$.

Condition (iii) of the above definition says $h \in H \cap K$ iff $\varphi(h) \in K$. We can locate $\mathcal{E}^{n+1}(A)$ decidable subgroups of G_φ (with normal form index) according to the following

Lemma 5.9. *Let G be an $\mathcal{E}^n(A)$ group for $n \geqslant 3$, φ an $\mathcal{E}^n(A)$ isomorphism in G, $K < G$ an $\mathcal{E}^n(A)$ decidable subgroup $\mathcal{E}^n(A)$ invariant under φ. Define H = domain φ and $\varphi' = \varphi|_{H \cap K}$. Then the embedding of the $\mathcal{E}^{n+1}(A)$ group $K_{\varphi'}$ into the $\mathcal{E}^{n+1}(A)$ group G_φ by $k \in K$, $k \mapsto k \in G$ and $t \mapsto t$ is an $\mathcal{E}^{n+1}(A)$ embedding relative to the indices given by Theorem 5.3.*

Proof. We proceed exactly as in the proof of Lemma 3.1 of [4]. Using the notation of that proof, we see $K_{\varphi'} < L'$ is an $\mathcal{E}^{n+1}(A)$ decidable subgroup of the $\mathcal{E}^{n+1}(A)$ group L' corresponding to some choice of $\mathcal{E}^n(A)$ coset representative system, $L' < L$ is $\mathcal{E}^{n+1}(A)$ decidable for an index on L given by a specific choice of coset representative system and by Corollary 4.9 the identity isomorphism on L is $\mathcal{E}^{n+1}$ for any choices of $\mathcal{E}^n(A)$ coset representative systems. Thus $K_{\varphi'} < L' < L \overset{\text{id}}{\cong} L > G_\varphi$ is an $\mathcal{E}^{n+1}(A)$ embedding. The verification of the appropriate compatibility conditions for $L' < L$ is identical to that given in the cited proof noting that the decisions involved are all $\mathcal{E}^n(A)$ decidable. □

We observe that when the conditions of Lemma 5.9 hold, $K_{\varphi'} = \{K, t\} < G_\varphi$ and, by Proposition 3.3 of [4], $\{K, t\} \cap G = K$ in G_φ.

Next, we consider a many-fold application of the strong Britton extension. In particular we allow countably many such extensions corresponding to isomorphisms $\varphi_1, \varphi_2, \ldots$ in G. Observe that $G_{\varphi_1, \varphi_2} \ldots$ may be finitely or (countably) infinitely generated. Here each φ_i is to be an $\mathcal{E}^n(A)$ isomorphism in G and not (more generally) in $G_{\varphi_1, \ldots, \varphi_{i-1}}$ so our result does not apply as it stands to the general case of "Britton towers".

Lemma 5.10. *Let G be an $\mathcal{E}^n(A)$ group for $n \geqslant 3$ and $\varphi_1, \varphi_2, \ldots$ be an ordered sequence of $\mathcal{E}^n(A)$ isomorphisms in G with domain $\varphi_k = H_k < G$. Define $G_\infty = G_{\varphi_1, \varphi_2, \ldots} = \langle G, t_1, t_2, \ldots; t_k h_k t_k^{-1} = \varphi_k(h_k)$ for all k, and $h_k \in H_k \rangle$. Then G_∞ is an $\mathcal{E}^{n+1}(A)$ group and the embedding $G < G_\infty$ is an $\mathcal{E}^{n+1}(A)$ embedding.*

Proof. Our proof is similar to that of Lemma 3.2 of [4] but with two modifications (note that the functions σ, σ_1, σ_2 of [4] are our J, K, L).

Let $G = G_0$, $G_{\varphi_1, ..., \varphi_m} = G_m$. We first show that each G_m is an $\mathcal{E}^{n+1}(A)$ group (clear for G_0). Let G' be a copy of G with $g' \in G'$, the copy of $g \in G$. Define $L_m = (G * \langle r_1, ..., r_m ;\rangle) *_\psi (G' * \langle s_1, ..., s_m ;\rangle)$ for $\psi : G * r_1 H_1 r_1^{-1} * ... * r_m H_m r_m^{-1} \to G' * s_1 \varphi_1 (H_1)' s_1^{-1} * ... * s_m \varphi_m (H_m)' s_m^{-1}$ by $g \mapsto g'$ and $r_i h r_i^{-1} \mapsto s_i \varphi_i (h)' s_i^{-1}$ for all $h \in H_i$ and $i = 1, ..., m$. The factors of L_m are $\mathcal{E}^n(A)$ groups by Corollary 4.7 and the domain and range of ψ are $\mathcal{E}^n(A)$ decidable by repeated application of the relativized version of Proposition 2.1 of [4]. It is clear that the individual homomorphisms $g \mapsto g'$ and $r_i h r_i^{-1} \mapsto s_i \varphi(h)' s_i^{-1}$ are $\mathcal{E}^n(A)$ computable (as homomorphisms from G or $r_i H_i r_i$ into $G' * \langle s_1, ..., s_m ;\rangle$). Moreover since the normal form of ψ(w) for $w \in G * \langle r_1, ..., r_n ;\rangle$ is the same as that for w with only the symbols changed, it is clear that the recursion used to define ψ as an extension of these individual homomorphisms by Corollary 4.8 can be bounded so ψ is $\mathcal{E}^n(A)$ computable. Similarly ψ^{-1} is $\mathcal{E}^n(A)$ computable so L_m is an $\mathcal{E}^{n+1}(A)$ group by Theorem 4.6. The homomorphism $\tau : L_m \to L_m$ given by $g \mapsto g$, $g' \mapsto g'$, $r_i \mapsto s_i^{-1} r_i$ and $s_i \mapsto \{1\}$ is $\mathcal{E}^{n+1}(A)$ computable, fixing only $G_m < L_m$ by an argument entirely analogous to that used in showing τ $\mathcal{E}^{n+1}$ computable in the proof of Theorem 3.1 of [4] using Lemma 2.6 to bound the recursion. Thus G_m is an $\mathcal{E}^{n+1}(A)$ decidable subgroup of L_m. Let L_m (and hence G_m) have index (i_m, m_m, j_m).

Next we show that for $k < m$, the embedding $G_k < G_m$ is an $\mathcal{E}^{n+1}(A)$ embedding. This is immediate for $k = 0$ and for $k > 0$ it suffices to show $L_k < L_m$ is an $\mathcal{E}^{n+1}(A)$ embedding. Observe that $G * r_1 H_1 r_1^{-1} * ... * r_k H_k r_k^{-1} < G * r_1 H_1 r_1^{-1} * ... * r_m H_m r^{-1}$ and $G' * s_1 \varphi_1 (H_1)' s_1^{-1} * ... * s_k \varphi_k (H_k)' s_k^{-1} < G' * s_1 \varphi_1 (H_1)' s_1^{-1} * ... * s_m \varphi_m (H_m)' s_m^{-1}$ are $\mathcal{E}^3$ decidable and so the conditions of Lemma 4.20 are satisfied and $L_k < L_m$ is an $\mathcal{E}^{n+1}(A)$ embedding.

To form $i(G_\infty)$ let $n(g)$ = minimum m such that $g \in G_m$ and set $i(g) = J(n(g), i_{n(g)}(g))$. Then

$$x \in i(G_\infty) \leftrightarrow L(x) \in i_{K(x)}(G_{K(x)}) \wedge (K(x) > 0 \to L(x) \notin i_{K(x)}(G_{K(x) \dot{-} 1}))$$

so $i(G_\infty)$ is $\mathcal{E}^{n+1}(A)$ decidable. The inverse operation on the group G_∞ is encoded by $j(x) = J(K(x), \bar{i}_{K(x)}(L(x)))$ so j is $\mathcal{E}^{n+1}(A)$ computable. To define the encoded group multiplication, let $\kappa_{k,m} : i_k(G_k) \to i_m(G_m)$ be the $\mathcal{E}^{n+1}(A)$ embedding for $k < m$. Since $g \in G_k$ and $\tilde{g} \in G_m$ for $k < m$ implies $g\tilde{g} \in G_m$ and $g\tilde{g} \notin G_{m-1}$ we define

$$m(x,y) = \begin{cases} J(K(x), m_{K(x)}(L(x), L(y))) & \text{if } K(x) = K(y), \\ J(K(x), m_{K(x)}(L(x), \kappa_{K(y),\, K(x)}L(y))) & \text{if } K(x) > K(y), \\ J(K(y), m_{K(y)}(\kappa_{K(x),\, K(y)}L(x), L(y))) & \text{if } K(x) < K(y), \end{cases}$$

so m is $\mathcal{E}^{n+1}(A)$ computable.

Clearly the embedding $G = G_0 < G_\infty$ is an $\mathcal{E}^{n+1}(A)$ embedding.□

In the above proof observe also that the embeddings $G_m < G_\infty$ are $\mathcal{E}^{n+1}(A)$ embeddings since the $G_k < G_m$ for $k < m$ are $\mathcal{E}^{n+1}(A)$ embeddings and for $x \in i_m(G_m)$ the computation of the minimal k such that $x \in i_k(G_k)$ can be performed by bounded minimalization and hence is $\mathcal{E}^{n+1}(A)$ computable.

We have the following extension to Lemma 5.9.

Lemma 5.11. *Under the assumptions of Lemma 5.10, assume also that $K < G$ is an $\mathcal{E}^n(A)$ decidable subgroup which is $\mathcal{E}^n(A)$ invariant under the φ_k for all k. Then*

i) $K_\infty = \langle K, t_1, \ldots; t_k h t_k^{-1} = \varphi_k(h)$ *for all* $k = 1, \ldots$ *and all* $h \in H_k \cap K \rangle = \{K, t_1, \ldots\} < G_\infty$ *is an* $\mathcal{E}^{n+1}(A)$ *embedding*

ii) $K_\infty \cap G = K.$

Proof. We use the notation in the proof of Lemma 5.10. It suffices to show the embedding $K_m = \langle K, t_1, \ldots, t_m ; t_k h t_k^{-1} = \varphi_k(h)$ for $k = 1, \ldots, m$ and all $h \in H_k \cap K \rangle < G_m$ is an $\mathcal{E}^{n+1}(A)$ embedding. Define $L'_m = (K * \langle r_1, \ldots, r_m ;\rangle) *_{\psi'} (K' * \langle s_1, \ldots, s_m ;\rangle)$ for K' a copy of K, by $k \mapsto k'$ and

$$\psi' : K * r_1 H_1 \cap K r_1^{-1} * \ldots * r_m H_m \cap K r_m^{-1} \to K' * s_1 \varphi_1(H_1 \cap K)' s_1^{-1} * \ldots * s_m \varphi_m(H_m \cap K)' s_m^{-1}$$

by $k \mapsto k'$ and $r_i h r_i^{-1} \mapsto s_i \varphi_i(h)' s_i^{-1}$ for $k \in K$ and $h \in H_i \cap K$. Then as in the case of L_m, L'_m is an $\mathcal{E}^{n+1}(A)$ group. It suffices to show the embedding $L'_m < L_m$ is an $\mathcal{E}^{n+1}(A)$ embedding. This is immediate from Lemma 4.20 observing that

$$K * r_1 H_1 \cap K r_1^{-1} * ... * r_m H_m \cap K r_m^{-1} < G * r_1 H_1 r_1^{-1} * ... * r_m H_m r_m^{-1}$$

and

$$K' * s_1 \varphi_1 (H_1 \cap K)' s_1^{-1} * ... * s_m \varphi_m (H_m \cap K)' s_m^{-1} < G' * s_1 \varphi_1 (H_1)' s_1^{-1} * ... * s_m \varphi_m (H_m)' s_m^{-1}$$

are $\mathcal{E}^n(A)$ decidable and ψ' is the restriction of ψ. It is obvious that $K_m = \{K, t_1, ..., t_m\} < G_m$ and so $K_\infty = \{K, t_1, ...\} < G_\infty$ proving (i).

To prove (ii) it suffices to show $K_m \cap G = K$ for all m. By lemma 4.20, $K_m \cap G < L'_m \cap G = L'_m \cap G \cap (G * \langle r_1, ..., r_m ;\rangle) = (K * \langle r_1, ..., r_m ;\rangle) \cap G = K$. Since $K_m > K$ and $G > K$, $K_m \cap G > K$ proving $K_m \cap G = K$ for all m. □

References

[1] F.B. Cannonito, Hierarchies of computable groups and the word problem, Journal of Symbolic Logic 31 (1966) 376-392.

[2] F.B. Cannonito, The algebraic invariance of the word problem in groups, this volume.

[3] M. Davis, Computability and unsolvability (McGraw-Hill, New York, 1957).

[4] R.W. Gatterdam, Embeddings of primitive recursive computable groups, doctoral dissertation, University of California, Irvine, 1970. Submitted for publication.

[5] A. Grzegorczyk, Some classes of recursive functions, Rozprawy Matematyczne 4 (1953) 46 pp.

[6] G. Higman, Subgroups of finitely presented groups, Proceedings of the Royal Society, A 262 (1961) 455-475.

[7] S.C. Kleene, Introduction to metamathematics (Van Nostrand, Princeton, New Jersey, 1952).

[8] W. Magnus, A. Karrass and D. Solitar, Combinatorial Group Theory (Interscience Publishers, New York, 1966).

[9] M.O. Rabin, Computable algebra, general theory and theory of computable fields, Transactions of the American Mathematical Society 95 (1960) 341-360.
[10] R.W. Ritchie, Classes of recursive functions based on Ackerman's function, mimeographed lecture notes, University of Washington, 1963.
[11] R.M. Robinson, Primitive recursive functions, Bulletin of the American Mathematical Society 53 (1947) 925-942.

THE WORD, POWER AND ORDER PROBLEMS IN FINITELY PRESENTED GROUPS

Donald J. COLLINS
Queen Mary College, London

§1. Introduction

We consider some generalizations of the word problem. In particular we consider their relationship to the word problem and also the relationship of any of these more general problems to another.

By the *power problem* for a group presentation G we shall mean the problem of determining of any two words u and v of G whether or not u is a power of v *. This is a true decision problem and when we say that the problem is (recursively) soluble we shall, of course, mean that the set of ordered pairs (u, v) of words of G such that u is a power of v is a recursive set. By the *order problem* for a group presentation G we shall mean the problem of computing, for any word w of G, the number $\text{ord}(w)$ where *ord* is the function from the set of words of G to the natural numbers defined by

$$\text{ord}(w) = \begin{cases} \text{order of } w \text{ if } w \text{ has finite order} \\ 0 \qquad \text{if } w \text{ has infinite order.} \end{cases}$$

This problem is not a decision problem in the usual sense of the phrase. We shall say that the problem is (recursively) soluble if the function *ord* is a recursive function. The solubility of either

* We adopt the convention that 1 is the 0th power of any word v, i.e., $1 = v^0$ in G. A perhaps more natural but rather clumsy name for this problem might be the *generalized word problem for cyclic subgroups.*

problem implies the solubility of the word problem. For w is a power of 1 if and only if $w = 1$ in G and $\mathrm{ord}(w) = 1$ if and only if $w = 1$ in G.

These problems have been studied by Lipschutz [6,7] and McCool [8] for various types of 'small cancellation' groups and have been shown to be soluble in these cases. Here we are more concerned with the question of the existence of groups for which these problems are not soluble. Granted the existence of groups with insoluble word problem, this question is trivially answered since any such group will have insoluble power and order problems. If, however, we require that the examples we desire must have soluble word problem, then the answer is less immediate, particularly if we also insist that the groups be finitely presented. As a further question, we might ask whether or not we can construct examples in which one of these two problems is soluble but the other is insoluble. Gathering these together we can formulate a more general question, which includes the previous ones, as follows: if **a**, **b** and **c** are any three recursively enumerable (r.e.) degrees of insolubility with $\mathbf{a} \leqslant \mathbf{b}$ and $\mathbf{a} \leqslant \mathbf{c}$, does there exist a finitely presented group whose word problem is of degree **a**, whose power problem is of degree **b** and whose order problem is of degree **c**.* Our Theorem A below gives an affirmative answer to this question.

The first step forward answering these questions was taken by McCool [8] who constructed two infinite presentations such that in one the power problem is soluble but not the order problem while in the other the reverse is the case. The existence of such examples is not unexpected since when we permit infinite presentations, we allow much more scope for establishing that certain things are not recursive. Interestingly it turns out that despite the scope thus afforded, the construction of the example mentioned first above, although requiring only a few lines, is quite delicate.

In attempting to move from an infinite presentation to a finite presentation, there is one obvious way for us to go. We can firstly embed in a finitely generated group using, say, the technique of

* The order problem is of degree **c** if the function ord is **c**-recursive and is not **d**-recursive for any degree $\mathbf{d} < \mathbf{c}$.

Higman, Neumann and Neumann [5] and then use Higman's remarkable method [4] to embed in a finitely presented group and hope that the degree of insolubility of the problems will not be raised as a result of the embeddings. Since Clapham [1,2] has shown that the Higman embedding preserves the degree of the word problem, it is reasonable to hope that this approach will be fruitful and indeed it turns out that this method gives exactly what we require.

We now state our results formally.

Theorem A. *Let* **a**, **b** *and* **c** *be any r.e. degrees with* $\mathbf{a} \leqslant \mathbf{b}$ *and* $\mathbf{a} \leqslant \mathbf{c}$. *Then there exists a finitely presented group L such that*

(i) *the word problem for L is of degree* **a**;
(ii) *the power problem for L is of degree* **b**;
(iii) *the order problem for L is of degree* **c**.

Theorem A is an immediate consequence of the following two theorems.

Theorem B. *Let* **a**, **b** *and* **c** *be any r.e. degrees with* $\mathbf{a} \leqslant \mathbf{b}$ *and* $\mathbf{a} \leqslant \mathbf{c}$. *Then there exists a finitely generated group G with a recursively enumerable set of defining relations such that*

(i) *the word problem for G is of degree* **a**;
(ii) *the power problem for G is of degree* **b**;
(iii) *the order problem for G is of degree* **c**.

Theorem C. *Let G be a finitely generated group with a recursively enumerable set of defining relations and suppose*

1. *the word problem for G is of degree* **a**;
2. *the power problem for G is of degree* **b**;
3. *the order problem for G is of degree* **c**.

Then there exists a finitely presented group L in which G can be (recursively) embedded such that

(i) *the word problem for L is of degree* **a**;
(ii) *the power problem for L is of degree* **b**;
(iii) *the order problem for L is of degree* **c**.

Theorem C is a generalization of Clapham's theorem in [1,2] concerning the Higman embedding. We prove Theorem C by simply following Clapham and verifying that the additional requirements are satisfied.

§ 1. Theorem C

It is convenient to begin with Theorem C. Unless otherwise specified our notation and terminology are taken directly from [1,2].

Initially we concentrate on the power problem. Clapham uses three main constructions and we therefore establish some lemmas concerning these constructions.

Lemma 1.1 *. *Let G and G' be groups with recursively enumerable sets of generators, let U be a subgroup of G which is $\mathcal{A}$-soluble in G, U' a subgroup of G' which is $\mathcal{A}$-soluble in U' and ϕ an $\mathcal{A}$-recursive isomorphism between U and U'. If G and G' have $\mathcal{A}$-soluble power problems with respect to these sets of generators, then $\{G, G'; u = \phi(u), u \text{ in } U\}$ has $\mathcal{A}$-soluble power problem with respect to the set of generators consisting of the union of the two given sets.*

Proof. Our hypothesis imply that we can $\mathcal{A}$-recursively compute reduced forms for elements of the generalized free product $\{G, G'\}$. Without loss of generality we may therefore assume that our arbitrary u and v are in reduced form. Let these reduced forms be $u_1 u_2 \dots u_m$ and $v_1 v_2 \dots v_n$ where successive u_i and v_i come alternately from G and G' but not from the amalgamated part (except when $m = 1$ or $n = 1$).

Case (i). Let us suppose that v is cyclically reduced, i.e., either $n = 1$ or v_1 and v_n belong to different factors of the generalized free product. If $n = 1$, then $(\exists r) u = v^r$ implies that u belongs to the same factor as v so that in this case we may use the $\mathcal{A}$-recursive algorithm for that factor. If $n > 1$, we use the fact that for any

* This is an analogue of Clapham's Lemmas 3.5 and 9.1. $\mathcal{A}$ denotes some set of natural numbers.

r, v^r is in reduced form and hence $u = v^r$ only if $|r| \leqslant m$. Thus the power problem reduces to the word problem which, by Clapham's Lemma 9.1, is $\mathscr{A}$-soluble.

Case (ii). Suppose that v is not cyclically reduced. Then there is an $\mathscr{A}$-recursive procedure to express v as $y^{-1} v_0 y$ where v_0 is cyclically reduced. Moreover, $(\exists r) u = v^r$ if and only if $(\exists r) yuy^{-1} = v_0{}^r$ Since we can put yuy^{-1} in reduced form we can apply the procedure of case (i).

Lemma 1.2*. *Let G be a finitely generated group with* $\mathscr{A}$*-soluble power problem and let* $\phi_1, \phi_2, \ldots,$ *be an* $\mathscr{A}$*-recursively enumerable set of* $\mathscr{A}$*-recursive isomorphisms of the* $\mathscr{A}$*-recursively enumerable set* $A_1, A_2, \ldots,$ *of subgroups of G, into G, such that for* $i = 1, 2, \ldots,$ A_i *and* $\phi_i(A_i)$ *are* $\mathscr{A}$*-soluble in G. Then*

$$\{ G, t_1, t_2, \ldots; t_i^{-1} a_i t_i = \phi_i(a_i), a_i \text{ in } A_i, i = 1, 2, \ldots \}$$

has $\mathscr{A}$*-soluble power problem with respect to this set of generators.*

Proof. Let $G_n = \{ G, t_1, t_2, \ldots, t_n; t_i^{-1} a_i t_i = \phi_i(a_i), a_i$ in A_i, $i = 1, 2, \ldots, n \}$; then $G = \bigcup_n G_n$. The lemma will follow if we can prove that each G_n has $\mathscr{A}$-soluble power problem and that the mapping from arbitrary G_n to the corresponding algorithm is $\mathscr{A}$-recursive.

The former can be established by a repetition of the argument given by Clapham for his Lemma 9.2 but replacing applications of his Lemma 3.5 by applications of our Lemma 1.1.

The $\mathscr{A}$-recursiveness of the mapping from a G_n to the corresponding algorithm follows obviously from the general character of the argument**.

In addition to the two constructions just considered, Clapham also makes use of direct products. For our purpose it would be

* This is an analogue of Clapham's Lemmas 3.6 and 9.2.

** In the proof of his Lemma 9.2, Clapham neglects to verify that the mapping from a G_n to the corresponding algorithm is $\mathscr{A}$-recursive. It is clear that this is indeed the case but the point should be mentioned since it is crucial to the argument.

most convenient if we could show that whenever G and G' have $\mathcal{A}$-soluble power problems, then $G \times G'$ has $\mathcal{A}$-soluble power problem. Unfortunately, we have been unable to establish this and it may well be false – we return to this question in §3. However, it turns out that the following lemma is sufficient for our present purpose.

Lemma 1.3. *Let G and G' be groups with $\mathcal{A}$-soluble power problems relative to some generators. If either G or G' is torsion-free, then the direct product $G \times G'$ has $\mathcal{A}$-soluble power problem relative to the union of the two sets of generators.*

Proof. Let us assume that G is torsion-free. If u and v are arbitrary elements of $G \times G'$, we may write $u = u_1 u_2$ and $v = v_1 v_2$ where $u_1, v_1 \in G$ and $u_2, v_2 \in G'$.

We consider two cases according as $u_1 = 1$ or $u_1 \neq 1$. Since the word problem for G is $\mathcal{A}$-soluble this distinction can be made $\mathcal{A}$-recursively. If $u_1 = 1$ then $u = v^r$ if and only if $v_1 = 1$ and $u_2 = v_2{}^r$. The problem thus reduces to the word problem for G and the power problem for G'.

Now suppose $u_1 \neq 1$; then $u = v^r$ implies $u_1 = v_1{}^r$ and there can be exactly one such value of r since G is torsion-free. We can $\mathcal{A}$-recursively determine whether or not $(\exists r) u_1 = v_1{}^r$ and if such an r exists we can $\mathcal{A}$-recursively determine what this unique value is. It follows that the problem is reducible to the power problem for G and the word problem for G'.

Initially we assumed that G was torsion-free and then gave a procedure for $\mathcal{A}$-recursively solving the power problem for $G \times G'$. The assumption that G' was torsion-free would also lead to a procedure. In general we may not know which of G and G' is torsion-free. From the point of view of recursive function theory, this is quite irrelevant – all that is required is that one establishes that a procedure exists. The lemma is therefore proved.

The following are well known; we state them as lemmas for ease of reference.

Lemma 1.4. *Let G, G', U, U' and ϕ be as in Lemma 1.1. If G and G' are torsion-free, then* $\{G, G'; u = \phi(u), u \text{ in } U\}$ *is torsion-free*.*

Lemma 1.5. *Let $G, \phi_1, \phi_2, \ldots, A_1, A_2, \ldots,$ be as in Lemma 1.2. If G is torsion-free, then*

$$\{G, t_1, t_2, \ldots; t_i^{-1} a_i t_i = \phi_i(a_i), a_i \text{ in } A_i, i = 1, 2, \ldots\}$$

is torsion-free.*

Lemma 1.6. *Let G and G' be torsion-free. Then $G \times G'$ is torsion-free.*

The fundamental aspect of Clapham's work is the relationship between the concept of an $\mathcal{A}$-recursive, recursively enumerable set and his notion of an $\mathcal{A}$-strongly benign subgroup of a finitely generated group. We parallel this by defining the notion of $\mathcal{A}$-completely benign based on the following lemma.

Lemma 1.7. *Let G be a f.g. torsion-free group and H a subgroup of G. Then the following are equivalent:*

(i) *G can be embedded in a f.p. torsion-free group K with $\mathcal{A}$-soluble power problem which has a f.g. subgroup L such that $G \cap L = H$ and G, L and $G \cdot L$ are $\mathcal{A}$-soluble in K;*

(ii) *the group $\{G, t; t^{-1}ht = h, h \text{ in } H\}$ can be embedded in a f.p. torsion-free group K' with $\mathcal{A}$-soluble power problem and G and $\{G, t\}$ are $\mathcal{A}$-soluble in K';*

(iii) *the group $\{G, G_1; h = h_1, h \text{ in } H\}$ can be embedded in a f.p. torsion-free group K'' with $\mathcal{A}$-soluble power problem and G, G_1 and $\{G, G_1\}$ are $\mathcal{A}$-soluble in K''.*

Proof. The lemma is, of course, very similar to Clapham's Lemma 4.1. Moreover, it is proved by a repetition of the argument given by Clapham for his lemma but replacing applications of his Lem-

* The assumptions concerning $\mathcal{A}$-recursiveness are of course redundant here.

mas 3.5 and 3.6 by applications of our Lemmas 1.1 and 1.2 and noting that by our Lemmas 1.4 and 1.5 all groups involved are torsion-free.

A subgroup H of a f.g. torsion-free group G is called *$\mathcal{A}$-completely benign* (*in* G) if any, and hence all, of conditions (i), (ii) and (iii) of Lemma 1.7 are satisfied. We emphasize that this definition applies only to torsion-free groups.

We now have our main technical result.

Theorem 1.8. *A subgroup of a f.g. free group is $\mathcal{A}$-completely benign if and only if it is $\mathcal{A}$-recursive and recursively enumerable*.*

Proof. Yet again we mimic Clapham's argument with some alterations and additions.

Throughout his paper, Clapham is concerned to show that certain word problems are $\mathcal{A}$-soluble. He always does this in one of three ways. These are:

I. by using Lemma 9.1 or a finite version of it;
II. by using Lemma 9.2 or a finite version of it;
III. by using the fact that if two groups have $\mathcal{A}$-soluble word problem, then so does their direct product.

It follows from this that our theorem will be proved if we repeat Clapham's paper from Corollary 4.2.1 to Lemma 11.2 with the following alterations and additions:

1. replace ' $\mathcal{A}$-soluble word problem' by ' $\mathcal{A}$-soluble power problem';
2. replace ' $\mathcal{A}$-strongly benign' by ' $\mathcal{A}$-completely benign';
3. whenever any group is given, add the hypothesis that the group is torsion-free;
4. replace any proof by I by applications of our Lemma 1.1;
5. replace any proof by II by applications of our Lemma 1.2;
6. replace any proof by III by applications of our Lemma 1.3;
7. use our Lemmas 1.4–1.6 to verify that any groups constructed are torsion-free.

* This is the analogue of Clapham's main technical result, his Lemma 11.2.

Although we have not yet mentioned the order problem, the bulk of the proof of Theorem C is now complete. For the final stages we introduce the order problem and, firstly, obtain another trio of lemmas.

Lemma 1.9. *Let G, G', U, U' and ϕ be as in Lemma 1.1 save that we assume the $\mathcal{A}$-solubility of the order problems rather than the power problems. Then $\{G, G', u = \phi(u), u \text{ in } U\}$ has $\mathcal{A}$-soluble order problem.*

Proof. It is well known that in a free product with amalgamation an element w has finite order if and only if it is a conjugate of an element x which belongs to a factor and is of finite order. We can $\mathcal{A}$-recursively determine if such an x exists and compute it if it does. Then $\operatorname{ord}(w) = \operatorname{ord}(x)$.

Lemma 1.10. *Let G be a f.g. group with $\mathcal{A}$-soluble order problem and let $\phi_1, \phi_2, \ldots,$ and $A_1, A_2, \ldots,$ be as in Lemma 1.2. Then*

$$\{G, t_1, t_2, \ldots; t_i^{-1} a_i t_i = \phi_i(a_i), a_i \text{ in } A_i, i = 1, 2, \ldots\}$$

has $\mathcal{A}$-soluble order problem.

Proof. This can be derived from Lemma 1.9 since this construction is essentially a special case of a free product with amalgamation.

Lemma 1.11. *Let G and G' have $\mathcal{A}$-soluble order problems. Then $G \times G'$ has $\mathcal{A}$-soluble order problem.*

Proof. Let $w = w_1 w_2$ where $w_1 \in G$ and $w_2 \in G'$. Then the order of w is the lowest common multiple of the orders of w_1 and w_2 (where l.c.m. $(0, n) = 0$ for any n).

Proof of Theorem C

We are given a f.g. group whose word, power and order problems are of degree **a**, **b** and **c** respectively. Let us suppose that $G = F/R$ where F is a f.g. free group and R is an $\mathcal{A}$-recursive, recursively enumerable normal subgroup (where $\mathscr{A}$ is some representative of **a**). Then of course R is $\mathscr{A}$-soluble in F and so by Theorem 1.8, is $\mathscr{A}$-completely benign in F, i.e., $H = \{F_1, F_2; r_1 = r_2, r \text{ in } R\}$ can be embedded in a f.p. torsion-free group K whose power problem is $\mathscr{A}$-soluble and F_1, F_2 and $\{F_1, F_2\}$ are $\mathscr{A}$-soluble in K.

The required f.p. group L is then obtained in the following manner. Firstly we construct the direct product $K \times G$. Now there is a mapping $\lambda : F_1 \to K \times G$ defined by sending $f_1 \overset{\lambda}{\to} (f_1, f_1 R)$ and a mapping $\nu : F_2 \to K \times G$ defined by sending $f_2 \overset{\nu}{\to} (f_2, 1)$ and clearly λ and ν agree on the amalgamated part of H. Hence there is a homomorphism $\phi : H \to K \times G$ which extends λ and ν and in fact ϕ is a monomorphism. Thus we may form $L = \{G, t; t^{-1}ht = \phi(h), h \text{ in } H\}$ which contains G and is in fact finitely presented. All assertions made in this paragraph are verified in Higman [4].

We claim further than L satisfies the conditions (i), (ii) and (iii) of the statement of Theorem C.

Part (i). The power problem for K is $\mathscr{A}$-soluble and hence the word problem for K is $\mathscr{A}$-soluble. The word problem for G is $\mathscr{A}$-soluble and thus the word problem for $K \times G$ is $\mathscr{A}$-soluble. Clapham verifies that H and $\phi(H)$ are $\mathscr{A}$-soluble in $K \times G$ whence L has $\mathscr{A}$-soluble word problem. Since G is (recursively) embedded in L the word problem for L is of degree **a**.

Part (ii). Let $\mathscr{B}$ be a representative of **b**. The power problem for K is $\mathscr{A}$-soluble and hence $\mathscr{B}$-soluble since $\mathscr{A} \leqslant \mathscr{B}$. By hypothesis the power problem for G is $\mathscr{B}$-soluble and, since K is torsion-free, the power problem for $K \times G$ is $\mathscr{B}$-soluble (by Lemma 1.3). Since H and $\phi(H)$ are $\mathscr{B}$-soluble in $K \times G$, Lemma 1.2 gives us that the power problem for L is $\mathscr{B}$-soluble. The fact that G is embedded in L ensures that the power problem for L is of degree **b**.

Part (iii). Since K is torsion-free, the order problem for K is equivalent to the word problem and thus is $\mathscr{C}$-soluble (where $\mathscr{C}$ is some representative of **c**). By Lemmas 1.11 and 1.10, the order problem for L is $\mathscr{C}$-soluble. Since G is embedded in L, the order problem for L is of degree **c**.

One final point in connection with Theorem C should be noted. We have shown that the degree of the order problem is preserved in a rather indirect sort of way by utilising the torsion-freeness of

K. It is clear, however, that if we were concerned solely with the order problem, then on the basis of Lemmas 1.9 – 1.11, an analogue of Theorem 1.8 could be established directly and then the remainder of the proof carried out as in part (iii) above.

§2. Theorem B

We construct group presentations with infinitely many generators and defining relations and with word, power and order problems of varying degrees of insolubility. Then we use the technique of Higman, Neumann and Neumann to obtain the f.g. group G of Theorem B. Throughout this section $\mathbf{N}$ denotes the natural numbers.

The following lemma is well known.

Lemma 2.1. *Let* $\mathbf{b}$ *be any r.e. degree. Then there exists a one-to-one recursive function f of one variable such that*

(i) $0 \notin f(N)$;

(ii) *$f(N)$ is of degree* $\mathbf{b}$.

Corollary 2.2. *Let f be as in the above lemma and let* $S_{\mathbf{b}} = \{(f(n), n+2); n \in \mathbf{N}\}$. *Then*

(i) *$S_{\mathbf{b}}$ is recursive;*

(ii) $\{m \in \mathbf{N}; (\exists r)\,(m, r) \in S_{\mathbf{b}})\}$ *is of degree* $\mathbf{b}$;

(iii) *if* $(m, r) \in S_{\mathbf{b}}$ *and* $(m, r') \in S_{\mathbf{b}}$, *then* $r = r'$;

(iv) *if* $(m, r) \in S_{\mathbf{b}}$ *then* $m > 0$ *and* $r > 1$*.

Proof. This follows from the lemma and the fact that f is one-to-one.

Let T be any r.e. set of degree $\mathbf{a}$. Let $M_1 = (a, b, c, d; a^m b a^m = c^m d c^m, m \in T)$.

Lemma 2.3. *The word, power and order problems for M_1 are of degree* $\mathbf{a}$.

* We require conditions (iii) and (iv) for later technical use. Condition (iii) is essential whereas (iv) is a mere convenience.

Proof. Clapham shows that the word problem is of degree **a** by observing that M_1 is the free product of two free groups amalgamating two subgroups which are **a**-soluble* in the free groups but are not **d**-soluble for any $\mathbf{d} < \mathbf{a}$. By Lemma 1.1, the power and order problems are also **a**-soluble and hence of degree **a**.

Let **b** be an r.e. degree and let $S_\mathbf{b}$ satisfy conditions (i)–(iv) of Corollary 2.2. Then let

$$M_2 = (x_1, x_2, ..., y_1, y_2, ...; x_i y_i = y_i x_i, x_m = y_m{}^r, \quad i = 1, 2, ..., (m, r) \in S_\mathbf{b})$$

Lemma 2.4. *The word and order problem for M_2 are soluble. The power problem for M_2 is of degree* **b**.

Proof. This group presentation is due, essentially, to McCool [8] who shows that the word and order problems are soluble and that the power problem is insoluble. For the sake of completeness we shall repeat his arguments.

If we let $G_m = (x_m, y_m; x_m y_m = y_m x_m)$ when for all r, $(m, r) \notin S_\mathbf{b}$ and let $G_m = (x_m, y_m; x_m y_m = y_m x_m, x_m = y_m{}^r)$ when $(m, r) \in S_\mathbf{b}$**, then M_2 is the free product of the G_m. Thus M_2 is certainly torsion-free and so the word problem and order problem are equivalent. We solve the former by showing that we can compute the normal form of an arbitrary word. It clearly suffices to consider a word w in the form $x_{m_1}^{\alpha_1} y_{m_1}^{\beta_1} x_{m_2}^{\alpha_2} y_{m_2}^{\beta_2} ... x_{m_t}^{\alpha_t} y_{m_t}^{\beta_t}$ where α_i and β_i, $i = 1, 2, ..., t$ are not both zero. To compute the normal form of w we must be able to tell of an arbitrary $x_m{}^\alpha y_m{}^\beta$, α and β not both zero, whether or not this equals 1 in G_m. Clearly $x_m{}^\alpha y_m{}^\beta = 1$ in G_m if and only if there exists r such that $(m, r) \in S_\mathbf{b}$ and $r\alpha + \beta = 0$, i.e., $x_m{}^\alpha y_m{}^\beta = 1$ in G_m if and only if $-\beta/\alpha$ is a natural number such that $(m, -\beta/\alpha) \in S_\mathbf{b}$. Since $S_\mathbf{b}$ is recursive we can test whether or not this is the case.

The insolubility of the power problem is obvious since $(\exists r) x_m =$

* Since we no longer follow Clapham's notation quite so closely, we write a-soluble directly instead of $\mathcal{A}$-soluble with $\mathcal{A}$ some representative of a.

** Here, and subsequently, we use condition (iii) of Corollary 2.2.

y_m^r if and only if $(\exists r)\,(m, r) \in S_b$. It remains for us to establish the degree of the power problem. Since we can compute normal forms, we may assume, possibly after some conjugation, that u and v are in normal form and that r is, in addition, cyclically reduced. If v has length greater than 1, then by the same length argument as used in Lemma 1.1, the problem reduces to the word problem. If v has length 1 we need only consider u if it also has length 1 and is in the same G_m and thus we have the situation where we are examining two words $x_m{}^\alpha y_m{}^\beta$ and $x_m{}^\gamma y_m{}^\delta$.

We can **b**-recursively decide whether or not $(\exists r)\,(m, r) \in S_b$. If not, then $u = v^t$ if and only if $\alpha = t\gamma$ and $\beta = t\delta$ and this is clearly decidable. If on the other hand $(m, r) \in S_b$ then $u = v^t$ if and only if there exists s such that $x_m^{\alpha - t\gamma}\, y_m^{\beta - t\delta} = x_m{}^{-s} y_m{}^{rs}$ in the free abelian group on x_m and y_m. The latter occurs if and only if $\alpha - t\gamma = -u$ and $\beta - t\delta = -rs$ and the existence of these two equations is equivalent to $t = (r\alpha + \beta)/(r\gamma + \delta)$. (Notice that since $x_m{}^\gamma y_m{}^\delta$ can be assumed not to be 1 in G_m, $r\gamma + \delta \neq 0$). The integer t is thus uniquely defined and computable (if it exists) and the problem reduces to the word problem.

Let S_c satisfy conditions (i)–(iv) of Corollary 2.2 but with degree **c** instead of degree **b**. Let

$$M_3 = (x_1, x_2, \ldots; x_m{}^{r!} = 1,\ (m, r) \in S_c).$$

Lemma 2.5. *The word and power problems for M_3 are soluble. The order problem for M_3 is of degree* **c**.

Proof. This presentation is again essentially due to McCool [8], and he solves the word and power problems and shows the order problem insoluble. For completeness we repeat his arguments.

If $G_m = (x_m; \phi)$ when for all r, $(m, r) \notin S_c$ and $G_m = (x_m; x_m{}^{r!} = 1)$ when $(m, r) \in S_c$, M_3 is the free product of the G_m. To solve the word problem it therefore suffices to determine if an arbitrary $x_m{}^\alpha$, $\alpha \neq 0$, whether or not this is 1 in G_m. Clearly, $x_m{}^\alpha = 1$ in G_m if and only if there exists r such that $(m, r) \in S_c$ and $r!$ divides α. Since S_c is recursive we can determine whether or not such an r exists.

The solution of the power problem is, in the present author's view, surprisingly delicate. We may assume that we are given two words u and v in normal form with v in addition cyclically reduced. If the length of v is greater than 1, we reduce to the word problem. It then suffices to consider two words $x_m{}^\alpha$ and $x_m{}^\beta$ where $\alpha, \beta \neq 0$. We want to test whether or not $x_m{}^\alpha = x_m{}^{t\beta}$ in G_m for some t. As a first test we check to see if $\alpha = t\beta$ for some t. If so then the problem is settled; so suppose that α is not an integral multiple of β.

If $x_m{}^\alpha = x_m{}^{t\beta}$ in G_m, then there exists r and k such that $\alpha = t\beta + kr!$ and $(m, r) \in S_{\mathbf{c}}$. We claim that this equation implies that $r < |\beta|$. For if $r \geqslant |\beta|$ then $t\beta + kr!$ is an integral multiple of β and we have assumed that this is not the case*.

The insolubility of the order problem follows from the fact that x_m has finite order if and only if $(\exists r)\ (m, r) \in S_{\mathbf{c}}$. To determine the degree of the order problem we argue as follows. It suffices to consider elements whose normal form is $x_m{}^\alpha$, $\alpha \neq 0$ since any element whose cyclically reduced normal form has length greater than 1 is not of finite order. We can $\mathbf{c}$-recursively determine whether or not $(\exists r)\ (m, r) \in S_{\mathbf{c}}$. If $(m, r) \in S_{\mathbf{c}}$, then the order of $x_m{}^\alpha$ can be computed from the order of x_m which is $r!$ If no such r exists, then $\operatorname{ord}(x_m{}^\alpha) = 0$. The order problem is therefore $\mathbf{c}$-soluble and hence of degree $\mathbf{c}$.

Now let $M = M_1 * M_2 * M_3$ be the free product of M_1, M_2 and M_3.

Lemma 2.6.

(i) *The word problem for M is of degree* **a**.

(ii) *The power problem for M is of degree* **b**.

(iii) *The order problem for M is of degree* **c**.

Proof. This is trivial from Lemmas 1.1 and 1.9.

Let $F(p, q)$ be the free group on the symbols p and q. Also let M be rewritten as $(g_1, g_2, \ldots \mid R_\lambda = 1, \lambda = 1, 2, \ldots)$ and let

* The delicacy of the argument is in the use of the factorial which seems to be quite crucial.

$E = M * F(p, q)$. If $F(s, t)$ is the free group on the symbols s and t then let

$$G = (E * F(s, t);\ g_i p^i q p^i = s^i t s^i,\ i = 1, 2, \ldots)\ .$$

Proof of Theorem B. It clearly suffices to prove that the groups $\langle g_i p^i q p^i \rangle$ and $\langle s^i t s^i \rangle$ where $i = 1, 2, \ldots$, are **a**-soluble in E and $F(s, t)$ respectively. This is easily shown using the standard Nielsen-style argument and the normal form theorem for free products. Details of the required argument are given in Lemma 2.1 of Collins [3] (note that no two generators of M are equal).

§3. Direct products

In § 1, we mentioned that certain difficulties seem to occur in connection with the power problem and direct products. We examine the matter a little more closely here. We firstly establish the following lemma.

Lemma 3.1. *Let G and G' each have soluble power problem and soluble order problem. Then $G \times G'$ has soluble power problem.*

Proof. Let u and v be arbitrary elements of $G \times G'$ and suppose $u = u_1 u_2$, $v = v_1 v_2$ where $u_1, v_1 \in G$ and $u_2, v_2 \in G'$.

As a first test we check whether or not there exists r_1 and r_2 such that $u_1 = v_1{}^{r_1}$ and $u_2 = v_2{}^{r_2}$. If no such r_1 and r_2 exist then u is not a power of v. Let us suppose that they do exist; we can certainly then recursively compute such a pair of integers. Also we can recursively compute $p_1 = \text{ord}(v_1)$ and $p_2 = \text{ord}(v_2)$. Then u is a power of v if and only if there exist integers k and l such that $r_1 + kp_1 = r_2 + lp_2$. Such a k and l can exist if and only if $r_1 - r_2$ is a multiple of the greatest common divisor of p_1 and p_2 (where g.c.d. $(0, t) = t =$ g.c.d. $(t, 0)$ for any t), and we can determine whether or not this is so.

The above argument breaks down if we remove the hypotheses that the order problems for G and G' be soluble. Without these hypotheses we run into difficulties if the values of r_1 and r_2 which we obtain are not equal. For it is conceivable that there could be other values satisfying the equations $u_1 = v_1^{r_1}$ and $u_2 = v_2^{r_2}$ which would be equal. This will not occur if either v_1 or v_2 is of infinite order but without the solution to the order problem we have, in general, no way of knowing when other values exist and what they are when they do exist.

All this leads to the conjecture that there must exist groups G and G', each with soluble power problem, but such that the power problem for $G \times G'$ is insoluble. By our lemmas neither of these two groups can be torsion-free and at least one must have insoluble order problem. Unfortunately the author is at present unable to settle this question and it therefore appears in the list of problems posed during this conference.

In attempting to deal with the problem the first piece of information that appears useful is knowing what kinds of groups exist which have soluble power problem and insoluble order problem. As matters stand at present, the only examples known are those which can be derived from McCool's fundamental example by embeddings. We seek to show that these examples are of no avail with regard to this question.

Example. Let S and T be two sets satisfying conditions (i)–(iv) of Corollary 2.2 (for the moment we are not concerned with degrees and insist only that in (ii) we have non-recursiveness). Let

$$M = (x_1, x_2, ...; x_m{}^{n!} = 1,\ \ (m, n) \in S)$$

and

$$N = (y_1, y_2, ...;\ y_p{}^{q!} = 1,\ \ (p, q) \in T)\,.$$

We consider the power problem in $M \times N$.

Let $u = u_1 u_2$ and $v = v_1 v_2$, as usual, where $u_1, v_1 \in M$ and $u_2, v_2 \in N$. If either v_1 or v_2 is not a conjugate of a power of a generator, then the problem is easily solved since in this case the v_1 or

v_2 must be of infinite order and we can apply the uniqueness argument of Lemma 1.3. It should, of course, be observed that we can determine when an element is a conjugate of a power of a generator.

In our present situation we can always remove conjugating elements without loss of generality and it therefore suffices to consider the situation in which $v_1 = x_m{}^{\beta}$, $v_2 = y_p{}^{\delta}$ and $u_1 = x_m{}^{\alpha}$, $u_2 = y_p{}^{\gamma}$. The first step is to test whether or not $(\exists r_1) x_m{}^{\alpha} = x_m{}^{r_1\beta}$ and $(\exists r_2) y_p{}^{\gamma} = y_p{}^{r_2\delta}$. A negative answer here means a negative answer overall. So suppose we have found r_1 and r_2 satisfying the equation. We now subdivide into cases according as the following conditions are satisfied:

1(a) $\alpha = \beta r_1$;

1(b) $(\forall t)\alpha \neq \beta t$ and $(\exists n, k)$ $(x_m{}^{n!} = 1$ and $\alpha = \beta r_1 + kn!)$;

2(a) $\gamma = \delta r_2$;

2(b) $(\forall t)\gamma \neq \delta t$ and $(\exists q, l)$ $(y_p^{q!} = 1$ and $\gamma = \delta r_2 + lq!)$

It is clear that 1(a) and 1(b) are mutually exclusive as are 2(a) and 2(b) and that 1 and 2 are quite independent. Moreover, our use of the algorithms for M and N will tell us exactly which of these hold – and when 1(b) or 2(b) holds, the algorithm will compute the value of n or q respectively.

Case 1(b), 2(b). Here we have v_1 and v_2 both of finite order and we can compute these orders since we know the values of n and q. The argument of Lemma 3.1 can therefore be applied.

Case 1(a), 2(a). On account of the very simple way in which $x_m{}^{\alpha} = x_m{}^{\beta r_1}$ and $y_p{}^{\gamma} = y_p{}^{\delta r_2}$ we have no information as to whether $x_m{}^{\beta}$ and $y_p{}^{\delta}$ are of finite order. As a result this case is more complex.

We firstly check whether any of the equations $r_1 = r_2$, $x_m{}^{\alpha} = x_m{}^{\beta r_2}$ in M and $y_p{}^{\gamma} = y_p{}^{\delta r_1}$ in N hold. If any one is true, then u is a power of v. So suppose all three fail. We want to know whether or not there exists r such $x_m{}^{\alpha} = x_m{}^{\beta r}$ and $y_p{}^{\gamma} = y_p{}^{\delta r}$. Since $r_1 \neq r_2$, this can happen only if there exist n, q, k and l such that $\alpha = \beta r + kn!$, $\gamma = \delta r + lq!$, $x_m{}^{n!} = 1$ and $y_p{}^{q!} = 1$. We want to examine what conditions are imposed on n and q if this does indeed occur.

Recalling that $\alpha = \beta r_1$ and $\gamma = \delta r_2$ we obtain, after a little manipulation, the equation

$$\beta\delta(r_1-r_2) = \delta k(n!) - \beta l(q!) \qquad (*)$$

Without loss of generality we may suppose that $n \leqslant q$. Then by McCool's divisibility argument (see Lemma 2.5) it follows from $*$ that $n! \leqslant |\beta\delta(r_1-r_2)|$ and hence a fortiori $n \leqslant |\beta\delta(r_1-r_2)|$. Now we assumed that $x_m{}^\alpha \neq x_m{}^{\beta r_2}$; this means that $\alpha - \beta r_2$ is not a multiple of $n!$, i.e., $\beta(r_1-r_2)$ is not a multiple of $n!$ Then clearly $\beta\delta(r_1-r_2)$ is not a multiple of $\delta n!$. Using divisibility again, we deduce from $*$ that $q < |\delta n!| \leqslant |\beta\delta^2(r_1-r_2)|$.

The net result of the above argument is that if r does exist then $x_m{}^\beta$ and $y_p{}^\delta$ are elements of finite order and that these orders have upper bounds which can be recursively computed from the values of β, δ, r_1 and r_2. This enables us to determine whether or not r exists.

Case 1(a), 2(b) and 1(b), 2(a). These two are obviously similar. We examine the latter in detail. Let $\alpha = \beta r_1 + k_1 n!$ and $\gamma = \delta r_2$. If $x_m{}^\alpha = x_m{}^{\beta r}$ and $y_p{}^\gamma = y_p{}^{\delta r}$ then there exist l, q, k, such that $\alpha = \beta r + kn!$, $\gamma = \delta r + lq!$ and $y_p{}^{q!} = 1$. This yields

$$\beta\delta(r_1-r_2) = \delta(k-k_1)n! - \beta l q!\,. \qquad (\dagger)$$

We can then argue exactly as in the previous case since it follows from $\dagger$ that $q \leqslant |\delta n!|$.

The above argument shows that the power problem for $M \times N$ is soluble. Although we have here considered only McCool's fundamental example rather than the supergroups in which they are embedded, it should be clear that for the supergroups, essentially the only time that we do not have an element of infinite order is when we have an element which is an image of some $x_m{}^\beta$ so that the situation is not altered in any way by considering these supergroups.

We conclude by proving a lemma which shows that groups with soluble power problem and insoluble order problem have an unexpected algebraic property.

Lemma 3.2. *Let G be a group with soluble power problem and insoluble order problem. Then G has elements of every order.*

Proof. Firstly we remark that G must have elements of infinite order. For if every element is of finite order then given any w we can compute ord(w) by successively comparing $w, w^2, w^3, \ldots,$ with 1 until we find the least n such that $w^n = 1$.

To complete the proof we shall establish the following claim from which the lemma then follows easily. The claim is: for every $n \geqslant 2$ there exists $w \in G$ such that ord(w) is a multiple of n*. Suppose not; then there exists $n_0 \geqslant 2$ such that for every $w \in G$, either ord$(w) = 0$ or there exist k and r such that ord$(w) = kn_0 + r$ with $0 < r < n_0$. In this situation we assert that an element w of G has finite order if and only if the following disjunction holds: $(\exists\, k) w^{-1} = w^{kn_0}$ or $(\exists\, k) w^{-2} = w^{kn_0}$ or ... or $(\exists\, k) w^{-(n_0-1)} = w^{kn_0}$. If the disjunction holds then certainly w has finite order. But the converse is immediate from our supposition.

This means that by a fixed finite number of applications of the algorithm which solves the power problem, we can determine whether or not w has finite order. If w has finite order we can then go on to compute ord(w) by comparing $w, w^2, w^3, \ldots,$ with 1. If w does not have finite order, then ord$(w) = 0$. We have thus established the existence of a procedure for solving the order problem which is contradictory**.

* The author is indebted to C.F. Miller, III, for pointing out that the apparently more sweeping assertion that G has elements of every finite order is a trivial consequence of the claim.

** We may not know what n_0 is, but this is irrelevant. It is enough that it exists in order for the existence of the procedure to be established.

References

[1] C.R.J. Clapham, Finitely presented groups with word problems of arbitrary degrees of insolvability, Proc. Lond. Math. Soc., Series 3, 14 (1964) 633-676.

[2] C.R.J. Clapham, An embedding theorem for finitely generated groups, Proc. Lond. Math. Soc., Series 3, 17 (1967) 419-430.

[3] D.J. Collins, On embedding groups and the conjugacy problem, to appear in J. Lond. Math. Soc.

[4] G. Higman, Subgroups of finitely presented groups, Proc. Roy. Soc., Series A, 262 (1961) 455-475.

[5] G. Higman, B.H. Neumann and Hanna Neumann, Embedding theorems for groups, J. Lond. Math. Soc., 25 (1949(247-254.

[6] S. Lipschutz, An extension of Greendlinger's results on the word problem, Proc. Amer. Math. Soc., 15 (1964) 37-43.
[7] S. Lipschutz, On *T*-fourth groups, in preparation.
[8] J. McCool, On free product sixth group, Proc. Glas. Math. Soc. 10 (1969) 1-15.

THE HIGMAN THEOREM FOR PRIMITIVE-RECURSIVE GROUPS – A PRELIMINARY REPORT

R.W. GATTERDAM*

University of Wisconsin, Parkside

An extensive study of $\mathcal{E}^n(A)$ groups can be found in this volume, [1]. A group is an A-p.r. group (primitive recursive relative to $A \subset N$) if it is $\mathcal{E}^n(A)$ for some (finite) n; it is a p.r. group if it is $\mathcal{E}^n(N)$ for some n. As mentioned in [1] the following version of the Higman theorem, [5], holds for A-p.r. groups:

Theorem. *Let $A \subset N$ be a recursively enumerable set. Then an A-p.r. group can be embedded as an A-p.r. decidable subgroup of a finitely presented A-p.r. group. The embedding and its inverse, where defined, are A-p.r. computable.*

The complete proof of this theorem is contained in [4]. Also observe the similarity to the Clapham result, [2] and [3], which can be viewed as the same statement replacing "A-p.r." by "A recursive". In [4] both the theorem and our version of the Clapham result are proved by a technique similar to that of Schoenfield [6] using the computability of the group constructions free product with amalgamation and strong Britton extension as discussed in [1]. The purpose of this report is to briefly outline the proof given in [4] considering two crucial facets in some detail. First we see how countably many applications of the strong Britton extension (as in Lemma 5.10 of [1]) are used in the proof. The proof is then completed by a direct induction based on the A-p.r. computable

* AFOSR-1321-67 and AFOSR-70-1870 (Grants).

decision function for the word problem of a finitely generated, A-p.r. group. This report should be viewed as preliminary in that it does not utilize the full information on the computability of the strong Britton construction as found in [1] but rather the weaker results of Chapters 2 and 3 of [4]. With these newer results it is hoped that the Higman construction can be shown to induce at most a jump of a few computability levels of the relativized Grzegorczyk hierarchy.

An embedding of an A-p.r. group into another A-p.r. group is an A-p.r. embedding if it is an $\mathcal{E}^n(A)$ embedding for some n as in [1]. Similarly an A-p.r. isomorphism φ in G is an $\mathcal{E}^n(A)$ isomorphism and we consider the A-p.r. group $G_\varphi = \langle G, t; tht^{-1} = \varphi(h), h \in H = \text{domain } \varphi \rangle$ as in [1]. In view of Theorem 5.5 of [1] we restrict our attention to finitely generated (f.g.), A-p.r. groups and following [6] say such a group is *A-p.r. Higman* if it can be A-p.r. embedded in a f.p., A-p.r. group. Again following [6] an A-p.r. isomorphism φ in an A-p.r. Higman group G is said to be *A-p.r. benign* if G_φ is A-p.r. Higman and, in particular, an A-p.r. decidable subgroup $H < G$ is *A-p.r. benign* if $G_H = G_{1|H}$ is A-p.r. Higman.

Then paralleling [6] the following are proved in [4] (replacing p.r. by A-p.r.):

1. *If $H < K < G$ for G A-p.r. Higman, K f.g. and A-p.r. decidable, then H is A-p.r. benign in G iff it is A-p.r. benign in K.*

2. *The intersection of A-p.r. benign subgroups is A-p.r. benign.*

3. *The (group theoretic) union of A-p.r. benign subgroups is A-p.r. benign if A-p.r. decidable.*

4. *The image of an A-p.r. benign subgroup under an A-p.r. computable homomorphism is A-p.r. benign if A-p.r. decidable.*

5. *The preimage of an A-p.r. benign subgroup under an A-p.r. computable homomorphism is A-p.r. benign.*

6. *The restriction of an A-p.r. isomorphism in G with domain G to an A-p.r. benign subgroup is an A-p.r. isomorphism in G.*

As might be expected the proofs of 1. through 6. given in [4] are similar to proofs in [6] with attention given to the decidability of the various subgroups involved and the computability of the homomorphisms (both in the statements and internal to the proofs). Using the above facts the link between A-p.r. Higman groups and A-p.r. benign subgroups is established by

7. *Let G be f.g., A-p.r. and let* $1 \to K \to F \to G \to 1$ *be a presentation of G for F free and f.g. Then G is A-p.r. Higman iff K is A-p.r. benign.*

In view of 7. we turn our attention to A-p.r. decidable subsets of a f.g. groups F. For $P \subset F$, an A-p.r. decidable subset we define the subgroup $E_p < F * \langle z ; \rangle$ to be that subgroup generated by all words of the form XzX^{-1} for $X \in P$. We say P is an *A-p.r. benign subset* of F if $E_P < F * \langle z ; \rangle$ is an A-p.r. benign subgroup. We are now in a position to prove the crucial lemma (Lemma 5.1 of [4]). It is the proof of this lemma which requires countably many applications of the strong Britton extension and we consider it in some detail.

Lemma. *If* $P < F$ *is an A-p.r. decidable subgroup of the f.g. free group F which is A-p.r. benign as a subset then it is A-p.r. benign as a subgroup.*

To prove the lemma, set $G = F * \langle c, d ; \rangle$ and for every word $X \in F$ define $\varphi_X : \{c\} \to \{dX\}$ an isomorphism of infinite cyclic subgroups of G. Let $X_1, X_2, ..$ enumerate the words of F and consider $G_{|F|} = \langle G, t_{X_1}, t_{X_2}, ...; t_X c t_X{}^{-1} = dX \ \forall\, X \in F \rangle$. The proof proceeds in the following steps:

1. $G_{|F|}$ is an A-p.r. group (by Lemma 5.10 of [1]; it is here that the strong form of Lemma 5.10 as compared to Lemma 3.2 of [4] may be used to strengthen the result).

2. $G_{|F|}$ is A-p.r. Higman (by a construction).

3. $G_{|P|} = \{ G, t_X$ for all $X \in P \} < G_{|F|}$ is A-p.r. decidable (by an inductive argument during the construction of $G_{|F|}$).

4. $\{c, d, t_X$ for all $X \in P\} = \{c, d, t_X$ for all $X \in P, P\} < G_{|P|}$ is A-p.r. decidable (an application of Lemma 5.11 of [1]).

5. $\{c, d, t_X$ for $X \in P\}$ is A-p.r. benign (as the image of $\{c, d, XzX^{-1}$ for all $X \in P\} < F * \langle c, d, z;\rangle$ under an A-p.r. homomorphism).

6. $P = F \cap \{c, d, t_X$ for all $X \in P\}$ completing the proof of the lemma.

Next we consider certain subsets of $F = \langle a, b;\rangle$. We say an A-p.r. decidable subset $B \subset N^k$ (k-tuples of natural numbers) is *A-p.r. benign* if the set of all words in F of the form, $a^{x_1} ba^{x_2} b \ldots ba^{x_k}$ for $(x_1, \ldots, x_k) \in B$ is an A-p.r. benign subset of F. An A-p.r. computable function $f: N^k \to N$ is *A-p.r. benign* if the subset of $k+1$-tuples of the form $(x_1, \ldots, x_k, f(x_1, \ldots, x_k))$ is A-p.r. benign. We then establish the following:

Lemma. *If $A \subset N$ is recursively enumerable and $f: N^k \to N$ is A-p.r. computable then f is A-p.r. benign.*

The proof of the above in [4] differs somewhat from the proof given in [6] in that the characterization of A-p.r. computable functions as the smallest class of functions containing the initial functions Z, $U_m{}^n$, f_0, E and c_A and closed under substitution and (unlimited) iteration (see [1]), is used directly. In particular we show that c_A is in the class of A-p.r. benign functions since $x \in A \leftrightarrow \exists y Q(x, y)$ for Q a p.r. (actually $\mathcal{E}^3$) predicate related to the Kleene T predicate. Also the closure of the class of A-p.r. benign functions under (unlimited) iteration is shown by a direct construction (Proposition 5.10 of [4]).

The proof of the theorem is now completed by proving

Lemma. *Let $A \subset N$ be recursively enumerable and $P \subset F$ be an A-p.r. decidable subset of a f.g. free group F. Then P is A-p.r. benign.*

The proof uses the previous lemma applied to the characteristic function of P. The construction given in [4] follows that of [6] closely verifying A-p.r. computability at each step.

References

[1] F.B. Cannonito and R.W. Gatterdam, The computability of group constructions, part 1, this volume.

[2] C.R.J. Clapham, Finitely presented groups with word problem of arbitrary degrees of insolubility, Proceedings of the London Mathematical Society 14 (1964) 633-676.

[3] C.R.J. Clapham, An embedding theorem for finitely generated groups, Proceedings of the London Mathematical Society 17 (1967) 419-430.

[4] R.W. Gatterdam, Embeddings of primitive recursive computable groups, doctoral dissertation, University of California, Irvine, 1970. Submitted for publication.

[5] G. Higman, Subgroups of finitely presented groups, Proceedings of the Royal Society, A 262 (1961) 455-475.

[6] J.R. Schoenfield, Mathematical Logic (Addison Wesley, 1967).

CONNECTIONS BETWEEN TOPOLOGICAL AND GROUP THEORETICAL DECISION PROBLEMS

Wolfgang HAKEN
University of Illinois, Urbana

The main purpose of this article is to provide the logician with a brief (and thus rather incomplete) survey concerning results and open questions on decision problems in topology and their relations to group theoretic decision problems.

§ 1. Description of some basic decision problems in topology

Usually topological decision problems are concerned with *finite simplicial complexes.* Such complexes can be described (up to isomorphism) by their *incidence matrices* which are finite matrices with integral entries. (For definitions see for instance [20] or [3].) If the questions of a decision problem deal with more general topological spaces, as for instance *topological manifolds,* or with spaces with more structure, as for instance *differentiable manifolds,* then a detailed discussion is required as to how such a question can be posed *in finite terms* (as is required for a decision problem). A method of presenting a differentiable manifold in finite terms is discussed in [3]. In this article we restrict ourselves to decision problems of *finite combinatorial topology*, i.e., the spaces considered are finite simplicial complexes, in particular, the n-manifolds are *compact combinatorial n-manifolds* (= finite complexes such that the simplices incident to a vertex form a *combinatorial n-ball,* i.e., a complex which is isomorphic to a piecewise linear subdivision of a standard n-dimensional simplex). All mappings are *simplicial* (= piecewise linear, continuous) mappings; thus all homeomorphisms are *combinatorial equivalences.* For dimensions $n \leqslant 3$ this means no restriction of generality since Moise [13] has proved that every topological 3-manifold can be triangulated into a combinatorial 3-manifold and that all its triangulations are combinatorially equivalent.

1.a. Homeomorphism problems

The (piecewise linear) *homeomorphism problem for n-dimensional complexes* is the problem to decide for any two given n-complexes K_1^n, K_2^n whether or not K_1^n and K_2^n are (piecewise linearly) homeomorphic.

The most interesting sub-problem is the *homeomorphism problem for n-manifolds* where the given complexes are assumed to be n-manifolds M_1^n, M_2^n. Usually one considers further subproblems by specifying the dimension n, e.g., $n = 2, n = 3, n = 4$.

An extremely specialized (but especially interesting and for $n > 2$ still open) sub-problem is the *homeomorphism problem with the n-sphere* where one of the given manifolds is fixed to be the n-sphere S^n.

A more general problem is the *homeomorphism problem for pairs of complexes* (one being a sub-complex of the other). This is to decide for any two given pairs $(K_1, L_1), (K_2, L_2)$ where $L_1 \subset K_1, L_2 \subset K_2$ whether or not there exists a (piecewise linear) homeomorphism f from K_1 onto K_2 which maps L_1 onto L_2.

A well publicized sub-problem of this is the *equivalence problem of (classical) knot theory,* where both K_1 and K_2 are fixed to be the 3-sphere S^3 and the given sub-complexes L_1, L_2 are polygonal 1-spheres ("knots"). What is sometimes called the *knot problem* is the equivalence problem with the *trivial knot,* i.e., the sub-problem of the above where L_2 is fixed to be the boundary of a plane triangle. In general, if K_1 and K_2 are homeomorphic to some fixed complex K, and L_1, L_2 are homeomorphic to a fixed complex L, we have the *knotting problem of L in K.*

1.b. Isotopy and homotopy problems

A modification of the knotting problem of L in K is the *isotopy problem of L in K.* Here K_1 and K_2 are not only homeomorphic but identical to K and the question is whether or not L_1 and L_2 are *isotopic in K* (i.e., L_1 can be deformed into L_2 within K in such a way that no self-intersections occur at any stage of the deformation). Another modification is the *homotopy problem of L in K* where instead of L_1, L_2 we consider two given (continuous, simplicial) mappings $l_1 : L \to K, l_2 : L \to K$, and the question is whether or not l_1 and l_2 are *homotopic* (i.e., l_1 can be deformed into l_2 where self-intersections of the image are permissible at all stages).

An interesting special case of the homotopy problem is obtained if L is chosen to be the 1-sphere S^1. We call this the *homotopy problem of closed curves in K.* An important sub-problem arises if one of the mappings, say $l_2 : S^1 \to K$, is fixed to be constant (i.e., maps all of S^1 into one point in K). Let us call this the *contractibility problem of closed curves in K* (since a closed curve $l_1 : S^1 \to K$ is called *contractible in K* if l_1 is homotopic to a constant map).

The importance of the two problems mentioned is mainly due to the fact that the homotopy problem of closed curves in a *connected* complex K is logically equivalent to the *conjugacy problem in the fundamental group* $\pi_1(K)$ of K whereas the contractibility problem of closed curves is equivalent to the *word problem* in $\pi_1(K)$. This can be seen as follows. The elements of $\pi_1(K)$ are the homotopy classes of loops in K which originate and terminate at some fixed base point O in K. (Two loops are in the same class if and only if they are homotopic in K relative to the restriction that the initial and terminal points must be kept fixed at O at all stages of the deformation.) The identity element of $\pi_1(K)$ is the class of all contractible loops. Now an oriented closed curve C in K corresponds to a conjugacy class $[c]$ of an element c of $\pi_1(K)$. In order to produce a loop $\tilde{C}$ that represents c we choose an auxiliary path A from O to a point P on C and we compose A with a path from P to P around C (in the direction of the orientation of C) followed by A in the direction from P to O. The freedom we have in choosing A allows us to obtain representatives of all conjugates of c in $\pi_1(K)$. Two closed curves correspond to the same conjugacy class if and only if they are homotopic in K. On the other hand a presentation $\mathfrak{P}$ of $\pi_1(K)$ can be obtained from K by an easy procedure (see for instance [20]) where the generators are represented by certain loops in the 1-skeleton of K and the defining relations are read from the triangles of K. For a given word c (in the generators of $\mathfrak{P}$) it is then easy to construct a closed curve C in K that corresponds to $[c]$; and to a given closed curve C one can obtain a corresponding word c. With this one can easily establish the asserted equivalences.

The equivalences mentioned in the above paragraph are so obvious that topologists usually *identify* the equivalent decision problems and thus, for instance talk about the word problem in $\pi_1(K)$ without ever mentioning a presentation $\mathfrak{P}$ of $\pi_1(K)$ but just dealing with closed curves in K.

If K is an n-manifold M^n with $n > 2$ then to every closed curve in M^n one can find a homotopic *simple* closed curve. Thus instead of presenting a closed curve by a mapping $l : S^1 \to M^n$ we can consider a homotopic simple closed curve in M^n, presented by its image, which is just a closed polygon L in M^n. Then the questions which make up our decision problems can be formulated as follows: Given two (polygonal) simple closed curves L_1, L_2 in M^n, do L_1 and L_2 bound a singular annulus in M^n? (where a "singular annulus" is the continuous, simplicial image of an annulus; the existence of such a singular annulus between L_1 and L_2 in M^n is obviously equivalent to the fact that L_1 and L_2 are homotopic in M^n). And: Given a simple closed curve L in M^n, does L bound a singular disk in M^n?

Remark: A *singular disk* means a disk that may have self-intersections (being just the continuous image of a disk). This definitely distinguishes the word problem in the fundamental group of a 3-manifold M^3 from the knot problem in M^3. For, L is a trivial knot in M^3 if and only if it bounds a *non-singular disk* in M^3.

1.c. Isomorphism problems of fundamental groups

The questions of the isomorphism problem are whether or not two given complexes K_1, K_2 have isomorphic fundamental groups. An important sub-problem (a solution of which would be implied by a solution of the word problem) is the *simply connectedness problem* where the questions are whether or not a given complex K has trivial fundamental group (i.e. is simply connected).

§2. Unsolvability results

All unsolvability results about topological decision problems which have been obtained so far have been derived from the unsolvability of the word problem [2,4,14], or triviality problem [1,11,18] of group theory.

2.a. n = 2

To a given presentation $\mathfrak{P}$ of a group G we can construct a 2-dimensional complex $K^2(\mathfrak{P})$ which corresponds to $\mathfrak{P}$ where the fundamental group $\pi_1(K^2(\mathfrak{P}))$ is isomorphic to G. We construct $K^2(\mathfrak{P})$ as a *cell-complex* (but it could be easily subdivided into a simplicial complex). A group presentation $\mathfrak{P} = (\{g_1, ..., g_r\}, \{\mathfrak{r}_1, ..., \mathfrak{r}_s\})$ consists of a set of r generators $g_1, ..., g_r$ and a set of s relators $\mathfrak{r}_1, ..., \mathfrak{r}_s$ where the relators $\mathfrak{r}_j$ are words in the symbols $g_i^{\pm 1}$. We do *not* demand that all the trivial relators $g_i g_i^{-1}$ and $g_i^{-1} g_i$ be included among the $\mathfrak{r}_j$'s. However, we allow that two (or more) of the relators be equal ($\mathfrak{r}_j = \mathfrak{r}_k$ although $j \neq k$) or that some of the relators be empty words (the empty word being denoted by $*$). In order to construct $K^2(\mathfrak{P})$ we choose a point O for the only vertex of $K^2(\mathfrak{P})$. Corresponding to each generator g_i in $\mathfrak{P}$ we choose a 1-dimensional cell G_i of $K^2(\mathfrak{P})$ which is an arc that originates and terminates at O. Except for their end points, the G_i are chosen pairwise disjoint. Thus $\bigcup_{i=1}^{r} G_i$ is a wedge of r loops with fundamental group the free group of rank r. Corresponding to each relator $\mathfrak{r}_j$ in $\mathfrak{P}$ we choose a 2-dimensional cell R_j of $K^2(\mathfrak{P})$ which is a disk whose boundary ∂R_j is identified to a closed curve in $\bigcup_{i=1}^{r} G_i$ which runs through the G_i in the same order in which the g_i occur in $\mathfrak{r}_j$ (such that one direction of G_i corresponds to g_i and the opposite direction to g_i^{-1}). We say that "the reading of ∂R_j in the G_i is equal to $\mathfrak{r}_j$". (If $\mathfrak{r}_j = *$ then we identify all of ∂R_j to Q.) Except for their boundaries, the R_j are chosen pairwise disjoint.

This finishes the construction of $K^2(\mathfrak{P})$. Obviously, if we add a trivial relator to $\mathfrak{P}$ we do not change the group G presented by $\mathfrak{P}$; however we change the homeomorphy type of $K^2(\mathfrak{P})$ by adding a 2-dimensional cell (in such a way that the 2nd Betti number of $K^2(\mathfrak{P})$ is increased by 1).

Now the unsolvability of the triviality problem of group theory implies immediately the *unsolvability of the simply connectedness problem for 2-complexes* (and thus of the word problem, the conjugacy problem, and the isomorphism problem of the fundamental groups for 2-complexes).

2.b. $n \geqslant 4$

In order to construct an n-dimensional manifold $M^n(\mathfrak{P})$ corresponding to a group presentation $\mathfrak{P}$ with $\pi_1(M^n(\mathfrak{P})) \cong G$ (where G is the group presented by $\mathfrak{P}$) we may embed the 2-complex $K^2(\mathfrak{P})$ into the $(n+1)$-dimensional Euclidean space E^{n+1} (provided that $n \geqslant 4$); then we choose a simplicial, regular neighborhood N of $K^2(\mathfrak{P})$ in E^{n+1} (which is an $(n+1)$-manifold with boundary). It is not difficult to show that not only the fundamental group of N but also the fundamental group of its boundary is isomorphic to G. Thus we may choose the boundary of N for $M^n(\mathfrak{P})$. This construction of $M^n(\mathfrak{P})$ is essentially the same as that indicated in the textbook of Seifert-Threlfall [20] as an exercise (see also [3]).

Again, we conclude immediately the *unsolvability of the triviality problem, word problem, conjugacy problem, isomorphism problem of the fundamental-groups for n-manifolds with* $n \geqslant 4$.

It was the great discovery of Markov [12] that one can also conclude the *unsolvability of the homeomorphism problem for n-manifolds with* $n \geqslant 4$. His proof is based on two observations (for details see also [3]):

i) If $\mathfrak{P}$ presents the trivial group then the presentation $\mathfrak{P}^*$ which is derived from $\mathfrak{P}$ by adding r times the empty word as a new relator can be transformed into the "standard presentation" $\mathfrak{S}_{r,r+s} = (\{g_1, ..., g_r\}, \{g_1, ..., g_r, *^s\})$ (where r of the relators are equal to the 1-letter-words $g_1, ..., g_r$ and s relators are empty) by certain especially simple Tietze transformations. We call them Markov operations.

ii) If a group presentation $\mathfrak{P}_2$ is derived from a group presentation $\mathfrak{P}_1$ by a Markov operation then $M^n(\mathfrak{P}_2)$ is homeomorphic to $M^n(\mathfrak{P}_1)$.

Consequently, if a given presentation $\mathfrak{P}$ presents the trivial group then $M^n(\mathfrak{P}^*)$ is homeomorphic to $M^n(\mathfrak{L}_{r,\, r+s})$. Thus a solution of the homeomorphism problem for n-manifolds (for some $n \geqslant 4$) would imply a solution of the triviality problem of group theory.

Moreover, since $M^n(\mathfrak{L}_{r,\, r+s})$ can be seen to be an "n-sphere with s handles of index 2", it is an *unsolvable problem to decide whether or not a given n-manifold with* $n \geqslant 4$ is *an n-sphere with handles of index* 2. However the homeomorphism problem with the n-sphere itself is still open (see Section 4.a).

2.c. n = 3

It seems to be rather difficult to obtain unsolvability results about 3-dimensional manifolds. Not every finitely presented group is isomorphic to the fundamental group of a 3-manifold. Moreover, some progress has been made in *solving* decision problems for 3-manifolds (see Section 3.b). In this section I shall consider an unsolvable decision problem on 3-manifolds the questions of which are somewhat more sophisticated then those of the basic decision problems described in Section 1. However, the questions of this unsolvable problem have a certain similarity to questions of solved decision problems which we shall discuss in Section 3.b. So it appears that in dimension 3 solvable and unsolvable decision problems come relatively close to each other.

Let G be a finitely presented group with unsolvable word problem and let $\mathfrak{P}$ be a presentation of G (as for instance exhibited in [2]). Now we construct a 3-manifold M_G^3 which is rather simple in so far as it is a subspace of the 3-sphere S^3 which is bounded by some tori. The questions of our decision problem will be concerned with (polygonal) simple closed curves in this fixed 3-manifold.

Corresponding to each generator g_i ($i = 1, ..., r$) in $\mathfrak{P}$ we choose a (non-singular, polyhedral) disk G_i^2 in the 3-sphere S^3 so that the G_i^2 are pairwise disjoint. We choose small (regular, polyhdedral) neighborhoods N_i of the boundary curves of the G_i in S^3 and we remove the interiors of the N_i from S^3. This yields a 3-manifold F^3 which is bounded by r tori T_i^2 (the boundaries of the N_i) and the fundamental group of which is the free group of rank r. Corresponding to each relator $\mathfrak{r}_j$ in $\mathfrak{P}$ ($j = 1, ..., s$) we choose a (polygonal) simple closed curve C_j in the interior of F^3 which pierces the disks G_i^2 in the same order in which the corresponding letters g_i occur in $\mathfrak{r}_j$ (so that to a symbol g_i^{-1} there corresponds a piercing in the opposite direction to a piercing corresponding to g_i). In order to obtain M_G^3 we remove from F^3 the interiors of small (regular, polyhedral) neighborhoods of the C_j. Then M_G^3 is bounded by the tori T_i^2 and by s tori, say U_j^2.

Now our decision problem consists of the following questions. Given a (polygonal) simple closed curve L in the interior of M_G^3, does L bound a singular disk with holes in M_G^3 such that the boundary curves of the holes lie in the tori U_j^2?

This decision problem is unsolvable since it is logically equivalent to the word problem in G. For each word c in the generators g_i one can construct a polygonal simple closed curve L in M_G^3 which pierces the G_i^2 corresponding to c (and from each polygonal simple closed curve in M_G^3 one can read the corresponding word up to cyclic permutations). Then c represents the group identity in G if and only if L is contractible in a complex K which is the union of M_G^3 and s cones C_j^3 over the tori U_j^2 (the interiors of the C_j^3 being pairwise disjoint and disjoint from M_G^3). K is not a 3-manifold since it contains precisely s exceptional points (the vertices of the cones C_j^3) the neighborhoods of which are not 3-cells. But L is contractible in K if and only if it bounds a singular disk in K, and this again is the case if and only if L bounds a singular disk with holes in M_G^3 as demanded.

§3. Solvability results

3.a. n = 2

The *homeomorphism problem for 2-dimensional complexes* has been solved by Papakyriakopoulos [17].

The *fundamental groups of* the *2-manifolds* without boundary are *one-relator groups* and thus the corresponding word problems are solved by Magnus [10]. The 2-manifolds with boundary have free fundamental groups.

3.b. n = 3

Solutions have been obtained for some special cases of the homeomorphism problem of 3-manifolds and of the word problem in their fundamental groups. These solutions apply only to 3-manifolds with or without boundary which contain so-called incompressible and boundary-incompressible surfaces. So it appears that the more complicated 3-manifolds are more easily accessible for solving the basic decision problems.

In a 3-manifold M^3 we consider surfaces (i.e., polyhedral 2-manifolds) M^2 so that the interior of M^2 lies in the interior of M^3 and the boundary of M^2 lies in the boundary of M^3 (but may be empty). The concepts of *incompressibility* and *boundary-incompressibility* for such a surface M^2 in M^3 were originally defined geometrically [5] by conditions that no handle of M^2 can be "cut off" by a (non-singular) disk in the complement of M^2 in M^3, etc. But in the case that M^2 is orientable the loop theorem and Dehn's lemma, as proved by Papakyriakopoulos [15,16], permit definitions in terms of homotopy as follows. "Incompressible" means, if a closed curve C in M^2 is contractible in M^3 then it is also contractible in M^2. "Boundary-incompressible" means, if a curve C in M^2 that originates and terminates at the boundary of M^2 can be deformed within M^3 into the boundary of M^3 without moving its end points, then such a deformation of C can also be carried out within M^2.

The class of 3-manifolds which contain incompressible and boundary-incompressible surfaces is rather large: i) If an orientable 3-manifold M^3 has a boundary which does not only consist of 2-spheres then it contains an incompressible and boundary-incompressible, orientable surface. ii) If an orientable 3-manifold without boundary has an infinite first homology group or has a fundamental group which is a free product with amalgamation then and only then it contains an orientable, incompressible surface. (The latter part of ii was observed by Waldhausen [23].)

We can solve (as indicated in [6]), *the special case of the homeomorphism problem for 3-manifolds* which we obtain by restricting the questions as follows.

(I) The given 3-manifolds M_1^3, M_2^3 are given in such a way that they are known to be "irreducible", i.e., that every polygonal 2-sphere in M_i^3 bounds a 3-ball in M_i^3 $(i = 1,2)$.

(II) M_1^3 is given in such a way that it is known to contain orientable incompressible and boundary-incompressible surfaces.

(III) M_1^3 does *not* contain a sub-manifold $\widetilde{M}^3$ with the following properties. i) $\widetilde{M}^3$ is a fibre bundle over S^1 with fibre a surface F^2 (with or without boundary). ii) The boundary of $\widetilde{M}^3$ (if not empty) consists of incompressible surfaces in M^3. iii) The first Betti number of $\widetilde{M}^3$ is 1.

The main tools for solving the problem described are algorithms for determining incompressible and boundary-incompressible surfaces of minimal genus in given 3-manifolds. The basic theory for deriving such algorithms has been developed in [5] (see also [19]); one of the simplest applications was an algorithm for the knot problem: to determine whether or not a given polygonal simple closed curve in a 3-manifold M^3 bounds a (non-singular) disk (= a surface of genus 0) in M^3. In general it is possible to determine for a given 3-manifold M^3 and for a given integer g whether or not M^3 contains an incompressible and boundary-incompressible surface of genus $\leqslant g$. If such a surface exists then the algorithm constructs one which is of minimal genus. Moreover, as indicated in [6], in many cases the algorithm yields *all possible types* of such surfaces in M^3.

Remark: If in the questions of the unsolvable decision problem stated in 2.c the word "singular" is replaced by "non-singular" then we get a decision problem that can be solved (by some modification of the algorithm for the knot problem). This shows how much easier it is to find algorithms for determining non-singular surfaces than for determining singular ones, or, in other words, how much easier it is to solve isotopy problems than homotopy problems.

The algorithms mentioned above can be applied to split the given 3-manifold M_1^3 by incompressible and boundary-incompressible surfaces of minimal genus into sub-manifolds, provided that M_1^3 contains such surfaces and that we know some upper bound for their genus. For this reason we need the restriction (II) above. Then we split the sub-manifolds again by simplest possible surfaces into sub-manifolds, and so on. We can continue this process as long as not all boundary surfaces of the sub-manifolds obtained are 2-spheres. It can be proved that this process must eventually terminate and thus yield a finite decomposition of M_1^3 the 3-dimensional pieces of which are bounded only by 2-spheres. Here we need the restriction (I) in order to conclude that these pieces are 3-balls, i.e., that we actually arrived at a *cell-decomposition* of M_1^3. We try to arrange this splitting procedure in such a way that at each stage there are only finitely many possibilities of splitting and that all the possible splitting surfaces can be algorithmically found. If this succeeds then there exist only finitely many (isomorphism types of) cell-decompositions of M_1^3 we can end up with, and all of them can be constructed. For this purpose we need the additional restriction (III) (which was not recognized in [6]). Then the other given 3-manifold, M_2^3, is homeomorphic to M_1^3 if and only if the same procedure can be applied to it and yields cell-decompositions which are pairwise isomorphic to those obtained for M_1^3.

Waldhausen [24] found a *solution for the word problem in the fundamental groups of 3-manifolds under the restrictions* (I) *and* (II) stated above.

The first part of his algorithm is a splitting procedure applied to the given 3-manifold M^3 as described above. However, it is sufficient to obtain some one of the special cell-decompositions of M^3 and it is not necessary that there are only finitely many of them. Therefore the restriction (III) is not required. Then the given closed curve L in M^3 is deformed into a nicest possible position relative to the special cell-decomposition which permits to decide whether or not L is contractible in M^3. (This solves the contractibility problem for closed curves in M^3 which is, according to our discussion in Section 1, equivalent to the word problem in $\pi_1(M^3)$.)

From another result of Waldhausen [23] it follows that *the homeomorphism problem for orientable, irreducible 3-manifolds without boundary that contain orientable, incompressible surfaces is equivalent to the isomorphism problem of their fundamental groups.*

§4. Some particular open decision problems

In this section we mention some topological decision problems which are still open but are so closely related to group theoretic decision problems that one might attack them with methods of combinatorial group theory as well as with topological methods.

4.a. The homeomorphism problem with the n-sphere for n ⩾ 5

We call a group presentation $\mathfrak{P}$ *balanced* if the number s of relators in $\mathfrak{P}$ equals the number r of generators in $\mathfrak{P}$. *If the triviality problem for balanced presentations could be proved unsolvable then this would imply the unsolvability of the homeomorphism problem with the n-sphere for each* $n \geqslant 5$.

This can be seen as follows. If a presentation $\mathfrak{P}$ of the trivial group is balanced then the n-manifold $M^n(\mathfrak{P})$, as constructed in Section 2.b for any $n \geqslant 4$, is a *homotopy n-sphere* (i.e., its fundamental group and its homology groups are isomorphic to those of the n-sphere S^n). But the generalized Poincaré conjecture for dimensions $\geqslant 5$, as proved by Smale [21] and Stallings [22], says that a homotopy n-sphere is an n-sphere if $n \geqslant 5$. Thus a given balanced group presentation $\mathfrak{P}$ presents the trivial group if and only if $M^n(\mathfrak{P})$ is homeomorphic to S^n for any $n \geqslant 5$. This would hold for $n = 4$ also, if the generalized Poincaré conjecture were proved for the (very special) case of Seifert-Threlfall-Markov-4-manifolds as constructed in Section 2.b.

4.b. The homeomorphism problem with the 3-sphere, the simply connectedness problem for 3-manifolds, and the Poincaré conjecture

The (still open) Poincaré conjecture states that a compact, simply connected 3-manifold without boundary is a 3-sphere. This does not constitute a decision problem. However, *the truth of the Poincaré conjecture would imply the equivalence of the two decision problems* mentioned in the headline.

Some partial results have been obtained regarding the two decision problems which has been discussed in Sections II.b and II.c of [8].

Recently Jaco [9] obtained the result that *the simply connectedness problem for 3-manifolds is equivalent to a decision problem* the questions of which may be formulated *as follows.* Given 2s elements $x_1, ..., x_s, y_1, ..., y_s$ in a direct product $F_s \times F_s$ of two free groups of rank s such that

(i) the projections of $x_1, ..., x_s, y_1, ..., y_s$ into the factors of $F_s \times F_s$ generate the factors,

(ii) the relation $x_1 y_1 x_1^{-1} y_1^{-1} x_2 y_2 x_2^{-1} y_2^{-1} ...... x_s y_s x_s^{-1} y_s^{-1} = 1$ holds.

Question: Do $x_1, ..., x_s, y_1, ..., y_s$ generate $F_s \times F_s$?

In Jaco's theory the $x_1, ..., x_s, y_1, ..., y_s$ are generators of the fundamental group $\pi_1(M^2)$ of a handle surface M^2 of genus s, where (ii) is a defining relation for $\pi_1(M^2)$. Given is a homomorphism $\pi_1(M^2) \to F_s \times F_s$, the image of which projects onto the factors, and the question is whether or not this homomorphism is onto. So we may call this an *epimorphism problem.*

Regarding this problem it is remarkable that C.F. Miller found a simple proof that *if the condition* (ii) *is omitted then the questions form an unsolvable problem.*

4.c. The sufficiently largeness problem for 3-manifolds

Waldhausen calls an orientable 3-manifold *sufficiently large* [23] if it contains an orientable, incompressible surface. The restriction (II) in Section 3.b would be weakened if we could algorithmically decide whether or not a given 3-manifold M^3 is sufficiently large. If the first homology group $H_1(M^3)$ is infinite then M^3 is known to be sufficiently large. However, if $H_1(M^3)$ is finite we have an open problem. This problem is equivalent to the problem to decide whether or not the fundamental group of M^3 is a free product with amalgamation.

4.d. The homeomorphism problem for fibre bundles over S^1 with fibre a 2-manifold

A fibre bundle over S^1 with fibre a 2-manifold M^2 (with or without boundary) is a 3-manifold obtained from the cartesian product $M^2 \times [0,1]$ of M^2 with the unit interval $[0,1]$ by identifying the top, $M^2 \times (1)$, to the bottom, $M^2 \times (0)$, according to a homeomorphism φ from $M^2 \times (1)$ onto $M^2 \times (0)$. Thus the 3-manifold obtained is determined by M^2 and φ. If $\alpha : M^2 \to M^2 \times (0)$ and $\beta : M^2 \to M^2 \times (1)$ map a point p of M^2 to $p \times (\sigma)$ and $p \times (1)$, respectively, then φ is determined by a self-homeomorphism $\tilde{\varphi} = \alpha^{-1} \circ \varphi \circ$ of M^2. Two 3-manifolds obtained in this way from the same M^2 by two self-homeomorphisms $\tilde{\varphi}_1$ and $\tilde{\varphi}_2$, respectively, are homeomorphic if and only if $\tilde{\varphi}_1$ and $\tilde{\varphi}_2$ are conjugate elements in the group of self-homeomorphisms of M^2. Thus *the homeomorphism problem for these 3-manifolds is equivalent to the conjugacy problem in the group of self-homeomorphisms of* M^2.

A solution of this problem would remove the restriction (III) in Section 3.b.

References

[1] S.I. Adjan, The algorithmic unsolvability of checking certain properties of groups, Dokl. Akad. Nauk SSSR 103 (1955) 533-535 (in Russian).

[2] W.W. Boone, The word problem, Ann. Math. 70 (1959) 207-265.

[3] W.W. Boone, W. Haken and V. Poénaru, On recursively unsolvable problems in topology and their classification, in: A. Schmidt et al. (eds.), Contributions to mathematical logic (North-Holland, Amsterdam, 1968) 37-74.

[4] J.L. Britton, The word problem, Ann. Math. 77 (1963) 16-32.

[5] W. Haken, Theorie der Normalflächen, ein Isotopiekriterium für den Kreisknoten, Acta Math. 105 (1961) 245-375.

[6] W. Haken, Über das Homöomorphieproblem der 3-Mannigfaltigkeiten, I, Math. Zeitschr. 80 (1962) 89-120.

[7] W. Haken, Some results on surfaces in 3-manifolds, in: Studies in modern topology, MAA Studies in Mathematics, vol. 5 (Prentice-Hall, Englewood Cliffs, 1968) 39-98.

[8] W. Haken, Various aspects of the 3-dimensional Poincare problem, in: Topology of manifolds, (Markham, Chicago, 1970) 140–152.
[9] W. Jaco, Stable equivalence of splitting homeomorphisms, in: Topology of manifolds, (Markham, Chicago, 1970) 153–156.
[10] W. Magnus, A. Karras and D. Solitar, Combinatorial group theory, (Wiley, New York, 1966).
[11] A.A. Markov, Impossibility of algorithms for recognizing some properties of associative systems, Dokl. Akad. Nauk SSSR 77 (1951) 953-956 (in Russian).
[12] A.A. Markov, Insolubility of the problem of homeomorphy, Proc. Intern. Congress of Mathematicians, 1958 (Cambridge University Press) 300-306.
[13] E.E. Moise, Affine structures in 3-manifolds, V, The triangulation theorem and Hauptvermutung, Ann. Math. 56 (1952) 96-114.
[14] Novikov, P.S. On the algorithmic unsolvability of the word problem in group theory, Trudy Mat. Inst. Steklov, no. 44, 1955 (in Russian).
[15] C.D. Papakyriakopoulos, On solid tori, Proc. London Math. Soc. (3), 7 (1957) 281-299.
[16] C.D. Papakyriakopoulos, On Dehn's lemma and the asphericity of knots, Ann. Math. 66 (1957) 1-26.
[17] C.D. Papakyriakopoulos, A new proof for the invariance of the homology groups of a complex, Bull. Soc. Math. Grèce 22 (1943) 1-154. (1946) (in Greek).
[18] M.O. Rabin, Recursive unsolvability of group theoretic problems, Ann. Math. 67 (1958) 172-194.
[19] H. Schubert, Bestimmung der Primfaktorzerlegung von Verkettungen, Math. Zeitschr. 76 (1961) 116-148.
[20] H. Seifert and W. Threlfall, Lehrbuch der Topologie, (Akademische Verlagsgesellschaft B.G. Teubner, Leipzig, 1934).
[21] S. Smale, Generalized Poincaré's conjecture in dimensions greater than four, Ann. Math. 74 (1961) 391-406.
[22] J.W. Stallings, Polyhedral homotopy spheres, Bull. Am. Math. Soc. 66 (1960) 485-488.
[23] F. Waldhausen, On irreducible 3-manifolds which are sufficiently large, Ann. Math. 87 (1968) 56-88.
[24] F. Waldhausen, The word problem in fundamental groups of sufficiently large irreducible 3-manifolds, Ann. Math. 88 (1968) 272-280.

ON THE WORD PROBLEM AND T-FOURTH-GROUPS

Seymour LIPSCHUTZ
Temple University, Philadelphia

§1. Introduction

Let G be a finitely presented group with generating elements $a_1, ..., a_\mu$ and defining relations $R_1 = 1, ..., R_\lambda = 1$. We assume without loss in generality that the R_i, called *relators*, form a *symmetric set,* i.e. that the R_i are cyclically reduced and are closed under the operations of taking inverses and cyclic transforms. We call G a *T-fourth-group* if the following two conditions hold:

(i) *One-fourth condition*: If $R_i \cong XY$ and $R_j \cong XZ$ are distinct relators, then the length of the common initial segment X is less than ¼ the length of either relator.

(ii) *Triangle condition*: If each of three relators R_i, R_j and R_k is written along one side of a triangle then cancellation cannot occur at all three vertices, i.e. at least one of the products R_iR_j, R_jR_k, R_kR_i is freely reduced without cancellation.

The class of T-fourth-groups is of interest because Greendlinger [4] solved the word and conjugacy problems for these groups. Also Lyndon [6] solved the word problem and Schupp [9] the conjugacy problem for a class of groups with similar conditions. Our main theorem follows.

Main Theorem. *Let K be the free product of T-fourth-groups with an infinite cyclic group amalgamated. Then K has a solvable word problem.*

Results identical to that above and others below have been proven by the author [5] for Greendlinger sixth-groups, groups which satisfy condition (i) above with $\frac{1}{4}$ replaced by $\frac{1}{6}$. We also note that McCool [8] has proven similar results for Britton free-product sixth-groups.

§2. Notation

Capital letters, $V, W, A, B, \ldots$ will denote freely reduced words unless otherwise stated or implied. We use the following notation for words V and W in the group G:

$l(W)$ for the length of W,

$V = W$ means V and W are the same element of the group G,

$V \cong W$ means V and W are identical words,

$V \wedge W$ means V does not react with W, i.e. $U \cong VW$ is freely reduced,

$V \subset W$ means V is a subword of W, i.e. $W \cong AVB$.

The letter R with or without subscripts shall always denote one of the relators $R_1, \ldots, R_\lambda$.

We say W is *fully reduced* if it is freely reduced and does not contain more than half of a relator. We say W is *cyclically fully reduced* if every cyclic transform of W is fully reduced.

§3. Preliminary lemmas

We shall require the following obvious consequence of Theorem 2 in [4]. (Also stated in [3].)

Lemma 1 (Greendlinger). *Let W be a freely reduced word in a T-fourth-group G, and suppose $W = 1$. Then W contains more than $\frac{3}{4}$ of a relator, or W contains disjointly two subwords, each containing more than $\frac{1}{2}$ of a relator.*

Remark. The generalized word problem for a group G modulo a subgroup H is to decide whether or not an arbitrary element of G is also in H. The generalized word problem reduces to the word problem when $H = 1$; hence is unsolvable in general (c.f. Boone [1, §35]).

However, the following result is easily shown using properties of free products with amalgamations (c.f. [5, Lemma 2] or [7, Section 4.2]):

Lemma 2. *Let K be the free product of groups G_i with a subgroup H amalgamated. If the word problem for the G_i and the generalized word problem for the G_i modulo H are solvable, then K has a solvable word problem.*

The next two lemmas are about T-fourth-groups G. As the proofs are relatively long and combinatorial, we give them in Sections 5 and 6.

Lemma 3. *Suppose $W = U$ where U is fully reduced. Then $l(U) \leqslant rl(W)$, where r is the length of the largest defining relation in G.*

Lemma 4. *Suppose W is cyclically fully reduced, and suppose there is no relator $R \cong U^t$ such that W is conjugate to a power of U. If W^2 is also cyclically fully reduced then W^n is fully reduced for all n. If W^2 is not cyclically fully reduced then there is a relator*

$$R \cong ABAT^{-1}$$

where AB is a cyclic transform of W, and $(TB)^n$ is fully reduced for all n. In either case, W has infinite order.

§4. Main results

Our first theorem is similar to one proved by Greendlinger [2, Theorem VIII] for sixth-groups.

Theorem 1. *An element W in a T-fourth-group G has finite order if and only if there is a relator $R \cong U^k$ and W is conjugate to a power of U.*

Proof. Clearly, if there is a relator $R \cong U^k$ and W is conjugate to a power of U, then W has finite order. On the other hand, suppose $W \neq 1$ and suppose there is no relator $R \cong U^k$ such that W is conjugate to a power of U. By taking cyclic tranforms of W and substituting whenever more than half of a relator appears, we can obtain a cyclically fully reduced word V which is conjugate to W. By Lemma 4, V has infinite order. Therefore W also has infinite order and the theorem is proved.

The *order problem* for a group G is to decide the order of any element in G. If k is the maximum positive integer such that there is a relator $R \cong U^k$ in a T-fourth-group G then, by Theorem 1, no element has finite order greater than k. This fact together with the solvability of the word problem for G imply

Corollary 1. *The order problem for a T-fourth-group G is solvable.*

Theorem 2. *Let G be a T-fourth-group. Then the generalized word problem for G modulo any cyclic subgroup H is solvable.*

Proof. Let V be any element in G. By Corollary 1, we can determine the order of the generator W of H. Suppose the order of W is finite. Then V belongs to H iff $V = W^n$ for $1 \leqslant m \leqslant$ order (W). Since the word problem for G is solvable, this case is decidable.

Now suppose W has infinite order. Note that V belongs to gp(W) if and only if $A^{-1}VA$ belongs to gp($A^{-1}WA$). Thus by taking cyclic transforms of W and substituting whenever more than half of a relator appears, we can reduce our problem to the case where W is cyclically fully reduced and has the properties of Lemma 4. In fact, we can assume in Lemma 4 that $W \cong AB$.

Suppose W^n is fully reduced for all n. By Lemma 3, $V = W^m$ only if $|m| \leqslant l(W^m) \leqslant rl(V)$ where r is the largest defining relation in G. Thus we can decide if V belongs to gp(W). On the other hand, suppose W^n is not fully reduced. Then, by Lemma 4,

$$W^n \cong (ABAB)^k W^\epsilon = (TB)^k W^\epsilon$$

where $\epsilon = 1$ or $\epsilon = 0$. Hence V is in gp(W) iff V or VW^{-1} is in gp(TB). But $(TB)^k$ is fully reduced for all k. So the theorem is true for this last case also.

Remark. Our Main Theorem now follows directly from Theorem 2 and Lemma 2. In fact, we can generalize even further. Let Z_1 denote the class of T-fourth-groups. Let Z_i, $i > 1$, be the class of groups which are the free product of groups in Z_{i-1} with an infinite cyclic group amalgamated. Let G belong to Z_i for some i. Then by Theorem 5 in [5], the word problem for G and the generalized word problem for G modulo any infinite cyclic group are solvable.

Theorem 3. *Consider a T-fourth-group*

$$G = \text{gp}(a_1, ..., a_n, b_1, ..., b_m ; R_1 = 1, .., R_k = 1)$$

such that every freely reduced word $W = W(a_\lambda)$ *is fully reduced. Then* $H = \text{gp}(a_1, ..., a_n)$ *is a free subgroup of G, and the generalized word problem for G modulo H is solvable.*

Proof. By Lemma 1, every non-trivial word $W(a_\lambda) \neq 1$; hence H is free. Let V belong to G. If V is in H, i.e. if $V = W(a_\lambda)$ then, by Lemma 3, we know the maximum length of W. Since there are only a finite number of words $W(a_\lambda)$ of any given length, and since the word problem for G is solvable, the theorem is true.

§5. Proof of Lemma 3

Suppose the lemma is not true and suppose W is a word of minimum length for which the lemma does not hold, say $l(W) = n$. Then there exists a fully reduced word V such that $W = V^{-1}$ and $l(V) > rn$. So

(1) $\quad WV = 1.$

The minimality of W guarantees that W is fully reduced, and that (1) is freely reduced. Thus (1) satisfies Lemma 1. Moreover, since W and V are fully reduced, WV must contain more than $\frac{3}{4}$ of a relator R. Say $W \cong AB$, $V \cong CD$, $S \cong BC$; where $R \cong SE^{-1} \cong BCE^{-1}$ and $l(S) > \frac{3}{4}\, l(R)$.

Now, substituting in (1), we have

$$WV \cong ABCD \cong ASD = AED = 1.$$

Furthermore,

$$D \text{ is fully reduced and } l(D) > r(n-1),$$
$$l(C) \leqslant \tfrac{1}{2}\, l(R) \text{ since } V \text{ is fully reduced},$$
$$l(B) > \tfrac{1}{4}\, l(R) \text{ since } l(S) > \tfrac{3}{4}\, l(R),$$
$$l(E) < \tfrac{1}{4}\, l(R) \text{ since } l(S) > \tfrac{3}{4}\, l(R).$$

Thus, $l(E) < l(B)$, which implies $l(AE) < l(AB) = l(W)$. Let $W' \cong AE$, $V' \cong D$. Then $W'V' = 1$ and V' is fully reduced. Moreover, $l(W') = l(AE) < l(W) = n$, so $l(W') \leqslant n-1$. But $l(V') = l(D) > r(n-1) \geqslant rl(W')$, so that W' and V' violate the lemma. This contradicts the minimality of W, so our lemma is true.

§6. Proof of Lemma 4

Suppose W^n contains more than half of a relator; say W contains S where $R \cong ST^{-1}$ and $l(S) > \frac{1}{2}\, l(R)$. Since $S \subset W^n$, we say $S \cong V^tA$ where $V \cong AB$, $B \not\cong 1$, and V is a cyclic transform of W. Note $t \neq 0$ because V and so A are fully reduced. We claim that $t = 1$. Suppose not, i.e. suppose $t > 1$. Then $V^{t-1}A$ is a common initial segment of the relators $R \cong V^tAT^{-1}$ and $R' \cong V^{t-1}AT^{-1}V$. But

$$l(V^{t-1}A) \geqslant \tfrac{1}{2}\, l(S) > \tfrac{1}{4}\, l(R);$$

hence, by the one-fourth condition, $R \cong R'$. Consequently, V and

$V^{t-1}AT^{-1}$ commute as words in a free group, and so are powers of the same word U. Therefore, W is conjugate to a power of U and $R \cong V(V^{t-1}AT^{-1})$ is a power of U. However, this contradicts our original hypothesis. Thus $t = 1$ and the first part of the lemma is proved.

We can now assume that $S \cong ABA$, $V \cong AB$ and $R \cong ABAT^{-1}$. We also have $B \not\cong 1$. Moreover, since $l(V) \leqslant \frac{1}{2} l(R)$, we have $A \not\cong 1$ and $T \not\cong 1$. Furthermore, by choosing the largest possible S, we can assume that $B \wedge T$ and $T \wedge B$. (That is, if say $T \cong T'x^{-1}$ and $B \cong xB'$, then $R \cong ST^{-1} \cong SxT'^{-1}$ and $V^2 \cong ABAxB'$ and so we could have chosen Sx in place of S.) We also have that $A \wedge B$ and $B \wedge A$ because W is cyclically fully reduced, and we have $A \wedge T^{-1}$ and $T^{-1} \wedge A$ because R is freely and cyclically reduced. In addition, we have $l(T) < \frac{1}{2} l(R)$ because $l(S) > \frac{1}{2} l(R)$, and we have $l(B) < \frac{1}{2} l(R)$ because $l(V) \leqslant \frac{1}{2} l(R)$ and $A \not\cong 1$.

We are now ready to show that $(TB)^n$ is fully reduced for all n. Suppose $(TB)^n$ contains Q where Q is more than half of a relator, say $R' \cong QP$ and $l(Q) > \frac{1}{2} l(R')$. There are seven possibilities.

Case 1. $Q \subset B$. Impossible because V and so B are fully reduced.

Case 2. $Q \subset T$. Say $T \cong MQN$. Then $R \cong ABAN^{-1}Q^{-1}M^{-1}$. By the one-fourth condition, the relators $R' \cong QP$ and $R^* \cong QNA^{-1}B^{-1}A^{-1}M$ are identical; hence $l(R') = l(R^*) = l(R)$. Accordingly,

$$\frac{1}{2} l(R) = \frac{1}{2} l(R') < l(Q) \leqslant l(T) < \frac{1}{2} l(R),$$

which is a contradiction. Hence Q is not contained in T.

Case 3. $Q \subset TB$ but $Q \not\subset T$ and $Q \not\subset B$. Say $T \cong T_2T_1$, $B \cong B_1B_2$ and $Q \cong T_1B_1$. Then $R \cong AB_1B_2AT_1^{-1}T_2^{-1}$ and $R' \cong T_1B_1P$. Note that $Q \subset T$ implies $B_1 \not\cong B$ and that $Q \not\subset B$ implies $T_1 \not\cong 1$. Since $l(Q) > \frac{1}{2} l(R')$, either $l(T_1) > \frac{1}{4} l(R')$ or $l(B_1) > \frac{1}{4} l(R')$.

(a) Suppose $l(T_1) > \frac{1}{4} l(R')$. By the one-fourth condition, the relators $R' \cong T_1B_1P$ and $R'' \cong T_1A^{-1}B^{-1}A^{-1}T_2$ are identical. Since $B_1 \not\cong 1$, this contradicts the fact that $A \wedge B \cong B_1B_2$.

(b) Suppose $l(B_1) > \frac{1}{4} l(R')$. Then the relators $R'' \cong B_1PT_1$ and $R^* \cong B_1B_2AT^{-1}A$ are identical. Since $T_1 \not\cong 1$, this contradicts the fact that $A \wedge T^{-1} \cong T_1^{-1}T_2^{-1}$.

Thus Case 3 cannot occur.

Case 4. $Q \subset BT$ but $Q \not\subset B$ and $Q \not\subset T$. Say $B \cong B_4 B_3$, $T \cong T_3 T_4$ and $Q \cong B_3 T_3$. Then $R \cong AB_4 B_3 A T_4^{-1} T_3^{-1}$ and $R' \cong B_3 T_3 P$. Note that $Q \not\subset B$ implies $T_3 \not\cong 1$ and that $Q \not\subset T$ implies $B_3 \not\cong 1$. Since $l(Q) > \frac{1}{2} l(R')$, either $l(B_3) > \frac{1}{4} l(R')$ or $l(T_3) > \frac{1}{4} l(R')$.

(a) Suppose $l(B_3) > \frac{1}{4} l(R')$. By the one-fourth condition, the relators $R' \cong B_3 T_3 P$ and $R'' \cong B_3 A T_4^{-1} T_3^{-1} A B_4$ are identical. Since $T_3 \not\cong 1$, this contradicts the fact that $T_4^{-1} T_3^{-1} \cong T^{-1} \wedge A$.

(b) Suppose $l(T_3) > \frac{1}{4} l(R')$. Then the relators $R'' \cong T_3 P B_3$ and $R^{-1} \cong T A^{-1} B^{-1} A^{-1}$ are identical. Since $B_3 \not\cong 1$, this contradicts the fact that $B_4 B_3 \cong B \wedge A$.

Thus this case cannot occur.

Case 5. $Q \subset TBT$ but $Q \not\subset TB$ and $Q \not\subset BT$. Say $T \cong EF \cong MN$ and $Q \cong FBM$. Then $R' \cong FBMP$ and $R \cong ABAF^{-1}E^{-1}$. Note that $F \not\cong 1$ because $Q \not\subset TB$. We also have that $A \not\cong 1$ and $B \not\cong 1$. Thus the three relators

$$R_i \cong BMPF, \quad R_j \cong F^{-1}E^{-1}ABA, \quad R_k \cong A^{-1}TA^{-1}B^{-1}$$

violate the triangle condition. Hence this case is impossible.

Case 6. $Q \subset BTB$ but $Q \not\subset BT$ and $Q \not\subset TB$. Say $B \cong EF \cong MN$ and $Q \cong FTM$. Then $R' \cong FTMP$ and $R \cong AMNAT^{-1}$. Note that $M \not\cong 1$ because $Q \not\subset BT$. We also have that $A \not\cong 1$ and $T \not\cong 1$. Thus the three relators

$$R_i \cong MPFT, \quad R_j \cong T^{-1}AMNA, \quad R_k \cong A^{-1}TA^{-1}N^{-1}M^{-1}$$

violate the triangle condition. Hence this case cannot occur.

Case 7. $Q \not\subset TBT$ and $Q \not\subset BTB$. Then $Q \cong EBF \cong MTN$ where F and T have a common nontrivial initial segment, and where N and B have a common nontrivial initial segment. Now the relators $R \cong ABAT^{-1}$ and $R^* \cong AT^{-1}AB$ are distinct because $B \wedge T$. Hence by the one-fourth condition, $l(A) < \frac{1}{4} l(R)$. Accordingly, either $l(B) > \frac{1}{4} l(R)$ or $l(T) > \frac{1}{4} l(R)$.

(a) Suppose $l(B) > \frac{1}{4} l(R)$. Now $R \cong ABAT^{-1}$ and $R' \cong QP \cong EBFP$. By the one-fourth condition, the relators $R_i \cong BAT^{-1}A$

and $R_j \cong BFPE$ are identical. Since F and T have a common nontrivial initial segment, this contradicts the fact that $T^{-1} \wedge A$.

(b) Suppose $l(T) > \frac{1}{4} l(R)$. Now $R \cong ABAT^{-1}$ and $R' \cong QP \cong MTNP$. BY the one-fourth condition, the relators $R_k \cong TNPM$ and $R^{-1} \cong TA^{-1}B^{-1}A^{-1}$ are identical. Since N and B have a common nontrivial initial segment, this contradicts the fact that $A \wedge B$.

We have shown that this last case is also impossible.

It remains to show that W has infinite order. By Lemma 1, a fully reduced nonempty word cannot be 1. Thus we have proved that $W^n \neq 1$ for all n or that W^2 is conjugate to $ABAB = TB$ such that $(TB)^n \neq 1$ for all n. In either case, W has infinite order. Thus the lemma is proved.

References

[1] W.W. Boone, The word problem, Ann. of Math. (2) 70 (1959) 207-265.

[2] M. Greendlinger, On Dehn's algorithm for the conjugacy and word problems, with applications, Comm. Pure Appl. Math. 13 (1960) 641-677.

[3] M. Greendlinger, Solutions of the word problems for a class of groups by means of Dehn's algorithm, and of the conjugacy problem by means of a generalization of Dehn's algorithm, Doklady Akad. Nauk SSSR 154 (1964) 507-509.

[4] M. Greendlinger, On the word problem and the conjugacy problem, Izv. Akad. Nauk SSSR Ser. Mat. 29 (1965) 245-268.

[5] S. Lipschutz, An extension of Greendlinger's results on the word problem, Proc. Amer. Math. Soc. 15 (1964) 37-43.

[6] R.C. Lyndon, On Dehn's algorithm, Math. Ann. 166 (1966) 208-228.

[7] W. Magnus, A. Karrass and D. Solitar, Combinatorial group theory (Wiley, New York, 1966).

[8] J. McCool, The order problem and the power problem for free product sixth-groups, Glasgow Math. J. 10 (1969) 1-9.

[9] P.E. Schupp, On Dehn's algorithm and the conjugacy problem, Math. Ann. 178 (1968) 119-130.

ON A CONJECTURE OF W. MAGNUS

James McCOOL and Alfred PIETROWSKI
University of Toronto

Let F_n be the free group with free generating set $X_1, \ldots, X_n$, and let G be a group with one defining relation given by the presentation $\langle X_1, \ldots, X_n ; R = 1 \rangle$. We denote by $P(G)$ the set of all presentations of G of the form $\langle X_1, \ldots, X_n ; S = 1 \rangle$. Two elements $\langle X_1, \ldots, X_n ; S = 1 \rangle$ and $\langle X_1, \ldots, X_n ; Q = 1 \rangle$ of $P(G)$ are said to be N-equivalent if there exists an automorphism ϕ of F_n such that $\phi S = Q^{\pm 1}$. This defines an equivalence relation on $P(G)$; the number of equivalence classes under this relation is denoted by $|P(G)|$. It has been conjectured by Magnus ([1], page 401) that $|P(G)| = 1$ for every group G with one defining relation. In theorem I we show that this conjecture is false.

It was shown by A. Shenitzer [6] that if G has presentation $\langle X_1, \ldots, X_n ; R = 1 \rangle$, and G is a non-trivial free product, then there exists an automorphism ϕ of F_n such that ϕR does not involve all of the free generators $X_1, \ldots, X_n$. By combining this result with a result due to J.H.C. Whitehead [7], Shenitzer gave a constructive method for deciding, given a group G with one defining relation, whether or not G is a non-trivial free product. In an attempt to generalize this result, one might conjecture that if G has presentation $\langle X_1, \ldots, X_n ; R = 1 \rangle$, and G is a non-trivial free product of two groups with an infinite cyclic subgroup amalgamated, then there exists an automorphism ϕ of F_n such that $\phi R = W_1 W_2$, where W_1 and W_2 are non-identity elements of F_n which have no free generator X_i in common. We show in Theorem 2 that this conjecture is false.

Let k and l be integers. We denote by $G_{k,l}$ the group with presentation $\langle X_1, X_2 ; X_1{}^k = X_2{}^l \rangle$.

Lemma 1. *Let $l = pt + 1$ for some integers p and t. Then $G_{k,l}$ has presentation $\langle X_1, X_2; X_2 = (X_1{}^k X_2{}^{-t})^p \rangle$.*

Proof.

$$\begin{aligned} G_{k,l} &\cong \langle X_1, Y; X_1{}^k = Y^{pt+1} \rangle \\ &\cong \langle X_1, Y, X_2; X_1{}^k Y^{-pt} = Y, X_2 = Y^p \rangle \\ &\cong \langle X_1, X_2, Y; X_2 = (X_1{}^k X_2{}^{-t})^p, Y = X_1{}^k X_2{}^{-t} \rangle \\ &\cong \langle X_1, X_2; X_2 = (X_1{}^k X_2{}^{-t})^p \rangle \end{aligned}$$

Definition. Two elements W_1 and W_2 of F_n are said to be *equivalent* if there exists an automorphism ϕ of F_n such that $\phi W_1 = W_2$. The element W of F_n is said to be *of minimal length* if the length of W is less than or equal to the length of any element equivalent to W.

Theorem 1. *Let k, l, p, t be positive integers such that $k \geqslant 2$, $p \geqslant 2$, $t \geqslant 2$ and $l = pt + 1$. Then $|P(G_{k,l})| > 1$.*

Proof. It follows immediately from Theorem 3 of [6] that the elements $X_1{}^k X_2{}^{-l}$ and $(X_1{}^k X_2{}^{-t})^p X_2{}^{-1}$ of F_2 are of minimal length. Now $X_1{}^k X_2{}^{-l}$ has length $k + l$, while $(X_1{}^k X_2{}^{-t})^p X_2{}^{-1}$ has length $kp + l$, and $k + l \neq kp + l$, since $p \geqslant 2$. Since these elements are of minimal length, and their lengths are different, it follows that $(X_1{}^k X_2{}^{-t})^p X_2{}^{-1}$ is not equivalent to $(X_1{}^k X_2{}^{-l})^{\pm 1}$. Hence the two presentations $\langle X_1, X_2; X_1{}^k X_2{}^{-l} = 1 \rangle$ and $\langle X_1, X_2; (X_1{}^k X_2{}^{-t})^p X_2{}^{-1} = 1 \rangle$ of $G_{k,l}$ are not N-equivalent, and so $|P(G_{k,l})| > 1$.

Corollary. *Let m be any positive integer. Then we can choose k and l so that $|P(G_{k,l})| \geqslant m$.*

Proof. Choose $k \geqslant 2$ and $l = p_1 p_2 \ldots p_m + 1$, where $p_1, \ldots, p_m$ are the first m positive primes. Let t_i be such that $l = p_i t_i + 1$ $(i = 1, 2, \ldots, m)$. Then $R_i = (X_1{}^k X_2{}^{-t_i})^{p_i} X_2{}^{-1}$ is of minimal length in F_2, and has length $k p_i + l$, so that R_i is not equivalent to $R_j^{\pm 1}$ if $i \neq j$, since $kp_i + l \neq kp_j + l$. Now $\langle X_1, X_2; R_i = 1 \rangle$ is a presentation of $G_{k,l}$ ($i = 1, 2, \ldots, m$), so that $G_{k,l}$ has at least m non-equivalent presentations.

A more complete list of $G_{k,l}$ groups for which $|P(G_{k,l})| > 1$ is given in [3], together with some additional results which may be useful in determining the number of non-equivalent presentations of such groups.

It is interesting to note that O. Schreier [5] has shown that every automorphism of $G_{k,l} = \langle X_1, X_2; X_1^k X_2^{-l} = 1 \rangle$ is induced by an automorphism of F_2; this may be contrasted with the results of E.S. Rapaport [4], that the group $L = \langle X_1, X_2; X_1^3 X_2 X_1^{-1} X_2^{-2} X_1^{-1} X_2 = 1 \rangle$ has an automorphism which is not induced by any automorphism of F_2, and that $|P(L)| = 1$.

We note that the method of the proof of Proposition 6.3 of [2] can be used to give the following result.

Lemma 2. *Let k, l, k_1 and l_1 be integers different from* $0, \pm 1$. *Then $G_{k,l}$ is isomorphic to G_{k_1,l_1} if, and only if, either $|k| = |k_1|$ and $|l| = |l_1|$, or $|k| = |l_1|$ and $|l| = |k_1|$.*

Theorem 2. *Let k, l, p, t be as in Theorem 1. Then $(X_1^k X_2^{-t})^p X_2^{-1}$ is not equivalent in F_2 to any element of the form $X_1^r X_2^{-s}$.*

Proof. Suppose $(X_1^k X_2^{-t})^p X_2^{-1}$ is equivalent to $X_1^r X_2^{-s}$. Then $G_{k,l}$ is isomorphic to $G_{r,s}$, and so, by Lemma 2, either $|r| = |k|$ and $|s| = |l|$, or $|r| = |l|$ and $|s| = |k|$. It follows from this that $X_1^k X_2^{-l}$, and so $(X_1^k X_2^{-t})^p X_2^{-1}$ is equivalent to $X_1^k X_2^{-l}$. This is a contradiction.

Thus Theorem 2 gives examples of groups $\langle X_1, \ldots, X_n; R = 1 \rangle$ which are non-trivial free products of two free groups with an infinite cyclic subgroup amalgamated, such that no automorphic image of R in F_n is of the form $W_1 W_2$, where W_1 and W_2 are non-identity elements of F_n with no free generator X_i in common.

References

[1] W. Magnus, A. Karrass and D. Solitar, Combinatorial Group Theory (Interscience, 1966).

[2] W.S. Massey, Algebraic Topology, An Introduction (Harcourt, 1967).

[3] J. McCool and A. Pietrowski, On free products with amalgamation of two infinite cyclic groups, to appear in Jour. Alg.

[4] E.S. Rapaport, Note on Nielsen transformations, Proc. Amer. Math. Soc. 10 (1959) 228-235.

[5] O. Schreier, Über die Gruppen $A^aB^b = 1$, Abh. Math. Sem. Univ., Hamburg.

[6] A. Shenitzer, Decomposition of a group with a single defining relation into a free product, Proc. Amer. Math. Soc. 6 (1955) 273-279.

[7] J.H.C. Whitehead, On equivalent sets of elements in a free group, Ann. of Math. 37 (1936) 782-800.

AN ELEMENTARY CONSTRUCTION OF UNSOLVABLE WORD PROBLEMS IN GROUP THEORY*

Ralph McKENZIE and Richard J. THOMPSON

University of California, Berkeley

In this paper we introduce a new approach to constructing finitely presented groups with unsolvable word problems. The argument has a concrete motivation which should make it easy to follow. A brief intuitive sketch of the procedure is the following:

Let ${}^{\omega}\omega$ denote the set of all numerical functions (i.e. infinite sequences of natural numbers) and let $\mathfrak{G}$ denote the group of all permutations of ${}^{\omega}\omega$. Behind our arguments is the consideration of a certain primitive recursive set of relations $\mathcal{R}$, involving altogether only a finite set of letters correlated with elements of $\mathfrak{G}$, which has the property that the subset $\mathcal{R}_{\mathfrak{G}} \subset \mathcal{R}$ consisting of the relations of $\mathcal{R}$ which are valid in $\mathfrak{G}$ is non-recursive. After some preliminaries we will be in a position to define $\mathcal{R}$ and to prove: there exists a finite set of relations $\mathcal{U}$, which are true in $\mathfrak{G}$ and which imply $\mathcal{R}_{\mathfrak{G}}$. Then $\mathcal{U}$ will be constructed from a set of letters L_0, consisting of the letters which appear in $\mathcal{R}$ and new letters correlated with additional elements of $\mathfrak{G}$. A member of $\mathcal{R}$ will then be implied by $\mathcal{U}$ if and only if it belongs to $\mathcal{R}_{\mathfrak{G}}$ – since the relations $\mathcal{U}$ are true in $\mathfrak{G}$ – hence the finitely presented group $(L_0; \mathcal{U})$ will have an unsolvable word problem.

The novelty (and the virtue) of this approach is that after one understands the underlying notions, the argument itself is nearly trivial, and combinatorial (in a broad sense) rather than group-theoretic in nature.

* This work was aided by National Science Foundation grants GP-7578 and GP-6232X3.

§0. Preliminaries

The basic notation we will use is as follows: The set of all natural numbers is denoted by ω. Given two sets X and Y, ${}^Y X$ denotes the set of all functions from Y into X. We consider every natural number n to be identical with the set $\{0,1, ..., n-1\}$ and therefore ${}^n\omega$ is a set of functions, which we call n-termed sequences: $\alpha = \langle \alpha_0, ..., \alpha_{n-1} \rangle$ is the function such that $\alpha(i) = \alpha_i$. ${}^\omega\omega$ is the set of all infinite sequences of natural numbers; for these we may use the notation $\alpha = \langle \alpha_0, \alpha_1, \alpha_2, ... \rangle$. We put ${}^{\underset{\smile}{\omega}}\omega = \bigcup_{n \in \omega} {}^n\omega$, i.e. the set of all sequences of natural numbers having an arbitrary finite length. Given $\sigma \in {}^{\underset{\smile}{\omega}}\omega$ and $\alpha \in {}^{\underset{\smile}{\omega}}\omega \cup {}^\omega\omega$, we can define, of course, the concatenation of the two sequences: this is the sequence $\sigma^\frown\alpha = \langle \sigma_0, ..., \sigma_{m-1}, \alpha_0, ..., \alpha_{n-1} \rangle$ if $\sigma = \langle \sigma_0, ..., \sigma_{m-1} \rangle$ and $\alpha = \langle \alpha_0, ..., \alpha_{n-1} \rangle$; or $\sigma^\frown\alpha = \langle \sigma_0, ..., \sigma_{m-1}, \alpha_0, \alpha_1, ... \rangle$ if $\alpha \in {}^\omega\omega$. The formula $\sigma \subset \beta$ asserts that σ is an initial part of β (or β begins with σ), i.e. $\exists\alpha\ \sigma^\frown\alpha = \beta$. We use the notation $\sigma^{(n)}$ for the sequence obtained by concatenating σ with itself n times (where $\sigma \in {}^{\underset{\smile}{\omega}}\omega$). Thus $\langle 0 \rangle^{(n)}$ is a sequence of n zeroes.

The *interval* in ${}^\omega\omega$ determined by σ, where $\sigma \in {}^{\underset{\smile}{\omega}}\omega$, is the set $I_\sigma = \{\alpha \in {}^\omega\omega \mid \sigma \subset \alpha\}$. It is easily seen that when $\sigma, \tau \in {}^{\underset{\smile}{\omega}}\omega$, $I_\sigma \subset I_\tau$ iff $\tau \subset \sigma$.

We shall sometimes denote the members of ${}^\omega\omega$ by $f, g, h, ... \in {}^\omega\omega$ if it seems desirable to emphasize their character as numerical functions rather than as sequences. Our functions are often written on the left so that, for instance, if $f, g \in {}^\omega\omega$ and $i \in \omega$ then $fg(i) = f(g(i))$; interchangeably, $g(i)$, gi and g_i then denote the same thing. Functions being considered as elements of permutation groups, however, will always be written on the right. Then of course FG will denote the function such that $xFG = ((x)F)G$.

Let x and y be two elements of a given group. Their commutator is the element $[x, y] = xyx^{-1}y^{-1}$. We have $[x, y] = 1$ if, and only if $y * x = y$, where we write $y * x$ for $x^{-1}yx$ (for reasons which will become apparent later). Here of course "1" denotes the identity element of a group. Elsewhere it may denote the second smallest natural number.

The most important concept of this paper is contained in

Definition 0.1. $\mathfrak{C}$ is the group of permutations of ${}^{\omega}\omega$ whose members are the permutations F satisfying

$$\forall\alpha \in {}^{\omega}\omega\ \exists\sigma, \tau \in {}^{\underline{\omega}}\omega\ [\sigma \subset \alpha\ \&\ \forall\beta \in {}^{\omega}\omega\ [(\sigma^\frown\beta)F = \tau^\frown\beta]\,].$$

We shall not bother to justify the definition; it is not hard to show that if F and G satisfy the defining condition then so do the permutations FG and F^{-1}. A more useful working definition of $\mathfrak{C}$ is the following. Let us say that a set $\Sigma \subset {}^{\underline{\omega}}\omega$ is a *partition set* for ${}^{\omega}\omega$ provided that every sequence $\alpha \in {}^{\omega}\omega$ has exactly one initial part belonging to Σ (in other words, provided the intervals I_σ, $\sigma \in \Sigma$, form a partition of the set of infinite sequences). For example, ${}^{2}\omega$ is a partition set. Every partition set (if it fails to contain the empty sequence) must clearly be denumerably infinite.

Theorem 0.2. *Let $\Sigma_0, \Sigma_1, \varphi$ be any triple where Σ_0 and Σ_1 are partition sets and φ is a one-one map from Σ_0 onto Σ_1. Then the formulas $(\sigma^\frown\beta)F = \varphi(\sigma)^\frown\beta$ (for $\sigma \in \Sigma_0$ and $\beta \in {}^{\omega}\omega$) define a permutation F which belongs to $\mathfrak{C}$. Moreover, every member of $\mathfrak{C}$ can be put in this form.*

Proof. The first statement is clear from the definitions. Let us suppose, on the other hand, that $F \in \mathfrak{C}$. Then we let $\varphi'(\sigma, \tau)$ (for $\sigma, \tau \in {}^{\underline{\omega}}\omega$) mean that $(\sigma^\frown\beta)F = \tau^\frown\beta$ whenever $\beta \in {}^{\omega}\omega$. We put

$$\Sigma_0 = \{\,\sigma \mid \forall\gamma(\,\exists\tau(\varphi'(\gamma, \tau) \wedge \gamma \subset \sigma) \leftrightarrow \gamma = \sigma)\}.$$

As φ' is a function (note that $\varphi'(\sigma) = \tau$ and $\varphi'(\sigma, \tau)$ have the same meaning), if we let φ be the restriction of φ' to Σ_0, and Σ_1 be the range of φ, $\Sigma_0, \Sigma_1, \varphi$ and F will be seen to be related as required by the theorem.

§1. The arrow notation

It now appears from Theorem 0.2 that the elements of $\mathfrak{C}$ are, in essence, certain transformations of finite sequences. We shall employ an abbreviated system of notation which reflects this fact and makes no mention of infinite sequences; however, the reader should always remember that the objects being discussed are really permutations of ${}^{\omega}\omega$.

The arrow notation is best defined by giving an example from the next section. There we shall write

$$\begin{aligned} E:\quad &\langle 0\rangle \to \langle 0,0\rangle,\\ &\langle 1,i\rangle \to \langle 0,i+1\rangle\\ &\langle i+2\rangle \to \langle i+1\rangle. \end{aligned}$$

This formula is understood to define the function $E \in \mathfrak{C}$ given by the partition sets $\Sigma_0 = \{\langle 0\rangle\} \cup \{\langle 1,i\rangle, \langle i+2\rangle \mid i \in \omega\}$ and $\Sigma_1 = \{\langle 0,0\rangle\} \cup \{\langle 0,i+1\rangle, \langle i+1\rangle \mid i \in \omega\}$, and the map $\varphi : \Sigma_0 \to \Sigma_1$ specified by the arrows; in other words, if $n_0, n_1, \ldots \in \omega$ then

$$\begin{aligned} &\langle 0, n_0, n_1, \ldots\rangle E = \langle 0,0, n_0, n_1, \ldots\rangle,\\ &\langle 1, n_0, n_1, \ldots\rangle E = \langle 0, n_0+1, n_1, \ldots\rangle,\\ &\langle n_0+2, n_1, \ldots\rangle E = \langle n_0+1, n_1, \ldots\rangle. \end{aligned}$$

The permutation E^2 $(= EE)$ is specified in this notation by

$$\begin{aligned} E^2 :\ &\langle 0\rangle \to \langle 0,0,0\rangle,\\ &\langle 1,i\rangle \to \langle 0,0,i+1\rangle,\\ &\langle 2,i\rangle \to \langle 0,i+1\rangle,\\ &\langle i+3\rangle \to \langle i+1\rangle. \end{aligned}$$

The reader may formulate for himself the rules for computing arrow notation for products and inverses of given permutations. We remark that, in general, many arrow notations define the same function: e.g., E above is also given by $\langle 0,j\rangle \to \langle 0,0,j\rangle$; $\langle 1,i,j\rangle \to \langle 0,i+1,j\rangle$; $\langle i+2\rangle \to \langle i+1\rangle$. However we shall always use the unique, "irredundant" notation stemming from the partition sets constructed in proving Theorem 0.2.

We will take $\sigma F = \tau$, where σ and τ are given finite sequences, to mean that F acts on any $\alpha \in {}^{\omega}\omega$ which begins with σ by deleting σ and replacing it by τ. Thus $\sigma F = \tau (\sigma, \tau \in {}^{\underset{\smile}{\omega}}\omega)$ implies $\sigma ^\frown \langle j \rangle F = \tau ^\frown \langle j \rangle$ for all $j \in \omega$.

§2. The groups $\mathfrak{H}$ and $\mathfrak{P}$

The elements of $\mathfrak{C}$ whose relations we shall study can be exhibited as the generators of a certain subgroup. In fact, we shall work with the subgroup $\mathfrak{H} \subset \mathfrak{C}$ generated by three basic permutations – D, E, R – and a system of permutations A_f, correlated with every function $f \in {}^{\omega}\omega$. These are defined below.

$$
\begin{aligned}
R:\quad & \langle 0 \rangle \to \langle 0 \rangle \\
& \langle 1, i, j, k \rangle \to \langle 1, j, k, i \rangle \\
& \langle i+2 \rangle \to \langle i+2 \rangle \\
E:\quad & \langle 0 \rangle \to \langle 0,0 \rangle \\
& \langle 1, i \rangle \to \langle 0, i+1 \rangle \\
& \langle i+2 \rangle \to \langle i+1 \rangle \\
D:\quad & \langle 0 \rangle \to \langle 0 \rangle \\
& \langle 1 \rangle \to \langle 1,0 \rangle \\
& \langle 2, i \rangle \to \langle 1, i+1 \rangle \\
& \langle i+3 \rangle \to \langle i+2 \rangle \\
A_f:\quad & \langle 0, i, j \rangle \to \langle 1, i, j \rangle \text{ when } 0 \leqslant j < f(i) \\
& \langle 0, i, j + f(i) \rangle \to \langle 0, i, j \rangle \\
& \langle 1, i, j \rangle \to \langle 1, i, j + f(i) \rangle \\
& \langle i+2 \rangle \to \langle i+2 \rangle
\end{aligned}
$$

Definition 2.1. (i) $\mathfrak{H} = \mathrm{Gp}(D, E, R, A_f(f \in {}^{\omega}\omega))$.
(ii) $\mathfrak{P} = \mathrm{Gp}(D, E)$.

It will be useful subsequently to remark that when we set $M = E^{-1}D$ and $Q = D^2RD^{-1}$ then $\langle 1, i \rangle M = \langle 1, i+1 \rangle$ and $\langle 1, i \rangle Q = \langle 1, i, 0 \rangle$.

§3. Letters, words and relations

Since our readers are assumed to know the basic facts about groups defined by generators and relations (see, for instance, [4]) we can be very brief here. Let $\mathfrak{F} = FG(L)$ be a free group freely generated by a set $L = \{\mathbf{D}, \mathbf{E}, \mathbf{R}, \mathbf{A}_f(f \in {}^{\omega}\omega)\}$ in one to one correspondence with the permutations defined in §2. The elements of L will be called the *basic letters*, and the elements of $\mathfrak{F}$ will be called *words*; expressions denoting words will be printed in **italics**. (Note that we do not distinguish between a word and its reduced form.) A *relation* is either an ordered pair of words, written $\mathbf{W}_0 \approx \mathbf{W}_1$, or a word $\mathbf{W}$, thought of as the pair $\mathbf{W} \approx \mathbf{1}$. The formula $\vdash_{\mathcal{S}} \mathcal{E}$ asserts that the relation $\mathcal{E}$ is a consequence of, or is derivable from, the set of relations $\mathcal{S}$.

Each group $\mathfrak{G}$ which we discuss will have a set of generators correlated with some set of basic letters L_0. Implicitly, we have an epimorphism $r^{\mathfrak{G}}$ from $FG(L_0)$ onto $\mathfrak{G}$ which maps the letters to the corresponding generators. In this context the expressions "$\mathbf{W}_0 = \mathbf{W}_1$ in $\mathfrak{G}$," "$\mathbf{W}_0 \approx \mathbf{W}_1$ is true (or valid) in $\mathfrak{G}$," "$r^{\mathfrak{G}}(\mathbf{W}_0) = r^{\mathfrak{G}}(\mathbf{W}_1)$" all have the same meaning (where $\mathbf{W}_0, \mathbf{W}_1 \in FG(L_0)$). Let $\mathcal{S}$ be a set of relations in the letters L_0. We say that $\mathfrak{G}$ has the *presentation* $(L_0; \mathcal{S})$ iff the kernel of the map $r^{\mathfrak{G}}$ is identical with the normal subgroup of $FG(L_0)$ generated by $\{\mathbf{W}_0\mathbf{W}_1^{-1} \mid \mathbf{W}_0 \approx \mathbf{W}_1 \in \mathcal{S}\}$. If this is true then we have, for $\mathbf{W} \in FG(L_0)$: $\mathbf{W} = \mathbf{1}$ in $\mathfrak{G}$ if and only if $\vdash_{\mathcal{S}} \mathbf{W}$.

Theorem 3.1. $\mathfrak{P}$ *has the presentation*

$$\text{(i) } (\mathbf{D}, \mathbf{E}; [\mathbf{E}^{-1}\mathbf{D}, \mathbf{E}\mathbf{D}\mathbf{E}^{-1}], [\mathbf{E}^{-1}\mathbf{D}, \mathbf{E}^2\mathbf{D}\mathbf{E}^{-2}]).$$

Proof. First we must show that the two commutator relations are valid in $\mathfrak{P}$ (check Definition 2.1(ii)). This is so because $\mathbf{E}^{-1}\mathbf{D}$ acts only on $I_{\langle 0\rangle} \cup I_{\langle 1\rangle}$ whereas $\mathbf{EDE}^{-1}$ and $\mathbf{E}^2\mathbf{DE}^{-2}$ leave the points of this set fixed.

Next we must show that every relation valid in $\mathfrak{P}$ is derivable from the two given relations. Let $\mathfrak{G}$ be the finitely presented group (i) and let us put $\mathbf{W}_0 \equiv \mathbf{E}$, $\mathbf{W}_{n+1} \equiv \mathbf{E}^n\mathbf{DE}^{-n}$. From the given relations we have that $\mathbf{E}^{-1}\mathbf{D} \cdot \mathbf{W}_\alpha = \mathbf{W}_\alpha \cdot \mathbf{E}^{-1}\mathbf{D}$ in $\mathfrak{G}$ for $\alpha = 2,3$. Using this we prove

(a) In $\mathfrak{G}$ we have, for $m \geqslant 0$ and $k \geqslant 1$:

(1) $\mathbf{W}_m \cdot \mathbf{W}_{m+k} = \mathbf{W}_{m+k+1} \cdot \mathbf{W}_m$;

(2) $[\mathbf{E}^{-1}\mathbf{D}, \mathbf{W}_{k+1}] = \mathbf{1}$.

In fact, (a1) is trivial for $m = 0$ and follows from (a2) for $m > 0$:

$$\begin{aligned}\mathbf{W}_m \cdot \mathbf{W}_{m+k} &= \mathbf{E}^{m-1}\mathbf{DE}^k\mathbf{DE}^{-k}\mathbf{E}^{-(m-1)}\\ &= \mathbf{E}^m(\mathbf{E}^{-1}\mathbf{D})\mathbf{W}_{k+1}\mathbf{E}^{-(m-1)}\\ &= \mathbf{E}^m\mathbf{W}_{k+1}\mathbf{E}^{-1}\mathbf{DE}^{-(m-1)}\\ &= \mathbf{W}_{m+k+1} \cdot \mathbf{W}_m .\end{aligned}$$

Thus we have (a1), (a2) for $k = 1,2$. But if (a1), (a2) are true for $k = l, l+1$ $(l \geqslant 1)$ then we have $\mathbf{W}_{l+3} = \mathbf{W}_{l+1} \cdot \mathbf{W}_{l+2} \cdot \mathbf{W}_{l+1}^{-1}$ by the case $k = 1$, and so (a2) for $k = l + 2$ follows, and then (a1). Hence the proof of (a) can be given by induction.

Now (a1) leads to a normal form for elements of $\mathfrak{G}$:

(b) Let $\mathbf{W} \in FG(\mathbf{D}, \mathbf{E})$. In $\mathfrak{G}$ we have

(b1) $\mathbf{W} = (\mathbf{W}_{i_k}^{p_k} \cdot \ldots \cdot \mathbf{W}_{i_0}^{p_0})^{-1} \cdot (\mathbf{W}_{j_l}^{q_l} \cdot \ldots \cdot \mathbf{W}_{j_0}^{q_0})$

for some $k, l \geqslant 0$, where $0 \leqslant i_0 < i_1 < \ldots < i_k$ and $0 \leqslant j_0 < j_1 < \ldots < j_l$ and the p_s and q_t are positive. (The proof of (b) is by induction on the reduced length of $\mathbf{W}$: $\mathbf{D} = \mathbf{W}_1^{-1} \cdot \mathbf{W}_1^2$, $\mathbf{E} = \mathbf{W}_0^{-1} \cdot \mathbf{W}_0^2$ can be put in this form, and (a1) implies that products and inverses of words in this form are equivalent in $\mathfrak{G}$ to words which have the form.)

To complete the proof, let $\mathbf{W} \in FG(\mathbf{D}, \mathbf{E})$ such that $\mathbf{W} = \mathbf{1}$ in $\mathfrak{P}$. Then the expression (b1) for $\mathbf{W}$ in $\mathfrak{G}$ will also hold in $\mathfrak{P}$ (by the first paragraph of the proof); therefore it is enough to show that if

$$\text{(b2)} \qquad \mathbf{W}_{i_k}^{p_k} \cdot \ldots \cdot \mathbf{W}_{i_0}^{p_0} = \mathbf{W}_{j_l}^{q_l} \cdot \ldots \cdot \mathbf{W}_{j_0}^{q_0} \text{ in } \mathfrak{P},$$

with positive exponents and strictly decreasing indices, then this relation is also valid in $\mathfrak{G}$. In fact, (b2) implies that $k = l$ and $i_s = j_s$ and $p_s = q_s$ for all $s \leqslant k$, as one will see by checking the definitions of the permutation: observe that the permutation denoted by $\mathbf{W}_i{}^p$ satisfies $\langle i \rangle \mathbf{W}_i{}^p = \langle i \rangle ^\frown \langle 0 \rangle^{(p)}$, and $\langle k \rangle \mathbf{W}_i{}^p = \langle k \rangle$ for all $k < i$. Thus (b2) implies $i_0 = j_0$ and $p_0 = q_0$. Then one can cancel $\mathbf{W}_{i_0}^{p_0}$ from both sides and repeat the argument.

§4. Deferred action; $\mathfrak{H}$-computable functions

Definition 4.1. (See Definition 0.1) Let $F \in \mathfrak{C}$ and let $\sigma \in {}^{\underline{\omega}}\omega$. We define F_σ, the *σ-deferment* of F, as follows: $(\sigma^\frown\beta)F_\sigma = \sigma^\frown(\beta F)$ for all $\beta \in {}^\omega\omega$; and $\alpha F_\sigma = \alpha$ if $\alpha \in {}^\omega\omega - I_\sigma$.

F_σ acts on I_σ exactly as F acts on $I_\emptyset = {}^\omega\omega$, and acts trivially elsewhere. Thus one can easily show that if $F_\sigma = G_\tau$ and neither $\sigma \subset \tau$ nor $\tau \subset \sigma$, then $F = G = 1$. We proceed to show that the group $\mathfrak{P}$ contains the deferments of all its members.

Definition 4.2. We correlate with every word $\mathbf{W}$ in the letters $\mathbf{D}$ and $\mathbf{E}$ and with every finite sequence of natural numbers $\sigma = \langle \sigma_0, \ldots, \sigma_k \rangle$ a new word $\mathbf{W}(\sigma)$, or $\mathbf{W}(\sigma_0, \ldots, \sigma_k)$:

(1) $\mathbf{W}(\emptyset) \equiv \mathbf{W}$; and for $n \in \omega$:

(2) $\mathbf{E}(0) \equiv \mathbf{E}^{-1}\mathbf{D}^{-1}\mathbf{E}^2, \mathbf{E}(1) \equiv \mathbf{D}^{-1}\mathbf{E}\mathbf{D}^{-1}\mathbf{E}^{-1}\mathbf{D}^2, \mathbf{E}(n+1) \equiv \mathbf{E}^n\mathbf{E}(1)\mathbf{E}^{-n}$;

(3) $\mathbf{D}(0) \equiv \mathbf{E}^{-1}\mathbf{E}(1)\mathbf{E}, \mathbf{D}(1) \equiv \mathbf{D}^{-1}\mathbf{E}(2)\mathbf{D}, \mathbf{D}(n+1) \equiv \mathbf{E}^n\mathbf{D}(1)\mathbf{E}^{-n}$;

(4) $\mathbf{W}(\sigma_0, \ldots, \sigma_k) \equiv \mu_{\sigma_0} \cdots \mu_{\sigma_k}(\mathbf{W})$ where μ_m is the endomorphism of $FG(\mathbf{D}, \mathbf{E})$ such that $\mu_m(\mathbf{D}) = \mathbf{D}(m)$ and $\mu_m(\mathbf{E}) = \mathbf{E}(m)$.

Theorem 4.3. *Suppose that* $\mathbf{W} \in FG(\mathbf{D}, \mathbf{E})$ *and* $\mathbf{W}$ *defines the element* $F \in \mathfrak{P}$. *Then for* $\sigma \in {}^{\underline{\omega}}\omega$, $\mathbf{W}(\sigma)$ *defines* F_σ. *Therefore* $F_\sigma \in \mathfrak{P}$.

Proof. Let μ_m be as in Definition 4.2, let r be the canonical epimorphism $FG(\mathbf{D}, \mathbf{E}) \to \mathfrak{P}$, and for each finite sequence σ let ν_σ denote the endomorphism of $\mathfrak{C}$ which maps $F \to F_\sigma$. The reader may check Definition 4.2 and § 2 to see that for each one-term sequence $\sigma = \langle m \rangle$, $\nu_\sigma r(\mathbf{E}) = r\mu_m(\mathbf{E})$; $\nu_\sigma r(\mathbf{D}) = r\mu_m(\mathbf{D})$; hence $\nu_\sigma r = r\mu_m$ on $FG(\mathbf{D}, \mathbf{E})$. Now if $\sigma = \langle \sigma_0, \ldots, \sigma_k \rangle$, we calculate

$$\begin{aligned} r\mu_{\sigma_0} \cdot \ldots \cdot \mu_{\sigma_k} &= \nu_{\langle \sigma_0 \rangle} r\mu_{\sigma_1} \cdot \ldots \cdot \mu_{\sigma_k} \\ &\;\;\vdots \\ &= \nu_{\langle \sigma_0 \rangle} \cdot \ldots \cdot \nu_{\langle \sigma_k \rangle} r \\ &= \nu_\sigma r, \end{aligned}$$

which is the desired result.

Our precise goal in the next two sections is now describable. If $F, G \in \mathfrak{C}$, then one can check that we have

$$F^{-1} G_\sigma F = G_\tau,$$

whenever $\sigma, \tau \in {}^{\underline{\omega}}\omega$ and $\sigma F = \tau$. If $F \in \mathfrak{H}$ and $G \in \mathfrak{P}$ then, by Theorem 4.3, these equations can be written as relations which are true in $\mathfrak{H}$. From now on we write $x * y$ for $y^{-1}xy$; and we now make a definition:

Definition 4.4. A function $f \in {}^\omega\omega$ will be called $\mathfrak{H}$*-computable* if there is a finite set of relations, $\mathcal{B}_f$, which are valid in $\mathfrak{H}$ and from which the relations

$$\mathbf{X}(1,i,j) * \mathbf{A}_f \approx \mathbf{X}(1,i,j+f(i)) \; (i,j \in \omega \text{ and } \mathbf{X} \in \{\mathbf{D},\mathbf{E}\})$$

can be derived. A set with these properties will be said to *compute* f.

Note that in this definition the relations $\mathfrak{B}_f$ are allowed to involve any finite subset of the letters $L = \{\mathbf{D},\mathbf{E},\mathbf{R},\mathbf{A}_g,\mathbf{A}_h, ...\}$; the relations assumed to be derivable contain, on the other hand, only the three letters $\mathbf{D}$, $\mathbf{E}$ and $\mathbf{A}_f$. Our immediate goal is to prove the following theorem (whose converse is obvious):

Theorem 4.5. *Every (general) recursive function is* $\mathfrak{H}$*-computable.*

§5. The basic relations $\mathfrak{B}$

We define a set of basic relations which will be useful in the proof of Theorem 4.5. Put $\mathbf{M} \equiv \mathbf{E}^{-1}\mathbf{D}$, $\mathbf{Q} \equiv \mathbf{D}^2\mathbf{R}\mathbf{D}^{-1}$, $\Gamma \equiv \mathbf{M} * \mathbf{R}$, $\Delta \equiv \mathbf{M} * \mathbf{R}^2$.

B1: $[\mathbf{M}, \mathbf{E}\mathbf{D}\mathbf{E}^{-1}]$

B2: $[\mathbf{M}, \mathbf{E}^2\mathbf{D}\mathbf{E}^{-2}]$

B3: $\mathbf{R}^3$

$$\left.\begin{array}{l} \text{B4: } [[\mathbf{M},\Delta],\mathbf{X}(1)] \\ \text{B5: } [\mathbf{D}\Delta\mathbf{D}^{-1}\mathbf{M}^{-1},\mathbf{X}(1)] \\ \text{B6: } [\mathbf{D}\Gamma\mathbf{D}^{-1}\Delta^{-1},\mathbf{X}(1)] \\ \text{B7: } [\mathbf{R},\mathbf{X}(1,0,0,0)] \end{array}\right\} \quad (\text{for } \mathbf{X} \in \{\mathbf{D},\mathbf{E}\})$$

This set of 11 relations we call $\mathfrak{B}$. The words $\mathbf{X}(1)$, $\mathbf{X}(1,0,0,0)$ were defined above (4.2).

Lemma 5.1. *The relations* $\mathfrak{B}$ *are valid in* $\mathfrak{H}$.

Proof. See Theorem 3.1 for B1 and B2, and the definitions in §2 for the others. Each of the commutator relations $[\mathbf{W}, \mathbf{X}(\sigma)]$ is true because $\mathbf{W}$ defines an element $F \in \mathfrak{H}$ such that $\sigma F = \sigma$.

Lemma 5.2. *Let* $\mathbf{W}$ *be an arbitrary word and let* $\rho, \sigma, \tau \in {}^{\omega}\omega$. *The two relations* $\mathbf{X}(\sigma) * \mathbf{W} \approx \mathbf{X}(\tau)$, $\mathbf{X} \in \{\mathbf{D}, \mathbf{E}\}$, *imply for every* $\mathbf{Y} \in FG(\mathbf{D}, \mathbf{E})$, *the relation* $\mathbf{Y}(\sigma^\frown\rho) * \mathbf{W} \approx \mathbf{Y}(\tau^\frown\rho)$.

Proof. By Definition 4.2, if $\mathbf{A}$ is any word in $\mathbf{D}$ and $\mathbf{E}$, say

$$\mathbf{A} = \mathbf{X}_0^{n_0} \cdot \ \dots \ \cdot \mathbf{X}_k^{n_k} \ (\mathbf{X}_0, \dots, \mathbf{X}_k \in \{\mathbf{D}, \mathbf{E}\})$$

then

$$\mathbf{A}(\sigma) = \mathbf{X}_0(\sigma)^{n_0} \cdot \ \dots \ \cdot \mathbf{X}_k(\sigma)^{n_k}; \text{ and}$$

$$\mathbf{A}(\tau) = \mathbf{X}_0(\tau)^{n_0} \cdot \ \dots \ \cdot \mathbf{X}_k(\tau)^{n_k}.$$

Hence obviously from $\mathbf{X}(\sigma) * \mathbf{W} \approx \mathbf{X}(\tau)$ $(\mathbf{X} = \mathbf{D}, \mathbf{E})$ one may derive $\mathbf{A}(\sigma) * \mathbf{W} \approx \mathbf{A}(\tau)$. Let us take $\mathbf{A} = \mathbf{Y}(\rho)$. Then $\mathbf{A}(\sigma) = \mathbf{Y}(\sigma^\frown\rho)$ and $\mathbf{A}(\tau) = \mathbf{Y}(\tau^\frown\rho)$, as follows from 4.2; hence we have the result.

We will frequently use in the following derivations various immediate consequences of this lemma, such as for instance the fact that relations B4 imply, for any finite sequence σ, $[[\mathbf{M}, \Delta], \mathbf{X}(\langle 1 \rangle^\frown\sigma)] \approx 1$, or what is equivalent, $\mathbf{X}(\langle 1 \rangle^\frown\sigma) * \mathbf{M}\Delta \approx \mathbf{X}(\langle 1 \rangle^\frown\sigma) * \Delta\mathbf{M}$.

Theorem 5.3. *Let* $\mathbf{X} \in \{\mathbf{D}, \mathbf{E}\}$. *For every triplet of natural numbers* i, j, k *we have*

(i) $\vdash_{\mathfrak{B}} \mathbf{X}(1) * \mathbf{D} \approx \mathbf{X}(1,0)$;

(ii) $\vdash_{\mathfrak{B}} \mathbf{X}(1, i) * \mathbf{M} \approx \mathbf{X}(1, i+1)$;

(iii) $\vdash_{\mathfrak{B}} \mathbf{X}(1, i, j, k) * \mathbf{R} \approx \mathbf{X}(1, j, k, i)$;

(iv) $\vdash_{\mathfrak{B}} \mathbf{X}(1, i) * \mathbf{Q} \approx \mathbf{X}(1, i, 0)$.

Proof. We write $\mathbf{W}_0 \sim \mathbf{W}_1$ to denote that $\vdash_{\mathfrak{B}} \mathbf{W}_0 \approx \mathbf{W}_1$. By 3.1 and 5.1, if $\mathbf{W}_0, \mathbf{W}_1$ contain only the letters $\mathbf{D}$ and $\mathbf{E}$ then $\mathbf{W}_0 \sim \mathbf{W}_1$ iff

$\mathbf{W}_0 \approx \mathbf{W}_1$ holds in $\mathfrak{H}$. Hence (i) and (ii) follow immediately, since $\mathbf{M} \equiv \mathbf{E}^{-1}\mathbf{D}$.

To prove (iii); by B4, B5, B6, for $\mathbf{X} \in \{\mathbf{D}, \mathbf{E}\}$ we have

(a) (1) $\mathbf{X}(1) * \mathbf{M}\Delta \sim \mathbf{X}(1) * \Delta\mathbf{M}$,

(2) $\mathbf{X}(1) * \mathbf{D}\Delta \sim \mathbf{X}(1) * \mathbf{MD}$,

(3) $\mathbf{X}(1) * \mathbf{D}\Gamma \sim \mathbf{X}(1) * \Delta\mathbf{D}$.

By 5.2 this remains true when we replace $\mathbf{X}(1)$ by $\mathbf{X}(\langle 1 \rangle^\frown \sigma)$.

From this we obtain

(b) (1) $\mathbf{X}(1, i, j) * \Delta \sim \mathbf{X}(1, i, j+1)$;

(2) $\mathbf{X}(1, 0, i, j) * \Gamma \sim \mathbf{X}(1, 0, i, j+1)$;

whenever $i, j \in \omega$ *and* $\mathbf{X} \in \{\mathbf{D}, \mathbf{E}\}$.

In fact, for $i = 0$ we have

$$\begin{aligned} \mathbf{X}(1,0,j) * \Delta &\sim \mathbf{X}(1,j) * \mathbf{D}\Delta; \text{ by (i) and Lemma 5.2} \\ &\sim \mathbf{X}(1,j) * \mathbf{MD}; \text{ by (a2) and 5.2} \\ &\sim \mathbf{X}(1,0,j+1); \text{ by (i), (ii) and 5.2.} \end{aligned}$$

If (b1) has been proved for $i = i_0$ then compute

$$\begin{aligned} \mathbf{X}(1, i_0+1, j) * \Delta &\sim \mathbf{X}(1, i_0, j) * \mathbf{M}\Delta; \text{ by (ii) and 5.2} \\ &\sim \mathbf{X}(1, i_0+1, j+1); \text{ by (a1), 5.2 and} \end{aligned}$$

the induction assumption. To get (b2), use (i), (a3) and (b1):

$$\begin{aligned} \mathbf{X}(1,0,i,j) * \Gamma &\sim \mathbf{X}(1,i,j) * \mathbf{D}\Gamma \\ &\sim \mathbf{X}(1,i,j) * \Delta\mathbf{D} \\ &\sim \mathbf{X}(1,0,i,j+1). \end{aligned}$$

Now (b), together with (ii) and 5.2, yields

$$\mathbf{X}(1, i, j, k) \sim \mathbf{X}(1,0,0,0) * \Gamma^k \mathbf{M}^i \Delta^j.$$

If we conjugate both sides by **R**, taking note of B3 and B7, we have that $\Gamma * \mathbf{R} \sim \Delta$, $\mathbf{M} * \mathbf{R} \sim \Gamma$, $\Delta * \mathbf{R} \sim \mathbf{M}$ and

$$\mathbf{X}(1, i, j, k) * \mathbf{R} \sim \mathbf{X}(1,0,0,0) * \Delta^k \Gamma^i \mathbf{M}^j$$
$$\sim \mathbf{X}(1, j, k, i); \text{ by (ii), (b) and 5.2.}$$

To prove (iv); just apply (i) and (iii).

§6. Proof of Theorem 4.5

The characterization of (general) recursive functions which we shall use is taken from J. Robinson [6]: We say that the function $f \in {}^\omega\omega$ is obtained by *general recursion* from $g, h, u, v \in {}^\omega\omega$ if
(i) $fg = u$, $fh = vf$ and
(ii) every natural number n belongs to the range of some one of the functions $h^k g$ $(k \geqslant 0)$.

Theorem 6.1. [6] *The class of (general) recursive functions is the smallest class of numerical functions which is closed under composition and general recursion and contains the zero function* $\theta(\theta(n) = 0)$ *and the successor function* $\pi(\pi(n) = n + 1)$.

The proof of Theorem 4.5 is contained in a series of lemmas, showing that the class of $\mathfrak{H}$-computable functions contains π and θ and is closed under composition and general recursion. We first extend Definition 4.4.

Definition 6.2. *Let* $f \in {}^\omega\omega$ and $n \in \omega$. We say that f is n-computable from a set of relations $\mathcal{R}$ iff the following are derivable from $\mathcal{R}$:

$$\mathbf{X}(1, n, k) * \mathbf{A}_f \approx \mathbf{X}(1, n, k + f(n)) \ (\mathbf{X} \in \{\mathbf{D}, \mathbf{E}\}, k \in \omega).$$

Lemma 6.3. π *and* θ *are* $\mathfrak{H}$*-computable.*

Proof. Referring back to Definition 4.4 we see that θ is computed by the set $\{ \mathbf{A}_\theta \approx \mathbf{1} \}$.

Let $\mathcal{B}_\pi$ be obtained from the set $\mathcal{B}$ (§5) by adding the four relations:

$$\left.\begin{array}{ll} (1) & [\mathbf{D}\mathbf{A}_\pi \mathbf{D}^{-1}\mathbf{M}^{-1}, \mathbf{X}(1)] \\ (2) & [[\mathbf{M}, \mathbf{A}_\pi] \cdot \Delta^{-1}, \mathbf{X}(1)] \end{array}\right\} \quad \mathbf{X} \in \{ \mathbf{D}, \mathbf{E} \}$$

Here, and in the future, we will not bother to show that the relations we write down are true in $\mathfrak{H}$ since these verifications are always routine. Moreover we always write $\sim$ to indicate equality provable from the set of relations being considered – in the present case $\mathcal{B}_\pi$. The applications of Lemma 5.2 will not be explicitly mentioned. We now show by induction on n that π is n-computable from $\mathcal{B}_\pi$ for every n.

For $n = 0$;

$$\begin{aligned} \mathbf{X}(1,0,k) * \mathbf{A}_\pi &\sim \mathbf{X}(1,k) * \mathbf{D}\mathbf{A}_\pi; \text{ by 5.3 (i)} \\ &\sim \mathbf{X}(1,k) * \mathbf{M}\mathbf{D}; \text{ by (1) above} \\ &\sim \mathbf{X}(1,0,k+1); \text{ by 5.3.} \end{aligned}$$

Assuming π is n-*computable from* $\mathcal{B}_\pi$;

$$\begin{aligned} \mathbf{X}(1,n+1,k) * \mathbf{A}_\pi &\sim \mathbf{X}(1,n,k) * \mathbf{M}\mathbf{A}_\pi; \text{ by 5.3} \\ &\sim \mathbf{X}(1,n,k) * \Delta\mathbf{A}_\pi\mathbf{M}; \text{ by (2) above} \\ &\sim \mathbf{X}(1,n+1,k+n+2); \end{aligned}$$

by 5.3 (ii, iii), definition of Δ and the induction assumption.

Thus π is n-computable from $\mathcal{B}_\pi$ for all n; hence it is $\mathfrak{H}$-computable.

Definition 6.4. For any two words $\mathbf{W}_0$ and $\mathbf{W}_1$ we put

$$\mathbf{W}_0 \,\square\, \mathbf{W}_1 \equiv \mathbf{W}_1 * (\mathbf{Q}\mathbf{W}_0\mathbf{R})^{-1}.$$

The two technical lemmas which follow considerably simplify the remaining discussion.

Lemma 6.5. *Let* W *be any word and* $i, j, k, l \in \omega$. *From the six relations (where* $\mathbf{X} \in \{\mathbf{D}, \mathbf{E}\}$*)*

(1) $[[\Gamma, \mathrm{W}], \mathbf{X}(1)]$,

(2) $[[\mathbf{DR}, \mathrm{W}], \mathbf{X}(1)]$,

(3) $\mathbf{X}(1, i, j, p) * \mathrm{W} \approx \mathbf{X}(1, k, l, p)$;

together with $\mathfrak{B}$, *one can derive*

(4) $\mathbf{X}(1, i, j) * \mathrm{W} \approx \mathbf{X}(1, k, l)$ $(\mathbf{X} \in \{\mathbf{D}, \mathbf{E}\})$.

Lemma 6.6. *Let* $\mathrm{W}_0, \mathrm{W}_1$ *be any two words and* $i, j, k, l \in \omega$. *Consider the five pairs of relations (where* $\mathbf{X} \in \{\mathbf{D}, \mathbf{E}\}$*):*

(a) $\mathbf{X}(1, i, 0) * \mathrm{W}_0 \approx \mathbf{X}(1, i, j)$,

(b) $\mathbf{X}(1, j, k) * \mathrm{W}_1 \approx \mathbf{X}(1, j, k+l)$,

(c) $\mathbf{X}(1, i, k) * (\mathrm{W}_0 \,\square\, \mathrm{W}_1) \approx \mathbf{X}(1, i, k+l)$,

(d) $[[\Gamma, \mathrm{W}_1], \mathbf{X}(1)]$,

(e) $[[\mathbf{DR}, \mathrm{W}_1], \mathbf{X}(1)]$.

In the presence of $\mathfrak{B}$ *one can derive I:* (c) *from* (a) *and* (b)*; II:* (b) *from* (a), (c), (d) *and* (e).

Proof of 6.5. We calculate

$$\begin{aligned} \mathbf{X}(1, i, j) * \mathbf{W}\mathbf{DR}\Gamma^p &\sim \mathbf{X}(1, i, j) * \mathbf{DR}\mathrm{W}\Gamma^p; \text{ 6.5 (2)} \\ &\sim \mathbf{X}(1, i, j, 0) * \mathrm{W}\Gamma^p; \text{ 5.3} \\ &\sim \mathbf{X}(1, i, j, 0) * \Gamma\mathrm{W}\Gamma^{p-1}; \text{ 6.5 (1)} \end{aligned}$$

$$\sim \mathbf{X}(1, i, j, 1) * \mathbf{W}\Gamma^{p-1}; \text{ 5.3 and def.of } \Gamma$$

$$\cdot \quad \cdot$$

$$\cdot \quad \cdot$$

$$\sim \mathbf{X}(1, i, j, p) * \mathbf{W};$$

$$\sim \mathbf{X}(1, k, l, p); \text{ 6.5 (3)}$$

$$\sim \mathbf{X}(1, k, l) * \mathbf{DR}\Gamma^{p}; \text{ 5.3};$$

and then we conjugate by $(\mathbf{DR}\Gamma^p)^{-1}$ to obtain 6.5 (4).

Proof of 6.6. The derivation of (c) from (a), (b) and $\mathfrak{B}$ is a very interesting but perfectly straightforward computation (using 6.4 and 5.3) which we leave to the reader.

To get (b) from (a), (c) – (e) and $\mathfrak{B}$; first transform (c) – using (a) and 5.3 – and conjugate both sides by $\mathbf{QW}_0\mathbf{R}$ to obtain the relations: $\mathbf{X}(1, j, k, i) * \mathbf{W}_1 \approx \mathbf{X}(1, j, k+l, i)$. Then apply Lemma 6.5 taking $\mathbf{W} = \mathbf{W}_1$.

Lemma 6.7. *If g and h are* $\mathfrak{H}$*-computable then gh is* $\mathfrak{H}$*-computable.*

Proof. Let $\mathfrak{B}_g$ and $\mathfrak{B}_h$ be finite sets of relations which compute g and h. Define $\mathfrak{B}_{gh}$ to be $\mathfrak{B} \cup \mathfrak{B}_g \cup \mathfrak{B}_h$ together with the two relations $[(\mathbf{A}_h \square \mathbf{A}_g) \cdot \mathbf{A}_{gh}^{-1}, \mathbf{X}(1)]$ where $\mathbf{X} = \mathbf{D}, \mathbf{E}$. Then apply Lemma 6.6.I.

Lemma 6.8. *If f is obtained by general recursion from* $\mathfrak{H}$*-computable functions g, h, u and v, then f is* $\mathfrak{H}$*-computable.*

Proof. To $\mathfrak{B} \cup \mathfrak{B}_g \cup \mathfrak{B}_h \cup \mathfrak{B}_u \cup \mathfrak{B}_v$ we add the following relations to form a set $\mathfrak{B}_f$ (where $\mathbf{X} \in \{\mathbf{D}, \mathbf{E}\}$):

(1) $[(\mathbf{A}_g \square \mathbf{A}_f) \cdot \mathbf{A}_u^{-1}, \mathbf{X}(1)]$,

(2) $[(\mathbf{A}_h \square \mathbf{A}_f) \cdot (\mathbf{A}_f \square \mathbf{A}_v)^{-1}, \mathbf{X}(1)]$,

(3) $[[\Gamma, \mathbf{A}_f], \mathbf{X}(1)]$,

(4) $[[\mathbf{DR}, \mathbf{A}_f], \mathbf{X}(1)]$.

We are assuming that $\mathcal{B}_g$, $\mathcal{B}_h$, ... compute $g, h, \ldots$, repectively; hence $\mathcal{B}_f$ computes each of g, h, u, v. By the definition of general recursion, $fg = u$ and $fh = vf$. To show that $\mathcal{B}_f$ computes f it will be enough to show, for each $p \geqslant 0$ and for each $m \in \operatorname{rng} h^p g$, that f is m-computable from $\mathcal{B}_f$. This we do by induction on p.

For $p = 0$; by (1) we have, for $n, k \in \omega$:

$$\mathbf{X}(1, n, k) * \mathbf{A}_g \square \mathbf{A}_f \sim \mathbf{X}(1, n, k + u(n));$$

also

$$\mathbf{X}(1, n, 0) * \mathbf{A}_g \sim \mathbf{X}(1, n, g(n)).$$

Hence by 6.6.II, (3) and (4) we obtain

$$\mathbf{X}(1, g(n), k) * \mathbf{A}_f \sim \mathbf{X}(1, g(n), k + u(n))$$

which is the desired result since $fg(n) = u(n)$.

The induction step; assuming f is m-computable from $\mathcal{B}_f$ for every $m \in \operatorname{rng} h^p g$: consider $m \in \operatorname{rng} h^{p+1} g$, say $m = h(n)$ where $n \in \operatorname{rng} h^p g$. By (2) we have, for any $k \in \omega$:

$$\begin{aligned} \mathbf{X}(1, n, k) * \mathbf{A}_h \square \mathbf{A}_f &\sim \mathbf{X}(1, n, k) * \mathbf{A}_f \square \mathbf{A}_v \\ &\sim \mathbf{X}(1, n, k + vf(n)) \end{aligned}$$

by 6.6.I and the induction assumption;

$$\sim \mathbf{X}(1, n, k + f(m))$$

since $vf = fh$. Since we also have $\mathbf{X}(1, n, 0) * \mathbf{A}_h \sim \mathbf{X}(1, n, m)$ the desired result, $\mathbf{X}(1, m, k) * \mathbf{A}_f \sim \mathbf{X}(1, m, k + f(m))$, follows from (3), (4) and 6.6.II.

§7. The Novikov-Boone theorem

Theorem 7.1. ([1], [5]) *There exists a finitely presented group whose word problem is not recursively solvable.*

Proof. Let f be a recursive function whose range is a non-recursive set. By Theorem 4.5 there exists a finite set of relations $\mathcal{B}_f$ which computes f as per Definition 4.4. We can assume that $\mathcal{B} \subset \mathcal{B}_f$.

Let h be the function such that $h(n) = 0$ if $n \in \operatorname{rng} f$ and $h(n) = 1$ if $n \notin \operatorname{rng} f$. Then the following relations are true in $\mathfrak{H}$:

$$[\mathbf{A}_f \,\square\, \mathbf{A}_h, \mathbf{X}(1)] \approx 1,$$

$$[[\Gamma, \mathbf{A}_h], \mathbf{X}(1)] \approx 1,$$

$$[[\mathbf{DR}, \mathbf{A}_h], \mathbf{X}(1)] \approx 1;$$

in which $\mathbf{X} = \mathbf{D}$ or $\mathbf{E}$ and $\mathbf{A}_f \,\square\, \mathbf{A}_h$ was defined in 6.4. Let $\mathcal{U}$ be the set of relations $\mathcal{B}_f$ together with the above relations.

Now it follows from Lemma 6.6.II that

$$\vdash_{\mathcal{U}} \mathbf{E}(1, n, 0) * \mathbf{A}_h \approx \mathbf{E}(1, n, 0), \text{ if } n \in \operatorname{rng} f.$$

On the other hand, if $n \notin \operatorname{rng} f$ then the above relation is not true in $\mathfrak{H}$, whence certainly

$$\nvdash_{\mathcal{U}} \mathbf{E}(1, n, 0) * \mathbf{A}_h \approx \mathbf{E}(1, n, 0), \text{ if } n \notin \operatorname{rng} f.$$

Combining these two results, we get that the relations in the three letters $\mathbf{D}, \mathbf{E}$ and $\mathbf{A}_h$ which are derivable from $\mathcal{U}$ are non-recursive. Hence the finitely presented group $(L_0; \mathcal{U})$, where L_0 consists of all the letters appearing in $\mathcal{U}$, has an unsolvable word problem.

§8. Concluding remarks

First, we should mention that by using Britton's Lemma (see [2]) and suitably modifying the finite presentations given in this paper the second author has succeeded in giving a proof for the existence of finitely presented groups with word problems of an arbitrary recursively enumerable truth-table degree. This proof will be published elsewhere by Thompson.

Second, we remark that our definition of recursive function

could have been taken from R.M. Robinson [7], instead of [6]. The proof, in that case, would have been simpler in some respects, since A_{f+g} and $A_f A_g$ have the same effect on the interval $I_{\langle 1 \rangle}$; but it would have required us to prove a variant of [7; Theorem 3, p. 940].

Finally, we should like to recount a short history of our proof which will introduce some interesting related notions.

We have dealt in this paper with a group $\mathfrak{C}$, whose members are homeomorphisms of the topological space ${}^{\omega}\omega$, provided with the product topology (so that it is homeomorphic to the space of irrational numbers). Let us consider now the Cantor space ${}^{\omega}2$,, whose members are the infinite 01-sequences. We define ${}^{\underline{\omega}}2$ to be the set of all finite 01-sequences. Our Definition 0.1 is now transformed into the definition of a group $\mathfrak{C}'$ of homeomorphisms of ${}^{\omega}2$. Theorem 0.2 is still true in this context but since ${}^{\omega}2$ is compact every partition set $\subseteq {}^{\underline{\omega}}2$ is finite. Therefore $\mathfrak{C}'$ is a denumerable group.

Actually $\mathfrak{C}'$ is a finitely generated, algebraically simple group which has a finite presentation (as Thompson will show in a forthcoming paper). The group $\mathfrak{P}$ which plays a leading role in the present paper is isomorphic to the subgroup of $\mathfrak{C}'$ composed of the functions which preserve the natural ordering of ${}^{\omega}2$ (see below).

Thompson discovered the groups $\mathfrak{C}'$ and $\mathfrak{P}'$ in connection with his research in logic about 1965. He studied them thoroughly and obtained results which led eventually to our joint work on the word problem. Among his discoveries at that time were the arrow notation, equivalents of Theorems 3.1 and 4.3, and a finite presentation for $\mathfrak{C}'$. The key observation which triggered our interest in the word problem was made by him: viz., that arbitrary numerical functions f can be coded as permutations B_f of ${}^{\omega}2$ (in various ways) so that f is uniquely determined by the relations between B_f and the elements of $\mathfrak{C}'$. Working together the two of us completed, in December 1967, a proof rather similar to the one given in this paper, which referred however to the Cantor space. Later, McKenzie observed that a much neater argument is obtained by translating into ${}^{\omega}\omega$.

The translation which was used is the following: Let $\Lambda \subset {}^{\omega}2$ be the set of 01-sequences which assume the value 0 infinitely many times. Λ is homeomorphic to ${}^{\omega}\omega$ under the map η defined by

$$\langle n_0, n_1, \ldots \rangle \eta = \langle 1 \rangle^{(n_0)\frown}\langle 0 \rangle^{\frown}\langle 1 \rangle^{(n_1)\frown}\langle 0 \rangle^{\frown}\ldots .$$

The set Λ is invariant under $\mathfrak{C}'$ and only the identity element of $\mathfrak{C}'$ acts trivially on Λ. Hence (one sees that) the map $F \to \eta F \eta^{-1}$ embeds $\mathfrak{C}'$ into $\mathfrak{C}$. Under this map the two functions $E': \langle 0 \rangle \to \langle 0,0 \rangle, \langle 1,0 \rangle \to \langle 0,1 \rangle, \langle 1,1 \rangle \to \langle 1 \rangle$; and $D': \langle 0 \rangle \to \langle 0 \rangle$, $\langle 1,0 \rangle \to \langle 1,0,0 \rangle, \langle 1,1,0 \rangle \to \langle 1,0,1 \rangle, \langle 1,1,1 \rangle \to \langle 1,1 \rangle$ are taken onto the E and D defined in §2 which generate $\mathfrak{P}$.

In order to see that E' and D' generate the subgroup of order-isomorphisms of $\mathfrak{C}'$ we shall now introduce Thompson's original arrow notation. (Which, incidentally, should give an idea of the logical origin of the group $\mathfrak{C}'$.) First we define an operation on finite subsets of ${}^{\underset{\smile}{\omega}}2$. Given Σ_0 and Σ_1, finite subsets of ${}^{\underset{\smile}{\omega}}2$, put

$$\Sigma_0 + \Sigma_1 = \{\langle 0 \rangle^{\frown}\sigma \mid \sigma \in \Sigma_0\} \cup \{\langle 1 \rangle^{\frown}\tau \mid \tau \in \Sigma_1\}.$$

Clearly the set of all partition sets for ${}^{\omega}2$ is generated by applying this operation repeatedly, beginning with the partition set $\{\emptyset\}$ ($\emptyset$ is the empty sequence).

We now introduce a free groupoid generated by an infinite set of symbols $X_0, X_1, X_2, \ldots$. (The word "groupoid" is taken here to mean an arbitrary algebraic system whose only fundamental operation is binary.) The elements of the groupoid we call *exps*, and we denote them by juxtaposition and the use of parentheses ($X_0, X_2X_1, X_0(X_1X_2)$, etc.). With each exp, we correlate the partition set Σ_s which it represents: (i) $X_0, X_1, \ldots$ all represent $\{\emptyset\}$; (ii) the product of two exps s and t (in the groupoid) represents $\Sigma_s + \Sigma_t$. For two exps s and t we now have $\Sigma_s = \Sigma_t$ iff s can be obtained from t by replacing some of the generating symbols in t, at some of their occurrences, by other symbols. Thus we need consider only reduced exps, in which no X_i occurs more than once. For each reduced exp s there is a canonical one-one mapping

g_s from Σ_s onto the set of symbols in s: if $\Sigma_s = \{\sigma_0, ..., \sigma_n\}$ where, lexicographically, $\sigma_0 < \sigma_1 < ... < \sigma_n$, then s has length $n + 1$; upon removing parentheses s looks like $X_{i_0} ... X_{i_n}$ – we define $g_s(\sigma_k) = X_{i_k}$. (For instance, if $s = X_0(X_1X_2)$ then $\Sigma_s = \{\langle 0\rangle, \langle 1,0\rangle, \langle 1,1\rangle\}$ and $g_s(\langle 0\rangle) = X_0$, $g_s(\langle 1,0\rangle) = X_1$, $g_s(\langle 1,1\rangle) = X_2$.)

Finally, take any expression $s \to t$ where s and t are reduced exps having the same symbols X_i occurring in them. We then consider the formal expression $s \to t$ to represent the element of $\mathfrak{C}'$ defined by the partition sets Σ_s and Σ_t, and the one-one map $g_t^{-1}g_s$ from Σ_s onto Σ_t. With this convention, we see that the function $E' \in \mathfrak{C}'$ is represented by the "associative law":

$$X_0(X_1X_2) \to (X_0X_1)X_2;$$

and that the function D' is represented by the associative law "deferred to the right":

$$X_0(X_1(X_2X_3)) \to X_0((X_1X_2)X_3).$$

The inverse of $s \to t$ is $t \to s$, so we find that $E'^{-1}D'$ is represented by:

$$(X_0X_1)\,(X_2X_3) \to X_0((X_1X_2)X_3).$$

The representation defined above is a many-one map from the set of expressions $\{s \to t\}$ onto $\mathfrak{C}'$. It is very useful in the study of the group. The order-preserving members of $\mathfrak{C}'$ are obviously those which simply rearrange parentheses, and it is intuitively quite clear that they are generated by the two simple rearrangements E' and D'.

References

[1] W.W. Boone, The word problem, Ann. of Math. (2) 70 (1959) 207–265.
[2] J.L. Britton, The word problem, Ann. of Math. (2) 77 (1963) 16–32.
[3] G. Higman, Subgroups of finitely presented groups, Proc. Roy. Soc. London, Ser. A, 262 (1961) 455–475.

[4] W. Magnus, A. Karras and D. Solitar, Combinatorial Group Theory (John Wiley, New York, 1966).
[5] P.S. Novikov, On the algorithmic unsolvability of the word problem in group theory, Trudy Mat. Inst. im. Steklov. No. 44, Izdat. Akad. Nauk SSSR, Moscow, 1955 [Russian].
[6] J. Robinson, Recursive functions of one variable, Proc. Amer. Math. Soc. 19 (1968) 815–820.
[7] R. M. Robinson, Primitive recursive functions, Bull. Amer. Math. Soc. 53 (1947) 925–942.

A NON-ENUMERABILITY THEOREM FOR INFINITE CLASSES OF FINITE STRUCTURES*

T.G. McLAUGHLIN
University of Illinois, Urbana

Let $\mathfrak{A}$ be a fixed, finitely axiomatized class of algebras with finitely many operations each of finite arity; e.g., let $\mathfrak{A}$ = the class of all *groups*. (Axiomatizability is understood to mean axiomatizability *in the lower predicate calculus with equality*.) Let $\mathfrak{P}_{\mathfrak{A}}$ denote the set of all finite presentations, relative to some fixed presentational alphabet A, for members of $\mathfrak{A}$; let $\mathfrak{F}_{\mathfrak{A}}$ denote the class of all *finite* members of $\mathfrak{A}$; and let $\mathfrak{P}_{\mathfrak{A}}^*$ denote the class of all presentations in $\mathfrak{P}_{\mathfrak{A}}$ which determine elements of $\mathfrak{F}_{\mathfrak{A}}$. We assume a fixed Gödel numbering g of the elements of $\mathfrak{P}_{\mathfrak{A}}$; g is understood to possess all of the usual effectivity and uniqueness properties demanded of such codings in recursion-theoretic contexts. The following easy lemma is used several times in our proof of Theorem 2.

Lemma 1. *The isomorphism problem is solvable uniformly for finite members of* $\mathfrak{A}$, *in the sense that there exists a uniform effective decision procedure (i.e., a partial recursive function) for determining from any pair* $\langle g(P_1), g(P_2)\rangle$ *such that* P_1 *and* P_2 *both belong to* $\mathfrak{P}_{\mathfrak{A}}^*$ *whether or not* P_1 *and* P_2 *present isomorphic structures.*

We impose upon $\mathfrak{A}$ the additional requirement that for all natural numbers k there exist members of $\mathfrak{F}_{\mathfrak{A}}$ having cardinality $\geqslant k$.

* The present paper summarizes the main features of some material which will appear in detailed form elsewhere. Research supported by the U.S. National Science Foundation under contract No. GP-7421.

Rogers' book [4] is our source for notation and terminology in connection with the "arithmetical hierarchy" of Kleene and Mostowski ([2, Chapter XI]). Thus, we denote by Σ_k (by Π_k) the class of all sets of natural numbers definable by a k-quantifier prenex arithmetical predicate having initial quantifier $\exists$(having initial quantifier $\forall$); and by a *complete* set in Σ_k (in Π_k) we mean a set which belongs to Σ_k (to Π_k) and has the property that any other set in Σ_k (in Π_k) is recursively 1-1 reducible to it.

Let Inf($\mathfrak{P}_{\mathfrak{A}}$) denote the family of all strongly infinite, recursively enumerable classes of elements of $\mathfrak{P}_{\mathfrak{A}}{}^*$; by a *strongly infinite* class of presentations we mean one whose associated class of presented structures contains representatives of infinitely many different isomorphism types. Let *Inf*($\mathfrak{P}_{\mathfrak{A}}{}^*$) denote the family of all classes $\mathfrak{C}$ such that $\mathfrak{C}$ is the class of structures presented by (the elements of) some class $I \in$ *Inf* ($\mathfrak{P}_{\mathfrak{A}}{}^*$). By an *index* of a member of *Inf* ($\mathfrak{P}_{\mathfrak{A}}^*$) we mean a Gödel number of a class $I \in$ *Inf*($\mathfrak{P}_{\mathfrak{A}}{}^*$); here our Gödel numbering of *classes* of presentations is one which is induced (in any of a variety of natural ways) by the numbering g.

Theorem 2. (a) *The set α of all indices of members of Inf($\mathfrak{P}_{\mathfrak{A}}^*$) is a complete set in Π_2.*

(b) *Inf*($\mathfrak{P}_{\mathfrak{A}}{}^*$) *is not "$\Sigma_2$-enumerable"; i.e., there is no Σ_2 set $\beta, \beta \subset \alpha$, such that* $(\forall \mathfrak{C})$ [$\mathfrak{C} \in$ *Inf*($\mathfrak{P}_{\mathfrak{A}}{}^*$) $\langle == \rangle$ $(\exists n)$ [$n \in \beta$ & n is an index of $\mathfrak{C}$]] .

Actually, we prove a theorem which is more "effective" than Theorem 2(b) as stated above, in that we establish a *productivity* property relative to Σ_2-enumerable subfamilies of *Inf*($\mathfrak{P}_{\mathfrak{A}}^*$). (Investigations of productivity properties relating to classes and families of classes have previously been carried out in [1] and [3].) The statement of the sharper form of the theorem, however, would involve technical terminology too cumbersome for the present note.

We mention some consequences of Theorem 2.

Corollary 3. *Let a finite presentation of a group be termed* strongly finite *in case the group presented by it is finite. Then: the* *indices for the various strongly infinite r.e. classes of strongly finite group presentations form a complete* Π_2 *set, no* Σ_2 *subset of which suffices to enumerate the family of all infinite r.e. classes of finite groups.*

Corollary 4. *The collection* α_0 *of all indices of infinite recursively enumerable sets of natural numbers is complete in* Π_2, *and* α_0 *has no* Σ_2 *subset sufficient to enumerate the class of all such r.e. sets. (cf.* [1, Theorems 3.3 and 6.4]; *using the sharper form of* Theorem 2, *we actually obtain productivity relative to* Σ_2*-enumerable subclasses.)*.

From Corollary 4 and an observation in [3], we can further show:

Corollary 5. *The collection* β_0 *of all indices of recursively enumerable classes of infinite r.e. sets is complete in* Π_2, *and* β_0 *has no* Σ_2 *subset sufficient to enumerate the family of all such classes.*

We would like to have some information on the "levels of enumerability" of families of not necessarily strongly infinite classes of finitely presented structures in $\mathfrak{F}_{\mathfrak{A}}$, for various $\mathfrak{A}$; and it can be demonstrated easily that Theorem 2 is not generally true if the strong infiniteness condition is dropped.

References

[1] J.C.E. Dekker and J. Myhill, Some theorems on classes of recursively enumerable sets, Trans. Amer. Math. Soc. 89 (1958) 25-59.

[2] S.C. Kleene, Introduction to Metamathematics (D. Van Nostrand, Princeton, 1952).

[3] T.G. McLaughlin, The family of all recursively enumerable classes of finite sets, to appear.

[4] H. Rogers, Theory of recursive functions and effective computability (McGraw-Hill, New York, 1967).

SOME CONNECTIONS BETWEEN HILBERT'S 10th PROBLEM AND THE THEORY OF GROUPS

Charles F. MILLER III*
The Institute for Advanced Study, Princeton

The tenth problem listed by Hilbert [9] in his famous address of 1900 is the following decision problem: to determine of an arbitrary polynomial equation $P(x_1, ..., x_n) = 0$, with integer coefficients, whether or not it has a solution in integers. Whether or not this problem is recursively solvable is as yet unknown.** The purpose of this article is to examine some decision problems in the theory of groups which are equivalent*** to Hilbert's 10th problem, or which, if they were unsolvable, would imply the unsolvability of Hilbert's 10th.

Although there is essentially no literature on the connections between these group-theoretic decision problems and Hilbert's 10th, some of this material seems to have been known to several people. Consequently, this article might best be regarded as a survey of the possibilities of settling Hilbert's problem via group-theoretic problems. An effort has been made to supply sufficient detail for this work to be accessible to both logicians and group theorists.

* Work supported by U.S. National Science Foundation under contract No. GP-7421.

** This article was written prior to the publication of Matejasevich's remarkable theorem [23].

*** Equivalent and reducible in this article generally mean Turing equivalent and Turing reducible as decision problems.

§ 1. The logical setting

Interesting work on Hilbert's 10th is discussed in Davis [7] and J. Robinson [16]. In contrast to their approach, we will pursue a development more oriented towards logic and model theory, and some familiarity with these subjects is assumed. First, we recall some definitions from universal algebra (see Cohn [5]):

Let $N = \{0, 1, 2, \ldots\}$ denote the natural numbers. An *operator domain* is a set Ω together with a map $a : \Omega \to N$. If $\sigma \in \Omega$ and $n = a(\sigma)$, then σ is called an n-ary operator. If A is a set, an Ω-algebra structure on A is an assignment to each $\sigma \in \Omega$ of a function from the cartesian product A^n to A where $n = a(\sigma)$. The set A with this structure is called an Ω-algebra. Note that if σ is a 0-ary operator, then σ is just a distinguished element (constant) of A.

For example, the ring of integers **Z** is an Ω-algebra where Ω consists of two 0-ary operators (for 0 and 1) and two binary operators (for + and $\cdot$). Also, a group G is an Ω'-algebra where Ω' consists of a 0-ary operator (for 1), a unary operator (for inverse) and a binary operator (for group multiplication).

The first order language associated with an operator domain Ω, written $L(\Omega)$, has as symbols the following:

(1) a (countable) set of variables $x, y, z, \ldots$;

(2) an n-ary function symbol for each n-ary operator of Ω and the symbol =;

(3) the usual logical symbols: $\neg, \vee, \wedge, \exists, \forall, (,)$.

A *term* of $L(\Omega)$ is defined inductively by

(i) a variable or a 0-ary function symbol (i.e. a constant) is a term

(ii) if $t_1, \ldots, t_n$ are terms and f is an n-ary function symbol, then $f(t_1, \ldots, t_n)$ is a term.

An *atomic formula* is an expression of the form $t_1 = t_2$ where t_1 and t_2 are terms.

Formulas are then defined inductively using atomic formulas and the logical symbols. An Ω-algebra A is just an $L(\Omega)$-structure in the usual terminology of logic. If B is a formula of $L(\Omega)$, then that B is *valid* (respectively *satisfiable*) in A is defined as usual. Intuitively, a formula B is valid in A if B is true for all possible interpretations in A of the free variables of B. The formula B is satisfiable in A if B is true for at least one interpretation in A of the free variables of B. See Shoenfield [20], Church [4], or Stoll [21] for details.

Let **A** be a class of Ω-algebras. By the *elementary theory* of **A** we mean the first order theory with language $L(\Omega)$ whose non-logical axioms are exactly those formulas of $L(\Omega)$ which are valid in every $A \in \mathbf{A}$. The elementary theory of **A** will be denoted by $\mathrm{Th}_\Omega(\mathbf{A})$ or simply Th(**A**).

If B is a formula, $\forall B$ will denote the universal closure of B. Recall that, in Th(**A**), B is a theorem if and only if $\forall B$ is a theorem. Also, B is a theorem if and only if $\neg, B$ is not satisfiable in any $A \in \mathbf{A}$. Moreover, B is satisfiable in some $A \in \mathbf{A}$ if and only if $(\exists x)B$ is satisfiable in some $A \in \mathbf{A}$.

An *open formula* is a formula which contains no quantifiers. A formula of the form $E_1 \vee E_2 \vee \ldots \vee E_k$ where each E_i is either an atomic formula or the negation of an atomic formula will be called a *basic formula*. The disjunctive normal form theorem implies that if B is an open formula then B is logically equivalent to a disjunction $D_1 \vee D_2 \vee \ldots \vee D_m$ where each D_i is a basic formula. The D_i can be effectively found from B and will be called the *disjunctive components* of B. Clearly, B is satisfiable in some $A \in \mathbf{A}$ if and only if at least one of the disjunctive components of B is satisfiable in some $A \in \mathbf{A}$.

We are particularly interested in the following three decision problems concerning Th(**A**):

DPOF(**A**) - the *decision problem for open formulas* of Th(**A**), i.e., to determine of an arbitrary open formula B whether or not B is a theorem of Th(**A**).

SPOF(**A**) - the *satisfaction problem for open formulas* of Th(**A**), i.e. to determine of an arbitrary open formula B whether or not B is satisfiable in some $A \in \mathbf{A}$.

SPBF(**A**) - the *satisfaction problem for basic formulas* of Th(**A**), i.e. to determine of an arbitrary basic formula B whether or not B is satisfiable in some $A \in \mathbf{A}$.

The following lemma is an immediate consequence of the above discussion:

Lemma 1. DPOF(A) *is equivalent to* SPOF(A) *and to* SPBF(A). *Moreover, the set of satisfiable open formulas is recursively enumerable if and only if the set of satisfiable basic formulas is recursively enumerable.*

§2. Regarding Hilbert's 10th

Recall that **Z** denotes the ring of integers. We now specialize the above discussion to **A** = { **Z** }, and consider Th(**Z**). An atomic formula of Th(**Z**) is simply an equation of the form $P_1 = P_2$ where P_1 and P_2 are polynomials with positive coefficients. Now $P_1 = P_2$ is (algebraically) equivalent to the polynomial equation $(P_1 - P_2) = 0$. Conversely, a polynomial equation $P = 0$ is (algebraically) equivalent to an atomic formula obtained by separating the parts of P occurring with positive and negative coefficients. Indeed, we could suppose that **Z** has an additional binary operation "-" for minus. Consequently, for convenience it will be assumed that every atomic formula of Th(**Z**) has the form $P = 0$ where P is a polynomial (allowing minus signs); the negation of an atomic formula will be written as $P \neq 0$.

Lemma 2. *Hilbert's 10th problem is equivalent to* SPBF(**Z**) *and hence to* DPOF(**Z**) *and to* SPOF(**Z**).

Proof. Since $P = 0$ is a basic formula, it is clear that Hilbert's 10th problem is reducible to SPBF(**Z**). An arbitrary basic formula D of Th(**Z**) has the form

$$P_1 = 0 \wedge P_2 = 0 \wedge \ldots \wedge P_k = 0 \wedge Q_1 \neq 0 \wedge \ldots \wedge Q_m \neq 0.$$

Let u, v, x, y be variables not occurring in any of the P_i or Q_j. In view of Lagrange's theorem that every non-negative integer is a sum of four squares, it follows that D is satisfiable (in **Z**) if and only if the formula D_1 having form

$$P_1 = 0 \wedge \dots \wedge P_k = 0$$
$$\wedge (Q_1)^2 - (1 + u^2 + v^2 + x^2 + y^2) = 0$$
$$\wedge Q_2 \neq 0 \wedge \dots \wedge Q_m \neq 0$$

is satisfiable (in **Z**). Continuing in this way, one obtains a basic formula D^* having the form

$$P_1 = 0 \wedge \dots \wedge P_k = 0 \wedge P_{k+1} = 0 \wedge \dots \wedge P_{k+m} = 0$$

such that D is satisfiable if and only if D^* is satisfiable. Now D^* is satisfiable if and only if the single equation E given by

$$(P_1)^2 + \dots + (P_{k+m})^2 = 0$$

is satisfiable. Note that E can be effectively found from D. Hence SPBF(**Z**) is reducible to Hilbert's 10th. This completes the proof.

Lemma 3. *The set of open formulas of* Th(**Z**) *which are satisfiable in* **Z** *is recursively enumerable.*

Proof. Let B be any open formula. Then B is logically equivalent to $D_1 \vee \dots \vee D_n$ where the D_i are basic formulas. As in the proof of Lemma 2, D_i is satisfiable if and only if an associated polynomial equation $P_i = 0$ is satisfiable. But then the polynomial equation

$$P(B) = P_1 \cdot P_2 \cdot \dots \cdot P_n = 0$$

is satisfiable if and only if B is satisfiable. Further, $P(B)$ can be effectively found from B. Now the set of polynomials Q such that $Q = 0$ has a solution in **Z** is easily seen to be recursively enumerable. Hence the set of satisfiable open formulas is recursively enumerable.

Let $R(x_1, \dots, x_n)$ be a predicate of n variables concerning **Z**. Then R is called *polynomial* if there is a polynomial $P(x_1, \dots, x_n)$ such that

$$R(x_1, ..., x_n) \Leftrightarrow P(x_1, ..., x_n) = 0$$

R is called *diophantine* (or existentially definable) if there is a polynomial $P(x_1, ..., x_n, y_1, ..., y_k)$ such that

$$R(x_1, ..., x_n) \Leftrightarrow (\exists y_1) ... (\exists y_k) P(x_1, ..., x_n, y_1, ..., y_k) = 0.$$

Proofs of the following facts can be found in Davis [7]:
(1) If R_1, R_2 are polynomial (resp. diophantine) then $R_1 \wedge R_2$ and $R_1 \vee R_2$ are polynomial (resp. diophantine).
(2) If R is diophantine, then $(\exists u)R$ is diophantine.
(3) Let m, n be fixed integers and let x, y be variables. Then the following predicates are diophantine:

(i) $x = y$	(v) $x \equiv y \bmod n$
(ii) $x \neq y$	(vi) $x \equiv m \bmod n$
(iii) $x = n$	(vii) x is even
(iv) $x \neq n$	(viii) x is odd.

(4) Every diophantine predicate R is recursively enumerable, i.e. the set of all n-tuples $(x_1, ..., x_n)$ such that $R(x_1, ..., x_n)$ is recursively enumerable.
(5) If there exists a diophantine predicate R which is not recursive, then Hilbert's 10th problem is unsolvable.

§3. Diophantine matrix groups

GL(n, **Z**) will denote the group of $n \times n$ matrices with coefficients in **Z** having determinant ± 1. SL(n, **Z**) denotes the subgroup of GL(n, **Z**) consisting of those matrices with determinant + 1. (SL(n, **Z**) has index 2 in GL(n, **Z**) – it is the kernel of the determinant map.)

Let G be an abstract group. G will be called a *diophantine group* if there exist an integer $n > 0$ and a faithful representation $\varphi : G \to \mathrm{GL}(n, \mathbf{Z})$ and a diopahntine predicate $R(x_{11}, ..., x_{nn})$ of n^2 variables such that:

$$R(x_{11}, ..., x_{nn}) \Leftrightarrow \begin{pmatrix} x_{11} & \cdots & x_{1n} \\ \cdot & & \cdot \\ \cdot & & \cdot \\ \cdot & & \cdot \\ x_{n1} & \cdots & x_{nn} \end{pmatrix} \in \varphi(G).$$

That is, the conditions which an arbitrary matrix with n^2 variable entries must satisfy in order to belong to $\varphi(G)$ are expressible as a diophantine predicate.

Example 1 (Sanov [17]). Let F_2 be the free group of rank 2. Then F_2 is isomorphic to the subgroup $\widetilde{F}_2$ of SL(2, **Z**) generated by the matrices

$$\begin{pmatrix} 1 & 2 \\ 0 & 1 \end{pmatrix}, \begin{pmatrix} 1 & 0 \\ 2 & 1 \end{pmatrix}.$$

Let

$$M = \begin{pmatrix} u & v \\ x & y \end{pmatrix}$$

be an arbitrary 2×2 matrix. Then $M \in \widetilde{F}_2$ if and only if the following conditions are satisfied:

(1) $uy - xv = 1$

(2) u and y are congruent to 1 mod 4

(3) x and v are even.

By our previous discussion (section 2), it follows that F_2 is a diophantine group.

Example 2. Let H be the free nilpotent group of class 2 and rank 2. Then H is isomorphic to the subgroup $\widetilde{H}$ of SL(3, **Z**) consisting of all matrices of the form

$$\begin{pmatrix} 1 & x & y \\ 0 & 1 & z \\ 0 & 0 & 1 \end{pmatrix}.$$

Again it is easy to see that this characterization implies that H is a diophantine group.

Example 3. GL(n, **Z**) and SL(n, **Z**) are diophantine groups. The conditions here are just the determinant conditions. Many arithmetic groups are diophantine (see Borel and Harish-Chandra [3]).

Remark. It is easy to show that the direct product of two diophantine groups is again a diophantine group.

Let G be an arbitrary group and let $\Delta = \{g_1, ..., g_k\}$ be a finite (possibly empty) set of fixed elements of G. The pair $\langle G, \Delta \rangle$ will be called a *group with distinguished elements.* $\langle G, \Delta \rangle$ can be regarded as an Ω'-algebra where Ω' is formed from the usual operator domain Ω for groups by adding new 0-ary operators in one-one correspondence with Δ. Formulas of $L(\Omega')$ then contain constants corresponding to elements of Δ and a formula B is valid in $\langle G, \Delta \rangle$ if and only if B is valid in G when the g_i are substituted for the corresponding constants (and similarly for satisfiable).

Theorem 1. *Let $\langle G, \Delta \rangle$ be a group with distinguished elements and assume that G is a diophantine group. Then* SPBF($\langle G, \Delta \rangle$) *is reducible to Hilbert's 10th problem. Hence,* DPOF($\langle G, \Delta \rangle$) *and* SPOF($\langle G, \Delta \rangle$) *are reducible to Hilbert's 10th problem. Moreover, the set of open formulas of* Th($\langle G, \Delta \rangle$) *which are satisfiable in $\langle G, \Delta \rangle$ is recursively enumerable.*

Remark. Theorem 2 will provide a converse to this result. Namely, there is a diophantine group H with a single distinguished element c such that Hilbert's 10th problem is reducible to SPBF($\langle H, \{c\} \rangle$). Indeed, H is just the free nilpotent group of class 2 and rank 2.

Proof. Theorem 1 is an immediate consequence of Lemmas 1, 2 and 3 together with the following:

Lemma 4. *Let $\langle G, \Delta \rangle$ be a group with distinguished elements and assume G is diophantine. Then there is an effective process which,* *when applied to any basic formula D of* Th($\langle G, \Delta \rangle$), *gives an open formula E_D of* Th(**Z**) *such that D is satisfiable in $\langle G, \Delta \rangle$ if and only if E_D is satisfiable in* **Z**.

Proof. By hypothesis G is diophantine so for some fixed n there is a monomorphism $\varphi : G \to \mathrm{GL}(n, \mathbf{Z})$ and a diophantine predicate $R(x_{11}, \ldots, x_m)$ of n^2 variables such that

$$M = \begin{pmatrix} x_{11} \cdots x_{1n} \\ \vdots \quad \vdots \\ x_{n1} \cdots x_{nn} \end{pmatrix} \in \varphi(G)$$

if and only if $R(x_{11}, \ldots, x_{nn})$ or for simplicity $R(M)$. Now $\Delta = \{g_1, \ldots, g_k\}$ is finite so we can assume that R together with the matrices $\varphi(g_1), \ldots, \varphi(g_k)$ are explicitly given (R is specified by a polynomial).

Let D be any basic formula of Th($\langle G, \Delta \rangle$), and let $u_1, \ldots, u_m$ be the variables occurring in D. Pick correspondingly m variables matrices $M_1, \ldots, M_m$ with different variables (for integers) as entries – a total of $m \cdot n^2$ variables.

Let D_1 be the formula concerning matrices formed from D as follows: replace u_i by M_i; replace each constant corresponding to g_i by the matrix $\varphi(g_i)$; replace 1 by the identity matrix; group multiplication is replaced by matrix multiplication u_i^{-1} is replaced by the formal inverse of M_i which can be written down in terms of cofactors and the determinant of M_i.

Now by simply carrying out all of the (matrix) multiplications in D_1 formally one obtains an open formula D_2 of Th(**Z**) of the following kind: D_2 is a conjunction of formulas B_i and $\neg B_i$ where each B_i is a conjunction of n^2 polynomial equations (coming from atomic parts of D_1).

Let D_3 denote the formula

$$D_2 \wedge R(M_1) \wedge \ldots \wedge R(M_m)$$

of Th(**Z**). Now D_3 is satisfiable in **Z** if and only if D is satisfiable in $\langle G, \Delta \rangle$, since φ is monic and R characterizes $\varphi(G)$. Let D_4 be the prenex form of D_3. Then D_4 has the form

$$(\exists v_1) \dots (\exists v_l)\ [D_2 \wedge P_1 = 0 \wedge \dots \wedge P_k = 0]$$

where the P_i are polynomials. Now D_3 is satisfiable if and only if D_4 is satisfiable. Finally, let E_D be the open formula

$$D_2 \wedge P_1 = 0 \wedge \dots \wedge P_k = 0.$$

E_D is satisfiable if and only if D_4 is satisfiable. Hence, E_D is satisfiable. Hence, E_D is satisfiable in **Z** if and only if D is satisfiable in $\langle G, \Delta \rangle$. Clearly, E_D can be effectively found from D. This completes the proof of the lemma and thereby Theorem 1.

§4. Applications to free groups

For any cardinal number α, let F_α denote the free group on α generators and let **F** denote the class of all free groups (including the trivial and infinite cyclic groups). Consider an arbitrary open formula B of Th(**F**). Since B contains only finitely many variables, B is satisfiable in some $F \in \mathbf{F}$ if and only if B is satisfiable in $F_{\aleph_0}$. Because subgroups of free groups are free and $F_{\aleph_0}$, it follows that B is satisfiable in some $F \in \mathbf{F}$ if and only if B is satisfiable in F_2. Hence SPOF(**F**) is equivalent to SPOF(F_2).

According to example 1 of section 2, F_2 is a diophantine group (with an empty set of distinguished elements). Applying Theorem 1 we conclude:

Lemma 5. DPOF(**F**), SPBF(**F**), *and* SPOF(**F**) *are reducible to Hilbert's 10th problem. Moreover, the set of open formulas of* Th(**F**) *which are satisfiable in some* $F \in \mathbf{F}$ *is recursively enumerable.*

Any basic formula D of Th(**F**) is logically equivalent to a basic formula of the form

$$R_1 = 1 \wedge \ldots \wedge R_m = 1 \wedge w_1 \neq 1 \wedge \ldots \wedge w_k \neq 1$$

where the R_i and w_j are words in the free group on the variables, say $x_1, \ldots, x_n$, which occur in D.

To give a useful interpretation to this formula, we introduce the finitely presented group $G_D = \langle x_1, \ldots, x_n; R_1 = 1, \ldots, R_m = 1 \rangle$. Observe that G_D does not depend on the inequalities which occur in D. Now D is satisfiable in some $F \in \mathbf{F}$ if and only if there is a homomorphism ψ from G_D into some $F \in \mathbf{F}$ such that $\psi(w_1) \neq 1 \wedge \ldots \wedge \psi(w_k) \neq 1$. (This follows from the definition of satisfiable). However, since subgroups of free groups are free,it follows that D is satisfiable in some $F \in \mathbf{F}$ if and only if G_D has a quotient group F' which is free and such that the images of $w_1, \ldots, w_k$ do not map to 1 in F' (take $F' = \psi(G_D)$ in the previous statement).

For any group G, define

$$K_G = \bigcap_K \{K \lhd G \mid G/K \in \mathbf{F}\}$$

(the intersection of all normal subgroups K of G such that G/K is free). Now G/K_G is a residually free group. Moreover, if G_1 is any other homomorphic image of G which is residually free, then G_1 is a homomorphic image of G/K_G. This property is in fact equivalent to the definition of K_G.

Lemma 6. *Let G be a finitely presented group. Then $L_G = G \backslash K_G =$ $\{w \mid w$ is a word in the generators of G and $w \notin K_G\}$ is recursively enumerable. Moreover, the method of enumerating L_G is uniform in G in the sense that given G we can effectively find the enumerating process.*

Proof. Suppose $G = \langle x_1, \ldots, x_n; R_1, \ldots, R_m \rangle$ and let D_w be the basic formula

$$R_1 = 1 \wedge \ldots \wedge R_m = 1 \wedge w \neq 1.$$

Then D_w is satisfiable in some $F \in \mathbf{F}$ if and only if $w \notin K_G$. The result now follows from Lemma 5.

Lemma 7. *If Hilbert's 10th problem is recursively solvable, then there is a uniform algorithm* $\mathbf{A}(G)$ *which, when applied to any finitely presented group* G, *solves the word problem for* G/K_G.

Lemma 7′. *If Hilbert's 10th problem is recursively solvable, then there is a uniform algorithm* $\mathbf{A}'(G)$ *which, when applied to any finitely presented group* G, *enumerated the words of* K_G.

Proof. In view of Lemma 6, Lemmas 7 and 7′ are (logically) equivalent. Let G and D_w be as in the proof of Lemma 6. Now D_w is satisfiable in some $F \in \mathbf{F}$ if and only if $w = 1$ in G/K_G. The result now follows by Lemma 5.

Problem 1. Does there exist a finitely presented group G such that G/K_G is *not* recursively presented?

The following problem was posed several years ago by M.O. Rabin (unpublished) who knew of its connection with Hilbert's 10th problem.

Problem 2 (Rabin's Problem). Does there exist a recursive class of finitely presented groups $G_1, G_2, ...$ such that there is no algorithm to tell of an arbitrary G_i whether or not G_i has a non-abelian free quotient?

Clearly if the answer to Problem 1 is affirmative, then by Lemma 7′ Hilbert's 10th problem would be unsolvable. Consider Problem 2. Let $G_i = \langle x_1, ..., x_n ; R_1 = 1, ..., R_m = 1 \rangle$. Now G_i has a non-abelian free quotient if and only if one of the finitely many commutators $[x_i, x_j] \neq 1$ in G/K_G. But this is reducible to the word problem for G/K_G. Hence, if there were such a class of groups, by Lemma 7, Hilbert's 10th problem would be unsolvable. Note that $\{ G_i | G_i$ has a non-abelian free quotient$\}$ is recursively enumerable by Lemma 6.

Problem 3. Does there exist a recursive class of finitely presented groups $G_1, G_2, ...$ with uniformly solvable word problem such that

$\{G_i | G_i$ is residually free$\}$ is recursively enumerable, but not recursive? Here "uniformly solvable word problem" means there is a uniform algorithm which, when applied to a G_i, solves the word problem for G_i.

If the answer to Problem 3 is affirmative, then Hilbert's 10th problem would be unsolvable. To see this assume that Hilbert's 10th were solvable and that $G_1, G_2, \ldots$ is such a class. Now G_i is *not* residually free if and only if $(\exists w \in G_i)$ $(w \neq 1$ in $G_i \wedge w \in K_{G_i})$. Since we are assuming Hilbert's 10th is solvable, by Lemma 7′ it follows that

$$S = \{(G_i, w) | w \in G_i \wedge w \in K_{G_i}\}$$

is recursively enumerable. Since the word problem for the G_i is uniformly solvable the subset

$$T = \{(G_i, w) | w \in G_i \wedge w \in K_{G_i} \wedge w \neq 1 \text{ in } G_i\}$$

of S is also recursively enumerable. But G_i appears as the first component in some member of T if and only if G_i is not residually free. Hence $\{G_i | G_i$ is *not* residually free$\}$ is recursively enumerable. But this contradicts our assumption on the class $G_1, G_2, \ldots$, and the claim is established.

Collins [6] has constructed a recursive class of finitely presented groups $G_1, G_2, \ldots$ such that each G_i has solvable word problem (note: *not* uniformly) such that $\{G_i | G_i$ is residually free$\}$ is recursively enumerable, but not recursive. This does not settle Problem 3 since the word problem is not uniformly solvable. Indeed, his construction works exactly because the word problem for the G_i is *not* uniformly solvable.

§5. A free nilpotent group

In this section a construction of Malcev [13] will be used to establish a "converse" to Theorem 1. H will denote the free nilpotent group of class 2 and rank 2. The commutator notation $[u, v] = uvu^{-1}v^{-1}$ will be used frequently.

Now H is generated by two elements, say a_1 and a_2, which freely generate H (in the sense of varieties). H satisfies the law $[[H, H], H] = 1$ and the center of H, written $C(H)$ is infinite cyclic with generator $c = [a_2, a_1] = a_2 a_1 a_2^{-1} a_1^{-1}$.

Recall that H has a faithful matrix representation $\widetilde{H}$ in $\mathrm{SL}(3, \mathbf{Z})$ where $\widetilde{H}$ consists of those matrices of the form

$$\begin{pmatrix} 1 & m & n \\ 0 & n & k \\ 0 & 0 & 1 \end{pmatrix} .$$

For convenience, H and $\widetilde{H}$ will often be identified. In terms of matrices, one can take

$$a_1 = \begin{pmatrix} 1 & 0 & 0 \\ 0 & 1 & 1 \\ 0 & 0 & 1 \end{pmatrix} , \quad a_2 = \begin{pmatrix} 1 & 1 & 0 \\ 0 & 1 & 0 \\ 0 & 0 & 1 \end{pmatrix}$$

and
$$c = \begin{pmatrix} 1 & 0 & 1 \\ 0 & 1 & 0 \\ 0 & 0 & 1 \end{pmatrix}$$

The equation

$$a_1{}^k a_2{}^m c^n = \begin{pmatrix} 1 & m & n \\ 0 & 1 & k \\ 0 & 0 & 1 \end{pmatrix}$$

is easily verified – hence every element of H is uniquely expressible in the form $a_1{}^k a_2{}^m c^n$. Elements of H multiply according to the equation:

$$(a_1{}^k a_2{}^m c^n) \cdot (a_1{}^h a_2{}^i c^j) = a_1{}^{k+h} a_2{}^{m+i} c^{mh+n+j} .$$

The general commutator is given by

$$[(a_1{}^h a_2{}^i c^j), (a_1{}^k a_2{}^m c^n)] = c^{ik-mh}.$$

Finally, while the above equations are given for two particular free generators a_1, a_2, if y_1, y_2 are any other two free generators of H, then the same statements hold for y_1, y_2 in place of a_1, a_2 except that y_1, y_2 correspond to different matrices.

Lemma 8. *Let $z_1 = c^k, z_2 = c^l$ be arbitrary elements of $C(H)$. Then there exist elements $x_1, x_2 \in H$ which satisfy the equations*

$$(*) \qquad \begin{cases} x_1 a_2 = a_2 x_1 \quad x_2 a_1 = a_1 x_2 \\ a_1 x_1 a_1^{-1} x_1 = z_1 \\ a_2 x_2 a_2^{-1} x_2^{-1} = z_2 \end{cases}$$

Moreover, if x_1, x_2 are any two elements of H satisfying (*), then $[x_2, x_1] = x_2 x_1 x_2^{-1} x_1^{-1} = c^{k\cdot l}$.

Proof. To see that such x_1, x_2 exist, put $x_1 = a_2^{-k}$ and $x_2 = a_1{}^l$. Then clearly $x_1 a_2 = a_2 x_1$ and $x_2 a_1 = a_1 x_2$. Moreover, by our general rule for commutators:

$$[a_1, x_1] = [a_1, a_2^{-k}] = c^{-1\cdot(-k)} = c^k = z_1$$

and

$$[a_2, x_2] = [a_2, a_1{}^l] = c^{1\cdot l} = c^l = z_2$$

as required. It also follows that $[x_2, x_1] = [a_1{}^l, a_2^{-k}] = c^{-l\cdot(-k)} = c^{k\cdot l}$.

Now let x_1, x_2 be arbitrary elements of H satisfying (*). Write $x_1 = a_1{}^h a_2{}^m c^n$ and consider the equation $x_1 a_2 = a_2 x$. Then $1 = [x_1, a_2] = [a_1{}^h a_2{}^m c^n, a_2] = c^{-1\cdot h}$ and so $h = 0$ and $x_1 = a_2{}^m c^n$. But now $c^k = z_1 = [x_1, a_1] = [a_2{}^m c^n, a_1] = c^{-m\cdot 1} = c^{-m}$. Hence $m = -k$ and $x_1 = a_2^{-k} c^n$. Similarly it follows that $x_2 = a_1{}^l c^s$. But then $[x_2, x_1] = [a_1{}^l c^s, a_2^{-k} c^n] = [a_1{}^l, a_2^{-k}] = c^{k\cdot l}$ as claimed.

Lemma 9. *Let z be any element of $C(H)$. Then there exist elements $u, v \in H$ such that $z = [u, v]$. Conversely, for any $u, v \in H$, $[u, v] \in C(H)$.*

Proof. That $[u, v] \in C(H)$ is trivial since H satisfies the law $[[H, H], H] = 1$. Let $z = c^k$ be an arbitrary element of $C(H)$. Then, taking $u = a_2$, $v = a_1{}^k$, it follows that $[u, v] = [a_2, a_1{}^k] = c^k = z$ as required.

Lemma 10. *Let y_1, y_2 be arbitrary elements of H such that $[y_2, y_1] = y_2 y_1 y_2^{-1} y_1^{-1} = c$. Then y_1, y_2 freely generate H. In particular, all of the above results hold with y_1, y_2 in place of a_1, a_2.*

Proof. We first show that $[y_2, y_1] = c$ implies y_1, y_2 generate H. Put $y_1 = a_1{}^k a_2{}^m c^n$ and $y_2 = a_1{}^h a_2{}^i c^j$. From the general formula for commutators, it follows that $[y_2, y_1] = c^{ik-mh} = c$, and hence $ik - mh = 1$. Let G be the subgroup of H generated by y_1, y_2. Clearly $C(H) \subset G$. Now $y_1{}^i = a_1{}^{ik} a_2{}^{mi} c^p$ for some p and $y_2{}^m = a_1{}^{mh} a_2{}^{mi} c^q$ for some q. Hence $a_1{}^{ik} a_2{}^{mi} \in G$ and $a_1{}^{mh} a_2{}^{mi} \in G$ and $a_1{}^{mh} a_2{}^{mi} \in G$. Thus $a_1{}^{ik-mh} = a_1 \in G$. Similarly $a_2 \in G$ and so $G = H$ as claimed.

H is residually finite and therefore Hopfian, so according to H. Neumann [15] (41.33) y_1, y_2 freely generate H (in the sense of varieties).

Next we consider $\mathrm{Th}(\langle H, \{c\}\rangle)$, where $\langle H, \{c\}\rangle$ is the group H with distinguished element c (the generator of $C(H)$). Let $R(z_1, z_2, x_1, x_2, y_1, y_2)$ be the conjunction of the following equations:

$$y_2 y_1 y_2^{-1} y_1^{-1} = c$$

$$x_1 y_2 = y_2 x_1 \qquad x_2 y_1 = y_1 x_2$$

$$y_1 x_1 y_1^{-1} x_1^{-1} = z \qquad y_2 x_2 y_2^{-1} x_2^{-1} = z_2 .$$

If the formula $(\exists y_1)(\exists y_2) R(z_1, z_2, x_1, x_2, y_1, y_2)$ is satisfied by a particular assignment of the variables z_1, z_2, x_1, x_2 to elements of H, by Lemmas 8, 9 and 10 it follows that $z_1, z_2 \in C(H)$ and $x_2 x_1 x_2^{-1} x_1^{-1} = c^{k \cdot l}$ where $z_1 = c^k$ and $z_2 = c^l$.

Our purpose is to interpret $\mathrm{Th}(\mathbf{Z})$ in $\mathrm{Th}(\langle H, \{c\}\rangle)$ via the following scheme:

$$1 \to c, \ 0 \to 1$$

$$k + l \to z_1 \cdot z_2 = c^{k+l}$$

$$k \cdot l \to x_2 x_1 x_2^{-1} x_1^{-1} = c^{k \cdot l}$$

where $z_1 = c^k$, $z_2 = c^l$ and $(\exists y_1)(\exists y_2) R(z_1, z_2, x_1, x_2, y_1, y_2)$. Although it is possible to interpret any formula of Th(**Z**) in Th($\langle H, \{c\}\rangle$) by suitably adapting Malcev [13], for our purposes it is only necessary to interpret polynomial equations.

Lemma 11. *There is an effective process which associates with any polynomial equation* $P = 0$ *an open formula* B_P *of* Th($\langle H, \{c\}\rangle$) *such that* $P = 0$ *is satisfiable in* **Z** *if and only if* B_P *is satisfiable in* $\langle H, \{c\}\rangle$.

Proof. The equation $P = 0$ is, as previously noted, algebraically equivalent to an atomic formula $P_1 = P_2$ of Th(**Z**) (no minus signs allowed) where P_1 and P_2 are the positive and negative parts of P. First we show that any atomic formula $P_1 = P_2$ is logically equivalent to a conjunction of atomic formulas

$$E_1 \wedge E_2 \wedge \ldots \wedge E_l$$

which can be effectively found from $P_1 = P_2$ where each E_i has one of the following forms:

$$t_1 = t_2, \quad t_1 + t_2 = t_3, \quad t_1 \cdot t_2 = t_3$$

where each t_i is 0, 1 or a variable of Th(**Z**). A formula of the type E_i will be called *elementary*. The proof is by induction on the number of occurrences of + and $\cdot$ in $P_1 = P_2$ (recall that P_1, P_2 are both terms). If there are either zero or one occurrences of + and $\cdot$ in $P_1 = P_2$, the result is obvious from the definition of term. Assume at least two + or $\cdot$ occur in $P_1 = P_2$. Then one of P_1 or P_2 has the form $Q_1 + Q_2$ or $Q_1 \cdot Q_2$ where Q_1, Q_2 are terms, say $P_1 = Q_1 + Q_2$. Let x, y, z be variables not occurring in P_1 or P_2. Then $P_1 = P_2$ is logically equivalent to

$$(x = Q_1) \wedge (y = Q_2) \wedge (z = P_2) \wedge (x + y = z).$$

Each of the conjuncts contains fewer occurrences of + and · than $P_1 = P_2$. Hence by induction each conjunct is logically equivalent to a conjunction of elementary formulas. Thus $P_1 = P_2$ is equivalent to a conjunction of elementary formulas. A similar argument applies in case P_1 has form $Q_1 \cdot Q_2$. Moreover, it is clear these instructions for obtaining the desired E_i's are effective. This proves our claim.

From $P_1 = P_2$ one can effectively find a logically equivalent formula D of the form $E_1 \wedge E_2 \wedge \ldots \wedge E_l$ where each E_i is elementary. Assume the variables of D are among $u_1, \ldots, u_k$. With each E_i we associate a formula S_i of Th($\langle H, \{c\}\rangle$) with free variables among $z_1, \ldots, z_k$ as follows:

(1) If E_i has form $u_1 = u_2$ take S_i to be

$$(z_1 = z_2) \wedge (\exists x_1)(\exists x_2)([x_2, x_1] = z_1) \wedge$$
$$(\exists x_1)(\exists x_2)([x_2, x_1] = z_2).$$

(2) If E_i has form $u_1 + u_2 = u_3$ take S_i to be

$$z_1 \cdot z_2 = z_3 \wedge (\exists x_1)(\exists x_2)([x_2, x_1] = z_1) \wedge$$
$$(\exists x_1)(\exists x_2)([x_2, x_1] = z_2).$$

(3) If E_i has form $u_1 \cdot u_2 = u_3$ take S_i to be

$$(\exists x_1)(\exists x_2)(\exists y_1)(\exists y_2)(z_3 = x_2 x_1 x_2^{-1} x_1^{-1} \wedge$$
$$R(z_1, z_2, x_1, x_2, y_1, y_2))$$

where R is as defined above.

(4) If E_i contains 0 or 1 instead of certain variables, find the corresponding S_i as in (1) – (3) except that 0 or 1 becomes 1 or c respectively. Let T be the prenex form of $S_1 \wedge \ldots \wedge S_l$. From Lemmas 8, 9, and 10 together with the remarks preceding the present lemma, it follows that $P = 0$ is satisfiable in **Z** if and only if

T is satisfiable in $\langle H, \{c\}\rangle$. But T has the form $(\exists v_1) \ldots (\exists v_m) B_P$ where B_P is an open formula. Hence $P = 0$ is satisfiable in $\mathbf{Z}$ if and only if B_P is satisfiable in $\langle H, \{c\}\rangle$. Moreover, B_P can be effectively found from $P = 0$, as required.

Since H is diophantine, Theorem 1 and Lemma 11 imply:

Theorem 2. *Let H be the free nilpotent group of rank 2 and class 2 and let $c \in H$ generate the center of H. Then Hilbert's 10th problem is Turing equivalent to* SPBF($\langle H, \{c\}\rangle$) *and hence also to* SPOF($\langle H, \{c\}\rangle$) *and* DPOF($\langle H, \{c\}\rangle$).

Remarks. We have actually proved somewhat more than Theorem 2. In particular, since the B_p found in Lemma 11 contains no negation signs, it follows that Hilbert's 10th problem and SPBF($\langle H, \{c\}\rangle$) are equivalent to the satisfaction problem for conjunctions of atomic formulas in $\langle H, \{c\}\rangle$. In a different direction, it would now be a simple matter to prove most of the results in Malcev [13]. In particular, it now follows easily that Th($\langle H, \{c\}\rangle$) and Th(H) are undecidable. See [13] for more details.

It would be desirable to replace SPBF($\langle H, \{c\}\rangle$) by SPBF(H) in Theorem 2. In order to do this, we would need an open formula $B(x, y_1, \ldots, y_n)$ not involving c such that $(\exists y_1) \ldots (\exists y_n) B(x, y_1, \ldots, y_n)$ holds in H if and only if $x = c$. The author has not been able to do this. There are more complicated formulas which define c, but they do not help for Theorem 2. Over in Th($\mathbf{Z}$), the formula $(x \cdot x = x) \wedge (x \neq 0)$ holds in $\mathbf{Z}$ if and only if $x = 1$. However, the translation of this formula into Th($\langle H, \{c\}\rangle$) involves the formula R and hence c.

Toh [22] has obtained some partial results concerning SPBF(H). Toh's paper contains, among other things, some interesting examples which depend on substantial results from number theory. In view of Theorem 2, this dependence might well be expected.

The following result is weaker than Theorem 2, and follows immediately from Theorems 1 and 2:

Theorem 3. *Hilbert's 10th problem is unsolvable if and only if there exists a diophantine group G with distinguished elements* Δ *such that* SPBF($\langle G, \Delta \rangle$) *is unsolvable.*

§6. Applications to the conjugacy problem

Recall that the *conjugacy problem* for a group G is the problem of deciding for an arbitrary pair u, v of words on the generators of G whether or not u and v are conjugate in G, i.e. whether or not $(\exists w \in G)(w^{-1}uw = v \text{ in } G)$.

A group L is a *diophantine group of matrices,* if for some integer $n > 0$, $L \subset \mathrm{GL}(n, \mathbf{Z})$ and there is a diophantine predicate R of n^2 variables such that for M any $n \times n$ matrix, $R(M) \Leftrightarrow M \in L$.

Lemma 12. *Let L be a diophantine group of matrices, as above. Then the conjugacy problem for L is reducibel to Hilbert's 10th problem.*

Proof. Let M_1, M_2 be arbitrary (but fixed) matrices belonging to L. Now M_1 and M_2 are conjugate in L if and only if the formula $M_1 M = MM_2$ concerning matrices is satisfiable by some $M \in L$. Call this formula D_1. As in the proof of Lemma 4,* there is an effective process which, when applied to D_1, gives an open formula E of Th($\mathbf{Z}$) such that E is satisfiable in $\mathbf{Z}$ if and only if D_1 is satisfiable in L, i.e. if and only if M_1 and M_2 are conjugate in L. By Lemma 2, the result follows.

Theorem 4. *Let G be a finitely generated diophantine group. Then the conjugacy problem for G is reducible to Hilbert's 10th problem.*

Proof. Let $g_1, ..., g_k$ denote a finite set of generators for G. Since G is diophantine, there is a faithful representation $\varphi : G \to \mathrm{GL}(n, \mathbf{Z})$ for some $n > 0$ so that $\varphi(G)$ is a diophantine group of matrices. Now put $l_1 = \varphi(g_1), ..., l_k = \varphi(g_k)$ and let $L = \varphi(G)$. The matrices l_i generate L. Clearly, there is an effective process which, when ap-

* The starting point in Lemma 4 was somewhat different. However, it is easy to see the proof of that Lemma also applies to these circumstances.

plied to any word $u(g_i)$ on the g_i in G, gives the matrix $\varphi(u(g_i))$ – namely $\varphi(u(g_i)) = u(l_i)$. Since two words u, v on the g_i are conjugate in G if and only if $\varphi(u)$, $\varphi(v)$ are conjugate in L, the result follows from Lemma 12.

Remark. Consider a diophantine group with finitely many distinguished elements $\langle G, \Delta \rangle$. If Δ generates G, it follows that the conjugacy problem for G is reducible to SPOF($\langle G, \Delta \rangle$) and hence to Hilbert's 10th problem. To see this, let u, v be arbitrary words in the generators Δ. Then $w^{-1}uw = v$ is satisfiable in $\langle G, \Delta \rangle$ if and only if u and v are conjugate in G. But $w^{-1}uw = v$ is an open formula of Th($\langle G, \Delta \rangle$), as claimed.

These considerations led W.W. Boone (unpublished) to ask the following question?

Problem 4 (Boone). Does there exist a finitely generated diophantine group with unsolvable conjugacy problem?

In view of Theorem 4, an affirmative answer to Problem 4 would show Hilbert's 10th problem unsolvable. In this connection, Miller [14] has given an example of a finitely generated subgroup of GL(4, **Z**) with an unsolvable conjugacy problem. Whether or not this subgroup is diophantine is not known.

§7. Concluding remarks

In view of the connections established above between diophantine groups and Hilbert's 10th problem, the following vague question seems appropriate:

Problem 5. Which groups are diophantine?

In trying to answer this question, one may as well consider groups of matrices and ask which are diophantine. Since a diophantine predicate is recursively enumerable (section 2), it follows that:

(1) A diophantine subgroup of GL(n, **Z**) is a recursively enumerable set of matrices. Some of the work of Davis, Putnam, and J. Robinson [8] on Hilbert's 10th problem is concerned with the following question which is as yet unanswered: Is every recursively enumerable predicate diophantine? An affirmative answer would of course show Hilbert's10th unsolvable (see section 2 and [7]) and would provide an answer to Problem 5. As Boone and Davis have pointed out, an appropriate answer to any of problems 1 through 4 could show Hilbert's 10th unsolvable without answering the above question.

Added in proof: Matijasevič [23] has now proved the remarkable theorem that every recursively enumerable predicate is diophantine. This answers problems 4 and 5. Problems 1, 2, and 3 still seem to be open.

Next we shall list several known results concerning groups of matrices which seem relevant to Problem 5. The following notation will be used:

Φ = arbitrary field of characteristic zero
C = field of complex numbers
Q = field of rational numbers
R = field of real numbers
GL(n, Φ) = invertible $n \times n$ matrices with coefficients in Φ.

Clearly, for any such Φ, one has

$$\mathrm{GL}(n, \mathbf{Z}) \subset \mathrm{GL}(n, \Phi).$$

It is known that GL(n, **Z**) is finitely presented (see [10] for instance).

(2) A finitely generated subgroup of GL(n, Φ) is residually finite (Malcev [12]). Since residual finiteness is heriditary, it follows that any diophantine group is residually finite.

(3) Let G be a finitely generated subgroup of GL(n, Φ). Then G contains a torsion free normal subgroup of finite index (Selberg [18]).

(4) A solvable subgroup of GL(n, **Z**) is polycyclic (Malcev [11]).

(5) The order of a finite subgroup of GL(n, **Q**) is bounded by a function of n (see for instance Serre [19]).

An important class of groups related to diophantine groups is the class of arithmetic groups [1]. For example, the free nilpotent group of section 4 is an arithmetic group. We will briefly indicate connections between diophantine and arithmetic groups.

A complex algebraic group is a subgroup G of GL(n, **C**) which consists of all invertible matrices $M = (m_{ij})$ whose coefficients annihilate some set of polynomials $\{P_\alpha(x_{11}, ..., x_{nn})\}$ with complex coefficients. G is said to be defined over Q if the P_α may be chosen to have rational coefficients. In view of the Hilbert basis theorem one can assume the number of P_α is finite.

Assume hereafter that G is an algebraic groups defined over Q by polynomials $P_1, ..., P_k$. Let $G_R = G \cap$ GL(n, **R**) and $G_Z =$ $G \cap$ GL(n, **Z**). The group G_Z is called an *arithmetic subgroup* (or *group of units*) of G_R. It is easy to show that G_Z consists of exactly those matrices in GL(n, **Z**) which satisfy a suitable polynomial equation with integer coefficients constructed from $P_1, ..., P_k$. In particular, G_Z is a diophantine group.

It is known that G_Z is finitely presented (see Borel [2]). G_Z is a discrete subgroup of the real Lie group G_R. Borel and Harish-Chandra [3] have given necessary and sufficient conditions under which G_R/G_Z is compact, or of finite invariant measure. A theory of fundamental sets for G_Z has been developed.

Of course the study of arithmetic groups relies on the study of algebraic groups. In particular, the methods of algebraic geometry and Lie theory are involved. To what extent similar techniques might be helpful in investigating diophantine groups is not clear to the author.

References

[1] A. Borel, Reduction theory for arithmetic groups, Proc. Sympos. Pure Math., vol. 9 (Amer. Math. Soc., Providence, R.I., 1966) 20-25.

[2] A. Borel, Arithmetic properties of linear algebraic groups, Proc. International Congress of Math. (Stockholm, 1962) 10-22.

[3] A. Borel and Harish-Chandra, Arithmetic subgroups of algebraic groups, Ann. of Math. (2) 75 (1962) 485-535.
[4] A. Church, Introduction to mathematical logic, vol. I (Princeton University Press, Princeton, N.J., 1956).
[5] P. M. Cohn, Universal Algebra (Harper and Row, New York, 1965).
[6] D.J. Collins, to be published.
[7] M. Davis, Computability and unsolvability (McGraw-Hill, New York, 1958).
[8] M. Davis, H. Putnam and J. Robinson, The decision problem for exponential diophantine equations, Ann. of Math. 74 (1961) 425-436.
[9] D. Hilbert, Mathematical problems, Bull. of Amer. Math. Soc. 8 (1901-2) 437-479).
[10] W. Magnus, A. Karras and D. Solitar, Combinatorial group theory (Wiley, New York, 1966).
[11] A.I. Malcev, On some classes of infinite soluble groups, Amer. Math. Soc. Translations, series 2, vol. 2 (1956) 1-22.
[12] A.I. Malcev, On the faithful representation of infinite groups by matrices, Amer. Math. Soc. Translations, series 2, vol. 45 (1965) 1-18.
[13] A.I. Malcev, On a correspondence between rings and groups, Amer. Math. Soc. Translations, series 2, vol. 45 (1965) 221-232.
[14] C.F. Miller III, Unsolvable problems in direct products of free groups and in unimodular groups, unpublished.
[15] H. Neumann, Varieties of Groups, Ergebnisse der Mathematik und ihrer Grenzgebiete, Band 37 (Springer, Berlin, 1967).
[16] J. Robinson, Diophantine decision problems, in: W.J. LeVeque, ed., MAA Studies in Math., Vol. 6 (Prentice-Hall, 1969).
[17] I.N. Sanov, A property of a certain representation of a free group (Russian), Doklady Akad. Nauk SSSR. vol. 57 (1947) 657-659.
[18] A. Selberg, On discontinuous groups in higher-dimensional symmetric spaces, Internat. Colloq. Function Theory, 147-164, Tata Inst. of Fund. Research, Bombay, 1960.
[19] J.-P. Serre, Lie Algebras and Lie groups (Benjamin, New York, 1965).
[20] J.R. Shoenfield, Mathematical logic (Addison-Wesley, Reading, Mass., 1967).
[21] R.R. Stoll, Set theory and logic (Freeman, San Francisco, 1963).
[22] K.H. Toh, Problems concerning residual finiteness in nilpotent groups, to be published.
[23] Ju.V. Matijasevič, Enumerable sets are diophantine (Russian), Dokl. Akad. Nauk. SSSR, Tom 191 (1970), No. 2 – English translation: Soviet Math. Dokl., Vol. 11 (1970), No. 2.

DECISION PROBLEMS IN ALGEBRAIC CLASSES OF GROUPS (A SURVEY)

Ch.F.MILLER, III*

The Institute for Advanced Study, Princeton

This is a survey of what is known concerning the world problem, conjugacy problem, and other decision problems in certain classes of finitely presented groups. Each of the decision problems is known to be unsolvable for finitely presented groups in general. Our aim is to investigate these decision problems in classes of finitely presented groups which are in some sense "elementary".

To methods of "measuring" the elementary nature of finitely presented groups will be considered: (1) algebraic constructions – how a group is built from free groups by algebraic constructions such as free products with amalgamation; and (2) algebraic classes – groups defined by algebraic properties such as nilpotence and residual finiteness.

Most of the results discussed are summarized in Tables 1 and 2, and the text might be regarded as an explanation of these tables. Section 1 consists mostly of terminology and has been included to make our survey accessible to a more diverse audience. Section 2 is concerned with Table 1, and Section 3 is concerned with Table 3. Section 4 contains an example relevant to both tables, and Section 5 consists of additional remarks and problems.

All groups are assumed to be finitely presented unless otherwise stated.

Regarding the tables: The notation "+WP" in a table means the word problem is known to be solvable for the corresponding class of groups; "¬WP" means examples are known for which the word

* Work supported by U.S.National Science Foundation under contract No. GP-7421.

Table 1

Construction / Iterations	free products with amalgamation	HNN-construction (= Britton extensions)	Split Extensions	Direct Products
Level 3	⋮	⋮ (Boone ¬WP) ⋮	⋮	⋮
Level 2	⋮ ¬WP	⋮ (Collins ¬CP) ¬WP	⋮	⋮
Level 1	+WP ¬CP ¬GWP	+WP ¬CP ¬GWP	+WP ¬CP ¬GWP	+WP +CP ¬GWP

f.g. free groups
+WP
+CP
+GWP

(all groups are assumed to be f.p.)

problem is unsolvable; and "?WP" means the status of the word problem is an open question. Similarly for the other problems considered.

1. Some terminology

By a presentation of a group we mean an ordered pair $\langle S;D\rangle$ where S is a set and D a collection of words on the elements of S and their inverses. By the group G presented by $\langle S;D\rangle$ we mean the quotient group of the free group on S by the normal closure of the words in D. (For further details see [12].) We usually write

Table 2

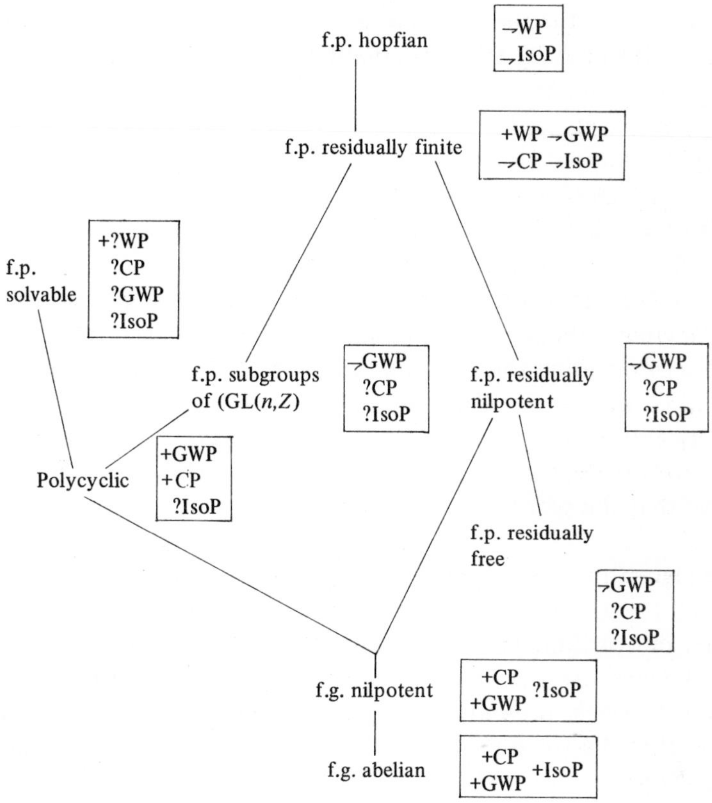

(all groups are assumed to be f.p.)

$G = \langle S;D \rangle$ in this situation. Except when necessary, we will not distinguish between a group (as an abstract algebraic object) and its presentation (as a notation in some logical system)*.

$G = \langle S;D \rangle$ is said to be *finitely generated* (f.g.) if S is finite and *finitely presented* (f.p.) when S and D are both finite. Note that being generated by a finite set of elements or having a finite presentation is an algebraic property of groups (preserved under iso-

* There are other possible viewpoints towards presentations. In particular, presentations can be viewed as logical systems – see Boone [6] for this approach.

morphism). Finitely presented groups arise naturally in topology: they are exactly the fundamental groups of finite simplicial complexes or alternatively the fundamental groups of closed differentiable n-manifolds ($n \geqslant 4$).

Viewing the group $G = \langle S;D \rangle$ as the quotient of the free group F on S by the normal closure of D, recall that a word $W = 1$ (the empty word) in G if and only if W is equal in F to a product of conjugates of words in D. Moreover, $W_1 = W_2$ in G if and only if there is a word $Z = 1$ in G such that $W_1 = ZW_2$ in F.

In case G is finitely generated, the *word problem* (WP) (or in some languages the identity problem) for G is the algorithmic problem of deciding for arbitrary words W of G whether or not $W = 1$ in G. In case $G = \langle S;D \rangle$ is such that S is finite and D is a recursively enumerable (r.e.) set of words, G is said to be *recursively presented**. In particular, f.p. groups are recursively presented. Observe that the word problem is r.e. for recursively presented groups.

The algorithmic problem of deciding for an arbitrary pair of words $U, V \in G$ whether or not U and V are conjugate in G (i.e. whether or not there exists $W \in G$ such that $U = W^{-1}VW$ in G) is called the *conjugacy problem* (CP) or *transformation problem* for G. For recursively presented groups, the conjugacy problem is r.e. Observe that the word problem for G is one-one reducible to the conjugacy problem for G since $U = 1$ in G if and only if U is conjugate to 1 in G.

Let $\Phi = \{\prod_i, i \geqslant 0\}$ be a recursive class of finite presentations of groups (say on some fixed alphabet). The *isomorphism problem* (IsoP) for Φ is the algorithmic problem of deciding whether or not $\prod_i \cong \prod_j$ for arbitrary i and j (i.e. whether or not $\prod_i$ and $\prod_j$ are presentations for the same abstract group). The class Φ is required to be recursive so that the algorithmic problem is well-posed.

In the context of topology, the word problem corresponds to the problem of deciding whether a closed path is contractible to a

* By a device due to Craig, a group is recursively presented (in our sense) if and only if it has some presentation $\langle S';D' \rangle$ where S' is finite and D' is a recursive set of words. See Boone [6] in this connection.

point. The conjugacy problem corresponds to the problem of deciding whether two closed paths are free homotopic. Since homotopy equivalent spaces have isomorphic fundamental groups, the isomorphism problem is related to the problem of homotopy equivalence.

Finally consider a f.g. group $G = \langle S;D \rangle$ and a finite set of words $W_1, \ldots, W_n$. Let H be the subgroup of G generated by $W_1, \ldots, W_n$. The *generalized word problem* (GWP) or *membership problem* for H in G is the algorithmic problem of deciding whether or not an arbitrary word $U \in G$ belongs to the subgroup H (i.e. whether or not there is a product of the W_i and their inverses which is equal to U in G). The GWP is r.e. provided G is recursively presented. Clearly, the word problem for G is just the GWP for the trivial subgroup in G. The GWP as formulated here is often called the *extended word problem*, but we follow [12] in our choice of terminology.

A property P of groups is called a *poly-property* if, whenever N and G/N have the property P, so has G. Here N is a normal subgroup of G.

Proposition 1. *The following are poly-properties:*
(1) *being finitely generated*
(2) *having a finite presentation*
(3) *satisfying the maximum condition for subgroups*
(4) *being finitely generated and having a solvable word problem.*

That (1) – (3) are poly-properties is shown in P. Hall [9], and that (4) is a poly-property is easily verified.

A group G is *poly-P* if it can be obtained from the trivial group by a finite succession of extensions by groups in P, i.e. if there is a finite series of subgroups

$$G = G_0 \supset G_1 \supset \ldots \supset G_n = \{1\}$$

such that G_{i+1} is normal in G_i and $G_i/G_{i+1} \in P$. As an application of Proposition 1, it follows that polycyclic groups are finitely presented and satisfy the maximum condition for subgroups. Observe that solvable groups are just the polyabelian groups.

Let P be a property of groups. A group G is said to be *residually* P if for every $1 \neq W \in G$ there is a normal subgroup N_W of G such that $W \notin N_W$ and G/N_W has the property P. Equivalently, G is residually P if the intersection of the normal subgroups N of G such that G/N has P is the identity. For instance, free groups are residually nilpotent and residually finite (see [12]). Baumslag [3] has shown that the automorphism group of a f.g. residually finite group is again residually finite.

A group G is called *hopfian* if $N \lhd G$ and $G/N \cong G$ imply that $N = \{1\}$. Equivalently, G is hopfian if every epic endomorphism of G is an automorphism. F.g. residually finite groups are hopfian (see [21] 41.44).

2. Algebraic constructions

This section concerns decision problems in f.p. groups which are "built" (allowing iterations) from free groups by the following constructions:

(1) free products with amalgamated subgroup
(2) the construction of Higman, Neumann and Neumann [10] (also called "Britton extensions")
(3) split extensions
(4) direct products.

From the viewpoint of decision problems, free groups are rather pleasant. Namely, the word problem, the conjugacy problem, and the generalized word problem for finitely generated subgroups are all recursively solvable (see [12]).

Each of the above constructions occurs quite frequently in infinite group theory. If a f.p. group is built from free groups by one application of these constructions, it is in a sense "elementary" – say it is Level 1 above free groups (with respect to that construction). If the same construction is applied to Level 1 groups, obtaining a Level 2 group, the resulting group may be more complicated and less elementary.

Next we inquire as to the status of the various decision problems at Level 1, Level 2, etc. It turns out that, for these four constructions, a rather complete picture is available. In general, the various decision problems are unsolvable at the lowest "reasonable" level.

First we recall the Higman-Neumann-Neumann (HNN) construction in the case of f.p. groups. Let E be a f.p. group and let $A_1, ..., A_n$ be f.g. subgroups of E. Assume that for each i there is an isomorphic mapping φ_i from A_i onto a subgroup B_i of E. Then the group E^* obtained from E by adding new generators $p_1, ..., p_n$ and new relations $p_i^{-1} a_i p_i = \varphi_i(a_i)$ for all $a_i \in A_i$ has the following properties: (i) E is embedded naturally in E^* and (ii) $p_1, ..., p_n$ freely generate a free subgroup of E^*. (For a proof see [15], [20], or [22].) From the proof of this result one can easily deduce the following: Assume E is a free group and let F_n denote the free group of rank n. Then $\bar{E} = E^* * F_n$ (ordinary free product) is the free product with amalgamation of two free groups. Clearly, the WP and CP for E^* and $\bar{E}$ are equivalent. Moreover, if E^* has unsolvable GWP, then so does $\bar{E}$. Note that E^* has Level 1 in the HNN-construction and $\bar{E}$ has Level 1 using free products with amalgamation.

Level 1. If a group G is the free product with amalgamation of two f.g. free groups and if G is f.p., then (i) (Baumslag [2]) the amalgamated subgroup must be f.g. and so (ii) ([12], p. 272) the WP for G is solvable. That is, at Level 1 using free products with amalgamation, the WP is solvable. Moreover, by the previous remarks, it follows that a Level 1 using the HNN-construction the WP is solvable.

Since having a solvable WP is a poly-property, the split extension of one f.g. free group by another must have solvable WP. (Observe this also follows from the fact that such a split extension can be viewed as an HNN-construction over a free group.) Clearly, the direct product of two f.g. free groups has solvable WP and solvable CP.

Miller [16] (see Section 4) has shown, there exists a split extension G of one f.g. free group by another having unsolvable CP and a f.g. subgroup $H \subset G$ having unsolvable GWP. Moreover, G is obtained from a free group by one application of the HNN-construction. Hence, by the above remarks, the CP and GWP can be unsolvable at Level 1 using any of the first three kinds of construction.

On the other hand, the direct product of two free groups has solvable CP. However, Mihailova [14] has shown there is a f.g. subgroup of the direct product of two free groups with unsolvable GWP.

Higher Levels. It is clear that if we take the direct product of groups at Level 1 in the direct product column, the result is still a direct product of (possibly many) free groups. Hence the status of the various decision problems at higher levels of direct products is the same as at Level 1.

Suppose G is the split extension of two Level 1 split extension groups. Then, since having solvable WP is a poly-property, G has solvable WP. Since the CP and GWP are already unsolvable at Level 1, the status of the various decision problems at higher levels of split extensions is the same as at Level 1.

Let E be a f.p. group at Level 1 in the HNN-construction column which has a f.g. subgroup A with unsolvable GWP in E (as noted before such A and E exist). Say A is generated by words $w_1, \ldots, w_n$. Now apply the HNN-construction to E as follows: E^* is obtained from E by adding a new generator p and n new relations $p^{-1} w_i p = w_i$ (i=1,...,n). Then E^* is at Level 2 in the HNN-construction column. But E^* has unsolvable WP since $p^{-1} u p u^{-1} = 1$ in E^* if and only if $u \in A$ where u is a word on the generators of E and the GWP for A in E is unsolvable (we are using Britton's Lemma, see [22]). Thus at Level 2 in the HNN-construction column, the WP can be unsolvable.

Similarly, let H be a f.p. group at Level 1 in the free product with amalgamation column which has a f.g. subgroup A with unsolvable GWP in H. Let H_1 be another copy of H with corresponding subgroup A_1. Now form the free product with amalgamation $G = (H * H_1; a=a_1$ for all $a \in A)$. Now if h is an arbitrary word of H and h_1 the corresponding word in H_1, then $hh_1^{-1} = 1$ in G if and only if $h \in A$. Since the GWP for A in H is unsolvable, this shows the WP for G is unsolvable. But G has Level 2 and can be f.p., so at Level 2 in the free product with amalgamation column, the WP can be unsolvable.

Recall that the WP is reducible to the CP and the GWP. Thus at Levels 2 and higher in the HNN-construction column and the free products with amalgamation column, any of the WP, CP, and GWP can be unsolvable.

Historical remark. The groups studied by Boone and Britton (see [22]) in showing the WP unsolvable for f.p. groups are located at Level 3 in the HNN-construction column. In [7], Collins exhibits a f.p. group with solvable WP and unsolvable CP. Collins' groups are located at Level 2 in the HNN-construction column. These are indicated in Table 1.

3. Algebraic classes

In Table 2 various classes of f.p. groups characterized by algebraic properties are listed, together with the known status of the various decision problems for groups in those classes. If two classes are connected by an upward line, then the "lower" class is contained in the "higher" class. Note that if one of the decision problems considered is solvable in a class of groups, then it is also solvable in any subclass. Accordingly, the solvability of WP has not been repeated for subclasses of the f.p. residually finite groups.

Hopfian groups. The largest class considered is the class of f.p. hopfian groups. Schupp and the author [18] have shown that any f.p. group G can be embedded in a f.p. hopfian group H in such a way that the WP for G is (Turing) equivalent to the WP for H. It follows that the WP and IsoP for f.p. hopfian groups can be unsolvable. Hence the CP and GWP can also be unsolvable.

Residually finite groups. Any f.g. residually finite group is hopfian (for instance, see [21] 41.44). Dyson [8] and Mostowski [19] have shown that f.p. residually finite groups have solvable WP. Briefly, the argument is as follows: Let G be f.p. residually finite, and let w be a word of G. Then $w \neq 1$ in G if and only if $w \neq 1$ in some finite quotient of G. Since G is f.p., one can construct a list of all finite quotients of G and check to see whether or not $w \neq 1$ in successive quotients. Hence, one can enumerate $\{w|w\neq 1 \text{ in } G\}$. On the other hand, since G is f.p., $\{w|w=1 \text{ in } G\}$ can be enumerated. Since every word of G appears in exactly one of these lists, this solves the word problem for G.

Miller [16] (see also Section 4 below) has given an example of a f.p. residually finite group with unsolvable CP and unsolvable GWP. Moreover, he shows that there exists a recursive class of f.p. residually finite groups for which IsoP is unsolvable.

Observe that the property of being residually finite is hereditary, i.e. subgroups of residually finite groups are residually finite.

Subgroups of GL(n,Z). Let GL(n,Z) denote the group of $n \times n$ matrices with integer entries and determinant ± 1. Let A_n denote the free abelian group of rank n. Then GL(n,Z) is isomorphic to the automorphism group of A_n. Now A_n is f.p. residually finite, so by Baumslag [3] it follows that GL(n,Z) is residually finite. Finite presentations for GL(n,Z) are known (see [12], p. 168).

Thus any f.p. subgroup of GL(n,Z) is residually finite and hence has solvable WP. Since GL(4,Z) contains a "nice" isomorphic copy of the direct product of two free groups, by Milailova [14] it follows that GL(4,Z) has unsolvable GWP.

The status of the CP and IsoP for f.p. subgroups of GL(n,Z) seems to be unknown. Even the CP for GL(n,Z) itself seems to be unsettled. In this connection, Miller [17] has shown the CP is unsolvable for finitely generated subgroups of GL(n,Z) and that there is no algorithm to tell of two finite sets of matrices whether or not they generate the same subgroup of GL(n,Z) (or even isomorphic subgroups).

Solvable groups. For f.p. solvable groups in general, there seems to be little known concerning the various decision problems. It is remored that several Polish mathematicians have shown the WP is solvable for f.p. solvable groups, but the author knows of no published proof.

Polycyclic groups. Polycyclic groups are certainly solvable, indeed a group is polycyclic if and only if it is solvable and satisfies the maximum condition for subgroups. As noted above, polycyclic groups are f.p. Malcev [13] has shown that a f.g. solvable subgroup of GL(n,Z) is polycyclic. Conversely, Auslander [1] has proved that any polycylic group can be embedded in GL(n,Z) for suitable n. It follows that polycyclic groups are residually finite, a result due to Hirsch which can be proved directly (see [21] 32.1).

Toh [24] has shown that polycyclic groups are *subgroup separable,* i.e. if H is a subgroup of the polycyclic group G and $w \notin H$, then there exists an epimorphism φ: $G \to F$ where F is a finite group such that $\varphi(w) \notin \varphi(H)$. Thus, by an argument similar to that showing the WP is solvable for f.p. residually finite groups, Toh has shown the GWP is solvable for polycyclic groups.

The CP and IsoP are open for polycyclic groups. Regarding the CP, it does not seem to be known whether or not polycyclic groups are *conjugacy separable.* (G is conjugacy separable means that if u, v are not conjugate in G then their images in some finite quotient of G are not conjugate. If G is conjugacy separable and f.p., then G has solvable CP.)

Added in proof: Remeslennikov [25] has now shown that polycyclic groups are conjugacy separable and hence have solvable conjugacy problem.

F.g. nilpotent groups. A f.g. nilpotent group is polycyclic and hence is f.p. (see [11], vol. II, p. 232, or [21]). In particular, f.g. nilpotent groups are residually finite and have solvable WP.

Blackburn [5] has shown that f.g. nilpotent groups are conjugacy separable and hence have solvable CP. Since polycylic groups are subgroup separable, so are f.g. nilpotent groups. Thus the GWP is solvable for f.g. nilpotent groups. (This was known before Toh's work – see Mostowski [19] for this and other decision problems.)

The IsoP for nilpotent groups seems to be open and very interesting because of connections with Hilbert's 10th problem.

F.g. abelian groups. That the WP, CP, GWP, IsoP are solvable for f.g. abelian groups follows easily from the "fundamental theorem for f.g. abelian groups" (see for instance [11], vol. I, p. 145). Clearly, f.g. abelian groups are f.p. and nilpotent.

Residually free groups. Since free groups are residually nilpotent (see [12], p. 311), it follows that residually free groups are residually nilpotent and hence are residually finite. In particular, f.p. residually free groups have solvable WP. However, the direct product of two free groups is residually free, so Mihailova [14] has shown the GWP can be unsolvable in f.p. residually free groups.

The CP and IsoP for f.p. residually free groups seem to be open. However, for f.g. recursively presented residually free groups, Miller [17] has shown the CP can be unsolvable.

Residually nilpotent groups. A f.p. residually nilpotent group is residually finite because f.g. nilpotent groups are residually finite. In particular, such groups have solvable WP. Beyond this, no more seems to be known other than the results cited above for the subclass of f.p. residually free groups.

We remark that a polycyclic, residually nilpotent group need not be nilpotent. For example the infinite dihedral group $D = \langle x,y; x^2=1, x^{-1}yx=y^{-1}\rangle$ is residually a finite 2-group and polycylic, but not nilpotent.

4. An example

In this section, the author's example [16] of a f.p. residually finite group G with unsolvable CP is presented. The group G is also the split extension of one free group by another and accounts for most of the unsolvability results for the CP discussed in the above classifications.

Let $H = \langle s_1, s_2, ..., s_n; R_1, ..., R_m\rangle$ be a f.p. group with unsolvable WP. Put $F = \langle q, s_1, ..., s_n\rangle$, a free group of rank $n+1$ on the listed generators. Finally define the f.p. group G as follows:

Generators:

$$q,\ s_1,\ ...,\ s_n,\ t_1,\ ...,\ t_m,\ d_1,\ ...,\ d_n$$

Relations:

$$\left.\begin{array}{ll} t_i^{-1} q t_i &= qR_i \\ t_i^{-1} s_\alpha t_i &= s_\alpha \end{array}\right\} \qquad \begin{array}{l} 1 \leqslant i \leqslant m \\ 1 \leqslant \alpha \leqslant n \end{array}$$

$$\left.\begin{array}{ll} d_\alpha^{-1} q d_\alpha &= s_\alpha^{-1} q s_\alpha \\ d_\alpha^{-1} s_\beta d_\alpha &= s_\beta \end{array}\right\} \qquad \begin{array}{l} 1 \leqslant \alpha \leqslant n \\ 1 \leqslant \beta \leqslant n \end{array}$$

where the R_i are the words in s_i which appear as the given defining relations for H.

Lemma 1. *G is obtained from F by the* HNN-*construction adding the letters* $\{t_1,...,t_m,d_1,...,d_n\}$.

Proof. Each of the subgroups in question is F itself. For example, for fixed i, the subgroups involved with t_i are generated by $\{q,s_1,...,s_n\}$ and $\{qR_i,s_1,...,s_n\}$. Since each of these sets consist of n+1 elements of F which generate F which has rank n+1, they freely generate F. Consequently, the map $q \to qR_i$, $s_\alpha \to s_\alpha$ generates an automorphism of F as was to be verified. The same argument applies to the subgroups associated with each d_α.

Let T be the subgroup of G generated by $\{t_1,...,t_m,d_1,...,d_n\}$. Then, as noted above regarding the HNN-construction, T is a free group freely generated by $\{t_1,...,t_m,d_1,...,d_n\}$. Observe that every element of T normalizes F in view of the defining relations. Hence

Lemma 2. *G is the split extension of the free group F by the free group T. In particular G has solvable* WP.

Lemma 3. *G is residually finite.*

This lemma follows immediately from the following result which seems to have been known to several people. The proof is omitted (see [16]):

Extension Lemma. *Suppose that* $1 \to A \to E \to B \to 1$ *is an exact sequence of groups (i.e. E is an extension of A by B) where A and B are residually finite and A is f.g. Assume that* any one *of the following conditions holds:*

(1) *A has trivial center.*

(2) *The sequence splits (i.e. E is a split extension or semi-direct product of A by B).*

(3) *B is free or A is non-abelian free. Then E is residually finite.*

Lemma 4. *Let X be an arbitrary word on the* s_α. *Then qX is conjugate in G to q if and only if* $X = 1$ *in H.*

From Lemmas 3 and 4, since H has unsolvable WP, it follows that:

Theorem. *The f.p. residually finite group G has unsolvable* CP.

Proof of Lemma 4. Assume that qX is conjugate in G to q. Define φ: $G \to H$ to be the epimorphism obtained by mapping $q \to 1$, $t_i \to 1$, $d_j \to 1$, $s_\alpha \to s_\alpha$. Then $1 = \varphi(q)$ and hence $\varphi(qX) = 1$ in H. But $\varphi(qX) = \varphi(q)\varphi(X) = \varphi(X) = X = 1$ in H as claimed.

Because of notational complexity, we only sketch the proof that if $X = 1$ in H then q is conjugate to qX in H. First notice that the t_i and d_j commute with the s_α. Let Y, Z be words on the s_α and consider YqZ. Conjugating YqZ by t_i gives YqR_iZ. Thus conjugating YqZ by t_i corresponds to inserting R_i between Y and Z in the formal product YZ, i.e. YZ becomes YR_iZ.

Now conjugating YqZ by d_α gives $Ys_\alpha^{-1}qs_\alpha Z$. This corresponds to inserting $s_\alpha^{-1}s_\alpha$ in the formal product YZ to get $Ys_\alpha^{-1}s_\alpha Z$. The q keeps track (in G) of the position at which we are operating in the formal product. Finally, conjugating $Ys_\alpha qZ$ by d_α gives $Ys_\alpha s_\alpha^{-1}qs_\alpha Z = Yqs_\alpha Z$. Thus, by conjugations with d_α, the position of the marking symbol q can be changed.

Since $X = 1$ in H, X is freely equal to a product of conjugates of the R_i. Thus X can be built up from the empty word by inserting inverse pairs of generators and inserting $R_i^{\pm 1}$, and finally keep freely reducing (after all insertions). But as argued above this corresponds to conjugating q by some sequence of t_i and d_α to give qX. This completes the sketch of the proof.

Finally we remark that the GWP the subgroup of G generated by T and q is unsolvable. Moreover, using the fact that T acting on F by conjugation induces an automorphism of F, it can be shown that the GWP for Aut(F) is unsolvable.

5. Remarks and problems

The above discussion leaves as open questions the status of certain decision problems in several algebraic classes of groups. Here we consider some additional open problems and make some remarks related to the above.

As noted in Proposition 1, "being finitely presented and having solvable WP" is a poly-property. The group G of Section 4 shows that "being f.p. and having solvable CP" is not a poly-property. Having solvable WP is hereditary in the sense that a f.p. subgroup of a group with solvable WP again has solvable WP. This is false for the CP. By modifying the example of Section 4, Miller [16] has constructed a f.p. residually finite group with solvable CP which contains a f.p. subgroup having unsolvable CP.

Related to this is the following:

Problem 1. (B.H.Neumann, Schupp, Miller). Let G be a f.p. group and H a subgroup of finite index in G. Is the conjugacy problem for G (Turing) equivalent to the conjugacy problem for H? If not, is one always reducible to the other?

It would suffice to prove the equivalence for H also normal. For suppose H has index n in G. Let K be the intersection of all subgroups of index n in G. Then K is contained in H. characteristic in G and normal in H, and has finite index in both.

Problem 2. Does there exist a f.p. group G such that (1) G has unsolvable WP and (2) G satisfies a non-trivial identity?

Note that identities are not allowed in defining G – only finitely many relations. For f.g. recursively presented groups, the answer is affirmative. An example may be easily derived from a group constructed in *P*. Hall [9].

For almost all of the unsolvability results given in Tables 1 and 2, an "arbitrary r.e. degree result" has been proved, in the sense that for suitable choice of the groups involved, the decision problem has preassigned r.e. degree of unsolvability.

Problem 3. (Boone, Collins, Miller). Given two r.e. degrees D_1, D_2 such that $D_1 \leqslant D_2$, does there exist a f.p. group G with WP of degree D_1 and CP of degree D_2?

For f.g. recursively presented groups Miller [16] has shown the answer is affirmative. Collins has made some progress on the f.p. case, but the problem remains open. For $D_1 = 0$, the answer is affirmative and due to Collins [7].

Added in proof: Collins [26] has now given an affirmative answer to Problem 3.

This survey has omitted several important topics – among them metabelian groups, single relator groups, and groups satisfying a cancellation condition. Fortunately, the latter are discussed by Schupp [23] in a survey in this volume.

References

[1] L.Auslander, On a problem of Philip Hall, Ann. of Math. 86 (1967) 112-116.
[2] G.Baumslag, A remark on generalized free products, Proc. of Amer. Math. Soc. 13 (1962) 53-54.
[3] G.Baumslag, Automorphism groups of residually finite groups, Journal London Math. Soc. 38 (1963) 117-118.
[4] G.Baumslag, Finitely presented groups, Proc. Internat. Conf. Theory of Groups, Australian National University, Canberra, August, 1965, pp. 37-50, Gordon and Breach, New York, 1967.
[5] N.Blackburn, Conjugacy in nilpotent groups, Proc. of Amer. Math. Soc. 16 (1965) 143-148.

[6] W.W.Boone, The word problem, Ann. of Math. 70 (1959) 207-265.

[7] D.J.Collins, Recursively enumerable degrees and the conjugacy problem, Acta mathematica 122 (1969) 115-160.

[8] V.H.Dyson, The word problem and residually finite gorups, Notices of Amer. Math. Soc. 11 (1964) 743.

[9] P.Hall, Finiteness conditions for soluble groups, Proc. London Math. Soc. (3) 4 (1954) 419-436.

[10] G.Higman, B.H.Neumann, and H.Neumann, Embedding theorems for groups, Journal London Math. Soc. 24 (1949) 247-254.

[11] A.G.Kurosh, The theory of groups, Vols. I and II (Chelsea, New York, 1956).

[12] W.Magnus, A.Karrass and D.Solitar, Combinatorial group theory (Wiley, New York, 1966).

[13] A.I.Malcev, On some classes of infinite solvable groups, A.M.S. Translations, series 2, vol. 2 (1956).

[14] K.A.Mihailova, The occurrence problem for direct products of groups, Dokl. Akad. Nauk SSSR 119 (1958) 1103-1105 (Russian).

[15] C.F.Miller, III, On Britton's theorem A, Proc. of Amer. Math. Soc. 19 (1968) 1151-1154.

[16] C.F.Miller, III, On group-theoretic decision problems and their classification, Annals of Math. Studies, No. 68, Princeton, Princeton University Press, 1971.

[17] C.F.Miller, III, Unsolvable problems in direct products of free groups and in unimodular groups, unpublished.

[18] C.F. Miller, III and P.E. Schupp, Embedding into hopfian groups, Journal of Algebra, 17 (1971), pp. 171–176.

[19] A.W.Mostowski, On the decidability of some problems in special classes of groups, Fund. Math. LIX (1966) 123-135.

[20] B.H.Neumann, An essay on free products of groups with amalgamations, Phil. Trans. Royal Soc. of London, No. 919, 246 (1954) 503-554.

[21] H.Neumann, Varieties of groups, Ergebnisse der Mathematik und ihrer Grenzgebiete 37 (Springer, Berlin, 1967).

[22] J.J.Rotman, The theory of groups (Allyn and Bacon, Boston, 1965).

[23] P.E.Schupp, A survey of small cancellation theory, this volume.

[24] K.H.Toh, Problems concerning residual finiteness in nilpotent groups, to be published.

[25] V.N. Remeslennikov, Conjugacy in polycyclic groups, Algebrai i Logika, Seminar, Vol. 8, No. 6 (1969) 712-725.

[26] D.J. Collins, Representation of Turing reducibility by word and conjugacy problems in finitely presented groups, to appear in Acta Mathematica.

UNIFORM ALGORITHMS FOR DECIDING GROUP-THEORETIC PROBLEMS

A. Wlodzimierz MOSTOWSKI
Mathematical Institute of the University of Gdańsk

§1. Introduction

This paper is devoted to an investigation of the concept of a uniform algorithm for deciding group-theoretic problems, for any group-theoretical predicate with a fixed number of variables, and invariant under isomorphism. Many problems such as the word problem, conjugacy problem and others are connected with such predicates, cf. Schupp [12]. For the word problem (w.p.), uniform algorithms were investigated in Boone and Rogers [3]. Collins [4] and also in Mostowski [10], but in the last paper the notion was not explicitly mentioned.

Given a recursively enumerable (r.e.) sequence

(1) $$P_{\alpha_1}, P_{\alpha_2}, \ldots$$

of finite presentations, a *uniform algorithm* is a decision method applicable to any member of (1).

Preliminaries are in Section 2. A precise definition of a uniform algorithm, in the language of recursive functions, is given in Section 3. Some reasons for limiting the notion to r.e. sets of presentations are explained. The main Theorem 5 concerning extensions of uniform algorithms to the closure of (1) under isomorphism is given in Section 4. The proof is based on Rabin [11]. Applications are given in Section 5. The extension of the theory to predicates meaningful for any finite number of variables is given in Section 6. An example of a problem described by such a predicate is

Magnus' extended word problem. A method of treating the isomorphism problem is suggested in Theorem 9 of Section 7. The last section, viz. 8, is devoted to a relativisation of the theory to the decidability problem with some oracle. Therein is to be found Theorem 11, in which we prove that grouptheoretic problems for isomorphic presentations have the same degree of unsolvability.

§2. Preliminaries

We shall deal with a fixed set $S = (s_1, s_2, ...)$ of symbols and an effective enumeration of the free group $F(S)$. For this fixed enumeration, we shall denote by $\overline{\tau}$ a number of $\tau \in F(S)$, and conversely by $\overline{k}$ a $\tau \in F(S)$ such that $\overline{\tau} = k$. A finite presentation of a group is a pair $P = (X; R)$ where the set X of generators is a finite subset of S, and the set R of relations is a finite subset of $F(X)$. We shall deal with a fixed Godel enumeration of the set $\mathcal{P}$ of all finite presentations. The number of $P \in \mathcal{P}$ will be denoted by $n(P)$. We shall use the notations $P = P_m$ and $n(P) = m$ as well.

An n-variable predicate $\Gamma(x_1, ..., x_n)$ where $n = 0, 1, 2, ...$ is a grouptheoretical predicate iff:

for any group G and $g_1, ..., g_n \in G$

$$\Gamma(g_1, ..., g_n) \text{ is true or false}$$

and Γ is invariant under any isomorphism of G.

The decision problem for Γ relative to a class $\mathcal{K} \subset \mathcal{P}$ is to decide for any $P = (X; R) \in \mathcal{K}$ and any $\tau_1, ..., \tau_n \in F(X)$ whether

$$\Gamma(g_1, ..., g_n) \text{ is true or false,}$$

where $g_1, ..., g_n$ are elements of the group $G(P)$ presented by the words $\tau_1, ..., \tau_n$. Note that because of the invariancy of Γ under isomorphism the problem is well posed, i.e., is independent on any isomorphic copy of a group $G(P)$ presented by P.

Now let us define a new predicate $\Gamma_P(\tau_1, ..., \tau_n)$ which is partial and connected with Γ. For $P = (X; R) \in \mathcal{P}$, and $\tau_1, ..., \tau_n \in F(X)$

presenting elements $g_1, \dots, g_n$ of a group $G(P)$ with a presentation P, we define

$$\Gamma_P(\tau_1, \dots, \tau_n) \quad \text{iff} \quad \Gamma(g_1, \dots, g_n).$$

If for some $i = 1, 2, \dots, n$, $\tau_i \notin F(X)$, i.e., τ_i contains a letter s_j or s_j^{-1} not belonging to X, then $\Gamma_P(\tau_1, \dots, \tau_n)$ is undefined.

Since we can effectively decide for any $P = (X; R) \in \mathcal{P}$ and any $\tau_1, \dots, \tau_n \in F(S)$ whether or not some of $\tau_1, \dots, \tau_n$ contains a letter out of X, then the decision problem for Γ is the same as the decision problem whether or not $\Gamma_P(\tau_1, \dots, \tau_n)$ is true or false. So the decidability of Γ is equivalent to the computability of the following function $f(m, k_1, \dots, k_n)$ on integers:

$$f(m, k_1, \dots, k_n) = \begin{cases} 1 \\ 0 \text{ for } \Gamma_{P_m}(\bar{k}_1, \dots, \bar{k}_n) \\ 2 \end{cases} \begin{cases} \text{true} \\ \text{false} \\ \text{undefined.} \end{cases}$$

§3. Decision functions and uniform algorithms

In the literature, see [3, 4], a notion of uniform algorithm for deciding a problem for some recursively enumerable set $\Sigma \subset \mathcal{P}$ is used. The idea of the notion is, that having effectively listed the presentations of Σ as $(P_{\alpha_1}, P_{\alpha_2}, \dots)$, we need one effective method for deciding the problem for $P_{\alpha_1}, P_{\alpha_2}$, and so on.

For some interesting classes of groups it is not known whether or not the sets of all finite presentations of these groups are r.e. sets. That is, nothing is known about the recursive enumerability of the following sets:

$\mathcal{R}$, all finite presentations of residually finite groups
$\mathcal{S}$, all finite presentations of simple nontrivial groups.

For both of these classes it is known that there is a uniform method of solving the word problem. For residually finite groups cf., e.g., [10]; for simple nontrivial groups see the known method presented in Section 4. According to the supposition that both R and S are neither Σ^1 nor Π^1 in Kleene–Mostowski hierarchy, both R and S are out of the notion of uniform algorithm for the w.p. despite the existence of uniform methods of solving the w.p. relative to this classes.

We shall look for a notion more extensive than the notion of uniform algorithm, to cover both these cases. It is given in the following definition based on our preliminaries.

Definition 1. A partial recursive function $f(m, k_1, ..., k_n)$ is called a *decision function* for a grouptheoretical predicate $\Gamma(x_1, ..., x_n)$ relative to a class $K \subset P$ iff for any integer m such that $P_m \in K$,

(a) the function $f_m(k_1, ..., k_n) = f(m, k_1, ..., k_n)$ is total;

(b) moreover,

$$f_m(k_1, ..., k_n) = \begin{cases} 1 \\ 0 \end{cases}$$

for

$$\Gamma_{P_m}(\bar{k}_1, ..., \bar{k}_n) \text{ defined and } \begin{cases} \text{true} \\ \text{false;} \end{cases}$$

(c) for $\Gamma_{P_m}(\bar{k}_1, ..., \bar{k}_n)$ undefined, the value $f_m(k_1, ..., k_n)$ can be any integer different from 0 and 1.

Note that the definition can be inessentially weakened since, according to our preliminaries, (a)–(c) can be replaced by the following: for any integer m such that $P_m \in K$,

(d) for

$$\Gamma_{P_m}(\bar{k}_1, ..., \bar{k}_n) \text{ being } \begin{cases} \text{true} \\ \text{false} \end{cases}$$

the value

$$f(m, k_1, ..., k_n) = \begin{cases} 1 \\ 0. \end{cases}$$

Using the definition of a decision function, we can give a following precise definition of a uniform algorithm.

Definition 2. A function $f(i, m_1, ..., m_n)$ is a uniform algorithm for deciding $\Gamma(x_1, ..., x_n)$ relative to a r.e. set $\Sigma \subset P$ iff there exists a decision function $f'(k, m_1, ..., m_n)$ for Γ relative to Σ such that for some enumeration $\alpha(i)$ of Σ

$$f(i, m_1, ..., m_n) = f'(\alpha(i), m_1, ..., m_n) \text{ for } i = 1, 2, ...$$

From the definition it follows that f is a computable function. Thus, from the definition, we have the following.

Theorem 1. *If a decision function $f'(k, m_1, ..., m_n)$ for Γ relative to* $K \subset P$ exists, *then for any r.e. subset Σ of K there exists a uniform algorithm for deciding Γ relative to Σ.*

Proof. Let $\alpha(i)$ be an enumeration of Σ, the $f(i, m_1, ..., m_n) = f'(\alpha(i), m_1, ..., m_n)$ is the desired uniform algorithm.

Now we shall prove the following.

Theorem 2. *Let $f(i, m_1, ..., m_n)$ and $h(s, m_1, ..., m_n)$ be two uniform algorithms for deciding Γ relative to a r.e. set, then there exists a computable function $t(s)$ such that*

$$f(i, m_1, ..., m_n) = h(t(i), m_1, ..., m_n).$$

Proof. Let $\alpha(i)$ and $\beta(s)$ be two enumerations of Σ and $f'(k, m_1, ..., m_n)$ and $h'(k, m_1, ..., m_n)$ be two decision functions relative to Σ such that

$$f(i, m_1, ..., m_n) = f'(\alpha(i), m_1, ..., m_n) \quad \text{for} \quad i = 1, 2, ...$$

$$h(s, m_1, ..., m_n) = h'(\beta(s), m_1, ..., m_n) \quad \text{for} \quad s = 1, 2, ...$$

We have for $k \in n(\Sigma)$

$$f'(k, m_1, ..., m_n) = h'(k, m_1, ..., m_n)$$

with both sides defined.

The function $t_0(k)$ is defined as follows: $t_0(k)$ = minimum i such that $\beta(i) = k$ is a partial recursive function. The function $t(x) = t_0(\alpha(x))$ is recursive and total, so computable.

Now $h(t_0(l), m_1, \ldots, m_k) = h'(l, m_1, \ldots, m_n) = f'(l, m_1, \ldots, m_n)$ for any $l \in n(\Sigma)$, according to $\beta(t_0(l) = l$. Then $h(t(i), m_1, \ldots, m_n) =$ $= h(t_0(\alpha(i)), m_1, \ldots, m_n) = h'(\alpha(i), m_1, \ldots, m_n) = f'(\alpha(i), m_1, \ldots, m_n) =$ $= f(i, m_1, \ldots, m_n)$. This proves that $t(x)$ has the desired property.

Now we shall prove:

Theorem 3. *If there exists uniform algorithms f_1 and f_2 for deciding Γ relatively to r.e. subsets Σ_1 and Σ_2 of P, then there exists a uniform algorithm f_0 for deciding Γ relative to $\Sigma_0 = \Sigma_1 \cup \Sigma_2$.*

Proof. Let α, β be enumeration of Σ_1 and Σ_2 such that for suitable decision functions f_1' and f_2'

$$f_1(i, m_1, \ldots, m_n) = f_1'(\alpha(i), m_1, \ldots, m_n)$$

$$f_2(i, m_1, \ldots, m_n) = f_2'(\beta(i), m_1, \ldots, m_n)$$

The function:

$$\delta(i) = \begin{cases} \alpha(j) & \text{for } i = 2j \\ \beta(j) & \text{for } i = 2j+1 \end{cases}$$

is an enumeration for $\Sigma_0 = \Sigma_1 \cup \Sigma_2$.

Note that for k such that $k = \delta(2j) = \delta(2j'+1)$ for some integers j, j', $f_1'(k, m_1, \ldots, m_n) = f_2'(k, m_1, \ldots, m_n)$ so the function

$$f_0'(k, m_1, \ldots, m_n) = \begin{cases} f_1'(k, m_1, \ldots, m_n) & \text{for } k = \delta(2j) \\ f_2'(k, m_1, \ldots, m_n) & \text{for } k = \delta(2j'+1) \end{cases}$$

is a well-determined partial recursive function. Evidently, f_0' is a decision function relative to Σ_0. Now

$$f_0(i, m_1, \ldots, m_n) = f_0'(\delta(i), m_1, \ldots, m_n)$$

is the desired uniform algorithm for Σ_0.

From Theorems 1 and 3 we obtain immediately.

Corollary. *The family of subsets of* $\mathcal{P}$ *with uniform algorithms for* Γ *forms an ideal of the lattice of all r.e. subsets of* $\mathcal{P}$.

The following example due to V. Dyson explains difficulties connected with the notion of a decision function, i.e., a decision process relative to a class $\mathcal{K}$ which is not a recursively enumerable set of presentations. The example shows that decision processes for two classes $\mathcal{K}_1$ and $\mathcal{K}_2$ of presentations can give different results when used for a $P \in \mathcal{K}_1 \cup \mathcal{K}_2$ even in case they both terminate.

Let $\mathcal{K}_1$ be the class $\mathcal{R}$ of all finite presentations of residually finite groups. The decision process described in Mostowski [10] for deciding the word problem is as follows:

For $P = (X;R)$, list all elements

(1) $$\tau_1, \tau_2, \tau_3, \ldots$$

of the normal subgroup N_R generated by R in $F(X)$. Similarly list all homomorphisms

$$\varphi_1, \varphi_2, \ldots$$

of the group $G(P)$ onto finite groups, and for a $\tau \in F(X)$ look for $\varphi_1(\tau), \varphi_2(\tau), \ldots$ step by step. Then either τ is in a sequence (1) and then $\tau = 1$ in $G(P)$ or for some i $\varphi_i(\tau) \neq 1$, and then $\tau \neq 1$ in $G(P)$. For details, see [10].

Let $\mathcal{K}_2$ be the class $\mathcal{S}$ of nontrivial simple groups. The decision process goes as follows:

For $P = (X;R)$ list all elements (1) of the normal subgroup N_R. Similarly list all presentations

(2) $$P_{\alpha_1}, P_{\alpha_2}, P_{\alpha_3}, \ldots$$

of trivial groups.

For a $\tau \in F(X)$, either $\tau \in N_R$ and is in the sequence (1). That is case $\tau = 1$ in $G(P)$. Or in the case $\tau \neq 1$ in $G(P)$, the presentation $P' = (X; R, \tau)$ is a presentation of a trivial group (since G is simple). Then P' is in the sequence (2).

For a presentation P of a trivial group, the first process always gives the answer $\tau = 1$ for any $\tau \in F(X)$. The second process can give the wrong answer $\tau \neq 1$, for some τ. That is the case when we first assert P' as being in the sequence (2), before asserting that τ is in the sequence (1).

§4. Extensions of uniform algorithms

In Rabin's paper [11] it is proved that there exists a (total) recursive function $e(n, k)$ such that in the sequence

$$P_{e(n,0)}, P_{e(n,1)}, \ldots$$

there are all finite presentations of groups isomorphic to a group presented by $P_n \in \mathrm{P}$. Denote by Σ^* a set obtained from $\Sigma \subset \mathrm{P}$ by adding all presentations $Q \in \mathrm{P}$ isomorphic to some $P \in \Sigma$. Then if Σ is a r.e. set and $\alpha(i)$ an enumeration for Σ, then for an enumeration $c(i, j)$ of pairs of integers, and projection functions $l(k)$ and $r(k)$ such that $k = c(l(k), r(k))$ the function

$$\beta(k) = e(\alpha(l(k)), r(k))$$

is an enumeration of Σ^*. Thus Σ^* is a r.e. set when Σ is a r.e. set.

Now let $P' = (S'; R')$ and $P'' = (S''; R'')$ be two finite presentations. A mapping ψ of S' into $F(S'')$ such that the extension ψ_0 of ψ to a homomorphism induces an isomorphism i of $G(P')$ with $G(P'')$ such that $\psi_0(\tau) = i(\tau)$ in $G(P'')$, for any $\tau \in F(S')$, which we shall call an *isomorphism of presentations*. Note that the definition is independent of the choice of $G(P')$ and $G(P'')$. In fact, let $i_1 \colon G(P') \to G_1$ and $i_2 \colon G(P'') \to G_2$ be two isomorphisms and $\psi_0(\tau) = i(\tau)$ where $\psi_0(\tau)$ is an element of $G(P'')$ and τ in $i(\tau)$ is an element of $G(P')$, then $i_2\,\psi_0(\tau) = i_2 i\, i_1^{-1}(i_1(\tau))$. Now $i_2 i\, i_1^{-1}$ is an isomorphism of $G_1 \to G_2$. And $i_2(\psi_0(\tau))$ is an element $\psi_0(\tau)$ in G_2, and $i_1(\tau)$ is an element τ in G_1. So $\psi_0(\tau) = i_2 i\, i_1^{-1}(\tau)$ for $\psi_0(\tau)$ in G_2 and τ in G_1.

Now we shall prove the following.

Theorem 4. *There exists a family $\psi_{n,k}$ of isomorphisms, each isomorphism between P_n and $P_{e(n,k)}$ which is described effectively; strictly speaking, there exists a (total) recursive function $t(n, k, m)$ such that*

$$t(n, k, \overleftarrow{\tau}) = \overline{\psi_{0\,n,k}(\tau)}$$

for τ being a word in the letters of the presentation P_n.

Proof. Let T be a pair $\langle s_j, \omega\rangle$ where $s_j \in S$ and $\omega \in F(S)$. For any finite presentation $Q = (X; R)$ and any pair $T = \langle s_j, \omega\rangle$, define a presentation $Q(T)$ as

$$Q(T) = \begin{cases} (X, s_j; R, s_j\omega^{-1}) & \text{for } s_j \notin X \text{ and } \omega \in F(X) \\ Q & \text{for either } s_j \in X \text{ or } \omega \notin F(X). \end{cases}$$

Let E be an equation, i.e. an expression of the form

(1) $$\tau_1^{-1}(S)\, y_1^{\epsilon_1}\, \tau_1(S) \cdots \tau_n^{-1}(S)\, y_n\, \tau_n(S) = w(S)$$

where $n > 0$, $\epsilon_i = \pm 1$, $\tau_1, \ldots, \tau_n$, $w \in F(S)$ and $y_1, \ldots, y_n$ are some letters out of S.

For a finite presentation $Q = (X; R)$ and an equation E of the form (1) we define a new presentation

$$Q(E) = \begin{cases} (X; R, w) & \text{in case } \tau_1, \ldots, \tau_n \in F(X) \text{ and for some } r_1, \ldots, r_n \in R \\ & \tau_1^{-1} r_1^{\epsilon_1} \tau_1 \cdots \tau_n^{-1} r_n^{\epsilon_n} \tau_n = w \text{ in } F(X), \\ Q & \text{in the opposite case.} \end{cases}$$

A sequence

(2) $$\Gamma = \langle A_0, B_0, A_1, \ldots, A_{s-1}, B_{s-1}, A_s\rangle$$

where $A_0, \ldots, A_s$ are finite presentations and each $B_0, \ldots, B_{s-1}$ is either a pair or an equation, is called a *proof of isomorphy* iff for $i = 0, \ldots, s-1$, either

$$A_i(B_i) = A_{i+1} \quad \text{or} \quad A_{i+1}(B_i) = A_i.$$

According to Tietze's theorem, for $P, Q \in \mathrm{P}$, the groups $G(P)$ and $G(Q)$ are isomorphic iff there exists a proof of isomorphy Γ such that $P = A_1$ and $Q = A_s$.

All sequences (2) can be effectively enumerated and for any sequence Γ it can be effectively checked whether or not Γ is a proof of isomorphy owing to the finiteness of all presentations $A_0, \ldots, A_s$ in Γ. Thus all proofs of isomorphy can be effectively enumerated as

$$\Gamma_1, \Gamma_2, \ldots, \Gamma_k, \ldots$$

The above investigations are completely Rabin's. The function $e(n, k)$ is defined by him as follows:

For $\Gamma_k = \langle A_0, B_0, \ldots, A_s \rangle$,

$$e(n, k) = \begin{cases} n(A_s) & \text{if } n = n(A_0) \\ n & \text{if } n \neq n(A_0) \end{cases}$$

Now we wish to define for $e(n, k)$ a family $\psi_{n,k}$ of isomorphisms and a function $t(n, k, m)$.

It does not matter whether we define a family of $\psi_{n,k}$ or $\psi_{0n,k}$. First for any proof of isomorphism Γ, we define a homomorphism ψ_0.

For $\Gamma = \langle A_0, B, A_1 \rangle$ being a proof of isomorphy, we define a homomorphism ψ_0 and thereby an isomorphism ψ of A_0 and A_1 as follows:

If B is an equation, ψ_0 is the identity: $\psi_0(\tau) = \tau$.

If B is a pair $\langle s_j, \omega \rangle$ and Γ a proof of isomorphy, such that $A_0(B) = A_1$, ψ_0 is the identity. If $A_1(B) = A_0$, two cases are to be distinguished:

(1) $A_0 = A_1(B) = A_1$ then ψ_0 is an identity.

(2) $A_0 = A_1(B) \neq A_1$, then we delete the letter s_j from A_0 containing A_1. In this case ψ_0 is defined as follows: for $\tau = \tau(x_1, \ldots, x_n, s_j)$,

$$\psi_0(\tau) = \tau(x_1, \ldots, x_n, \omega(x_1, \ldots, x_n)).$$

Then for a proof of isomorphy $\Gamma = \langle A_0, B_0, \ldots, A_s \rangle$ the homomorphism $\psi_0(\Gamma)$ is a product of homomorphisms for proofs of isomorphy $\langle A_0, B_0, A_1 \rangle, \ldots, \langle A_{s-1}, B_{s-1} A_s \rangle$.

Then $\psi_0(\Gamma)$ induces an isomorphisms i of groups $G(A_0)$ and $G(A_s)$ such that $\psi_0(\Gamma)(\tau) = i(\tau)$. For $\Gamma_k = \langle A_0, B_0, \ldots, A_s \rangle$, $\psi(n, k)$ is defined as follows:

$$\psi(n, k) = \begin{cases} \psi_0(\Gamma_k) & \text{for } \; n = n(A_0) \\ \text{identity} & \text{for } \; n \neq n(A_0). \end{cases}$$

Then $\psi(n, k)$ is an isomorphism of P_n and $P_{e(n,k)}$.

To determine the function $t(n, k, m)$ such that

$$t(n, k, m) = \overline{\psi_{0\,n,k}(\overline{m})}$$

for m such that $\overline{m} \in F(X)$ where $P_n = (X; R)$ we shall first notice that the function leading from a triple x, y, j of natural numbers to a number n of a word obtained from $\overline{x}$ by the substitution of $\overline{y}$ to a letter s_j, is a total recursive function.

Now let all pairs and equations be effectively enumerated as

$$T_1, T_2, T_3, \ldots .$$

We shall define a function $z(x, m)$ as

$$\begin{cases} x & \text{for } T_m \text{ being an equation,} \\ x & \text{for } T_m \text{ being a pair } \langle s_j, \omega \rangle \text{ and in} \\ & \overline{x} = \tau(x_1, \ldots, x_n), \text{ all } x_1, \ldots, x_n \text{ different} \\ & \text{from } s_j. \\ \overline{\tau(x_1, \ldots, x_n, \omega)} & \text{for } T_m \text{ being a pair } \langle s_j, \omega \rangle \text{ and} \\ & \overline{x} = \tau(x_1, \ldots, x_n, s_j). \end{cases}$$

According to the definition $z(x, m)$ is total recursive.

From the definition of ψ_0 for a proof of isomorphy $\Gamma = \langle A_0, T_m, A_1 \rangle$ it follows exactly that

$$z(x, m) = \overline{\psi_0(\overline{x})}$$

for x such that $\overline{x} \in F(X)$ where $A_0 = (X; R)$. Now for $\Gamma_k = \langle A_0, B_0, ..., A_s \rangle$ we define

$$t(n, k, x) = \begin{cases} x \text{ for } & n \neq n(A_0) \\ z(z(\cdots z(z(x_1 m_0), m_1) \ldots, m_{s-2}) m_{s-1}) \\ \text{where } B_0 = T_{m_0}, \ldots, B_{s-1} = T_{m_{s-1}} \end{cases}$$

The function $t(n, k, x)$ is total recursive and

$$t(n, k, x) = \overline{\psi_{0\, n,k}(\overline{x})} \text{ for } \overline{x} \in F(X)$$

where $A_0 = (X; R)$. The proof of the theorem is thereby finished.

Now we return to uniform algorithms. Suppose that for a predicate $\Gamma(x_1, ..., x_n)$ there exists an uniform algorithm $f(i, m_1, ..., m_n)$ relative to a r.e. set Σ. Suppose $\alpha(i)$ is an enumeration for Σ. Then the set Σ^* is r.e. and suppose $\alpha^*(s)$ is s its enumeration. Our aim is to define an uniform algorithm $f^*(s, l_1, ..., l_n)$ relative to Σ^*.

When defining $f^*(s, l_1, ..., l_n)$, two cases must be distinguished. Suppose $P_{\alpha^*(s)} = (S'; R')$.

The first case: Some $\overline{l}_1, ..., \overline{l}_n$ contains a letter $s_p \notin S'$. We define

$$f^*(s, l_1, ..., l_n) = 2$$

(or any fixed integer different from 0 and 1).

The second case: All letters in $\overline{l}_1, ..., \overline{l}_n$ are in S'. Then we find $i, x, m_1, ..., m_n$ such that

$$\begin{aligned} &\alpha^*(s) = e(\alpha(i), x) \\ (3) \qquad &t(\alpha(i), x, m_j) = l_j \quad \text{for} \quad j = 1, ..., n \\ &f(i, m_1, ..., m_n) \text{ is either 0 or 1} \end{aligned}$$

and define

$$f^*(s, l_1, \ldots, l_n) = f(i, m_1, \ldots, m_n)$$

To show the correctness of the definition it is to be proven that such $i, x, m_1, \ldots, m_n$ exist, and that for any other $i', x', m'_1, \ldots, m'_n$ such that (3), the equality

$$(4) \qquad f(i, m_1, \ldots, m_n) = f(i', m'_1, \ldots, m'_n)$$

holds.

Since $P_{\alpha^*(s)} \in \Sigma^*$, there exists $P_y \in \Sigma$ isomorphic to $P_{\alpha^*(s)}$ so for some i, j, $e(y, j) = \alpha^*(s)$ and $\alpha(i) = y$. Now we consider the isomorphism $\psi_{0\,\alpha(i),x}$. When the second case of the definition takes place, then $l_1, \ldots, \bar{l}_n$ are in letters of $P_{\alpha^*(s)}$ and some words $\tau_1, \ldots, \tau_n$ in letters of $P_{\alpha(i)}$ exist such that

$$\psi_{0\,\alpha(i),x}(\tau_j) = \bar{l}_j \quad \text{for} \quad j = 1, \ldots, n.$$

For $m_1 = \bar{\tau}_1, \ldots, m_n = \bar{\tau}_n$, we have

$$t(\alpha(i), x, m_j) = l_j.$$

Since $\Gamma_{P_{\alpha(i)}}(\tau_1, \ldots, \tau_n)$ is determined, the values of $f(i, m_1, \ldots, m_n)$ are zeros and ones.

Now suppose that for some $i', x', m'_1, \ldots, m'_n$ the equalities (3) are satisfied. Then according to $e(\alpha(i), x) = \alpha^*(s) = e(\alpha(i'), x')$ we have isomorphisms $\psi_1 = \psi_{0\,\alpha(i),x}$, $\psi_2 = \psi_{0\,\alpha(i),x}$ onto $P_{\alpha^*(s)}$ such that $\psi_{0\,\alpha(i),x}(\overline{m}_j) = \bar{l}_j$ and $\psi_{0\,\alpha(i'),x}(\overline{m}'_j) = l_j$. Then $\psi_2^{-1}\psi_1$ is an isomorphism of $P_{\alpha(i)}$ and $P_{\alpha(i')}$ and moreover $\psi_2^{-1}\psi_1(\overline{m}_j) = \overline{m}'_j$ for $j = 1, \ldots, m$. According to the invariance of Γ we have

$$\Gamma_{P_{\alpha(i)}}(\overline{m}_1, \ldots, \overline{m}_n) \text{ iff } \quad \Gamma_{P_{\alpha(i')}}(\overline{m}'_1, \ldots, \overline{m}'_n)$$

This proves the equality (4).

The function $f^*(s, l_1, \ldots, l_n)$ is well determined and by definition is a computable function.

Now according to the definitions of uniform algorithm and decision function, the following holds:

$$f(i, m_1, ..., m_n) = \begin{cases} 0 \\ 1 \\ \neq 0, 1 \end{cases} \quad \text{iff} \quad \Gamma_{P_{\alpha(i)}}(\overline{m}_1, ..., \overline{m}_n) \begin{cases} \text{false} \\ \text{true} \\ \text{undefined;} \end{cases}$$

but due to the invariance of Γ under the isomorphism $\psi_{0\,\alpha(i),x}$

$$\Gamma_{P_{\alpha(i)}}(\overline{m}_1, ..., \overline{m}_n) \begin{cases} \text{false} \\ \text{true} \\ \text{undefined} \end{cases} \quad \text{iff} \quad \Gamma_{P_{\alpha^*(s)}}(\overline{l}_1, ..., \overline{l}_n) \begin{cases} \text{false} \\ \text{true} \\ \text{undefined.} \end{cases}$$

This proves that $f^*(s, l_1, ..., l_n)$ is a uniform algorithm for deciding Γ relatively to Σ^*.

So we have proved the following main theorem in this section.

Theorem 5. *If for a predicate Γ there exists a uniform algorithm f relative to a r.e. set Σ then there exists a uniform algorithm f^* for Γ relative to a necessarily r.e. closure Σ^* of Σ under isomorphism.*

Note: When we deal with a decision function relative to a set $K \subset P$ which is not r.e., the question of whether or not the decision function relative to K^* exists is open. In the proof of the latter theorem the recursive enumerability of Σ was used very strongly.

§5. Applications of the main theorem to the word problem and similar problems

In the sequel we shall give some applications of Theorem 5. First we consider applications to the word problem. We obtain the word problem when we set the predicate Γ as

$$\Gamma(x_1) \text{ iff } x_1 = 1.$$

For the word problem, the set $\mathcal{D}$ of all finite presentations with decidable word problem is neither a r.e. set nor the complement of a r.e. set. See Boone and Rogers [3]. It is also proved there that there exists a r.e. set $\Sigma \subset \mathcal{D}$ such that no uniform algorithm for the w.p. relative to Σ exists. From this it follows that no decision function for the w.p. relative to $\mathcal{D}$ exists.

For some interesting classes of presentations such as $\mathcal{R}$ (residually finite groups) and $\mathcal{S}$ (nontrivial simple groups), decision functions for the w.p. exist; cf. the investigations of Section 3. Note that for both $\mathcal{R}$ and $\mathcal{S}$ it is not known whether or not they are r.e. or complements of r.e. sets.

Now we shall give some examples of r.e. sets with uniform algorithms for the word problem

Let $\Sigma \subset \mathcal{D}$ be a r.e. set with uniform algorithm. Then according to Theorem 5 the uniform algorithm for Σ^* exists.

The first interesting case is when $\Sigma = (P)$, i.e., Σ consists of one given presentation with decidable w.p. Then Σ^* is the class of all finite presentations of the group $G(P)$. The result can be stated that when P has decidable word problem, then there is one uniform method to decide the word problem for any presentation of $G(P)$.

The second interesting case is the class $\mathcal{M}$ of all finite presentations with a single defining relation. It is known (see Magnus [6]) that there exists a uniform algorithm for deciding the word problem relative to the (evidently r.e.) set $\mathcal{M}$. The class $\mathcal{M}^*$ is the class of all finite presentations of groups which can be defined by a single defining relation. Then according to Theorem 5, there exists a uniform algorithm for deciding the w.p. relative to $\mathcal{M}^*$.

The third case is the class $\mathcal{F}$ of all finite presentations of finite groups. It is known that there exists an enumeration $\alpha(i)$ such that in the sequence

$$(1) \qquad \Sigma = (P_{\alpha(1)}, P_{\alpha(2)}, \ldots)$$

there is exactly one presentation of each finite group (meaning different groups as nonisomorphic). In Mostowski [10] Eq. (1) is given as the set of group tables. Then $\Sigma^* = \mathcal{F}$ is an r.e. set. From Theorem 5 follows the well-known result that the uniform algorithm relative to $\mathcal{F}^*$ exists.

The fourth application is to nilpotent groups. Let N_c be the class of all finite presentations of nilpotent groups of nilpotency $\leqslant c$, and $\mathsf{N} = \bigcup_{c=1}^{\infty} \mathsf{N}_c$ be the class of all finite presentations of nilpotent groups. We shall prove the following.

Lemma. *The sets* N *and* N_c *for* $c = 1, 2, \ldots$ *are r.e. sets.*

Proof. A group $G = \{g_1, \ldots, g_n\}$ is nilpotent of nilpotency $\leqslant c$, iff for any $\alpha_1, \ldots, \alpha_{c+1} \in (1, \ldots, n)$

$$(g_{\alpha_1}, \ldots, g_{\alpha_{c+1}}) = 1,$$

where $(x_1, x_2) = x_1^{-1} x_2^{-1} x_1 x_2$, $(x_1, \ldots, x_k, x_{k+1}) = ((x_1, \ldots, x_k), x_{k+1})$. See, e.g., Mostowski [9], Lemma 3.

Now for a presentation $P = (s_1, \ldots, s_n; R)$ of a group we form a new presentation $P^{(c)}$ by adding new relations

$$(s_{\alpha_1}, \ldots, s_{\alpha_{c+1}}) = 1 : \alpha_1, \ldots, \alpha_{c+1} \in (1, \ldots, n).$$

Then $P^{(c)}$ is a presentation of a nilpotent group of nilpotency $\leqslant c$. The function $P \to P^{(c)}$ maps P into N_c. Moreover, $P \in \mathsf{N}_c$ iff P and $P^{(c)}$ are presentations of isomorphic groups. It is easy to see that the two argument function

$$g(k, c)\colon k \to n(P_k^{(c)})$$

is a total recursive function. This proves that $\mathsf{P}^{(c)}$ and $\bigcup_{c=1}^{\infty} \mathsf{P}^{(c)}$ are r.e. sets. So $\mathsf{N}_c = \mathsf{P}^{(c)*}$ and $\mathsf{N} = (\bigcup_{c=1}^{\infty} \mathsf{P}^{(c)})^*$ are r.e. sets.

The proof of the lemma is finished. Since $\mathsf{N}, \mathsf{N}_c \subset \mathsf{R}$ $c = 1, 2, \ldots$ according to the lemma just proved and to Theorem 1, there exist uniform algorithms for deciding the w.p. relative to N and N_c for $c = 1, 2, \ldots$.

The next example is an application of Theorem 5 to the conjugacy problem, i.e., a problem for a predicate

$$\Gamma(x_1, x_2) \text{ iff there exists } z \text{ such that } zx_1 = x_2 z.$$

It is known that for the conjugacy problem there exists a decision function relative the class C of all finite presentations of conjugacy separable groups. It is known (see Blackburn [2]) that $N \subset C$. Owing to the recursive enumerability of N, there exists a uniform algorithm for the conjugacy problem relative to N.

All that we can obtain as a corollary to Theorem 5 is the following: If there exists a uniform algorithm for the conjugacy problem relative to an r.e. set set Σ, then there exists a unfiorm algorithm relative to the set Σ^*.

There is a long list of other applications of Theorem 5. We shall describe a general idea. Let A be an algebraic property, i.e., a property of groups shared by isomorphisms. The first case: Define a zero argument predicate Γ as

Γ true for a group having A,

Γ false for a group having non A.

The interesting examples of A are: triviality, finiteness, abelianity, nilpotency, solvability, having decidable word problem, and so on. A very long list of such properties is described in [1].

The other case is when we define an n-argument predicate Γ as follows

$\Gamma(g_1, ..., g_n)$ iff an n-generator subgroup $\{g_1, ..., g_n\}$ of the group G has a property A.

For both cases, Theorem 5 can be restated as the following.

Theorem 6. *If there exists a uniform algorithm for deciding whether the group has an algebraic property A, relative to a r.e. set Σ of finite presentations, then there exists a uniform algorithm relative to Σ^*. If there exists a uniform algorithm for deciding whether an n-generator subgroup of the group has a property A relative to a r.e. set Σ, then there exists a uniform algorithm relative to Σ^*.*

But note that all except the last of the examples of the properties A listed above, are strong Markov properties, hence there exists a r.e. set Σ (depending of A) such that there is no uniform algorithm for deciding whether the group has the property A relative to Σ. For details see Collins [4].

Other possible applications of the second part of Theorem 6 arise when we take as A the following subgroup properties: being of finite index, being normal, being isomorphic with a given group, being a finitely related group, and so on; cf. also [1].

At the end of this section we shall mention the inclusion problem, sometimes called Magnus' extended word problem.

For a single presentation $P = (X; R)$, there are in fact three inclusion problems; some of them surely different; c.f. Michailowa [7] and [8]:

(1) *Weak inclusion problem*: For a given $\tau_1, \ldots, \tau_n \in F(X)$ decide for any $\tau_{n+1} \in F(X)$ whether or not

$$\tau_{n+1} \in \{\tau_1, \ldots, \tau_n\} \quad \text{in } G(P)$$

(2) *n-generator inclusion problem.* The n is fixed. The problem is to decide for any $\tau_1, \ldots, \tau_n, \tau_{n+1} \in F(X)$ whether or not

$$\tau_{n+1} \in \{\tau_1, \ldots, \tau_n\} \quad \text{in } G(P).$$

(3) *The inclusion problem* (*the strong inclusion problem*). This is the union of all n-generator inclusion problems for all n. The weak inclusion problem does not make sense for a class of presentations.

The n-generators inclusion problem allows a treatment by our theory, since the $n + 1$ variable predicate Γ

$$\Gamma(x_1, \ldots, x_{n+1}) \text{ iff } x_{n+1} \in \{x_1, \ldots, x_n\} \quad \text{in } G(P)$$

is algebraic.

So from Theorem 5 we obtain a corollary.

Corollary. *If there exists a uniform algorithm for solving the n-generator inclusion problem relative to a r.e. set Σ then the uniform algorithm relative to Σ^* exists.*

For a discussion of the (strong) inclusion problem see the next section.

§6. Families of predicates

In the previous section, we were dealing with decision problems such as:

the n-generator subgroup has algebraic property A or

the $(n-1)$-generator inclusion problem.

But these problems can be treated in another sense as follows:

a finitely generated subgroup has an algebraic property A or

a strong inclusion problem.

In both cases we have a family Γ_n, $n = 1, 2, 3, \ldots$ of algebraic predicates $\Gamma_n = \Gamma_n(x_1, \ldots, x_n)$ such that

$$\Gamma_n(x_1, \ldots, x_n) \quad \text{iff} \quad \{x_1, \ldots, x_n\} \text{ has a property } A$$

or

$$(1) \qquad \Gamma_n(x_1, \ldots, x_n) \quad \text{iff} \quad x_n \in \{x_1, \ldots, x_{n-1}\}$$

and we wish to decide the problems for all predicates from the family Γ_n relative to the same class $\mathcal{K}$ of finite presentations. This can be translated into the language of recursive functions as follows:

Let us denote by $\Pi_n(X)$ a set of n-tuples

$$u = \langle \tau_1, \ldots, \tau_n \rangle \text{ where } \tau_1, \ldots, \tau_n \in F(X)$$

for $X \subset S$ put

$$\Pi(X) = \bigcup_{n=1}^{\infty} \Pi_n(X).$$

We fix some Gödel enumeration of the set $\Pi(S)$ and denote by $\bar{u}$ a number of $u \in \Pi(S)$, and for a natural number k, we denote by $\bar{k}$ an $n(k)$-tuple $u \in \Pi(S)$ such that $\bar{u} = k$.

For a family $\Gamma_n = \Gamma_n(x_1, \ldots, x_n)$ of algebraic predicates, define a new partial predicate $\Gamma_P(u)$ where $u \in \Pi(S)$ and $P = (X; R)$ is a presentation $P \subset \mathcal{P}$ as follows.

For $u = \langle \tau_1, \ldots, \tau_n \rangle \in \Pi_n(X)$,

$$\Gamma_P(u) \quad \text{iff} \quad \Gamma_n(\tau_1, \ldots, \tau_n) \quad \text{in the group } G(P)$$

and for $u \in \Pi(S)\backslash\Pi(X)$, the predicate $\Gamma_P(u)$ is undefined.

Then a notion of a decision function and a uniform algorithm can be extended to the family Γ_n as follows:

Definition 1′. A partial recursive function $f(m, k)$ is called a decision function for a family Γ_n; $n = 1, 2, ...,$ of predicates relatively to a class $K \subset P$ iff for any integer m such that $P_m \in K$

$$(d')\text{ for } \Gamma_{P_m}(\bar{k}) \text{ being } \begin{cases}\text{true}\\ \text{false}\end{cases} \text{ the value } f(m, k) = \begin{cases}1\\ 0.\end{cases}$$

Definition 2′. A function $f(i, m)$ is a uniform algorithm for deciding a family Γ_n; $n = 1, 2, ...,$ of algebraic predicates relative to a r.e. set $\Sigma \subset P$ iff there exists a decision function $f'(k, m)$ for the family such that for some enumeration $\alpha(i)$ of Σ

$$f(i, m) = f'(\alpha(i), m)$$

Theorems 1, 2, and 3 can be easily proved for a family Γ_n of algebraic predicates (instead of for an algebraic predicate). The analog of the main Theorem 5 can be proved also.

Theorem 5′. *If for a family Γ_n, $n = 1, 2, ...,$ of algebraic predicates there exists a uniform algorithm f relatively to a r.e. set $\Sigma \subset P$, then there exists a uniform algorithm f^* for the family, relative to the necessarily r.e. closure Σ^* of Σ under isomorphism.*

The proof needs only small technical changes in the proof of Theorem 5, given in Section 4.

As an application of Theorem 5′ or families (1) we obtain immediately the following corollaries.

Corollary. *If there exists a uniform algorithm for deciding whether a finitely generated subgroup has an algebraic property A, relatively to a r.e. set $\Sigma \subset P$, then there exists a uniform algorithm relative to Σ^*.*

Corollary. *If there exists a uniform algorithm for deciding the (strong) inclusion problem relative to a r.e. set $\Sigma \subset \mathcal{P}$, then there exists a uniform algorithm relative to Σ^*.*

§7. Isomorphism problem

Let G be a fixed group. Define a zero argument predicate Γ as follows. For a $P \in \mathcal{P}$ defining a group isomorphic to G

Γ is true

and for a $P \in \mathcal{P}$ defining a group nonisomorphic to G

Γ is false.

Then evidently Γ is algebraic.

The problem of deciding Γ relative to a class $\mathcal{K} \subset \mathcal{P}$ is simply a weak isomorphism problem, i.e. the problem of isomorphism with a fixed group. The decidability of a weak isomorphism means simply the existence of a decision function for Γ (or a uniform algorithm in the case when $\mathcal{K}$ is an r.e. set) relative to $\mathcal{K}$. So the following theorem holds.

Theorem 7. *If a weak isomorphism problem is decidable relatively to a r.e. set Σ, then it is decidable relative to the set Σ^*.*

The isomorphism problem is the most interesting ; i.e., the problem of whether for presentations P and Q the groups $G(P)$ and $G(Q)$ are isomorphic. The theory developed above does not fit exactly in this case. We now develop some modifications. First, the decidability of the problem must be stated in terms of recursive functions.

Definition. The isomorphism problem is decidable relative to a class $\mathcal{K}$ of presentations if there exists a recursive function $f(n, k)$ such that

$$(1) \quad f(n,k) = \begin{cases} 0 \text{ for } P_n, P_k \in K \text{ and } G(P_n) \text{ and } G(P_k) \text{ non isomorphic} \\ 1 \text{ for } P_n, P_k \in K \text{ and } G(P_n) \text{ and } G(P_k) \text{ isomorphic.} \end{cases}$$

The value of $f(n,k)$, in the case $P_n \notin K$ or $P_k \notin K$, need not be defined.

Now we shall prove the following

Theorem 8. *If the isomorphism problem is decidable relative to a r.e. set Σ of presentations, then the problem is decidable relative to the set Σ^*.*

For the proof, we need to define a recursive function $f'(n,k)$ satisfying (1) relative to Σ^*. Let $g(i)$ be an enumeration for Σ, i.e., $\Sigma = (P_{g(1)}, P_{g(2)}, \ldots)$. For $P_n, P_k \in \Sigma^*$, there exists some s, t, i, j such that

$$(2) \qquad P_{e(n,s)} = P_{g(i)} \quad \text{and} \quad P_{e(k,t)} = P_{g(j)}$$

Set $f'(n,k) = f(g(i), g(j))$ in case there exist s, t, i, j such that (2) and undefined in the other case. The function $f'(n,k)$ is well defined. Suppose there exist s', t', i', j' such that

$$P_{e(n,s')} = P_{g(i')} \quad \text{and} \quad P_{e(k,t')} = P_{g(j')}.$$

Then groups $G(P_n)$, $G(P_{g(i)})$ and $G(P_{g(i')})$ are isomorphic since $G(P_x)$ is isomorphic to $G(P_{e(x,y)})$. Similarly the groups $G(P_k)$, $G(P_{g(j)})$ and $G(P_{g(j')})$ are isomorphic. This proves that $f(g(i), g(j)) = f(g(i'), g(j'))$.

The function $f'(n,k)$ is recursive by definition, and evidently satisfies (1).

From the theorem just proven immediately follows

Theorem 9. *If there exists a recursive set Σ of finite presentations such that different presentations give nonisomorphic groups, then for Σ^* the isomorphism problem is decidable.*

Proof. In the case of the theorem the function

$$f(x, y) = \begin{cases} 0 & x \neq y \\ 1 & x = y \end{cases}$$

satisfies (1) relatively to Σ.

Now we shall give some well-known facts which are corollaries from this theorem.

Let A_0 be a set of presentations of the following form

$$P = (s_1, ..., s_r; s_1^{n_1}, ..., s_k^{n_k}, (s_i, s_j) \text{ for all } 1 \leqslant i < j \leqslant r)$$

where $k \leqslant r$, and $n_1 \leqslant \cdots \leqslant n_k$ are all primes or powers of primes. Evidently A_0 is recursively enumerable; different presentations from A_0 present nonisomorphic groups.

Moreover, since any finitely generated abelian group is a direct sum of cyclic groups, A_0^* is the set of all finite presentations of abelian groups. According to our theorem, the isomorphism problem for abelian groups is decidable (cf. also Rabin [11]).

In Mostowski [10] it is proved that there exists a recursive sequence Σ of presentations of finite groups (given by tables) such that different tables present nonisomorphic groups and any finite group has a presentation by tables. From this follows the well-known result: the isomorphism problem is decidable relative to finite groups.

§8. Relativisation and degrees of unsolvability

Let F be a set of total functions from integers to integers and $T(F)$ be the smallest class of functions containing $0, x + 1$, $I_1(x_1, x_2) = x_1$, $I_2(x_1, x_2) = x_2$, and the set F, and closed under substitution, recursion scheme and minimization.

Note that $F_1 \subset F_2$ implies $T(F_1) \subset T(F_2)$, and $TT(F) = T(F)$. If F is empty, $T(F) = T$ is the class of all (total) recursive functions. If F consists of a single function g, we shall set $T(g) = T(F)$. The equality $T(g) = T$ means g is total and recursive.

Call $T(g)$ the degree of unsolvability of the function g. The degree of g will be denoted as $\deg(g)$ and $\deg(g) \leqslant \deg(h)$ will mean that $T(g) \subset T(h)$. All total recursive functions have the lowest de= gree of unsolvability. For f being a characteristic function of a set A of integers, $\deg(f)$ is simply the degree of unsolvability of A (strictly speaking the degree of unsolvability of the problem $x \in A$); cf. [5].

Let $\Gamma(x_1, ..., x_n)$ be an algebraic predicate and $\Sigma \subset P$ a r.e. set with an enumeration $\alpha(i)$.

Define a total function $f(i, m_1, ..., m_n)$ as follows:

(a) For $P_{\alpha(i)} = (X; R)$ and $\overline{m}_1, ..., \overline{m}_n \in F(X)$

$$(1)\quad f(i, m_1, ..., m_n) = \begin{cases} 0 & \Gamma(\overline{m}_1, ..., \overline{m}_n) \text{ false in } G(P_{\alpha(i)}) \\ & \text{iff} \\ 1 & \Gamma(\overline{m}_1, ..., \overline{m}_n) \text{ true in } G(P_{\alpha(i)}) \end{cases}$$

(b) For some of $\overline{m}_1, ..., \overline{m}_n$ not belonging to $F(X)$, set

$$f(i, m_1, ..., m_n) = 2.$$

Note that since the set U of $(n + 1)$-tuples $\langle i, m_1, ..., m_n \rangle$ such that some of $\overline{m}_1, ..., \overline{m}_n$ does not belong to $F(X)$ is a computable set (cf. Section 3), then for any total function $g(i, m_1, ..., m_n)$ satisfying (a) we have $T(f) \subset T(g)$.

From the proof of Theorem 2, it follows that if, for any other enumeration $\beta(i)$ of the set Σ we have a total function g satisfying (a), (b), then there exists a function $t(x) \in T$ such that

$$f(i, m_1, ..., m_n) = g(t(i), m_1, ..., m_n).$$

This proves that $T(f) \subset T(g)$ and interchanging the roles of f and g, $T(g) \subset T(f)$, and therefore $T(f) = T(g)$. Then we shall call $\deg(f)$, the degree of unsolvability of Σ for the predicate Γ and we shall write

$$(2)\qquad \deg_\Gamma \Sigma = \deg f.$$

For a fixed Γ, we shall write $\deg \Sigma = \deg_\Gamma \Sigma$. Now let $\Gamma = \Gamma(x_1, \ldots, x_n)$ be fixed. If f_1 is a function relative a r.e. set Σ_1 satisfying (a), (b) and Σ is a r.e. subset of Σ_1. Then there exists a function f satisfying (a), (b) relative to Σ such that (cf. proof of Theorem 1)

$$f \in T(f_1). \tag{3}$$

For f_1 and f_2 being functions satisfying (a), (b) relative to r.e. sets Σ_1 and Σ_2, there exists a function f satisfying (a), (b) relative to the set $\Sigma = \Sigma_1 \cup \Sigma_2$, and such that (cf. proof of Theorem 3)

$$f \in T(f_1, f_2). \tag{4}$$

So we have the following.

Theorem 10. *The function* $\deg(\Sigma)$ *is defined on an ideal of r.e. subsets of* P, *and moreover*

$$\Sigma \subset \Sigma_1 \quad \textit{implies} \quad \deg \Sigma \leqslant \deg \Sigma_1,$$

and

$$\deg(\Sigma_1) \cup \deg(\Sigma_2) = \deg(\Sigma_1 \cup \Sigma_2). \tag{5}$$

Proof. From (3) we have $T(f) \leqslant T(f_1)$ and thereby $\deg(\Sigma) \leqslant \deg(\Sigma_1)$. Now from (4)

$$T(f) \subset T(f_1, f_2)$$

and since

$$T(f_1) \subset T(f) \text{ and } T(f_2) \subset T(f)$$

we have

$$T(f) = T(f_1, f_2).$$

Now, by relativisation of the proof of Theorem 5 to $T(f)$ (instead of T) we obtain the following: If there exists a function f

satisfying (a), (b) relative to Σ, then there exists a function f^* satisfying (a), (b) relative to a (necessarily r.e.) closure Σ^* of Σ under isomorphism, and such that $f^* \in T(f)$. So the theorem.

Theorem 11. *For an algebraic predicate Γ and any r.e. subset $\Sigma \subset P$ and its closure Σ^* under isomorphism*

$$deg_\Gamma \Sigma = \deg_\Gamma \Sigma^*.$$

Proof. Since $\Sigma \subset \Sigma^*$, $\deg_\Gamma \Sigma \leqslant \deg_\Gamma \Sigma^*$ and since $f^* \in T(f)$, then $T(f^*) \subset T(f)$ and thereby $\deg_\Gamma \Sigma^* \leqslant \deg_\Gamma \Sigma$.

The most interesting application of Theorem 11 is a case when $\Sigma = (P)$ is a class containing a single presentation. Then $\deg_\Gamma \Sigma$ is simply the degree of unsolvability of Γ for P. Now when Q is a presentation of a group isomorphic with that for P, then $(Q)^* = \Sigma^*$, and

$$\deg_\Gamma (P) = \deg_\Gamma \Sigma^* = \deg_\Gamma (Q).$$

So we have the following

Corollary. *For an algebraic predicate Γ all presentations of the same group have the same degree of unsolvability.*

As an interesting special cases of this corollary we obtain the following result:

All presentations of the same group have the same degree of unsolvability of the word problem, the conjugacy problem, and the n-generator inclusion problem.

References

[1] G. Baumslag, W.W. Boone and B.H. Neumann, Some unsolvable problems about elements and subgroups of groups, Math. Scand. I (1959) 191–201.
[2] N. Blackburn, Conjugacy in nilpotent groups, Proc. Amer. Math. Soc. (1965) 143–148.

[3] W.W. Boone and H. Rogers, On a problem of J.H.C. Whitehead and a problem of Alonzo Church, Math. Scand. 19 (1966) 185–192.
[4] D.J. Collins, On recognising properties of groups which have solvable word problem, Archiv d Math. 25 (1970) 31–39.
[5] S.C. Kleene and E.L. Post, The uppersemilattice of degrees of unsolvability, Ann. Math. (2) 59 (1954) 379–407.
[6] W. Magnus, Die Identiäts problem für Gruppen mit einer definirenden Relationen, Math. Ann. 106 (1932) 295–307.
[7] K.A. Mihailova, Inclusion problem for a direct product of groups, D.A.N. 119 No. 6 (1958) 1103–1105.
[8] K.A. Mihailova, Inclusion problem for a free product of groups DAN 127 No. 4 (1959) 746–748.
[9] A.W. Mostowski, Computational algorithms for deciding some problems for nilpotent groups, F.M. 59 (1966) 137–152.
[10] A.W. Mostowski, On decidability of some problems in special classes of groups, F.M. 59 (1966) 123–135.
[11] M.O. Rabin, Recursive unsolvability in group theoretic problems, Ann. Math. 67 (1958) 172–194.
[12] P.E. Schupp, A note on recursively enumerable predicates in groups, F.M. LXVI (1969) 61–63.

THE ISOMORPHISM PROBLEM FOR ALGEBRAICALLY CLOSED GROUPS

B.H.NEUMANN

Vanderbilt University, The University of Cambridge, and The Australian National University

1. Known facts and a conjecture

Let $\mathfrak{C}$ be a class of algebraic systems (briefly: algebras), all of the same species; for example the class of groups; and let A be an algebra in $\mathfrak{C}$. Consider a set Σ of sentences made from elements of A, element variables $x_1, x_2, \ldots, y, z, \ldots$, the algebraic operations of the species of $\mathfrak{C}$, equality, and the operations of the lower predicate calculus. We call Σ *consistent* over A if A can be embedded in an algebra A' in $\mathfrak{C}$ such that Σ is satisfied, that is to say, all sentences of Σ are valid in A'.

Next let $\mathfrak{S}$ be a class of such sets Σ of sentences. Then A is called *algebraically closed* (or, if necessary, $\mathfrak{C}$-$\mathfrak{S}$-algebraically closed) if every set $\Sigma \in \mathfrak{S}$ that is consistent over A is satisfied in A.

Take, for example, $\mathfrak{C}$ to be the class of (commutative) fields, and $\mathfrak{S}$ the class of singletons $\Sigma = \{\sigma\}$, where each sentence σ is a polynomial equation in a single variable prefaced by an existential quantifier binding the variable: then we get the usual notion of an algebraically closed field.

Again, let $\mathfrak{S}$ consist of singletons $\Sigma = \{\sigma\}$, where now each sentence σ is a finite conjunction of equations

$$f(x_1,x_2,\ldots,x_n) = f'(x_1,x_2,\ldots,x_n)$$

and negations of such equations,

$$g(x_1,x_2,...,x_n) \neq g'(x_1,x_2,...,x_n) ,$$

prefaced by existential quantifiers binding all the variables; here f, f', g, g' are *words* formed with the relevant algebraic operations from variables and constants, that is elements of the algebra A under consideration. When $\mathfrak{C}$ is the class of all groups, the $\mathfrak{C}$-$\mathfrak{S}$-algebraically closed algebras are the algebraically closed groups introduced by Scott [7].

Other classes of algebraic systems will also serve instead of groups, and in particular the facts that will be stated and the conjecture to be discussed about algebraically closed groups apply also, *mutatis mutandis,* to algebraically closed semigroups [6].

The following theorem and corollary were proved by Scott [7]:

Theorem 1.1. *Every group can be embedded in an algebraically closed group.*

Corollary 1.2. *Every countable group can be embedded in a countable algebraically closed group.*

This latter fact is a corollary of the proof method rather than of the theorem; and the proof applies equally to other classes of algebras, provided that they admit direct limits and their species is defined by finitely many finitary operations: it is in essence the proof devised by Steinitz [8] to prove the corresponding statement for fields.

We deduce an easy consequence:

Theorem 1.3. *There are* $2^{\aleph_0}$ *mutually non-isomorphic countable algebraically closed groups.*

To see this we need only remember that there are $2^{\aleph_0}$ mutually non-isomorphic 2-generator groups [4], and each can, by Corollary 1.2, be embedded in a countable algebraically closed group; but each countable group contains only countably many pairs of elements and thus only countably many 2-generator groups. Hence $2^{\aleph_0}$ countable algebraically closed groups are needed to accommodate all the 2-generator groups. On the other hand there can not be more than $2^{\aleph_0}$ isomorphism classes of countable algebraically closed groups, because there are no more than $2^{\aleph_0}$ isomorphism classes of countable groups altogether.

We note in passing that the same argument will show the same fact for semigroups instead of groups; but not for fields: there are only countably many countable algebraically closed fields, one for each combination of characteristic (a positive prime number or zero) and transcendence degree (a cardinal number, in this case $\leqslant \aleph_0$).

However, no algebraically closed group is *explicitly* known, the existence proof being highly non-constructive. This stems in part from the fact that there is no useful criterion known that tells one what sentences are or are not consistent over a given group.

But assume now that we are given the knowledge that two countable groups, say A and B, are algebraically closed. We then ask how one can tell whether or not they are isomorphic. Now I conjecture that this problem, *the isomorphism problem for countable algebraically closed groups, is algorithmically unsolvable*. The aim of this note is to give some reasons for entertaining this conjecture.

2. Absolute presentations of groups

A proof of the conjecture would first of all require a more precise formulation of it; in particular we would need to know just how the groups A and B are given, and what algorithms are admitted. But as I am not proving the conjecture, I shall allow myself much imprecision.

We may think of A and B as given, for example, in some presentation, that is by defining relations in some set of generators; and we note a fact that is not really relevant:

Theorem 2.1. *An algebraically closed group can not be finitely generated, nor finitely related.*

Proof. If $g_1, g_2, ..., g_n$ are finitely many elements in an arbitrary group G, then the sentence

$$\exists z : g_1 z = z g_1 \ .\&.\ g_2 z = z g_2 \ .\&. \ \dots \ .\&.\ g_n z = z g_n \ .\&.\ z \neq 1$$

is consistent over G, as it is satisfied in any direct product of G and a non-trivial group. It follows that in an algebraically closed group every finitely generated subgroup has a non-trivial centralizer; but it is known [5] that every algebraically closed group is simple, and thus has trivial centre: hence it can not be finitely generated. Now in a finitely related group in infinitely many generators infinitely many of these generators occur in none of the defining relations and thus generate freely a free group that is a free factor of the whole group: take any two of these free generators, and observe that the centralizer of the subgroup they generate is trivial: but in an algebraically closed group this can not happen, as we have seen, and the theorem follows.

How could we hope to prove that B is not isomorphic to A? We know that B can not contain (isomorphic copies of) all possible 2-generator groups as subgroups; so we might try to spot one that is not in B, but is in A. Thus we pick two elements, say a_1 and a_2, of A and try to show that the subgroup H, say, they generate can not be matched in B.

To this end we need to know what this subgroup H is. A set of defining relations of H in terms of a_1 and a_2, that is a presentation of H, suffices if we are given that it is a presentation, that is to say, if we know that the only relations between a_1 and a_2 are the consequences of the given relations. We observe that to be able to decide whether an arbitrary word in a_1 and a_2 equals the unit element in consequence of the given defining relations, we should be able to solve the word problem for this presentation of H; and to be able to decide whether H is isomorphic to the subgroup K, say, of B generated by a pair b_1, b_2 of elements picked from B, we should be able to solve the isomorphism problem for our presentation of H.

If H is finitely presented, say by

$$r_1(a_1,a_2) = 1,\ r_2(a_1,a_2) = 1,\ \dots,\ r_m(a_1,a_2) = 1\ ,$$

then the sentence

$$\exists x_1 . \exists x_2 : r_1(x_1,x_2) = 1 .\&. \, r_2(x_1,x_2) = 1 .\&. \, \dots .\&.$$
$$r_m(x_1,x_2) = 1 .\&. \, x_1 \neq 1$$

is consistent over every group G, as it is valid in $G \times H$, where we have assumed, without loss of generality as we shall presently see (Lemma 2.5), that H is not cyclic; hence this sentence is satisfied in B. This does not quite mean that H is matched in B and thus useless for telling B from A: if b_1, b_2 in B satisfy this sentence, the group K they generate might still satisfy further relations that are not satisfied by a_1 and a_2 in H.

This leads us to think of H as specified more precisely by what we shall call an *absolute presentation*, that is a set of relations

(2.2) $\quad r_1(a_1,a_2) = 1, r_2(a_1,a_2) = 1, \dots$

together with a set of "irrelations"

(2.3) $\quad s_1(a_1,a_2) \neq 1, s_2(a_1,a_2) \neq 1, \dots$

that jointly ensure that if b_1, b_2 are two group elements that satisfy them in place of a_1, a_2, then mapping a_1 on b_1 and a_2 on b_2 necessarily generates an isomorphism of H onto the group K generated by b_1 and b_2. If a presentation by generators and defining relations only is to be distinguished from an absolute presentation, it will be called a *relative presentation.*

We note in passing that all the classical problems on relative group presentations, such as the word problem, the conjugacy problem, the isomorphism problem, also make sense for absolute presentations, and are still in need of solving, or unsolving.

It is easy to see that if H is finitely absolutely presented, then it is matched in B and thus useless for distinguishing B from A:

Lemma 2.4. *Every algebraically closed group contains an isomorphic copy of every finitely absolutely presented group.*

We need only observe that the sentence obtained by conjunction of the defining relations and irrelations, written in variables x_1, x_2 that are then bound by existential quantifiers, is consistent over an arbitrary group, hence satisfied in every algebraically closed group.

We have already, in claiming that no generality is lost by assuming $a_1 \neq 1$, made use of a fact that is almost a corollary of this lemma:

Lemma 2.5. *Every algebraically closed group contains cyclic groups of every order.*

Cyclic groups of finite order are, like all finite groups, finitely absolutely presented, hence contained in every algebraically closed group by Lemma 2.4. For an infinite cyclic group we use the sentence

$$\exists x_1 . \exists x_2 . \exists x_3 . \exists x_4 : x_1^{x_2} = x_1^2 \;.\&.\; x_2^{x_3} = x_2^2 \;.\&.\; x_3^{x_4} = x_3^2 \;.\&.$$

$$x_4^{x_1} = x_4^2 \;.\&.\; x_1 \neq 1 \;,$$

which is satisfied in a group introduced by Graham Higman [1], and which has the property that in every non-trivial epimorph the generators have infinite orders: this sentence must be satisfied in every algebraically closed group, and the lemma follows.

Corollary 2.6. *An algebraically closed group can not be periodic.*

3. Infinite presentations

It follows from what has been said that our hope of picking two elements a_1, a_2 from the algebraically closed group A such that the subgroup H they generate is not matched in the algebraically closed group B, so as to distinguish B from A, now rests with an H that can not be finitely absolutely presented, and is not infinite cyclic either. However, as we seek an effective procedure that distinguishes our groups, it seems to me that we need to assume at least that H is *recursively* absolutely presented, in the sense that in some effective enumeration of the elements of the free group on a_1, a_2 the relators $r_1(a_1,a_2)$, $r_2(a_1,a_2)$, $r_3(a_1,a_2)$, ... are recursively enumerated, and so are the irrelators $s_1(a_1,a_2)$, $s_2(a_1,a_2)$, $s_3(a_1,a_2)$, ..., where the absolute presentation is given by (2.2), (2.3). Note that one and the same group can have absolute presentations that are recursive and others that are not: for example the infinite cyclic group generated by an element a can be given by an arbitrary infinite sequence of irrelations of the form

$$a^{n(1)!} \neq 1,\ a^{n(2)!} \neq 1,\ a^{n(3)!} \neq 1, ...,\ n(1) < n(2) < (3) < ... ,$$

and the recursiveness or otherwise of this absolute presentation depends on the sequence of numbers $n(i)$ chosen.

We need a procedure to reduce a countable set of irrelations to a single irrelation; the method is adapted from [5]. For notational convenience we describe it here only for a 2-generator group: it is clear that it applies in the same way to groups without this restriction.

Lemma 3.1. *Let the group H be generated by a_1, a_2 with the defining relations*

(3.1.1) $\quad r_i(a_1,a_2) = 1,\ i = 1, 2, ...,$

and assume that H also satisfies the irrelations

(3.1.2) $\quad a_1 \neq 1$,

(3.1.3) $\quad s_j(a_1,a_2) \neq 1,\ j = 1, 2, ...$;

here i and j range over some or all positive integers. Then H can be embedded in the group H_1 generated by H and three elements c, c', d with the defining relations of H and

(3.1.4) $\quad c^2 = 1,\ (cc')^2 = 1,\ c'^2 = a_1$,

(3.1.5) $(d^{-j}cd^js_j(a_1,a_2))^3 = 1, j = 1, 2, \ldots,$

with j ranging over the same integers as in (3.1.3).

Corollary 3.2. *With the same notation, if H is absolutely presented by* (3.1.1), (3.1.2), (3.1.3), *then every homomorphism of* H_1 *under which the image of* a_1 *is not* 1 *is a monomorphism on H.*

Corollary 3.3. *If H is recursively absolutely presented by* (3.1.1), (3.1.2), (3.1.3), *then* H_1 *is recursively relatively presented by* (3.1.1), (3.1.4), (3.1.5).

Proof of the lemma. Denote the order of a_1 by α, the order of $s_j(a_1,a_2)$ by $\alpha(j)$. Form the group C generated by c, c' and a_1 with the defining relations

$$c^2 = 1, (cc')^2 = 1, c'^2 = a_1, a_1^\alpha = 1 ;$$

and for each relevant j the group C_j generated by c_j and $s_j(a_1,a_2)$ with the defining relations

$$c_j^2 = 1, (c_js_j(a_1,a_2))^3 = 1, s_j(a_1,a_2)^{\alpha(j)} = 1 .$$

It is known that the order of a_1 in C is α, and the order of $s_j(a_1,a_2)$ in C_j is $\alpha(j)$: hence we can form the generalized free product of H and C, C_1, C_2, ..., amalgamating a_1 in H and C, then $s_1(a_1,a_2)$ in the product so obtained and C_1, and so on. In the resulting group the elements c, c_1, c_2, ... all have order 2 and are otherwise not related among one another: that is to say, they generate the free product of their cyclic groups of order 2. The partial automorphism of this free product generated by mapping c on c_1, c_1 on c_2, ... can be obtained by conjugation [3] by an element d that we now adjoin: the resulting group is the H_1 of the lemma. Note that now $c_j = d^{-j}cd^j$, so that the relations $c_j^2 = 1$ follow from $c^2 = 1$ and may be omitted. This completes the proof of the lemma. The corollaries are obvious.

Note that if H is embedded in a simple group, this would serve equally well for Corollary 3.2; but we need an embedding that is constructive, so as to retain recursiveness; and it is convenient to have H_1 finitely generated

We now use Graham Higman's famous embedding theorem [2] to make our final embedding.

Theorem 3.4. *If H is recursively absolutely presented by* (3.1.1), (3.1.2), (3.1.3), *then H can be so embedded in a finitely relatively presented group H^* that every homomorphism of H^* under which the image of a_1 is not* 1 *is a monomorphism on H.*

We need only embed H_1, which is recursively relatively presented, in a finitely relatively presented group: the rest follows from the lemma and its corollaries.

Corollary 3.5. *Every algebraically closed group contains isomorphic copies of every recursively absolutely presented 2-generator group.*

This is hardly surprising, as there are only a countable infinity of such groups; but it seems to me to leave no hope of distinguishing two algebraically closed groups by picking out pairs of elements and looking at the subgroups they generate.

References

[1] G. Higman, A finitely generated infinite simple group, J. London Math. Soc. 26 (1951) 61–64.

[2] G.Higman, Subgroups of finitely presented groups, Proc. Roy. Soc. London (A) 262 (1961) 455-475.

[3] G. Higman, B.H. Neumann and H. Neumann, Embedding theorems for groups, J. London Math. Soc. 24 (1949) 247–254.

[4] B.H.Neumann, Some remarks on infinite groups, J. London Math. Soc. 12 (1937) 122-127.

[5] B.H.Neumann, A note on algebraically closed groups, J. London Math. Soc. 27 (1952) 247-249.

[6] B.H. Neumann, Algebraically closed semigroups, Studies in Pure Mathematics (ed. L. Mirsky) (Academic Press, New York, London, 1971) 185-194.

[7] W.R.Scott, Algebraically closed groups, Proc. Amer. Math. Soc. 2 (1951) 118-121.
[8] E. Steinitz, Algebraische Theorie der Körper, J. reine angew. Math. 137 (1910) 167-309, especially §21.

EQUATIONS OVER GROUPS

Helmut SCHIEK

University of Bonn

Dedicated to W. Krull on his 70th birthday

Introduction

Let G be any group and $X = \langle x \rangle$ an infinite cyclic group, let $G[x]$ be the free product $G * X$ and $R = R(x) = a_0 x^{i_1} a_1 \ldots a_{n-1} x^{i_n} a_n$ with $a_0, \ldots, a_n \in G$ be any element of $G[x]$. We say, that the equation $R(x) = 1$ has a solution over G if and only if $R(h) = 1$ for some element h of a group H containing G. A solution exists if and only if the normal closure N of R in $G[x]$ has trivial intersection with G.

The study of equations over groups was initiated by B.H. Neumann [4]. Levin [3] showed that a solution always exists provided $n \geqslant 1$ and $i_1, \ldots, i_n > 0$. Gerstenhaber and Rothaus [1] showed that a solution always exists provided the exponent sum $e(R) = i_1 + \ldots + i_n$ does not vanish and that G can be embedded in a compact connected Lie group. If $e(R) = 0$ we don't have always a solution. For example, the equation $ax^{-1}bx = 1$ can have a solution only if a and b have the same order, while it is a special case of a result of Higman, Neumann and Neumann [2] that if a and b do have the same order a solution always exists. The case $e(R) = 0$ has been studied by the author in several papers [6,7]. Here we use results from these earlier papers to obtain the following new result.

Theorem. *Let $R(x) = aE^{-1}bE$ where a and b are elements of G of the same order and where E has the form $E = x^{-s_1} a_1 x^{s_1} \ldots x^{-s_n} a_n x^{s_n}$ with $n \geqslant 1$, all non-trivial $a_1, \ldots, a_n \in G$, $E \neq 1$ and all of $s_1, \ldots, s_n \neq 0$. Then the equation $R(x) = 1$ has a solution.*

Proof of the theorem. We first state a result from Schiek [6]. Let H be any group. Assume we have isomorphic copies $H_i \cong H$ for every integer $i \in \mathbf{Z}$, $H_i \cap H_j = 1$ for $i \neq j$. Let σ_i be the isomorphism with $\sigma_i(H) = H_i$ and let H^* be the free product $H^* = * H_i$ ($i \in \mathbf{Z}$); in special cases we can have $H_i = H^{x^i}$ for a subgroup H of a group G and x an element of G. We define $l(W)$ for an element W of H^* to be the least s such that for some i, $W \in H_{i+1} * ... * H_{i+s}$. We define $R \in H^*$ to be *incompressible* with respect to the factorization of H^* into the factors H_i if for every non-trivial element S in the normal closure of R in H^* we have $l(S) \geqslant l(R)$; if $H_i = H^{x^i}$ we also say: R is incompressible with respect to H and x. In H^* we have an automorphism σ, defined by $\sigma[\sigma_j(a)] = \sigma_{j+1}(a)$ (for every $a \in H$, every $j \in \mathbf{Z}$) and the homomorphy property. Let $R_i = \sigma^i(R)$ for every integer i; then for every non-trivial T in the normal closure of the set of all R_i ($i \in \mathbf{Z}$) we have: $l(T) \geqslant l(R)$; for $H_i = H^{x^i}$ every non-trivial T in the normal closure of all R^{x^i} has this property.

Define a series of subgroups of $G[x]$ as follows. Let $K_0 = G[x]$ and, for all $i \geqslant 0$, let K_{i+1} be the normal closure of G in K_i. This series was studied in Schiek [5]. Evidently K_1 consists of all W such that $e(W) = 0$, and $K_1 = * G_i$ where $G_i = G^{x^i}$. Also $K_2 = * G^E$, where E runs over all elements of K_1 with normal form $E = E_1 ... E_n$, $n \geqslant 0$, where every $E_i (1 \leqslant i \leqslant n)$ has the form $x^{-s_i} a_i x^{s_i}$, $a_i \in G$, $a_i \neq 1$, $s_i \neq 0$, $s_i \neq s_{i+1}$ for $1 \leqslant i \leqslant n-1$ (for $n = 0$ we have $E = 1$, $G^E = G$).

Under the hypothesis of the theorem it is clear that $R \in G * G^E$, whence $l(R) \leqslant 2$. Imposing the relation $R = 1$ on $G * G^E$ amounts to passing from $G * G^E$ to a free product with amalgamation of the cyclic group generated by a^{-1} in G and that generated by b^E in G^E. This extends to a homomorphism from $H = * G^{E^s}$ with kernel the normal closure of R in H. It follows that no non-trivial element in the normal closure of R in H lies in any G^{E^s}, hence that R is incompressible with respect to G and E (of course we can assume $l(R) = 2$, otherwise a, b and R would be trivial). From the result cited above it follows that the normal closure M of the set of R^{E^s} in H has trivial intersection with G. Now $K_2 = H * P$, where P is the free product of all factors G^F of K_2 with $F \neq E^s$ (for every integer s). The normal closure N_1 in K_2 of the set of R^F, one for each factor G^F of K_2, has intersection M with H, hence trivial intersection with G. Because N_1 is the normal closure of R in K_1, we have shown that the normal closure of R in K_1 has trivial intersection with G; we want to show the same is true of the normal closure N of R in $K_0 = G[x]$.

We treat first the special case that $E = x^{-j}cx^j$ with $1 \neq c \in G$ and $j \neq 0$. For each i let p_i be the natural projection of $K_1 = *G_i$ onto G_i. If $i \neq 0$, we evidently have $p_i(R) = 1$, whence $p_i(N_1) = 1$ and it follows that N_1 has no non-trivial element in G_i. We have seen above that N_1 has no non-trivial element in $G_0 = G$. Thus N_1 has no non-trivial element of length less than $l(R) = 2$, and R is incompressible with respect to G and x. It now follows from the result cited earlier that the normal closure in K_1 of the set of R_i has no non-trivial element of length 1. This normal closure is N and we have shown that N has trivial intersection with G.

For the general case we use induction. Let $E = E_1 \ldots E_n$, $E_i = x^{-s_i}a_i x^{s_i}$, $1 \neq a_i \in G$, $s_i \neq s_{i+1}$ $(1 \leqslant i \leqslant n)$, $n \geqslant 1$. We say that $R = aE^{-1}bE$ has a normal form of type n. For $n = 1$ we have already proved that $R(x) = 1$ has a solution. We now assume $n \geqslant 2$ and we make the induction hypothesis that every R^* with a normal form of type $t < n$ has a solution. E_1 is an element of G_{s_1}; for $n \geqslant 2$ not all E_i are elements of G_{s_1} (because of the condition $s_i \neq s_{i+1}$). K_1 is the free product of G_{s_1} and $K_1^* = *G_i (i \neq s_1)$. E^* shall be the projection of E into K_1^*, $E^* = E_{j_1} \ldots E_{j_t}$, $j_1 < \ldots < j_t$, where $\{j_1, \ldots, j_t\}$ is the set of all i $(1 \leqslant i \leqslant n)$ for which $s_i \neq s_1$. Surely we have $t \geqslant 1$ $(s_2 \neq s_1)$.

$R^* = aE^{*-1}bE^*$ has a normal form of type $t < n$. According to the induction hypothesis $R^* = 1$ has a solution. Let N^* be the normal closure of R^* in $G[x]$, $H = G[x]/N^*$, then we have $H \geqslant G$. Now we get R^* from R if we introduce the commutator relations $[G_{s_1}, K_1^*] = 1$ in K_1; that means: $R^{-1}R^*$ is in the normal closure of $[G_{s_1}, K_1^*]$ in K_1. We will prove that the normal closure Q of the set theoretical union of $[G_{s_1}, K_1^*]$ and R^* in $G[x]$ has trivial intersection with G; then we have $N \subseteq Q$, $N \cap G = 1$ and $R(x) = 1$ has a solution.

Let $y \notin H$, $Y = \langle y \rangle$ an infinite cyclic group, and let $H_i = Hy^i$, $L_1 = *H_i$, $L_1^* = *H_i$ $(i \neq s_1)$. We can introduce the relations

$[H_{s_1}, L_1^*] = 1$ without changing $H \geqslant G$; we take the *direct* product $L^* = H_{s_1} \times L_1^*$, $L^* \supseteq H \supseteq G$.

Let $z \notin L^*$, $Z = \langle z \rangle$ an infinite cyclic group, and $L[z] = L * Z$. In $L[z]$ we can introduce relations:

(1) $\qquad (a^z)^{-1} a^y = 1$ for all $a \in G$

(2) $\qquad (a^z)^{-1} a^x = 1$ for all $a \in G$.

The projection of $(a^z)^{-1} a^y = z^{-1} a^{-1} z y^{-1} a y$ into G is trivial, also the projection of $(a^z)^{-1} a^x$ into G. Let $\tilde{N}$ be the normal closure of the union of all $(a^z)^{-1} a^y$ and all $(a^z)^{-1} a^x$ with $a \in G$ in $L^*[z]$, then we have: $\tilde{N} \cap G = 1$.

In this group we have the relations: $a^x = a^z = a^y$, $G_{s_1} = H_{s_1}$, $K_1^* = L_1^*$, $R^{-1} R^* = 1$, $R^* = 1$ and therefore $R = 1$. The group $L^*[z]/\tilde{N}$ contains G. Therefore we have proved that $R(x) = 1$ has a solution.

Remarks. $R(x)$ is an element of $K_2 - K_3$. We can ask if we can prove the existence of a solution of an equation $R_i(x) = 1$, $R_i(x)$ being an element of $K_i - K_{i+1}$, of a similar structure. In [5] the structure of K_i is given by induction on i. For K_i we have a system of generators: $S_i = T_i \cup G$. The set T_1 is defined by $T_1 = \{a^{x^i} \mid 1 \neq a \in G, i \neq 0\}$. If T_i is given, we define Λ_i as the set of all products $E = E_1 \ldots E_t$ (for some $t \geqslant 1$), all $E_1, \ldots, E_t \in T_i$. We define: $T_{i+1} = \{a^E \mid 1 \neq a \in G, E \in \Lambda_i\}$.

Let $R_i(x) = aF$, $F = b^E \in T_i$; then we can ask if $R_i(x) = 1$ has a solution. $R_i(x)$ is an element of $K_i - K_{i+1}$. For every $R \in G[x] - G$ there is an integer i such that $R \in K_i - K_{i+1}$ (in [5] it is proved that $\cap K_i = G$). It might be possible that one can use induction on i in order to get more general results; this would need quite a lot of technique.

References

[1] M.Gerstenhaber and O.S.Rothaus, The solutions of sets of equations in groups, Proc. Nat. Acad. Sci. U.S. 48 (1962) 1951-1953.

[2] G. Higman, B.H. Neumann and Hanna Neumann, Embedding theorems for groups, Journal London Math. Soc. 24 (1949) 247–254.

[3] F.Levin, Solutions of equations over groups, Bulletin American Math. Soc. 68 (1962) 603-604.

[4] B.H.Neumann, Adjunction of elements to groups, Journal London Math. Soc. 18 (1943) 4-11.

[5] H.Schiek, Über eine spezielle Reihe von Normalteilern, Archiv der Mathematik 9 (1958) 236-240.

[6] H.Schiek, Adjunktionsproblem und inkompressible Relationen, Math. Annalen 146 (1962) 314-320.

[7] H.Schiek, Adjunktionsproblem und inkompressible Relationen II, Math. Annalen 161 (1965) 163-170.

A SURVEY OF SMALL CANCELLATION THEORY

Paul E.SCHUPP*

University of Illinois, Urbana

1. Introduction

1.1. A preview

In 1911 M.Dehn [5] posed the word and conjugacy problems for groups in general and provided algorithms which solved these problems for the fundamental groups of closed orientable two-dimensional manifolds. A crucial feature of these groups is that (with trivial exceptions) they are defined by a single relator r with the property that if s is any cyclic conjugate of r or r^{-1}, $s \neq r^{-1}$, there is very little cancellation in forming the product rs. Dehn's algorithms have been extended to large classes of groups possessing presentations in which the defining relations have a similar "small cancellation" property. At first, investigations were concerned with the solution of the word problem for groups G presented as a "small cancellation" quotient of a free group F. The theory was subsequently extended to the case where F is a free product or a free product with amalgamation. Moreover, strong results were obtained about algebraic properties; for example, one can classify torsion elements and commuting elements in "small cancellation" quotients.

Dehn's methods were geometric, making use of regular tessellations of the hyperbolic plane. The first extensions of Dehn's re-

* This manuscript was prepared while the author was a visiting member at the Courant Institute of Mathematical Sciences, New York University, and was supported in part by the New York State Science and Technology Foundation, Grant SSF-(8)-8. The author thanks Bruce Chandler and Roger Lyndon for making many valuable suggestions.

sults to larger classes of groups were obtained using cancellation arguments of combinatorial group theory, independent of any geometric considerations. More recently, the geometric character of Dehn's argument has been restored in the form of elementary combinatorial geometry. "Small cancellation" theory is now emerging as a unified and powerful theory. In what follows, we shall outline the central ideas of this theory and present some important and typical results.

1.2. The hypotheses

We now turn to the formulation of the conditions which, for suitable values of the parameters, allow one to "do" small cancellation theory. In order to fix our notation and terminology, let F be a free group on a set X of generators. A *letter* is an element of the set Y of generators and inverses of generators. A *word* w is a finite string of letters, $w = y_1 ... y_m$. We shall not distinguish between w and the element of F that it denotes. We denote the identity of F by 1. Each element of F other than the identity has a unique representation as a *reduced word* $w = y_1 ... y_n$ in which no two successive letters $y_j y_{j+1}$ form an inverse pair $x_i x_i^{-1}$ or $x_i^{-1} x_i$. The integer n is the *length* of w, which we denote by $|w|$. A reduced word w is called *cyclically reduced* if y_n is not the inverse of y_1. If there is no cancellation in forming the product $z = u_1 ... u_n$ we write $z \equiv u_1 ... u_n$.

A subset R of F is called *symmetrized* if all elements of R are cyclically reduced and, for each r in R, all cyclically reduced conjugates of both r and r^{-1} also belong to R.

Suppose that r_1 and r_2 are distinct elements of R with $r_1 \equiv bc_1$ and $r_2 \equiv bc_2$. Then b is called a *piece* relative to the set R. (Since we only work with one symmetrized set at a time, we will omit the phrase "relative to R" and simply say that b is a piece.) Since b is cancelled in the product $r_1^{-1} r_2$, and R is symmetrized, a piece is simply a subword of an element of R which can be cancelled by the multiplication of two non-inverse elements of R.

The hypotheses of "small cancellation" assert that pieces are relatively small parts of elements of R. The most usual condition takes a metric form, $C'(\lambda)$, where λ is a positive real number.

Condition $C'(\lambda)$: If $r \in R$, $r \equiv bc$ where b is a piece, then $|b| < \lambda|r|$.

A closely related, non-metric, condition is $C(p)$ where p is a natural number.

Condition $C(p)$: No element of R is a product of fewer than p pieces.

Observe that $C'(\lambda)$ implies $C(p)$ for $\lambda \leqslant 1/(p-1)$. As illustration, the fundamental group of a closed, orientable 2-manifold of genus g has a presentation

$$G = \langle a_1, b_1, \ldots, a_g, b_g ; a_1 b_1 a_1^{-1} b_1^{-1} \ldots a_g b_g a_g^{-1} b_g^{-1} \rangle .$$

In this case R consists of all cyclic permutations of r and r^{-1} where r is $a_1 b_1 a_1^{-1} b_1^{-1} \ldots a_g b_g a_g^{-1} b_g^{-1}$. Clearly, pieces are single letters and R satisfies $C'(1/4g-1)$ and $C(4g)$. Groups which have a presentation $G = \langle X, R \rangle$ where R satisfies $C'(1/6)$ are sometimes called sixth-groups. Analogously, we have eight-groups, etc.

We shall sometimes need a condition $T(q)$, for q a natural number, whose intuitive meaning will be clarified shortly.

Condition $T(q)$: Let $3 \leqslant h < q$. Suppose $r_1, \ldots, r_h$ are elements of R with no successive elements r_i, r_{i+1} an inverse pair. Then at least one of the pairs $r_1 r_2, \ldots, r_{h-1} r_h, r_h r_1$ is reduced without cancellation.

The cancellation conditions just introduced can be extended naturally to the case where F is a free product or a free product with amalgamation by using the appropriate normal forms and associated length functions. The definitions are then essentially the same as for F a free group. One must be careful on a few points, however, so we defer precise definitions of the cancellation conditions in the more general cases to Section 5.

As an illustration of the power of small cancellation theory over free products we note that the theory applies to "most" Fuchsian groups

$$G = \langle a_1, b_1, ..., a_g, b_g, x_1, ..., x_n, f_1, ..., f_k ;$$

$$x_1^{m_1} = 1, ..., x_n^{m_n} = 1, f_1 ... f_k x_1 ... x_m \prod_{i=1}^{g} [a_i, b_i] = 1 \rangle$$

We shall consistently use the notation introduced above. F will be a free group on generators X, or a free product, or a free product with amalgamation. R will be a symmetrized subset of F with N the normal closure of R in F. G will be the quotient group F/N. If F is a free group, G has a presentation $\langle X;R \rangle$. The natural map will be denoted by $\nu : F \to F/N$.

1.3. A brief historical sketch

Dehn's methods were geometric [5,6]. He used the fact that with the fundamental group G of an orientable closed 2-manifold there is an associated regular tessellation of the hyperbolic plane which is composed of transforms of a fundamental region for G. Using the hyperbolic metric, Dehn inferred that a non-trivial word w equal to 1 in G contained more than half of an element of R. Reidemeister [37] pointed out that Dehn's conclusion followed from the combinatorial properties of the tessellation, without metric considerations.

In 1949, V.A.Tartakovskii [47,48,49] initiated the algebraic study of small cancellation theory. Tartakovskii solved the word problem for finitely presented quotients of free products of cyclic groups by symmetrized R satisfying $C(7)$. J.Britton [3], in 1957, independently investigated quotient groups of arbitrary free products by R satisfying $C'(1/6)$. The triangle condition, Condition $T(4)$, was introduced in 1956 by Schiek [38], who solved the word problem for R satisfying $C'(1/4)$ and $T(4)$. Greendlinger [11,12], in 1960, solved the conjugacy problem for $C'(1/8)$, gave a new proof of the solvability of the word problem for $C'(1/6)$, and obtained several other important results. Greendlinger [15] subsequently also investigated the $C'(1/4)$ and $T(4)$ hypothesis.

Very few group presentations have Cayley diagrams which are embeddable in the plane. However, it turns out that for any group $G = \langle X;R \rangle$, if w is in N, there exists a finite planar diagram M, each edge of which is labelled by an element of F, and such that each region (face) D of M has as label on its boundary an element of R, while the label on the boundary of the entire diagram M is the reduced word w.

The existence of such a diagram M was observed by Van Kampen [50] in 1933. Van Kampen's paper seems to have been totally ignored until Weinbaum [51], 1966, used the ideas to prove some of the results of Greendlinger. The above ideas were rediscovered independently by R.C.Lyndon [29] in his 1966 paper "On Dehn's Algorithm", which provided a unification, simplification, and generalization of many previous results.

Lyndon observed that the Condition $C(p)$ asserts that every interior region of the diagram M borders on at least p other regions. Condition $T(q)$ expresses the dual condition that each interior vertex of M (excluding vertices of degree two) has at least q incident edges. Lyndon solved the word problem for finite R satisfying one of the hypotheses $C(p)$ and $T(q)$ where (p,q) is one of the pairs (6,3), (4,4), or (3,6). (Condition $T(3)$ is vacuous.) These hypotheses correspond naturally to the three regular tessellations of the Euclidean plane. For example, the hypothesis $C(4)$ and $T(4)$ corresponds to the regular tessellation of the plane by squares. In the regular tesselation, all vertices and all regions have degree four. In the diagrams considered under the hypothesis $C(4)$ and $T(4)$, all interior regions and interior vertices have degree greater than or equal to four.

2. Decision problems

2.1. Dehn's algorithm and Greendlinger's lemma

In his study of the word problem for fundamental groups of orientable 2-manifolds, Dehn concluded that if a freely reduced non-trivial word v is equal to 1 in the fundamental group, then v contains more than half of some cyclic permutation of the defining relator or its inverse.

This conclusion gives *Dehn's algorithm* for the word problem. Suppose a group G has a presentation $G = \langle x_1, \ldots, x_n; R \rangle$ where R is a finite symmetrized set of defining relators and it has been established that freely reduced non-trivial words which are equal to 1 in G contain more than half of some element of R. Let w be a non-trivial word of G. If $w = 1$ in G, then w has some factorization $w \equiv bcd$ where, for some r in R, $r \equiv ct$ with $|t| < |c|$. In G then, $w = bt^{-1}d$, a word of shorter length. A finite number of such reductions either leads to 1, giving a "proof" that $w = 1$ in G, or to a word w^* which cannot be so shortened, establishing $w \neq 1$ in G.

The most fundamental result of small cancellation theory is that Dehn's algorithm is valid for R satisfying one of the metric hypotheses $C'(1/6)$, or $C'(1/4)$ and $T(4)$. Actually, as first discovered by Greendlinger [11], considerably more is true. In order to state the sharper results we need to single out certain "large" subwords of elements of R.

Definition. A word s is called a *j-remnant* (with respect to R) if some $r \in R$ has the form $r \equiv sb_1...b_j$ where $b_1,...,b_j$ are pieces.

Theorem 1 (Greendlinger's lemma). *Let F be a free group, free product, or free product with amalgamation. Let R be a symmetrized subset of F with N the normal closure of R. Assume that R satisfies the hypothesis C(p) and T(q) where (p,q) is one of the pairs* (6,3), (4,4), *or* (3,6).

If $w \in N$, $w \neq 1$, *then* $w \in R$ *or some cyclically reduced conjugate* w^* *of* w *has the form* $w^* \equiv u_1 s_1 ... u_m s_m$ *where each* s_k *is an* $i(s_k)$*-remnant. The number m of the* s_k *and the numbers* $i(s_k)$ *satisfy the relation*

$$\sum_{k=1}^{m} \left[\frac{p}{q} + 2 - i(s_k) \right] \geqslant p \, .$$

If the hypothesis is $C(6)$ we have

$$\sum_{k=1}^{m} [4 - i(s_k)] \geqslant 6 \, .$$

For this inequality to hold w must contain a j-remnant with $j \leqslant 3$. If the hypothesis is strengthened to $C'(1/6)$ we conclude that w contains more than half of an element of R.

The conclusion that an element w of N contains more than half of an element of R was obtained under varying hypotheses by Britton [3] and Schiek [38]. Theorem 1 was proved by Greendlinger [11] for F free and R satisfying $C'(1/6)$. He subsequently [15] proved the theorem for R satisfying $C'(1/4)$ and $T(4)$. Weinbaum [51] gave a geometric proof for the $C'(1/6)$ case. The general formula

$$\sum_{k=1}^{m} \left[\frac{p}{q}+2-i(s_k)\right] \geqslant p$$

is due to Lyndon [29]. In the geometric approach, the nature of this formula is revealed as a combinatorial "curvature formula". The relationship of the curvature formula to Greendlinger's Lemma is discussed by Schupp [42].

2.2. The word problem

Greendlinger's Lemma is independent of any cardinality or effectiveness assumptions on either F or R. The conclusion that a non-trivial element of N contains more than half of an element of R allows one to use Dehn's algorithm to solve the word problem in many cases where F is a free product and R is infinite. We consider the following effectiveness condition on F and R.

Condition $E(k)$: Assume that F is a finitely generated free group or free product. Suppose that R is a symmetrized subset of F satisfying $C'(1/6)$ or $C'(1/4)$ and $T(4)$. Assume that there is an algorithm which decides, given $w \in F$, whether or not there exist $r \in R$ such that $|r| \leqslant k|w|$ and $r \equiv ws$, and, if so, produces all such r. (In particular, there are only finitely many such r. In most cases of interest, if such an r exists it will be unique by the cancellation condition.)

The following corollary is then immediate from Theorem 1.

Corollary 1. *Let F and R satisfy Condition $E(2)$. Then the word* *problem for $G = F/N$ is reducible to the word problem for F and conversely. (In the language of recursive function theory, the word problems for F and G have the same Turing degree.)*

Under the non-metric hypothesis $C(p)$ and $T(q)$ a rather different argument is needed to solve the word problem. The following theorem is due to Lyndon [29].

Theorem 2. *Let F be a free group, and let R be a finite symmetrized subset of F satisfying the hypothesis $C(p)$ and $T(q)$ for (p,q) one of* (6,3), (4,4), *or* (3,6). *Let N be the normal closure of R in F. Then F/N has solvable word problem.*

The proof of Theorem 2 involves a combinatorial "area formula" which shows that if $w \in N$, one can calculate from w a bound on the number of conjugates of elements of R necessary to form a product $(u_1 r_1 u_1^{-1})...(u_m r_m u_m^{-1}) = w$. The lengths of the conjugating elements u_i are also bounded. One could then try the finite number of possible products to see if w can be so formed.

2.3. The conjugacy problem

Our discussion of conjugacy begins with *Dehn's algorithm for the conjugacy problem* which Dehn [6] discovered for fundamental groups G of 2-manifolds. Let w and v be two words of G. We can effectively replace w and v by words w' and v' which are conjugate in G to w and v respectively, which are cyclically reduced, and which are "cyclically R-reduced" in the sense that no cyclic permutation of w' or v' contains more than half of an element of R.

If one of w', v' is the identity 1 and the other is not, then w and v are not conjugate in G. If both w' and v' are 1 they are certainly conjugate in G. Assuming that both w' and v' are non-trivial, then w' and v' are conjugate in G if and only if for some cyclic permutations w^*, v^* of w' and v' respectively, the equation $w^* = uv^*u^{-1}$ holds in G for u a subword of some $r \in R$. There are only finitely many such u, and, in view of the solvability of the word problem for G, one can decide if the above equation holds. Dehn also obtained a bound on $|u|$ in terms of $|r|$, namely $|u| < |r|/8$.

Greendlinger [12] showed that Dehn's algorithm solves the conjugacy problem if F is a free group and R is finite satisfying $C'(1/8)$. Greendlinger subsequently showed that generalizations of Dehn's algorithm solve the conjugacy problem for R satisfying $C'(1/6)$ [17], and $C'(1/4)$ and $T(4)$ [15]. A geometric proof extending these results essentially unchanged to free products is given by Schupp [41].

To investigate conjugacy when F is a free product we need a mild additional assumption which is automatically true if F is a free group.

Condition J: If $r \in R$, then r is not conjugate to r^{-1} in F.

Theorem 3. *Let F be a free group or free product, and let R be a symmetrized subset of F satisfying* $C'(1/6)$, *or* $C'(1/4)$ *and* $T(4)$. *Assume that R satisfies Condition J and that R and F satisfy Condition* $E(2q)$ (q *is* 3 *and* 4 *for* $C'(1/6)$, *and* $C'(1/4)$ *and* $T(4)$ *respectively). Then the conjugacy problem for* $G = F/N$ *and F have the same Turing degree.*

As with the word problem, a somewhat different argument is needed to solve the conjugacy problem under the non-metric hypotheses. This was done by Schupp [40].

Theorem 4. *Let F be a free group, and let R be a finite symmetrized subset of F satisfying the hypothesis* $C(p)$ *and* $T(q)$ *for* (p,q) *one of* (6,3), (4,4), *or* (3,6). *Then* $G = F/N$ *has solvable conjugacy problem.*

3. Algebraic applications

3.1. Small cancellation products

In this section we discuss some of the algebraic applications of small cancellation theory.

Definition. A group G is a *product of the groups* X_i, $i \in I$, if $G \cong$

F/P where F is the free product of the X_i, and P is a normal subgroup of F such that the natural map $\nu : F \to F/P$ embeds each of the groups X_i. (Viewing each X_i as a subgroup of F, we require $\nu(X_i) \cong X_i$.)

We see immediately from Greendlinger's Lemma that if F is a free product of groups X_i and R satisfies one of the cancellation hypotheses $C'(1/6)$, or $C'(1/4)$ and $T(4)$, then N contains no elements of length one. Hence, F/N is a product of the X_i.

Thus we have some new products of groups at our disposal, which we will call *small cancellation products.* We shall see that these products have several "nice" properties and are powerful tools in dealing with embedding problems and adjunctions of solutions to equations over groups.

The theorem of Higman-Neumann-Neumann [19] that a countable group can be embedded in a two-generator group is now well-known. As an illustration of the use of small cancellation products we will prove a much stronger result. The idea of using small cancellation theory to prove embedding theorems is originally due to J.L.Britton (unpublished) and has been used by McCool [33], Levin [21], and Miller and Schupp [35]. The following proof is Britton's.

Definition. A countable group K is called *S-Q universal* if every countable group can be embedded in a quotient group of K.

Theorem 5. *Let F be any non-trivial free product, $F = X * Y$, with the single exception of the free product of two copies of C_2, the cyclic group of order two. Then F is S-Q universal.*

Proof. Since we have excluded the case of $C_2 * C_2$, we can pick distinct elements x_1, x_2, neither the identity, in one group, X say, and an element $y \neq 1$ in Y. Let H be a countable group with presentation $H = \langle h_1, \ldots ; S\rangle$. S is a set of defining relations among the h_i.

Let $F' = H * X * Y$. Let

$$r_1 = h_1 x_1 y x_2 y (x_1 y)^2 x_2 y (x_1 y)^3 x_2 y ... (x_1 y)^{80} x_2 y$$

and, in general, let

$$r_i = h_i (x_1 y)^{80(i-1)+1} x_2 y ... (x_1 y)^{80i} x_2 y \ .$$

Let R be the symmetrized set generated by the r_i. It is not difficult to see that R satisfies the cancellation condition $C'(1/10)$. This follows from the fact that no piece can contain a subword of the form $[x_2 y (x_1 y)^k x_2 y (x_1 y)^{k+1} x_2 y]^{\pm 1}$. Let N be the normal closure of R in F'. $G = F'/N$ is a small cancellation product and hence embeds all the factors of F', in particular H.

Now G is certainly generated by the images of X and Y, since each $r_i = 1$ in G and each h_i is thus equal to a word on x_1, x_2, and y. Indeed, since each r_i contains precisely one h_i, we can eliminate the relations R by Tietze transformations, rewriting the relations S in terms of x_1, x_2, and y. Hence G is actually a quotient group of $X * Y$ which embeds H. This completes the proof.

3.2. Torsion elements

If R satisfies the metric hypothesis $C'(1/6)$ one can completely classify the elements of finite order in F/N.

Theorem 6. *Let F be a free group or a free product, and let R be a symmetrized subset of F satisfying $C'(1/6)$. Let $\nu : F \to F/N$ be the natural map.*

If w has finite order in $G = F/N$ then either
(i) *$w = \nu(w')$ where w' is an element of finite order in F, or*
(ii) *there is an $r \in R$ which is a proper power in F, say $r = v^n$, $n > 1$, and w is conjugate in G to a power of v.*

Theorem 6 says that the only elements of finite order in G are the "obvious" ones. In particular, if R contains no proper powers and F is torsion free (certainly the case if F is free) then F/N is torsion free.

Theorem 6 was proved for F free by Greendlinger [12], and for F a free product by McCool [31]. The same result was proved for F free and $C'(1/4)$ and T by Soldatova [45].

3.3. Commuting elements

Like elements of finite order, commuting elements can be successfully characterized in suitable small cancellation quotients.

Theorem 7. *Let F be a free group or a free product. Let R be a symmetrized subset of F satisfying Condition J, and which also satisfies one of $C'(1/6)$, or $C'(1/4)$ and $T(4)$. Let $\nu : F \to F/N$ be the natural map.*

Then two elements u and v of $G = F/N$ commute if and only if $u = \nu(u')$ and $v = \nu(v')$ for elements u', v' which commute in F.

Greendlinger [12] and Lipschutz [22] first investigated commuting elements for the case F free and R satisfying $C'(1/6)$. Greendlinger [14] then showed that for F free and R satisfying $C'(1/8)$, two elements of G commute if and only if they are powers of a common element. Greendlinger [17] then extended the result to $C'(1/6)$. Schupp [44] gave a geometric proof which extends the results to free products and to the $C'(1/4)$ and $T(4)$ hypothesis.

3.4. Endomorphisms and hopficity

It turns out that endomorphisms of certain small cancellation products may be severely limited. We need to recall some definitions. A group K is *hopfian* if every endomorphism of K onto K is an automorphism. Dually, K is *co-hopfian* if all 1-1 endomorphisms of K are onto. K is *complete* if all automorphisms of K are inner and K has trivial center.

Let $F' = H * C_m * C_n$ where H is an arbitrary countable group, and C_m and C_n are cyclic of orders $m \geqslant 3$ and $n \geqslant 2$. Miller and Schupp [35] investigate the endomorphisms of $G = F'/N$ where N is the normal closure of the set R defined in the proof of Theorem 5. They prove that G is always complete and hopfian. In addition, if H has no elements of order m, or no elements of order n, then G is co-hopfian. In particular, we have

Theorem 8. *Any countable group H can be embedded in a complete, hopfian quotient G_H of the modular group $C_2 * C_3$. If H is finitely presented then so is G_H.*

3.5. A uniqueness theorem

Theorem 9. *Let F be a free group or a free product. Let R and R' be symmetrized sets of F which satisfy $C'(1/6)$ or $C'(1/4)$ and $T(4)$. If R and R' have the same normal closure, then $R = R'$.*

Theorem 9 was discovered by Greendlinger [13] for F free and $C'(1/6)$. The theorem is essentially a direct consequence of Greendlinger's Lemma. Schupp [42] observed that the above generalization holds.

3.6. Small cancellation theory over free products with amalgamation

It turns out that small cancellation theory can be done over free products with amalgamation. For precise definitions of how the hypotheses are to be interpreted see Section 5. Greendlinger's Lemma, when properly interpreted, and the uniqueness theorem, Theorem 9, continue to be true for F a free product with amalgamation and R satisfying $C'(1/6)$.

One can use the existence of small cancellation theory over free products with amalgamation to prove that many free products with amalgamation are *S-Q* universal. The idea is to attempt to imitate the proof of Theorem 5. In order to do so, we need a definition.

Definition. Let H be a group with subgroup A. Let $\{x_1,x_2\}$ be a pair of distinct elements of H, neither of which are in A. The pair $\{x_1,x_2\}$ is called a *blocking pair for A in H* if the following two conditions hold:

(i) $x_i^\epsilon x_j^\delta \notin A$, $1 \leqslant i, j \leqslant 2$, $\epsilon = \pm 1$, $\delta = \pm 1$ unless the product $x_i^\epsilon x_j^\delta = 1$.

(ii) If $a \in A$, $a \neq 1$, then $x_i^\epsilon a x_j^\delta \notin A$, $1 \leqslant i, j \leqslant 2$, $\epsilon = \pm 1$, $\delta = \pm 1$.

The existence of a blocking pair for a subgroup is not an unreasonable condition in groups which have a fair amount of "freeness". For instance, if H is free and A is a finitely generated subgroup of infinite index in H, then there does exist a blocking pair for A in H. (See Burns [4].) Schupp [43] proves

Theorem 10. *Let $F = (H*J;A=B)$ be a free product with amalgamation, where A and B are proper subgroups of H and J respectively If there exists a blocking pair for A in H then F is S-Q universal.*

Perhaps the most surprising group to which Theorem 10 applies is the group

$$G = \langle a,b,c,d; b^{-1}ab=a^2, c^{-1}bc=b^2, d^{-1}cd=c^2, a^{-1}da=d^2 \rangle$$

which G.Higman [20] used to prove the existence of a finitely generated infinite simple group.

4. Some open questions

There are some interesting open questions in small cancellation theory. For this section, a "*suitable* small cancellation group" will be a finitely presented group $G = \langle X;R \rangle$ where R satisfies $C'(\lambda)$ for some "sufficiently small" λ and R also satisfies "suitable" additional hypotheses. What additional "suitable" hypotheses are necessary to make progress on the questions listed is not known, but it is assumed that they eliminate the obvious counterexamples not thought of by the author. Unfortunately, all the questions are probably quite difficult, with the possible exception of question 4.

1. Does a suitable small cancellation group have solvable generalized word problem?

The generalized word problem asks for an algorithm which, when given a word w of G and a finitely generated subgroup H of G, decides whether or not $w \in H$. Asking for a solvable generalized word problem is a very strong condition. It is known (Mikhailova

[34]) that even the direct product, $F_2 \times F_2$, of two free groups of rank two has unsolvable generalized word problem! D.J.Collins has observed that the obvious presentation of $F_2 \times F_2$ satisfies $C(4)$ and $T(4)$.

On the positive side, Lipschutz [25,28] has shown that for R satisfying either $C'(1/6)$, or $C'(1/4)$ and T, the generalized word problem restricted to cyclic subgroups is solvable. A tempting case in which to hope for a positive answer is that of the Fuchsian groups.

2. (C.F.Miller). What can we say about the algebraic properties of subgroups of suitable small cancellation groups?

For example, is the intersection of two finitely generated subgroups again finitely generated? (This latter property is sometimes called the Howson property.)

3. What can one say about the endomorphisms of suitable small cancellation groups?

It is reasonable to expect that endomorphisms should be quite restricted. We remarked in Section 3.4 that certain very special small cancellation groups are hopfian. Are suitable small cancellation groups always hopfian? Perhaps suitable small cancellation groups are even residually finite. This is true, of course, in the Fuchsian case.

What about the automorphisms of small cancellation groups? A possible conjecture is that all automorphisms of G are induced by automorphisms of the free group F which fix the set R. This was proved by Nielsen [36] for the fundamental groups of 2-manifolds. (The conjecture is not quite true for free products of cyclic groups, so among the "suitable" additional hypotheses should be one saying that R has elements involving all the generators of G.)

4. (Lipschutz). If G is a suitable small cancellation group, is the centralizer of every element cyclic?

We already know that, assuming only $C'(1/6)$, two elements commute if and only if they are powers of a common element. What is desired here is to establish that if u^n and v^m commute then u and v are powers of a common element. This should be true and perhaps even approachable.

5. What "is" a small cancellation group?

What is desired here is a geometric characterization of small cancellation groups. For example, the Fuchsian groups are essentially *the* groups with planar Cayley diagrams. The geometric approach to small cancellation theory suggests that there should be a characterization of small cancellation groups by means of "natural" geometric properties of their Cayley diagrams, or in terms of their possible action on other complexes. Such a characterization would bring us full circle back to Dehn.

5. The hypothesis revisited

In this section we want to clarify exactly what the small cancellation hypotheses are in case F is a free product or a free product with amalgamation. If F is the free product of non-trivial groups X_j then each non-identity element w of F has a unique representation in *normal form* as $w = y_1 ... y_n$ where each of the *letters* y_i is a non-trivial element of one of the factors X_j, and where no adjacent y_i, y_{i+1} come from the same factor. The integer n is the *length* of w, written $|w|$.

If $u = y_1 ... y_k c_1 ... c_t$ and $v = c_t^{-1} ... c_1^{-1} d_1 ... d_s$ in normal form where $d_1 \neq y_k^{-1}$, we say that the letters $c_1, ..., c_t$ are *cancelled* in forming the product uv. If y_k and d_1 are in different factors of F, then $w = uv$ has normal form $y_1 ... y_k d_1 ... d_s$. It is possible that d_1 and y_k are in the same factor of F with $d_1 \neq y_k^{-1}$. Let $a = y_k d_1$. Then $w = uv$ has normal form $y_1 ... y_{k-1} a d_2 ... d_t$. We say that y_k and d_1 have been *consolidated* to give the single letter a in the normal form of uv.

We say that a word w has *semi-reduced form* uv if there is no cancellation in forming the product uv, and write $w \equiv uv$. Consolidation is expressly allowed. (Our notation $w \equiv uv$ in this context is somewhat a departure from usual usage since $\equiv$ often means "identically equal".) More generally, we write $w \equiv u_1 ... u_n$ if there is no cancellation in bringing the product $u_1 ... u_n$ into normal form.

There are two reasonable definitions of "cyclically reduced" for elements of free products. The most common is to mean that $|w| = 1$ or if $w = y_1 ... y_n$ in normal form, then y_n and y_1 are in different factors of F. This is equivalent to asserting that w is not conjugate in F to an element of shorter length. We shall say that w is *strictly cyclically reduced* in this situation. We shall say that w is *cyclically reduced* if $|w| = 1$ or if $w = y_1 ... y_n$ in normal form and $y_n \neq y_1^{-1}$. Thus there is no cancellation between y_n and y_1 although consolidation is allowed.

A subset R of F is called *symmetrized* if every $r \in R$ is cyclically reduced and every cyclically reduced conjugate of r and r^{-1} is also in R.

A word b is called a *piece* if there are distinct elements r_1 and r_2 with *semi-reduced forms* $r_1 \equiv bc_1$ and $r_2 \equiv bc_2$. Note that the last letter of b does not have to be a letter of the normal form of r_1 or r_2.

Condition $C'(\lambda)$: If $r \in R$, $r = bc$ in semi-reduced form where b is a piece, then $|b| < \lambda |r|$. To avoid pathological cases, we further require that if $r \in R$ then $|r| > 1/\lambda$.

Condition $C(k)$: No element of R is a product in semi-reduced form of fewer than k pieces. If $r \in R$ then $|r| \geqslant k$.

The triangle condition, *Condition* $T(4)$, splits into two parts:

(i) If r, s, t are strictly cyclically reduced elements of R then at least one of the products rs, st, tr is reduced as stands without cancellation or consolidation.

(ii) If each of y_1, y_2, y_3 is a letter occurring in the normal forms of strictly cyclically reduced elements r, s, t of R, then $y_1 y_2 y_3 \neq 1$.

We now turn to the case of F a free product with amalgamation. Let X_i, $i \in I$, be a collection of groups with proper subgroups $A_i \subseteq X_i$, each isomorphic to a fixed group A. Let $\psi_i : A \to A_i$, $i \in I$, be isomorphisms. Let $F = (*X_i; \psi_i(A) = \psi_j(A))$.

In discussing a normal form for elements of F it is usual to choose coset representatives for each A_i in X_i. We specifically do not want to do this. We thus depart from the usual usage of "normal form".

Definition. An element $w \neq 1$ if F is said to be in *normal form* if w is $y_1 \ldots y_n$ where the successive y_i come from different factors of F and no y_i is in the amalgamated part A unless $n = 1$.

Under this definition, an element may have infinitely many normal forms. It is well known, however, that if w also has a normal form $y'_1 \ldots y'_m$, then $m = n$. The integer n is the length of w and we again write $|w| = n$.

Suppose that u and v are elements of F with normal forms $u = y_1 \ldots y_n$ and $v = x_1 \ldots x_m$ respectively. If $y_n x_1$ is in the amalgamated part A we say that there is *cancellation* between u and v in forming the product $w = uv$. If y_n and x_1 are in the same factor of F but $y_n x_1 \notin A$ we say that y_n and x_1 are *consolidated* in forming a normal form of uv.

A word w is said to have *semi-reduced* form $u_1 \ldots u_m$ if there is no cancellation in bringing the product $u_1 \ldots u_m$ into a normal form. Again, we write $w \equiv u_1 \ldots u_m$.

As for free products, if $w = y_1 \ldots y_n$ in normal form, we say that w is *strictly cyclically reduced* if y_n and y_1 are in different factors of F. w is *cyclically reduced* if $y_n y_1$ is not in the amalgamated part.

The definition of "symmetrized" is as before.

A word b is said to be a *piece* if there exist distinct elements r_1, r_2 of R (i.e., $r_1 \neq r_2$ in F) such that $r_1 \equiv bc_1$ and $r_2 \equiv bc_2$ in semi-reduced form.

The definition of the cancellation conditions is now exactly as for F a free product. The cancellation conditions are quite strong in the case of free products with amalgamation since *all* normal forms of r_1 and r_2 must be used to determine whether or not "subwords" b are pieces. A great deal thus depends on the amalgamated subgroup.

As an illustration of what can happen, let us consider the following example. In the case of an ordinary free product, no amalgamation, if x and y are elements from different factors, then the symmetrized set generated by $(xy)^n$ will, for sufficiently large n, satisfy whatever cancellation condition $C'(\lambda)$ we desire.

Now let $F = (\langle x\rangle * \langle y\rangle ; x^2 = y^2)$, the free product of two infinite cyclic groups with the squares of the generators identified. Consider the symmetrized set R generated by $(xy)^n$. Now $(xy)^n \neq (x^{-1}y^{-1})^n$ in F, but

$$(x^{-1}y^{-1})^n = x^{-1}y^{-1}(xy)^{n-1} = (x^{-1}x^2)(y^{-2}y^{-1})(xy)^{n-1}$$

$$= xy^{-3}(xy)^{n-1} = (xy)^{n-1}xy^{-(4n-1)} .$$

Thus $(xy)^{n-1}x$ is a piece relative to R and any cancellation condition fails badly.

References

[1] C.Blanc, Une interprétation élémentaire des théorèms fondamentaux de M.Nevanlinna, Comm. Math. Helv. 12 (1940) 153-163.

[2] C.Blanc, Les réseaux Riemanniens, Comm. Math. Helv. 13 (1941) 54-67.

[3] J.L.Britton, Solution of the word problem for certain types of groups, I, II, Proc. Glasgow Math. Assoc. 3 (1956) 45-54 (1958) 68-90.

[4] R.G.Burns, A note on free groups, Proc. Amer. Math. Soc. 23 (1969) 14-17.

[5] M.Dehn, Über unendliche diskontinuierliche Gruppen, Math. Ann. 71 (1911) 116-144.

[6] M.Dehn, Transformation der Kurven auf zweiseitigen Flächen, Math. Ann. 72 (1912) 413-421.

[7] F.Fiala, Sur les polyèdres à faces triangulaires, Comm. Math. Helv. 19 (1946) 83-90.

[8] A.V.Gladkii, On the nilpotency classes of groups with δ-bases, Doklady Akad. Nauk SSSR 125 (1959) 963-965.

[9] A.V.Gladkii, On simple Dyck words, Sib. Mat. Zh. 2 (1961) 36-45.

[10] A.V.Gladkii, On groups with k-reducible bases, Sib. Mat. Zh. 2 (1961) 366-383. These results were announed in Doklady Akad. Nauk SSSR 134, 16-18.

[11] M.Greendlinger, On Dehn's algorithm for the word problem, Comm. Pure Appl. Math. 13 (1960) 67-83.

[12] M.Greendlinger, On Dehn's algorithms for the word and conjugacy problems with applications, Comm. Pure Appl. Math. 13 (1960) 641-677.

[13] M.Greendlinger, An analogue of a theorem of Magnus, Archiv d. Math. 12 (1961) 94-96.

[14] M.Greendlinger, A class of groups all of whose elements have trivial centralizers, Math. Z. 78 (1962) 91-96.

[15] M.Greendlinger, On the word problem and the conjugacy problem, Izv. Akad. Nauk SSSR Ser. Mat. 29 (1965) 245-268. These results were announced in Doklady Akad. Nauk SSSR 154 (1964) 507-509.

[16] M.Greendlinger, Strengthened forms of two theorems for one class of groups, Sib. Mat. Zh. 6 (1965) 972-985.

[17] M.Greendlinger, Problem of conjugacy and coincidence with the anticenter in group theory, Sib. Mat. Zh. 7 (1966) 785-803. English translation, Siberian Math. J. 7 (1966) 626-640. These results were announced in Doklady Akad. Nauk SSSR 158 (1964) 1254-1257.

[18] M.Hall, Jr., Generators and relations in groups – the Burnside Problem, in: Lectures in Modern Mathematics, Vol. II, ed. T.Saaty (New York, 1964).

[19] G.Higman, B.H.Neumann and H.Neumann, Embedding theorems for groups, J. London Math. Soc. 24 (1949) 247-254.

[20] G.Higman, A finitely generated infinite simple group, J. London Math. Soc. 26 (1951) 61-64.

[21] F.Levin, Factor groups of the modular group, J. London Math. Soc. 43 (1968) 195-203.

[22] S.Lipschutz, Elements in *S*-groups with trivial centralizers, Comm. Pure Appl. Math. 13 (1960) 679-683.

[23] S.Lipschutz, On powers of elements in *S*-groups, Proc. Amer. Soc. 13 (1962) 181-186.

[24] S.Lipschutz, On square roots in eight-groups, Comm. Pure Appl. Math. 15 (1962) 39-43.

[25] S.Lipschutz, An extension of Greendlinger's results on the word problem, Proc. Amer. Math. Soc. 15 (1964) 37-43.

[26] S.Lipschutz, Powers in eight-groups, Proc. Amer. Math. Soc. 16 (1965) 1105-1106.

[27] S.Lipschutz, On the conjugacy problem and Greendlinger's eight groups, Proc. Amer. Math. Soc. 23 (1969) 101-106.

[28] S. Lipschutz, On the word problem and T-fourth-groups, this volume.

[29] R.C.Lyndon, On Dehn's algorithm, Math. Ann. 166 (1966) 208-228.

[30] R.C.Lyndon, A maximum principle for graphs, J. Combinatorial Theory 3 (1967) 34-37.

[31] J.McCool, Elements of finite order in free product sixth-groups, Glasgow Math. J. 9 (1968) 128-145.

[32] J.McCool, The order problem and the power problem for free product sixth-groups, Glasgow Math. J. 10 (1969) 1-9.

[33] J.McCool, Embedding theorems for countable groups, to appear.

[34] K.A.Mikhailova, The occurrence problem for direct products of groups, Doklady Akad. Nauk SSSR 119 (1958) 1103-1105.

[35] C.F. Miller and P.E. Schupp, Embeddings into hopfian groups, J. Algebra, 17 (1971) 171–176.

[36] J.Nielsen, Untersuchungen zur Topologie der geschlossenen zweiseitigen Flächen, I, II, III, Acta Math. 50 (1927) 189-358; 53 (1929) 1-76; 58 (1931) 87-167.

[37] K.Reidemeister, Einführung in die kombinatorische Topologie (Braunschweig, 1932).
[38] H.Schiek, Ähnlichkeitsanalyse von Gruppenrelationen, Acta Math. 96 (1956) 157-251.
[39] H.Schiek, Das Adjunktionsproblem der Gruppentheorie, Math. Ann. 147 (1962) 159-165.
[40] P.E.Schupp, On Dehn's algorithm and the conjugacy problem, Math. Ann. 178 (1968) 119-130.
[41] P.E.Schupp, On the conjugacy problem in certain quotient groups of free products, Math. Ann. 186 (1970) 123–129.
[42] P.E.Schupp, On Greendlinger's Lemma, Comm. Pure Appl. Math. 23 (1970) 233-240.
[43] P.E.Schupp, Small cancellation theory over free products with amalgamation, to Math. Ann., to appear.
[44] P.E.Schupp, Commuting elements in small cancellation groups, to appear.
[45] V.V.Soldatova, On groups with a δ-basis, for $\delta < \frac{1}{4}$, and a single additional condition, Sib. Mat. Zh. 7 (1966) 627-637. English translation, Siberian Math. J 7 (1966) 504-511.
[46] P.Stender, On the application of the sieve method to the solution of the word problem for certain groups with a denumerable set of generating elements and a denumerable set of defining relations, Mat. Sbornik (N.S.) 32 (1953) 97-108.
[47] V.A.Tartakovskii, The sieve method in group theory, Mat. Sbornik (N.S.) 25 (1949) 3-50.
[48] V.A.Tartakovskii, Application of the sieve method to the solution of the word problem for certain types of groups, Mat. Sbornik (N.S.) 25 (1949) 251-274.
[49] V.A.Tartakovskii, Solution of the word problem for groups with a k-reduced basis for $k > 6$, Izv. Akad. Nauk SSSR, Ser. Mat. 13 (1949) 483-494. English translation of [47,48,49], Amer. Math. Soc. Translations 60 (1952), reprint 1 (1962).
[50] E.R.Van Kampen, On some lemmas in the theory of groups, Amer. J. Math. 55 (1933) 268-273.
[51] C.M.Weinbaum, Visualizing the word problem, with an application to sixth groups, Pacific J. Math. 16 (1966) 557-578.
[52] C.M. Weinbaum, The word and conjugacy problems for the knot group of any tame, prime, alternating knot, Proc. Amer. Math. Soc., to appear.

I would like to mention two interesting results obtained since this survey article was written. C.M. Weinbaum [52] has shown that if G is the group of a prime alternating knot, then $G * \langle x \rangle$ (the free product of G and an infinite cyclic group) has a presentation satisfying $C(4)$ and $T(4)$. Thus both the word and conjugacy problems for G are solvable.

We have not touched on small cancellation theory for semigroups in this article. John Remmers (Thesis, Univ. of Michigan, 1971) has recently developed a geometric version of small cancellation theory for semigroups and extended the previous results.

THE ASSOCIATIVITY PROBLEM FOR MONOIDS AND THE WORD PROBLEM FOR SEMIGROUPS AND GROUPS

Dov TAMARI

Department of Mathematics, State University of New York at Buffalo

§1. Associativity of monoids

Definition of associativity

A *word* with formally correct binary bracketing is a *monomial.* Parentheses are a priori indispensable for giving words of length > 2 an interpretation by computation in terms of the given, possibly partial, binary operation. Associativity of such a multiplicative system $(M, \cdot)$ is usually intended to mean the independence of the "value" or "sense" of words (if there is one!) from the distribution of the parentheses. The complexity and, in spite of it, insufficiency of such a concept of general associativity are well hidden for *complete*, i.e. everywhere defined, operations $\cdot : M \times M \to M$. We call these systems *monads* *, although they are more often referred to as groupoids *; associative monads are semigroups.

General systems with a partial operation $\cdot : P_M \to M$, where $P_M \subset M \times M$ is an arbitrary binary relation over M, will be called

* *Terminological remark*: The legitimate use of the term "groupoid" refers to Brandt groupoids or relatively slight generalizations, synonymous to "partial groups", "group germs" or "local groups" and "categories of morphisms". "Partial groupoids" is a pleonasm and compounded abuse of language. To correct the misused "groupoid" to "Ore groupoid" is unwanted; why should one honor the unlucky choice of terminology? It was unfortunate that Ore ignored the legitimate use of "groupoid"; but it was fortunate for category theory (or should one say for category theorists?) that, as a consequence, the limited rights of Brandt could be ignored in good conscience. Confusion is compounded by the use of "monoid" for "semigroup with 1" under the influence of Bourbaki (who probably intended to abolish the term "semigroup" for reasons of linguistic taste).

*monoids**, although they are more often referred to as partial groupoids*. One is inevitably led to a problem of definition: What does or should associativity of monoids mean? This question is reduced to that of monads as follows: *M is associative if it can be completed to,* i.e. *embedded in, a semigroup*; in other words, *if there is an injective homomorphism (monomorphism) $M \to S$, S a semigroup*. One can choose $S = S_M$, the semigroup *generated* by M (universal object in the category of morphisms of M into semigroups). M is therefore associative iff its distinct elements, considered as the *generators* or one-letter words of S, represent mutually distinct elements of S: $s_1 \neq s_2$ in $M \Rightarrow s_1 \neq s_2$ in S. The canonical homomorphism ϕ: $M \to S$ maps each $s \in M$ into the equivalence class s^ϕ containing the one-letter word s. M is associative, or what is the same, ϕ is a monomorphism, iff s is the only one-letter word in s^ϕ.

Immediate consequences

M associative $\Rightarrow$ card M = card $M^\varphi \leqslant$ card S;

M associative & card $M > 1 \Rightarrow$ card $S > 1$, i.e. S is not trivial;

M non-associative $\Rightarrow$ card $M >$ card $M^\phi \Rightarrow \exists$ a couple $s_1 \neq s_2$ in M but $s_1 = s_2$ in S. *M^ϕ is the greatest associative homomorphic image of M* (up to isomorphism).

Why has this simple conceptual definition of associativity not been used before? Its meaning in explicit, constructive terms, while simple and definite for monads, is combinatorially complicated for general monoids and cannot be written down in ordinary finite terms. For *symmetrical monoids*, or *partial inverse property loops* [9, 33] the definition in constructive terms is still infinite, but simplified by the "one-mountain theorem" which can be traced back to Newman [25].

Construction of $S_M \simeq F_M/E_M$

F_M is the free semigroup over the set (support) $|M|$ of M, E_M the congruence generated over F_M by the relations $ab = c$ in M, acting as elementary substitutions: $ab \to c$ (products or contrac-

* See Terminological remark (preceding page).

tions) and $c \to ab$ (factorizations or expansions). For $A, B \in F_M$, $A = B$ in $S_M \Leftrightarrow A \sim B \pmod{E_M} \Leftrightarrow \exists$ a finite chain of substitutions

$$C_A^B: A \equiv W_0 \to W_1 \to \cdots \to W_{i-1} \to W_i \to \cdots \to W_n \equiv B,$$

where either $W_{i-1} \equiv U_i a_i b_i V_i$ and $W_i \equiv U_i c_i V_i$, or vice versa. "M associative" means that a chain contains at most one one-letter word, although possibly several times. Call chains starting and ending with one-letter words a, b *special* chains C_a^b; then: "M associative" $\Leftrightarrow$ "$C_a^b \Rightarrow a = b$". An inner piece of a special chain is a general chain. Conversely, every general chain can be extended to, or embedded in, a special chain over a suitable extension monoid.

Non-validity of the one-mountain theorem for general chains

Arranging words in levels by length, chains consist of alternating sequences of expansions or "ascents" and contractions or "descents". The one-mountain theorem, valid for symmetrical monoids, permits limitation to chains with a single ascent followed by a descent. The monoid

$$M = \{a, b, c, d, f, f'; ab = ac = bc = cb = d, dc = f, db = f'\}$$

shows that the one-mountain theorem is not valid in general:

$$f \to dc \to abc \to ad \to acb \to db \to f';$$

i.e., $f = f'$ in S_M but not in M (M non-associative). Yet, there is no word contracting over M to f as well as to f'.*

Universal and partial algebras

A class A of algebras is equationally defined by basic identities (laws, axioms); what is the natural generalization to partial or incomplete A-algebras? *A partial algebra P with the "same" operations as A is an incomplete A-algebra if it can be completed to*, i.e.

* In this example it is essential that M is not cancellative as (left) cancellation would imply $b = c$ and thus $f = f'$; however, one can easily construct somewhat larger cancellation monoids similarly exhibiting non-validity of the one-mountain theorem.

embedded in, an A*-algebra A*. As before one can choose the universal object in the category of morphisms of P into A-algebras, i.e. $A = A_P$, the A-algebra *generated* by P. Unconditional equations, i.e. identities

$$I: f(x_1, \dots, x_n) = g(x_1, \dots, x_n), \quad f, g \text{ polynomials},$$

are replaced, or "translated", by implications from chains

$$I^*: C_y^z \Rightarrow y = z,$$

where C_y^z is a chain of elementary substitution relations $o_i(u_1, \dots, u_{n_i}) = v_i$, the o_i basic operations, the v_i and u_j intermediary variables, computing $f(x_1, \dots, x_n) = y$ and $g(x_1, \dots, x_n) = z$. Denote: C the axiom of complete operations; I_0 the set of basic laws of A; $I = \bigcup_{i=0}^{\infty} I_i$ the r.e. set of all valid identities of A in some convenient graduation I_i; I_0^*, resp. $I^* = \bigcup_{i=0}^{\infty} I_i^*$, the corresponding "translation" to implicational chains. In A

$$I_0 \;\&\; C \Rightarrow I;$$

but in the corresponding class of incomplete A-algebras *no finite subset* $I_{(m)}^* = \bigcup_{i=0}^{m} I_i^*$ *needs imply all of* I^*. Therefore the original, finite set of laws I_0 of A must be replaced by the infinity of implications I^*, or, at least, by an infinite subset $D^* \subset I^*$ such that $D^* \Rightarrow I^*$.

Associativity chains

For A being the class of semigroups, one single binary operation and one single identity $(xy)z = x(yz)$ suffice as a set of basic operations and laws. For the class of *incomplete semigroups*, or *associative monoids*, one needs an infinity of axioms*, a r.e. system of conditional identities in the form of implications from finite chains of ever-increasing length and complexity. These associativity chains

* This insight had been obtained independently and indirectly by Evans [12]. From the fact that the word problem for semigroups is unsolvable he concludes that "no finite set of axioms for semigroups will admit an embedding theorem".

are similar to the Malcev conditions for the embeddability of semigroups into groups. They are most conveniently pictured by literally the same graphs which are geometrical models of Malcev chains as introduced by Tamari [32]. Their deduction and algebraic interpretation are even simpler and more general. To some degree this more basic meaning is already present in the older work of the early 50's. The interesting theory of these associativity chains will appear in detail elsewhere.*

§2. Presentations

The proof of the equivalence of the associativity and the word problem is based on conceptual work, in particular on wider interpretation of the concept of presentations as partial algebras.

Preliminaries

A word is a shorthand for all its monomials and the *associativity relations equating them.* These relations are a kind of a priori implied *"trivial" defining relations.* With the postulated *existence* of a monomial, the existence of all its *components* or *submonomials* is implied; with the existence of a word, also that of the associativity relations in its subwords.**

The concept of presentation of semigroups (G, R), G an alphabet or set of generators, R a set of defining relations, is extended to *presentation of monoids* (Q, R), Q a set of existing monomials (letters included) and R a set of defining relations among members of Q. Monoids themselves are presentations of themselves, as well

* For a first informal account see "Le probleme de l'associativite des monoides et le probleme des mots pour les demi-groupes: Algebres partielles et chaines elementaires", Sem. Dubreil-Pisot 24 (1970/71) 8.01–15.

** These preliminary conventions are not equivalent to consideration of semigroups in the category of monads. While abelian groups are singled out in the category of groups by the "trivial" commutativity relations of couples of generators, $g_ig_j = g_jg_i$, and groups in the category of semigroups by adjoining an identity, formal inverses, and the "trivial" defining relations $g_ig_i^{-1} = g_i^{-1}g_i = 1$, the associativity relations and new generators of the standard presentation below will not make the generated monad a semigroup. Thus associativity is a *deeper* property.

as of other structures, in particular of the semigroups *generated* by them. Conversely, *every presentation of a semigroup can be standardized to a monoid generating the same semigroup* (up to isomorphism); *finite presentations standardize to finite monoids.*

Extensions

A presentation of a semigroup, standardized or not, can be extended by adjunction to Q of monomials, in particular of words. The extended standard presentations are extension monoids of the original standard presentation: The alphabet and the set of defining relations are extended by new letters for each new subword and the implied associativity relations.

Construction of standard presentation

Let M be a monoid, $A = s_{i_1} s_{i_2} \cdots s_{i_n}$ an incontractible word over M. Construct first an extension $M(A)$ of M with a chain $\mathcal{C}^A_{s_A}$, $s_A \in M(A)$, $S_{M(A)} = S_M$: Screen and name the *distinct* two-letter subwords in A among the $n-1$ neighbor couples $s_{i_j} s_{i_{j+1}} = s_{i_j i_{j+1}}$; the *distinct* three-letter subwords among its $n-2$ neighbor triples introducing new symbols and couples of elementary relations

$$s_{i_j} s_{i_{j+1} i_{j+2}} = s_{i_j i_{j+1} i_{j+2}} = s_{i_j i_{j+1}} s_{i_{j+2}};$$

the *distinct* k-letter words, $k < n$, among the $n-k+1$ neighbor k-tuples, introducing new symbols and sets of $k-1$ new elementary relations; finally,

$$s_A \equiv s_{i_1 \cdots i_n} = s_{i_1} s_{i_2 \cdots i_n} = s_{i_1 i_2} s_{i_3 \cdots i_n} = s_{i_1 i_2 \cdots i_{n-1}} s_{i_n}$$

with one new symbol and $n-1$ new relations. Thus M is extended bya finite number of new letters and relations to $M(A)$ in which the word A is contractible to one letter s_A. Similarly, one can adjoin to M a set of words $\{A, B, \ldots\}$. Assume all words incontractible over M, or replace them by incontractible contractions.* Screen

* A contractible word may admit several incontractible contractions; any one will do.

and name their *distinct* neighbor couples, triples, ..., k-tuples, until the exhaustion of the longest words, introducing new symbols and sets of new elementary relations. One obtains an extension monoid $M(A, B, \ldots)$ generating the same S_M because its elements are words over M and its relations anyhow valid in S_M. Any C_A^B can be embedded in a special chain $C_{s_A}^{s_B} \equiv C_{s_A}^A \, C_A^B \, C_B^{s_B}$. *Prescribed relations, say $A = B$, are taken in account by putting $s_A \equiv s_B$ and considering them one and the same symbol.*

Given any ordinary presentation of a semigroup (G, R), start with its set of generators and no relations; i.e., consider the abstract set G as a monoid with empty multiplication table. Adjoin the set of all words appearing in R and "take account" of all defining relations of R. This standardizes the given presentation to a monoid. The result of the construction is unique up to isomorphism.

As each (monoidal) presentation can be extended by adjunction of any word, *one can thus associate with each finite semigroup presentation r.e. classes of finite monoids, all presenting the same semigroup, for any r.e. set of words, in particular the r.e. set of all words.*

Non-associativity

Non-associativity of finite or countably infinite *symmetrical* monoids is a readily assertable property. It suffices to enumerate the words and to subject each to a finite number of finite computations. *A word is non-associative* if it can be contracted to at least two distinct elements of M. If M is non-associative, one must come across a first non-associative word, indeed across any non-associative word, in a finite number of steps. Therefore the set of non-associative words over M – if not empty – is effectively constructible.

For general, finite or countably infinite monoids for which the one-mountain theorem is not valid, non-associativity is still readily assertable. For finite monoids, the number of chains of all types of given length is finite because a word admits only a finite number of elementary contractions and expansions. This is still so for infinite monoids with the property that each element appears only a finite number of times in the multiplication table, i.e., admits only a finite number of binary factorizations. Even without such a property, one can enumerate all chains by *rank* $r = l + i$, l the length of the chain, i the highest index of letters in any of its words. *A chain is non-associative* if it connects two distinct letters. Thus, given an enumeration of all chains in order of increasing rank, if a monoid is non-associative one can find the first, and ultimately, any non-associative chain. Therefore the set of non-associative chains is either empty or effectively constructible. It follows that *as long as one can decide if a finite monoid is associative or not, one can effectively construct its greatest associative homomorphic image* $M^{\phi} \simeq M/E_M$ by constructing the congruence generated by its finitely many "*non-associative*" *couples of elements* connected by non-associative chains.

Homomorphic images

A presenting monoid may or may not be associative. If it is non-associative all its extensions are so, too. One can make sure that associative monoids are among the standard presentations of a semigroup by including homomorphic images, i.e., closing the class for homomorphisms. Among them must be associative homomorphic images. The greatest associative homomorphic image still presents S and is identifiable with a generating subset of S and an extract of its multiplication table. A finite monoid admits only a finite number of congruences and corresponding homomorphic images. Thus a r.e. class of finite monoids remains r.e. by homomorphic closure. On the other hand, homomorphic closure introduces presentations of new semigroups which are proper homomorphic images of the originally presented semigroup. If one has an *associativity test* for the class of finite monoids under consideration, this can be avoided by limiting oneself to congruences induced by non-associative couples only. Iterating these quotient constructions a finite number of times, one is stopped at the first, i.e. greatest associative homomorphic image. For monoids presenting S with a *word comparison test* one admits only congruences generated by $a \neq b$ in M, but $a =_s b$ in S. The meaning of "homomorphic image" is thus restricted from now on.

§3. Reduction of the word problem of semigroups to the associativity problem for monoids

Theorem. *To any given finite presentation of a semigroup one can construct a r.e. class of finite monoids such that an algorithmic solution of the associativity problem for this class implies a solution of the word problem for the semigroup, and vice versa.*

Corollary. *The known existence of finitely presented semigroups with unsolvable word problem* [27, 34, 10] *implies unsolvability of the associativity problem for the corresponding classes of finite monoids and for the class of finite monoids in general.*

Remark. Checking associativity of finite monads is a finite routine task; passing to monoids makes holes in the multiplication table; therefore, the "devil of unsolvability" slips in through these holes at the passage to partial systems.

Proof. The hypothetical solution of the associativity problem by an assocativity test AsT for a r.e. class M_S of finite monoids associated with S leads to a solution of the word problem for S. AsT can be imagined as an oracle, black box or machine putting out the right answer to the question "associative or not" for every monoid of the class fed into its input. Narrowing down the class apparently strengthens the result; but it is convenient to start with the whole class M of finite monoids; in fact, considerably smaller classes will do.* A priori the class is not r.e., even not a set. One means the finite monoids in standard notation with elements $s_1, s_2, ..., s_m$ from an infinite standard set of symbols $s_1, s_2, ...$. For each m exist $(m + 1)^{m^2}$ partial multiplication tables; therefore the standard class of finite monoids is r.e.; the isomorphism type of each finite monoid is represented in it.**

* The more restricted classes actually used are implicit in the construction of the proof. Explicit definition of r.e. classes, good for both reductions, seems difficult and is itself a problem (see discussion of unsolvability degrees).

** One can constructively assert the *first* appearance of an isomorphism type in a recursive enumeration of finite multiplication tables.

Let M be a standard presentation of S. AsT either asserts M as associative, or it enables one to pick out its greatest associative homorphic image. Without loss of generality one can assume $M = M_0$ associative. One constructs an ascending chain of finite associative monoids exhausting S and solving its word problem:

$$M_0 \subset M_1 \subset M_2 \subset \cdots \subset S \equiv \lim_{n \to \infty} M_n = \bigcup_{n=0}^{\infty} M_n.$$

Assume M_n with elements $s_1, \ldots, s_{m_n}$ and its partial multiplication table already constructed. Arrange the incontractible words over M_n in an order θ_n progressing first by length, then by highest indexed letter in a word, and last by alphabet; e.g.:

$$aa; ab, ba; bb; ac, bc, ca, cb, cc; ad, \ldots; aaa; aab, aba, abb, baa, bab, bba, bbb; aac, \ldots .$$

In this way the last couples contain the last generator and are of form $s_i s_{m_n}, s_{m_n} s_i$; similarly for triples, etc. Denote $a_n b_n = c_n$, the first incontractible two-letter word over M_n. Consider c_n a new letter and $a_n b_n = c_n$ a new relation yielding an extension $M'_n = M_n(c_n)$. (c_n is of course the letter s_{m_n+1}.) Apply AsT to M'_n and distinguish two cases

(a) *M'_n is associative*: Put $M'_n = M_{n+1}$. Words containing $a_n b_n$ which were incontractible become contractible: $a_n b_n$ is replaced everywhere by c_n. One gets a new enumeration θ_{n+1} of incontractible words over M_{n+1}: $a_n b_n$ disappears from the head of the two-letter word sequence, while new incontractible words of form $s_i c_n, c_n s_i$ appear at its end.*

(b) *M'_n non-associative*: $\exists! d_n \in M_n \mid C_{c_n}^{d_n}$ in M'_n, i.e. $d_n \sim c_n$ (mod $E_{M'_n}$). $M_{n+1} \simeq M'^{\phi}_n$. $E_{M'_n}$ is a trivial congruence; its equivalence class are singletons except for $\{d_n, c_n\}$. As convened, standardize by replacing c_n by $d_n \equiv s_k$ ($k \leqslant m_n$). The supports of M_n and M_{n+1} are the same, but M_{n+1} has an additional relation, $a_n b_n = d_n$. In the new order θ_{n+1} of incontractible words over M_{n+1} again the first two-letter word of θ_n disappears, but *no new two-letter words*

* Possibly also $c_n c_n$ if $b_n a_n$ is incontractible over M_n. Words heading the three- and more letter word sequences disappear and words with c_n appear at the ends.

are added: formerly incontractible three- or more letter words containing $a_n b_n$ shrink by one or more letters.

(c) From M_{n+1} proceed as from M_n. If the non-associative case b) repeats itself a sufficient number of times, one will be stopped at M_{n+p_n} with $|M_{n+p_n}| = |M_n|$ and completely filled multiplication table, if M_n has empty places. $M_{n+p_n} \simeq S$ *is a finite semigroup*; its word problem is solved.*

(d) Otherwise one must encounter the associative case (a) sufficiently often, replenishing the supply of new incontractible two-letter words. One keeps going ad infinitum. At each step a new relation is added; words shrink in length to become replaced by existing or new incontractibles; from time to time a new generator is added. One exhausts all original two-letter incontractibles over M_n and reaches M_{n+p_n}. All incontractibles over M_n have disappeared and are replaced by shorter words consisting mostly of new generators.

One continues filling the $q_n \leqslant p_n$ new lines and columns corresponding to the q_n new letters $s_{m_n+1}, s_{m_n+2}, ..., s_{m_n+q_n}$, etc. Any given, originally incontractible word of arbitrary length will eventually be reached, shrunk to one letter at some stage of sufficiently high index M_t. Each word contracts finally to one and only one letter representing an element of $S \simeq \lim M_n$; its word problem is solved. One can express the result as follows: The unsolvability degree of the associativity problem for finite monoids, in particular also for the r.e. classes constructively associated to a given semigroup, is greater or equal the unsolvability degree of the word problem of any finitely presented semigroup.

§4. The converse reduction

In order to prove the equivalence of the word and the associativity problems one needs a converse reducing the associativity problem for sufficiently large r.e. classes M of finite monoids to the word problem of finitely presented semigroups S_{M} associated

* This solves the *finiteness* problem as a special case of the associativity problem.

with this class. Assumption of an algorithmic solution for the word problem of S_M leads to the solution of the associativity problem for M.

The algorithmic solution of the word problem can be imagined as a *word comparison test* WcT (oracle, black box or machine) admitting as input couples of words in the generators and yielding as output correct answers "$=_S$" or "$\neq_S$" (equal or not equal in S). Each monoid known to be a presentation of S has elements expressible as words over the original presentation. Apply WcT to the finite number of couples of distinct elements written as such words. Getting each time the same answer "$\neq_S$" means "M is associative", getting once "$=_S$" "non-associative". Doing this for the standard class of all finite monoids presenting S would solve the associativity problem for this class. The trouble is that this class is not r.e.; there is no decision procedure for the isomorphism problem of finitely presented semigroups. One has therefore to be satisfied with a smaller, less elegantly defined, r.e. class of finite monoids, e.g., that defined by the construction of the former reduction. There the argument can evidently be inversed. This r.e. class M_M starts with the finite monoid M and includes all its modifications by adjunction of words in a suitable enumeration and by restricted homomorphic image. M generates the semigroup S_M for which the WcT is assumed. A finite number of applications of WcT decides for each monoid of the class if it has distinct elements a, b with $a =_S b$ or not, i.e., if it is non-associative or not, if it admits a homomorphic image or not in the class, solving its associativity problem.

One has thus proven that the degree of unsolvability of the word problem for $S = S_M$ is at least as great as that of the associativity problem for the implicitly and vaguely defined r.e. class M_M of finite monoids. As the classes M_M used in both reductions are the same, one has thus proven that the word problem and the associativity problem are equivalent in this sense.

§5. Unsolvability degrees, open questions

Contrary to what was said for the first reduction, the converse theorem gets stronger if the r.e. class of finite monoids, for which the associativity problem is reduced to the word problem for a certain ad hoc constructed finitely presented semigroup, becomes wider. Clearly, classes which are too small can not do. Consider the extreme case of classes reduced to a single monoid; there is no genuine, i.e., algorithmic associativity problem. For an individual monoid the problem is decidable: it is either associative or not. One may say that its unsolvability degree is 0. True, the solution of the word problem for the semigroup generated by M solves the "associativity problem" for M as its generator or one-letter word problem; but to use an algorithmic solution for the word problem of S_M for its solution seems an abuse.*

Denote $\alpha\mathsf{M}$ the unsolvability degree of the associativity problem for a r.e. class M of finite monoids, βS that of the word problem of a finitely presented semigroup S. One has:

I (*first reduction*): For all finitely presented semigroups $S = S_M$ there exist r.e. classes M_S of finite monoids such that $\alpha\mathsf{M}_S \geqslant \beta S_M$.

II (*second reduction*): Any finite monoid M generates a r.e. class of finite monoids M_M and a semigroup $S = S_M$ such that $\beta S_M \geqslant \alpha\mathsf{M}_M$.

I becomes stronger if M_S gets smaller, II becomes stronger if M_M gets larger. It is therefore desirable that M_S be as small and M_M as large as possible, $\mathsf{M}_M \supset \mathsf{M}_S$, and $S = S_M$. This, certainly, would justify the conclusion $\alpha\mathsf{M} = \beta S$, i.e., a strong equivalence of both problems; but $\mathsf{M}_M = \mathsf{M}_S$ suffices. As a matter of fact the definition of $\mathsf{M}_M = \mathsf{M}_S$ is not very satisfactory. One can doubt that they are well defined r.e. classes, or even suspect circular definition. More to the point is the recognition of a "grey" area (instead of a clear borderline) containing r.e. M_M, M_S, which needs further re-

* "Like driving a small nail with a sledge hammer" (Boone), shooting birds with cannons, or, quite usual today, using universal computers for a few bills.

search. Does an ideal borderline exist? If so, do there exist r.e. M on it? Is there a "principle of continuity" assuring existence of r.e. classes on it? Can M_M be replaced by larger and simpler r.e. classes? More generally, can one construct to *any* r.e. class of finite monoids a *finitely* generated semigroup such that the solution of its word problem solves the associativity problem for the class? If so, this would in particular apply to the standard class of *all* finite monoids and imply the existence of a "universal" finitely presented semigroup with word problem of unsolvability degree *equal* to that of the associativity problem for the class of all finite monoids. Presently one can only assert that the unsolvability degree of this associativity problem is at least equal that of the word problem for finitely presented semigroups.

Another open question

A direct proof of the unsolvability of the associativity problem for finite monoids (for certain r.e. classes), say by a diagonal method, is desirable. It would immediately yield a more significant proof for the unsolvability of the word problem for semigroups and groups. Originally the search for such a method delayed publication of this work in the early 60's, until it was lost.* At that time the author believed himself near his aim; the question is still open.

§6. Symmetrical monoids and groups

Symmetrical monoids (= partial inverse property loops; for definition see [9, 33]) generate by themselves semigroups which are groups. Symmetrical presentations standardize by themselves to symmetrical monoids. For arbitrary monoids and presentations one needs a theory of symmetrization and symmetrizability. In this way methods and results transfer from semigroups to groups. One has the advantage of the one-mountain theorem simplifying associativity, but the disadvantage of the more difficult word problem for groups. This method with emphasis on groups was pre-

* A handbag containing manuscripts and documents of the author disappeared in the harbor of Haifa, Israel, September 9, 1962. In spite of rewards offered it was never recovered.

valent in the earlier work of the author in the beginnings of the 60's. An informal, partial account of some of its results is given in [33b]. Among others, it is proven there that *a cancellation semigroup is embeddable in a group if and only if its symmetrization is associative.*

Acknowledgements

As already mentioned the first ideas of this paper go back to the beginnings of the 50's [32a, b]. The author feels deeply obliged to Professors H. Cartan, A. Châtelet and P. Dubreil, who made this earlier work in France possible. He further owes a great intellectual debt to Malcev; also a moral one for encouragement when he met him personally at the occasion of the I.C.M. Stockholm 1962.

This paper owes its origin to an attempt to freely reconstruct and to update, as far as feasible, work lost in 1962.* It is part of this attempt of reconstruction, which has perhaps partially succeeded. The author takes this occasion to thank those French, Dutch and American Mathematicians who helped him in his need by offering shelter, interest and generosity of mind; in particular in direct connection with this work thanks are due to Professor Gödel who invited him a second time to the Institute for Advanced Study at Princeton 1967/68; and to the Organizers and Participants of these Proceedings, in particular to Professor Boone, for hospitality and critical discussions.

Updating too is only partial, due to the limited capacity and difficult circumstances of the author. During the last decade progress in the fields of Algebra, Universal and Quasi-Universal Algebra, and of Decision Problems has been great, generating a vast volume of literature. New, powerful tools like Category, Automata and Language Theories have become readily available. There are some near approaches in the direction of our subject, but the most relevant one, Evans' papers [12], were already available in the early 50's. Yet, to the best of the author's knowledge, and fortunately for this paper, and others to follow, the point of view of this work and its results have not been covered elsewhere.

* See historical remark in footnote preceding page.

Bibliography and References

[1] S.I. Adjan, Algorithmic undecidability of certain group properties, Doklady Akad. Nauk. SSSR 103 (1955) 533–535 (Russian).

[2] S.I. Adjan, Defining relations and algorithmic problems for groups and semigroups, Trud Mat. Inst. Steklov 85 (1966). Translated A.M.S. 1967, IV + 152.

[3] R. Baer, Free sums of groups and their generalizations. An analysis of the associative law. I–III. Am. J. Math. 71 (1949) 706–742; 72 (1950) 625–646; 647–670.

[4] G.E. Bates, Free loops and nets and their generalizations, Am. J. Math. 69 (1947) 499–550.

[5] W.W. Boone, The word problem, Ann. Math. 70 (1959) 207–265.

[6] W.W. Boone, Word problems and recursively enumerable degrees of unsolvability, A sequel on finitely presented groups, Ann. Math. 84 (1966) 49–84.

[7] J.L. Britton, The word problem for groups; Proc. London Math. Soc. 8 (1958) 493–506.

[8] G.C. Bush, The embedding theorems of Malcev and Lambek; Canad. J. Math. 15 (1963) 49–58.

[9] Carvalho and Tamari, Sur l'associativité partielle des symétrisations de semigroupes; Portug. Mat. 21 (1962) 157–169.

[10] G.S. Cejtin, An associative calculus with an unsolvable problem of equivalence; Trudi Mat. Inst. Steklov 52 (1958) 172–189 (Russian).

[11] R. Doss, Sur l'immersion d'un semi-groupe dans un group. Bull. Sci. Math. 72 (1948) 139–150.

[12] T. Evans, (a) The word problem for abstract algebras, J. London Math. Soc. 26 (1951) 64–71; (b) Embeddability and the word problem, ibid. 28 (1953) 76–80.

[13] Freedman and Tamari, Problèmes d'associativité: Une structure de treillis finis induite par une loi demi-associative; J. Comb. Th. 2 (1967) 215–242.

[14] A.A. Fridman, Degrees of Unsolvability of Word Problems for Groups (Nauka, Moscow, 1967) 190 pp. (Russian).

[15] A. Ginzburg and D. Tamari, Representation of binary systems and generalized groups by families of binary relations, Israel J. Math. 7 (1969) 21–45.

[16] G. Higman, Subgroups of finitely presented groups, Proc. Roy. Soc. Ser. A 262 (1961) 455–475.

[17] J. Lambek, The immersibility of a semigroup into a group, Canad. J. Math. 3 (1951) 34–43.

[18] A. Malcev, On the immersion of an algebraic ring into a field, Math. Ann. 113 (1937) 686–691.

[19] A. Malcev, On embedding associative systems into groups I, Mat. Sbornik 6 (1939) 331–336; II, ibid. 8 (1940) 251–263.

[20] A. Malcev, Sur les groupes topologiques locaux et complets; Comptes Rendus (Dokl.) 32 (1941) 606–608.

[21] A. Malcev, Quasiprimitive classes of abstract algebras (Russian); Doklady Akad. Nauk SSSR 108 (1956) 187–189.

[22] A. Malcev, Some questions bordering on algebra and mathematical logic; AMS transl.

[23] K.A. Mihailova, The occurrence problems for direct and free products of groups; Doklady. Akad. Nauk SSSR 119 (1958) 1103–05; ibid. 127 (1959) 746–748.

[24] H. Neumann, Generalized free products with amalgamated subgroups, Am. J. Math. 70 (1948) 590–625.
[25] M.H.A. Newman, On theories with a combinatorial definition of "equivalence"; Ann. Math. 43 (1942) 223–243.
[26] N.S. Novikov, On the algorithmic unsolvability of the word problem in groups; Trudi Mat. Inst. Steklov, 44 (1955) 1–143.
[27] E.L. Post, Recursive unsolvability of a problem of THUE, J. Symbol. Logic 12 (1947) 1–11.
[28] V. Ptak, Immersibility of semigroups; Acta Fac. Nat. Un. Gar. Prague 192 (1949) 16.
[29] M. Rabin, Recursive unsolvability of group theoretic problems, Ann. Math. 67 (1958) 172–194.
[30] S. Swierczkowski, Some examples of non-associative monoids (symmetric or not) of low order; refers to Malcev [20]; a private letter communicated by J. Mycielski.
[31] D. Tamari, Les images homomorphes des groupoïdes de Brandt, Comptes Rendus (Paris) 229 (1949) 1291–1293; Representations isomorphes par des systemes de relations; ibid. 232 (1951) 1332–1334.
[32] D. Tamari, (a) Malcev chains and generalized Malcev conditions, Commun. Int. Congr. Math. Cambridge 1950; (b) Monoïdes préordonnés et chaînes de Malcev; Thèse, Paris 1951; (c) part of it in Bull. Soc. Math. France 82 (1954) 53–96.
[33] D. Tamari, (a) Imbeddings of partial (incomplete) multisystems (monoids), associativity and word problem, Abstract, Amer. Math. Soc. Notices, 7 (1960) 760. (b) Problèmes d'associativité des monoides et problèmes des mots pour les groupes; Sem. Dubreil-Pisot 16 (1962/63) 7.01–30.
[34] A.M. Turing, The word problem in semigroups with cancellation; Ann. Math. 52 (1950) 491–505.

MAXIMAL MODELS AND REFUTATION COMPLETENESS: SEMIDECISION PROCEDURES IN AUTOMATIC THEOREM PROVING*

Lawrence WOS
Argonne National Laboratory, Argonne, Illinois

George ROBINSON
Stanford Linear Accelerator, Stanford, California

§1. Introduction

In recent years the idea of using electronic computers to search for proofs of theorems of quantification theory has drawn considerable attention. One of the more successful methods of attack on the problem has stemmed from the work of Quine [12], Gilmore [5], Davis and Putnam [4] and J. Alan Robinson [16]. This paper is concerned with a portion of the theory underlying an extension of this line of development to systems – first-order theories with equality – in which there is a distinguished relation symbol for equality. The field of mathematics upon which we have concentrated our computer experiments in order to study various properties of our procedure is first-order group theory.

To say that a particular property is decidable for some class of statements means that there is a single uniform procedure which will correctly determine whether the property holds when presented with any given statement from the class. Church showed [2] that, for any fixed procedure, there exists a statement of first-order predicate calculus for which the procedure will not be able to answer correctly the question: is this statement a theorem? For group theory, the question of theoremhood is also known to be undecidable as proved by Tarski [17]. The situation is, however, far from hopeless for first-order theories at least. There do

* Work performed in part under the auspices of the United States Atomic Energy Commission.

exist procedures which will, if presented with a set of first-order axioms for a theory and a first-order statement that happens to be a theorem thereof, correctly identify the statement as a theorem. Such a procedure is called a *semidecision procedure for theoremhood*. The basis for such a procedure is often a set of inference rules (rules for reasoning from the axioms of the theory to the statement whose theoremhood is in question), in which case it is natural to call the procedure a *proof procedure* since the procedure generates a proof of the chosen statement if it is a theorem. In order for a procedure to be of interest from the computational viewpoint, it must also be reasonably efficient. The indications are that the inference rules given in this paper may provide the basis for an efficient semidecision procedure for theoremhood not only for first-order group theory but for other first-order theories with equality as well. The reasons for expecting efficiency are: equality is treated as a special logical symbol distinguished from the relation symbols of the mathematical theory, the inferences deduced with the rules have a certain important property of generality that causes us to call the rules *conservative,* and finally a number of inferences which are ordinarily obtained immediately in more classical systems are not deducible with these rules. This last property of nondeducibility is referred to as the lack of deduction completeness. For example, not all theorems of the familiar first-order predicate calculus can be deduced from the proposed set of rules. Contrary to intuition, this is often an advantage computationally.

The set, Π, of inference rules to be studied in this paper consists of resolution, factoring, and paramodulation. The first two are generalizations of well known inference rules for the propositional calculus. Paramodulation is a generalization of a substitution rule for equality. The formulation of resolution and factoring is essentially that of J. Robinson [16] while paramodulation originated with the present authors [13,15].

It is desirable that the underlying set of inference rules have certain logical properties related to semidecidability. The key property is that of R-refutation completeness. A refutation is a proof of contradiction. It is proved herein that in the presence of functional reflexivity Π is R-refutation complete, i.e., a refutation employing just the rules from Π exists for any set of statements (each of which is in the appropriate logical form) which possesses no equality model. In other words, if one starts with a statement which should lead to a contradiction with the usual meaning of equality, Π is strong enough to yield a contradiction in the presence of functional reflexivity.

The general idea is as follows. To prove that a statement is a theorem of some first-order theory with equality, one proceeds by first denying the statement. Then this denial and the axioms of the theory in question are converted to a particular logical form which may be said to be, loosely speaking, a conjunction of statements each of which is a disjunction. The existential quantifiers have been replaced by Skolem functions and the remaining variables are considered to be universally quantified precisely over the entire conjunct in which they occur. The resulting statements are called clauses. A contradiction is deducible (using the rules of Π) from the set of clauses if and only if the original statement was true in all equality models of the theory. The sufficiency requirement leads to a definition of R-refutation completeness. The necessity leads to a corresponding soundness concept.

Several examples from first-order group theory are discussed in Appendix I, and refutations (within Π) for two of them are given.

It is not within the scope of this paper to discuss the strategies which lead to a promising semidecision procedure. Intuitively, "strategy" may refer to the order in which the inference rules are applied or to certain constraints placed on their application. In order to have an efficient semidecision procedure, one needs, in addition to good inference rules, various strategies. The procedure used in Corollary 2 is not one which is recommended. One of the most successful approaches is that which allows only those inferences which are in part traceable to the special hypothesis of the theorem or to the denial of its conclusion. This strategy is a special case of the set of support strategy. The set of support strategy is known to be R-refutation complete when coupled with Π in the presence of functional reflexivity [21].

To illustrate the general approach, consider the theorem: if $x^2 = e$ for all x in a group G, then G is commutative. In the notation of first-order predicate calculus the axiom for the existence of an identity element (two-sided) is, $(\exists y)(x)[Pyxx \wedge Pxyx]$, where P denotes product. Replacing the existential quantifier by a Skolem function and then putting the results into the desired form (a set of clauses), one has the clauses $Pexx$ and $Pxex$. Since the computer attempts to find contradiction from assuming the theorem false, one has, in addition to the clauses corresponding to the other axioms, the clauses $Pxxe$ and $Pabc$ and $\bar{P}bac$. The last two clauses correspond to the assumption that the group is not commutative, i.e., that there exists a pair of elements which do not commute.

The system Π uses no logical axioms nor does it contain any of the more familiar classical rules of inference such as universal instantiation. In the example above, the procedure would consist of applying the various rules of Π, resolution, factoring, and paramodulation, in some order until a contradiction was found. Thus we are vitally interested in the property of R-refutation completeness.

The use of Herbrand models throughout the paper rather than the more familiar concrete models provides a convenient tool for proving the theorems underlying the theory of the approach discussed herein. The theorem of logic which permits use of Herbrand models states in effect that a set has an Herbrand model if and only if it has a concrete model.

In Section 2 we give definitions of a number of concepts, such as *interpretation* and *model*, which are necessary for our study. In Section 3 we prove the *Maximal Model Theorem* which states that, given a particular interpretation for a given satisfiable set of disjunctions, there is a model for that set of disjunctions, whose set-theoretic intersection with the particular interpretation is as small as it can be without losing the property of modelhood. In Sections 4 and 5 we give the inference rules. In Section 6 we introduce the terms, *refutation complete, deduction complete*, and *conservative* and prove that the inference system (set of inference rules) Π under study has the desired logical property of R-refutation completeness for *functionally reflexive* sets. The usual axiom set for basic group theory [13] and also that for basic ring theory, for example, are functionally reflexive relative to Π.

Although many classical inference systems have the property of refutation completeness, they are also deduction complete and not conservative. The system, Π, on the other hand, is conservative but not deduction complete. In the light of experience with existing computer programs, this seems to be advantageous in using computers to search for proofs of theorems.

§2. Definitions

We shall deal with a language having as primitive symbols *individual variables* $x_1, x_2, ..., x_n, ...$; *individual constants* $a_1, a_2, ..., a_n, ...$; *predicate constants* (relation symbols) $P_1^1, P_1^2, ..., P_k^j, ...$; and *function letters* $f_1^1, f_1^2, ..., f_k^j, ...$ (where superscripts indicate the number of arguments). (The *equality predicate* P_1^2 will frequently be abbreviated to R.) The remaining primitive symbol is a bar for negation. Individual constants and variables are terms; a function letter f_i^n applied to n terms $t_1, ..., t_n$ is also a *term*. A predicate letter P_i^n applied to n terms is an *atomic formula* or *atom*. A *literal* is an atom q or the negation $\bar{q}$ thereof. If q is an atom, the negation $\bar{\bar{q}}$ of $\bar{q}$ is just taken to be q. The *absolute value* $|h|$ of a literal h is the atom q such that either h is q or h is $\bar{q}$. A *clause* is a finite set of literals. Intuitively, one may view a clause as the disjunction of its literals, universally quantified (over the entire disjunction) on its individual variables. Also intuitively, existential quantification is (in effect) accomplished through the use of (Skolem) function letters. A *ground clause* (*literal, term*) is one that has no variables occurring in it.

The result $C\theta$ of a *substitution* $\theta = [t_1/u_1, ..., t_n/u_n]$ (where the u_i are all distinct) on a clause C (literal, term) is the clause (literal, term) obtained by replacing uniformly and simultaneously each occurrence (if any) of each variable u_i ($i = 1, ..., n$) by the corresponding term t_i. Here the clause (literal, term) $C\theta$ is called an *instance* of C. Suppose that a substitution θ transforms S, some set of literals, into a set $S\theta$. Then if for all literals h_1 and h_2 in $S\theta$, $|h_1| = |h_2|$, we shall call θ a *unifier* (or *match*) of S. The set

S is said to be *unifiable* if such a θ exists.* If θ transforms all members of a set T of terms into a single term, θ is likewise called a *unifier* or *match* of T. The process of finding and applying a unifier or match is called *unification* or *matching*.

If a substitution θ can be obtained by composing two substitutions θ_1 and θ_2 in that order, then θ is an *instance* of θ_1. If one clause (literal, term, substitution) is an instance of a second and the second is also an instance of the first, then each is a *variant* of the other. A finite, non-empty set of clauses (literals, terms, substitutions) that have an instance in common have a *most general common instance* (not necessarily unique), i.e., one that itself has as instances all common instances of the originals. Similarly, if a finite, non-empty set S of literals (terms) is unifiable, then it has a *most general unifier* (*match*), i.e., a substitution θ that unifies S and such that every unifier of S is an instance of θ.

If a term t occurs as the i_k-th argument of the i_{k-1}-st argument of ... of the i_1-st argument of a literal h, then the ordered k-tuple $(i_1, ..., i_k)$ is called the *position vector* of that occurrence of t in h. Two terms are *in the same position* in their respective literals if they have the same position vectors.

The *Herbrand universe* H_s of a set S of clauses (literals, terms) is the set of all ground terms that can be constructed from the symbols occurring in S. Here if no individual constant occurs in S, a_1 is to be added to the vocabulary. An *Herbrand atom* for a set S of clauses is a predicate letter P_i^n from S applied to n terms from H_S. An *interpretation I of S* is a set of ground *literals* whose absolute values are all the Herbrand atoms for S, such that for each Herbrand atom j exactly one of j and $\bar{j}$ is in I. An interpretation I *satisfies the ground clause* C' if at least one of the literals of C' is in I, i.e., if $I \cap C' \neq \emptyset$. The interpretation I *satisfies the* (possibly non-ground) *clause* C if it satisfies every ground instance of C (over the Herbrand universe under consideration); it *satisfies the set S of clauses* if it satisfies each clause in S; it *condemns the ground clause* C' if it contains the negation of every literal of C';

* When a finite set S of literals is to be unified, we do not usually find it particularly instructive to view that set as a clause, although it is not set-theoretically distinguishable from a clause.

it *condemns a clause* C if it condemns some ground instance of C (over the Herbrand universe under consideration); and it *condemns the set* S *of clauses* if it condemns some clause in S. When the empty set of literals is viewed as a clause *false*, it can be satisfied by no interpretation and is vacuously condemned by every interpretation. Note however that the same set-theoretic object $\emptyset$, when viewed as a set of clauses, is vacuously satisfied by all interpretations.

A *model* M *of the set* S *of clauses* is an interpretation of S that satisfies S; it is an *R-model* of S if, in addition, the set $\{(t_1, t_2) | Rt_1 t_2 \in M\}$ (i.e., the set $\{(t_1, t_2) | P_1^2 t_1 t_1 \in M\}$) is an equality relation for the terms occurring in M, i.e., if each of the following is true for all terms s and t in H_S and all function letters f_k^n:

(i) $Rtt \in M$.

(ii′) If the literal $k \in M$ and $Rst \in M$ and if h is obtained by replacing in k some one occurrence of s by an occurrence of t, then $h \in M$.

It is often more convenient to replace (ii′) by the condition

(ii) If an *atom* $k \in M$ and $Rst \in M$ and if h is obtained, etc.

since (i) and (ii) are equivalent to (i) and (ii′).

The other familiar properties follows easily from (i) and (ii′):

(iii) If $Rst \in M$, then $Rts \in M$.

(iv) If $Rst \in M$ and $Rtu \in M$, then $Rsu \in M$.

(v) For all $t_0, t_1, ..., t_n$ in H_S, if $Rt_j t_o \in M$ and f occurs in some literal in M, then

$$Rf(t_1, ..., t_{j-1}, t_j, ..., t_n)\, f(t_1, ..., t_{j-1}, t_0, ..., t_n) \in M.$$

§3. The Maximal Model Theorem

One can associate with each interpretation T of a set S of clauses a partial ordering relative to T of the set of all interpretations of S, and hence of the set of models of S, given by:

$M_1 \leqslant_T M_2$ if and only if $M_1 \cap T \supset M_2 \cap T$.

The maximal model theorem states that, given a set S of clauses possessing a model and an interpretation T of S, among the models of S there exists one, say M, such that no other model of S has a strictly smaller set-theoretic intersection with T.

Maximal model theorem. *If a set S of clauses has a model, then, given any interpretation T of S, S has a maximal model M_0 relative to T.*

Proof. Let $T' = \{x \mid \bar{x} \in T\}$, or equivalently the set of the negations of the literals in T. Then T' is an interpretation of S, and $T' \cap T = \emptyset$.

If we show that every simply ordered (relative to T) set of models of S has an upper bound which is itself a model, then an application of Zorn's lemma will suffice. Let the set $\{H_\alpha\}_{\alpha \in \Lambda}$ be a simply ordered (relative to T) set of models of S. Let $M = \bigcup_{\alpha \in \Lambda}(H_\alpha \cap T') \cup (\bigcap_{\alpha \in \Lambda}(H_\alpha \cap T))$. M will be shown to be both an upper bound for $\{H_\alpha\}$ and a model of S.

That M is an upper bound (if it is an interpretation) follows from considering H_β for an arbitrary β in Λ and noting that $H_\beta \cap T' \subset \bigcup_\alpha (H_\alpha \cap T') = M \cap T'$, since $T \cap T'$ is empty. Thus $H_\beta \cap T' \subset M \cap T'$, hence $H_\beta \cap T \supset M \cap T$ and $H_\beta \leqslant_T M$.

To see that M is an interpretation, consider an arbitrary literal b in T since every Herbrand atom is represented in T either by itself or its negation. If b is in H_α for all α in Λ, then b is in $\bigcap_\alpha (H_\alpha \cap T)$ so b is in M. If for some α, the literal b is not in H_α, then for that α, $\bar{b}$ is in H_α since each H_α is an interpretation. But $\bar{b} \in T'$, so $\bar{b} \in \bigcup_\alpha (H_\alpha \cap T') \subset M$. This shows that at least one of b and $\bar{b}$ is in M. To see that only one of b and $\bar{b}$ is in M assume that b is in M. Then since b is not in T' it cannot be in $\bigcup_\alpha (H_\alpha \cap T')$ and so must be in $\bigcap_\alpha (H_\alpha \cap T)$ and hence in $\bigcap_\alpha H_\alpha$. But with b in every H_α, $\bar{b}$ cannot be in any H_α and hence not in $\bigcup_\alpha (H_\alpha \cap T')$ and hence not in M.

To see that M is a model of S, consider some arbitrarily chosen ground instance D of some clause in S.

Case 1. $D \subset T'$. Let H_β be arbitrarily chosen. Since H_β is a model of S, $H_\beta \cap D$ is not empty and so contains some literal, say c. Since D is contained in T' by assumption, c is in T', so $c \in T' \cap H_\beta \subset \bigcup_\alpha (H_\alpha \cap T') \subset M$. Therefore, $c \in M \cap D$.

Case 2. $D \not\subset T'$. Then $D \cap T = \{c_1, c_2, ..., c_k\}$ for some finite $k > 0$, since D has only a finite number of literals and since all literals of all ground instances of all clauses of S are in $T \cup T'$.

If some c_i, say c, is in all H_α, then c is in $\bigcap_\alpha (H_\alpha \cap T) \subset M$, and $M \cap D$ is not empty. Otherwise assume that for $1 \leqslant i \leqslant k$ no c_i is in all H_α. Therefore, for each c_i, there is an H_α, say H_{α_i}, with $c_i \notin H_{\alpha_i}$. But the H_α are models and hence interpretations, so $\bar{c}_i \in H_{\alpha_i}$ for $1 \leqslant i \leqslant k$. Since $c_1, ..., c_k$ are in T, $\bar{c}_1, ..., \bar{c}_k$ are in T'. Since the family $\{H_\alpha\}_{\alpha \in \Lambda}$ is simply ordered relative to T, there is among $H_{\alpha_1}, ..., H_{\alpha_k}$ a largest (relative to T) H_{α_i}, say H. From the definition of $\leqslant_T$, if a literal is in some $H_{\alpha_i} \cap T'$ $(1 \leqslant i \leqslant k)$ then it is in $H \cap T'$.

Therefore, $\{\bar{c}_1, ..., \bar{c}_k\} \subset H \cap T'$. Since H is a model and hence an interpretation of S, no c_i for $1 \leqslant i \leqslant k$ is in H. But $H \cap D \neq \emptyset$ since H is a model of S, so $H \cap D$ contains some element a. Since no c_i is in H, $a \neq c_i$ for each i. Therefore, $a \notin T$ since $T \cap D = \{c_1, c_2, ..., c_k\}$. So a is in T', and therefore $a \in H \cap T'$. But $H \cap T' \subset \bigcup_\alpha (H_\alpha \cap T') \subset M$, so again $M \cap D \neq \emptyset$.

Since the cases for D are exhausted, and since D was arbitrarily chosen, for each ground instance D of a clause in S, $M \cap D \neq \emptyset$. M is thus a model of S.

The conditions for Zorn's Lemma being satisfied, there exists a maximal (relative to the ordering associated with T) model M_0 of S. Q.E.D.

In a similar fashion one can, by applying the maximal model theorem to T', show that there exists a minimal model of S with respect to T.

In a later section we apply the theorem to the case where T is the set of Herbrand atoms for S. In this case, the key point is that if M_0 is a maximal model relative to T, for each atom k in M_0 we can find a ground instance D of a clause of S with $D \cap M_0 = \{k\}$.

This in turn allows us certain applications of paramodulation, which we shall define in a later section.

Corollary A. *If S is a set of clauses, T an interpretation of S, and M a maximal (relative to T) model of S, then, for each literal b in $M \cap T$, there exists a clause in S having a ground instance (over H_S) D with $D \cap M = \{b\}$.*

Proof. Let S be a set of clauses, T be an arbitrary but fixed interpretation, and M be a maximal (relative to T) model of S. Assume by way of contradiction that there exists a literal b in $M \cap T$ falsifying the theorem. Since M is a model of S we can conclude, therefore, that, for any ground instance (over H_S) C of any clause in S, $(M \cap C) - \{b\}$ is not empty. Let M^* be obtained from M by replacing b by $\bar{b}$. M^*, therefore, has a non-empty intersection with all ground instances (over H_S) of all clauses in S. For each Herbrand atom a, M^* will still contain exactly one of a and $\bar{a}$. M^* is, therefore, an interpretation of S and, hence, a model of S. Since b is in T by assumption and T is an interpretation, $\bar{b}$ is not in T. Therefore, $M \cap T$ contains $M^* \cap T$ as a proper subset. This contradicts the maximality of M, and the proof is complete.

Corollary A and the maximal model theorem establish that there are models (the maximal models) which possess the intersection property (given in the conclusion of corollary A) upon which the proof of R-refutation completeness rests. That this intersection property, however, does not characterize maximality can be seen from the following example.

Let $S = \{A, B\}$, where $A = \{\overline{Pa}, Qa\}$ and $B = \{Pa, \overline{Qa}\}$. Let $T = \{Pa, Qa\}$. Let $M^* = \{Pa, Qa\}$. M^* is not maximal relative to T, for the model $M = \{\overline{Pa}, \overline{Qa}\}$ is such that $M^* \cap T = \{Pa, Qa\}$ contains $M \cap T = \emptyset$ as a proper subset. M is itself a maximal model relative to T. M^*, however, has the property that, for every literal b in $M^* \cap T$, there exists a clause in S one of whose ground instances intersects M^* in exactly b. This example shows that the property of maximality for models is not equivalent to the intersection property.

Depending on S and T, one cannot be assured of the existence of a unique maximal model. In the example just given, M is the only maximal model. But, if one replaces S by $S^* = \{A^*, B^*\}$, where $A^* = \{Pa, Qa\}$ and $B^* = \{\bar{P}a, \bar{Q}a\}$, there are two maximal models. $M_1 = \{Pa, \bar{Q}a\}$ and $M_2 = \{\bar{P}a, Qa\}$ are both maximal models relative to T.

By definition, a model M is maximal relative to T if and only if, whenever $M \cap T$ contains $M^* \cap T$ for a model M^*, $M \cap T = M^* \cap T$. Using the definition of interpretation one can easily show that a model M is maximal relative to T if and only if, whenever $M \cap T$ contains $M^* \cap T$ for a model M^*, M is equal to M^* set-theoretically.

§4. Resolution and Factoring

In this section we give a brief account of an inference rule called *resolution*. This rule may be viewed as a generalization of *modus ponens* and hypothetical syllogism.

Definition (Resolution). If A and B are clauses (with no variables in common) with literals k and h respectively such that k and h are opposite in sign (i.e., exactly one of them is an atom) but $|k|$ and $|h|$ have a most general common instance m, and if σ and τ are most general substitutions with $m = |k|\sigma = |h|\tau$, then infer from (any variants A^* and B^* of) A and B the clause $C = (A - \{k\})\sigma \cup (B - \{h\})\tau$. C is called a *resolvent* of A^* and B^* and is inferred by *resolution* [14,16].*

Example 1. Premisses $\{Pabc\}$ and $\{\bar{P}xyz, Pyxz\}$ yield by resolution $\{Pbac\}$. In this example resolution has the effect of first instantiating the second premiss (which might be thought of as the commutative axiom in a group theory problem) so that one has two premisses to which *modus ponens* can be applied, and then ap-

* Note that the premisses A^* and B^* to which the rule of inference is applied, are not themselves subject to the restriction of having no variables in common. Given an arbitrary A^* and B^*, in practice, one need merely re-letter B^* to obtain a variant B having no variable in common with A^* before constructing the resolvent C from A^* and B.

plying *modus ponens*. The clauses which serve as premisses for the *modus ponens* applications are most general instances of the given premisses, most general with respect to permitting application of *modus ponens*.

Example 2. Premisses $\{Pax, Qx\}$ and $\{Pyb, Qy\}$ yield by resolution $\{Qa, Qb\}$. Syllogistic inference is not directly applicable to the given pair of clauses. The clause $\{Qa, Qb\}$, however, can be inferred by instantiation followed by a syllogistic inference. By definition, resolution, in effect, seeks most general instances of the clauses which will permit syllogistic inference.

Example 1 shows that resolution yields in the spirit of *modus ponens* an inference from pairs of premisses even though *modus ponens* does not apply directly to those premisses. Example 2 illustrates the corresponding relationship to syllogistic inference. By placing no restriction on the (finite) number of literals in either premiss, resolution extends both *modus ponens* and syllogistic inference in yet another way.

Example 3. From $\{\overline{N}x, Px\}$ and $\{Py, Qy\}$ infer by resolution $\{\overline{N}y, Qy\}$. This is the familiar categorical syllogism, all F are G, all G are H, so all F are H.

Definition (Factoring). If A is a clause with literals k and h such that k and h have a most general common instance m, and if σ is a most general substitution with $k\sigma = h\sigma = m$, then infer the clause $A' = (A - \{k\})\sigma$ from A. A' is called an *immediate factor* of A. The *factors* of A are given by: A is a factor of A, and a immediate factor of a factor of A is a factor of A.

Example 4. From $\{Qax, Qyb, Quz\}$ infer (among other clauses) as an immediate factor $\{Qab, Quz\}$, which in turn has an immediate factor $\{Qab\}$.

§5. Paramodulation

The concept of equality is ordinarily handled in first-order theories in one of two ways: 1) a set of explicit first-order axioms (or axiom schemata) is given for the equality predicate; 2) a substitution rule is supplied together with the axiom of reflexivity. It is in the context of the second approach that the inference rule paramodulation is to be understood.

With the appropriate constraints on the variables in the terms s and t, one version of the substitution rule for equality permits inference of the formula $\rho(t)$ from the formulae Rst (i.e., $s = t$) and $\rho(s)$. Paramodulation [13,15] extends this substitution rule by permitting inference from the formulae Rst and $\rho(u)$ when the terms s and u, though not identical, have a common instance. (Since the only formulae of interest throughout this paper are clauses, paramodulation is defined only for clauses.)

Definition of Paramodulation. Let A and B be clauses (with no variables in common) such that a literal Rst (or Rts) occurs in A and a term u occurs in (a particular position in) B. Further assume that s and u have a most general common instance $s' = s\sigma = u\tau$ where σ and τ are most general substitutions such that $s\sigma = u\tau$. Where $\hat{B}$ is obtained by replacing by $t\sigma$ the occurrence of $u\tau$ in the position in $B\tau$ corresponding to the particular position of the occurrence of u in B, infer from any variants A^* and B^* of A and B respectively the clause $C = \hat{B} \cup (A - \{Rst\})\sigma$ (or $C = \hat{B} \cup (A - \{Rts\})\sigma$). C is called a *paramodulant of A^* and B^** (and also of B^* and A^*) and is said to be *inferred by paramodulation from A^* on the variant of Rst* (or Rts) *into B^* on* (*the occurrence in the particular position in B^* of*) the variant of u. The variant of the literal Rst (or Rts) is called the *literal of paramodulation* and the occurrence of the variant of u is called the *term of paramodulation.* *

Before giving examples to aid in one's understanding of paramodulation, we remark that examination of the definition shows

* See footnote on page 000 for a comment on how separation of variables works out in practice.

that the property of symmetry is inherent in the inference rule. It is also to be noted that, although in much of what follows we assume the presence of the reflexivity axiom, the definition of paramodulation could have been easily extended to obviate the need for this axiom. The extension does not, however, seem to be of practical value.

Example 1. Premisses $\{Rab\}$ and $\{Qa\}$ yield $\{Qb\}$ by paramodulation or by the usual substitution rule for equality.

Example 2. Premisses $\{Rab\}$ and $\{Qx\}$ yield $\{Qb\}$ by paramodulation. They also yield by paramodulation $\{Qa\}$. The substitution rule does not apply in its usual form since neither term, a nor b, occurs in the premiss $\{Qx\}$. In many systems from $\{Qx\}$ one could infer $\{Qa\}$, which could then have been used as one premiss together with $\{Rab\}$ and the substitution rule to yield $\{Qb\}$.

Example 3. Premisses $\{Rab\}$ and $\{Qx, Px\}$ yield among others the clause $\{Qb, Pa\}$ by paramodulation. Again in many systems $\{Qa, Pa\}$ could have been inferred from $\{Qx, Px\}$, providing a premiss to which the substitution rule for equality would apply.

Example 4. Premisses $\{Rxh(x)\}$ and $\{Qg(y)\}$ yield by paramodulation $\{Qh(g(y))\}$. Note that the presence among the constants occurring in the set of clauses under consideration of the individual constants a and b does not lead by paramodulation to either the inference $\{Qh(g(a))\}$ or $\{Qh(g(b))\}$. Paramodulation (in effect) first finds most general instances of the two premisses which permit straightforward application of the substitution rule for equality and then makes the equality substitution.

Example 5. Consider the premisses $\{Rf(xg(x))e\}$ and $\{Pyf(g(y)z)z\}$. For intuitive purposes think of P as product, f as product, and g as inverse. The functions f and g are frequently Skolem functions introduced in place of existential quantifiers. For this application of paramodulation let s be $f(xg(x))$ and u be $f(g(y)z)$. A most general common instance of u and s is $f(g(y)g(g(y)))$. The inference thus made is $\{Pyeg(g(y))\}$. Note that both premisses required non-trivial instantiation in order to apply the substitution rule.

Example 6. Premisses $\{Rf(xg(x))e, Qx\}$ and $\{Pyf(g(y)z)z, Qz\}$ yield by paramodulation with s and u as in Example 5 $\{Pyeg(g(y)), Qg(y), Qg(g(y))\}$. This example illustrates another way in which paramodulation extends the substitution rule, as both premisses in this example contain more than one literal. The substitution rule applies to pairs of formulae one of which is of the form $\{Rst\}$.

The extension in the direction illustrated by Example 6 is needed for R-refutation completeness as can be seen by examining the following set of clauses: $\{\{Rab, Qc\}, \{\bar{R}g(a)g(b), Qc\}, \{Rcd, \bar{Q}c\}, \{\bar{R}g(c)g(d), \bar{Q}c\}, \{Rxx\}\}$.

A most crucial property of paramodulation was illustrated by Example 4. From a theorem-proving viewpoint it is important that the inference rules have the property of being conservative (see Section 6).

§6. Refutation Completeness

A set S of clauses *implies* (*R-implies*) a clause C if no model (R-model) of S condemns C, and S *implies* (*R-implies*) a set T of clauses if it implies (R-implies) each clause in T. We write $S \models C$, $S \models_R C$, $S \models T$, $S \models_R T$ respectively to express these relationships. If for clauses C and D, $\{C\} \models D$, we also write $C \models D$ and say that C *implies* D (and similarly for $\models_R$). If A implies (R-implies) B, then B is a *consequence* (*R-consequence*) of A. If S has no model (R-model), then S is *unsatisfiable* (*R-unsatisfiable*).

A *deduction D of a clause C_n from S in an inference system* Ω is a finite sequence of pairs (C_i, J_i) $i = 1,2, ..., n$ where C_i is a clause and J_i is a "justification" of C_i in terms of one of the rules of inference of Ω and previous steps of D or in terms of membership of C_i in S. (By *inference system* we mean here nothing more than a set of rules of inference.) If such a deduction exists we write $S \vdash_\Omega C_n$. Except when the justifications J_i are of particular interest [21], D will be identified with the sequence $C_1, C_2, \ldots, C_n$. A *refutation* of S in Ω is a deduction of *false* from S in Ω.

Henceforth Π will be the inference system whose rules of inference are paramodulation, resolution, and factoring and Σ will be that whose rules are just resolution and factoring.

A system Ω is *deduction complete* (*R-deduction-complete*) if for any set S of clauses and any clause C, $S \vdash_\Omega C$ whenever $S \models C$ ($S \vdash_\Omega C$ if $S \models_R C$). It is (R-) *refutation complete* if for any (R-) unsatisfiable S, $S \vdash_\Omega$ *false.*

The system Σ is well known to be refutation-complete [16]. (An adaptation to the present formalism of the simple but rather elegant proof in [16] is given in Appendix 1.) If Σ were deduction complete as well, one could easily prove (see next paragraph) that Π is R-refutation complete (in the presence of the usual reflexivity axiom). (More generally, any deduction complete system Ω, when augmented by paramodulation, becomes R-refutation complete in the presence of reflexivity.) Both Σ and Π would then, however, be virtually worthless for automatic theorem proving algorithms of the type in use today!

Let Ω be any deduction complete system including the inference rule paramodulation and let S be any R-unsatisfiable set of clauses including $\{Rxx\}$. Consider the tautology $\{\bar{R}xy, Rxy\}$, a trivial logical consequence of S. Deduction completeness of Ω would give $S \vdash_\Omega \{\bar{R}xy, Rxy\}$. By paramodulating from this tautology into $\{Rxx\}$, one could show that $S \vdash_\Omega \{\bar{R}xy, Ryx\}$. By paramodulating from this clause into itself, one can deduce $\{\bar{R}xy, \bar{R}zy, Rxz\}$. Since $\{\{\bar{R}xy, Ryx\}, \{\bar{R}xy, \bar{R}zy, Rxz\}\} \models \{\bar{R}xy, \bar{R}yz, Rxz\}$, it follows by deduction completeness of Ω that $S \vdash_\Omega \{\bar{R}xy, \bar{R}yz, Rxz\}$, completing the equivalence relation properties for R. For the predicate substitution property, consider $P_i^j x_1 \ldots x_k \ldots x_j$, for arbitrary choice of i, j, and k. From the deduction completeness of Ω, $S \vdash_\Omega \{\bar{P}_i^j x_1 \ldots x_k \ldots x_j, P_i^j x_1 \ldots x_k \ldots x_j\}$. Paramodulating from symmetry into this clause on x_k in $\bar{P}_i^j x_1 \ldots x_k \ldots x_j$, one could show that $S \vdash_\Omega \{\bar{R}x_{j+1}x_k, \bar{P}_i^j x_1 \ldots x_{j+1} \ldots x_j, P_i^j x_1 \ldots x_k \ldots x_j\}$. For the function substitution property, since $\{Rxx\} \models \{Rf_i^i(x_1 \ldots x_k \ldots x_j)\, f_i^j(x_1 \ldots x_k \ldots x_j)\}$, deduction completeness of Ω gives $\{Rxx\} \vdash_\Omega \{Rf_i^j(x_1 \ldots x_k \ldots x_j)$ $f_i^j(x_1 \ldots x_k \ldots x_j)$. Paramodulating from symmetry into this clause on x_k would give $\{Rxx\} \vdash_\Omega \{\bar{R}x_{j+1}x_k, Rf_i^j(x_1 \ldots x_{j+1} \ldots x_j) \times$ $f_i^j(x_1 \ldots x_k \ldots x_j)\}$. Hence $S \vdash_\Omega S \cup E$ where E is a set of equality axioms strong enough to guarantee that any model of $S \cup E$ would also be an R-model of S. Suppose then, that S is R-unsatisfiable and includes $\{Rxx\}$. $S \cup E$ could then have no model; hence $S \cup E \models$ *false* (vacuously). Again applying the hypothesis of deduction completeness of Ω, one could show $S \cup E \vdash_\Omega$ *false* and hence $S \vdash_\Omega$ *false*. Since S was assumed only to be R-unsatisfiable and to include $\{Rxx\}$, this shows that Ω is R-refutation complete in the presence of $\{Rxx\}$.

To assess the problems involved in proving refutation completeness of systems that, unlike the hypothetical Ω above, are useful for (present techniques of) automatic theorem proving, one must take care to note that *one of the principal reasons that the systems under study here are useful is that they are devoid of the familiar deduction-completeness properties* of many of the customary formulations of quantification theory, hypothesized for the system Ω above. (So that, for example, the Gödel completeness theorem fails to hold for Σ and Π.) The simplest difference is that universal instantiation is not usually possible in Σ or Π. For example, $\{Rxa\} \models \{Raa\}$, but $\{Raa\}$ is not deducible from $\{Rxa\}$ in either Σ or Π. The property that makes systems such as Σ and Π valuable for automatic theorem proving is that they are *conservative* in the following sense: Loosely speaking, an inference rule is called *conservative* if it does not allow a *proper* instance C' of a clause C to be inferred from, say A and B, when C itself could have been inferred [13]. (A *proper* instance C' of C is an instance for which C is not an instance of C'.)

Resolution is a conservative inference rule. From $\{Pf(x), Qa\}$ and $\{\bar{P}y, Qy\}$ one infers by resolution $\{Qf(x), Qa\}$ and not, for example, $\{Qf(a), Qa\}$. The latter, of course, can be inferred by instantiation followed by a syllogistic inference. The instantiation involved therein is, however, not most general among those permitting syllogistic inference. Factoring is also a conservative inference rule, permitting, in effect, only the most general instantiations that allow application of the equivalence of $p \vee p \vee q$ with $p \vee q$.

Paramodulation, in effect, seeks instantiations that are most general among those that permit equality substitutions. In addition to extending substitution in the two ways already discussed, it is important that certain inferences are avoided. (See Example 4 in Section 5.) From a pair of premisses paramodulation yields an inference which could be made after some number (possibly zero) of applications of instantiation to either or both premisses, followed by an application of an equality substitution rule. It combines instantiation with equality substitution in a way such that the property of being conservative is present. Among the possible instantiations which permit application of equality substitution, paramodulation, in effect, allows only the most general instantiations.

The intuitive comparison above of resolution and paramodulation with syllogism (or *modus ponens*) and substitution, respectively, is intended as motivation for, not as a precise characterization of, resolution and paramodulation. For the latter one must refer to the definitions. The intuitive discussion might, for example lead one to take $C = (A\sigma - \{k\sigma\}) \cup (B\tau - \{h\tau\})$ as the resolvent instead of $C = (A - \{k\}\sigma \cup (B - \{h\})\tau$ as we intend in the present paper. If $A = \{Pa, Px\}$ and $B = \{\bar{P}a, Qb\}$, we intend the resolvent obtained from choosing Px and $\bar{P}a$ as focal literals to be $C = \{Pa, Qb\}$, not $C = \{Qb\}$.

A set S of clauses is said to be *functionally reflexive* if $\{Rxx\}$ is in S, and for each function letter f_k^j occurring in S, $\{Rf_k^j(x_1 ... x_j) f_k^j(x_1 ... x_j)\}$ is in S. S is said to be *functionally reflexive relative to* an inference system Ω if $S \vdash_\Omega \{Rxx\}$ and for each f_k^j occurring in S, $S \vdash_\Omega \{Rf_k^j(x_1 ... x_j) f_k^j(x_1 ... x_j)\}$. The principal result of this section is that the inference system Π is R-refutation complete for sets S that are functionally reflexive (or equivalently, for S functionally reflexive relative to Π). For deduction-complete systems Ω, the condition that S be functionally reflexive relative to Ω reduces to the condition that $S \models \{Rxx\}$. Fortunately for efficiency in automatic theorem proving, Π is not deduction complete; consequently, the additional functional reflexivity properties are not so easily eliminated. (When paramodulation was first introduced a few years ago, we conjectured that $S \vdash_{\Pi} \{Rxx\}$ was sufficient for R-refutation completeness of Π. No counterexample has yet been discovered to this conjecture, but neither has a proof been given that the completeness theorem proved in this section can be replaced by the simpler and somewhat more attrative conjecture. An earlier, weaker version of the completeness theorem was proved in [15]. There it was required that $S \vdash_{\Pi} \{Rtt\}$ for all terms $t \in H_S$, a much stronger restriction since there are infinitely many such clauses Rtt, but there are only $n + 1$ functional-reflexivity clauses where n is the number of distinct function symbols occurring in S. Several systems, such as basic group theory [13], satisfy the hypothesis of the present completeness theorem (Corollary 1) but fail to satisfy the hypothesis of the completeness theorem in [15].)

Lemma 1. *Let $A \in S$ have a ground instance (over H_S) $A' = A\tau$ such that in A and A' respectively the terms u^* in the literal m^* and u' in the literal $m' = m^*\tau$ are in the corresponding positions. Then there exists a factor E of A and a substitution μ such that* (1) *$A' = E\mu$ is an instance of E,* (2) *E and A' have the same number of literals, and* (3) *there exists a literal m in E and a term u in m with $m\mu = m'$ and with u in m in the corresponding position to u' in m'.*

Proof. The substitution τ induces a partition of the clause A given by j equivalent to k if and only if $j\tau = k\tau$, where j and k are literals in A. The sets A_i of literals of the partition are each unifiable. There exists, therefore, a most general single substitution θ which simultaneously unifies each of the A_i. Since τ unifies each set A_i, there exists a substitution μ with $\tau = \theta\mu$. Let $E = A\theta$. Let $m = m^*\theta$, and let $u = u^*\theta$. The clause E has the desired properties.

Lemma 2. *Let A' and B' be respectively ground instances (over H_S) of A and B in S, and let C' be a paramodulant of A' and B'. Further assume that the literal, $Rs't'$, of paramodulation is in A' and that the term, u', of paramodulation is in the literal m' of B'. Let the position of u' in m' be given by the vector $(q_1, q_2, ..., q_n)$. Let the substitution ρ and the literal m^* in B be such that $B\rho = B'$*

and $m^*\rho = m'$. *Then if there exists a term* u^* *in* m^* *such that* u^* *and* u' *are in the corresponding position respectively in* m^* *and* m', *then there exists a clause* C *having* C' *as an instance such that* C *is a paramodulant of some factors* E *and* F *of* A *and* B *respectively.*

Proof. From Lemma 1 we can conclude the existence of factors E and F of A and B respectively such that E and F have no variables in common, E has A' as an instance, E and A' have the same number of literals, F has B' as an instance, F and B' have the same number of literals, and F contains a literal m containing a term u with u in m and u' in m' in corresponding positions. Furthermore, we can conclude the existence of substitutions τ and θ with $A' = E\tau$, $B' = F\theta$, $m' = m\theta$, and $u' = u\theta$.

E contains a literal Rst with $Rst\tau = Rs't'$, where $Rs't'$ is the literal of paramodulation in A'. Assume without loss of generality that s' is the argument involved in the inference by paramodulation of C'.

Then s and u have a common instance. Let λ and μ be most general substitutions such that $s\lambda = u\mu$ is a most general common instance of s and u.

Let $\hat{F}$ be obtained from $F\mu$ by replacing $u\mu$ in the position $(q_1, q_2, ..., q_n)$ in $m\mu$ by $t\lambda$. Let $C = (E - \{Rst\})\lambda \cup \hat{F}$. C is a paramodulant of E and F having C' as an instance, and the proof is complete.

Lemma 3. *Let* S *be a functionally relfexive set of clauses closed under both paramodulation and factoring. If* A' *and* B' *are respectively ground instances (over* H_S*) of* A *and* B *in* S *and if* C' *is a paramodulant of* A' *and* B', *then there exists a* C *in* S *having* C' *as an instance.*

Proof. Assume without loss of generality that the paramodulation is from A' into B'. Let $Rs't'$ in A' be the literal of paramodulation, and let the literal m' in B' contain the term u' of paramodulation in the position $(q_1, q_2, ..., q_n)$. Let ρ be such that $B' = B\rho$.

Case 1. There exists a literal m^* in B with $m^*\rho = m'$ and with m^* containing a term u^* in the position $(q_1, q_2, ..., q_n)$. Lemma 2 applies to A, B, A', and B' to yield a clause C having C' as an instance. C is in S since S is closed under both paramodulation and factoring.

Case 2. No such m^* exists in B. Let m in B be such that $m\rho = m'$. There exists, therefore, a term (a variable) x in m in the position $(q_1, q_2, ..., q_k)$ with $k < n$. Thus there exist $p = n-k$ functions f_i such that ρ contains the element $f_1(... f_2(...(... f_p(... u' ...))))/x$.

Let $G_1, G_2, ..., G_p$ be the functional reflexivity axioms corresponding to $f_1, f_2, ..., f_p$. G_i are in S since S is functionally reflexive.

A set containing $G_1, G_2, ..., G_p$ and B and closed under paramodulation contains a clause B_p with the following properties: $B_p\theta = B'$ for some substitution θ, B_p contains a literal m_p with $m_p\theta = m'$, and m_p contains a term u_p with u_p in m_p and u' in m' in corresponding position. By applying the argument of Case 1 to A, B_p, A', and B', we can complete the proof.

Theorem 1. *If a functionally reflexive set S is closed under paramodulation and factoring, and if S is R-unsatisfiable, then S is unsatisfiable.*

Proof. Assume by way of contradiction that S is satisfiable, and hence, has a model. Let P be the set of atoms over H_S. By the maximal model theorem S has a maximal model M relative to P. We shall show that M is an R-model of S thus contradicting the R-unsatisfiability of S.

Since S is functionally reflexive, $Rxx \in S$. For arbitrary $t \in H_S$, therefore, $Rtt \in M$ since M is a model of S.

Consider arbitrary s and t in H_S, and assume Rst in M. Let k and h be atoms (in P) such that h is obtained from k by replacing some one occurrence of s by t. Assume $k \in M$. By Corollary A there exist ground clauses A' and B' such that $A' \cap M = \{Rst\}$ and $B' \cap M = \{k\}$. Let A and B in S be such that A' and B' are respectively instances of A and B.

Let $C' = (A' - \{Rst\}) \cup (B' - \{k\}) \cup \{h\}$. C' is a paramodulant of A' and B'. By Lemma 3 we can conclude that there exists a $C \in S$ having C' as an instance.

Since M is a model of S, the intersection of M and C' is not empty. This intersection, therefore, equals $\{h\}$, and so $h \in M$. Thus M has been shown to be an R-model, and a contradiction is reached. The proof is complete.

The desired proof, in the presence of functional reflexivity, of the R-refutation completeness of Π follows as a corollary from Theorem 1.

Corollary 1. *For sets (consisting of clauses) functionally reflexive relative to* Π, Π *is R-refutation complete.*

Proof. Let S be a functionally reflexive relative to Π, R-unsatisfiable set of clauses and let S^* be its closure under paramodulation, resolution, and factoring. S^* is therefore functionally reflexive. By Theorem 1, S^* is unsatisfiable. By the Resolution Theorem (see Appendix 1), *false* must then be in S^*. But every clause C in S^* is the last line of a deduction $D_1, ..., D_n$ of C from S in Π with $D_i \in S^*$ for $i = 1, ..., n$. Hence S^* contains a refutation of S in Π.

Corollary 2. *For finite or effectively enumerable sets S that are functionally reflexive relative to* Π, *there is a semi-decision procedure for R-unsatisfiability.*

The proof of this corollary is a form of a well known argument, whose details are supplied for the reader who may be unfamiliar with the formalism of paramodulation and resolution. The semi-decision procedure exhibited in the proof is chosen to make the argument transparent. It is *not* recommended as an efficient theorem-proving procedure; efficient procedures generally require a number of "strategies" for determining the sequence in which inferences are to be made and for suppressing unwise applications of the rules of inference.

Proof. Let S be a finite or effectively enumerable R-unsatisfiable set of clauses functionally reflexive relative to Π. Consider an effective enumeration $C_1, C_2, ...$ of S. Let $W^0 = \emptyset$. For $i > 0$, let $W^i = W^{i-1} \cup \{C_i\} \cup Q^i$, where Q^i is the set of all clauses C such that C is a paramodulant, resolvent, or factor of clauses in W^{i-1} and no lexically earlier variant of C has that property.* Each W^i is finite and can be effectively generated from W^{i-1}. For each clause C in the set S^* defined in the proof of Corollary 1, $\bigcup_i W^i$ contains a variant of C. Hence $false \in \bigcup_i W^i$ and thus is in some W^n.

To complete the proof, note that for any clause D, $S \vdash_\Pi D$ only if $S \models_R D$, so that $S \vdash_\Pi false$ only if S is R-unsatisfiable.

By examining the proofs of Theorem 1 and Corollaries 1 and 2, we can easily obtain parallel results about an inference system in which paramodulation is subject to the following constraint: allow paramodulation only when the term of paramodulation is contained in a positive literal. Thus, in the clause $\{Pb, \bar{Q}b\}$, only the first occurrence of b would be admitted as a term of paramodulation. One can then prove, for example: if a functionally reflexive set S is closed under paramodulation (restricted by the constraint above) and factoring, and if S is R-unsatisfiable, then S is unsatisfiable. The proof is identical to that given in Theorem 1. The parallel to Corollary 1 also holds when paramodulation is so constrained. If the semidecision procedure of Corollary 2 is modified by constraining paramodulation as above, the property of semidecidability is not lost.

* From any pair of clauses that have at least one non-ground resolvent, one can obtain infinitely many resolvents since any non-ground clause has infinitely many variants and any variant of a resolvent of A and B is also a resolvent of A and B. Hence the need to select a single representative in order that Q^i and W^i remain finite. A similar situation exists for paramodulation and factoring.

Appendix 1

Refutation Completeness Theorem for Σ. *If S is any unsatisfiable set of clauses, then* $S \vdash_\Sigma$ *false.*

Proof. (Adapted from [16].) Let Y be the closure of S under resolution and factoring. Since for each clause C in Y, Y contains (all the steps of) a deduction of C from S in Σ, we need only show that *false* $\in Y$. Consider an ordering $a_1, a_2, \ldots$ of the Herbrand atoms for S. Let $I_0 = \emptyset$, and for $j > 0$, let $I_j = I_{j-1} \cup \{a_j\}$ if $I_j \cup \{\bar{a}_j\}$ condemns some clause in Y, otherwise let $I_j = I_{j-1} \cup \{\bar{a}_j\}$. Let $I = \bigcup_j I_j$. I is an interpretation of Y, but cannot be a model of Y, since it would then be a model of S, but S is unsatisfiable. Hence I condemns some clause C in Y, that is, for some ground instance C' of C, the negation of each literal of C' is in I. Since there are only finitely many literals in C', C' must be condemned by I_j for some finite j. Let j be the smallest non-negative integer such that I_j condemns (some ground instance C' of) some clause C in Y. The following establishes that $j = 0$: Suppose that $j > 0$. Then I_{j-1} condemns no clause in Y while I_j condemns some clause C. It must be that $I_j = I_{j-1} \cup \{a_j\}$ since $I_j = I_{j-1} \cup \{a_j\}$ only if $I_{j-1} \cup \{a_j\}$ condemns no clause of Y, and by hypothesis I_j condemns C. But $I_{j-1} \cup \{a_j\}$ must also condemn some (ground instance D' of some) clause D in Y, since otherwise I_j would be $I_{j-1} \cup \{\bar{a}_j\}$. I_{j-1} by hypothesis condemns neither C' nor D'. The literal $\bar{a}_j$ must therefore occur in C' and a_j must occur in D'. Hence there must be a set C^* of literals contained in C with a substitution σ such that $C^*\sigma = \{\bar{a}_j\}$ and $(C - C^*)\sigma = C' - \{\bar{a}_j\}$ and a set $D^* \subset D$ with a substitution τ such that $D^*\tau = \{a_j\}$ and $(D - D^*)\tau = D' - \{a_j\}$. Let σ^* and τ^* be most general unifiers for C^* and D^* respectively. Then $C_1 = C\sigma^*$ and $D_1 = D\tau^*$ are factors of C and D respectively and must thus be in Y. The factors C_1 and D_1 then resolve on the literals $C^*\sigma^*$ and $D^*\tau^*$ to give a clause E also in Y. The clause $E' = (C' - \{\bar{a}_j\}) \cup (D' - \{a_j\})$ is a ground instance of E. Since $I_{j-1} \cup \{a_j\}$ condemns C', I_{j-1} must condemn $C' - \{\bar{a}_j\}$; and similarly since $I_{j-1} \cup \{\bar{a}_j\}$ condemns D', I_{j-1} must condemn $D' - \{a_j\}$. Hence I_{j-1} must condemn E, contrary to hypothesis. Hence $j \ngtr 0$, so j must be zero.

Thus $I_o = \emptyset$ condemns some clause C_o in Y. But this is possible only if C_o = *false*. Hence *false* $\in Y$, completing the proof.

To see that factoring is needed for refutation completeness one need only consider the example where S is composed of the clauses $\{Qa\}$, $\{Px, Py\}$, and $\{\bar{P}w, \bar{P}z\}$. Then the closure under resolution alone includes only variants of clauses in this unsatisfiable set S and of $\{\bar{P}u, Pt\}$, but does not include *false*, so no refutation can be obtained.

Resolution is sometimes defined in such a way as to subsume factoring. While this appears to simplify a few definitions and theorems, it leads, we feel, to undesirable consequences when applied to automatic theorem proving. Hence we prefer the older formulation in terms of separate rules for resolution and factoring.

Alternate Proof of the Maximal Model Theorem:

Since the proof of the maximal model theorem appearing in Section 3 was first given [19], a number of other proofs have been found. The simplest we have been able to work out to date is obtained by adapting an idea found in the proof of Lindenbaum's Lemma that is given in [11]. Briefly, the alternative proof thus obtained runs as follows: Consider an enumeration $a_1, a_2, \ldots$ of the interpretation T. Let $b_j = \bar{a}_j$ if $S \cup \{\{b_1\}, \ldots, \{b_{j-1}\}, \{\bar{a}_j\}\}$ is satisfiable, otherwise let $b_j = a_j$. Let $Q_o = \emptyset$ and for $j > 0$ let $Q_j = \{\{b_1\}, \ldots, \{b_j\}\}$. Let $M = \{b_1, b_2, \ldots\}$. Then M is an interpretation, and furthermore if M is a model from the manner of construction, it must be a maximal one relative to T. To see that M is a model, suppose that it is not. Then, since M is an interpretation, it must be that M condemns some clause in S. Hence the set $W = S \cup \{\{b_1\}, \{b_2\}, \ldots\}$ must be unsatisfiable. Some finite subset W' of W must then be unsatisfiable. Without loss of generality let W' be such that no proper subset of W' is unsatisfiable. Then let $W' \cap \{\{b_1\}, \{b_2\}, \ldots\} = \{\{b_{j_1}\}, \ldots \{b_{j_k}\}\}$ and let n be the largest subscript on b appearing therein. Now for $j > 0$, $S \cup Q_j$ must be satisfiable if $S \cup Q_{j-1}$ is, since otherwise both $S \cup Q_{j-1} \cup \{a_j\}$ and $S \cup Q_{j-1} \cup \{\bar{a}_j\}$ would be unsatisfiable. Hence, since S is satisfiable, $S \cup Q_n$ must by induction be satisfiable, which contradicts the unsatisfiability of its subset W'. Thus M condemns no clause in S and must be a model of S, hence a maximal one relative to T.

The construction given in the proof of the Refutation Completeness theorem for Σ can also be made to yield a maximal model. Failure to recognize the importance of closure of Y under resolution and factoring to the modelhood of I was, however, led to a number of spurious "constructive" proofs of the maximal model theorem. Indeed, if S is finite and the closure restriction is satisfied by S itself, I is a model, and an effective means of enumerating I is provided by the technique given in the refutation completeness proof for Σ. Luckham has successfully applied a similar technique [9] to a useful special case involving a finite set of ground clauses. It appears [10] that attempts to give a proof of the general Maximal Model theorem embodying an effective enumeration of the maximal model may be foredoomed by the existence in first-order logic of sentences for which there exist no effectively enumerable models, hence no such maximal ones.

Appendix 2

In this section we list a number of theorems from group theory and some from ring theory whose proofs have been obtained by one of our theorem-proving programs. One should not draw conclusions from comparing the times given to obtain proof, since the times obtained were affected materially by the fact that a number of different programs were employed. We shall also include the proofs of two theorems from group theory to illustrate the inference rule, paramodulation.

1. In a group, if $xy = e$, then $yx = e$. The time required to obtain a proof was 815 milliseconds.
2. If $xy = e$ and $zy = e$, $x = z$. 5.3 sec.
3. The right inverse of the right inverse of $x = x$. 326 milliseconds.
4. $(x^{-1})^{-1} = x$. 998 milliseconds.

Example 3 differs from example 4, obviously, by not treating inverse as known to be two sided.

5. A non-empty subset H of a group is a subgroup if and only if, for every x and y in H, xy^{-1} is in H. The necessity was proved in 1.4 sec., assuming that identity and inverse of the subgroup were that of the group as a whole. The sufficiency was proved as a series of lemmas. That H contains e required 89 milliseconds to prove; for every x in H, H contains x^{-1}, 222 milliseconds; H is closed under multiplication, 32 sec.

6. Exponent 2 implies commutativity. 8 sec.

7. The axioms of right identity and right inverse are dependent axioms. The proofs were obtained respectively in 2 seconds and 3+ seconds.

8. Exponent 2 implies commutativity, but using only those axioms sufficient to obtain a proof. 460 milliseconds. The addition of various lemmas, just as the presence of a full axiom set, can help or hurt the proof-search. In problems of some depth, lemmas will almost certainly be needed.

9. Subgroups of index 2 are normal. The theorem was proved by dividing it into two cases, as is often done in the standard proof. The proof for invariance with respect to elements of the subgroup was obtained in 4.9 sec., for elements outside the subgroup in 27.5 sec.

10. Boolean rings have characteristic 2. 41.7 sec.

11. Boolean rings are commutative, assuming the lemma of characteristic 2. 12+ min.

12. If a set is closed under an associative operation, contains an element e with $e^2 = e$, and every element has (with respect to e) at least one left inverse and at most one right inverse, then the set is a group. A proof was obtained in 5.8 sec.

Of the examples given, example 12 is clearly the most difficult, mathematically speaking. The theorem is not easy to prove for the average student of group theory.

On the other hand, 18.2.8 on page 322 of [7] has not yet been provable by a computer. The lemma states that exponent 3 implies $((a, b), b) = e$ for all a and b in the group. The proof is shortened considerably (see appendix of [13]) by the addition of paramodulation to the most successful of the previous theorem-proving systems. That previous system was based on resolution.

We now give a proof, employing paramodulation, of the theorem that the axiom of right inverse is dependent on the set consisting of left identity, left inverse, and associativity. In the following, R is read as equals, f as product, g as inverse, and e as the left identity. Assume by way of contradiction that there exists an element a lacking a right inverse. The resulting clause is $\{\bar{R}f(ay)e\}$. Braces and commas will be omitted in the following.

Proof.

1. $Rf(ex)x$. Left identity
2. $Rf(g(x)x)e$. Left inverse
3. $Rf(xf(yz))f(f(xy)z)$. Associativity
4. $\bar{R}f(ay)e$.
5. $\bar{R}f(f(ea)y)e$, paramodulate 1 on x into 4 on a.
6. $\bar{R}f(f(f(g(w)w)a)y)e$, 2 into 5 on first occurrence of e.
7. $\bar{R}f(f(g(w)f(wa))y)e$, 3 into 6 on $f(f(g(w)w)a)$.
8. $\bar{R}f(f(g(g(a))e)y)e$, 2 into 7 on $f(wa)$.
9. $\bar{R}f(g(g(a))f(ey))e$, 3 into 8 on $f(f(g(g(a))e)y)$.
10. $\bar{R}f(g(g(a))y)e$, 1 into 9 on $f(ey)$.
11. *false*, resolve 10 and 2.

For the 12th example, the following clauses were input to the theorem-proving program, PG5 [20], and the proof which follows the input (in essence) was obtained in 5.8 sec.

1. $Pxyf(xy)$. Closure
2. $Peee$.
3. $Pg(x)xe$. Left inverse
4. $\bar{P}xyu\ \bar{P}yzv\ \bar{P}uzw\ Pxvw$. } Associativity
5. $\bar{P}xyu\ \bar{P}yzv\ \bar{P}xvw\ Puzw$. } Associativity
6. $\bar{P}xye\ \bar{P}xze\ Ryz$.
7. Rxx. Reflexivity
8. $\bar{P}xyu\ \bar{P}xyv\ Ruv$. Well-definedness
9. $\bar{R}xy\ \bar{R}yz\ Rxz$. Transitivity

10. $\bar{R}uv\ \bar{P}uxy\ Pvxy.$	}	Substitutivity of equals
11. $\bar{R}uv\ \bar{P}xuy\ Pxvy.$		
12. $\bar{R}uv\ \bar{P}xyu\ Pxyv.$		
13. $\bar{R}uv\ Rf(ux)f(vx).$		
14. $\bar{R}uv\ Rf(xu)f(xv).$		
15. $\bar{R}uv\ Rg(u)g(v).$		
16. $\bar{P}eaa.$		*e* is not a left identity

Proof.

1. $Peee.$
2. $\bar{P}xyu\ \bar{P}yzv\ \bar{P}uzw\ Pxvw.$
3. $\bar{P}xye\ \bar{P}xze\ Ryz.$
4. $\bar{R}uv\ \bar{P}xyu\ Pxyv.$
5. $\bar{P}eaa.$
6. $Pg(x)xe.$
7. $Pxyf(xy).$
8. $\bar{R}f(xy)v\ Pxyv,$ resolve clause 4 on second literal against clause 7.
9. $\bar{R}f(ea)a$, 8_2 (i.e., second literal of clause 8) vs. 5.
10. $\bar{P}yzv\ \bar{P}f(xy)zw\ Pxvw$, 2_1 vs. 7.
11. $\bar{P}f(xe)ew\ Pxew$, 10_1 vs. 1.
12. $Pxef(f(xe)e)$. 11_1 vs. 7.
13. $\bar{P}g(z)ye\ Ryz$, 3_2 vs. 6.
14. $\bar{P}g(a)f(ea)e$, 13_2 vs. 9.
15. $\bar{P}yzv\ \bar{P}ezw\ Pg(y)vw$, 2_1 vs. 6.
16. $\bar{P}ezw\ Pg(y)f(yz)w$, 15_1 vs. 7.
17. $Pg(y)f(ye)e$, 16_1 vs. 1.
18. $Rf(xe)x$, 13_1 vs. 17.
19. $\bar{P}ezv\ \bar{P}f(f(xe)e)zw\ Pxvw$, 2_1 vs. 12.
    $\bar{P}ezv\ \bar{P}xzw\ Pxvw$, demodulate [Reference 20] second literal of 19 with 18.
20. $\bar{P}xzw\ Pxf(ez)w$, $19'_1$ vs. 7.
21. $Pg(x)f(ex)e$, 20_1 vs. 6.
22. *false*, 21 vs. 14.

PG5 gave as output only 1 thru 5, 9, 12, 14, and 21.

The following proof is also of example 12 but employing paramodulation as an additional inference rule.

Proof.

1. $Rf(ee)e$.
2. $Rf(xf(yz))f(f(xy)z)$.
3. $\bar{R}f(xy)e\ \bar{R}f(xz)e\ Ryz$.
4. $Rf(g(x)x)e$.
5. $\bar{R}f(ea)a$.
6. $\bar{R}f(g(z)y)e\ Ryz$, resolve 3_2 vs. 4.
7. $\bar{R}f(g(a)f(ea))e$, resolve 6_2 and 5.
8. $Rf(g(y)f(yz))f(ez)$, paramodulate 4 into 2 on $f(xy)$.
9. $Rf(g(y)f(ye))e$, paramodulate 1 into 8 on $f(ez)$.
10. $Rf(ye)y$, resolve 6_1 and 9.
11. $Rf(xf(ez))f(xz)$, paramodulate 10 into 2 on $f(xy)$.
12. $Rf(g(x)f(ex))e$, paramodulate 4 into 11 on $f(xz)$.
13. *false*, resolve 7 and 12.

Acknowledgements

The authors are deeply indebted to W.F. Miller for his support and encouragement of the automatic theorem-proving effort over the years; to D.L. Luckham and C.C. Green; and most particularly to W.W. Boone for his extensive and invaluable assistance in the preparation and criticism of this paper. An important portion of the fundamental research underlying the results in this paper was supported by the Computer Science Department of the University of Wisconsin and the University of Wisconsin Computing Center during 1966-67. The current work is supported by the U.S. Atomic Energy Commission.

References

[1] A. Church, Introduction to Mathematical Logic I, Princeton (1956).
[2] A. Church, A note on the Entscheidungsproblem, J. Symb. Logic 1, (1936) 40-41, 101-102.
[3] J. Darlington, Automatic theorem proving with equality substitutions and mathematical induction, in: D. Michie (ed.), Machine Intelligence (Edinburgh Univ. Press and Oliver and Boyd, 1968).

[4] M. Davis and H. Putnam, A computing procedure for quantification theory, J. Assn. Comput. Mach. 7 (1960) 201-215.
[5] P.C. Gilmore, A proof method for quantification theory: its justification and realization, IBM Journal (Jan. 1960) 28-35.
[6] C. Green, Theorem-proving by Resolution as a Basis for Question-answering Systems, in: B. Meltzer and D. Michie (eds.), Machine Intelligence 4 (Edinburgh Univ. Press and American Elsevier, N.Y., 1969).
[7] M. Hall, The Theory of Groups (Macmillan, New York, 1959), p. 322.
[8] R. Kowalski and P. Hayes, Semantic trees in automatic theorem proving, in: B. Meltzer and D. Michie (eds.), Machine Intelligence 4 (Edinburgh Univ. Press and American Elsevier, N.Y., 1969).
[9] D. Luckham, Some tree-paring strategies for theorem-proving, in: D. Michie (ed.), Machine Intelligence 3 (Edinburgh Univ. Press, 1968).
[10] D. Luckham, Personal communication (1969).
[11] E. Mendelson, Introduction to Mathematical Logic (Van Nostrand, 1964).
[12] W. Quine, A proof procedure for quantification theory, J. Symb. Logic 20 (1955) 141-149.
[13] G. Robinson and L. Wos, Paramodulation and theorem-proving in first-order theories with equality, in: B. Meltzer and D. Michie (eds.), Machine Intelligence 4 (Edinburgh Univ. Press and American Elsevier, N.Y., 1969).
[14] G. Robinson, L. Wos and D. Carson, Some theorem-proving strategies and their implementation. AMD Tech. Memo No. 72, Argonne National Laboratory (1964).
[15] G. Robinson and L. Wos, Completeness of paramodulation, J. Symb. Logic 34 (1969) 160 (abstract).
[16] J.A. Robinson, A machine-oriented logic based on the resolution principle, J. Assn. Comput. Mach. 12 (1965) 23-41.
[17] A. Tarski, A. Mostowski and R. Robinson, Undecidable theories (North Holland, Amsterdam, 1953).
[18] L. Wos, D. Carson and G. Robinson, The unit-preference strategy in theorem proving, AFIPS Conference Proceedings 26 (Spartan Books, Washington, D.C., 1964) 615-621.
[19] L. Wos and G. Robinson, The maximal model theorem, J. Symb. Logic 34 (1969) 159-160 (abstract).
[20] L. Wos, G. Robinson, D. Carson and L. Shalla, The concept of demodulation in theorem proving, J. Assn. Comput. Mach. 14 (1967) 698-709.
[21] L. Wos and G. Robinson, Paramodulation and Set of Support, Symposium on Automatic Demonstration, Lecture Notes in Mathematics 145 (Springer, Berlin, 1970).

Problems

1. Let Π be the inference system consisting of the inference rules paramodulation, resolution, and factoring. For sets of clauses which contain a reflexivity axiom for equality but are not necessarily functionally reflexive, is Π R-refutation complete? In other words, if the set S of clauses is R-unsatisfiable (has no equality model) and contains a reflexivity axiom, is Π strong enough to guarantee the existence of a refutation of S solely employing the inference rules in Π? Roughly, the question amounts to asking whether Π is sufficiently strong as to provide the basis for a semidecision procedure for theoremhood for first-order theories with equality. (The natural approach to this problem would be to apply the so-called "lifting lemma". Contrary to intuition, the standard type of "lifting lemma" does not hold, and the modified forms of the lemma that are known to hold have not yielded a proof of the desired completeness theorem.)

2. Find a "direct" proof of the unsolvability of the associativity problem for finite monoids by a diagonal procedure.

3. Solve the same question for the class of finite associative symmetrical monoids.

4. Or, as next best, consider proofs reducing directly to Turing machines as recursive functions, without the intervention of groups (or semi-groups).

5. Determine the existence of (Turing) degrees of unsolvability of associativity problems of monoids, symmetrical monoids, etc. (Conjecture: there are associativity problems of any given degree of unsolvability.)

6. Reduce Miller's Tableau on special group classes with solvable and unsolvable problems to symmetrical monoids, in particular, for the corresponding embedding problems.

7. The non-associative finite monoids, in particular the symmetrical ones, are effectively enumerable (in principle). Establish a reasonable practical method of enumeration giving those of lowest orders without gap, and in particular, establish an enumeration for strongly non-associative monoids, i.e., having no non-trivial associative homomorphic image.

8. Extend the method of group and semi-group diagrams (Dehn's "Gruppenbilder") to monoids.

9. Is the word problem for Thue systems with one relation solvable?

In problems 10–14 below let G be a Greendlinger group, (or a small cancellation group in general).

10. Does G have unique roots? That is, does $x^n = y^n$ imply $x = y$?

11. Does G contain a free nonabelian subgroup?

12. Does G satisfy the ascending chain condition for n-generator subgroups?

13. Is G conjugacy separable?

14. Does G have a solvable commutator problem, i.e., given an arbitrary word $w \in G$ can we decide if w is a commutator?

15. Is the conjugacy problem solvable for free products of Greendlinger groups with a cyclic amalgamation?

16. Are small cancellation groups Hopfian (residually finite)?

The *generalized word problem for the group H relative to the finitely generated subgroup K* is the problem to determine for an arbitrary word $w \in H$ whether or not $w \in K$. (Here, regard K as given by a finite set of words which generate K.) Then the *generalized word problem for H* is the union of the generalized word problems for H relative to each K, as K ranges over all finitely generated subgroups of H.

17. Do small cancellation groups have solvable generalized word problems?

18. Are the conjugacy problem and the generalized word problem solvable for one-relator groups?

19. Is the isomorphism problem solvable for one-relator groups?

20. Is there a one-relator group whose word problem is not solvable by a primitive recursive function?

21. When are one-relator groups Hopfian (residually finite), (SQ-universal), (orderable)?

22. Is the conjugacy problem solvable for GL(n, Z)?

23. Does the automorphism group of a finitely generated free group have solvable conjugacy problem?

24. Is the conjugacy problem solvable for finitely presented residually nilpotent or residually free groups?

25. Is the word problem, conjugacy problem, and generalized word problem for finitely presented solvable groups solvable? (One probably wants to consider "almost finitely presented" groups, i.e., finitely generated groups defined by a solvability law and a finite number of additional relations.)

26. Does there exist a finitely presented group which satisfies a non-trivial identity and has unsolvable word problem?

27. Does the free group of a finitely based variety of groups have solvable word problem? Or, to frame the question in a more combinatorial fashion: for any finite set L of laws defining a variety $\mathfrak{V}(L)$ of groups, the set $\mathfrak{C}(L)$ of all consequences of L, i.e., the set of all laws holding in $\mathfrak{V}(L)$, is obviously recursively enumerable. But is there a choice of L for which $\mathfrak{C}(L)$ is not recursive?

28. Does there exist a finitely presented infinite periodic group? Does there exist one of bounded exponent?

29. If G is a sixth group, $u^m \neq 1$, $v^n \neq 1$ in G and the commutator $[u^m, v^n] = 1$ in G, must u and v be powers of a common element?

30. (Michael Rabin) Is there a decision procedure to decide if a finitely presented group has a non-abelian free quotient? (Negative

answer gives an independent proof of the unsolvability of Hilbert's Tenth Problem.)

31. Same as above replacing "free" by "finite".

32. Let G be a finitely presented group. Let K be the intersection of all normal subgroups of finite index in G. Can G/K have unsolvable word problem? (A positive answer would show the open sentence problem for finite groups undecidable.)

33. Same as above with "finite" replaced by "free". (A positive answer would show the open sentence problem for free groups, and hence Hilbert's Tenth Problem, undecidable.)

34. Does there exist a recursive class Ω of groups with uniformly solvable word problem such that the set of groups in Ω which are residually free is recursively enumerable but not recursive. (A positive answer would give an independent proof of the unsolvability of Hilbert's Tenth Problem.)

35. Same as above with "free" replaced by "finite".

36. Do there exist finitely presented groups G_1, G_2 with solvable power problem such that $G_1 \times G_2$ has unsolvable power problem?

37. Let F_1, F_2 be two free groups of the same finite rank with $\theta: F_1 \to F_2$ an isomorphism. Let H_1 be a finitely generated subgroup of F_1. Does the free product with amalgamation $G = F_1 *_{\theta|H_1} F_2$ have solvable conjugacy problem?

38. Is the conjugacy problem solvable for knot groups?

39. Can one solve the word problem, conjugacy problem, and isomorphism problem for fundamental groups of 3-manifolds?

40. Does the automorphism group of a fundamental group of a 2-manifold have solvable conjugacy problem?

41. Call a finite presentation of a group *balanced* if it has the same number of generators as defining relators. Is there an algorithm to decide if a balanced presentation is trivial?

42. Consider the class W of groups which can be built up from free groups by taking Britton extensions and free products with